W9-BBP-150

Multivariate Statistics
and
Probability

Essays in Memory
of
Paruchuri R. Krishnaiah

P. R. Krishnaiah
1932–1987

The articles in this volume are dedicated to the memory of the founding editor, P. R. Krishnaiah

Multivariate Statistics and Probability

Essays in Memory of Paruchuri R. Krishnaiah

Edited by

C.R. Rao

Center for Multivariate Analysis
Pennsylvania State University
University Park, Pennsylvania

M.M. Rao

Department of Mathematics
University of California
Riverside, California

ACADEMIC PRESS, INC.

Harcourt Brace Jovanovich, Publishers

Boston San Diego New York
Berkeley London Sydney
Tokyo Toronto

ACADEMIC PRESS, INC.
1250 Sixth Avenue, San Diego, CA 92101

United Kingdom Edition published by
ACADEMIC PRESS INC. (LONDON) LTD.
24–28 Oval Road, London NW1 7DX

Library of Congress Cataloging-in-Publication Data
Multivariate statistics and probability.

Includes bibliography and index.
1. Multivariate analysis. 2. Probabilities.
3. Krishnaiah, Paruchuri R. I. Krishnaiah,
Paruchuri R. II. Rao, C. Radhakrishna (Calyampudi
Radhakrishna), Date- . III. Rao, M.M. (Malempati
Madhusudana), Date- .
QA278.M863 1989 519.5'35 89-14868
ISBN 0-12-580205-6 (alk. paper)

Printed in the United States of America
89 90 91 92 9 8 7 6 5 4 3 2 1

Contents

Contributors

Numbers in parentheses refer to the pages on which the authors' contributions begin.

G.J. Babu (15), *Department of Statistics, Pennsylvania State University, University Park, Pennsylvania 16802*

Z.D. Bai (24, 40), *Department of Mathematics, University of Pittsburgh, Pittsburgh, Pennsylvania 15260*

J.K. Baksalary (53), *Academy of Agriculture, 60-769 Poznan, Poland*

R.N. Bhattacharya (68, 80), *Department of Mathematics, Indiana University, Bloomington, Indiana 47405*

H.W. Block (91), *Department of Mathematics and Statistics, University of Pittsburgh, Pittsburgh, Pennsylvania 15260*

C.S. Chen (105), *Department of Mathematics and Statistics, University of Pittsburgh, Pittsburgh, Pennsylvania 15260*

X.R. Chen (116), *Department of Mathematics and Statistics, University of Pittsburgh, Pittsburgh, Pennsylvania 15260*

A. Cohen (131), *Department of Statistics, Rutgers University, New Brunswick, New Jersey 08903*

M. Csörgő (151), *Department of Mathematics and Statistics, Carleton University, Ottawa, Ontario, Canada K1S5B6*

C.M. Cuadras (447), *Department of Estadistica, Universitat de Barcelona, Gram Via 585, 08007, Barcelona, Spain*

M.H. DeGroot (539), *Department of Statistics, Carnegie-Mellon University, Pittsburgh, Pennsylvania 15213*

P. Erdös (169), *Mathematical Institute, Hungarian Academy of Sciences, Budapest, Hungary*

D.A.S. Fraser (181), *Department of Mathematics, York University, North York, Ontario, Canada M3J1P3*

Y. Fujikoshi (194), *Department of Mathematics, Hiroshima University, Hiroshima 730, Japan*

M.M. Gabr (484), *Department of Mathematics, University of Manchester, Institute of Science and Technology, Manchester M60lQP, England*

J.K. Ghosh (68), *Indian Statistical Institute, Calcutta, India*

M. Ghosh (206), *Department of Statistics, University of Florida, Gainesville, Florida 32611*

N.C. Giri (270), *Department of Mathematics and Statistics, University of Montreal, Montreal PQ, Canada H3C3J7*

P. Hall (228), *Department of Statistics, Australian National University, Cambera ACT 2601, Australia*

T. Hida (225), *Department of Mathematics, Nagoya University, Nagoya 464, Japan*

L. Horvath (151), *Department of Mathematics, Szeged University, Szeged 6720, Hungary*

H.P. Hucke (261), *Department of Mathematics, Fern University, D-4040 Neuss 1, Federal Republic of Germany*

G. Kallainpur (261), *Department of Statistics, University of North Carolina, Chapel Hill, North Carolina 27514*

R.L. Karandikar (261), *Indian Statistical Institute, New Delhi 110029, India*

T. Kariya (270), *Institute of Economic Research, Hitotubashi University, Kunitachi Tokyo 186, Japan*

S. Karlin (284), *Department of Mathematics, Stanford University, Stanford, California 94305*

A.R. Karmous (300), *Department of Statistics, Zagozig University, Zagozig, Egypt*

C.G. Khatri (319), *Department of Statistics, Gujarat University, Ahmedabad 380009, India*

T.L. Lai (334), *Department of Statistics, Stanford University, Stanford, California 94305*

O. Lee (80), *Department of Mathematics, Ewha Women's University, Seoul, South Korea*

K.-S. Lii (359), *Department of Statistics, University of California, Riverside, California 92521*

M.E. Mack (539), *Stuart Pharmaceutical Division 1Cl Americas, Inc., Wilmington, Delaware 19897*

J.I. Marden (131), *Department of Statistics, University of Illinois, Champaign, Illinois 61820*

T. Mathew (53), *Department of Mathematics and Statistics, University of Maryland, Catonsville, Maryland*

B.Q. Miao (375), *Department of Mathematics, University of Science and Technology of China, Hefei, Peoples Republic of China*

R. Nishii (392), *Faculty of Integrated Arts and Sciences, Hiroshima University, Hiroshima 730, Japan*

F. Perron (270), *Department of Mathematics and Statistics, University of Montreal, Montreal PQ, Canada H3C3J7*

J. Pfanzagl (404), *Mathematics Institute, Universität za Köln, D-5000 Köln 41, Federal Republic of Germany*

T.M. Pukkila (422), *Department of Mathematical Sciences, University of Tampere, SF-33101, Tampere, Finland*

C.R. Rao (15, 24), *Department of Statistics, Pennsylvania State University, University Park, Pennsylvania 16802*

M.B. Rao (505), *Department of Mathematical Science, North Dakota State University, Fargo, North Dakota 58105*

M.M. Rao (343), *Department of Mathematics, University of California, Riverside, California 92521*

P. Révész (169), *Institut für Statistics and Mathematics, Technische Universität Wien, A-1040 Wien, Austria*

Y. Rinott (284), *Department of Statistics, Hebrew University, Jerusalem 91905, Israel*

M. Rosenblatt (359), *Department of Mathematics, University of California, San Diego, La Jolla, California 92093*

C. Ruiz-Rivas (447), *Department de Matematicas, Universidad Autonoma de Madrid, Madrid, Spain*

A.R. Samson (91), *Department of Mathematics and Statistics, University of Pittsburgh, Pittsburgh, Pennsylvania 15260*

T.H. Savits (105), *Department of Mathematics and Statistics, University of Pittsburgh, Pittsburgh, Pennsylvania 15260*

V.V. Sazonov (457), *Steklov Mathematical Institute, Academy of Sciences of the USSR, Moscow B-333, USSR*

P.K. Sen (300), *Department of Biostatistics, School of Public Health, University of North Carolina, Chapel Hill, North Carolina 27514*

B.K. Sinha (206), *Department of Mathematics and Statistics, University of Maryland, Baltimore, Maryland 21228*

T. Subba Rao (484), *Department of Mathematics, University of Manchester, Institute of Science and Technology, Manchester M601QD, England*

K. Subramanyam (505), *Department of Mathematics and Statistics, University of Pittsburgh, Pittsburgh, Pennsylvania 15260*

M. Tainguchi (521), *Department of Mathematics, Hiroshima University, Hiroshima 730, Japan*

D.M. Titterington (228), *Department of Statistics, University of Glasgow, Glasgow G128QQ, United Kingdom*

V.V. Ulyanov (457), *Stecklov Mathematical Institute, Moscow B-333, USSR*

J.S. Verducci (539), *Department of Statistics, Ohio State University, Columbus, Ohio 43210*

Y.H. Wu (116), *Department of Mathematics and Statistics, University of Pittsburgh, Pittsburgh, Pennsylvania 15260*

Z. Ying (334), *Department of Statistics, University of Illinois, Champaign, Illinois 61820*

B.A. Zaleszkii (457), *Stecklov Mathematical Institute, Moscow B-333, USSR*

L.C. Zhao (24), *Department of Mathematics, University of Science and Technology of China, Hefei, Anhui, Peoples' Republic of China*

Preface

The contributions included in the following pages were originally planned for a presentation on the 55th birthday of Professor P.R. Krishnaiah. Unfortunately his illness became severe and soon after claimed his life. Hence this book is dedicated to honor his memory. The articles, initially intended for a standard journal, are based on original research by active and leading scientists in the areas of their current interests in the multivariate field. The authors were all associated with Krishnaiah professionally in his research and development of multivariate statistical analysis and stochastic theory, and many of them also knew him personally.

The chapters of this volume cover the main areas of multivariate statistical theory and its applications, as well as aspects of probability and stochastic analysis. They cover both finite sampling and asymptotic results, including aspects of decision theory, Bayesian analysis, classical estimation, and regression, as well as time-series problems. There are discussions of practical applications and computational solutions. The works on probability include results on the (vector) central limit theory for dependent random variables, the rates of convergence and asymptotic expansions, Markov processes, and foundational problems. The material covered in the volume should be of considerable interest to researchers as well as to graduate students working in theoretical and applied statistics, multivariate analysis, and random processes.

We wish to express our appreciation to the contributors who responded to our invitations and compiled their chapters within the time constraints. All the articles were refereed; and, as a result, several underwent revisions and alterations. We are deeply indebted to the many referees, whose names cannot be listed here, but whose assistance was essential. Also our special thanks go to the staff of Academic Press, Inc. for bringing out this volume expeditiously and accommodating all our requests. Finally, we wish to acknowledge our home institutions for providing the secretarial assistance to complete this project on schedule.

C.R. Rao
University Park, Pennsylvania
M.M. Rao
Riverside, California
May 1989

In Memoriam

P. R. Krishnaiah

(1932–1987)

Paruchuri Rama Krishnaiah was born in a suburb of Repalle in Andhra Pardesh, India, in 1932 into a respected middle class Kisan (or farming) family. He was one of the brightest students of the local high school and, by his parents and teachers noticing this fact, he was sent to the well-known Loyola College in Madras for higher education. Krishnaiah passed the 2-year intermediate course in a high first class in 1950 and was admitted to the then newly started and highly competitive (and difficult to gain admission) statistics honors program in the Presidency College, also of Madras University. Coincidentally, I joined the same school that year as a (mathematics) graduate student (for a master's degree) and although we saw each other, we met formally only at the end of the first quarter on a trip home for a vacation. To our mutual surprise we found that we came from adjacent districts, separated by the river Krishna, and our homes were no more than 30 miles apart. From then on, we maintained a close friendship, and thus it was a rude shock to me to learn in late 1986 that he had become a victim of a cruel disease (cancer). Indeed he was a teetotaler and a nonsmoker, had always been careful in whatever he ate, and had no bad habits. He fought the ailment with great courage and was optimistic of overcoming it, which he so expressed on his birthday, July 15, 1987. Finally, he succumbed to the dreadful disease on August 1, 1987, leaving his friends, loved ones, relatives, and colleagues in great sorrow. He is survived by his wife, Indira, two young sons, Raghu and Niranjan, five brothers, and two sisters. I shall now briefly describe his educational, professional, organizational, humanitarian, and research accomplishments.

Soon after finishing his B. Sc. (Honors) at Presidency College, Madras, in 1954, Krishnaiah went to the United States and joined the University of Minnesota in Minneapolis to continue his graduate work in statistics. He was always interested in both the theory and the applications of this subject. Even as a student he assisted in statistical methodology at the Bureau of Educational Research in Minnesota, and this resulted in some publications with his colleagues there. He took an M. S. degree in 1957 while searching for a suitable area of specialization for his dissertation.

Multivariate Statistics and Probability
ISBN 0-12-580205-6

Reprinted from *J. Mult. Anal.* **27**(1).

During the summer term of 1956 the late Professor S. N. Roy of the University of North Carolina visited Minnesota and offered a course in multivariate statistical analysis. This was the first substantial account of that subject given there; Krishnaiah took it and became interested in it immediately. He spent the summer of 1957 at the IMS Summer Institute in Boulder, Colorado, as a student member, where he was exposed to the analysis of variance and related problems from the seminars of Professors Bose, Kempthorne, Kruskal, Scheffé, and several other visiting scholars. Another such session operated at the University of Minnesota a year later, concentrating on ranking and selection problems, to which Krishnaiah was again attracted. These three general areas of statistical theory became his main research subjects for all his later work, consultation, and publications as reviewed below.

There was no central location for statistics at Minnesota until 1960, and students had to find the faculty whose interests (and appointments) were combined with other areas. For a while, Krishnaiah traveled to discuss the subjects of his interest, and in 1959–1960 he spent the year at Chapel Hill with Professor Roy for this purpose. From 1960 on he worked as a senior statistician at Remington Rand Univac in Blue Bell, Pennsylvania, until 1963 when he joined the Wright–Patterson Air Force Base in Dayton, Ohio, as a mathematical statistician. He was also awarded the Ph. D. degree by the University of Minnesota during the same year. Krishnaiah remained at Wright–Patterson until 1976 when he joined the University of Pittsburgh as a Professor of Statistics. In 1982 he became the founder and director of the Center for Multivariate Analysis and also took a joint appointment as a professor in the Graduate School of Business to reflect his interests in substantiative applications. Before turning to his research, it is appropriate to consider his professional and humanitarian work at this point.

Krishnaiah organized six international symposia on Multivariate Analysis which were held in June of 1965, 1968, 1972, 1975, and 1978 and in July of 1983. It is of interest to note that he dedicated the published "Proceedings" of the first conference to the memory of S. N. Roy and some of the other proceedings volumes to H. Hotelling, P. C. Mahalanobis, and H. Scheffé, paying his respects to these scholars from whose works and contacts he had learned the subject. He also organized a symposium on Applications of Statistics in June 1976, edited its proceedings, and gave short courses on multivariate data analysis. He served as a member of the council of the American Statistical Association for 1968–1969, was on various committees of the IMS, received the Statistician of the Year award in 1982 from the Pittsburgh chapter of the ASA, and was a member of the technical committee on statistical pattern recognition of the International Association of Pattern Recognition. Krishnaiah is a fellow of the ASA,

IMS, and AAAS, as well as an elected member of the International Statistical Institute. He was the founder and editor of the *Journal of Multivariate Analysis*, as well as the founder and editor of the series "Developments in Statistics," published by Academic Press (four volumes appeared), and the general editor of the "Handbook of Statistics," published by North-Holland (seven, of a dozen proposed volumes appeared). Krishnaiah served as a member of the editorial board of the *Journal of Statistical Planning and Inference* and was a coordinating editor of the North-Holland series "Statistics and Probability." He presented invited papers at several professional meetings, including the first, second, and fourth international conferences on Probability and Mathematical Statistics held at Vilnius, USSR, he visited the People's Republic of China in 1981 for 3 weeks at the invitation of some universities in that country, and at the time of his death he was in receipt of a fellowship by the Japan Society for Promotion of Science to visit Japan for a month in 1986–1987. Earlier, he was a visiting scientist at the Indian Statistical Institute in 1966 and had been at the Banach center in Warsaw, as well as at the Department of Mathematics and Statistics and the Academy of Agriculture in Poznán, under an exchange visitor program between the Polish and the U.S. National Academies of Sciences. In 1985, the Telegu Association of North America conferred on him a distinguished scientist award.

Moreover, Krishnaiah played a major role in developing the statistics program in the Department of Mathematics and Statistics at the University of Pittsburgh. This was rated as the most improved program in statistics in the nation. For instance, he was instrumental in bringing Professor C. R. Rao to this department when several other schools were trying to get him. Krishnaiah worked in different areas such as theoretical and computational statistics, signal processing, pattern recognition, medical statistics, and econometrics so that he was able to assist scientists in various disciplines during the last 30 years, and he derived great satisfaction from it. At the time of his death, he was serving as president of SHARE, a nonprofit organization devoted to scientific, health, and allied research education, helping mostly the third world countries.

With regard to his research activities, it should be observed that Krishnaiah edited (or coedited) 19 books and monographs and authored two more (one jointly) reference books which are in press. He also was the principal (or coprincipal) investigator of research contracts and grants at the University of Pittsburgh continuously from 1976 until his death.

Although by training he was inclined toward theoretical statistics, applications of statistics were always kept in view. Indeed, his initial papers (1959a, 1960a, 1962a) are on such applications. The early paper (1961a), the only joint work we published, deals with some aspects of multivariate gamma distribution which later played a key role in his theoretical work for many years. It was used in his thesis (1963a), was analyzed further in (1963b, 1964a), and played a role in several papers on simultaneous tests of hypotheses. It was generalized for use in tests involving multivariate F-statistics (in addition to multivariate χ^2-problems), distributional problems with Wishart matrices, and other sample covariance matrices. See (1984j) for an account of the work on these topics. Also the well-known "union–intersection" principle of S. N. Roy played an important part in Krishnaiah's work. This led to the "finite intersection tests" formulated by him and are now known by his name. Selection of the best, or a prescribed subset, of a collection of multiple populations was the topic of his research for several papers. Many of these results are surveyed in (1976b, 1978a, 1979a). These test procedures led Krishnaiah to consider the distributions of eigenvalues of various types of sample covariance matrices, complementing the works of S. N. Roy, H. Hotelling, and others. He also extended some of these results if the errors were correlated in some way, or if they formed a simple Markov process.

Since the exact distribution of the above types of statistics is quite involved, Krishnaiah was led to approximations and asymptotic expansions of distribution functions. These questions occupied a major part of his work in the last 10 years and are reflected in several publications (1977e; 1979b; 1980a, c; 1981a; 1982a, b; 1983a; 1986d). He was also studying the limit behavior of the distributions of the eigenvalues of sample matrices, as seen from the papers (1983e, f; 1984g, h; 1985g; 1986b, c, d). It is clear that his research has touched most areas of multivariate statistical analysis and made some inroads into time series (he was deeply interested in signal detection problems in the last 3 years), as well as some nonparametric estimation, multiple comparisons, and considerable work on the multivariate analysis of variance. In order to apply the latter results to practical problems, Krishnaiah expended much effort in constructing several types of statistical tables for significance tests.

A better idea of his research interests can be obtained by reading the titles of his extensive publication list, which is included below. He was very prolific in the last years. It reveals another fact. Krishnaiah interacted freely with different types of scientists, and this is why more than three-quarters of his publications involve at least one joint author. This collaborative effort helped widen his interests and also contributed to a broad and balanced view of the subjects for publication in the editorial work of the *Journal of Multivariate Analysis* as well as his inviting people of different backgrounds to participate in the symposia that he organized.

Until the end, Krishnaiah had a positive attitude toward life and was hopeful that he could beat the illness. He was participating in works even from his sick bed until almost the end. But adversity took over, and it was cruel. We all miss him.

In peparing this sketch and compiling the complete publication list, I am indebted to Professor C. R. Rao and to Mrs. Indira Krishnaiah for much help.

M. M. Rao

Riverside, California

Publications

Books Edited

1. (1966). *Multivariate Analysis.* Academic Press, New York.
2. (1969). *Multivariate Analysis-II.* Academic Press, New York.
3. (1973). *Multivariate Analysis-III.* Academic Press, New York.
4. (1977). *Multivariate Analysis-IV.* North-Holland, Amsterdam.
5. (1979). *Multivariate Analysis-V.* North-Holland, Amsterdam.
6. (1985). *Multivariate Analysis-VI.* North-Holland, Amsterdam.
7. (1977). *Applications of Statistics.* North-Holland, Amsterdam.
8. (1978). *Developments in Statistics, Vol.* 1. Academic Press, New York.
9. (1979). *Developments in Statistics, Vol.* 2. Academic Press, New York.
10. (1980). *Developments in Statistics, Vol.* 3. Academic Press, New York.
11. (1983). *Developments in Statistics, Vol.* 4. Academic Press, New York.
12. (with Kallianpur, G., and Ghosh, J. K.) (1981). *Statistics and Probability: Essays in Honor of C. R. Rao*, North-Holland, Amsterdam.
13. (1980). *Handbook of Statistics, Vol.* 1: *Analysis of Variance.* North-Holland, Amsterdam.
14. (with Kanal, L.) (1982). *Handbook of Statistics, Vol.* 2: *Classification, Pattern Recognition and Reduction of Dimensionality.* North-Holland, Amsterdam.
15. (with Brillinger, D. R.) (1983). *Handbook of Statistics, Vol.* 3: *Time Series in Frequency Domain.* North-Holland, Amsterdam.
16. (with Sen, P. K.) (1984). *Handbook of Statistics, Vol.* 4: *Nonparametric Methods.* North-Holland, Amsterdam.
17. (with Hannan, E. J., and Rao, M. M.) (1985). *Handbook of Statistics, Vol.* 5: *Time Series in the Time Domain.* North-Holland, Amsterdam.
18. (with Rao, C. R.) (1988). *Handbook of Statistics, Vol.* 6: *Sampling.* North-Holland, Amsterdam.
19. (with Rao, C. R.) (1988). *Handbook of Statistics, Vol.* 7: *Quality Control and Reliability.* North-Holland, Amsterdam.
20. (with Schuurmann, F. J.) (In press). *Computations of Complex Multivariate Distributions.* North-Holland, Amsterdam. (In press).
21. *Simultaneous Test Procedures.* Dekker, New York. (In press).

Papers

1959

a. (with Hoyt, C. J. and Torrance, E. P.) Analysis of complex contingency data. *J. Experimental Ed.* **27** 187–194.

1960

a. (with Hoyt, C. J.) Estimation of test reliability by analysis of variance technique. *J. Experimental Ed.* **28**.

1961

a. (with Rao, M. M.) Remarks on a multivariate Gamma distribution. *Amer. Math. Monthly* **79** 342–346.

1962

a. (with Kaskey, G.) Statistical routines for computers. In *Proceedings, Middle Atlantic Conference of American Society for Quality Control.*

b. (with Kaskey, G., and Azzari, A.) Cluster formation and diagnostic significance in psychiatric symptom evaluation. In *Proceedings of Fall Joint Computer Conference, Philadelphia*, pp. 285–302. Spartan, Washington, DC.

1963

a. *Simultaneous Tests and the Efficiency of Generalized Balanced Incomplete Block Designs.* Technical Report ARL 63–174, Wright-Patterson Air Force Base, OH.

b. (with Hagis, P., and Steinberg, L.) A note on the bivariate Chi distribution. *SIAM, Rev.* **5** 140–144.

1964

a. *Distribution of the Studentized Smallest Chi-Square with Tables and Applications.* Technical Report ARL 64–218. Wright-Patterson Air Force Base, OH.

b. (with Armitage, J. V.) *Tables for the Studentized Largest Chi-Square Distribution and Their Applications.* Technical Report ARL 64–188, Wright-Patterson Air Force Base, OH.

1965

a. On a multivariate generalization of the simultaneous analysis of variance test. *Ann. Inst. Stastist. Math.* **17** 167–173.

b. On the simultaneous ANOVA and MANOVA tests. *Ann. Inst. Statist. Math.* **17** 35–53.

c. Simultaneous tests for the equality of variances against certain alternatives. *Austral. J. Statist.* **7** 105–109. Correction (1968). **10** 43.

d. Multiple comparison tests in multi-response experiments. *Sankhyā, Ser. A* **27** 65–72.

e. (with Armitage, J. V.) Tables for the distribution of the maximum of correlated Chi-square variates with one degree of freedom. *Trabajos Estadíst.* **17** 91–96; this is a part of ARL 65–136.

f. (with Hagis, P., Jr., and Steinberg, L.) Tests for the equality of standard deviations in a bivariate normal population. *Trabajos Estadíst.* **17**, 3–15; this is a part of ARL 65–199.

1966

a. (with Armitage, J. V.) Tables for multivariate t distribution. *Sankhyā Ser. B* **28** 31–56; this is a part of ARL 65–199.

b. (with Murthy, V. K.) Simultaneous tests for trend and serial correlations for Gaussian Markov residuals. *Econometrica* **34** 472–480. Erratum. *Econometrica* **34**.

c. (with Rizvi, M. H.) Some procedures for selection of the multivariate normal populations better than a control. In *Multivariate Analysis* (P. R. Krishnaiah, Ed.), pp. 477–492. Academic Press, New York.

d. (with Rizvi, M. H.) A note on recurrence relations between expected values of functions of order statistics. *Ann. Math. Statist.* **37** 733–734.

1967

a. (with Breiter, M. C.) Tables for the moments of Gamma order statistics. *Sankhyā Ser. B* **30** 59–72.

b. Selection procedures based on covariance matrices of multivariate normal populations. *Trabajos Estadíst.* **19** 11–24.

c (with PATHAK, P. K.) Tests for the equality of covariance matrices under the intraclass correlation model. *Ann. Math. Statist.* **38** 1286–1288.

d. (with RIZVI, M. H.) A note on the moments of Gamma order statistics. *Technometrics* **9** 315–318.

1968

a. Simultaneous tests for the equality of covariance matrices against certain alternatives. *Ann. Math. Statist.* **39** 1303–1309.

b. (with PATHAK, P. K.) A note on confidence bounds for certain ratios of characteristic roots of covariance matrices. *Austral. J. Statist.* **10** 116–119.

1969

a. Simultaneous test procedures under general MANOVA models. In *Multivariate Analysis-II* (P. R. Krishnaiah, Ed.), pp. 121–143. Academic Press, New York.

b. Further results on simultaneous tests under general MANOVA models. *Bull. Internat. Statist. Inst.* **43**, 288–289.

c. (with ARMITAGE, J. V., AND BREITER, M. C.) *Tables for the Probability Integrals of the Bivariate t Distribution.* Technical Report ARL 69–0060, Wright-Patterson Air Force Base, OH.

d. (with ARMITAGE, J. V., AND BREITER, M. C.) *Tables for bivariate t distribution.* Technical Report ARL 69–0210, Wright-Patterson Air Force Base, OH.

1970

a. (with ARMITAGE, J. V.) On a multivariate F distribution. In *Essays in Probability and Statistics* (R. C. Bose *et al.*, Eds.), pp. 439–468. Univ. of North Carolina Press, Chapel Hill, NC.

1971

a. (with CHANG, T. C.) On the exact distribution of the extreme roots of the Wishart and MANOVA matrices. *J. Multivariate Anal.* **1** 108–117.

b. (with CHANG, T. C.) On the exact distribution of smallest root of the Wishart matrix using zonal polynomials. *Ann. Inst. Statist. Math.* **23** 293–295.

c. (with WAIKAR, V. B.) Simultaneous tests for equality of latent roots against certain alternatives–I. *Ann. Inst. Statist. Math.* **23** 451–468.

d. (with WAIKAR, V. B.) Exact joint distributions of any few ordered roots of a a class of random matrices. *J. Multivariate Anal.* **1** 308–315.

e. (with SEN, P. K.) Some asymptotic simultaneous tests for multivariate moving average processes. *Sankhyā Ser. A* **33** 81–90.

1972

a. (with CHANG, T. C.) On the exact distributions of the traces of $S_1(S_1+S_2)^{-1}$ and $S_1S_2^{-1}$. *Sankhyā Ser. A* **34** 153–160.

b. (with WAIKAR, V. B.) Simultaneous tests for equality of latent roots against certain alternatives—II. *Ann. Inst. Statist. Math.* **24** 81–85.

c. (with WAIKAR, V. B., AND CHANG, T. C.) Distributions of few roots of noncentral complex random matrices. *Austral. J. Statist.* **14** 84–88.

1973

a. (with CLEMM, D. S., AND CHATTOPADHYAY, A. K.) Upper percentage points of the individual roots of the Wishart matrix. *Sankhyā Ser. B* **35** 325–338.

b. (with CLEMM, D. S., AND WAIKAR, V. B.) Tables for the extreme roots of the Wishart matrix. *J. Statist. Comput. Simulation* **2** 65–92.

c. (with SCHUURMANN, F. J., AND WAIKAR, V. B.) Upper percentage points of the intermediate roots of the MANOVA matrix. *Sankhyā, Ser. B* **35** 339–358.

d. (with WAIKAR, V. B.) On the distribution of a linear combination of correlated quadratic forms. *Commun. Statist.* **1** 371–380.

e. (with SCHUURMANN, F. J., AND CHATTOPADHYAY, A. K.) On the distribution of the ratios of the extreme roots to the trace of the Wishart matrix. *J. Multivariate Anal.* **3** 445–453.

f. (with SCHUURMANN, F. J., AND WAIKAR, V. B.) Percentage points of the joint distribution of the extreme roots of the random matrix $S_1(S_1+S_2)^{-1}$. *J. Statist. Comput. Simulation* **2** 17–38.

1974

a. (with SCHUURMANN, F. J: *On the Exact Distributions of the Trace of a Complex Multivariate Beta Matrix.* Technical Report ARL 74–0107, Wright-Patterson Air Force Base, OH.

b. (with SCHUURMANN, F. J.) *On the Exact Distributions of the Individual Roots of the Complex Wishart and Multivariate Beta Matrices.* ARL 74–0111, Wright-Patterson Air Force Base, OH.

c. (with SCHUURMANN, F. J.) *On the Distributions of the Ratios of the Extreme Roots of the Real and Complex Multivariate Beta Matrices.* Technical Report ARL 74–0122, Wright-Patterson Air Force Base, OH.

d. (with SCHUURMANN, F. J.) *Approximations to the Distributions of the Traces of Complex Multivariate Beta and F Matrices.* Technical Report ARL 74–0123, Wright-Patterson Air Force Base, OH.

e. (with SCHUURMANN, F. J.) On the evaluation of some distributions that arise in simultaneous tests for the equality of the latent roots of the covariance matrix. *J. Multivariate Anal.* **4** 265–282.

f. (with SEN, P. K.) On a class of simultaneous rank order tests in MANOCOVA. *Ann. Inst. Statist. Math.* **26** 135–145.

g. (with LEE, J. C.) On covariance structures. *Sankhyā Ser. A* **38** 357–371.

1975

a. Tests for the equality of the covariance matrices of correlated multivariate normal populations. In *A Survey of Statistical Design and Linear Models* (J. N. Srivastava, Ed.). North-Holland, Amsterdam.

b. (with GUPTA, A. K., AND CHATTOPADHYAY, A. K.) Asymptotic distributions of the determinants of certain random matrices. *Commun. Statist.* **4** 33–47.

c. Simultaneous tests for multiple comparisons of growth curves. In *Proceedings, Meeting of the International Statistical Institute, Warsaw, Poland.*, 1975.

d. (with CHATTOPADHYAY, A. K.) On some noncentral distributions in multivariatate analysis. *South African Statist. J.* **9** 37–46.

e. (with SCHUURMANN, F. J., AND CHATTOPADHYAY, A. K.) Exact percentage points of the distributions of the trace of a multivariate Beta matrix. *J. Statist. Comput. Simulation* **3** 331–343.

f. (with SCHUURMANN, F. J., AND CHATTOPADHYAY, A. K.) Tables for a multivariate *F* distribution. *Sankhyā Ser. B* **37** 308–331.

1976

a. (with CHANG, T. C., AND LEE, J. C.) On the distribution of the likelihood ratio test statistic for compound symmetry. *South African Statist. J.* **10** 49–62.

b. Some recent developments on complex multivariate distributions. *J. Multivariate Anal.* **6** 1–30.

c. (with LEE, J. C., AND CHANG, T. C.) Approximations to the distributions of the likelihood ratio statistics for tests of certain covariance structures of complex muiltivariate normal populations. *Biometrika* **63** 543–549.

d. (with KIM, D. G., DUBRO, G. A., PHADIA, E., AND SCHUURMANN, F. J.) The measurement of flow velocity by transit time techniques. Unpublished.

1977

a. (with CHANG, T. C., AND LEE, J. C.) Approximations to the distributions of the likelihood ratio statistics for testing the hypotheses on covriance matrices and mean vectors simultaneously. In *Applications of Statistics* (P. R. Krishnaiah, Ed.), pp. 97–103. North-Holland, Amsterdam.

b. (with KHATRI, C. G. AND SEN, P. K.) A note on the joint distribution of correlated quadratic forms. *J. Statist. Planning Inference* **1** 299–307.

c. (with LEE, J. C.) Infeence on the eigenvalues of the covariance matrices of real and complex multivariate normal populations. In *Multivariate Analysis-IV* (P. R. Krishnaiah, Ed.), pp. 95–103. North-Holland, Amsterdam.

d. On generalized multivariate Gamma type distributions and their applications in reliability. In *Proceedings, Conference on the Theory and Applications of Reliability with Emphasis on Bayesian and Nonparametric Methods, Vol.* 1 (I. N. Shimi and C. P. Tsokos, Eds.), pp. 475–494. Academic Press, New York.

e. (with LEE, J. C., AND CHANG, T. C.) Approximations to the distributions of the likelihood ratio statistics for testing certain structures on the covariance matrices of real multivariate normal populations. In *Multivariate Analysis-IV* (P. R. Krishnaiah, Ed.), pp. 105–118. North-Holland, Amsterdam.

f. (with LEE, J. C., AND CHANG, J. C.) Approximations to the distributions of the determinants of real and complex multivariate Beta matrices. *South African Statist. J.* **11** 13–26.

1978

a. Some recent developments on real multivariate distributions. In *Developments in Statistics, Vol.* 1 (P. R. Krishnaiah, Ed.), pp. 135–169. Academic Press, New York.

1979

a. Some developments on simultaneous test procedures. In *Developments in Statistics, Vol.* 2 (P. R. Krishnaiah, Ed.), pp. 157–201. Academic Press, New York.

b. (with LEE, J. C.) On the asymptotic joint distributions of certain functions of the eigenvalues of four random matrices. *J. Multivariate Anal.* **9** 248–258.

1980

a. (with LEE, J. C.) On the asymptotic distributions of certain functions of the eigenvalues of correlation matrices. *Banach Center Publ.* **6** 229–237.

b. (with COX, C. M., LEE, J. C., REISING, J., AND SCHUURMANN, F. J.) A study on finite intersection tests for multiple comparions of means. In *Multivariate Analysis-V* (P. R. Krishnaiah, Ed.), pp. 435–466. North-Holland, Amsterdam.

c. (with FANG, C.) Asymptotic joint distributions of functions of the elements of sample covariance matrix. In *Statistics and Probability: Essays in Honor of C. R. Rao*, pp. 249–262. North-Holland, Amsterdam.

d. (with KASKEY, G., KOLMAN, B., AND STEINBERG, L.) Transformations to normality. In *Handbook of Statistics, Vol.* 1 (P. R. Krishnaiah, Ed.), pp. 321–342. North-Holland, Amsterdam.

e. Computations of some multivariate distributions. In *Handbook of Statistics, Vol.* 1 (P. R. Krishnaiah, Ed.), pp. 745–971. North-Holland, Amsterdam.

f. (with LEE, J. C.) Likelihood ratio tests for the mean vectors and covariance matrices of multivariate normal populations. In *Handbook of Statistics, Vol.* 1 (P. R. Krishnaiah, Ed.), pp. 513–570. North-Holland, Amsterdam.

g. (with YOCHMOWITZ, M. G.) Inference on the structure of interaction in two-way classification model. In *Handbook of Statistics*, Vol. 1 (P. R. Krishnaiah, Ed.), pp. 973–994. North-Holland, Amsterdam.

h. (with MUDHOLKAR, G. S., AND SUBBAIAH, P.) Simultaneous test procedures for mean vectors and covariance matrices. In *Handbook of Statistics*, Vol. 1 (P. R. Krishnaiah, Ed.), pp. 631–671. North-Holland, Amsterdam.

1981

a. (with FANG, C.) Asymptotic distributions of functions of the eigenvalues of the real and complex noncentral Wishart matrices. In *Statistics and Related Topics* (M. Csorgo *et al.*, Eds.), pp. 89–108. North-Holland, Amsterdam.

b. (with KARIYA, T., AND SINHA, B. K.) *Some Properties of Left Orthogonally Invariant Distributions* Technical Report No. 81–28, Institute for Statistics and Applications, University of Pittsburgh.

1982

a. (with FANG, C., AND NAGARSENKER, B. N.) Asymptotic distributions of the likelihood ratio test statistics for covriance structures of the complex multivariate normal distribution. *J. Multivariate Anal.* **12** 597–611.

b. (with FANG, C.) Asymptotic distributions of functions of the eigenvalues of some random matrices for non-normal populations. *J. Multivariate Anal.* **12** 39–63.

c. Selection of variables under univariate regression models. In *Handbook of Statistics*, Vol. 2 (L. Kanal and P. R. Krishnaiah, Eds.), pp. 805–820. North-Holland, Amsterdam.

d. Selection of variables in discriminant analysis. In *Handbook of Statistics*, Vol. 2 (L. Kanal and P. R. Krishnaiah, Eds.), pp. 883–892. North-Holland, Amsterdam.

1983

a. (with FANG, C.) On asymptotic distributions of test statistics for covariance matrices and correlation matrices. In *Studies in Econometrics, Time Series, and Multivariate Statistics* (S. Karlin, T. Amemiya, and L. Goodman, Eds.), pp. 419–433. Academic Press, New York.

b. (with KARIYA, T., AND RAO, C. R.) Inference on parameters of multivariate normal population when some data is missing. In *Developments in Statistics*, Vol. 4 (P. R. Krishnaiah, Ed.), pp. 137–184. Academic Press, New York.

c. (with LEE, J. C., AND CHANG, T. C.) Likelihood ratio tests on covariance matrices and mean vectors of complex multivariate normal populations and their applications in time series. In *Handbook of Statistics*, Vol. 3 (D. R. Brillinger and P. R. Krishnaiah, Eds.), pp. 439–476. North-Holland, Amsterdam.

d. (with SARKAR, S. K., AND SINHA, B. K.) Some tests with unbalanced data from a bivariate normal population. *Ann. Inst. Statist. Math.* **35** 63–75.

e. (with YIN, Y. Q., AND BAI, Z. D.) The limiting behavior of the eigenvalues of a multivariate F matrix. *J. Multivariate Anal.* **13** 508–516.

f. (with YIN, Y. Q.) Limit theorems for the eigenvalues of product of two random matrices. *J. Multivariate Anal.* **13**, 489–507.

g. (with FANG, C., AND NAGARSENKER, B. N.) Asymptotic distributions of sphericity test in a complex multivariate normal distribution. *Comm. Statist. A—Theory Methods* **12** 273–288.

1984

a. (with FANG, C.) Asymptotic distributions of functions of the eigenvalues of the doubly noncentral F matrix. In *Proceeding, Indian Statistical Institute Golden Jubilee International Conference on Statistics: Applications and New Directions*, pp. 254–269.

b. (with SEN, P. K.) Tables for order statistics. In *Handbook of Statistics*, Vol. 4 (P. R. Krishnaiah and P. K. Sen, Eds.), pp. 873–935. North-Holland, Amsterdam.

c. (with SARKAR, SHAKUNTALA) Nonparametric estimation of multivariate density using Laguerre and Hermite polynomials. In *Multivariate Analysis-VI* (P. R. Krishnaiah, Ed.), pp. 361–373. North-Holland, Amsterdam.

d. (with LIANG, W.-Q.) Multi-stage nonparametric estimation of density function using orthonormal systems. *J. Multivariate Anal.* **15** 228–241.

e. (with LIANG, W.-Q.) An optimum rearrangment of terms in estimation of density using orthonormal systems. Technical Report No. 84–23. *J. Statist. Plann. Inference*, in press.

f. (with SEN, P. K.) Selected tables for nonparametric statistics. In *Handbook of Statistics*, Vol. 4 (P. R. Krishnaiah and P. K. Sen Eds.). North-Holland, Amsterdam.

g. (with BAI, Z. D., AND YIN, Y. Q.) On Limiting empirical distribution function of the eigenvalues of a multivariate F matrix. *Teor. Veroyatnost. i Primenen.* (*Theory Probab. Appl.*) **29**.

h. (with YIN, Y. Q., AND BAI, Z. D.) On limit of the largest eigenvalue of the large dimensional sample covariance matrix. Technical Report No. 84–44. *Probab. Theory Related Fields*, in press.

i. (with KARIYA, T., AND FUJIKOSHI, Y.) Test for independence of two multivariate regression equations with different design matrices. *J. Multivariate Anal.* **15** 383–407.

j. Multivariate Gamma distributions and their applications in reliability. In *Developments in Statistics and Its Applications; Proceedings, First Saudi Symposium on Statistics and Its Applications.*

1985

a. Multivariate Gamma distribution. In *Encyclopedia of Statistical Sciences*, Vol. 6, pp. 63–66. Wiley, New York.

b. Multivariate multiple comparisons. In *Encyclopedia of Statistical Sciences*, Vol. 6, pp. 88–95. Wiley, New York.

c. (with LIANG, W. Q.) Nonparametric iterative estimation of multivariate binary density. *J. Multivariate Anal.* **16** 162–172.

d. (with LIN, J., AND WANG, L.) *Inference on the Ranks of the Canonical Correlation matrices for Elliptically Symmetric Populations.* Technical Report No. 85–14, Center for Multivariate Analysis, University of Pittsburgh.

e. (with LIN, J., AND WANG, L.) *Tests for the Dimensionality of the Regression Matrices when the Underlying Distributions Are Elliptically Symmetric.* Technical Report No. 85–36, Center for Multivariate Analysis, University of Pittsburgh.

f. (with ZHAO, L. C., AND BAI, Z. D.) *On Rates of Converence of Efficient Detection Criteria in Signal Processing with White Noise.* Technical Report No. 85–45, Center for Multivariate Analysis, University of Pittsburgh. *IEEE Trans. Inform. Theory*, in press.

g. (with Yin, Y. Q.) Limit theorems for the eigenvalues of the sample covariance matrix when the underlying distributions is isotropic. *Teor. Veroyatnost. i Primenen.* (*Theory Probab. Appl.* **30** 810–816.

1986

a. (with Rao, M. B., and Subramanyam, K.) *Extreme Point Methods in the Determination of the Structure of a Class of Bivariate Distributions and Some Applications to Contingency Tables.* Technical Report No. 86–01, Center for Multivariate Analysis, Univesity of Pittsburgh.

b. (with Yin, Y. Q.) Limit theorems for the eigenvalues of product of large dimensional random matrices when the underlying distribution is isotropic. *Teor. Veroyatnost. i Primenen.* (*Theory Probab. Appl.* **31** 394–398.

c. (with Bai, Z. D., and Yin, Y. Q.) On limiting spectral distribution of product of two random matrices when the underlying distribution is isotropic. *J. Multivariate Anal.* **19** 189–210.

d. (with Bai, Z. D., and Liang, W. Q.) On asymptotic joint distribution of the eigenvalues of the noncentral MANOVA matrix for nonnormal populations. *Sankhyā Ser. B* **48** 153–162.

e. (with Fang, C.) On asymptotic distribution of the test statistic for the mean of the non-isotropic principal component, *Commun. Statist.* **15** 1163–1168.

f. (with Zhao, L. C., and Bai, Z. D.) On detection of number of signals in presence of white noise. *J. Multivariate Anal.* **20** 1–25.

g. (with Zhao, L. C., and Bai, Z. D.) On detection of number of signals when the covariance matrix is arbitrary. *J. Multivariate Anal.* **20** 26–49.

h. (with Yin, Y. Q.) On some nonparametric methods for detection of the number of signals. *IEEE Trans. Acoustics Speech Signal Process.* **ASSP-35** 1533–1538.

i. (with Lin, J.) Complex elliptic distributions. *Comm. Statist. A—Theory Methods* **15** 3693–3718.

j. (with Sarkar, S.) Principal component analysis under correlated multivariate regression equations model. In *Proceedings, Fourth International Conference on Probability Theory and Mathematical Statistics*, Vol. 2, pp. 103–119. VNU Science Press, Vilnius, USSR.

k. (with Bai, Z. D., and Zhao, L. C.) Signal processing using model selection methods. Technical Report No. 86–03. *Inform. Sci.*, in press.

l. (with Taniguchi, M., and Chao, R.) *Normalizing Transformations of Some Statistics of Gaussian ARMA Processes.* Technical Report No. 86–05, Center for Multivariate Analysis, University of Pittsburgh.

m. (with Taniguchi, M.) *Asymptotic Distributions of Functions of Eigenvalues of the Sample Covariance Matrix and Canonical Correlation Matrix in Multivariate Time Series.* Technical Report No. 86–08, Center for Multivariate Analysis, *J. Multivar. Anal.* **22** 156–176.

n. (with Pukkila, T. M.) *On the Use of Autoregressive Order Determination Criteria in Univariate White Noise Tests.* Technical Report No. 86–15, Center for Multivariate Analysis, University of Pittsburgh. *IEEE Trans.*, in press.

o. (with Pukkila, T. M.) *On the Use of Autoregressive Order Determination Criteria on Multivariate White Noise Tests.* Technical Report No. 86–16, Center for Multivariate Analysis, University of Pittsburgh. *IEEE Trans.*, in press.

p. (with Bai, Z. D., and Yin, Y. Q.) *Inference on the Occurrence/Exposure Rate and Simple Risk Rate.* Technical Report No. 86–18, Center for Multivariate Analysis, University of Pittsburgh. *Ann. Inst. Statist. Math.*, in press.

q. (with Nishii, R.) *On the Moments of Classical Estimates of Explanatory Variables under a Multivariate Calibration Model.* Technical Report No. 86–27, Center for Multivariate Analysis, University of Pittsburgh.

r. (with BAI, Z. D., RAO, C. R., REDDY, P. S., SUN, Y. N., AND ZHAO, L. C.) *Reconstruction of the Left Ventricle from Two Orthogonal Projections.* Technical Report No. 86–33, Center for Multivariate Analysis, University of Pittsburgh. *Computer Vision Graphics Image Process.* in press.

s. (with BAI, Z. D., AND ZHAO, L. C.) *On Simultaneous Estimation of the Number of Signals and Frequencies under a Model with Sinusoids.* Technical Report No. 86–37, Center for Multivariate Analysis, University of Pittsburgh. *IEEE Trans.*, in press.

t. (with BAI, Z. D., AND ZHAO, L. C.) *On the Rate of Convergence of Equivariation Linear Prediction Estimation of the Number of Signals and Frequencies of Multiple Sinusoids.* Technical Report No. 86–38, Center for Multivariate Analysis, University of Pittsburgh. *IEEE Trans.*, in press.

u. (with TAUXE, W. N., KLEIN, H. A., BAGCHI, A., KUNDU, D., AND TEPE, P.) *Clinical Evaluation of the Filtration Fraction: A Multivariate Statistical Analysis.* Technical Report No. 86–41, Center for Multivariate Analysis, University of Pittsburgh.

v. (with NISHI, R., AND BAI, Z. D.) *Strong Consistency of Certain Information Theoretic Criteria for Model Selection in Calibration, Discriminant Analysis and Canonical Correlation Analysis.* Technical Report No. 86–42, Center for Multivariate Analysis, University of Pittsburgh.

w. (with MIAO, B. Q., AND ZHAO, L. C.) *On Detection of Change Points using Mean Vector.* Technical Report No. 86–47, Center for Multivariate Analysis, University of Pittsburgh. In *Handbook of Statistics*, Vol. 7 (P. R. Krishnaiah and C. R. Rao, Eds.). Academic Press, New York.

x. (with CHEN, X. R., AND LIANG, W. Q.) *Estimation of Multivariate Binary Density Using Orthonormal Functions.* Technical Report No. 86–48. Center for Multivariate Analysis, University of Pittsburgh.

y. (with CHEN, X. R.) *Test of Linearity in General Regression Models.* Technical Report No. 86–49, Center for Multivariate Analysis, University of Pittsburgh. *J. Multivariate Anal.*, in press.

z. (with CHEN, X. R.) *Estimation and Testing in Truncated and Nontruncated Linear Median-regression Models.* Technical Report No. 86–50. Center for Multivariate Analysis, University of Pittsburgh.

1987

a. (with KARIYA, T., AND FUJIKOSHI, Y.) On tests for selection of of variables and independence under multivariate regression models. *J. Multivariate Anal.* **21** 207–237.

b. (with SARKAR, S.) Tests for sphericity under correlated mulrivariate regression equations model. *Ann. Inst. Statist. Math.* **39** 163–175.

c. (with ZHAO, L. C., AND BAI, Z. D.) Remarks on certain criteria for detection of number of signals. *IEEE Trans. Acoustics Speech Signal Process.* **ASSP-35** 129–132.

d. (with TEPE, P. G., TAUXE, W. N., BAGCHI, A., AND REZENDA, P.) Comparison of measurement of glomerular filtration rate by single sample, plasma disappearance slope/intercept and other methods, *European J. Nucl. Med.* **13** 28–31.

e. Reduction of dimensionality. In *Encyclopedia of Physical Science and Technol.* Vol. 12, pp. 61–78. Academic Press, New York.

f. (with FUJIKOSHI, Y., AND SCHMIDHAMMER, J.) Effect of additional variables in principal component analysis, discriminant analysis and canonical correlation analysis. *Advances in Multivariate Statistical Analysis*, 45–61.

g. (with RAO, M. B., AND SUBRAMANYAM, K.) A structure theorem on bivariate positive quadrant dependent distributions and tests for independence in two-way contingency tables. *J. Multivariate Anal.* **23** 93–118.

h. (with BAI, Z. D., AND ZHAO. L. C.) *Multivariate Components of Covariance Model in*

Unbalanced Case. Technical Report No. 87–03, Center for Multivaiate Analysis, University of Pittsburgh. *Comm. Statist. A*, in press.

i. (with BAI, Z. D., RAO, C. R., SUN, Y. N., AND ZHAO, L. C.) *Reconstruction of the Shape and Size of Objects from Two Orthogonal Projections.* Technical Reprt No. 87–08, Center for Multivariate Analysis, University of Pittsburgh.

j. (with MIAO, B. Q.) *Control Charts when the Observations Are Correlated.* Technical Report No. 87–09, Center for Multivariate Analysis, University of Pittsburgh.

k. (with MIAO, B. Q.) *Detecting and Interval Estimation about a Slope Change Point.* Technical Report No. 87–11, Center for Multivariate Analysis, University of Pittsburgh.

l. (with BAI, Z. D., AND ZHAO, L. C.) *On the Direction of Arrival Estimation.* Technical Report No. 87–12, Center for Multivariate Analysis, University of Pittsburgh.

m. (with BAI, Z. D., AND ZHAO, L. C.) *On the Asymptotic Joint Distributions of the Eigenvalues of Random Matrices Which Arise under Components of Covariance Model.* Technical Report No. 87–16, Center for Multivariate Analysis, University of Pittsburgh.

n. (with BAI, Z. D., CHEN, X. R., WU, Y., AND ZHAO, L. C.) *Strong Consistency of Maximum Likelihood Parameter Estimation of Superimposed Exponential Signals in Noise.* Technical Report No. 87–17, Center for Multivariate Analysis, University of Pittsburgh.

o. (with BAI, Z. D., CHEN, X. R., AND ZHAO, L. C.) *Asymptotic Property on the EVLP Estimation for Superimposed Exponential Signals in Noise.* Technical Report No. 87–19, Center for Multivariate Analysis, University of Pittsburgh.

p. (with MIAO, B. Q., AND ZHAO, L. C.) *Local Likelihood Method in the Problems Related to Change Points.* Technical Report No. 87–22, Center for Multivariate Analysis, University of Pittsburgh.

q. (with BAI, Z. D., AND ZHAO, L. C.) *Variable Selection in Logistic Regression.* Technical Report No. 87–23, Center for Multivariate Analysis, University of Pittsburgh.

r. (with ZHAO, L. C., AND CHEN, X.R.) *Almost Sure L_r-Norm Convergence for Data-Based Histogram Density Estimates.* Technical Report No. 87–30, Center for Multivariate Analysis, University of Pittsburgh.

s. (with MIAO, B. Q., AND WONG, H.) *Multidimensional Control Chart for Correlated Data.* Technical Report No. 87–33, Center for Multivariate Analysis, University of Pittsburgh.

t. (with TANIGUCHI, M., ZHAO, L. C., AND BAI, Z. D.) *Statistical Analysis of Dyadic Stationary Processes.* Technical Report No. 87–40, Center for Multivariate Analysis, University of Pittsburgh.

u. (with BAI, Z. D., AND ZHAO, L. C.) *On Multiplicities of the Eigenvalues of Components of Covariance with Applications in Signal Processing.* Technical Report No. 87–41, Center for Multivariate Analysis, University of Pittsburgh.

v. (with MIAO, B. Q.) *Review about Estimation of Change Point.* Technical Report No. 87–48, Center for Multivariate Analysis, University of Pittsburgh.

Joint Asymptotic Distribution of Marginal Quantiles and Quantile Functions in Samples from a Multivariate Population*

G. JOGESH BABU

The Pennsylvania State University

AND

C. RADHAKRISHNA RAO

University of Pittsburgh

The joint asymptotic distributions of the marginal quantiles and quantile functions in samples from a p-variate population are derived. Of particular interest is the joint asymptotic distribution of the marginal sample medians, on the basis of which tests of significance for population medians are developed. Methods of estimating unknown nuisance parameters are discussed. The approach is completely nonparametric.

1. INTRODUCTION

Let $X=(x_1, ..., x_p)$ be a random vector with joint d.f. (distribution function) F, ith marginal d.f. F_i, (i, j)th marginal d.f. F_{ij} and ith marginal density function f_i. We denote the ith marginal quantile function by

$$\xi_i(q)=F_i^{-1}(q)=\inf\{x: F_i(x) \geqslant q\}, \qquad 0<q<1 \tag{1.1}$$

and, for convenience, a specific quantile say the q_ith of F_i by

$$\theta_i=\xi_i(q_i). \tag{1.2}$$

* Research sponsored by the Air Force Office of Scientific Research under Contract F49620-85-C-0008. The U.S. Government's right to retain a nonexclusive royalty-free license in and to the copyright covering this paper, for governmental purposes, is acknowledged.

Multivariate Statistics and Probability
ISBN 0-12-580205-6

Reprinted from *J. Mult. Anal.* **27**(1).

Further, let

$$\eta_{ij}(q, r) = F_{ij}(\xi_i(q), \xi_j(r)) \tag{1.3}$$

and denote for given q_i and q_j,

$$\sigma_{ij} = \eta_{ij}(q_i, q_j) - q_i q_j = F_{ij}(\theta_i, \theta_j) - q_i q_j. \tag{1.4}$$

The parameters (1.1)–(1.4) defined above refer to the d.f. of X.

Now let

$$X_i = (x_{1i}, ..., x_{pi}), \qquad i, ..., n \tag{1.5}$$

be n independent copies of X and denote the empirical d.f. of $\{X_i, i = 1, ..., n\}$ by $F^{(n)}$ and the corresponding ith and (i, j)th marginal distributions by $F_i^{(n)}$ and $F_{ij}^{(n)}$, respectively. We denote the quantities (1.1)–(1.4) defined in terms of $F^{(n)}$, $F_i^{(n)}$, and $F_{ij}^{(n)}$ by

$$\xi_i^{(n)}(q), \qquad \theta_i^{(n)}, \qquad \text{and} \qquad \sigma_{ij}^{(n)} \tag{1.6}$$

or simply as

$$\hat{\xi}_i(q), \qquad \hat{\theta}_i, \qquad \text{and} \qquad \hat{\sigma}_{ij} \tag{1.7}$$

as estimates of $\xi_i(q)$, θ_i, and σ_{ij}, respectively.

In this paper, we derive the asymptotic distribution of

$$\hat{\boldsymbol{\theta}}' = (\hat{\theta}_1, ..., \hat{\theta}_p) = (\hat{\xi}_1(q_1), ..., \hat{\xi}_p(q_p)) \tag{1.8}$$

for given $q_1, ..., q_p$ and also the joint distribution of the marginal quantile processes

$$\hat{\xi}_i(q), \qquad 0 < q < 1, i = 1, ..., p. \tag{1.9}$$

The asymptotic distributions of the empirical quantle process (Csörgö and Révész [6]) and of a fixed set of specified quantiles (Mosteller [11]) in one dimension are well known.

Of particular interest is the joint asymptotic distribution of the marginal sample medians

$$(\hat{\xi}_1(\tfrac{1}{2}), ..., \hat{\xi}_p(\tfrac{1}{2})) \tag{1.10}$$

using which we develop tests of significance for the population medians analogous to tests for the means in the multivariate case (see Rao [12, pp. 543–573]). An early work on the joint asymptotic distribution of the sample medians is due to Mood [10]; see also Kuan and Ali [8], where they assume the existence of the density function for the vector variable X.

We obtain the dsitribution in the general case in a form convenient for practical applications.

2. Distribution of the Marginal Sample Quantiles

We prove the following theorem concerning the joint asymptotic distribution of

$$(\hat{\theta}_1, \ldots, \hat{\theta}_p) = (\hat{\xi}_1(q_1), \ldots, \hat{\xi}_p(q_p)), \tag{2.1}$$

the sample q_1th, ..., q_pth quantiles of the marginal empirical distributions of $x_1, \ldots, x_p$, respectively.

Theorem 2.1. *Let F_i be continuously twice differentiable in a neighborhood of θ_i and $\delta_i = f_i(\xi_i(q_i)) = f_i(\theta_i) > 0$, $i = 1, \ldots, p$, where f_i denotes the derivative of F_i. Then the asymptotic distribution of*

$$y_n = \sqrt{n}(\hat{\theta}_1 - \theta_1, \ldots, \hat{\theta}_p - \theta_p) \tag{2.2}$$

is p-variate normal with mean vector zero, and variance-covariance matrix

$$\Sigma = \begin{pmatrix} \dfrac{q_1(1-q_1)}{\delta_1^2} & \dfrac{\sigma_{12}}{\delta_1 \delta_2} & \cdots & \dfrac{\sigma_{1p}}{\delta_1 \delta_p} \\ \vdots & \vdots & \cdots & \vdots \\ \dfrac{\sigma_{p1}}{\delta_p \delta_1} & \dfrac{\sigma_{p2}}{\delta_p \delta_2} & \cdots & \dfrac{q_p(1-q_p)}{\delta_p^2} \end{pmatrix}, \tag{2.3}$$

where σ_{ij} are as defined in (1.4).

Proof. By Bahadur's representation of the sample quantiles (see Bahadur [4]),

$$(\log n)^{-1} n^{3/4} \, |(\hat{\theta}_i - \theta_i) - \delta_i^{-1}(r_i - q_i)| \xrightarrow{p} 0, \qquad i = 1, \ldots, p, \tag{2.4}$$

where $r_i = F_i^{(n)}(\theta_i)$. Then, it follows that

$$y_n = \sqrt{n}(\hat{\theta}_1 - \theta_1, \ldots, \hat{\theta}_p - \theta_p) \tag{2.5}$$

and

$$z_n = \sqrt{n}(\delta_1^{-1}(r_1 - q_1), \ldots, \delta_p^{-1}(r_p - q_p)) \tag{2.6}$$

have the same asymptotic distribution. By the multivariate central limit theorem, z_n weakly converges to a p-variate normal distribution with mean vector zero and covariance matrix as given in (2.3). This proves Theorem 2.1.

For practical applications we need a consistent estimate of Σ as defined in (2.3). There are two sets of unknown $\{\sigma_{ij}\}$ and $\{\delta_i^{-1}\}$ in Σ. A consistent estimate of σ_{ij} is provided by $\hat{\sigma}_{ij}$ as shown in Theorem 2.2.

THEOREM 2.2. *Let F_{ij} be continuous at $(\theta_i, \theta_j) = (\xi_i(q_i), \xi_j(q_j))$. Then*

$$\hat{\sigma}_{ij} = F_{ij}^{(n)}(\xi_i^{(n)}(q_i), \xi_j^{(n)}(q_j)) = F_{ij}^{(n)}(\hat{\theta}_i, \hat{\theta}_j) \to \sigma_{ij} = F_{ij}(\theta_i, \theta_j) \quad \text{a.e. as} \quad n \to \infty. \tag{2.7}$$

Proof.

$$|F_{ij}(\theta_i, \theta_j) - F_{ij}^{(n)}(\hat{\theta}_i, \hat{\theta}_j)| \leqslant |F_{ij}(\theta_i, \theta_j) - F_{ij}(\hat{\theta}_i, \hat{\theta}_j)| + \sup_{x, y} |F_{ij}(x, y) - F_{ij}^{(n)}(x, y)|. \tag{2.8}$$

Since F_{ij} is continuous at (θ_i, θ_j) and

$$\sup |F_{ij}(x, y) - F_{ij}^{(n)}(x, y)| \to 0 \qquad \text{a.e.} \tag{2.9}$$

it follows that the expression on the left-hand side of (2.8) $\to 0$ a.e. which establishes the result (2.7) of Theorem 2.2. Equation (2.9) is a consequence of Theorem 7.2 of Rao [13].

The result (2.7) implies that σ_{ij} in (2.3) can be consistently estimated by its sample equivalent $\hat{\sigma}_{ij}$.

There exist several methods for the estimation of δ_i (see Krieger and Pickards, III [7] and the references therein). Recently, a consistent and efficient estimator of δ_i^{-1} based on a sample of size n has been proposed by Bahu [2] under the assumption that f_i is continuously differentiable at $\xi_i(q_i)$. There is a possibility of this estimate taking negative values, and when this happens some modification of the estimate may have to be made. Using consistent estimates of $\hat{\sigma}_{ij}$ and δ_i^{-1}, a consistent estimate of $\sigma_{ij}/\delta_i\delta_j$, the (i, j)th element of Σ, can be obtained as $\hat{\sigma}_{ij}/\hat{\delta}_i\hat{\delta}_j$.

Another possibility is to obtain a direct estimate of $\sigma_{ij}/\delta_i\delta_j$ by the bootstrap method

$$\hat{\sigma}_{ij}/\hat{\delta}_i\hat{\delta}_j = E^*[n(\theta_i^* - \hat{\theta}_i)(\theta_j^* - \hat{\theta}_j)] \tag{2.10}$$

where E^* is the expectation under the bootstrap distribution function. The consistency of the estimator (2.10) can be proved on the same lines as those given by Babu [3] for the bootstrap estimate of the variance of the sample median.

3. Tests of Significance Based on Medians

Let

$$\hat{\boldsymbol{\theta}}_i' = (\hat{\theta}_{1i}, ..., \hat{\theta}_{pi}), \hat{\Sigma}_i \tag{3.1}$$

be the marginal sample medians and an estimate of Σ (as defined in (2.3)) obtained from a sample of size n_i from a p-variate population Π_i, $i=1, ..., k$. Further let $\boldsymbol{\theta}_i = (\theta_{1i}, ..., \theta_{pi})$ be the true value of the marginal medians for Π_i. To test the hypothesis

$$\boldsymbol{\theta}_1 = \cdots = \boldsymbol{\theta}_p \tag{3.2}$$

we can use the statistic

$$\chi^2 = \text{trace}\left[\sum_{i-1}^{k} n_i \Sigma_i^{-1} \hat{\boldsymbol{\theta}}_i \hat{\boldsymbol{\theta}}_i' - \left(\sum_{i=1}^{k} n_i \Sigma_i^{-1}\right) \bar{\boldsymbol{\theta}} \bar{\boldsymbol{\theta}}'\right], \tag{3.3}$$

where

$$\bar{\boldsymbol{\theta}} = \left(\sum_{i=1}^{k} n_i \Sigma_i^{-1}\right)^{-1} \sum_{i=1}^{k} n_i \Sigma_i^{-1} \hat{\boldsymbol{\theta}}_i \tag{3.4}$$

as chi-square on $p(k-1)$ degrees of freedom, provided the individual sample sizes $n_1, ..., n_k$ are large.

In cases where a common Σ for the k populations can be assumed, we have the problem of estimating Σ from the combined sample. For this purpose we consider the residual vectors by replacing each observed vector by its difference from the sample median vector computed from the sample to which the observed vector belongs. There are altogether $n = (n_1 + \cdots + n_k)$ residual vectors, arising out of the k different samples, from which we construct a p-dimensional empirical distribution function E with the marginal medians as zeros. Then σ_{ij} can be estimated from E_{ij}, the (i, j)th marginal d.f. of E as indicated in (2.7) and δ_i from E_i, the ith marginal d.f. of E using any of the methods described at the end of Section 2. If we denote a common estimate of Σ by $\hat{\Sigma}$, then we can develop tests of significance concerning the structure of the median vectors $\boldsymbol{\theta}_i$, $i=1, ..., k$, as in the case of mean values (see Rao [12, p. 556]). For this purpose we compute the "between populations" matrix

$$S = \sum_{i=1}^{k} n_i \hat{\boldsymbol{\theta}}_i \hat{\boldsymbol{\theta}}_i' - n\bar{\boldsymbol{\theta}}\bar{\boldsymbol{\theta}}' \tag{3.5}$$

where $n\bar{\boldsymbol{\theta}} = n_1 \hat{\boldsymbol{\theta}}_1 + \cdots + n_k \hat{\boldsymbol{\theta}}_k$, and set up the determinental equation

$$|S - \lambda \hat{\Sigma}| = 0. \tag{3.6}$$

The roots of Eq. (3.6) can be used as in the table on p. 558 of Rao [12] to test the dimensionality of the configuration of median values.

4. Joint Distribution of the Marginal Quantile Processes

In Section 2 of the paper, we derived the joint asymptotic distribution of specified marginal quantiles. We now derive the weak limits of the entire marginal quantile processes after suitable scaling. More specifically we consider the processes $\{Z_n\}$ indexed by $(q_1, ..., q_p) \in (0, 1)^p$, where

$$Z_n(q_1, ..., q_p) = \sqrt{n}[f_1(\xi_1(q_1))(\xi_1^{(n)}(q_1) - \xi_1(q_1)), ..., f_p(\xi_p(q_p))(\xi_p^{(n)}(q_p) - \xi_p(q_p))]. \quad (4.1)$$

We first simplify the problem using the following result which is essentially a restatement of Theorem 5.2.2 of Csörgö and Révész [6].

Theorem 4.1. *Suppose that for $i = 1, ..., p$, the marginal d.f. F_i is twice differentiable on (a_i, b_i), where*

$$-\infty \leqslant a_i = \sup\{x: F_i(x) = 0\}$$
$$\infty \geqslant b_i = \inf\{x: F_i(x) = 1\}$$

and $F_i' = f_i \neq 0$ on $(a_i b_i)$. Further assume that

$$\max_i \sup_{a_i < x < b_i} F_i(x)[1 - F_i(x)] \frac{|f_i'(x)|}{f_i^2(x)} < \infty$$

and f_i is non-decreasing (non-increasing) on an interval to the right of a_i (to the left of b_i). Let

$$Y_n^*(q_1, ..., q_p) = \sqrt{n}(V_1^{(n)}(q_1) - q_1, ..., V_p^{(n)}(q_p) - q_p),$$

where $V_i^{(n)}$ is the empirical d.f. of the uniform variables

$$u_{ij} = F_i(x_{ij}), \qquad j = 1, ..., n.$$

Then

$$\sup_{\mathbf{q} \in (0,1)^p} \|Y_n^*(\mathbf{q}) - Z_n(\mathbf{q})\| \to 0 \qquad a.e. \quad (4.2)$$

Hence $\{Y_n^*\}$ and $\{Z_n\}$ have the same limit.

Note that the marginals of $\{Y_n^*\}$ converge weakly to a Brownian bridge on $C[0, 1]$ (see Billingsley [5, p. 105]). Since the paths of the limiting process are continuous, we define a new process Y_n close to Y_n^* as follows. Let $D_1^{(n)}(t)$ be, as a function of $t \in [0, 1]$, the d.f. corresponding to a uniform distribution of mass $(n+1)^{-1}$ over each of the $(n+1)$ intervals $[d_{j-1}, d_j]$, $j = 1, ..., n+1$, where $d_0 = 0$, $d_{n+1} = 1$, and $d_1, ..., d_n$ are the values of $u_{i1}, ..., u_{in}$ arranged in increasing order. Clearly

$$|V_i^{(n)}(t) - D_i^{(n)}| \leqslant \frac{1}{n}, \qquad 0 \leqslant t \leqslant 1 \text{ a.e.}$$

So if

$$Y_n(\mathbf{q}) = \sqrt{n}(D_1^{(n)}(q_1) - q_1, ..., D_p^{(n)}(q_p) - q_p)$$

then

$$\| Y_n(\mathbf{q}) - Y_n^*(\mathbf{q}) \| \leqslant n^{-1/2} \qquad \forall \mathbf{q} \in [0, 1]^p \text{ a.e.}$$

As a consequence, $\{Y_n\}$ and $\{Z_n\}$ have the same weak limits and the marginals of Y_n are continuous functions. Note that

$$Y_n \in B = \{h \colon h(\mathbf{q}) = (h_1(q_1), ..., h_p(q_p)), h_i$$
$$\text{is a continuous function on } [0, 1], i = 1, ..., p\}.$$

Clearly B is a separable closed linear subspace of the Banach space C_p of continuous functions on $[0, 1]^p$ into $\mathbb{R}^p$.

We shall show that $\{Y_n\}$ converges weakly to a Gaussian measure on B. A probability measure μ on B is called Gaussian if for every $H \in B^*$, the space of real continuous linear functionals on B, μH^{-1} is Gaussian on the line (see Aranjo and Giné [1, pp. 140–142, 28, and problem 2 on p. 33]).

To characterize B^*, let H be a real continuous linear functional on B. Then

$$\begin{aligned} H(h_1, ..., h_p) &= H(h_1, 0, ..., 0) + \cdots + H(0, 0, ..., h_p) \\ &= H_1(h_1) + \cdots + H_p(h_p), \qquad \text{say.} \end{aligned} \tag{4.3}$$

The zeroes in the first line of (4.3) refer to the zero function. Clearly, each H_i is a real continuous linear functional on $C[0, 1]$. It then follows that B^* is the k-fold direct sum of the dual space C^* of $C[0, 1]$. By Riesz's representation theorem, for any $L \in C^*$, there exists a signed measure v on $[0, 1]$ such that

$$L(f) = \int_0^1 f(x)\, dv(x)$$

for any $f \in C[0, 1]$ (see Dunford and Schwartz [9]). Thus for every $H \in B^*$, there exist signed measures $\nu_1, ..., \nu_p$ on $[0, 1]$ such that for $f = (f_1, ..., f_p) \in B$,

$$H(f) = \sum_{i=1}^{p} \int_0^1 f_i(x)\, d\nu_i(x).$$

Now let

$$A = \left\{ \sum_{j=1}^{r} \alpha_j \varepsilon_{x_j} : 0 \leqslant x_j \leqslant 1,\ x_j,\ \alpha_j \text{ rational},\ j = 1, ..., r,\ r = 1, 2, ... \right\},$$

where ε_x is the probability measure putting all its mass at x. It is easily seen that A is dense in C^* and is countable. We now state the main result.

THEOREM 4.2. *$\{Y_n\}$ converges weakly to a Gaussian random element $W = (W_1, ..., W_k)$ in B, where W_i is a Brownian bridge for each i and*

$$E(W_i(t)\, W_j(s)) = P(F_i(x_{i1}) \leqslant t,\ F_j(x_{j1}) \leqslant s) - ts \tag{4.4}$$

for all i, j and $0 \leqslant t, s \leqslant 1$.

Proof. Since $\{\sqrt{n}(D_i^{(n)}(t) - t) : 0 \leqslant t \leqslant 1\}$ is tight for each i in $C[0, 1]$, it follows that $\{Y_n\}$ is tight in B. Since A is dense in C^*, in order to show that $\{Y_n\}$ has a weak limit it is enough to show that for any $q_{11}, ..., q_{1r}, ..., q_{p1}, ..., q_{pr}$ in $[0, 1]$ and α_{ij} real

$$\sum_{i=1}^{p} \sum_{j=1}^{r} \alpha_{ij} \sqrt{n}(D_i^{(n)}(q_{ij}) - q_{ij})$$

converges weakly. This holds because of the central limit theorem and the fact that

$$\sup_{0 \leqslant t \leqslant 1} |V_i^{(n)}(t) - D_i^{(n)}(t)| \leqslant \frac{1}{n} \quad \text{a.e.}$$

To complete the proof it is enough to show the existence of W satisfying (4.4).

Since $\{Y_n\}$ is tight, there exists a random element Y on B and a subsequence $\{Y_{n'}\}$ such that $Y_{n'}$ converges weakly to $Y = (Y^{(1)}, ..., Y^{(p)})$. Further, from the above arguments

$$\sum_{i=1}^{p} \sum_{j=1}^{r} \alpha_{ij} Y^{(i)}(q_{ij}) \quad \text{and} \quad \sum_{i=1}^{p} \sum_{j=1}^{r} \alpha_{ij} W_i(q_{ij})$$

have the same distribution as that of normal random variables. So it follows that Y satisfies the properties of W mentioned in (4.4) and Y is Gaussian. Thus Y_n converges weakly to W, and in view of Theorem 4.1, $\{Z_n\}$ converges to W.

References

[1] ARANJO, A., AND GINÉ, E. (1980). *The Central Limit Theorem for Real and Banach Valued Random Variables.* Wiley, New York.

[2] BABU, G. J. (1986). Efficient estimation of the reciprocal of the density quantile function at a point. *Statist. Probab. Lett.* **4** 133–139.

[3] BABU, G. J. (1986). A note on bootstrapping the variance of sample quantiles. *Ann. Inst. Statist. Math.* **38** 439–443.

[4] BAHADUR, R. R. (1966). A note on quantiles in large samples. *Ann Math. Statist.* **37** 577–580.

[5] BILLINGSLEY, P. (1968). *Convergence of Probability Measures.* Wiley, New York.

[6] CSÖRGŐ, M., AND RÉVÉSZ, P. (1981). *Strong Approximations in Probability and Statistics.* Academic Press, New York.

[7] KRIEGER, A. M., AND PICKANDS, III, J. (1981). Weak convergence and efficient density estimation at a point. *Ann. Statist.* **9** 1066–1078.

[8] KUAN, K. S., AND ALI, M. M. (1980). Asymptotic distribution of quantiles from a multivariate distribution. *Multivariate Statistical Analysis, Proceedings, Conference at Dalhousie University, Halifax, Nova Scotia, 1979*, pp. 109–120, North-Holland, Amsterdam.

[9] DUNFORD, N., AND SCHWARTZ, J. T. (1958). *Linear Operators*, Part I. Interscience, New York.

[10] MOOD, A. M. (1941). On the joint distribution of the medians in samples from a multivariate population. *Ann. Math. Statist.* **12** 268–278.

[11] MOSTELLER, F. (1946). On some useful "inefficient statistics." *Ann. Math. Statist.* **17** 377–408.

[12] RAO, C. R. (1973). *Linear Statistical Inference and Its Application*, 2nd ed., Wiley, New York.

[13] RAO, R. R. (1962). Relations between weak and uniform convergence of measures with applications. *Ann. Mat. Statist.* **33** 659–680.

Kernel Estimators of Density Function of Directional Data

Z. D. BAI, C. RADHAKRISHNA RAO, AND L. C. ZHAO

University of Pittsburgh

Let X be a unit vector random variable taking values on a k-dimensional sphere Ω with probability density function $f(x)$. The problem considered is one of estimating $f(x)$ based on n independent observation $X_1, ..., X_n$ on X. The proposed estimator is of the form $f_n(x) = (nh^{k-1})^{-1} C(h) \sum_{i=1}^{n} K[(1 - x'X_i)/h^2]$, $x \in \Omega$, where K is a kernel function defined on R_+. Conditions are imposed on K and f to prove pointwise strong consistency, uniform strong consistency, and strong L_1-norm consistency of f_n as an estimator of f.

1. INTRODUCTION

There is considerable literature on non-parametric estimation of the probability density function (pdf) of a random variable taking values in R^k through kernel functions. If $X_1, ..., X_n$ is a sequence of random k-vectors with f as the common pdf, then the Rosenblatt–Parzen kernel estimator is of the form

$$f_n(x) = (nh_n^k)^{-1} \sum_{1}^{n} K[(x - X_i)/h_n], \qquad x \in R^k, \tag{1.1}$$

where K is a bounded pdf on R^k and $\{h_n\}$ is a sequence of positive numbers. The object of the present paper is to develop a suitable theory of kernel density estimation for random variables taking values on a k-dimensional unit sphere Ω_k, which we denote simply by Ω dropping the suffix throughout the paper.

* Research is sponsored by the Air Force Office of Scientific Research (AFSC) under Contract F49620-85-C-0008. The United States Government is authorized to produce and distribute reprints for governmental purposes notwithstanding any copyright notation hereon.

Multivariate Statistics and Probability
ISBN 0-12-580205-6

Reprinted from *J. Mult. Anal.* **27**(1).

The subject is of some practical interest as there are many situations where observed data are in the form of direction cosines or in the form of vectors scaled by an unknown positive scalar so that only the direction is known. Problems of inference based on such data are discussed under various parametric models for the pdf on Ω (for a review of the literature on the subject see books by Batschelet [1, 2], Mardia [7], and Watson [11], the review paper by J. S. Rao [9], and a recent paper by Pukkila and Rao [8] for derivation of particular parametric models for directional data).

Let $X_1, ..., X_n$ be i.i.d. unit vectors with f as the common pdf on Ω such that

$$\int_{\Omega} f(x)\, d\omega(x) = 1 \tag{1.2}$$

where ω is the Lebesgue measure on Ω.

Theoretically speaking, to estimate the density $f(x)$ on Ω, we can proceed as follows. First select a one-to-one mapping ϕ from Ω onto or into R^{k-1} (which may be chosen as continuous or even arbitrarily differentiable). Then based on the transformed data $\phi(X_1), ..., \phi(X_n)$, by using the usual (kernel, nearest neighbor, or orthogonal series, etc.) density estimation, we can construct an estimate of the density of $\phi(X)$. Finally, by the inverse transform, we get an estimate of $f(X)$. However, two kinds of difficulties arise in practice. First, the transform and its inverse may be complicated and difficult to compute, especially for large k. Second, whatever transformation is used, there is at least one point at which the density cannot be estimated. This happens even for $k=2$. If we consider the density function $f(x)$ on the unit circle as that on the interval $[-\pi, \pi]$ when $f(\pi)=f(-\pi)>0$, then $f(x)$ is not a continuous function on R^1 (assuming $f(x)=0$ outside this interval). Hence there is no kernel density estimate of $f(x)$ which is uniformly consistent (even in the sense of weak convergence). Therefore, we have to choose a mapping to transform the unit circle onto R^1. In this case, the transform and its inverse may be complicated and the value of the density at the point $(-1, 0)$ cannot be estimated since this point corresponds to infinity by the transform. The main purpose of this paper is to propose a method by which we directly estimate the density on Ω, and to investigate the limiting properties of this estimate.

When $k \geqslant 2$, we propose the following kernel estimator of $f(x)$ based on $X_1, ..., X_n$,

$$f_n(x) = (nh^{k-1})^{-1} C(h) \sum_{i=1}^{n} K[(1 - x'X_i)/h^2], \qquad x \in \Omega, \tag{1.3}$$

where $h = h_n > 0$, $K(\cdot)$ is a non-negative function defined on $R_+ = [0, \infty)$ such that

$$0 < \int_0^\infty K(v)\, v^{(k-3)/2}\, dv < \infty \tag{1.4}$$

and $C(h)$ is a positive number such that

$$h^{k-1}[C(h)]^{-1} = \int_\Omega K[(1 - x'y)/h^2]\, d\omega(y). \tag{1.5}$$

Here the above integral is obviously independent of x.

Using the result (2.2.2) given in Watson [11, p. 44], the integral (1.5) can be written as

$$\begin{aligned}[C(h)]^{-1} &= \frac{2\pi^{(k-1)/2}}{h^{k-1}\Gamma[(k-1)/2]} \int_{-1}^{+1} K[(1-z)/h^2](1-z^2)^{(k-3/2}\, dz \\ &= \frac{2\pi^{(k-1)/2}}{\Gamma[(k-1)/2]} \int_0^{2/h^2} K(v)\, v^{(k-3)/2}(2 - vh^2)^{(k-3)/2}\, dv. \end{aligned} \tag{1.6}$$

We note that if $\{h_n\}$ is such that $h_n \to 0$ as $n \to \infty$, then by (1.4) and the dominated convergence theorem

$$\lim_{n\to\infty} [C(h_n)]^{-1} = \frac{(2\pi)^{(k-1)/2}}{\Gamma[(k-1)/2]} \int_0^\infty K(v)\, v^{(k-3)/2}\, dv = \lambda \qquad \text{(say)}. \tag{1.7}$$

Some examples of the choice of the kernel function are as follows:

$K(v) = e^{-v}$	(Longevin–Von Mises–Fisher distribution)
$= 1$	if $v < 1$, $= 0$ otherwise (uniform distribution on a cup).

In this paper, we study the various conditions under which $f_n(x) \to f(x)$ a.s. pointwise, uniformly, and in L_1-norm.

We quote some lemmas which will be used in the proofs of theorems in later sections.

LEMMA 1. *Let $\xi_1, ..., \xi_n$ be independent random variables such that $E(\xi_i) = 0$ and $V(\xi_i) = \sigma_i^2$, $i = 1, ..., n$. Further let there exist a finite constant b such that $P(|\xi_i| \leqslant b) = 1$, $i = 1, ..., n$. Then for any $\varepsilon > 0$ and all n, we have*

$$P\left(\left|n^{-1} \sum_{i=1}^n \xi_i\right| \geqslant \varepsilon\right) \leqslant 2 \exp[-n\varepsilon^2/(2\sigma^2 + b\varepsilon)], \tag{1.8}$$

where $\sigma^2 = n^{-1}(\sigma_1^2 + \cdots + \sigma_n^2)$.

For a proof, see Hoeffding [6].

In order to state Lemmas 2 and 3, we introduce some concepts and notations. Let $x_1, ..., x_r$ be r points in R^k, and $\mathscr{A}$ be a class of Borel sets in R_k. Denote by $\Delta^{\mathscr{A}}(x_1, ..., x_r)$ the number of distinct sets in $\{F \cap A : A \in \mathscr{A}\}$, where $F = \{x_1, ..., x_r\}$. Define

$$m^{\mathscr{A}}(r) = \max_{F} \Delta^{\mathscr{A}}(x_1, ..., x_r). \tag{1.9}$$

Vapnik and Chervonenkis [10] showed that $m^{\mathscr{A}}(r) = 2^r$ for any positive integer r, or $m^{\mathscr{A}}(r) \leqslant r^{s+1}$, where s is the smallest integer j such that $m^{\mathscr{A}}(j) \neq 2^j$. A class of sets $\mathscr{A}$ for which the latter case holds will be called a V-C class with index s.

Let $X_1, X_2, ...,$ be a sequence of i.i.d. random vectors in R^k with a common distribution μ and μ_n be the empirical distribution of $X_1, ..., X_n$. Denote a "distance measure" between μ_n and μ by

$$D_n(\mathscr{A}, \mu) = \operatorname*{Sup}_{A \in \mathscr{A}} |\mu_n(A) - \mu(A)|. \tag{1.10}$$

Further, assume that

$$D_n(\mathscr{A}, \mu), \qquad \sup_{A \in \mathscr{A}} |\mu_n(A) - \mu_{2n}(A)|, \qquad \sup_{A \in \mathscr{A}} \mu_n(A) \tag{1.11}$$

are all random variables. We have the following lemma.

LEMMA 2. *Let $\mathscr{A}$ be a V-C class with index s such that*

$$\sup_{A \subset \mathscr{A}} \mu(A) \leqslant \delta \leqslant \tfrac{1}{8}. \tag{1.12}$$

Then for any $\varepsilon > 0$;

$$\begin{aligned} P\{D_n(\mathscr{A}, \mu) > \varepsilon\} \leqslant\ & 5(2n)^s \exp[-n\varepsilon^2/(91\delta + 4\varepsilon)] \\ & + 7(2n)^s \exp(-\delta n/68) \\ & + 2^{2+s} n^{1+2s} \exp(-\delta n/8) \end{aligned} \tag{1.13}$$

provided

$$n \geqslant \max(12\delta/\varepsilon^2, [68(1+s)\log 2]/\delta).$$

For a proof of Lemma 2, the reader is referred to Zhao [13].

Denote by $\|\cdot\|$ the Euclidean norm on R^k. Write

$$B(x, \rho) = \{y : \|y - x\| < \rho\}, \qquad x \in R^k, \rho > 0,$$

$$\bar{B}(x, \rho) = \{y : \|y - x\| \leq \rho\}, \qquad x \in R^k, \rho > 0.$$

We have the following lemma.

LEMMA 3. *If $B(k)$ denotes the set of all open balls $B(x, \rho)$ and $\bar{B}(k)$ denotes the set of all closed balls $\bar{B}(x, \rho)$, then $B(k)$ and $\bar{B}(k)$ both belong to the V-C class with the same index $s = k + 2$ for all $k = 1, 2, \ldots$.*

For a proof see Wenoeur and Dudley [12] (1981).

LEMMA 4 (A multinomial distribution inequality). *Let $n_1, \ldots, n_m$ be the frequencies in m classes of a multinomial distribution in $n = n_1 + \cdots + n_m$ independent trials. Then for all $\varepsilon \in (0, 1)$ and all m such that $(m/n) \leq \varepsilon^2/20$, we have*

$$P\left(\sum_{i=1}^{m} |n_i - En_i| > n\varepsilon\right) \leq 3 \exp(-n\varepsilon^2/25). \tag{1.14}$$

For a proof, see Devroye [3].

2. POINTWISE STRONG CONSISTENCY

We prove the following theorem on pointwise strong consistency of $\hat{f}_n(x)$ as defined in (1.3) as an estimator of $f(x)$.

THEOREM 1. *Let $K(\cdot)$ and $\{h_n\}$ satisfy the following conditions:*

(a) *K is bounded on R_+,*

(b) $0 < \int_0^\infty K(b)\, v^{(k-3)/2}\, dv < \infty$,

(c_1) $\lim_{v \to \infty} v^{(k-1)/2} K(v) = 0$ *or*

(c_2) *f is bounded on Ω,*

(d) $\lim_{n \to \infty} h_n = 0$, *and*

(e) $\lim_{n \to \infty} (nh_n^{k-1}/\log n) = \infty$.

Then at any continuity point x of f,

$$\lim_{n \to \infty} f_n(x) = f(x) \qquad a.s. \tag{2.1}$$

We need the following lemma to prove Theorem 1. For convenience of notation we write h for h_n throughout the paper except in the statements of theorems.

LEMMA 5. *Suppose that the conditions* (a) – (d) *of Theorem* 1 *hold. Then at any continuity point* x *of* f

$$|Ef_n(x) - f(x)| \to 0 \qquad as \quad n \to \infty. \tag{2.2}$$

Further, if f *is continuous on* Ω, *then*

$$\lim_{n \to \infty} \operatorname{Sup}_{x} |Ef_n(x) - f(x)| = 0. \tag{2.3}$$

Proof. Using (1.5),

$$\begin{aligned}
|Ef_n(x) - f(x)| &= C(h)\, h^{1-k} \left| \int_\Omega K[(1 - x'y)/h^2][f(y) - f(x)]\, d\omega(y) \right| \\
&\leqslant C(h)\, h^{1-k} \int_{1 - x'y \leqslant \delta} K[(1 - x'y)/h^2]\, |f(y) - f(x)|\, d\omega(y) \\
&\quad + C(h)\, h^{1-k} f(x) \int_{1 - x'y > \delta} K[(1 - x'y)/h^2]\, d\omega(y) \\
&\quad + C(h)\, h^{1-k} \int_{1 - x'y > \delta} K[1 - x'y)/h^2]\, f(y)\, d\omega(y) \\
&= I_1 + I_2 + I_3 \qquad \text{(say)}.
\end{aligned} \tag{2.4}$$

By continuity of f at x, we can find $\delta > 0$ for any given $\varepsilon > 0$ such that $|f(y) - f(x)| < \varepsilon$ for $1 - x'y \leqslant \delta$. Thus, by (1.5),

$$I_1 \leqslant \varepsilon C(h)\, h^{1-k} \int_\Omega K[1 - x'y)/h^2]\, d\omega(y) = \varepsilon. \tag{2.5}$$

Now, let condition (c_1) of Theorem 1 hold. Then

$$I_3 \leqslant C(h)\, \delta^{(1-k)/2} \sup_{v > \delta/h^2} K(v)\, v^{(k-1)/2} \int_\Omega f(y)\, d\omega(y) \to 0 \qquad \text{as} \quad n \to \infty \tag{2.6}$$

by (1.2), (1.7), and conditions (c_1) and (d) of Theorem 1. Further, we have

$$\begin{aligned}
I_2 &= \frac{2C(h)\, \pi^{(k-1)/2}}{\Gamma[(k-1)/2]} f(x) \int_{\delta/h^2}^{2/h^2} K(v)[v(2 - h^2 v)]^{(k-3)/2}\, dv \\
&\leqslant \frac{(2\pi)^{(k-1)/2} C(h)}{\Gamma[(k-1)/2]} f(x) \int_{\delta/h^2}^{\infty} K(v)\, v^{(k-3)/2}\, dv \to 0 \qquad \text{as} \quad n \to \infty
\end{aligned} \tag{2.7}$$

by (1.4) and (1.7). Equations (2.5)–(2.7) imply (2.2). The results (2.2) when (c_2) is true and (2.3) can be proved in a similar manner.

Proof of Theorem 1. By Lemma 5, we have

$$\lim_{n \to \infty} Ef_n(x) = f(x); \tag{2.8}$$

also we shall prove that

$$\lim_{n \to \infty} [f_n(x) - Ef_n(x)] = 0, \qquad \text{a.s.}, \tag{2.9}$$

so that (2.8) and (2.9) imply that $f_n(x) \to f(x)$ a.s., which is the desired result.

Put

$$\xi_i = h^{1-k} C(h)(K[(1 - x'X_i)/h^2] - EK[(1 - x'X_i)/h^2]).$$

Then $\xi_1, ..., \xi_n$ are i.i.d. and

$$E(\xi_1) = 0, |\xi_1| \leqslant 2h^{1-k} C(h) M$$

$$E(\xi_1^2) \leqslant h^{2(1-k)} C^2(h) \int_\Omega K^2 [(1 - x'y)/h^2] f(y)\, d\omega(y)$$

$$\leqslant Mh^{2(1-k)} C^2(h) \int_\Omega K[(1 - x'y)/h^2] f(y)\, d\omega(y), \tag{2.10}$$

where M is an upper bound of K on R_+. By (1.7) and Lemma 5, there exist constants $a > 0$ and $a(x) > 0$ such that

$$|\xi_1| \leqslant ah^{1-k}, \qquad E\xi_1^2 \leqslant a(x)\, h^{1-k}.$$

By Lemma 1,

$$\begin{aligned} P[|f_n(x) - f(x)| \geqslant \varepsilon] &= P\left[n^{-1} \left|\sum_1^n \xi_i\right| \geqslant \varepsilon\right] \\ &\leqslant 2 \exp[-n\varepsilon^2/(a\varepsilon h^{1-k} + 2a(x)\, h^{1-k})] \\ &= 2 \exp[-nh^{k-1}\varepsilon^2/(2a(x) + a\varepsilon)]. \end{aligned}$$

By condition (e) of Theorem 1,

$$\sum_n P[|f_n(x) - Ef_n(x)| \geqslant \varepsilon] < \infty \Rightarrow f_n(x) - Ef_n(x) = 0 \qquad \text{a.s.},$$

i.e., (2.9) holds, which together with (2.8) implies (2.1), the result of Theorem 1.

3. UNIFORM STRONG CONSISTENCY

In the following we assume that μ is a measure on Ω with $f(x)$ as the pdf and μ_n is the empirical measure based on the sample $X_1, ..., X_n$. We have the following theorem which is parallel to that for the standard case given by Bertrand–Retali [4]:

THEOREM 2. *Suppose that f is continuous on Ω and K is bounded on R_+ and Riemann integrable on any finite interval in R_+ with*

$$\int_0^{\infty} \operatorname{Sup}\{K(u): |\sqrt{u}-\sqrt{v}|<1\}\, v^{(k-3)/2}\, dv < \infty. \tag{3.1}$$

If $h_n \to 0$ and

$$(nh_n^{k-1}/\log n) \to \infty \tag{3.2}$$

as $n \to \infty$, then

$$\sup_x |f_n(x) - f(x)| \to 0 \qquad a.s. \tag{3.3}$$

Proof. The proof of Theorem 2 is similar to that of Theorem 1 in Devroye and Wagner [5]. Here we give only a sketch of the proof. By Lemma 3 of Devroye and Wagner [5], for each η, δ small and ρ large we can find a function

$$K^*(v) = \sum_i^{N_0} \alpha_i I_{A_i}(v),$$

where I_{A_i} is the indicator function:

(i) $\alpha_1, ..., \alpha_{N_0}$ are non-negative numbers,

(ii) $A_1, ..., A_{N_0}$ are disjoint intervals contained in $[0, \rho]$,

(iii) $|K^*(v) - K(v)| < \eta$ on $[0, \rho]$ except on a set D,

(iv) $D \subseteq B = \bigcup_1^{N^*} B_i$, where $B_1, ..., B_{N^*}$ are intervals in $[0, \rho]$ whose union has Lebesgue measure less than δ, and

(v) $\max_{1 \leqslant i \leqslant N_0} \alpha_i \leqslant \sup_v K(v) = M$ (say).

We note that continuity of f on Ω implies that it is uniformly continuous and $f(x) \leqslant M_f$ (some constant) on Ω. By Lemma 5,

$$\sup_x |Ef_n(x) - f(x)| \to 0 \qquad \text{a.s. as} \quad n \to \infty. \tag{3.4}$$

Putting

$$U_{1n}(x)=h^{1-k}C(h)\int_{\Omega}|K[(1-x'y)/h^2]-K^*[(1-x'y)/h^2]|\,f(y)\,d\omega(y),$$

$$U_{2n}(x)=h^{1-k}C(h)\left|\int_{\Omega}K^*[(1-x'y)/h^2]\,d[\mu_n(y)-\mu(y)]\right|,$$

$$U_{3n}(x)=h^{1-k}C(h)\int_{\Omega}|K^*[(1-x'y)/h^2]-K[(1-x'y)/h^2]|\,d\mu_n(y),$$

we have

$$\sup_x |f_n(x)-Ef_n(x)|\leqslant \sum_{i=1}^{3}\sup_x U_{in}(x). \tag{3.5}$$

Following Devoye and Wagner [5], we can prove that

$$\operatorname{Sup}_x U_{1n}(x) \qquad \text{and} \qquad \operatorname{Sup}_x U_{3n}(x) \tag{3.6}$$

can be made arbitrarily small by choosing η, δ small and ρ large enough. Let

$$A_i^*(x)=\{y\in\Omega: [(1-x'y)/h^2]\in A_i\}.$$

Then

$$\begin{aligned}\mu(A_i^*(x))&=\int_{A_i^*(x)}f(y)\,d\omega(y)\leqslant M_f\frac{(2\pi h^2)^{(k-1)/2}}{\Gamma[(k-1)/2]}\int_0^{\rho}v^{(k-3)/2}\,dv\\&=ch^{k-1}\qquad\text{(say).}\end{aligned} \tag{3.7}$$

Hereafter, c denotes a positive constant but may take different values at different appearances, even in the same expression.

If we choose $A_i=[a_i, b_i)$, $i=1, ..., N_0$, then

$$A_i^*(x)=\{y\in\Omega: \sqrt{2a_i}h\leqslant\|y-x\|<\sqrt{2b_i}h\}.$$

Writing

$$\mathscr{A}=\{A_i^*(x): x\in\Omega, i=1, ..., N_0\}$$

we have by Lemma 3,

$$m^{\mathscr{A}}(n)\leqslant 2(n^{k+2}+1)^2 \qquad \text{for any } n. \tag{3.8}$$

Hence $\mathscr{A}$ is a V-C class with some index s as defined in the text following (1.9). Then using Lemma 2 quoted in Section 1, we have

$$P\{\sup_x U_{2n}(x) \geqslant \varepsilon\} \leqslant cn^{1+2s} \exp(-cnh^{k-1}), \tag{3.9}$$

By (3.2), we have

$$\sum_n P[\sup_x U_{2n}(x) \geqslant \varepsilon] < \infty \qquad \text{for any} \quad \varepsilon > 0.$$

Then, by the Borel–Cantelli lemma,

$$\sup_x U_{2n}(x) \to 0 \qquad \text{a.s.} \tag{3.10}$$

(3.5), (3.6), and (3.10) complete the proof of Theorem 2.

4. Strong L_1-Norm Consistency

In the following, we establish under some conditions the strong L_1-norm consistency of $f_n(x)$, i.e.,

$$\int_\Omega |f_n(x) - f(x)| \, d\omega(x) \to 0 \qquad \text{a.s.} \tag{4.1}$$

The precise statement is given in Theorem 3.

Theorem 3. *Suppose that*

$$\text{(a)} \qquad \int_0^\infty v^{(k-3)/2} K(v) \, dv < \infty, \tag{4.2}$$

$$\text{(b)} \qquad h_n \to 0 \textit{ and } nh_n^{k-1} \to \infty \textit{ as } n \to \infty. \tag{4.3}$$

Then, for any given $\varepsilon > 0$, *there exists a constant* $c > 0$ *such that*

$$P\left\{\int_\Omega |f_n(x) - f(x)| \, d\omega(x) \geqslant \varepsilon\right\} \leqslant e^{-cn}. \tag{4.4}$$

Proof. By (1.5),

$$V_n = \int_\Omega |Ef_n(x) - f(x)| \, d\omega(x)$$

$$\leqslant h^{1-k} C(h) \int_\Omega d\omega(x) \int_\Omega K[(1 - x'y)/h^2] \, |f(y) - f(x)| \, d\omega(y).$$

Given $\varepsilon > 0$, we can find a non-negative continuous function $g(x)$ on Ω such that

$$\int_{\Omega} |f(x) - g(x)| \, d\omega(x) < \varepsilon/6. \tag{4.5}$$

Then

$$\begin{aligned} V_n \leqslant & \int_{\Omega} h^{1-k} C(h) \, d\omega(x) \int_{\Omega} K[(1 - x'y)/h^2] \, |f(y) - g(y)| \, d\omega(y) \\ & + \int_{\Omega} h^{1-k} C(h) \, d\omega(x) \int_{\Omega} K[(1 - x'y)/h^2] \, |f(x) - g(x)| \, d\omega(y) \\ & + \int_{\Omega} h^{1-k} C(h) \, d\omega(x) \int_{\Omega} K[(1 - x'y)/h^2] \, |g(y) - g(x)| \, d\omega(y) \\ = & \, J_{1n} + J_{2n} + J_{3n}, \qquad \text{say.} \end{aligned} \tag{4.6}$$

By (1.5) and (4.5),

$$\begin{aligned} J_{1n} &= h^{1-k} C(h) \int_{\Omega} |f(y) - g(y)| \, d\omega(y) \int_{\Omega} K[(1 - x'y)/h^2] \, d\omega(x) \\ &= \int_{\Omega} |f(y) - g(y)| \, d\omega(y) < \varepsilon/6. \end{aligned} \tag{4.7}$$

In the same way,

$$J_{2n} < \varepsilon/6. \tag{4.8}$$

Let us denote $M_g = \sup\{ g(x), \ x \in \Omega \}$ and $\Omega_1(x) = \{ y \in \Omega \colon 1 - x'y > \rho h^2 \}$. As in (3.9), we can take ρ sufficiently large such that

$$\begin{aligned} & \int_{\Omega} h^{1-k} C(h) \, d\omega(x) \int_{\Omega_1(x)} K[(1 - x'y)/h^2] \, |g(y) - g(x)| \, d\omega(y) \\ & \quad \leqslant M_g h^{1-k} C(h) \int_{\Omega} d\omega(x) \int_{\Omega_1(x)} K[(1 - x'y)/h^2] \, d\omega(y) \\ & \quad \leqslant M_g \frac{h^{1-k} C(h) (2\pi)^{(k-1)/2}}{\Gamma[(k-1)/2]} \int_{\Omega} d\omega(x) \int_{\rho}^{\infty} v^{(k-3)/2} K(v) \, dv < \varepsilon/12. \end{aligned} \tag{4.9}$$

By uniform continuity of $g(x)$ on Ω and (1.5), we see that for large n,

$$\begin{aligned} & \int_{\Omega} h^{1-k} C(h) \, d\omega(x) \int_{\Omega - \Omega_1(x)} K[(1 - x'y)/h^2] \, |g(y) - g(x)| \, d\omega(y) \\ & \quad \leqslant \frac{\varepsilon}{12 \omega(\Omega)} \int_{\Omega} d\omega(x) \int_{\Omega - \Omega_1(x)} h^{1-k} C(h) \, K[(1 - x'y)/h^2] \, d\omega(y) < \varepsilon/12. \end{aligned} \tag{4.10}$$

By (4.6)–(4.10), for large n,

$$V_n < \varepsilon/2. \tag{4.11}$$

Take $K^*(v) \geqslant 0$ such that

$$\frac{C(h)(2\pi)^{(k-1)/2}}{\Gamma[(k-1)/2]} \int_0^\infty |K(v) - K^*(v)| \, v^{(k-3)/2} \, dv < \varepsilon/6, \tag{4.12}$$

and put

$$f_n^*(x) = n^{-1} h^{1-k} C(h) \sum_{i=1}^n K^*[(1 - x'X_i)/h^2]. \tag{4.13}$$

As in (1.6) we have

$$\begin{aligned}
&\int_\Omega |f_n(x) - f_n^*(x)| \, d\omega(x) \\
&\quad \leqslant n^{-1} h^{1-k} C(h) \sum_{i=1}^n \int_\Omega |K[(1 - x'X_i)/h^2] - K^*[(1 - x'X_i)/h^2]| \, d\omega(x) \\
&\quad \leqslant \frac{C(h)(2\pi)^{(k-1)/2}}{\Gamma[(k-1)/2]} \int_\Omega |K(v) - K^*(v)| \, v^{(k-3)/2} \, dv < \varepsilon/6,
\end{aligned} \tag{4.14}$$

and

$$\int_\Omega |Ef_n(x) - Ef_n^*(x)| \, d\omega(x) < \varepsilon/6. \tag{4.15}$$

We can take

$$K^*(v) = \sum_{j=1}^N \alpha_j I_{A_j}(v),$$

where $A_1, ..., A_N$ are disjoint finite intervals on R_+. By (4.11), (4.14), and (4.15), in order that (4.4) holds, it is enough ro prove that for any $\varepsilon_1 > 0$, there exists a positive constant c such that

$$P\left\{\int_\Omega |f_n^*(x) - Ef_n^*(x)| \, d\omega(x) \geqslant \varepsilon_1\right\} \leqslant e^{-cn}. \tag{4.16}$$

Here we can take $K^*(v) = I_{[a,b)}(v)$.

For $x=(x_1, ..., x_k)' \in \Omega$, we can represent x in polar coordinates

$$\begin{aligned} x_1 &= \cos\theta_1 \\ x_2 &= \sin\theta_1 \cos\theta_2 \\ &\cdots \\ x_{k-1} &= \sin\theta_1 \cdots \sin\theta_{k-2} \cos\theta_{k-1} \\ x_k &= \sin\theta_1 \cdots \sin\theta_{k-1} \end{aligned} \tag{4.17}$$

with $0 \leqslant \theta_i \leqslant \pi$, $i=1, ..., k-2$ and $0 \leqslant \theta_{k-1} \leqslant 2\pi$. Such a representation is unique except for a Lebesgue null set $H \subset \Omega$. Take $L>0$, and put

$$\begin{aligned} J_{i_j}^{(j)} &= \{x = x(\theta_1, ..., \theta_{k-1}) \in \Omega - H: L^{-1}h(i_j - 1) \leqslant \theta_j < L^{-1}hi_j\}, \\ &\quad i_j = 1, 2, ..., u-1 = [h^{-1}L\pi], \qquad j = 1, ..., k-2, \\ &\quad i_{k-1} = 1, 2, ..., v-1 = [h^{-1}2L\pi], \\ J_u^{(j)} &= \{x = x(\theta_1, ..., \theta_{k-1}) \in \Omega - H: (u-1)\, L^{-1}h \leqslant \theta_j \leqslant \pi\}, \\ &\quad j = 1, ..., k-2, \\ J_v^{(k-1)} &= \{x = x(\theta_1, ..., \theta_{k-1}) \in \Omega - H: (v-1)\, L^{-1}h \leqslant \theta_{k-1} \leqslant 2\pi\} \end{aligned}$$

and

$$J_{i_1 \cdots i_{k-1}} = \bigcap_{j=1}^{k-1} J_{i_j}^{(j)}, \qquad i_1, ..., i_{k-2} = 1, 2, ..., u;\ i_{k-1} = 1, 2, ..., v.$$

All these $J_{i_1 \cdots i_{k-1}}$ constitute a pertition Ψ of $\Omega - H$.

Take c and L such that

$$c > \max\{\sqrt{2b}\, k^{3/2}, \sqrt{2a}\, k^{3/2} + (2L)^{-1}k^3\} \qquad \text{and} \qquad 2L^{-1}c < b - a.$$

Put

$$\begin{aligned} A &= [a, b), \qquad B = [a + L^{-1}c, b - L^{-1}c] \\ A^*(x) &= \{y \in \Omega - H: a \leqslant (1 - x'y)/h^2 < b\}, \qquad x \in \Omega - H, \\ B^*(x) &= \{y \in \Omega - H: a + cL^{-1} \leqslant (1 - x'y)/h^2 < b - L^{-1}c\}, \qquad x \in \Omega - H, \\ D(x) &= \bigcup_{J \in \Psi, J \subset A^*(x)} J, \qquad x \in \Omega - H. \end{aligned}$$

Now we proceed to prove that for $x \in \Omega - H$,

$$G(x) = A^*(x) - D(x) \subset A^*(x) - B^*(x) = G^*(x). \tag{4.18}$$

Assume that $y = y(\theta_1', \ldots, \theta_{k-1}') \in G(x)$. Then $y \notin D(x)$, and there exists a set $J_{i_1 \cdots i_{k-1}}$ and a point $\omega = \omega(\theta_1'', \ldots, \theta_{k-1}'') \in J_{i_1 \cdots i_{k-1}}$ such that

$$y \in J_{i_1 \cdots i_{k-1}}, \qquad \omega \in J_{i_1 \cdots i_{k-1}} \qquad \text{but} \quad \omega \notin A^*(x). \tag{4.19}$$

Thus $|\theta_j' - \theta_j''| < L^{-1}h$, and by (4.17), $|y_j - \omega_j| < jhL^{-1}$, where y_j and ω_j are the components of y and ω, respectively. Hence

$$\|y - \omega\| < k^{3/2}hL^{-1}. \tag{4.20}$$

But $\omega \notin A^*(x) \Rightarrow \|x - \omega\| \geqslant \sqrt{2b}\,h$ or $\|x - \omega\| < \sqrt{2a}\,h$, which in turn implies that

$$\|x - y\| > (\sqrt{2b} - k^{3/2}L^{-1})h \qquad \text{or} \qquad \|x - y\| < (\sqrt{2a} + k^{3/2}L^{-1})h$$

i.e.,

$$1 - x'y > (b - cL^{-1})h^2 \qquad \text{or} \qquad 1 - x'y < (a + cL^{-1})h^2. \tag{4.21}$$

Thus $y \in A^*(x) - B^*(x)$, and (4.18) is proved.

Since $K^*(v) = I_A(v)$, we have

$$\begin{aligned}
&\int_\Omega |f_n^*(x) - Ef_n^*(x)|\, d\omega(x) \\
&\quad = h^{1-k}C(h) \int_\Omega |\mu_n(A^*(x)) - \mu(A^*(x)|\, d\omega(x) \\
&\quad \leqslant h^{1-k}C(h) \int_\Omega \sum_{J \in \Psi, J \subset A^*(x)} |\mu_n(J) - \mu(J)|\, d\omega(x) \\
&\qquad + h^{1-k}C(h) \int_\Omega [\mu(G^*(x)) + \mu_n(G^*(x))]\, d\omega(x) \\
&\quad = Z_{1n} + Z_{2n} \qquad \text{(say).}
\end{aligned} \tag{4.22}$$

For any probability measure v on Ω, we have

$$\begin{aligned}
&h^{1-k}C(h) \int_\Omega v[G^*(x)]\, d\omega(x) \\
&\quad = \int_\Omega dv(y) \int_\Omega h^{1-k}C(h)\, I_{A-B}[(1 - x'y)/h^2]\, d\omega(x) \\
&\quad \leqslant \frac{C(h)(2\pi)^{(k-1)/2}}{\Gamma[(k-1)/2]} \int_{v \in A-B} v^{(k-3)/2}\, dv < \varepsilon_1/3
\end{aligned} \tag{4.23}$$

by taking L sufficiently large. Thus

$$Z_{2n} < 2\varepsilon_1/3. \tag{4.24}$$

If $J \in \Psi$, $y \in J \subset A^*(x)$, $x \in \Omega - H$, then $(1 - x'y) < bh^4$. Hence

$$\omega\{x \in \Omega - H : J \subset A^*(x)\} \leqslant \int_\Omega I_{[0,b)}[(1 - x'y)/h^2]\, d\omega(x) \leqslant ch^{k-1}, \tag{4.25}$$

where c is a positive constant. Thus by (4.22) and (4.25), we have

$$Z_{1n} \leqslant cC(h) \sum_{J \in \Psi} |\mu_n(J) - \mu(J)| \leqslant c \sum_{J \in \Psi} |\mu_n(J) - \mu(J)|. \tag{4.26}$$

Since $\#(\Psi) \leqslant ch^{1-k} = o(n)$ by (4.3), Lemma 4 can be used. Thus by (4.22), (4.24), and (4.26), we have

$$P\left\{\int_\Omega |f_n^*(x) - Ef_n^*(x)|\, d\omega(x) \geqslant \varepsilon_1\right\} \leqslant P\left\{\sum_{J \in \Psi} |\mu_n(J) - \mu(J)| \geqslant (\varepsilon_1/3c)\right\} < e^{-cn},$$

where $c > 0$ is a positive constant, which proves (4.16) the desired result.

References

[1] Batschelet, E. (1971). Recent statistical methods for orientation. *Amer. Inst. Biol. Sci. Bull.*

[2] Batschelet, E. (1981). *Circular Statistics in Biology*. Academic Press, London.

[3] Devroye, L. (1983). The equivalence of weak, strong and complete convergence in L_1 for kernel density estimates. *Ann. Statist.* **11** 896–904.

[4] Bertrand–Retali, R. (1978). Convergence uniforme d'une estimateur de la densité par la méthod du Noyau, *Rev. Roumaine Math. Pures Appl.* **23** 361–385.

[5] Devroye, L. P., and Wagner, T. J. (1980). The strong uniform consistency of kernel density estimates. In *J. Multivariate Anal.* Vol. 5 (P. R. Krishnaiah, Ed.), pp. 59–77, North-Holland, Amsterdam.

[6] Hoeffding, W. (1963). Probability inequalities for sums of bounded random variables. *J. Amer. Statist. Assoc.* **58** 13–30.

[7] Mardia, K. V. (1972). *Statistics of Directional Data*, Academic Press, New York.

[8] PUKKILA, T., AND RAO, C. R. (1986). *Pattern Recognition Based on Scale Invariant Discriminant Functions.* Technical Report 86-09. Center for Multivariate Analysis, University of Pittsburgh.

[9] RAO, J. S. (1984). Nonparametric methods in directional data analysis. In *Handbook of Statistics*, Vol. 4, pp. 757–770. Elsevier, Amsterdam/New York.

[10] VAPNIK, V. N., AND CHERVONENKIS, A. YA. (1971). On the uniform convergence of relative frequencies of events to their probabilities. *Theory Probab. Appl.* **16** 264–280 (English).

[11] WATSON, G. S. (1983). *Statistics on Spheres*, Wiley, New York.

[12] WENOCUR, R. S., AND DUDLEY, R. M. (1981). Some special Vapnik–Chervonenkis classes, *Discrete Math.* **33** 313318.

[13] ZHAO, L. C. (1985). *An Inequality Concerning the Deviation between Theoretical and Empirical Distributions.* Technical Report 85-30, Center for Multivariate Analysis, University of Pittsburgh.

On Determination of the Order of an Autoregressive Model

Z. D. BAI, K. SUBRAMANYAM, AND L. C. ZHAO*

University of Pittsburgh

To determine the order of an autoregressive model, a new method based on information theoretic criterion is proposed. This method is shown to be strongly consistent and the convergence rate of the probability of wrong determination is established.

1. INTRODUCTION

Consider an autoregressive (AR) model of order p ($p \geqslant 1$, unknown) generated by a purely random process $e(n)$ given by

$$\sum_{j=0}^{p} \alpha(j)\, X(n-j) = e(n), \qquad \alpha(0) = 1. \tag{1.1}$$

Assume that $\{e(n)\}$ is a sequence of i.i.d. random variables with $Ee(1) = 0$, $Ee^2(1) = \sigma^2$ and $0 < \operatorname{Var}(e^2(1)) < \infty$. Suppose the coefficients in the model $\alpha(0), \alpha(1), ..., \alpha(p)$ satisfy

$$g(z) = \sum_{j=0}^{p} \alpha(j) z^j \neq 0 \qquad \text{for} \quad |z| \leqslant 1. \tag{1.2}$$

In time series analysis, AR models play an important role. An interesting problem in the analysis of AR models is the determination of the order p of the model. There is a considerable amount of research work done on this topic. To name a few, the reader is referred to Akaike [1], Hannan [3], Hannan and Quinn [4], and Shibata [6].

* Present address: Dept. of Mathematics, University of Science & Technology of China, Hefei, Anhui, Peoples Republic of China.

Multivariate Statistics and Probability
ISBN 0-12-580205-6

Reprinted from *J. Mult. Anal.* **27**(1).

Let $X(1), X(2), ..., X(N)$ denote a random sample drawn from an AR model of order p. Assume that the order p is known a priori to be $p \leqslant K < \infty$. Using Yule–Walker equations and a recursive computing procedure, Hannan and Quinn [4] obtained an estimate $\hat{\sigma}_p^2$ of σ^2. To estimate p, the following criterion based on $\hat{\sigma}_p^2$ is proposed,

$$\psi(p) = \log \hat{\sigma}_p^2 + 2p\, CN^{-1} \log\log N, \tag{1.3}$$

where $C > 1$ is a constant. An estimate $\hat{p}$ of p is chosen as that one which minimises $\psi(p)$. Under weaker conditions than mentioned above, strong consistency of $\hat{p}$ is obtained.

In this article a new criterion to estimate the order of the AR model is proposed. Strong consistency as well as the convergence rate of the estimate $\hat{p}$ is established.

The paper is organized as follows. In Section 2, a new method to determine the order AR model is described. In Section 3, convergence rates of $P\{\hat{p} \neq p\}$ is derived. Some general remarks, including the strong consistency of $\hat{p}$, are made in Section 4.

2. Determination of the Order p

Let $X(1), X(2), ..., X(N)$ be a random sample from an AR series. Define

$$L_p(\boldsymbol{\alpha}_p) = \sum_{n=p+1}^{N} \left(X(n) + \sum_{i=1}^{p} \alpha(i)\, X(n-i) \right)^2, \tag{2.1}$$

where $\boldsymbol{\alpha}_p = (\alpha(1), ..., \alpha(p))'$. The true order p of the model and the true regression coefficients $\alpha(1), ..., \alpha(p)$ will be denoted as $p_0, \alpha_0(1), ..., \alpha_0(p_0)$, respectively.

For each $p \leqslant K$ choose $\hat{\boldsymbol{\alpha}}_p = (\hat{\alpha}(1), ..., \hat{\alpha}(p))'$ such that

$$L_p(\hat{\boldsymbol{\alpha}}_p) = \min_{\boldsymbol{\alpha}_p} L_p(\hat{\boldsymbol{\sigma}}_p^2) \triangleq N\hat{\sigma}_p^2. \tag{2.2}$$

Since L_p is a quadratic form of $\boldsymbol{\alpha}_p$, it is easy to compute $\hat{\boldsymbol{\alpha}}_p$ and $L_p(\hat{\alpha}_p)$. Define

$$\phi(p) = N \log \left[\frac{1}{N} L_p(\hat{\boldsymbol{\alpha}}_p) \right] + pC_N, \tag{2.3}$$

where constants C_N will be chosen suitably. Then any $\hat{p}$ minimizing

$$\phi(\hat{p}) = \min_{p \leqslant K} \phi(p) \tag{2.4}$$

will be taken as the estimate of the order p of the AR series.

Remark 2.1. In fact, $(1/N)\, L_p(\hat{\boldsymbol{\alpha}}_p)$ is an estimate of σ^2, which is slightly different from that used by Hannan and Quinn [4]. When N is not very large, $(1/(N-p))\, L_p(\hat{\alpha}_p)$ is a better estimate of σ^2 as compared to $(1/N)\, L_p(\hat{\alpha}_p)$. Since we are interested in the large sample properties, there is no harm in using $(1/N)\, L_p(\hat{\alpha}_p)$ as an estimate of σ^2.

Define

$$\begin{aligned} \hat{q}_p(i, j) &= \frac{1}{N} \sum_{n=p+1}^{N} X_{n-i} X_{n-j}, \quad i, j = 0, 1, 2, ..., p. \\ \hat{Q}_p &= (\hat{q}_p(i, j))_{i, j = 1, 2, ..., p} \\ \hat{\beta}_p &= (\hat{q}_p(0, 1), ..., \hat{q}_p(0, p))'. \end{aligned} \tag{2.5}$$

By differentiating $L_p(\hat{\boldsymbol{\alpha}}_p)$, we get

$$\hat{Q}_p\, \hat{\boldsymbol{\alpha}}_p = -\hat{\boldsymbol{\beta}}_p$$

or, equivalently,

$$\hat{\boldsymbol{\alpha}}_p = -\hat{Q}_p^{-1}\, \hat{\boldsymbol{\beta}}_p \tag{2.6}$$

provided $\hat{Q}_p$ is nonsingular. In the proof of our main result, it is shown that with probability one, for large N, $\hat{Q}_p$ is nonsingular. Hence we can use (2.6).

Using the above notation, the main theorems are stated below. Proofs are given in the next section.

THEOREM 2.1. *Suppose*

$$E \exp\{te(1)^2\} < \infty \qquad \textit{for some} \quad t > 0, \tag{2.7}$$

and choose C_N such that

$$C_N/N \to 0, \qquad C_N \to \infty. \tag{2.8}$$

Then

$$P(\hat{p} \neq p_0) \leqslant C_1 \exp\{-C_2 C_N\}, \tag{2.9}$$

where C_1, C_2 are two positive constants independent of N.

THEOREM 2.2. *Suppose* (2.8) *holds and*

$$E\, |e(1)|^{2t} < \infty, \qquad \textit{for some} \quad t \geqslant 2. \tag{2.10}$$

Then

$$P(\hat{p} \neq p_0) \leqslant C_1/(N^{t/2-1} C_N^{t/2}) + C_2 e^{-C_3 C_N}, \tag{2.11}$$

where C_1, C_2, C_3 are positive constants independent of N.

3. Proof of the Theorems

Lemma 3.1. *Let $y_1, \ldots, y_n$ be independent random variables with $Ey_i = 0$ and $E|y_i|^t < \infty$, $i = 1, \ldots, n$, for some $t \geqslant 2$. Denote*

$$S_n = \sum_{i=1}^{n} y_i, \quad B_n^2 = \sum_{i=1}^{n} \operatorname{Var}(y_i), \quad A_{t,n} = \sum_{i=1}^{n} E|y_i|^t.$$

Then for any $a > 0$,

$$P\{S_n \geqslant a\} \leqslant C_t^{(1)} A_{t,n} a^{-t} + \exp\{-C_t^{(2)} a^2 / B_n^2\},$$

where

$$C_t^{(1)} = (1 + 2/t)^t \qquad \text{and} \qquad C_t^{(2)} = 2(t+2)^{-2} e^{-t}.$$

Proof. Refer to Corollary 4 of Fuk and Nagaev [2].

Let $\boldsymbol{\alpha}_{p_0} = (\alpha_0(1), \ldots, \alpha_0(p_0))'$ and σ^2 be the true parameters of the model. Let

$$\begin{aligned} &\gamma(i-j) = E(X(n-i)\,X(n-j)), \\ &\Gamma_p = ((r(i-j)))_{i,j=1,\ldots,p,} \boldsymbol{\gamma}_p = (\gamma(1), \ldots, \gamma(p)),\ p \leqslant K. \end{aligned} \tag{3.1}$$

Suppose $p \geqslant p_0$, then from

$$\sum_{i=0}^{p_0} \alpha_0(i)\,X(n-i) = e(n),$$

it follows that

$$\sum_{i=0}^{p_0} \alpha_0(i)\,\gamma(i-j) = \delta_{0,j}\sigma^2, \qquad j = 0, 1, 2, \ldots, p, \tag{3.2}$$

where $\delta_{i,j}$ is Kronecker's delta. Thus, if we take $\boldsymbol{\alpha}_p^* = (\alpha_0(1), \ldots, \alpha_0(p_0), 0, \ldots, 0)'$, then $\boldsymbol{\alpha}_p = \boldsymbol{\alpha}_p^*$ is a unique solution of the equation

$$\Gamma_p \boldsymbol{\alpha}_p = -\boldsymbol{\gamma}_p. \tag{3.3}$$

It is well known that, under the conditions (1.1) and (1.2), for $0 \leqslant p \leqslant K$,

$$\lim_{N \to \infty} \hat{Q}_p = \Gamma_p \text{ a.s.}, \qquad \lim_{N \to \infty} \hat{\boldsymbol{\beta}}_p = \boldsymbol{\gamma}_p \text{ a.s.} \tag{3.4}$$

and

$$\lim_{N \to \infty} \hat{\boldsymbol{\alpha}}_p \overset{\text{a.s.}}{=} -\Gamma_p^{-1} \boldsymbol{\gamma}_p \triangleq \boldsymbol{\alpha}_p^* = (\alpha^*(1), \ldots, \alpha^*(p))'. \tag{3.5}$$

Note that for $p_0 \leqslant p \leqslant K$,

$$\boldsymbol{\alpha}_p^* = (\alpha_0(1), ..., \alpha_0(p_0), 0, ..., 0)', \tag{3.6}$$

and that

$$\lim_{N \to \infty} \hat{\sigma}_p^2 \stackrel{\text{a.s.}}{=} \gamma(0) - \boldsymbol{\alpha}_p^{*\prime} \Gamma_p \boldsymbol{\alpha}_p^* = \begin{cases} \sigma^{*2} > \sigma^2, & \text{if } p < p_0, \\ \gamma(0) - \boldsymbol{\alpha}_{p_0}' \Gamma_{p_0} \boldsymbol{\alpha}_{p_0} = \sigma^2, & \text{if } p \geqslant p_0. \end{cases} \tag{3.7}$$

It is easily seen that,

$$\begin{aligned} \hat{\sigma}_p^2 &= \min_{\boldsymbol{\alpha}_p} \frac{1}{N} \sum_{n=p+1}^{N} \left(X(n) + \sum_{i=1}^{p} \alpha(i)\, X(n-i) \right)^2 \\ &\geqslant \min_{\boldsymbol{\alpha}_{p+1}: \alpha(p+1)=0} \frac{1}{N} \sum_{n=p+2}^{N} \left(X(n) + \sum_{i=1}^{p+1} \alpha(i)\, X(n-i) \right)^2 \\ &\geqslant \min_{\boldsymbol{\alpha}_{p+1}} \frac{1}{N} \sum_{n=p+2}^{N} \left(X(n) + \sum_{i=1}^{p+1} \alpha(i)\, X(n-i) \right)^2 = \hat{\sigma}_{p+1}^2. \end{aligned} \tag{3.8}$$

First we establish the following proposition which will be used to prove our main theorems.

PROPOSITION 3.1. *Under conditions* (1.1), (1.2), *and* (2.8), *there exists a constant* $\varepsilon > 0$ *such that for large* N,

$$P\{\hat{p} \neq p_0\} \leqslant P_1 + P_2 + P_3 + P_4,$$

where

$$\begin{aligned} P_1 &= \sum_{i,j=0}^{K} P\{|\hat{q}_K(i, j) - \gamma(i, j)| > \varepsilon \sqrt{C_N/N}\} \\ P_2 &= \sum_{i=1}^{K} P\left\{ \left| \frac{1}{N} \sum_{n=K+1}^{N} e(n)\, X(n-i) \right| > \varepsilon \sqrt{C_N/N} \right\} \\ P_3 &= 2KP\{|e(0)| > \varepsilon \sqrt{C_N}\} \end{aligned} \tag{3.9}$$

and

$$P_4 = 2KP\{|X(0)| > \varepsilon \sqrt{C_N}\}.$$

Proof. Denote

$$\begin{aligned} A_1(\varepsilon) &= \{|\hat{q}_K(i, j) - \gamma(i-j)| \leqslant \varepsilon \sqrt{C_N/N} \quad \text{for all} \quad i, j \leqslant K\} \\ A_2(\varepsilon) &= \left\{ \left| \frac{1}{N} \sum_{n=K+1}^{N} e(n)\, X(n-i) \right| \leqslant \varepsilon \sqrt{C_N/N} \quad \text{for all} \quad 1 \leqslant i \leqslant K \right\} \\ A_3(\varepsilon) &= \{|e(n)| \leqslant \varepsilon \sqrt{C_N} \quad \text{for all} \quad n \leqslant 2K\} \\ A_4(\varepsilon) &= \{|X(n)| \leqslant \varepsilon \sqrt{C_N} \quad \text{for all} \quad n \leqslant 2K\}. \end{aligned}$$

For $p < p_0$, since $\hat{\sigma}_p^2$, as a function of $\hat{q}_K(i, j)$'s and $X(n)\,X(n-l)$, is continuously differentiable, we have

$$\begin{aligned}\hat{\sigma}_p^2/\hat{\sigma}_{p_0}^2 &\geq \hat{\sigma}_{p_0-1}^2/\hat{\sigma}_{p_0}^2\\ &= \frac{1}{N}\sum_{i=p_0}^{N}\left(X(n)+\sum_{i=1}^{p_0-1}\hat{\alpha}_{p_0-1}(i)\,X(n-i)\right)^2\Big/\hat{\sigma}_{p_0}^2\\ &= \{\hat{q}_{p_0-1}(0,0)-\hat{\boldsymbol{\alpha}}'_{p_0-1}\hat{Q}_{p_0-1}\hat{\boldsymbol{\alpha}}_{p_0-1}\}/\hat{\sigma}_{p_0}^2\\ &\geq \{\gamma(0)-\boldsymbol{\alpha}^{*\prime}_{p_0-1}\Gamma_{p_0-1}\boldsymbol{\alpha}^*_{p_0-1}\}/\sigma^2\\ &\quad - C\left\{\sum_{i,j=0}^{K}|\hat{q}_K(i,j)-\gamma(i,j)|+\frac{1}{N}\sum_{n=1}^{2K}X^2(n)\right\}. \qquad (3.10)\end{aligned}$$

Hereafter, C denotes a constant independent of N, but may take a different value at each appearance even in the same expression.

From (3.7), noting (3.10), there exists $\varepsilon > 0$ such that if $A_1(\varepsilon)\cap A_4(\varepsilon)$ holds then for any $p < p_0$ and large N,

$$\begin{aligned}\log(\hat{\sigma}_p^2/\hat{\sigma}_{p_0}^2) &\geq \log(\hat{\sigma}_{p_0-1}^2/\sigma_{p_0}^2)\\ &\geq \log(\sigma^{*2}/\sigma^2) - C\varepsilon\sqrt{C_N/N} > (p_0-p)\,C_N/N. \qquad (3.11)\end{aligned}$$

Now assume that $p_0 < p \leq K$. Put $\Delta\hat{\alpha}_p(i) = \hat{\alpha}_p(i) - \alpha^*(i)$, $\Delta\boldsymbol{\alpha}_p = \hat{\boldsymbol{\alpha}}_p - \boldsymbol{\alpha}_p^*$. By (2.2) and (3.8),

$$\begin{aligned}0 &\geq \hat{\sigma}_p^2 - \hat{\sigma}_{p_0}^2 \geq \hat{\sigma}_K^2 - \hat{\sigma}_{p_0}^2\\ &\geq \frac{1}{N}\sum_{n=K+1}^{N}\left(X(n)+\sum_{i=1}^{K}\hat{\alpha}_K(i)\,X(n-i)\right)^2\\ &\quad -\frac{1}{N}\sum_{n=p_0+1}^{N}\left(X(n)+\sum_{i=1}^{p_0}\alpha_0(i)\,X(n-i)\right)^2\\ &= \frac{1}{N}\sum_{n=K+1}^{N}\left(e(n)+\sum_{i=1}^{K}\Delta\hat{\alpha}_K(i)\,X(n-i)\right)^2\\ &\quad -\frac{1}{N}\sum_{n=p_0+1}^{N}e(n)^2\\ &\geq -\frac{1}{N}\sum_{n=1}^{K}e(n)^2 - \hat{\psi}'\hat{Q}_K^{-1}\hat{\psi}, \qquad (3.12)\end{aligned}$$

where $\hat{Q}_K$ is defined in (2.5) and $\hat{\psi} = (\hat{\psi}_1, ..., \hat{\psi}_K)'$, $\hat{\psi}_j = (1/N)\sum_{n=K+1}^{N} e(n)\,X(n-j)$, $j = 1, 2, ..., K$.

From this, one can see that, there exists $\varepsilon > 0$ such that for large N, if $A_1(\varepsilon)\cap A_2(\varepsilon)\cap A_3(\varepsilon)$ holds then for any $p_0 < p \leq K$,

$$\frac{\hat{\sigma}_{p_0}^2 - \hat{\sigma}_p^2}{\hat{\sigma}_p^2} < \frac{C_N}{2N} \qquad (3.13)$$

which in turn implies that for any $p_0 < p \leqslant K$,

$$\log(\hat{\sigma}_p^2/\hat{\sigma}_{p_0}^2) > -C_N/N \geqslant -(p-p_0)\, C_N/N. \tag{3.14}$$

From (3.11) and (3.14), Proposition 3.1 follows.

Proof of Theorem 2.1. Hereafter, C is a positive constant independent of N which can be assigned as large as you wish, but may take a different value at each appearance. To prove Theorem 2.1, it is enough to show that

$$P_\eta < C\exp\{-CC_N\}, \quad \eta = 1, 2, 3, 4, \tag{3.15}$$

where P_η's are defined in (3.9). It is easy to see that (3.15) is true for $\eta = 3, 4$ using (2.7). By (2.4),

$$X(n) = \sum_{j=0}^{\infty} a_j e(n-j), \quad |a_j| \leqslant M\rho^j, \quad j = 0, 1, 2, ..., \tag{3.16}$$

where $\rho \in (0, 1)$ and $M > 0$ are constants. In order to prove (3.15) for $\eta = 1, 2$, it is enough to show that for any $\varepsilon > 0$,

$$P\left\{\left|\frac{1}{N}\sum_{n=1}^{N} X(n)\, X(n-l) - \gamma(l)\right| > \varepsilon\sqrt{C_N/N}\right\} \leqslant C\exp\{-CC_N\}, \quad l = 0, 1, 2, ..., K, \tag{3.17}$$

and

$$P\left\{\left|\frac{1}{N}\sum_{n=1}^{N} e(n)\, X(n-l)\right| > \varepsilon\sqrt{C_N/N}\right\} \leqslant C\exp\{-CC_N\}, \quad l = 1, 2, ..., K. \tag{3.18}$$

By (3.16), $\gamma(l) = \sigma^2 \sum_{j=0}^{\infty} a_j a_{l+j}$, and

$$\frac{1}{N}\sum_{n=1}^{N} X(n)\, X(n-l) = \sum_{i=l}^{\infty} a_i a_{i-l} \frac{1}{N}\sum_{n=1}^{N} e(n-i)^2 + \sum_{(i,j)\, i \neq l+j} a_i a_j \times \frac{1}{N}\sum_{n=1}^{N} e(n-i)\, e(n-j-l). \tag{3.19}$$

Fix $l \leqslant K$. Take $\rho_1 \in (\rho, 1)$ and set

$$B(\varepsilon_1) = \left\{ \left| \frac{1}{N} \sum_{n=1}^{N} e(n-i)^2 - \sigma^2 \right| < (\rho_1/\rho)^{2i-l} \varepsilon_1 \sqrt{C_N/N} \right.$$
$$\left. \text{for } i = l, l+1, \ldots \right\}$$

$$D(\varepsilon_1) = \left\{ \left| \frac{1}{N} \sum_{n=1}^{N} e(n-i)\, e(n-j-l) \right| \leqslant (\rho_1/\rho)^{i+j} \varepsilon_1 \sqrt{C_N/N} \right.$$
$$\left. \text{for any } i \neq l+j,\ i, j = 0, 1, 2, \ldots \right\}.$$

Take $\varepsilon_1 < \varepsilon M^{-2}(1-\rho_1)^2$. If $B(\varepsilon_1) \cap D(\varepsilon_1)$ occurs, using (3.19) we get

$$\begin{aligned} &\left| \frac{1}{N} \sum_{n=1}^{N} X(n)\, X(n-l) - \gamma(l) \right| \\ &\quad \leqslant M^2 \sum_{i=l}^{\infty} \rho^{2i-l} \varepsilon_1 \sqrt{C_N/N}\, (\rho_1/\rho)^{2i-l} \\ &\qquad + M^2 \sum_{(i,j): i \neq l+j} \rho^{i+j} \varepsilon_1 \sqrt{C_N/N}\, (\rho_1/\rho)^{i+j} \\ &\quad \leqslant M^2 \left(\sum_{i=0}^{\infty} \rho_1^i \right)^2 \varepsilon_1 \sqrt{C_N/N} < \varepsilon \sqrt{C_N/N}. \end{aligned}$$

Thus, taking $\lambda = \rho_1/\rho (>1)$ and $\varepsilon_2 = (\rho_1/\rho)^{-l} \varepsilon_1$ we get

$$\begin{aligned} &P\left\{ \left| \frac{1}{N} \sum_{n-1}^{N} X(n)\, X(n-l) - \gamma(l) \right| > \varepsilon \sqrt{C_N/N} \right\} \\ &\quad \leqslant \sum_{i=0}^{\infty} P\left\{ \left| \frac{1}{N} \sum_{n=1}^{N} e(n)^2 - \sigma^2 \right| \geqslant \lambda^{2i} \varepsilon_2 \sqrt{C_N/N} \right\} \\ &\qquad + \sum_{i \neq j} P\left\{ \left| \frac{1}{N} \sum_{n=1}^{N} e(n)\, e(n-j+i) \right| \geqslant \lambda^{i+j} \varepsilon_2 \sqrt{C_N/N} \right\}. \end{aligned} \tag{3.20}$$

Setting $f(\tau) = E \exp\{\tau(e(i)^2 - \sigma^2)\}$, $\tau \in (0, t)$, we have $f(\tau) = 1 + f'(0)\tau + \frac{1}{2} f''(\tau_1)\tau^2$, where $\tau_1 \in (0, \tau)$. Hence

$$f(\tau) \leqslant 1 + C\tau^2 \leqslant \exp\{C\tau^2\}. \tag{3.21}$$

Thus,

$$P\left\{\frac{1}{N}\sum_{n=1}^{N}(e(n)^2-\sigma^2)\geqslant\lambda^{2i}\varepsilon_2\sqrt{C_N/N}\right\}$$
$$\leqslant\exp(-\tau\lambda^{2i}\varepsilon_2\sqrt{NC_N})\,f(\tau)^N$$
$$\leqslant\exp\{-\tau\lambda^{2i}\varepsilon_2\sqrt{NC_N}+C\tau^2N\}. \tag{3.22}$$

Taking $\tau=\delta\sqrt{C_N/N}\,\lambda^{2i}$, where $\delta>0$ is small, one can see that

$$P\left\{\frac{1}{N}\sum_{n=1}^{N}(e(n)^2-\sigma^2)\geqslant\lambda^{2i}\varepsilon_2\sqrt{C_N/N}\right\}$$
$$\leqslant\exp\{-C\lambda^{4i}C_N\}\leqslant C\lambda^{-2i}\exp(-CC_N). \tag{3.23}$$

In the same way,

$$P\left\{\frac{1}{N}\sum_{n=1}^{N}(e(n)^2-\sigma^2)\leqslant-\lambda^{2i}\varepsilon_2\sqrt{C_N/N}\right\}$$
$$\leqslant C\lambda^{-2i}\exp(-CC_N). \tag{3.24}$$

In a similar fashion it follows that if $\tau\in(0,t)$

$$E\exp(\tau e(0)\,e(i-j))\leqslant\exp(C\tau^2). \tag{3.25}$$

For $i>j$, by (3.25),

$$P\left\{\sum_{n=1}^{N}e(n)\,e(n+i-j)\geqslant\lambda^{i+j}\varepsilon_2\sqrt{NC_N}\right\}$$
$$\leqslant\sum_{m=0}^{i-j}P\left\{\sum_{n\leqslant N,n\equiv m(\mathrm{mod}(i-j+1))}e(n)\,e(n+i-j)\right.$$
$$\left.\geqslant\frac{\varepsilon_2}{i-j+1}\lambda^{i+j}\sqrt{NC_N}\right\}$$
$$\leqslant(i-j+1)\exp\left\{-\tau\frac{\varepsilon_2}{i-j+1}\lambda^{i+j}\sqrt{NC_N}\right\}$$
$$\times\exp\left(C\frac{N}{i-j+1}\tau^2\right). \tag{3.26}$$

Taking $\tau=\delta\sqrt{C_N/N}\,\lambda^{i+j}$, where $\delta>0$ is small enough to get

$$P\left\{\sum_{n=1}^{N}e(n)\,e(n+i-j)\geqslant\lambda^{i+j}\varepsilon_2\sqrt{NC_N}\right\}$$
$$\leqslant C(i+j)\exp\left\{-C\frac{1}{i-j+1}\lambda^{i+j}C_N\right\}$$
$$\leqslant C\exp\{-CC_N\}\lambda^{-i-j}. \tag{3.27}$$

Similarly,

$$P\left\{\sum_{n=1}^{N} e(n)\, e(n+i-j) \leqslant -\lambda^{i+j}\varepsilon_2 \sqrt{NC_N}\right\}$$
$$\leqslant C \exp\{-CC_N\} \lambda^{-i-j}. \tag{3.28}$$

Note that (3.27), (3.28) hold for $i<j$. Thus, by (3.20), (3.23), (3.24), (3.27), and (3.28),

$$P\left\{\left|\frac{1}{N}\sum_{n=1}^{N} X(n)\, X(n-l) - \gamma(l)\right| \geqslant \varepsilon \sqrt{C_N/N}\right\}$$
$$\leqslant 2C \sum_{i=0}^{\infty} \lambda^{-2i} \exp(-CC_N) + 2C \exp(-CC_N) \sum_{i,j=0}^{\infty} \lambda^{-i-j}$$
$$\leqslant C \exp(-CC_N), \tag{3.29}$$

which is (3.17). The proof of (3.18) is similar. That completes the proof of Theorem 2.1.

Proof of Theorem 2.2. The line of proof is similar to that of Theorem 2.1. Here Lemma 3.1 is used. For example, in order to prove

$$P\left\{\left|\frac{1}{N}\sum_{n=1}^{N} X(n)^2 - \gamma(0)\right| > \varepsilon \sqrt{C_N/N}\right\}$$
$$\leqslant CN^{-t/2+1}(C_N)^{-t/2} + C \exp(-CC_N), \tag{3.30}$$

we use $\gamma(0) = \sigma^2 \sum_{j=0}^{\infty} a_j^2$ and

$$\frac{1}{N}\sum_{n=1}^{N} X(n)^2 = \sum_{j=0}^{\infty} a_j^2 \frac{1}{N}\sum_{n=1}^{N} e(n-j)^2$$
$$+ \sum_{i\neq j} a_i a_j \frac{1}{N}\sum_{n=1}^{N} e(n-i)\, e(n-j). \tag{3.31}$$

Take $\rho_1 \in (\rho, 1)$ and set

$$B(\varepsilon_1) = \left\{\left|\frac{1}{N}\sum_{n=1}^{N} e(n-j)^2 - \sigma^2\right| < (\rho_1/\rho)^{2j}\right.$$
$$\left.\times \varepsilon_1 \sqrt{C_N/N} \text{ for } j=0, 1, 2, ...\right\},$$

$$D(\varepsilon_1) = \left\{\left|\frac{1}{N}\sum_{n=1}^{N} e(n-i)\, e(n-j)\right| \leqslant (\rho_1/\rho)^{i+j}\right.$$
$$\left.\times \varepsilon_1 \sqrt{C_n/N} \text{ for any } i\neq j,\ i,\ j=0, 1, 2, ...\right\}.$$

As before, by taking $\varepsilon_1 < \varepsilon M^{-2}(1-\rho_1)^2$, we get, when $B(\varepsilon_1) \cap D(\varepsilon_1)$ occurs

$$\left| \frac{1}{N} \sum_{n=1}^{N} X(n)^2 - \gamma(0) \right| < \varepsilon \sqrt{C_N/N}.$$

Thus, with $\lambda = \rho_1/\rho$ (>1), we have

$$\begin{aligned} P\left\{ \left| \frac{1}{N} \sum_{n=1}^{N} X(n)^2 - \gamma(0) \right| > \varepsilon \sqrt{C_N/N} \right\} \\ \leqslant \sum_{i=0}^{\infty} P\left\{ \left| \frac{1}{N} \sum_{n=1}^{N} e(n)^2 - \sigma^2 \right| \geqslant \lambda^{2i} \varepsilon_1 \sqrt{C_N/N} \right\} \\ + \sum_{i,j=0,\, i \neq j}^{\infty} P\left\{ \left| \frac{1}{N} \sum_{n=1}^{N} e(n)\, e(n-j+i) \right| \geqslant \lambda^{i+j} \varepsilon_1 \sqrt{C_N/N} \right\}. \end{aligned}$$

By Lemma 3.1,

$$\begin{aligned} \sum_{j=0}^{\infty} P\left\{ \left| \frac{1}{N} \sum_{n=1}^{N} (e(n)^2 - \sigma^2) \right| \geqslant \lambda^{2j} \varepsilon_1 \sqrt{C_N/N} \right\} \\ \leqslant (1+2/t)^t \sum_{j=0}^{\infty} NE\, |e(1)^2 - \sigma^2|^t\, \varepsilon_1^{-t} \lambda^{-2jt} (NC_N)^{-t/2} \\ + \sum_{j=0}^{\infty} \exp\{-2(t+2)^{-2}\, e^{-t} \varepsilon^2 \lambda^{4j} NC_N/(N \operatorname{Var} e(1)^2)\} \\ \leqslant C \sum_{j=0}^{\infty} \lambda^{-2j} N^{-t/2+1} C_N^{-t/2} + C \sum_{j=0}^{\infty} \lambda^{-j} \exp(-CC_N) \\ \leqslant CN^{-t/2+1} C_N^{-t/2} + C \exp(-CC_N). \end{aligned} \tag{3.33}$$

For the last term of the right-hand side of (3.32), we can obtain the same bound. The proof of the rest is similar to that Theorem 2.1. This completes the proof of the theorem.

4. Some Remarks

From Theorem 2.1 and Theorem 2.2, it is easily seen that, under the restriction $C_n = o(N)$, the larger the magnitude of C_N, the better the detection is in the large sample cases. By the same way, if (1.1), (1.2), and (2.8) hold then the detection is weakly consistent.

Now we point out that, if (1.1), (1.2) hold and

$$\lim_{N \to \infty} C_N/N = 0 \qquad \text{and} \qquad \lim_{N \to \infty} C_N/\log\log N = \infty, \tag{4.1}$$

then $\hat{p}$ determined by (2.4) is a strongly consistent estimate of p_0. In fact, if $p < p_0$, then by (2.8), (3.7), and $\lim_{N \to \infty} C_N/N = 0$,

$$\lim_{N \to \infty} [\phi(p) - \phi(p_0)]/N \geqslant \log(\sigma^{*2}/\sigma^2) > 0.$$

It follows that, with probability one for N large,

$$\phi(p_0) < \phi(p), \qquad \text{for} \quad p < p_0. \tag{4.2}$$

Now we assume $p_0 < p \leqslant K$. Under the conditions (1.1) and (1.2), by the law of the iterated logarithm,

$$|\hat{q}_K(i, j) - \gamma(i - j)| = O\left(\sqrt{\frac{\log\log N}{N}}\right) \quad \text{a.s.,}$$

$$\left|\frac{1}{N} \sum_{n=K+1}^{N} e(n)\, X(n-i)\right| = O\left(\sqrt{\frac{\log\log N}{N}}\right) \quad \text{a.s.}$$

for $i, j = 0, 1, ..., K$. Thus, by (3.12),

$$0 \geqslant \hat{\sigma}_p^2 - \hat{\sigma}_{p_0}^2 = O\left(\frac{\log\log N}{N}\right) \quad \text{a.s.} \tag{4.3}$$

By (3.7), (3.12), and $\lim_{N \to \infty} C_N/\log\log N = \infty$, with probability one for large N,

$$\begin{aligned} \phi(p) - \phi(p_0) &\geqslant N \log \hat{\sigma}_K^2/\hat{\sigma}_{p_0}^2 + (p - p_0) C_N \\ &= N \log\{1 + (\hat{\sigma}_K^2 - \hat{\sigma}_{p_0}^2)/\hat{\sigma}_{p_0}^2\} + (p - p_0) C_N \\ &= O(\log\log N) + (p - p_0) C_N > 0, \quad p_0 < p \leqslant K. \end{aligned} \tag{4.4}$$

From (4.2) and (4.4), it follows that with probability one for N large,

$$\hat{p} = p_0. \tag{4.5}$$

This shows strong consistency of $\hat{p}$.

Note that for strong consistency of $\hat{p}$, the last condition of (4.1) can be weakened as

$$C_N \geqslant 2C \log\log N \qquad \text{with} \quad C > 1. \tag{4.6}$$

But this needs more accurate calculations.

REFERENCES

[1] AKAIKE, H. (1969). Fitting autoregressive models for prediction. *Ann. Inst. Statist. Math.* **21** 243–247.

[2] FUK, D. K. H., AND NAGAEV, S. V. (1971). Probability inequalities for sums of independent random variables. *Theory Probab. Appl.* **16**, No. 4, 643–660.

[3] HANNAN, E. J. (1970). *Multiple Time Series.* Wiley, New York.

[4] HANNAN, E. J., AND QUINN, B. G. (1979). The determination of the order of an autoregression. *J. Roy. Statist. Soc. Ser. B* **41** 190–195.

[5] HANNAN, E. J. (1980). The estimation of the order of an ARMA process. *Ann. Statist.* **8** 1071–1081.

[6] SHIBATA, R. (1976). Selection of the order of an autoregressive model by Akaike's information criterion. *Biometrika* **63** 117–126.

Admissible Linear Estimation in a General Gauss–Markov Model with an Incorrectly Specified Dispersion Matrix

JERZY K. BAKSALARY*

Academy of Agriculture in Poznan, Poznan, Poland

AND

THOMAS MATHEW†

University of Maryland Baltimore County

Necessary and sufficient conditions are established for the set of all admissible linear estimators under $\mathbf{M}_0$ to be contained in the corresponding set of estimators under $\mathbf{M}$, where $\mathbf{M}_0$ and $\mathbf{M}$ are general Gauss–Markov models with identical model matrices but different dispersion matrices. As preliminary results, certain new characterizations of admissible linear estimators are derived, including explicit expressions for the general representations of such estimators and extensions of the admissibility criteria given by Rao (*Ann. Statist.* **4** (1976), 1023–1037) and Klonecki and Zontek (*J. Multivariate Anal.* **24** (1988), 11–30).

1. INTRODUCTION AND PRELIMINARIES

Throughout this article $\mathcal{M}_{m,n}$, $\mathcal{M}^s_m$, $\mathcal{M}^{\geqslant}_m$, and $\mathcal{M}^{>}_m$ will denote the set of all $m \times n$ matrices, the subset of $\mathcal{M}_{m,m}$ consisting of symmetric matrices, the subset of $\mathcal{M}^s_m$ consisting of non-negative definite matrices, and the subset of $\mathcal{M}^{\geqslant}_m$ consisting of positive definite matrices, respectively. Given $L \in \mathcal{M}_{m,n}$, the symbols L', L^-, L^+, $R(L)$, and $r(L)$ will stand for the transpose, an arbitrary generalized inverse, the Moore–Penrose inverse, the range, and the rank, respectively, of L, while I_m will stand for the identity matrix of

* Research partly supported by the Polish Academy of Sciences Grant MR I.1-2/1.

† The paper was started while the author was visiting the Department of Mathematical and Statistical Methods of the Academy of Agriculture in Poznan.

Multivariate Statistics and Probability
ISBN 0-12-580205-6

Reprinted from *J. Mult. Anal.* **27**(1).

order m. Further, $P_L = LL^+$ and $Q_L = I_m - P_L$ will denote the orthogonal projectors onto $R(L)$ and $R^{\perp}(L)$, respectively, where $R^{\perp}(L)$ stands for the orthogonal complement to $R(L)$ with respect to the standard inner product. Finally, $\text{tr}(L)$ and $\tau(L)$ will denote the trace and spectrum, respectively, of an $L \in \mathcal{M}_{m,m}$, while $L \geqslant K$ will mean that $L \in \mathcal{M}_m^s$ is a successor of $K \in \mathcal{M}_m^s$ with respect to the Loewner partial ordering, that is (cf. Marshall and Olkin [5, p. 462]), $L - K \in \mathcal{M}_m^{\geqslant}$.

Consider a general Gauss–Markov model,

$$\mathbf{M} = \{Y, X\beta, \sigma^2 V\}, \tag{1.1}$$

in which $Y \in \mathcal{M}_{n,1}$ has $E(Y) = X\beta$ as its expectation and $D(Y) = \sigma^2 V$ as its dispersion matrix, where $0 \neq X \in \mathcal{M}_{n,p}$ and $V \in \mathcal{M}_n^{\geqslant}$ are known, while $\beta \in \mathcal{M}_{p,1}$ and $\sigma^2 > 0$ are unknown parameters. Rao [9] pointed out that an important tool in analyzing the model (1.1) is a matrix of the form

$$T = V + XGX', \tag{1.2}$$

with any $G \in \mathcal{M}_p^{\geqslant}$ such that $R(T) = R(X : V)$. Now suppose that instead of the model $\mathbf{M}$, as defined in (1.1), we have the model $\mathbf{M}_0 = \{Y, X\beta, \sigma^2 V_0\}$ with an incorrectly specified dispersion matrix $V_0 \neq V$. Further, let $\mathcal{L}_0$ be a class of all statistics with certain property under $\mathbf{M}_0$, let $\mathcal{L}$ be the class of all statistics with the same property, but corresponding to the correct model $\mathbf{M}$, and let the problem consist in determining conditions under which the class $\mathcal{L}_0$ remains valid under $\mathbf{M}$ in the sense that $\mathcal{L}_0 \subseteq \mathcal{L}$. The validity problem so defined has thoroughly been discussed in the literature in the context of best linear unbiased estimation; see, e.g., Rao [9], Rao and Mitra [12, Chap. 8], Mitra and Moore [8], Kala [3], Mathew and Bhimasankaram [6]. One of the results concerning the validity of best linear unbiased estimators is restated here as the following lemma.

LEMMA 1. *Let $\mathbf{M}_0 = \{Y, X\beta, \sigma^2 V_0\}$ and $\mathbf{M} = \{Y, X\beta, \sigma^2 V\}$ be general Gauss–Markov models, and let $\mathcal{B}_0$ and $\mathcal{B}$ be the sets of all possible representations of the best linear unbiased estimator of $X\beta$ under $\mathbf{M}_0$ and $\mathbf{M}$, respectively. Then $\mathcal{B}_0 \subseteq \mathcal{B}$ if and only if $R(VZ) \subseteq R(V_0 Z)$, where Z is any matrix such that $R(Z) = R^{\perp}(X)$.*

The purpose of the present paper is to investigate the validity problem with reference to the sets $\mathcal{A}_0$ and $\mathcal{A}$, comprising all linear estimators that are admissible for $X\beta$ among

$$\mathcal{F} = \{FY : F \in \mathcal{M}_{n,n}\} \tag{1.3}$$

under $\mathbf{M}_0$ and $\mathbf{M}$, respectively, where admissibility is understood according to the following.

DEFINITION. Let $\mathbf{M}=\{Y, X\beta, \sigma^2 V\}$ be a general Gauss–Markov model, let $\Theta=\mathcal{M}_{p,1}\times(0,\infty)$, and let $W\in\mathcal{M}_n^{>}$. Then an estimator AY is said to be admissible for $X\beta$; among $\mathcal{F}=\{FY: F\in\mathcal{M}_{n,n}\}$ under $\mathbf{M}$ if there does not exist $FY\in\mathcal{F}$ such that the inequality

$$\rho_W(FY; X\beta)\leqslant\rho_W(AY; X\beta)$$

holds for every pair $(\beta,\sigma^2)\in\Theta$ and is strict for at least one such pair, where

$$\begin{aligned}\rho_W(FY; X\beta)&=E[(FY-X\beta)'W(FY-X\beta)]\\&=\sigma^2\operatorname{tr}(FVF'W)+\beta'X'(F-I_n)'W(F-I_n)X\beta.\end{aligned}\tag{1.4}$$

This definition is to be supplemented by pointing out that the choice of the weight matrix W is immaterial for the problem, for, as shown by Shinozaki [13] and Rao [10], if an estimator AY is admissible for $X\beta$ with respect to the risk function (1.4), then it is admissible for $X\beta$ with respect to any quadratic risk function of the form (1.4), with W replaced by any member of $\mathcal{M}_n^{>}$. Consequently, no loss in generality arises by restricting attention to the unweighted quadratic risk function, defined as in (1.4) with $W=I_n$ and denoted by the unsubscripted ρ. Moreover, the admissibility of AY for $X\beta$ among the set $\mathcal{F}$ of all homogeneous linear estimators of $X\beta$, specified in (1.3), will henceforth be denoted by the symbol $AY\sim X\beta$.

A solution to the problem of the validity of admissible linear estimators of $X\beta$ in the case where the dispersion matrix of the model is incorrectly specified is given in Section 3. It is preceded by certain results concerning the characterization of admissible linear estimators of $X\beta$ under a general Gauss–Markov model. These results include extensions of the admissibility criteria given by Rao [10] and Klonecki and Zontek [4] and also explicit expressions for the general representations of admisssible linear estimators of $X\beta$.

2. CHARACTERIZATION OF ADMISSIBLE LINEAR ESTIMATORS

The problem of the admissibility of linear estimators was investigated first by Cohen [2] in the context of a simple location model $\{Y, \xi, \sigma^2 I_n\}$. Ten years later, an exhaustive study of the problem under a Gauss–Markov model $\{Y, X\beta, \sigma^2 V \mid V\in\mathcal{M}_n^{>}\}$ was given by Rao [10]. In particular, the following characterization of admissible linear estimators of $X\beta$ under this model is immediately obtainable from his Theorem 6.6.

LEMMA 2. *$AY\sim X\beta$ under $\{Y, X\beta, \sigma^2 V\mid V\in\mathcal{M}_n^{>}\}$ if and only if*

$$R(A)\subseteq R(X),\qquad AV=VA',\qquad \text{and}\qquad AV\geqslant AVA'.\tag{2.1}$$

Rao's work stimulated further research in this area. Mathew, Rao, and Sinha [7], Klonecki and Zontek [4], and Baksalary and Markiewicz [1] extended Rao's work by relaxing the rank conditions on the design and dispersion matrices. In particular, Klonecki and Zontek [4] extended the result of Lemma 2 to the case where, instead of $r(V)=n$, the additional assumption on the model is

$$r(X:V)=n. \tag{2.2}$$

LEMMA 3. *$AY \sim X\beta$ under $\{Y, X\beta, \sigma^2 V \mid R(X:V)=\mathcal{M}_{n,1}\}$ if and only if*

$$R(A) \subseteq R(X), \quad R(A-I_n)=R[(A-I_n)V], \quad AV=VA', \quad AV \geqslant AVA'. \tag{2.3}$$

Commenting on another result, also derived under the condition (2.2), Klonecki and Zontek [4] remarked that if (2.2) is not fulfilled, then a general solution can be obtained from the solution valid under (2.2) via appropriately modifying the latter by P_T, where T is defined in (1.2). The same is adopted below in developing a characterization of admissible linear estimators under a general Gauss–Markov model.

THEOREM 1. *$AY \sim X\beta$ under a general Gauss–Markov model $\mathbf{M}=\{Y, X\beta, \sigma^2 V\}$ if and only if*

$$R[A(X:V)] \subseteq R(X), \tag{2.4}$$

$$R[(A-I_n)X] \subseteq R[(A-I_n)V], \tag{2.5}$$

$$AV=VA', \tag{2.6}$$

and

$$AV \geqslant AVA'. \tag{2.7}$$

Proof. Using the definition (1.4), with $W=I_n$, and the equalities

$$P_T X = X \quad \text{and} \quad P_T V = P_T(V+Q_T)=V, \tag{2.8}$$

in which T is any matrix of the form (1.2), it is fairly straightforward to observe that $AY \sim X\beta$ under $\mathbf{M}$ if and only if $AP_T Y \sim X\beta$ under $\mathbf{M}$, and also that $AP_T Y \sim X\beta$ under $\mathbf{M}$ if and only if $AP_T Y \sim X\beta$ under $\bar{\mathbf{M}}=\{Y, X\beta, \sigma^2(V+Q_T)\}$. Since $R(X: V+Q_T)=\mathcal{M}_{n,1}$, Lemma 2 is applicable to the model $\bar{\mathbf{M}}$, and hence $AY \sim X\beta$ under $\mathbf{M}$ if and only if

$$R(AP_T) \subseteq R(X), \tag{2.9}$$

$$R(AP_T - I_n) = R[(AP_T - I_n)(V+Q_T)], \tag{2.10}$$

$$AP_T(V+Q_T) = (V+Q_T)P_T A', \tag{2.11}$$

and

$$AP_T(V+Q_T) \geqslant AP_T(V+Q_T)\, P_T A'. \tag{2.12}$$

The equivalence of (2.9) to (2.4) follows from the definition of T, while the equivalences of (2.11) to (2.6) and of (2.12) to (2.7) are obvious by (2.8). It remains to prove, therefore, that (2.10) may be replaced by (2.5). From (2.8) it is clear that an alternative form of (2.10) is

$$R[(A-I_n)\, P_T - Q_T] = R[(A-I_n)\, V - Q_T], \tag{2.13}$$

while from (2.6) and (2.9) it is clear that

$$R[(A-I_n)\, V] \subseteq R(V) \qquad \text{and} \qquad R[(A-I_n)\, P_T] \subseteq R(T). \tag{2.14}$$

Consequently, in view of (2.14) and (2.8), premultiplying (2.13) by P_T yields

$$R[(A-I_n)\, P_T] = R[(A-I_n)\, V]. \tag{2.15}$$

On the other hand, since

$$R^{\perp}[(A-I_n)\, P_T - Q_T] = R^{\perp}[(A-I_n)\, P_T] \cap R(T)$$

and, similarly,

$$R^{\perp}[(A-I_n)\, V - Q_T] = R^{\perp}[(A-I_n)\, V] \cap R(T),$$

it is clear that (2.15) entails (2.13). This establishes the equivalence of (2.10) to (2.15) and actually concludes the proof, since the equivalence of (2.15) to (2.5) is obvious in view of the definition of T. ∎

It can be easily shown that if $R(X\colon V) = \mathcal{M}_{n,1}$, then the conditions (2.4) through (2.7) are replaceable by those given in (2.3), while in another particular case of the model **M**, specified by the inclusion $R(X) \subseteq R(V)$, Theorem 1 simplifies to the following extension of Lemma 2.

COROLLARY 1. *$AY \sim X\beta$ under $\{Y, X\beta, \sigma^2 V \mid R(X) \subseteq R(V)\}$ if and only if*

$$R(AV) \subseteq R(X), \qquad AV = VA' \qquad \textit{and} \qquad AV \geqslant AVA'.$$

An alternative characterization of admissible linear estimators (in the set-up of Theorem 1) has been obtained by Baksalary and Markiewicz [1, Corollary 3].

THEOREM 2. *$AY \sim X\beta$ under $\mathbf{M} = \{Y, X\beta, \sigma^2 V\}$ if and only if* (i) $R(VA') \subseteq R(x)$, (ii) $AV = VA'$, (iii) $AV \geqslant AVA'$, *and* (iv) $R[(A - I_n)X] = R[(A - I_n)H]$, *where H is any matrix such that $R(H) = R(x) \cap R(V)$.*

It is clear that the condition (iv) of Theorem 2 may be replaced by $r[(A - I_n)X] = r[(A - I_n)H]$. For the particular choice of H, viz. $H = X(X'T^+X)^- X'T^+V$ with $T = V + XX'$, the result of Theorem 2 was also proved independently by Zhu [14] using the results in Rao [10].

Consider now again the model $\{Y, X\beta, \sigma^2 V \mid V \in \mathcal{M}_n^>\}$, and let $L \in \mathcal{M}_{n,n}$ be such that $L'VL = I_n$. Then it is easily verified, using the conditions (2.1), that $AY \sim X\beta$ under this model if and only if

$$A = L'^{-1} P_{L'X} S P_{L'X} L' \tag{2.16}$$

with an arbitrary $S \in \mathcal{M}_n^s$ satisfying the condition $\tau(P_{L'X}S) \subset [0, 1]$. A similar representation of admissible linear estimators under a Gauss–Markov model with a singular dispersion matrix is given in the following.

THEOREM 3. *Let $\mathbf{M} = \{Y, X\beta, \sigma^2 V\}$ be a Gauss–Markov model in which $r(V) = v < n$, and let $L = (L_1 : L_2) \in \mathcal{M}_{n,n}$ be nonsingular and such that*

$$L'VL = \operatorname{diag}(I_v, 0). \tag{2.17}$$

Further, let

$$L'X = (X_1' : X_2')' \quad \text{and} \quad L^{-1}Z = (Z_1' : Z_2')', \tag{2.18}$$

where Z is any matrix such that $R(Z) = R^\perp(X)$. Then $AY \sim X\beta$ under $\mathbf{M}$ if and only if

$$A = L'^{-1} \begin{pmatrix} A_{11} & A_{12} \\ 0 & A_{22} \end{pmatrix} L', \tag{2.19}$$

with

$$A_{11} = Q_{Z_1} S Q_{Z_1}, \tag{2.20}$$

$$A_{12} = P_{Z_1} X_1 X_2^+ + (Q_{Z_1} S Q_{Z_1} - Q_{Z_1}) K_1 X_2^+ + K_2 Q_{X_2}, \tag{2.21}$$

$$A_{22} = P_{X_2} + K_3 Q_{X_2}, \tag{2.22}$$

where $K_1 \in \mathcal{M}_{v,p}$, $K_2 \in \mathcal{M}_{v,n-v}$, $K_3 \in \mathcal{M}_{n-v,n-v}$, and $S \in \mathcal{M}_v^s$ are all arbitrary except only for the condition $\tau(Q_{Z_1}S) \subset [0, 1]$.

Proof. It is clear that every $A \in \mathcal{M}_{n,n}$ may be represented as in (2.19), but with the zero matrix in the southwest corner replaced by some A_{21}.

However, in view of (2.17), the conditions (2.6) and (2.7) are satisfied if and only if $A_{21}=0$ holds along with

$$A_{11}=A'_{11} \quad \text{and} \quad \tau(A_{11})\subset[0,1]. \tag{2.23}$$

Further, on account of (2.17) and (2.18), it follows that the condition (2.4), which is alternatively expressible as the pair of equations $Z'AV=0$ and $Z'AX=0$, is fulfilled if and only if

$$Z'_1A_{11}=0 \tag{2.24}$$

and

$$(Z'_1A_{12}+Z'_2A_{22})\,X_2=0, \tag{2.25}$$

while the condition (2.5) is fulfilled if and only if

$$A_{22}X_2=X_2 \tag{2.26}$$

and

$$R(A_{12}X_2)\subseteq R(A_{11}-I_v). \tag{2.27}$$

Hence, observing that the equalities (2.26) and

$$Z'_1X_1+Z'_2X_2=0 \tag{2.28}$$

enable (2.25) to be re-expressed as

$$Z'_1(A_{12}X_2-X_1)=0, \tag{2.29}$$

it follows that $AY\sim X\beta$ under $\mathbf{M}$ if and only if A is of the form (2.19) with A_{11} satisfying (2.23) and (2.24), with A_{22} satisfying (2.26), and with A_{12} satisfying (2.27), with a given A_{11}, and (2.29). Consequently, the representation (2.20) is obtainable similarly as that in (2.16); further, (2.22) is the general solution to Eq. (2.26); and finally, (2.21) can be established as follows.

First notice that (2.27) is alternatively expressible in the form

$$A_{12}X_2=(A_{11}-I_v)\,K_0, \tag{2.30}$$

where $K_0\in\mathcal{M}_{v,p}$ is arbitrary but such that, given A_{11}, Eq. (2.30) is solvable with respect to A_{12}, for which it is necessary and sufficient that

$$(A_{11}-I_v)\,K_0=(A_{11}-I_v)\,K_0P_{X'_2}. \tag{2.31}$$

On account of (2.24) and (2.30) modified by (2.31), Eq. (2.29) transforms to

$$Z_1' K_0 P_{X_2'} = Z_1' X_1. \tag{2.32}$$

From (2.28) it is clear that

$$Z_1' X_1 P_{X_2'} = Z_1' X_1. \tag{2.33}$$

Hence

$$P_{Z_1'} Z_1' X_1 P_{X_2'} = Z_1' X_1$$

which implies, according to Theorem 2.3.2 in Rao and Mitra [12], that (2.32) is solvable with respect to K_0, having as the general solution

$$K_0 = P_{Z_1} X_1 P_{X_2'} + K_1 - P_{Z_1} K_1 P_{X_2'}, \tag{2.34}$$

with an arbitrary $K_1 \in \mathcal{M}_{v,p}$. The desired formula (2.21) now follows by substituting (2.34) into (2.30) modified by (2.31), and then solving the equation so obtained with respect to A_{12} and replacing A_{11} by its representation given in (2.20). ▮

In the particular cases of the model **M**, in which admissibility criteria were given in Lemma 2 and Corollary 1, the general representation of admissible linear estimators of $X\beta$ simplifies accordingly.

COROLLARY 2. *Let* $\{Y, X\beta, \sigma^2 V\}$ *be a Gauss–Markov model in which* $r(V) = v < n$, *but* $R(X : V) = \mathcal{M}_{n,1}$, *and let a non-singular* $L \in \mathcal{M}_{n,n}$ *satisfy* (2.17) *and* (2.18). *Then* $AY \sim X\beta$ *under this model if and only if* A *is of the form* (2.19), *with*

$$A_{11} = Q_{Z_1} S Q_{Z_1}, \quad A_{12} = P_{Z_1} X_1 X_2^+ + (Q_{Z_1} S Q_{Z_1} - Q_{Z_1}) K_1 X_2^+, \quad A_{22} = I_{n-v},$$

where K_1 *and* S *are specified in Theorem* 3.

Proof. The result follows from Theorem 3 by noting that $r(X : V) = n$ if and only if $r(X_2) = n - v$, in which case $P_{X_2} = I_{n-v}$. ▮

COROLLARY 3. *Let* $\{Y, X\beta, \sigma^2 V\}$ *be a Gauss–Markov model in which* $r(V) = v < n$, *but* $R(X) \subseteq R(V)$, *and let a nonsingular* $L \in \mathcal{M}_{n,n}$ *satisfy* (2.17) *and* (2.18). *Then* $AY \sim X\beta$ *under this model if and only if* A *is of the form* (2.19), *with*

$$A_{11} = Q_{Z_1} S Q_{Z_1}, \qquad A_{12} = K_2, \qquad A_{22} = K_3,$$

where K_2, K_3, *and* S *are specified in Theorem* 3.

Proof. The result follows from Theorem 3 by noting that $R(X) \subseteq R(V)$ if and only if $X_2 = 0$. ∎

Now, let $\mathscr{A}$ and $\mathscr{B}$ denote the set of all admissible linear estimators and the set of all possible representations of the best linear unbiased estimator, respectively, of $X\beta$ under the model (1.1). Then $\mathscr{A}$ is characterized by the conditions (2.4) through (2.7), or equivalently, by the formulae (2.19) through (2.22), while (cf. Rao [11])

$$\mathscr{B} = \{BY : BX = X, BVZ = 0\} \tag{2.35}$$

$$= \{BY : B = X(X'T^+X)^+X'T^+ + KQ_T\}, \tag{2.36}$$

where Z is any matrix such that $R(Z) = R^\perp(X)$, T is defined in (1.2), and $K \in \mathscr{M}_{n,n}$ is arbitrary. The result below specifies those Gauss–Markov models for which the equality

$$\mathscr{U} \cap \mathscr{A} = \mathscr{B}, \tag{2.37}$$

where $\mathscr{U}$ stands for the set of all linear unbiased estimators of $X\beta$ under the model (1.1), takes the form $\mathscr{A} = \mathscr{B}$.

Corollary 4. *For a general Gauss–Markov model* $\mathbf{M} = \{Y, X\beta, \sigma^2 V\}$ *the following statements are equivalent*:

(i) $\mathscr{A} = \mathscr{B}$

(ii) $R(X) \cap R(V) = \{0\}$.

Proof. We note that if $\mathscr{A} = \mathscr{B}$, then every $A \in \mathscr{A}$ must satisfy $AX = X$ and $AVZ = 0$. The latter condition is always satisfied, since $AV = VA'$ and $R(AV) \subseteq R(X)$ by Theorem 1. Thus $\mathscr{A} = \mathscr{B}$ if and only if $A \in \mathscr{A}$ satisfies $AX = X$. Using (2.19)–(2.21), we see that $A \in \mathscr{A}$ satisfies $AX = X$ if and only if

$$Q_{Z_1}SQ_{Z_1}X_1 + P_{Z_1}X_1X_2^+X_2 + (Q_{Z_1}SQ_{Z_1} - Q_{Z_1})K_1X_2^+X_2 = X_1 \tag{2.38}$$

for any $S \in \mathscr{M}_v^s$ and $K_1 \in \mathscr{M}_{v,p}$. Eq. (2.38) holding for all such S and K_1 is equivalent to

$$Q_{Z_1} = 0 \quad \text{and} \quad P_{Z_1}X_1X_2^+X_2 = X_1,$$

or, equivalently,

$$P_{Z_1} = I \quad \text{and} \quad R(X_1') \subseteq R(X_2'). \tag{2.39}$$

Note that in view of (2.17) and (2.18), $R(X_1') \subseteq R(X_2')$ is equivalent to $R(X) \cap R(V) = \{0\}$. Also, $R(X_1') \subseteq R(X_2')$ is equivalent to $R(Z_1) = \mathscr{M}_{v,1}$ or, equivalently, $P_{Z_1} = I_v$. This completes the proof of Corollary 4. ∎

3. Validity of Admissible Linear Estimators

A necessary and sufficient condition for a nonnegative definite matrix to commute with every nonnegative definite matrix having its range contained in a given subspace is derived below as an auxiliary result for the proof of Theorem 4, providing a solution to the problem of the validity of admissible linear estimators of the expectation vector in the case where the dispersion matrix of a Gauss–Markov model is incorrectly specified.

Lemma 4. *Given $A \in \mathcal{M}_m^{\geq}$ and $B \in \mathcal{M}_{m,n}$ such that $AB \neq 0$, let*

$$\mathscr{C} = \{C \in \mathcal{M}_m^{\geq} : R(C) \subseteq R(B)\}. \tag{3.1}$$

Then $AC = CA$ for every $C \in \mathscr{C}$ if and only if $AB = dB$ for some $d > 0$.

Proof. Only the necessity is to be proved. Assume that $r(B) = b > 0$, and let $U \in \mathcal{M}_{m,b}$ be such that $R(U) = R(B)$ and $U'U = I_b$. Then the set $\mathscr{C}$ defined in (3.1) may be represented as

$$\mathscr{C} = \{C = UNU' : N \in \mathcal{M}_b^{\geq}\}. \tag{3.2}$$

In view of (3.2), the requirement that $AC = CA$ for every $C \in \mathscr{C}$ means that

$$AUNU' = UNU'A \qquad \text{for every} \quad N \in \mathcal{M}_b^{\geq}, \tag{3.3}$$

and hence

$$U'AUN = NU'AU \qquad \text{for every} \quad N \in \mathcal{M}_b^{\geq}. \tag{3.4}$$

From the assumptions that $A \in \mathcal{M}_m^{\geq}$ and $AB \neq 0$, it follows that $U'AU \neq 0$, and thus it is clear that (3.4) holds if and only if $U'AU = dI_b$ or, equivalently,

$$UU'AU = dU \tag{3.5}$$

for some $d > 0$. But the choice of $N = I_b$ in (3.3) yields $UU'A = AUU'$, and thus it follows from (3.5) that $AU = dU$, which gives $AB = dB$. ∎

Theorem 4. *Let $\mathbf{M}_0 = \{Y, X\beta, \sigma^2 V_0\}$ and $\mathbf{M} = \{Y, X\beta, \sigma^2 V\}$ be general Gauss–Markov models, and let $\mathscr{A}_0$ and $\mathscr{A}$ be the sets of all admissible linear estimators of $X\beta$ under $\mathbf{M}_0$ and $\mathbf{M}$, respectively. Then in the case where*

$$R(X) \cap R(V_0) = \{0\}, \tag{3.6}$$

the inclusion $\mathscr{A}_0 \subseteq \mathscr{A}$ holds if and only if

$$R(VZ) \subseteq R(V_0), \tag{3.7}$$

where Z is any matrix such that $R(Z)=R^{\perp}(X)$, while in the case where

$$R(X)\cap R(V_0)\neq\{0\}, \tag{3.8}$$

the inclusion $\mathscr{A}_0\subseteq\mathscr{A}$ holds if and only if

$$R(V)\subseteq R(V_0) \tag{3.9}$$

and

$$VV_0^{-}H=dH \qquad \text{for some} \quad d>0, \tag{3.10}$$

where H is any matrix such that $R(H)=R(X)\cap R(V_0)$.

Proof. Let $\mathscr{U}$ denote the set of all linear unbiased estimators of $X\beta$ under $\mathbf{M}_0$ and $\mathbf{M}$, and let $\mathscr{B}_0$ and $\mathscr{B}$ denote the sets of all possible representations of the best linear unbiased estimators of $X\beta$ under $\mathbf{M}_0$ and $\mathbf{M}$, respectively. In view of (2.37), it is clear that $\mathscr{B}_0\subseteq\mathscr{A}$ is equivalent to $\mathscr{B}_0\subseteq\mathscr{B}$. But Corollary 4 asserts that if (3.6) holds, then $\mathscr{B}_0=\mathscr{A}_0$, and consequently, $\mathscr{A}_0\subseteq\mathscr{A}$ if and only if $\mathscr{B}_0\subseteq\mathscr{B}$. Hence the first part of Theorem 4 follows immediately from Lemma 1 by observing that, under (3.6), $R(V_0Z)=R(V_0)$. To prove the second part first notice that, on account of Theorem 6.2.3 in Rao and Mitra [12], there exists a nonsingular $L\in\mathscr{M}_{n,n}$ such that if $r(V_0)=n$, then

$$L'V_0L=I_n \qquad \text{and} \qquad L'VL=D, \tag{3.11}$$

while if $r(V_0)=v<n$, then

$$L'V_0L=\operatorname{diag}(I_v,0) \qquad \text{and} \qquad L'VL=\operatorname{diag}(D_1,D_2), \tag{3.12}$$

where $D=\operatorname{diag}(D_1,D_2)$ is a member of $\mathscr{M}_n^{\geqslant}$. It is clear that the conditions (2.4) through (2.7) may equivalently be expressed by replacing V, X, and A by $L'VL$, $L'X$, and $L'AL'^{-1}$. Hence, for proving the theorem, we assume without loss of generality that

$$V_0=\operatorname{diag}(I_v,0) \qquad \text{and} \qquad V=\operatorname{diag}(D_1,D_2). \tag{3.13}$$

We shall only consider the case $v<n$; the case $v=n$ is treated similarly. First, we establish the necessity of (3.9) and (3.10) when V_0 in (3.13) satisfies (3.8). From Theorem 3 it follows that if $\mathscr{A}_0\subseteq\mathscr{A}$, then (2.6) leads to the conditions

$$Q_{Z_1}S_1Q_{Z_1}D_1=D_1Q_{Z_1}S_1Q_{Z_1} \qquad \text{for every} \quad S_1\in\mathscr{S}_1 \tag{3.14}$$

and

$$[P_{Z_1}X_1X_2^{+}+(Q_{Z_1}S_1Q_{Z_1}-Q_{Z_1})K_1X_2^{+}+K_2Q_{X_2}]D_2=0 \tag{3.15}$$

for every $K_1 \in \mathcal{M}_{v,p}$, $K_2 \in \mathcal{M}_{v,n-v}$, and $S_1 \in \mathcal{S}_1$, where X_1, X_2, and Z_1 are as defined in (2.18) while

$$\mathcal{S}_1 = \{S_1 \in \mathcal{M}_v^s : \tau(Q_{Z_1} S_1) \subset [0, 1]\}.$$

Note that $Q_{Z_1} \neq 0$, since if $Q_{Z_1} = 0$, then $P_{Z_1} = I$ and, in view of the last part of the proof of Corollary 4, this contradicts (3.8). Since K_1, K_2, and S_1 vary independently and since $Q_{Z_1} \neq 0$, (3.15) gives $X_2^+ D_2 = 0$ and $Q_{X_2} D_2 = 0$. These two together give $D_2 = 0$ which is (3.9) in view of (3.13). Applying Lemma 4 to (3.14), we get

$$D_1 Q_{Z_1} = d Q_{Z_1} \qquad \text{for some} \quad d > 0, \tag{3.16}$$

provided $D_1 Q_{Z_1} \neq 0$. But this is always the case, since, when $\mathcal{A}_0 \subseteq \mathcal{A}$, (2.5) must hold for V in (2.13) and if $D_1 Q_{Z_1} = 0$, one can exhibit $A \in \mathcal{A}_0$, not satisfying (2.5). To conclude the proof of necessity, it remains to show the equivalence of (3.16) and (3.10). For V_0 in (3.13), we note that $R(V_0) \cap R(X) = R(X_1 Q_{X_2'})$. Also, since $Z_1' X_1 + Z_2' X_2 = 0$, $Z_1' X_1 Q_{X_2'} = 0$, and, consequently, $R(Q_{Z_1}) = R(X_1 Q_{X_2'}) = R(V_0) \cap R(X) = R(H)$ which concludes the proof of the necessity.

To prove the sufficiency of the pair (3.9), (3.10), let $AY \sim X\beta$ under $\mathbf{M}_0$. Then, according to Theorem 1,

$$R[A(X : V_0)] \subseteq R(X), \tag{3.17}$$

$$R[(A - I_n)X] \subseteq R[(A - I_n) V_0], \tag{3.18}$$

$$A V_0 = V_0 A', \tag{3.19}$$

$$A V_0 \geqslant A V_0 A'. \tag{3.20}$$

The conditions (3.9) and (3.17) entail

$$R[A(X : V)] \subseteq R[A(X : V_0)] \subseteq R(X),$$

which is (2.4). Further, (3.17) and (3.19) imply that

$$R(A V_0) \subseteq R(X) \cap R(V_0) = R(H).$$

Consequently, in view of (3.9), (3.19), and (3.10), it follows that

$$VA' = VV_0^- V_0 A' = VV_0^- A V_0 = dA V_0 \qquad \text{for some} \quad d > 0,$$

and hence it is clear that (3.6) and (3.7) are immediate consequences of (3.19) and (3.20), respectively. Finally, (3.18) implies that

$$R[VV_0^-(A - I_n)X] \subseteq R[VV_0^-(A - I_n) V_0]. \tag{3.21}$$

But, on account of (3.19), (3.9), and (2.6),

$$R[VV_0^-(A-I_n)\,V_0]=R[(A-I_n)\,V], \tag{3.22}$$

while on account of (3.17), (3.18), and (3.19),

$$R[(A-I_n)\,X]\subseteq R(X)\cap R(V_0)=R(H);$$

hence, according to (3.10),

$$R[VV_0^-(A-I_n)\,X]=R[(A-I_n)\,X]. \tag{3.23}$$

Applying (3.22) and (3.23) to (3.21) yields (3.25), thus completing the proof. ∎

Two corollaries will be given to conclude the paper. The first of them compares the criterion for the validity of the set of all admissible linear estimators of $X\beta$, given in Theorem 4, with the criterion for the validity of the set of all possible representations of the best linear unbiased estimator of $X\beta$, given in Lemma 1, while the second corollary establishes a necessary and sufficient condition for the equivalence of the models $\mathbf{M}_0$ and $\mathbf{M}$ with respect to admissible linear estimators of $X\beta$.

COROLLARY 5. *Let* $\mathbf{M}_0=\{Y, X\beta, \sigma^2 V_0\}$ *and* $\mathbf{M}=\{Y, X\beta, \sigma^2 V\}$ *be general Gauss–Markov models, and let* $\mathscr{A}_0$, $\mathscr{A}$ *and* $\mathscr{B}_0$, $\mathscr{B}$ *be the sets of all admissible linear estimators of* $X\beta$ *and the sets of all possible representations of the best linear unbiased estimator of* $X\beta$ *under* $\mathbf{M}_0$ *and* $\mathbf{M}$, *respectively. Then* $\mathscr{A}_0\subseteq\mathscr{A}$ *implies* $\mathscr{B}_0\subseteq\mathscr{B}$.

Proof. The result is a direct consequence of the equalities $\mathscr{B}_0=\mathscr{U}\cap\mathscr{A}_0$ and $\mathscr{B}=\mathscr{U}\cap\mathscr{A}$, where $\mathscr{U}$ is the set of all linear unbiased estimators of $X\beta$ under both $\mathbf{M}_0$ and $\mathbf{M}$. ∎

COROLLARY 6. *Let* $\mathbf{M}_0=\{Y, X\beta, \sigma^2 V_0\}$ *and* $\mathbf{M}=\{Y, X\beta, \sigma^2 V\}$ *be general Gauss–Markov models, and let* $\mathscr{A}_0$ *and* $\mathscr{A}$ *be the sets of all admissible linear estimators of* $X\beta$, *respectively. Then* $\mathscr{A}_0=\mathscr{A}$ *if and only if*

$$R(V_0)=R(V) \tag{3.24}$$

and

$$V_0^+H=dV^+H \qquad \text{for some} \quad d>0, \tag{3.25}$$

where H *is any matrix such that* $R(H)=R(X)\cap R(V)$.

Proof. First observe that if $\mathscr{A}_0 = \mathscr{A}$, then either

$$R(X) \cap R(V_0) = \{0\} \quad \text{and} \quad R(X) \cap R(V) = \{0\} \tag{3.26}$$

or

$$R(X) \cap R(V_0) \neq \{0\} \quad \text{and} \quad R(X) \cap R(V) \neq \{0\}. \tag{3.27}$$

In fact, if $R(X) \cap R(V) = \{0\}$ and $R(X) \cap R(V_0) \neq \{0\}$, then in view of Corollary 4, the former condition means that $\mathscr{A} = \mathscr{B}$, and consequently, $\mathscr{A}_0 \subseteq \mathscr{A}$ entails $\mathscr{A}_0 \subseteq \mathscr{U}$. Hence, on account of (2.37), it follows that $\mathscr{A}_0 = \mathscr{B}_0$, which in view of Corollary 4, constitutes a contradiction with $R(X) \cap R(V_0) \neq \{0\}$.

Now it is clear that in the case characterized by (3.26) the equality $\mathscr{A}_0 = \mathscr{A}$ reduces to $\mathscr{B}_0 = \mathscr{B}$, and also that (3.24) can be reformulated as $R(V_0 Z) = R(VZ)$, while (3.25) is trivially fulfilled. Consequently, the required result is an immediate consequence of Lemma 1. In the case characterized by (3.27), the necessity and sufficiency of the conditions (3.24) and (3.25) follow by Theorem 4, in view of the equalities $VV_0^- H = VV_0^+ H$ and $V_0 V^- H = V_0 V^+ H$ valid for any generalized inverses V_0^- and V^-. ∎

ACKNOWLEDGMENTS

The authors are grateful to Professor C. R. Rao for drawing their attention to the result of Zhu [14]. Thanks are also due to a referee whose suggestions simplified the proofs and improved the presentation.

REFERENCES

[1] BAKSALARY, J. K., AND MARKIEWICZ, A. (1988). Admissible linear estimators in the general Gauss–Markov model. *J. Statist. Plann. Inference* **19** 349–359.

[2] COHEN, A. (1966). All admissible linear estimates of the mean vector. *Ann. Math. Statist.* **37**, 458–463.

[3] KALA, R. (1981). Projectors and linear estimation in general linear models. *Comm. Statist. A–Theory Methods* **10** 849–873.

[4] KLONECKI, W., AND ZONTEK, S. (1988). On the structure of admissible linear estimators. *J. Multivariate Anal.* **24** 11–30.

[5] MARSHALL, A. W., AND OLKIN, I. (1979). *Inequalities: Theory of Majorization and Its Applications.* Academic Press, New York.

[6] MATHEW, T., AND BHIMASANKARAM, P. (1983). Optimality of BLUE's in a general linear model with an incorrect design matrix. *J. Statist. Plann. Inference* **8** 315–329.

[7] MATHEW, T., RAO, C. R., AND SINHA, B. K. (1984). Admissible linear estimation in singular linear models. *Comm. Statist. A–Theory Methods* **13** 3033–3045.

[8] MITRA, S. K., AND MOORE, B. J. (1973). Gauss–Markov estimation with an incorrect dispersion matrix. *Sankhyā Ser. A* **35** 139–152.

[9] RAO, C. R. (1971). Unified theory of linear estimation. *Sankhyā Ser. A* **33** 371–394.

[10] RAO, C. R. (1976). Estimation of parameters in a linear model. *Ann. Statist.* **4** 1023–1037.

[11] RAO, C. R. (1979). Estimation of parameters in the singular Gauss–Markov model. *Comm. Statist. A–Theory Methods* **8** 1353–1358.

[12] RAO, C. R., AND MITRA, S. K. (1971). *Generalized Inverse of Matrices and Its Applications.* Wiley, New York.

[13] SHINOZAKI, N. (1975). *A Study of Generalized Inverse of a Matrix and Estimation with Quadratic Loss.* Ph.D. thesis, submitted to Keio University.

[14] ZHU, X. F. (1987). A note on the sufficient and necessary conditions for an admissible linear estimator, unpublished manuscript.

On Moment Conditions for Valid Formal Edgeworth Expansions

RABI N. BHATTACHARYA*

Indiana University

AND

J. K. GHOSH

The Indian Statistical Institute, Calcutta, India

The validity of formal Edgeworth expansions for statistics which are functions of sample averages was established in R. N. Bhattacharya and J. K. Ghosh (1978, *Ann. Statist.* **6** 434–451) under a moment condition which is sometimes too severe. In this article this moment condition is relaxed. Two examples of P. Hall (1983, *Ann. Probab.* **11** 1028–1036; 1987, *Ann. Probab.* **15** 920–931) are discussed in this context. © 1988 Academic Press, Inc.

INTRODUCTION

The validity of formal Edgeworth expansions for classical statistics was established in Bhattacharya and Ghosh [2] under moment conditions which cannot be relaxed in general, but turn out to be too severe in some cases. Two such examples are considered in Hall [6, 7]. In these examples and many others the highest order of moments involved in the actual expansion is much smaller than the order of moments assumed finite in our earlier work [2], and special methods were used by Hall [6, 7] to relax this moment condition. Attempts to find minimal moment restrictions for the general case run into unexpected analytical difficulties.

Suppose that the statistic may be expressed as (or approximated by) $H(\bar{Z})$, where $\bar{Z} = (1/n) \sum_{j=1}^{n} Z_j$ is a mean of i.i.d. vectors and H is a smooth function in a neighborhood of $\mu = EZ_j$. If all the components of grad $H(\mu)$

Multivariate Statistics and Probability
ISBN 0-12-580205-6

Reprinted from *J. Mult. Anal.* **27**(1).

are nonzero then one cannot significantly weaken the earlier moment assumptions. In this article we provide a relaxation of the moment condition in case grad $H(\mu)$ has some zero components, as is true in both examples of Hall. Apart from the method we present in detail here, another method using conditioning with respect to some coordinates of Z_j (namely coordinates $Z_j^{(i)}$ for which $(\partial H(z)/\partial z^{(i)})(\mu)=0$) is sketched as Remark 5 in Section 7. This last method generalizes some ideas of Hall [7] dealing with Student's statistic.

1. The Main Result

Many classical statistics are (or, may be approximated by statistics) of the form $H(\bar{Z})$, where $\bar{Z}=(1/n)\sum_1^n Z_j$ is a k-dimensional mean vector of sample characteristics and H is smooth in a neighborhood of $\mu=E\bar{Z}$. If grad $H(\mu)\neq 0$, and $E|Z_j|^2<\infty$, then the normalized statistic $W_n=\sqrt{n}\,(H(\bar{Z})-H(\mu))$ is asymptotically normal. This follows from the Taylor expansion

$$W_n=\sqrt{n}\,(\bar{Z}-\mu)\cdot \text{grad } H(\mu)+o_p(1)). \tag{1.1}$$

If $E|Z_j|^s<\infty$ for some integer $s\geqslant 3$ and H is s-times continuously differentiable in a neighborhood of μ, then one may approximate W_n better by

$$W_n''=n^{1/2}\left\{\sum_{i=1}^{k} l_i(\bar{Z}^{(i)}-\mu^{(i)})+\frac{1}{2!}\sum_{i_1,i_2=1}^{k} l_{i_1,i_2}(\bar{Z}^{(i_1)}-\mu^{(i_1)})(\bar{Z}^{(i_2)}-\mu^{(i_2)})\right.$$
$$\left.+\cdots+\frac{1}{(s-1)!}\sum_{i_1,\dots,i_{s-1}=1}^{k} l_{i_1,\dots,i_{s-1}}(\bar{Z}^{(i_1)}-\mu^{(i_1)})\cdots(\bar{Z}^{(i_{s-1})}-\mu^{(i_{s-1})})\right\}. \tag{1.2}$$

Here superscripts denote coordinates and $l_i=(D_iH)(\mu)$, $l_{i_1,i_2}=(D_{i_1}D_{i_2}H)(\mu)$, etc., with D_i denoting differentiation with respect to the ith coordinate. One may compute the jth cumulant $K_{j,n}$ of W_n'' algebraically $(1\leqslant j\leqslant s)$, and keep only terms up to order $O(n^{-(s-2)/2})$:

$$K_{j,n}=\tilde{K}_{j,n}+o(n^{-(s-2)/2})\qquad (1\leqslant j\leqslant s), \tag{1.3}$$

$\tilde{K}_{j,n)}$ being a polynomial in $n^{-1/2}$ with coefficients determined by the moments of Z_j and the derivatives $l_i, k_{i_1,i_2},\dots,l_{i_1,\dots,i_{s-1}}$. One has $\tilde{K}_{1,n}=O(n^{-1/2})$, $\tilde{K}_{2,n}=\sigma^2+o(n^{-1/2})$, $\tilde{K}_{j,n}=O(n^{-(j-2)/2})$ $(j\geqslant 3)$, where

$$\begin{aligned}\sigma^2&=\text{grad } H(\mu)\cdot V\,\text{grad } H(\mu),\\ V&\equiv \text{cov } Z_j.\end{aligned} \tag{1.4}$$

The characteristic function of W_n is now approximated by

$$\begin{aligned}
&\exp\left\{\sum_{j=1}^{s-2}\frac{(i\xi)^j}{j!}\tilde{K}_{j,n}\right\} \\
&=\exp\left\{-\frac{\sigma^2\xi^2}{2}\right\}\exp\left\{i\xi\tilde{K}_{1,n}-\frac{\xi^2}{2}(\tilde{K}_{2,n}-\sigma^2)+\sum_{j=3}^{s-2}\frac{(i\xi)^j}{j!}\tilde{K}_{j,n}\right\} \\
&=\exp\left\{-\frac{\sigma^2\xi^2}{2}\right\}\left[1+\sum_{j=1}^{s-2}n^{-j/2}\pi_j(i\xi)\right] \\
&\quad+o(n^{-(s-2)/2})=\hat{\psi}_{s,n}(\xi)+o(n^{-(s-2)/2}),
\end{aligned} \tag{1.5}$$

say. For the second equality in (1.5) one expands in powers of $n^{-1/2}$. Here $\pi_j(i\xi)$ is a polynomial (in $i\xi$) whose coefficients depend on the moments of Z_j and the derivatives of H at μ. Now $\hat{\psi}_{s,n}$ is the Fourier transform of the *density* $\psi_{s,n}$ *of the formal Edgeworth expansion* of the distribution of W_n, obtained by inversion:

$$\begin{aligned}
\psi_{s,n}(x)&=\left[1+\sum_{j=1}^{s-2}n^{-j/2}\pi_j\left(-\frac{d}{dx}\right)\right]\phi_{\sigma^2}(x), \\
\phi_{\sigma^2}(x)&\equiv\frac{1}{\sqrt{2\pi\sigma^2}}e^{-x^2/2\sigma^2}.
\end{aligned} \tag{1.6}$$

Suppose that the observations Y_j $(j=1, 2, ...)$ are i.i.d. m-dimensional with common distribution G and that

$$Z_j=(f_1(Y_j), f_2(Y_j), ..., f_k(Y_j))=(Z_j^{(1)}, Z_j^{(2)}, ..., Z_j^{(k)}), \tag{1.7}$$

where f_r $(1\leqslant r\leqslant k)$ are real-valued Borel measurable functions on $\mathbb{R}^m$. Let Q_1 denote the (common) distribution of $Z_j-\mu$. The following assumptions were made in Bhattacharya and Ghosh [2], Bhattacharya [1], to prove the validity of the formal expansion (1.6) (i.e., to establish $\text{Prob}(W_n\in B)=\int_B\psi_{s,n}(x)\,dx+o(n^{-(s-2)/2})$ uniformly for all Borel sets B):

(B_1) *H is $(s-1)$-times continuously differentiable in a neighborhood of μ.*

(B_2) *grad $H(\mu)\neq 0$.*

(B_3) *$E\,|f_r(Y_j)|^s<\infty$ for $1\leqslant r\leqslant k$.*

(B_4) *There exists a nonempty open subset U of $\mathbb{R}^m$ with the properties: (i) G has a nonzero absolutely continuous component (with respect to Lebesgue measure on $\mathbb{R}^m$) with a positive density on U; (ii) f_r $(1\leqslant r\leqslant k)$ are continuously differentiable on U; (iii) $1, f_1, ..., f_k$ are linearly independent as elements of the vector space of real valued continuous functions on U.*

Let us now assume, instead of (B_1), (B_2), (B_3),

(B_1') *H is s-times continuously differentiable in a neighborhood of μ.*

(B_2') (i) *$l_i \neq 0$ for $1 \leqslant i \leqslant k_1$;* (ii) *$l_i = 0$ for $k_1 < i \leqslant k$, where k_1 is an integer satisfying $1 \leqslant k_1 < k$.*

(B_3') (i) *$E|f_r(Y_j)|^s < \infty$ for $1 \leqslant r \leqslant k_1$;* (ii) *$E|f_r(Y_j)|^{s-1} < \infty$ for $k_1 < r \leqslant k$, for some positive integer $s \geqslant 3$.*

Our main result relaxing earlier moment conditions is the following.

THEOREM. *Under the assumptions (B_1'), (B_2'), (B_3'), (B_4) one has*

$$\sup_{u \in \mathbb{R}^1} \left| \operatorname{Prob}(W_n \leqslant u) - \int_{-\infty}^{u} \psi_{s,n}(x)\, dx \right| = o(n^{-(s-2)/2}). \tag{1.8}$$

Proof. Recall the notation $W_n = \sqrt{n}(H(\bar{Z}) - H(\mu))$. Let

$$\begin{aligned} W_n' = & \sum_{1 \leqslant i \leqslant k_1} l_i \sqrt{n}\,(\bar{Z}^{(i)} - \mu^{(i)}) \\ & + \frac{n^{-1/2}}{2!} \sum_{1 \leqslant i_1, i_2 \leqslant k} l_{i_1, i_2} \sqrt{n}\,(\bar{Z}^{(i_1)} - \mu^{(i_1)}) \sqrt{n}\,(\bar{Z}^{(i_2)} - \mu^{(i_2)}) \\ & + \cdots + \frac{n^{-(s-1)/2}}{s!} \sum_{1 \leqslant i_1, i_2, \ldots, i_s \leqslant k} l_{i_1, i_2, \ldots, i_s} \sqrt{n}\,(\bar{Z}^{(i_1)} - \mu^{(i_1)}) \\ & \cdots \sqrt{n}\,(\bar{Z}^{(i_s)} - \mu^{(i_s)}). \end{aligned} \tag{1.9}$$

We first prove (1.8) with W_n replaced by W_n'. By Lemma 2.2 in Bhattacharya and Ghosh [2], Q^{*k} (i.e., the distribution of $\sum_1^k (Z_j - \mu)$) has a nonzero absolutely continuous component. Hence the distribution Q_n of $\sqrt{n}\,(\bar{Z} - \mu)$ has a nonzero absolutely continuous component for $n \geqslant k$. Write

$$\begin{aligned} h(z, \varepsilon) = & \sum_{1 \leqslant i \leqslant k_1} l_i z^{(i)} + \frac{\varepsilon}{2!} \sum_{1 \leqslant i_1, i_2 \leqslant k} l_{i_1, i_2} z^{(i_1)} z^{(i_2)} \\ & + \cdots + \frac{\varepsilon^{s-1}}{s!} \sum_{1 \leqslant i_1, \ldots, i_s \leqslant k} l_{i_1, \ldots, i_s} z^{(i_1)} \cdots z^{(i_s)}, \\ h(z, 0) = & \sum_{1 \leqslant i \leqslant k_1} l_i z^{(i)}. \end{aligned} \tag{1.10}$$

Now it is shown in Bhattacharya and Ranga Rao [3] (see the proof of Theorem 19.5 and the remark on p. 207) that there exists a part q_n' of the density (component) of Q_n which has the properties

$$\sup_B \left| \int_B q_n'(z)\, dz - Q_n(B) \right| = o(n^{-(s-)/2}) \qquad (B \text{ a Borel subset of } \mathbb{R}^k) \tag{1.11}$$

and

$$|q'_n(z) - \xi_{s-1,n}(z)| \leqslant c\delta_n n^{-(s-3)/2}(1 + |z|^{s+k}), \qquad [z \in \mathbb{R}^k], \quad (1.12)$$

where $\xi_{s-1,n}(z)$ *is the density of the* $(s-2)$*-term Cramér–Edgeworth expansion of* Q_n, c is a positive constant, and $\delta_n \to 0$ as $n \to \infty$. Note that (1.11) holds under the assumptions (B'_3), (B_4); i.e., $E|Z_j|^{s-1} < \infty$ suffices. Indeed the right side in (1.11) is $o(n^{-m})$ for *every* positive integer m (see relations (19.73), (19.76), (19.77) in Bhattacharya and Ranga Rao [3]).

By (1.11) the following holds uniformly for all u:

$$\begin{aligned}
\operatorname{Prob}(W'_n \leqslant u) &= \operatorname{Prob}\left(\sum_{1 \leqslant i \leqslant k_1} l_i \sqrt{n}\,(\bar{Z}^{(i)} - \mu^{(i)}) \leqslant u\right) \\
&\quad + \operatorname{Prob}\left(\{W'_n \leqslant u\} \Big\backslash \left\{\sum_{1 \leqslant i \leqslant k_1} l_i \sqrt{n}\,(\bar{Z}^{(i)} - \mu^{(i)}) \leqslant u\right\}\right) \\
&\quad - \operatorname{Prob}\left(\left\{\sum_{1 \leqslant i \leqslant k_1} l_i \sqrt{n}\,(\bar{Z}^{(i)} - \mu^{(i)}) \leqslant u\right\} \Big\backslash \{W'_n \leqslant u\}\right) \\
&= \operatorname{Prob}\left(\sum_{1 \leqslant i \leqslant k_1} l_i \sqrt{n}\,(\bar{Z}^{(i)} - \mu^{(i)}) \leqslant u\right) \\
&\quad + \int_{\{h(z,\varepsilon) \leqslant u\} \backslash \{h(z,0) \leqslant u\}} q'_n(z)\, dz \\
&\quad - \int_{\{h(z,0) \leqslant u\} \backslash \{h(z,\varepsilon) \leqslant u\}} + o(n^{-(s-2)/2}). \qquad (1.13)
\end{aligned}$$

But in view of $(B'_3)(i)$ (and (B_4)) one has, unformly for all u,

$$\begin{aligned}
&\operatorname{Prob}\left(\sum_{1 \leqslant i \leqslant k_1} l_i \sqrt{n}\,(\bar{Z}^{(i)} - \mu^{(i)}) \leqslant u\right) \\
&\quad = \int_{\{z \in \mathbb{R}^{k_1} : \sum_1^{k_1} l_i z^{(i)} \leqslant u\}} {}^1\xi_{s,n}(z)\, dz + o(n^{-(s-2)/2}), \qquad (1.14)
\end{aligned}$$

where ${}^1\xi_{s,n}$ *is the density of the* $(s-1)$*-term Cramér–Edgeworth expansion of the distribution of* $\sqrt{n}\,(\bar{Z}^{(1)} - \mu^{(1)}, ..., \bar{Z}^{(k_1)} - \mu^{(k_1)})$.

On the other hand,

$$\begin{aligned}
&\int_{\{h(z,\varepsilon) \leqslant u\} \backslash \{h(z,0) \leqslant u\}} q'_n(z)\, dz - \int_{\{h(z,0) \leqslant u\} \backslash \{h(z,\varepsilon) \leqslant u\}} q'_n(z)\, dz \\
&\quad = \int_{\{h(z,\varepsilon) \leqslant u\} \backslash \{h(z,0) \leqslant u\}} \xi_{s-1,n}(z)\, dz - \int_{\{h(z,0) \leqslant u\} \backslash \{h(z,\varepsilon) \leqslant u\}} \xi_{s-1,n}(z)\, dz + \eta_n, \\
&\hfill (1.15)
\end{aligned}$$

where, by (1.12),

$$\eta_n \leqslant \left(\int_{\{h(z,\varepsilon) \leqslant u\} \Delta \{h(z,0) \leqslant u\}} (1 + |z|^{s+k})^{-1} \, dz \right) c\delta_n n^{-(s-3)/2}. \tag{1.16}$$

Here Δ denotes *symmetric difference*: $B \Delta C = (B \setminus C) \cup (C \setminus B)$. Note that for z in $\{|z| < 1/\varepsilon^{1/(s-1)}\}$ there are positive constants c_1, d_1 *such that*

$$h(z, \varepsilon) - c_1 \varepsilon \, |z||^2 - d_1 \varepsilon \leqslant h(z, 0) \leqslant h(z, \varepsilon) + c_1 \varepsilon \, |z|^2 + d_1 \varepsilon. \tag{1.17}$$

Write, for given u satisfying $|u| < 2 \, |l|/\varepsilon^{1/(s-1)} (|l|^2 = \sum_{1 \leqslant i \leqslant k_1} l_i^2)$,

$$A_\varepsilon = (\{h(z, \varepsilon) \leqslant u\} \, \Delta \{h(z, 0) \leqslant u\} \cap \{|z| < 1/\varepsilon^{1/(s-1)}\}. \tag{1.18}$$

Then

$$\begin{aligned} &A_\varepsilon \subset A_{\varepsilon 1} \cup A_{\varepsilon 2}, \\ &A_{\varepsilon 1} = \{u - c_1 \varepsilon \, |z|^2 - d_1 \varepsilon \leqslant h(z, 0) \leqslant u\} \cap \{|z| < 1/\varepsilon^{1/(s-1)}\}, \\ &A_{\varepsilon 2} = \{u < h(z, 0) \leqslant u + c_1 \, |z|^2 + d_1 \varepsilon\} \cap \{|z| < 1/\varepsilon^{1/(s-1)}\}. \end{aligned} \tag{1.19}$$

Now make an orthogonal transformation $z \to y$ with $y^{(1)} = h(z, 0)/|l| = \sum_{1 \leqslant i \leqslant k_1} l_i z^{(i)} / (\sum l_i^2)^{1/2}$. Then

$$\begin{aligned} &\int_{A_{\varepsilon 1}} (1 + |z|^{s+k})^{-1} \, dz \\ &\quad = \int_{\{(u - c_1 \varepsilon |y|^2 - d_1 \varepsilon)/|l| \leqslant y^{(1)} \leqslant u/|l|\} \cap \{|y| < 1/\varepsilon^{1/(s-1)}\}} (1 + |y|^{s+k})^{-1} \, dy. \end{aligned} \tag{1.20}$$

Write $|y|^2 = (y^{(1)})^2 + \sum_2^k (y^{(i)})^2 = (y^{(1)})^2 + r^2$ and solve the quadratic equation (in $y^{(1)}$): $y^{(1)} = (u - c_1 \varepsilon (y^{(1)})^2 - c_1 \varepsilon r^2 - d_1 \varepsilon) // |l|$, to derive from (1.20) the inequality

$$\begin{aligned} &\int_{A_{\varepsilon 1}} (1 + |z|^{s+k})^{-1} \, dz \\ &\quad \leqslant \int_{\{(u/|l|) - c_2 \varepsilon \leqslant y^{(1)} \leqslant u/|l|\} \cap \{|y| < 1/\varepsilon^{1/(s-1)}\}} (1 + |y|^{s+k})^{-1} \, dy \leqslant c_3 \varepsilon, \end{aligned} \tag{1.21}$$

which holds for some positive constants c_2, c_3 and for all sufficiently small $\varepsilon > 0$. Similarly, one has

$$\int_{A_{\varepsilon 2}} (1 + |z|^{s+k})^{-1} \, dz \leqslant c_4 \varepsilon \tag{1.22}$$

for some positive constant c_4 and all sufficiently small $\varepsilon > 0$. Also,

$$\int_{\{|z| > 1/\varepsilon^{1/(s-1)}\}} (1+|z|^{s+k})^{-1}\,dz = \omega_k \int_{1/\varepsilon^{1/(s-1)}}^{\infty} x^{k-1}(1+x^{s+k})^{-1}\,dx$$

$$\leqslant \omega_k \int_{1/\varepsilon^{1/(s-1)}} \frac{1}{x^{s+1}}\,dx \leqslant c_5 \varepsilon, \qquad [0<\varepsilon<1], \tag{1.23}$$

where ω_k, c_5 are suitable positive constants.

Combining (1.16)–(1.23) one gets, with $\varepsilon = n^{-1/2}$,

$$\eta_n = o(n^{-(s-2)/2}), \tag{1.24}$$

uniformly for all u satisfying $|u| < 2\,|l|/\varepsilon^{1/(s-1)}$. For $u \geqslant 2\,|l|/\varepsilon^{1(s-1)}$, $A_{\varepsilon 1}$ is empty for all sufficiently small ε (see (1.20)). For $u \leqslant -2\,|l|/\varepsilon^{1/(s-1)}$,

$$\int_{A_{\varepsilon 1}} (1+|z|^{s+k})^{-1}\,dz$$

$$\leqslant \int_{\{y^{(1)} \leqslant -2/\varepsilon^{1/(s-1)}\}} (1+|y|^{s+k})^{-1}\,dy$$

$$\leqslant c_6 \int_0^{\infty} r^{k-2} \left\{ \int_{\{y^{(1)} \leqslant -2/\varepsilon^{1/(s-1)}\}} (|y^{(1)}|+r)^{-s-k}\,dy^{(1)} \right\} dr$$

$$= \frac{c_6}{s+k-1} \int_0^{\infty} r^{k-2} \left(\frac{2}{\varepsilon^{1/(s-1)}} + r \right)^{-s-k+1} dr$$

$$\leqslant \frac{c_6}{s+k} \int_{2/\varepsilon^{1/(s-1)}}^{\infty} v^{-s-1}\,dv \leqslant c_7 \varepsilon, \tag{1.25}$$

for appropriate constants c_6, c_7. Similarly, one shows that

$$\int_{A_{\varepsilon 2}} (1+|z|^{s+k})^{-1}\,dz = O(\varepsilon) \qquad \text{as} \quad \varepsilon \downarrow 0, \tag{1.26}$$

in case $u \leqslant -2\,|l|/\varepsilon^{1/(s-1)}$. In exactly the same manner one shows that for $u \geqslant 2\,|l|/\varepsilon^{1/(s-1)}$, the integrals of $(1+|z|)^{-s-k}$ over $A_{\varepsilon 1}$ and $A_{\varepsilon 2}$ are $O(\varepsilon)$. Hence (1.24) holds uniformly for *all* u. Now use (1.24), (1.13)–(1.15) to get

$$\sup_{u \in \mathbb{R}^1} \Bigg| \operatorname{Prob}(W'_n \leqslant u) - \Bigg[\int_{\{z \in \mathbb{R}^{k_1} : \sum l_i z^{(i)} \leqslant u\}} {}^1\xi_{s,n}(z)\,dz$$

$$+ \int_{\{h(z,\varepsilon) \leqslant u\} \setminus \{h(z,0) \leqslant u\}} \xi_{s-1,n}(z)\,dz$$

$$- \int_{\{h(z,0) \leqslant u\} \setminus \{h(z,\varepsilon) \leqslant u\}} \xi_{s-1,n}(z)\,dz \Bigg] \Bigg| = o(n^{-(s-2)/2}). \tag{1.27}$$

The reduction of the above integrals is now carried out exactly as in Bhattacharya and Ghosh [2] to yield

$$\sup_{u \in \mathbb{R}^1} \left| \operatorname{Prob}(W_n' \leqslant u) - \int_{-\infty}^{u} \psi_{s,n}(x)\, dx \right| = o(n^{-(s-2)/2}). \tag{1.28}$$

Finally note that there exists a constant c_8 such that

$$|W_n - W_n'| \leqslant C_8 n^{\;\;s/2} / \sqrt{n}\,(\bar{Z} - \mu)|^{s+1}. \tag{1.29}$$

Now, by Corollary 17.12 in Bhattacharya and Ranga Rao [3] one has, for every $\varepsilon > 0$,

$$\begin{aligned} \operatorname{Prob}(\sqrt{n}\,|\bar{Z} - \mu| > \varepsilon n^{1/(s+1)}) &= o(n^{-(s-3)/2} n^{-(s-1)/(s+1)}) \\ &= o(n^{-(s-2)/2}) \qquad (s \geqslant 3). \end{aligned} \tag{1.30}$$

Since $\psi_{s,n}$ is bounded (uniformly in n), (1.28)–(1.30) imply (1.8). ∎

Remark 1. The proof essentially shows that one may replace the assumption (B_3') by (B_3''): $E\,|Z_1^{(i)}|^{s-r} < \infty$ *for all* i *which appear in the expression* (1.9) *for the first time in the sum* $n^{-r/2} \sum l_{i_1}, ..., {}_{i_{r+1}} \sqrt{n}\,(\bar{Z}^{(i_1)} - \mu^{(i_1)}) \cdots \sqrt{n}\,(\bar{Z}^{(i_{r+1})} - \mu^{(i_{r+1})})$ $(0 \leqslant r \leqslant s-2)$.

Remark 2. The proof goes over to the case of vector-valued statistics $\sqrt{n}\,(H(\bar{Z}) - H(\mu))$ (or, more generally, vector-valued statistics which may be adequately approximated, coordinate wise, in the form (1.9)).

Remark 3. In Bhattacharya and Ghosh [2], (also see Bhattacharya [1]) it is proved under the assumptions (B_1)–(B_4) that

$$\sup_{B} \left| \operatorname{Prob}(W_n \in B) - \int_B \psi_{s,n}(x)\, dx \right| = o(n^{-(s-2)/2}), \tag{1.31}$$

where the supremum is over the class of *all Borel subsets* B of $\mathbb{R}^1$. Our proof above, under the moment relaxation (B_3') (or (B_3'')), only provides an approximation of the *distribution function.* Although this proof may be extended to carry over to the case of probabilities of sets with *smooth boundaries* (e.g., Borel measurable convex sets), it does not yield (1.31). We do not know if (1.31) is valid under the hupothesis of the present theorem. (Of course, (1.31) holds in this case if the right side is replaced by $o(n^{-(s-3)/2})$.)

Remark 4. An entirely analogous result holds for statistics $H(\bar{Z})$ for which $l_i = 0$ for all i, while $l_{i_1, i_2} \neq 0$ for some i_1, i_2. Thus for statistics $n(H(\bar{Z}) - H(\mu))$ arising in testing statistical hypotheses (See Chandra and Ghosh [4]) moment conditions may be relaxed for those coordinates which do not appear in the principal term of the Taylor expansion around μ.

Remark 5 (Conditioning argument). We write $Z'_j = (Z_j^{(1)}, ..., Z_j^{(k_1)})$, $Z''_j = (Z_j^{(k_1+1)}, ..., Z_j^{(k)})$, $EZ'_j = \mu'$, $EZ''_j = \mu''$. Under (B_4), $(\sum_1^k Z'_j, \sum_1^k Z''_j)$ has a joint density and, therefore, $\sum_1^k Z'_j$ has a conditional density given $\sum_1^k Z''_j$. Dividing up $\sum_1^n Z'_j$, $\sum_1^n Z''_j$ into consecutive blocks of k summands each, one may first obtain an asymptotic expansion of the conditional distribution of the first sum (centered around its conditional expectation) given block sums of Z''_j. The successive block sums of Z'_j are still *independent* under this conditioning, but *not identically distributed.* However, by restricting $\bar{Z}''$ close to μ'' (the complementary event having small probability), one may often justify an asmptotic expansion of the above conditional distribution (see, e.g., Bhattacharya and Ranga Rao [3, Theorem (9.3)]). Under this conditioning regard $H(\bar{Z})$ as a function of $\bar{Z}'$ with (block sums of) Z''_j as parameters, center $H(\bar{Z})$ around its conditional expectation, rewrite $\sqrt{n}\,(H(\bar{Z}) - H(\mu))$ in terms of this new centering, and proceed as in Bhattacharya and Ghosh [2] to obtain an asymptotic expansion of its conditional distribution. Finally expand the expectation of this expansion, this time dealing with (sample) means of i.i.d. summands. Such a procedure sometimes also succeeds in relaxing moment conditions. See Hall [7] for a similar procedure applied to the *Student's statistic.* Clearly, for the expansion of the conditional distribution of the statistic up to an error $o(n^{-(s-2)/2})$ one only needs $E\,|\,Z'_j|^s < \infty$, together with an appropriate moment condition on Z''_j to ensure that $\bar{Z}''$ remains sufficiently close to μ'' with probability $1 - o(n^{-(s-2)/2})$. However, higher moments may be needed in carrying out the expansion of the expectation of the conditional expansion mentioned above. See Example 2 in Section 2 for an additional comment on this.

2. Examples

Example 1 (Hall [6]). Let Y_j $(j = 1, 2, ...)$ be a sequence of i.i.d. radom variables having zero mean, unit standard deviation and a nonzero third moment μ_3, say $\mu_3 > 0$. One may expect that the $100(1-\alpha)\%$ point of the distribution of $\sqrt{n}\,\bar{Y} = (Y_1 + \cdots + Y_n)/n^{1/2}$ is better approximated (than the $100(1-\alpha)\%$ point $z = z(\alpha)$ of the standard normal) by that of the normalized chisquare χ_N^2 having N degrees of freedom, where N is chosen so that the third moment (namely, $(8/N)^{1/2}$) of $T_N \equiv (2N)^{-1/2}(\chi_N^2 - N)$ equals that of $\sqrt{n}\,\bar{Y}$ (namely, $\mu_3/n^{1/2}$); i.e.,

$$N = 8n/\mu_3^2. \tag{2.1}$$

One may use the gamma tables to find $z_N = z_N(\alpha)$ such that

$$\text{Prob}(T_N \leqslant z_N) = 1 - \alpha. \tag{2.2}$$

Hall [6] shows that z_N is indeed a better approximation of the $100(1-\alpha)\%$ point for $\sqrt{n}\,\bar{Y}$ than usual estimates, under Cramér's condition as well as in the lattice case. In case μ_3 is unknown, replace it by the sample third moment $\hat{\mu}_3$ and write

$$\hat{N} = 8n/\hat{\mu}_3^2. \tag{2.3}$$

Hall [6, Theorem 5] provides an asymptotic expansion of $\text{Prob}(\sqrt{n}\,\bar{Y} \leqslant z_{\hat{N}})$ up to order $o(n^{-1})$, uniformly for $\alpha \in [\varepsilon, 1-\varepsilon]$ for every $\varepsilon > 0$, under the assumptions (i) $EY_1^6 < \infty$ and (ii) (Y_1, Y_1^3) satisfies Cramér's condition. He correctly points out that this expansion may be derived from Bhattacharya and Ghosh [2] only if (i) is strengthened to (i)′ $EY_1^{12} < \infty$. Let us show that our present results may be used to derive Hall's expansion under the conditions (i) $EY_1^6 < \infty$ and (ii)″ (B_4) *holds with* $m=1$, $k=2$; $f_1(y) = y$, $f_2(y) = y^3$.

By Lemma 1 of Hall [6], obtained by equating the asymptotic expansion of $\text{Prob}(T_N \leqslant y)$ with $1-\alpha$, one has

$$z_N = z + N^{-1/2} P_1(z) + N^{-1} P_2(z) + o(N^{-1}), \tag{2.4}$$

uniformly for $\alpha \in [\varepsilon, 1-\varepsilon]$ (for every fixed positive ε). Here P_1, f_2 are polynomials. Thus it is enough to expand $\text{Prob}(\sqrt{n}\,\bar{Y} \leqslant z')$, where

$$\begin{aligned} z' &= z + \hat{N}^{-1/2} P_1(z) + \hat{N}^{-1} P_2(z) \\ &= z + \frac{\hat{\mu}_3}{\sqrt{8n}} P_1(z) + \frac{\hat{\mu}_3^2}{8n} P_2(z) \\ &= z + \frac{\mu_3 P_1(z)}{\sqrt{8n}} + \frac{\mu_3^2 P_2(z)}{8n} \\ &\quad + n^{-1} \left\{ \sqrt{n}\,(\hat{\mu}_3 - \mu_3) \left(\frac{P_1(z)}{\sqrt{8}} + \frac{2\mu_3}{8n^{1/2}} P_2(z) \right) \right\} \\ &\quad + n^{-2} (\sqrt{n}\,(\hat{\mu}_3 - \mu_3))^2 \frac{P_2(z)}{8}. \end{aligned} \tag{2.5}$$

Expressing $\sqrt{n}\,\bar{Y} \leqslant z'$ in the form (1.9), one may now apply Remark 1 with $s=4$. Note that $\sqrt{n}\,(\bar{Z}^{(2)} - \mu^{(2)}) = \sqrt{n}\,(\hat{\mu}_3 - \mu_3)$ appears the first time with coefficient n^{-1}, so that (B_3'') becomes

$$EY_1^4 < \infty, \qquad E\,|Y_1^3|^2 \equiv EY_1^6 < \infty. \tag{2.6}$$

We have taken $\hat{\mu}_3 = n^{-1} \sum_{j=1}^n Y_j^3$ above. One may modify the calculations a little in case $\hat{\mu}_3 = n^{-1} \sum_{j=1}^n (Y_j - \bar{Y})^3$, to prove that (2.6) suffices along with (B_4) (with $k=3$, $f_i(y) = y^i$ for $i = 1, 2, 3$).

The expansion of $\mathrm{Prob}(\sqrt{n}\,\bar{Y} \leqslant z')$ in terms up to order n^{-1} involves EY_1^4 (see Hall [6, p. 1032]). It may be shown by complicated algebra that the coefficient of $n^{-3/2}$ in the formal expansion involves EY_1^6. Also, looking at (2.5) one would not expect a valid asymptotic expansion with error $o(n^{-1})$ unless $\sqrt{n}\,(\hat{\mu}_3 - \mu_3)$ converges in distribution. Thus it is unlikely that the desired expansion holds in general under the condition $E\,|Y_1|^r < \infty$ for some $r < 6$.

EXAMPLE 2 (Studentized statistics). Consider the Student's statistic $t = \bar{Y}/\hat{\sigma}$, where $\hat{\sigma}^2 = (1/n)\sum_{j=1}^n Y_j^2 - \bar{Y}^2$. Here $m = 1$, $k = 2$; $Z_j^{(1)} = Y_j$, $Z_j^{(2)} = Y_j^2$, $EY_j = 0$. According to the theorem in Section 1, under (B_4) the distribution of $n^{1/2}t$ has an asymptotic expansion with error $o(n^{-(s-2)/2})$ if

$$EY_j^{2(s-1)} < \infty, \tag{2.7}$$

instead of the earlier requirement: $EY_j^{2s} < \infty$. Thus for an error $o(n^{-1/2})$ one needs finite fourth moments. By a conditioning argument, similar to the one sketched in Remark 5, Hall [7] proves that for an error $o(n^{-1/2})$, $E\,|Y_1^3| < \infty$ is enough. He also shows that for a higher order expansion of the conditional distribution of t, given $\{Y_j^2, 1 \leqslant j \leqslant n\}$, $E\,|Y_j|^s < \infty$ suffices; but we are unable to obtain the appropriate expansion of the expectation of the conditional expansion under this moment condition.

Consider now the asmptotic expansion of the Studentized sample moment $\hat{\mu}_r = n^{-1}\sum_{j=1}^n Y_j^r$ (r is a positive integer). The studentized statistic is $T = (\hat{\mu}_r - \mu_r)/\hat{\sigma}_r$, where $\hat{\sigma}_r^2$ is obtained by replacing population moments by sample moments in the expression $\mathrm{var}(\hat{\mu}_r)$ calculated at least approximately keeping the principal terms (i.e., terms of order n^{-1}). For an expansion with an error term $o(n^{-(s-2)/2})$, the theorem in Section 1 requires $E\,|Y_j|^{2r(s-1)} < \infty$ instead of the older moment condition $E\,|Y_j|^{2rs} < \infty$.

ACKNOWLEDGMENTS

Professor W. van Zwet has kindly brought to our attention the articles by Chibishov [5] in which moment conditions are relaxed much further for *polynomial statistics*. It is not clear to us if Chibishov's results lead in general to better moment conditions for nonpolynomial statistics. Also our method is different and much simpler than that of Chibishov.

REFERENCES

[1] BHATTACHARYA, R. N. (1985). Some recent results on Cramér–Edgeworth expansions with applications. *Multivariate Analysis–VI* (P. R. Krishnaiah, Ed.) pp. 57–75. Elsevier, New York.

[2] BHATTACHARYA, R. N., AND GHOSH, J. K. (1978). On the validity of the formal Edgeworth expansion. *Ann. Statist.* **6** 434–451.

[3] BHATTACHARYA, R. N., AND RANGA RAO, R. (1986). *Normal Approximation and Asymptotic Expansions.* Krieger, Melbourne, Fl.

[4] CHANDRA, T. K., AND GHOSH, J. K. (1979). Valid asymptotic expansions for the likelihood ratio statistic and other perturbed chi-square variables. *Sankhyā Ser. A* **41** 22–47.

[5] CHIVISHOV, D. M. (1980, 1981). An asymptotic expansion for the distribution of a statistic admitting a stochastic expansion I, II. *Theory Probab. Appl.* **15** 732–744; **16** 1–12.

[6] HALL, P. (1983). Chi squared approximations to the distribution of a sum of independent random variables. *Ann. Probab.* **11** 1028–1036.

[7] HALL, P. (1987). Edgeworth expansion for student's *t* statistic under minimal moment conditions. *Ann. Probab.* **15** 920–931.

Ergodicity and Central Limit Theorems for a Class of Markov Processes

RABI N. BHATTACHARYA*

Indiana University

AND

OESOOK LEE

Ewha Women's University, Seoul, South Korea

We consider a class of discrete parameter Markov processes on a complete separable metric space S arising from successive compositions of i.i.d. random maps on S into itself, the compositions becoming contractions eventually. A sufficient condition for ergodicity is found, extending a result of Dubins and Freedman [8] for compact S. By identifying a broad subset of the range of the generator, a functional central limit theorem is proved for arbitrary Lipschitzian functions on S, without requiring any mixing type condition or irreducibility.

1. INTRODUCTION

Recent work has shown that the Billingsley–Ibragimov martingale central limit theorem (Billingsley [6, Theorem 23.1]) is the right tool for deriving functional central limit theorems for general ergodic Markov processes (Gordin and Lifsic [10], Bhattacharya [2]). There are several reasons for this. First, no mixing type condition is needed. Computations of mixing rates are often virtually impossible, and there are many important ergodic Markov processes for which none of the usual mixing rates goes to zero. Second, the martingale central limit theorem is applicable to each

* Research of this author is partially supported by NSF Grant DMS 8503358.

Multivariate Statistics and Probability
ISBN 0-12-580205-6

Reprinted from *J. Mult. Anal.* **27**(1).

centered function belonging to the range of the generator of the Markov process. The class of such functions is dense in the $\mathbf{L}^2$-space with respect to the invariant probability. Last, but not least, an analytical expression for the variance parameter of the limiting Brownian motion is automatically provided. Some illustrations of these different aspects of the theory may be found in Bhattacharya and Gupta [4], Bhattacharya [3], and Bhattacharya and Lee [5]. The present article provides another class of such processes. The nontrivial tasks in these applications are (1) the derivation of a *criterion for ergodicity* and (2) the *identification of* (a large subset of) *the range of the generator.*

In this article, we consider a discrete parameter Markov process $\{X_n\}$ on a complete separable metric space (S, ρ), represented as $X_n = \alpha_n \alpha_{n-1} \cdots \alpha_1 X_0$, where X_0 is a given random variable with values in S and $\{\alpha_n\}$ is an independent and identically distributed (i.i.d.) sequence of continuous random maps on S into itself. Also, X_0 and $\{\alpha_n\}$ are independent. It is assumed that there exists a positive integer m_0 such that with probability one, $\alpha_m \cdots \alpha_1$ is a *contraction* for each $m \geqslant m_0$. Under two additional assumptions (see (A_1), (A_2) in Section 2) it is shown that there exists a unique invariant probability π, and that the n-step transition probability $p^{(n)}(x, dy)$ converges weakly to $\pi(dy)$, as $n \to \infty$, for every $x \in S$ (Theorem 2.2). This extends to noncompact spaces an earlier result of Dubins and Freedman [8, Corollary 2.3]. What is novel about such a result is that the transition probability $p(x, dy)$ need not be *irreducible.* Recall that p is said by *φ-irreducible* with respect to a non-zero sigma finite measure φ if $\varphi(B) > 0$ implies, for each x, the existence of an integer $n = n(x, B)$ such that $p^{(n)}(x, B) > 0$ (Orey [13]). Typically, irreducibility is violated when the distribution of α_1 has a finite or discrete support. Such examples arise even in the case of linear autoregressive models of order one. See Bradley [7, Example 6.2] for a discussion of a example originally due to Rosenblatt [14].

Under an additional assumption (see (2.19)) it is shown that all centered Lipschitzian function f in $\mathbf{L}^2(S, \pi)$ belong to the range of $T - I$, where $(Tg)(x) = E(g(\alpha_1 x)) = \int g(y)\, p(x, dy)$, and I is the identity operator. It then follows from Gordin and Lifsic [10] and Bhattacharya [2] that the functional central limit theorem holds for such functions f (Theorem 2.5).

2. Main Results

Let S be a complete separable metric space with metric ρ and $\mathscr{B}(S)$ its Borel sigma field. Let Γ be a set of continuous maps on S into S. Endow Γ with the topology of uniform convergence on bounded sets and let $\mathscr{B}(\Gamma)$ be the Borel sigma field on Γ. Let P be a probability measure on $(\Gamma, \mathscr{B}(\Gamma))$. Consider a probability space $(\Omega, \mathscr{F}, Q)$ on which are defined an i.i.d. sequence of random maps $\alpha_1, \alpha_2, \ldots$ with common distribution P, and a random variable X_0 with values in S independent of the sequence $\{\alpha_n\}$. Then the following sequence $\{X_n\}$ is a Markov process on S,

$$X_0, X_n := \alpha_n \cdots \alpha_1 X_0 \qquad (n \geqslant 1). \tag{2.1}$$

Here, we write γx for the value of the map $\gamma \in \Gamma$ at x, and $\gamma_n \cdots \gamma_1$ for the composition of the maps $\gamma_1, \gamma_2, \ldots, \gamma_n$. It is well known (Kifer [12, Theorem 1.1, p. 8]) that every discrete parameter Markov process on S may be constructed in this manner, although Γ need not be a set of continuous maps.

Write Γ^m for the usual Cartesian product $\Gamma \times \cdots \times \Gamma$, and $\Gamma^{(m)}$ for the set of all compositions $\gamma_1 \gamma_2 \cdots \gamma_m$ of elements $\gamma_i \in \Gamma$ $(i = 1, \ldots, m)$. Let P^m denote the product probability on $(\Gamma^m, \mathscr{B}(\Gamma^m))$.

The following assumptions are made:

(A_0) *There exists m_0 such that for all $m \geqslant m_0$ every element of $\Gamma^{(m)}$ is a contraction, i.e., $\rho(\gamma x, \gamma y) \leqslant \rho(x, y)$ for $\gamma \in \Gamma^{(m)}$.*

(A_1) *Let m_0 be as in (A_0). For every $\varepsilon > 0$ there exists $\beta_\varepsilon < 1$ such that $P^{m_0}(\{(\gamma_1, \ldots, \gamma_{m_0}) \in \Gamma^{m_0}: \rho(\gamma_{m_0} \cdots \gamma_1 x, \gamma_{m_0} \cdots \gamma_1 y) \leqslant \max(\beta_\varepsilon \rho(x, y), \varepsilon)\ \forall x, y\}) > 0$.*

Write $\operatorname{diam}(C)$ for the *diameter* of $C \subset S$, $\operatorname{diam}(C) = \sup\{\rho(x, y): x, y \in C\}$. Also, γC denotes the set $\{\gamma x : x \in C\}$.

LEMMA 2.1. *Under the assumptions* (A_0), (A_1), $\operatorname{diam}(\alpha_n \cdots \alpha_1 C) \to 0$ *almost surely for every bounded $C \subset S$, as $n \to \infty$.*

Proof. Fix a bounded set C. For each $\varepsilon > 0$ and positive integer N define the sequence $\{F_j\}$ of events (in $(\Omega, \mathscr{F}, Q)$) by

$$\begin{aligned} F_j = [&\rho(\alpha_{mm_0} \cdots \alpha_{(m-1)m_0+1} x, \alpha_{mm_0} \cdots \alpha_{(m-1)m_0+1} y) \\ &\leqslant \max\{\beta_\varepsilon \rho(x, y), \varepsilon\}\ \forall x, y, \text{ and } \forall m \text{ satisfying } (j-1)N < m \leqslant jN] \\ &\qquad\qquad (j = 2, 3, \ldots). \end{aligned} \tag{2.2}$$

Then $Q(F_j) = Q(F_2) > 0$, each F_j being the intersection of N independent events each with the probability appearing in (A_1). Also, $\{F_j\}$ are independent. Therefore, by the Borel–Cantelli lemma, with Q-probability one, infinitely many F_j occur. Now if F_j occurs (for some $j \geqslant 2$) then for all $n \geqslant (jN+1)m_0$ one has, for every pair x, y in C,

$$\begin{aligned}
&\rho(\alpha_n \cdots \alpha_1 x, \alpha_n \cdots \alpha_1 y) \\
&\quad \leqslant \rho(\alpha_{jNm_0} \cdots \alpha_1 x, \alpha_{jNm_0} \cdots \alpha_1 y) \\
&\quad \leqslant \max\{\varepsilon, \beta_\varepsilon \rho(\alpha_{(jN-1)m_0} \cdots \alpha_1 x, \alpha_{(jN-1)m_0} \cdots \alpha_1 y)\} \\
&\quad \leqslant \max\{\varepsilon, \beta_\varepsilon^2 \rho(\alpha_{(jN-2)m_0} \cdots \alpha_1 x, \alpha_{(jN-2)m_0} \cdots \alpha_1 y)\} \\
&\quad \leqslant \cdots \leqslant \max\{\varepsilon, \beta_\varepsilon^N \rho(\alpha_{(j-1)Nm_0} \cdots \alpha_1 x, \alpha_{(j-1)Nm_0} \cdots \alpha_1 y)\} \\
&\quad \leqslant \max\{\varepsilon, \beta_\varepsilon^N \rho(x, y)\} \leqslant \max\{\varepsilon, \beta_\varepsilon^N \operatorname{diam}(C)\}.
\end{aligned} \tag{2.3}$$

Now find N such that $\beta_\varepsilon^N \operatorname{diam}(C) < \varepsilon$. Then for all sufficiently large n (depending on $\omega \in \Omega$) one has for all $x, y \in C$,

$$\rho(\alpha_n \cdots \alpha_1 x, \alpha_n \cdots \alpha_1 y) \leqslant \varepsilon. \quad \blacksquare$$

Let $p^{(n)}(x, dy)$ denote the *n-step transition probability* for the Markov chain $\{X_n\}$, where $p^{(1)}(x, dy) = p(x, dy)$. Note that $p^{(n)}(x, dy)$ is the distribution of $\alpha_n \cdots \alpha_1 x$.

On the *set* $\mathscr{P}(S)$ *of all probability measures on* $(S, \mathscr{B}(S))$ define the *bounded Lipschitzian distance*

$$d_{\mathrm{BL}}(\mu, \nu) = \sup\left\{\left|\int f\, d\mu - \int f\, d\nu\right| : \|f\|_\infty \leqslant 1, \|f\|_{\mathrm{L}} \leqslant 1\right\} \qquad (\mu, \nu \in \mathscr{P}(S)), \tag{2.4}$$

where $\|f\|_\infty = \sup\{|f(x)| : x \in S\}$, $\|f\|_{\mathrm{L}} = \sup\{|f(x) - f(y)|/\rho(x, y) : x \neq y \in S\}$. It is known that d_{BL} metrizes the weak-star topology on $\mathscr{P}(S)$ (Dudley [9]).

For the next result, we need the following additional assumption.

(A_2) *For some* $x_0 \in S$, $p^{(n)}(x_0, dy)$ *has the following property: for every* $\varepsilon > 0$ *there exists* $M_\varepsilon, n_\varepsilon$ *finite such that* $p^{(n)}(x_0, \{x : \rho(x, x_0) \geqslant M_\varepsilon\}) < \varepsilon$ $\forall n \geqslant n_\varepsilon$.

THEOREM 2.2. *Assume* (A_0), (A_1), (A_2). *There exists a unique invariant probability* $\pi(dy)$ *for* $p(x, dy)$, *and*

$$\sup\{d_{\mathrm{BL}}(p^{(n)}(x, dy), \pi(dy)) : x \in C\} \to 0, \qquad \text{as} \quad n \to \infty, \tag{2.5}$$

for every bounded set $C \subset S$.

Proof. Fix a bounded set C. For all $x_1, x_2 \in C$ one has

$$\begin{aligned}
&d_{\mathrm{BL}}(p^{(n)}(x_1, dy), p^{(n)}(x_2, dy)) \\
&\quad = \sup\{|Ef(\alpha_n \cdots \alpha_1 x_1) - Ef(\alpha_n \cdots \alpha_1 x_2)| : \|f\|_\infty \leqslant 1, \|f\|_{\mathrm{L}} \leqslant 1\} \\
&\quad \leqslant E(\rho(\alpha_n \cdots \alpha_1 x_1, \alpha_n \cdots \alpha_1 x_2) \wedge 1) \\
&\quad \leqslant E(\mathrm{diam}(\alpha_n \cdots \alpha_1 C) \wedge 1) \to 0 \quad \text{as} \quad n \to \infty, \qquad (2.6)
\end{aligned}$$

by Lemma 2.1. Similarly, writing $B(x_0 : M)$ for the ball of radius M centered at x_0, for all f satisfying $\|f\|_\infty \leqslant 1$, $\|f\|_L \leqslant 1$, one has

$$\begin{aligned}
&|Ef(\alpha_{n+m} \cdots \alpha_1 x_0) - Ef(\alpha_n \cdots \alpha_1 x_0)| \\
&\quad = |Ef(\alpha_1 \cdots \alpha_n \cdots \alpha_{n+m} x_0) - Ef(\alpha_1 \cdots \alpha_n x_0)| \\
&\quad \leqslant E(\rho(\alpha_1 \cdots \alpha_n \alpha_{n+1} \cdots \alpha_{n+m} x_0, \alpha_1 \cdots \alpha_n x_0) \wedge 1) \\
&\quad \leqslant Q(\{\rho(\alpha_{n+1} \cdots \alpha_{n+m} x_0, x_0) \geqslant M\}) \\
&\qquad + Q(\{\mathrm{diam}(\alpha_1 \cdots \alpha_n B(x_0 : M)) > \delta\}) + \delta, \qquad (2.7)
\end{aligned}$$

for every $M > 0$, $\delta > 0$. Given $\varepsilon > 0$, let $\delta = \varepsilon/3$ and choose $M = M'_\varepsilon$ such that

$$Q(\{\rho(\alpha_1 \cdots \alpha_m x_0, x_0) \geqslant M'_\varepsilon\}) < \varepsilon/3 \qquad \forall m = 1, 2, \ldots. \qquad (2.8)$$

This is possible since the family of distributions of $\rho(\alpha_1 \cdots \alpha_m x_0, x_0)$, $m \geqslant 1$, is relatively weak-star compact, by (A_2). By Lemma 2.1, $Q(\{\mathrm{diam}(\alpha_1 \cdots \alpha_n B(x_0 : M'_\varepsilon)) > \varepsilon/3\}) \to 0$ as $n \to \infty$. Hence, by (2.7) and (2.8), for all sufficiently large n, say $n > n_1(\varepsilon)$,

$$d_{\mathrm{BL}}(p^{(n+m)}(x_0, dy), p^{(n)}(x_0\, dy)) < \varepsilon \qquad \forall m = 1, 2, \ldots. \qquad (2.9)$$

Since $(\mathscr{P}(S), d_{\mathrm{BL}})$ is a complete metric space (Dudley [9]), it follows that there exists a probability measure π such that

$$d_{\mathrm{BL}}(p^{(n)}(x_0, dy), \pi(dy)) \to 0 \qquad \text{as} \quad n \to \infty. \qquad (2.10)$$

Now (2.6), (2.10) imply the uniform convergence of $p^{(n)}(x, dy)$ to $\pi(dy)$, in the d_{BL} metric, for $x \in C$. Since $x \to p(x, dy)$ is weak-star continuous, it is easily checked that π is the unique invariant probability. ∎

Theorem 2.2 extends Theorem 4.4 of Dubins and Freedman [8]. We state their result as a corollary.

COROLLARY 2.3 (Dubins and Freedman [8]). *Let S be a compact metric space, Γ a set of contractions on S, and P a probability measure on*

$(\Gamma, \mathscr{B}(\Gamma))$. If there exists a strict contraction γ_0 in the support of P, then there exists a unique unvariant probability π, and $p^{(n)}(x, dy)$ converges weakly to $\pi(dy)$, as $n \to \infty$, for each $x \in S$.

Proof. Assumptions (A_0), (A_2) are trivially satisfied in this case. It is enough to check (A_1) with $m_0 = 1$. For each $\varepsilon > 0$ define $\beta'_\varepsilon = \sup\{\rho(\gamma_0 x, \gamma_0 y)/\rho(x, y): x, y \text{ such that } \rho(x, y) \geqslant \varepsilon\}$. Then $\beta'_\varepsilon < 1$. For each $\delta > 0$ let $\Gamma_\delta = \{\gamma \in \Gamma: \rho(\gamma x, \gamma_0 x) < \delta\ \forall x\}$. Then $P(\Gamma_\delta) > 0$. Now if $\gamma \in \Gamma_\delta$ then

$$\begin{aligned}
\rho(\gamma x, \gamma y) &\leqslant \rho(\gamma x, \gamma_0 x) + \rho(\gamma_0 x, \gamma_0 y) + \rho(\gamma_0 y, \gamma y) \\
&< 2\delta + \rho(\gamma_0 x, \gamma_0 y) \\
&\leqslant 2\delta + \beta'_{\varepsilon/2} \rho(x, y)\, \chi_{\{\rho(x, y) \geqslant \varepsilon/2\}} + \frac{\varepsilon}{2} \chi_{\{\rho(x, y) < \varepsilon/2\}} \\
&\leqslant (\beta'_{\varepsilon/2} \rho(x, y) + 2\delta)\, \chi_{\{\rho(x, y) \geqslant \varepsilon/2\}} + (2\delta + \varepsilon/2)\, \chi_{\{\rho(x, y) < \varepsilon/2\}} \\
&\leqslant \left(\beta'_{\varepsilon/2} + \frac{4\delta}{\varepsilon}\right) \rho(x, y)\, \chi_{\{\rho(x, y) \geqslant \varepsilon/2\}} + \left(2\delta + \frac{\varepsilon}{2}\right) \chi_{\{\rho(x, y) < \varepsilon/2\}}.
\end{aligned} \tag{2.11}$$

Choose $\delta < \varepsilon/4$ such that $\beta_\varepsilon := \beta'_{\varepsilon/2} + 4\delta/\varepsilon < 1$. Then (2.11) becomes

$$\begin{aligned}
\rho(\gamma x, \gamma y) &\leqslant \beta_\varepsilon \rho(x, y)\, \chi_{\{\rho(x, y) \geqslant \varepsilon/2\}} + \varepsilon \chi_{\{\rho(x, y) < \varepsilon/2\}} \\
&\leqslant \max\{\beta_\varepsilon \rho(x, y), \varepsilon\} \qquad \forall \gamma \in \Gamma_\delta.
\end{aligned} \tag{2.12}$$

□

Remark 2.3.1. Assumption (A_2) is obviously necessary. It may be violated even for linear autoregressive models,

$$X_{n+1} = aX_n + \varepsilon_{n+1} \tag{2.13}$$

with $|a| < 1$, $\{\varepsilon_n\}$ an i.i.d. sequence. Here $S = \mathbb{R}^1$, $\Gamma = \{\gamma_\varepsilon : \varepsilon \in \mathbb{R}^1\}$ with $\gamma_\varepsilon(x) = ax + \varepsilon$, so that P is determined by the distribution G, say, of ε_1. It is easy to check that a unique invariant probability exists if and only if $\sum_{n=1}^{\infty} a^n \varepsilon_n$ converges almost surely or, equivalently, in distribution. For example, if $\sum_{n=1}^{\infty} G(\{\varepsilon: |a^n \varepsilon| > \delta\}) = \infty$ for some $\delta > 0$ then an invariant probability does not exist.

Remark 2.3.2. It is not difficult to check that Theorem 2.2 holds if the hypothesis (A_1) is replaced by the following alternative (A'_1). A contraction γ_0 will be said to be *asymptotically uniform on bounded sets* abbreviated as $a \cdot u \cdot b$, if

$$\lim_{m \to \infty} \sup_{\{x, y: \rho(x, y) \leqslant r\}} \rho(\gamma_0^m x, \gamma_0^m y) = 0 \qquad \forall r > 0.$$

(A_1') *There exists an* $a \cdot u \cdot b$ *contraction* γ_0 *such that for all* $\varepsilon > 0$ *and all* $m \geqslant m_0$ *one has*

$$P^m(\{(\gamma_1, ..., \gamma_m) \in \Gamma^m : \rho(\gamma_m \cdots \gamma_1 x, \gamma_0 x) \leqslant \varepsilon \; \forall x\}) > 0.$$

Assume that the hypothesis of Theorem 2.2 holds. Let T be the *transition operator* on $L^2(S, \pi)$,

$$(Tf)(x) := \int f(y) \, p(x, dy), \qquad f \in \mathbf{L}^2(S, \pi). \tag{2.14}$$

Then $(T^n f)(x) = \int f(y) \, p^{(n)}(x, dy)$. We will denote the $\mathbf{L}^2$-norm on $\mathbf{L}^2(S, \pi)$ by $\| \; \|_2$. Let I denote the identity operator. Write

$$\bar{f} = \int f \, d\pi. \tag{2.15}$$

LEMMA 2.4. *Let* $f \in \mathbf{L}^2(S, \pi)$. *If* $\sum_{n=0}^{\infty} \|T^n(f - \bar{f})\|_2 < \infty$, *then* $f - \bar{f}$ *belongs to the range of* $T - I$; *indeed,* $(T - I) g = f - \bar{f}$, *where*

$$g = - \sum_{n=0}^{\infty} T^n(f - \bar{f}). \tag{2.16}$$

Proof. Apply T to both sides of (2.16). ∎

It will be convenient to denote the sequence (2.1) as $\{X_n(x)\}$ if $X_0 \equiv x$,

$$X_0(x) := x, \; X_n(x) := \gamma_n \cdots \gamma_1 x \qquad (n \geqslant 1). \tag{2.17}$$

In order to state the functional central limit theorem, fix $f \in \mathbf{L}^2(S, \pi)$. For each positive integer n, write

$$Y_n(t) := n^{-1/2} \left[\sum_{j=0}^{[nt]} (f(X_j) - \bar{f}) + \left(t - \frac{[nt]}{n} \right) (f(X_{[nt]+1}) - \bar{f}) \right], \qquad (t \geqslant 0), \tag{2.18}$$

where $[nt]$ is the integer part of nt.

THEOREM 2.5. *Let the assumptions* (A_0), (A_1), (A_2) *hold. In addition, assume*

$$\sum_{n=0}^{\infty} \left(\int \left[\int E\rho(X_n(x), X_n(y)) \, \pi(dy) \right]^2 \pi(dx) \right)^{1/2} < \infty. \tag{2.19}$$

(a) *If the initial distribution is* π, *then for every Lipschitzian* f *in* $\mathbf{L}^2(S, \pi)$ *the function* $f - \bar{f}$ *belongs to the range of* $T - I$, *and for every such*

f the processes $Y_n(\cdot)$ converge in distribution to a Brownian motion with mean zero and variance parameter $\|g\|_2^2 - \|Tg\|_2^2$, where $(T-I)g = f - \bar{f}$.

(b) *If, further,*

$$n^{-1/2} \sum_{k=0}^{n} \left[\int E\rho(X_j(x), X_j(y))\, \pi(dy) \right] \to 0, \tag{2.20}$$

as $n \to \infty$, then the convergence in (a) *holds when $X_0 \equiv x$.*

Proof. (a) Let f be Lipschitzian on S, $|f(x) - f(y)| \leqslant M\rho(x, y)$ for all x, y. Then

$$\begin{aligned} |T^n(f-\bar{f})(x)|^2 &= \left(\int [Ef(X_n(x)) - Ef(X_n(y))]\, \pi(dy) \right)^2 \\ &\leqslant M^2 \left[\int E\rho(X_n(x), X_n(y))\, \pi(dy) \right]^2. \end{aligned} \tag{2.21}$$

Therefore,

$$\|T^n(f-\bar{f})\|_2^2 \leqslant M^2 \int \left[\int E\rho(X_n(x), X_n(y))\, \pi(dy) \right]^2 \pi(dx). \tag{2.22}$$

Hence if (2.19) holds, $f - \bar{f}$ belongs to the range of $T - I$ by Lemma 2.4. To prove the functional central limit theorem under the initial distribution π, let g be given by (2.16). Consider the representation

$$\begin{aligned} \sum_{j=0}^{n-1} (f(X_j) - \bar{f}) &= \sum_{j=0}^{n-1} (Tg(X_j) - g(X_j)) \\ &= \sum_{j=1}^{n} (Tg(X_{j-1}) - g(X_j)) + (g(X_n) - g(X_0)). \end{aligned} \tag{2.23}$$

Since $Tg(X_{j-1}) - g(X_j)$ $(j \geqslant 0)$ is, under the initial distribution π, a stationary ergodic sequence of martingale differences the functional central limit theorem follows (see Billingsley [6, Theorem 23.1], Gordin and Lifsic [10], Bhattacharya [2, Theorem 2.1]). In this case, the variance parameter of the limiting Brownian motion is $E(Tg(X_{j-1}) - g(X_j))^2 = \|g\|_2^2 - \|Tg\|_2^2$.

(b) Suppose (2.20) holds for some x (By (2.19) this is true for almost all $(\pi)x$.) Then, if f is as in (a),

$$\begin{aligned} &E\left(\max_{0 \leqslant j' \leqslant n} \left| n^{-1/2} \sum_{j=0}^{j'} (f(X_j(x)) - \bar{f}) - n^{-1/2} \sum_{j=0}^{j'} (f(X_j(y)) - \bar{f}) \right| \right) \\ &\leqslant Mn^{-1/2} \sum_{j=0}^{n} E\rho(X_j(x), X_j(y)). \end{aligned} \tag{2.24}$$

Let X_0 have distribution π and be independent of the sequence $\{\alpha_n\}$. Denoting $X_j = \alpha_j \cdots \alpha_1 X_0$, and letting $Y_n(\cdot)$ be the process defined by (2.18) and $Y_n^x(\cdot)$ the corresponding process with X_j replaced by $X_j(x)$ $(j \geqslant 0)$, one gets

$$E(\max_{0 \leqslant t \leqslant 1} |Y_n^x(t) - Y_n(t)|) \leqslant Mn^{-1/2} \left(\sum_{j=0}^{n} \int E\rho(X_j(x), X_j(y))\, \pi(dy) \right),$$

which goes to zero, as $n \to \infty$, by (2.20). ∎

Remark 2.5.1. By Hölder's inequality, (2.19) implies

$$\int \left[\sum_{n=0}^{\infty} \int E\rho(X_n(x), X_n(y))\, \pi(dy) \right] \pi(dx) < \infty. \tag{2.25}$$

Therefore, (2.20) is a mild extra condition and holds for all x outside a set of π-measure zero.

Remark 2.5.2. It is simple to check that every Lipschitzian f is in $\mathbf{L}^2(S, \pi)$ if, for some $z \in S$,

$$\int \rho^2(x, z)\, \pi(dx) < \infty. \tag{2.26}$$

EXAMPLE 2.5.3 (Linear time series models). Let $S = \mathbb{R}^k$, $\gamma_\varepsilon x = Ax + \varepsilon$, where A is a $k \times k$ matrix and $\Gamma = \{\gamma_\varepsilon : \varepsilon \in \mathbb{R}^k\}$ is endowed with the Euclidean topology on the set of labels ε. Let P be a probability measure on $(\Gamma, \mathscr{B}(\Gamma))$, i.e., on $(\mathbb{R}^k, \mathscr{B}(\mathbb{R}^k))$ such that $\int |\varepsilon|^2 P(d\varepsilon) < \infty$. Assume that the *eigenvalues* of A are all *less than one in magnitude*. Since the *spectral radius* $r(A)$, i.e., the largest magnitude of the eigenvalues, equals $\lim \|A^n\|^{1/n}$ (see Halmos [11, p. 182]), there exists m_0 such that $\|A^n\| < \delta^n$ for some $\delta < 1$ and for all $n \geqslant m_0$. The hypotheses (A_0), (A_1), (A_2) of Theorem 2.5 are satisfied with $\beta_\varepsilon = \delta$, and $x_0 = 0$, since $|X_n(x) - X_n(y)| = |AX_{n-1}(x) - AX_{n-1}(y)| = \cdots = |A^n(x-y)| \leqslant \|A^n\|\, |x-y|$. Also, the invariant distribution π is the distribution of $\sum_{n=0}^{\infty} A^n \varepsilon_n$, where ε_n are i.i.d. with common distribution P.

It is easy to check now that (2.19) holds, and (2.20) holds for all x. Hence the functional central limit theorem holds for $Y_n(\cdot)$ with f Lipschitzian, whatever the initial distribution is. In particular, $Z_n \equiv n^{-1/2} \sum_{j=0}^{n-1} (X_j - (I-A)^{-1} E\varepsilon_1)$ converges in distribution to a Gaussian law on R^k with mean zero. To calculate the *dispersion matrix* of this limiting Gaussian, check that $g(x) = -c'(I-A)^{-1}(x - (I-A)^{-1} E\varepsilon_1)$

solves $(T-I)\,g(x)=c'(x-(I-A)^{-1}E\varepsilon_1)$ for every $c\in\mathbb{R}^k$. Hence the variance of the limiting distribution of $c'Z_n$ is $\|g\|_2^2-\|Tg\|_2^2=c'Dc$, where

$$\begin{aligned} D&=(I-A)^{-1}V(I-A')^{-1},\\ V&:=\text{dispersion matrix of } \varepsilon_j \text{ under } P. \end{aligned} \tag{2.27}$$

This D is then the desired dispersion matrix.

One may treat the so-called $AR(q)$ or *linear autoregressive models of order k, and ARMA(k, q)* or *autoregressive-moving average models* of order (k, q) as special cases of the above example.

An ARMA(k, q) model is given by

$$U_{n+k}=\sum_{i=1}^{k}\beta_i U_{n+k-i}+\sum_{i=1}^{q}\delta_i\eta_{n+k-i}+\eta_{n+k}, \tag{2.28}$$

where η_n are i.i.d. real-valued and $\beta_1, ..., \beta_k, \delta_1, ..., \delta_q$ are real constants. Write $X_n=(U_n, ..., U_{n+k-1},\ \eta_{n+k-q}, ..., \eta_{n+k-1})'$, $\varepsilon_n=(0, ..., 0,\ 0, ..., 0, \eta_{n+k})'$. Then (2.28) may be expressed as

$$X_{n+1}=AX_n+\varepsilon_{n+1}, \tag{2.29}$$

where A is the $(k+q)\times(k+q)$ matrix

$$A=\begin{bmatrix} 0 & 1 & 0 & \cdot & \cdot & 0 & 0 & 0 & \cdot & \cdot & \cdot & 0 & 0\\ 0 & 0 & 1 & 0 & \cdot & 0 & 0 & 0 & \cdot & \cdot & \cdot & 0 & 0\\ \cdot & \cdot & \cdot & \cdot & \cdot & \cdot & \cdot & \cdot & \cdot & \cdot & \cdot & \cdot & \cdot\\ 0 & 0 & 0 & \cdot & \cdot & 1 & 0 & 0 & \cdot & \cdot & \cdot & 0 & 0\\ \beta_k & \beta_{k-1} & \cdot & \cdot & \cdot & \beta_1 & \delta_q & \delta_{q-1} & \cdot & \cdot & \cdot & \cdot & \delta_1\\ 0 & 0 & \cdot & \cdot & \cdot & 0 & 0 & 1 & 0 & \cdot & \cdot & \cdot & 0\\ 0 & 0 & \cdot & \cdot & \cdot & 0 & 0 & 0 & 1 & 0 & \cdot & \cdot & 0\\ \cdot & \cdot & \cdot & \cdot & \cdot & \cdot & \cdot & \cdot & \cdot & \cdot & \cdot & \cdot & \cdot\\ 0 & 0 & \cdot & \cdot & \cdot & 0 & 0 & 0 & 0 & 0 & \cdot & \cdot & 0 \end{bmatrix}. \tag{2.30}$$

Since $\text{Det}(A-\lambda I)=\text{Det}(B-\lambda I)\cdot(-\lambda)^q$, where B comprises the first k rows and columns of A, the nonzero roots of the characteristic polynomial equation for A are those of $\text{Det}(B-\lambda I)=0$. This last equation may be expressed as

$$-\lambda^k+\sum_{i=1}^{k}\beta_i\lambda^{k-i}=0. \tag{2.31}$$

As a special case of Example 2.5.3, therefore, there exists in this ARMA(k, q) model a unique invariant probability for X_n on $(\mathbb{R}^{k+q}, \mathscr{B}^{k+q})$

if the roots of (2.31) all lie within the unit circle and if $E\eta_n^2 < \infty$, and then the central limit theorem also applies.

A comprehensive account for the traditional treatment of the AR and ARMA models may be found in Anderson [1, Chaps. 5, 8]. By making use of Theorem 2.5 one may, however, prove central limit theorems for a broad class of nonlinear functions of X_n, and therefore of U_n, not provided by the classical treatment.

Remark 2.5.4. One may let U_{n+k} in (2.28) depend on all U_j, $-\infty < j < n+k$. In this case $S = \mathbb{R}^\infty$ and, given appropriate convergence of the coefficients, one may again derive conditions under which Theorems 2.2 and 2.5 apply. However, applications to nonlinear models of the form $X_{n+1} = \phi(X_n) + \varepsilon_{n+1}$ promise to be of greater significance.

REFERENCES

[1] ANDERSON, T. W. (1971). *The Statistical Analysis of Time Series.* Wiley, New York.

[2] BHATTACHARYA, R. N. (1982). On the functional central limit theorem and the law of iterated logarithm for Markov Processes. *Z. Wahrsch. Verw. Gebiete* **60** 185–201.

[3] BHATTACHARYA, R. N. (1985). A central limit theorem for diffusions with periodic coefficients. *Ann. Probab.* **13** 385–396.

[4] BHATTACHARYA, R. N., AND GUPTA, V. K. (1984). On the Taylor–Aris theory of solute transport in a capillary. *SIAM J. Appl. Math.* **44**, 33–39.

[5] BHATTACHARYA, R. N., AND LEE, O. (1988). Asymptotics of a class of Markov processes which are not in general irreducible. *Ann. Probab.*, in press.

[6] BILLINGSLEY, P. (1968). *Convergence of Probability Measures.* Wiley, New York.

[7] BRADLEY, R. C. (1986). Basic properties of strong mixing conditions. In *Dependence in Probability and Statistics: A Survey of Recent Results* (E. Eberlein, M. S. Taqqu, Eds.), pp. 165–192. Birkhauser, Boston.

[8] DUBINS, L. E., AND FREEDMAN, D. A. (1966). Invariant probabilities for certain Markov processes. *Ann. Math. Statist.* **37**, 837–847.

[9] DUDLEY, R. M. (1968). Distance of probability measures and random variables. *Ann. Math. Statist.* **39** 1563–1572.

[10] GORDIN, M. I., AND LIFSIC, B. A. (1978). The central limit theorem for stationary ergodic Markov process. *Dokl. Akad. Nauk. SSSR* **19** 392–393.

[11] HALMOS, P. R. (1958). *Finite-Dimensional Vector Spaces.* Van Nostrand, New York.

[12] KIFER, YU. (1986). *Ergodic Theory of Random Transformations.* Birkhäuser, Boston.

[13] OREY, S. (1971). *Limit Theorems for Markov Chain Transition Probabilities.* Van Nostrand, New York.

[14] ROSENBLATT, M. (1984). Linear processes and bispectra. *J. Appl. Probab.* **17**, 79–84.

Conditionally Ordered Distributions

HENRY W. BLOCK* AND ALLAN R. SAMPSON†

University of Pittsburgh

The concepts of conditionally more positively quadrant dependent, and conditionally more dispersed are introduced and studied. Based on these two concepts, new conditions are given for multivariate cdfs F and G so that $E_F h(\mathbf{X}) \geqslant E_G h(\mathbf{X})$ for suitable $h(\mathbf{X})$. Special cases include the multivariate normal distribution and elliptically contoured distributions. Conditional positive and negative dependence concepts as well as applications to the Farlie–Gumbel–Morgenstern distribution are also considered.

1. INTRODUCTION

Joag-dev, Perlman, and Pitt [6] study a type of pairwise condition on a function of n variables which implies monotonicity of the expected value of the function in the covariance matrix of a multivariate normal distribution. A related condition has been used by Cambanis and Simons [3] in obtaining a similar result. Both sets of authors also consider extensions to elliptically contoured distributions.

In this paper, we make the observation that the pairwise conditions of Joag-dev *et al.* actually represent conditions of two different types: (a) a condition related to pairwise dependence and (b) a condition related to dispersion orderings. Second, we demonstrate that the monotonicity result of Joag-dev *et al.* applies to any distributions which are conditionally pairwise dependence ordered or to distributions which are conditionally dispersion ordered.

In Section 2 we consider results for distributions which are conditionally positively quadrant-dependent ordered and in Section 3 we examine distributions which are conditionally dispersion ordered. In both sections, we derive the results of Joag-dev *et al.* [6] and Cambanis and Simon [3]

* Supported by ONR Contract N00014-84-K-0084 and AFOSR Grant AFOSR-84-0113.
† Supported by AFOSR Grant AFOSR-84-0113.

Multivariate Statistics and Probability
ISBN 0-12-580205-6

Reprinted from *J. Mult. Anal.* **27**(1).

as special cases. We also discuss in these two sections some improvements of the results of Joag-dev *et al.* under weaker regularity assumptions. In Section 4, the concepts of conditional positive and negative dependence are examined, and in Section 5 another example is considered.

Some notation which is used follows. For a given vector $\mathbf{a} = (a_1, ..., a_p)'$, define for each pair of integers $1 \leqslant i \leqslant j \leqslant p$ the corresponding vector

$$\mathbf{a}^{(i,j)} \equiv (a_1, ..., a_{i-1}, a_{i+1}, ..., a_{j-1}, a_{j+1}, ..., a_p)'.$$

In the case $i=j$ we write $\mathbf{a}^{(i)}$. For a given pair of integers $1 \leqslant i \leqslant j \leqslant p$ let $R(i,j) = \{1, ..., i-1, i+1, ..., j-1, j+1, ..., p\}$. (In the case $i=j$, we write $R(i)$.)

For a given cdf. $F(x_1, ..., x_p)$ and pair of integers $1 \leqslant i \leqslant j \leqslant p$ let

$$F(x_i, x_j \mid \mathbf{X}^{(i,j)} = \mathbf{t})$$

denote the conditional cdf of X_i, X_j given $\mathbf{X}^{(i,j)} = \mathbf{t}$. (In the case $i=j$, we write $F(x_i \mid \mathbf{X}^{(i)} = \mathbf{t})$.) Let $F_i(x_i)$ and $F_{R(i,j)}(x^{(i,j)})$, $1 \leqslant i \leqslant j \leqslant p$, denote the marginal cdfs, respectively, of X_i and $\mathbf{X}^{(i,j)}$. When densities exist, the following notations are used: $f(x_i, x_j \mid \mathbf{X}^{(i,j)} = \mathbf{t})$, $f_i(x_i)$ and $f_{R(i,j)}(\mathbf{x}^{(i,j)})$.

Let $a(x)$ be a function defined on $\mathbb{R}^1$. The number of sign changes of a, denoted by $S^-(a(x))$ is defined as $\sup S^-(a(x_1), ..., a(x_n))$ (over all) sequences $x_1 < \cdots < x_n$, $n = 1, 2, ...,$ where $S^-(\alpha_1, ..., \alpha_n)$ denotes the number of sign changes in $\alpha_1, ..., \alpha_n$, zero terms being ignored.

Let $\mathbf{I}_{ij}$ be the matrix whose every entry is zero, except for the (i,j)th entry which is 1. The dimension of $\mathbf{I}_{ij}$ is to be appropriate to the usage. Occasionally, we require a symmetrized version of the matrix, namely $\mathbf{I}_{ij} + \mathbf{I}_{ji}$, which we denote by $\mathbf{I}_{ij}^S$.

We follow the notation of Cambanis, Huang, and Simons [2] and say $\mathbf{X}$: $(p \times 1)$ is an elliptically contoured distribution with parameters $\boldsymbol{\mu}$, $\boldsymbol{\Sigma}$, ϕ, where $\boldsymbol{\Sigma}$ is nonnegative definite, if the characteristic function of $\mathbf{X} - \boldsymbol{\mu}$ has the form

$$\phi_{\mathbf{X}-\boldsymbol{\mu}}(\mathbf{t}) = \phi(\mathbf{t}'\boldsymbol{\Sigma}\mathbf{t}).$$

This is denoted by $\mathbf{X} \sim EC_p(\boldsymbol{\mu}, \boldsymbol{\Sigma}, \phi)$.

2. Conditionally More Positively Quadrant Dependent

In this section, we introduce our conditional positive quadrant dependence (PQD) ordering. We show that this ordering is preserved under a function with a pairwise condition, and then we obtain various special cases.

DEFINITION 2.1. *Let* $F(x_1, ..., x_p)$ and $G(x_1, ..., x_p)$ be two cdfs. Fix $1 \leqslant i < j \leqslant p$ and suppose that the following conditions are satisfied:

(a) $F_{R(i,j)}(\mathbf{t}) = G_{R(i,j)}(\mathbf{t})$, for all $\mathbf{t}$,

(b) (i) $F(x_i, \infty \mid \mathbf{X}^{(i,j)} = \mathbf{t}) = G(x_i, \infty \mid \mathbf{X}^{(i,j)} = \mathbf{t})$, for all x_i and $\mathbf{t}$,

(ii) $F(\infty, x_j \mid \mathbf{X}^{(i,j)} = \mathbf{t}) = G(\infty, x_j \mid \mathbf{X}^{(i,j)} = \mathbf{t})$, for all x_j and $\mathbf{t}$,

(c) $F(x_i, x_j \mid \mathbf{X}^{(i,j)} = \mathbf{t}) \geqslant G(x_i, x_j \mid \mathbf{X}^{(i,j)} = \mathbf{t})$, for all x_i, x_j, and $\mathbf{t}$.

Then F is said to be *conditionally more* (i, j)*-positively quadrant dependent than* G, written as $F \rightarrow^{P(i,j)} G$.

Sometimes for notational ease, if $\mathbf{X} \sim F$ and $\mathbf{Y} \sim G$, we write $\mathbf{X} \rightarrow^{P(i,j)} \mathbf{Y}$ instead of $F \rightarrow^{P(i,j)} G$.

Note 2.1. Conditions (a) and (b) of Definition 2.1 together are equivalent to both

(a′) $F_{R(i)}(\mathbf{s}) = G_{R(i)}(\mathbf{s})$ for all $\mathbf{s}$, and

(b′) $F_{R(j)}(\mathbf{s}) = G_{R(j)}(\mathbf{s})$ for all $\mathbf{s}$.

We subsequently show that under certain conditions the elliptically symmetrical distributions can be (i, j)-PQD ordered and, hence, so can the multivariate normal distribution. In Section 4, we provide some general techniques for obtaining (i, j)–PQD ordered distributions and also apply these techniques to obtaining inequalities for the generalized Farlie–Gumbel–Morgenstern family of distributions in Section 5.

A function $h(x, y)$ is called *quasi-monotone* if for all $x_1 \leqslant x_2$, $y_1 \leqslant y_2$,

$$h(x_1, y_1) + h(x_2, y_2) - h(x_1, y_2) - h(x_2, y_1) \geqslant 0.$$

Note 2.2. (i) Quasi-monotone is sometimes termed superadditive.

(ii) $h(x, y)$ is quasi-monotone if and only if e^h is TP_2.

(iii) If $h(x, y)$ is absolutely continuous, then $h(x, y)$ is quasi-monotone if and only if $(\partial^2/\partial x \partial y)\, h(x, y) \geqslant 0$ for almost all (x, y) in R^2.

DEFINITION 2.2. A function $h(x_1, ..., x_p)$ is (i, j)*-quasi-monotone* if $h(x_1, ..., x_i, ..., x_j, ..., x_p)$ is quasi-monotone in x_i, x_j for all possible fixed values of $\mathbf{x}^{(i,j)}$. We say $h(x_1, ..., x_p)$ is *quasi-monotone in pairs* if it is (i, j)-quasi-monotone for all $1 \leqslant i \leqslant j \leqslant p$. (Tchen [15, p. 824] calls functions that are quasi-monotone in pairs superadditive.)

Note 2.3. (i) When viewing $h(x_1, ..., x_i, ..., x_j, ..., x_p)$ as a function of x_i, x_j for fixed $\mathbf{x}^{(i,j)}$, we sometimes employ the notation $h(x_i, x_j; \mathbf{x}^{(i,j)})$ or $h_{\mathbf{x}^{(i,j)}}(x_i, x_j)$.

(ii) Observe that $h(x_1, ..., x_p)$ is quasi-monotone in pairs if and only if

$$h(\mathbf{x} \vee \mathbf{y}) + h(\mathbf{x} \wedge \mathbf{y}) \geqslant h(\mathbf{x}) + h(\mathbf{y}) \qquad \text{for all } \mathbf{x}, \mathbf{y}.$$

This follows from Kemperman [8, p. 329(i)], since $e^h > 0$.

One of our two main theorems is given next. Although it holds under a variety of assumptions, we give it in a form with conditions on the function h which are easy to state. More general conditions on h under which the theorem is true are given following the theorem.

THEOREM 2.1. *Let $F(x_1, ..., x_p)$ and $G(x_1, ..., x_p)$ be cdfs and fix $1 \leqslant i < j \leqslant p$. Suppose that $h(x_1, ..., x_p)$ is bounded, right-continuous, and (i, j)-quasi-monotone. If $F \to^{P(i,j)} G$, then $E_F h(\mathbf{X}) \geqslant E_G h(\mathbf{X})$.*

Proof. Consider $h(x_i, x_j; \mathbf{x}^{(i,j)})$ for any fixed $\mathbf{x}^{(i,j)}$. This function is bounded, right-continuous, and quasi-monotone in (x_i, x_j). Consequently, since $F \to^{P(i,j)} G$ from Tchen [15, Theorem 2, $n=2$] we have

$$\iint h(x_i, x_j; \mathbf{x}^{(i,j)}) \, dF\{x_j \mid \mathbf{X}^{(i,j)} = \mathbf{x}^{(i,j)}\}$$

$$\geqslant \iint h(x_i, x_j; \mathbf{x}^{(i,j)}) \, dG\{x_i, x_j \mid \mathbf{X}^{(i,j)} = \mathbf{x}^{(i,j)}\}.$$

The conclusion follows by integration.

Note 2.4. Notice that to apply Tchen's result we only need that $h(\mathbf{x})$ is bounded and right-continuous in (x_i, x_j) for fixed $\mathbf{x}^{(i,j)}$ and so the assumptions above can be weakened. (See also Corollary 2.1 of Tchen [15].)

Theorem 2.1 holds for many other classes of h's than those specified in the theorem. We state several other sets of conditions. The first set is due to Cambanis, Simons, and Stout [4] and various refinements of it can be found following Theorem 1 in that paper. The second set is due to Ruschendorf [11]. A comment similar to Note 2.4 above also applies to these conditions:

(1) For fixed $1 \leqslant i < j \leqslant p$, $h(\mathbf{x})$ is right-continuous, (i, j)-quasi-monotone, and either of the following is satisfied:

(i) $h_{\mathbf{x}^{(i,j)}}(x_i, x_j)$ is symmetric in x_i, x_j for almost all $\mathbf{x}^{(i,j)}$ and $\int h_{\mathbf{x}^{(i,j)}}(x_i, x_i) \, dF(x_i, \infty \mid \mathbf{X}^{(i,j)} = \mathbf{x}^{(i,j)})$ and $\int h_{\mathbf{x}^{(i,j)}}(x_j, x_j) \, dF(\infty, x_j \mid \mathbf{X}^{(i,j)} = \mathbf{x}^{(i,j)})$ are finite for almost all $\mathbf{x}^{(i,j)}$; or

(ii) there exist x_i^* and x_j^* such that $\int h_{\mathbf{x}^{(i,j)}}(x_i, x_j^*)\, dF(x_i, \infty \mid \mathbf{X}^{(i,j)} = \mathbf{x}^{(i,j)})$ and $\int h_{\mathbf{x}^{(i,j)}}(x_j^*, x_j)\, dF(\infty, x_j \mid \mathbf{X}^{(i,j)} = \mathbf{x}^{(i,j)})$ are finite for almost all $\mathbf{x}^{(i,j)}$.

(2) For fixed $1 \leqslant i < j \leqslant p$, $h(\mathbf{x})$ is right-continuous, (i,j)-quasi-monotone, $\int h_{\mathbf{x}^{(i,j)}}(x_i, x_j)\ dF(x_i, x_j \mid \mathbf{X}^{(i,j)} = \mathbf{x}^{(i,j)})$ and $\int h_{\mathbf{x}^{(i,j)}}(x_i, x_j)\ dG(x_i, x_j \mid \mathbf{X}^{(i,j)} = \mathbf{x}^{(i,j)})$ are finite for almost all $\mathbf{x}^{(i,j)}$, and either of the following are satisfied:

(i) $h_{\mathbf{x}^{(i,j)}}(x_i, x_j)$ is nondecreasing in x_i and x_j for almost all $\mathbf{x}^{(i,j)}$ or

(ii) $h_{\mathbf{x}^{(i,j)}}(x_i, x_j) \to 0$ as $x_i \to -\infty$ or as $x_j \to -\infty$ for almost all $\mathbf{x}^{(i,j)}$.

We now give a situation in which quasi-monotonicity is naturally satisfied.

Corollary 2.1. *Let $F(x_1, ..., x_p)$ and $G(x_1, ..., x_p)$ be cdf's and fix $1 \leqslant i \leqslant k < j \leqslant p$. Suppose $h(x_1, ..., x_p) = f(x_1, ..., x_k)\, g(x_{k+1}, ..., x_p)$, where f and g are both decreasing or both increasing, are bounded, and right-continuous. If $F \to^{P(i,j)} G$, then $E_F(h(\mathbf{X})) \geqslant E_G(h((\mathbf{X}))$.*

Proof. This follows directly from Theorem 2.1, since $f(x_1, ..., x_k)\, g(x_{k+1}, ..., x_p)$ is (i,j)-quasi-monotone for $1 \leqslant i \leqslant k < j \leqslant p$.

We now remove the regularity assumptions on f and g, i.e., we assume only that f and g are both decreasing or both increasing.

Corollary 2.2. *Suppose $h(x_1, ..., x_p) = f(x_1, ..., x_k)\, g(x_{k+1}, ..., x_p)$, where f and g are both increasing or both decreasing and are Borel measurable. If $F \to^{P(i,j)} G$, then $E_F(h(\mathbf{X})) \geqslant E_G(h(\mathbf{X}))$, provided the expectations exist.*

Proof. The proof is divided into five steps.

Step 1. Let $f = I_{C_1}, g = I_{C_2}$, where C_1 and C_2 are closed upper sets. The result follows immediately from Corollary 2.1. Similarly if C_2 is an open lower set, $-g$ is increasing and right-continuous so that

$$E_F(-h) \geqslant E_G(-h) \qquad \text{or} \qquad E_F(h) \leqslant E_G(h).$$

If C_1 and C_2 are both open lower sets then $E_F(h) \geqslant E_G(h)$.

Step 2. Let $f = I_{D_1}, g = I_{D_2}$ be Borel measurable upper sets. Then as in Block and Savits [1] we can approximate the D_i by closed upper sets C_i and apply Step 1. If either of the D_i are Borel measurable lower sets we can approximate by an open lower set. We have $E_F(h) \geqslant E_G(h)$ for both upper or both lower and $E_f(h) \leqslant E_G(h)$ for one upper and one lower.

Step 3. Let $f \geqslant 0$, $g \geqslant 0$ be nondecreasing Borel measurable. Then as in Block and Savits [1] we can find $f = (1/2^k) \sum_{i=1}^{k2^k} I_{D_{ik}}$ which converges upward to f, where D_{ik} are Borel measurable upper sets. A similar comment for g and the monotone convergence theorem gives the result. Similar comments apply if f and g are both nondecreasing Borel measurable functions or one is nondecreasing and one is nonincreasing.

Step 4. Let f and g be nondecreasing Borel measurable functions. Then f^+ and g^+ are nondecreasing and f^- and g^- are nonincreasing nonnegative Borel measurable functions. Thus from Step 3,

$$E_F(f^\pm g^\pm) \geqslant E_G(f^\pm g^\pm)$$

and

$$E_G(f^\pm g^\mp) \leqslant E_G(f^\pm g^\mp).$$

Under the assumptions that $E_F(h(\mathbf{X}))$ and $E_G(h(\mathbf{X}))$ exist (but are not necessarily finite) it is not hard to show that

$$E_F(f \cdot g) \geqslant E_G(f \cdot g).$$

Step 5. Let f and g be nonincreasing Borel measurable functions. The proof is similar to Step 4.

Conditional positive quadrant ordering is a concept which follows from covariance conditions in the multivariate normal case and its generalizations. We state as lemmas some of the results where covariance conditions imply orderings.

LEMMA 2.1. *Let $\mathbf{Y} \sim N(0, \mathbf{\Sigma})$ and $\mathbf{\Sigma} \sim N(0, \mathbf{\Sigma} + \delta \mathbf{I}_{ij}^S)$ and fix $1 \leqslant i < j \leqslant p$. Assume $\delta > 0$ and that $\mathbf{\Sigma} + \delta \mathbf{I}_{ij}^S$ is nonnegative definite. Then $\mathbf{X} \rightarrow^{P(i,j)} \mathbf{Y}$.*

Proof. Without loss of generality assume $i = 1$ and $j = 2$, and partition $\mathbf{\Sigma}$ accordingly into dimensions 2 and $p - 2$. Denote the cdfs of $\mathbf{X}$ and $\mathbf{Y}$ by F and G, respectively. Then $F(x_1, x_2 \mid \mathbf{X}^{(1,2)} = \mathbf{t})$ corresponds to $N(\mathbf{\Sigma}_{12} \mathbf{\Sigma}_{22}^- \mathbf{t}, \mathbf{\Sigma}_{1.2} + \delta \mathbf{I}_{12}^S)$ and $G(x_1, x_2 \mid \mathbf{X}^{(1,2)} = \mathbf{t})$ corresponds to $N(\mathbf{\Sigma}_{12} \mathbf{\Sigma}_{22}^- \mathbf{t}, \mathbf{\Sigma}_{1.2})$, where $\mathbf{\Sigma}_{1.2} = \mathbf{\Sigma}_{11} - \mathbf{\Sigma}_{12} \mathbf{\Sigma}_{22}^- \mathbf{\Sigma}_{21}$, and $\mathbf{\Sigma}_{22}^-$ is a generalized inverse of Σ_{22}. Clearly (a) and (b) of Definition 2.1 are satisfied. For every t, (c) of Definition 2.1 follows from Slepian's inequality (Slepian [14] or see Tong [16, Theorem 2.1.1]).

The following result gives the conditional orderings for elliptically contoured distributions. We use the notation of the paper by Cambais, Huang, and Simons [2] throughout.

LEMMA 2.2. *Let* $\mathbf{Y} \sim EC_p(\mathbf{0}, \boldsymbol{\Sigma}, \phi)$, $\mathbf{X} \sim EC_p(\mathbf{0}, \boldsymbol{\Sigma} + \delta \mathbf{I}_{ij}^S, \phi)$, *and fix* $1 \leqslant i < j \leqslant p$. *Assume* $\delta > 0$ *and that* $\boldsymbol{\Sigma} + \delta \mathbf{I}_{ij}^S$ *is nonnegative definite. Then* $\mathbf{X} \to^{P(i,j)} \mathbf{Y}$.

Proof. Without loss of generality assume $i = 1$ and $j = 2$, and partition $\boldsymbol{\Sigma}$ accordingly into dimensions 2 and $p - 2$. Denote the cdfs of $\mathbf{X}$ and $\mathbf{Y}$ by F and G, respectively. Suppose $\mathbf{t}$ is in $\mathscr{L}(\boldsymbol{\Sigma}_{22})$, the row space of $\boldsymbol{\Sigma}_{22}$. Then by Cambanis, Huang, and Simons [2, Corollary 5], $F(x_1, x_2 \mid \mathbf{X}^{(1,2)} = \mathbf{t})$ corresponds to $EC_2(\boldsymbol{\Sigma}_{12} \boldsymbol{\Sigma}_{22}^- \mathbf{t}, \boldsymbol{\Sigma}_{1.2} + \delta \mathbf{I}_{12}^S, \phi_{q(\mathbf{t})}$ and $G(x_1, x_2 \mid \mathbf{X}^{(1,2)} = \mathbf{t})$ corresponds to $EC_2(\boldsymbol{\Sigma}_{12} \boldsymbol{\Sigma}_{22}^- \mathbf{t}, \boldsymbol{\Sigma}_{1.2}, \phi_{q(t)})$, where $\boldsymbol{\Sigma}_{1.2} = \boldsymbol{\Sigma}_{11} - \boldsymbol{\Sigma}_{12} \boldsymbol{\Sigma}_{22}^- \boldsymbol{\Sigma}_{21}$, and $\phi_{q(t)}$ depends on ϕ and $q(\mathbf{t}) = \mathbf{t}' \boldsymbol{\Sigma}_{22}^- \mathbf{t}$. When $\mathbf{t} \notin \mathscr{L}(\Sigma_{22})$, the conditional distributions puts mass on $\mathbf{0}$ (Cambanis, Huang, and Simons [2, (17b)]. In the case $\mathbf{t} \in \mathscr{L}(\boldsymbol{\Sigma}_{22})$, parts (a) and (b) of Definition 2.1 follow from the fact that if $(\mathbf{W}_1' : \mathbf{W}_2')' \sim EC_{p_1 + p_2}((\boldsymbol{\mu}_1' : \boldsymbol{\mu}_2')'; \boldsymbol{\Sigma}, \phi)$, then $\mathbf{W}_1 \sim EC_{p_1}(\boldsymbol{\mu}_1, \boldsymbol{\Sigma}_{11}, \phi)$. For every $\mathbf{t}$, part (c) follows from Cambanis and Simon [3, Theorem 3.2]. For the case $\mathbf{t} \notin \mathscr{L}(\boldsymbol{\Sigma}_{22})$, the result is obvious.

We now give the general result for elliptically contoured distributions. It holds under weaker regularity conditions on h as pointed out in the note following the corollary.

COROLLARY 2.3. *Let* $\mathbf{X} \sim EC_p(\mathbf{0}, \boldsymbol{\Sigma}, \phi)$ *and let* $h(\mathbf{x})$ *be a bounded, right-continuous function which is quasi-monotone in pairs. Then* $E_{\boldsymbol{\Sigma}}(h(\mathbf{X}))$ *is increasing in the off-diagonal elements of* $\boldsymbol{\Sigma}$.

Proof. Apply Theorem 2.1 and the previous lemma iteratively.

Note 2.5. (a) If $h(\mathbf{x})$ is absolutely continuous in x_i and x_j for all $1 \leqslant i < j \leqslant n$ we can replace the quasi-monotone assumption above with the condition $\partial^2 h(x)/\partial x_i \partial x_j \geqslant 0$ for all $\mathbf{x}$.

(b) As mentioned in the note following Theorem 2.1 the corollary above holds under a variety of conditions. One strengthening of the above is to assume $h(\mathbf{x})$ is right-continuous, quasi-monotone in pairs, and that there exist $x_1, x_2, ..., x_p$ such that $E_{\boldsymbol{\Sigma}}(h(x_i; \mathbf{X}^{(i)}))$ are finite for $i = 1, 2, ..., p$.

(c) The normal case of the above corollary corresponds to the $i \neq j$ part of Proposition 1 of Joag-dev, Perlman, and Pitt [6]. Because of notes (a) and (b) above the conditions on h are somewhat weaker than those in the proposition cited.

3. CONDITIONALLY MORE DISPERSED

We now examine a concept of one distribution being conditionally more dispersed than another. Our main result of this section shows that if a

p-variate function is convex in its relevant argument then it preserves this ordering. Normal and elliptically contoured cases are then examined.

DEFINITION 3.1. Let $F(x_1, ..., x_p)$ and $G(x_1, ..., x_p)$ be two cdfs. Fix $1 \leqslant i \leqslant p$ and suppose the following conditions are satisfied:

(a) $F_{R(i)}(\mathbf{t}) = G_{R(i)}(\mathbf{t})$ for all $\mathbf{t}$,

(b) $E_F(X_i \mid \mathbf{X}^{(i)} = \mathbf{t}) = E_G(X_i \mid \mathbf{X}^{(i)} = \mathbf{t})$ for all $\mathbf{t}$,

(c) for all $\mathbf{t}$, both conditional distributions are degenerate, or

(i) $S^-(F(x_i \mid \mathbf{X}^{(i)} = \mathbf{t}) - G(x_i \mid \mathbf{X}^{(i)} = \mathbf{t})) = 1$, and

(ii) the sign sequence in (i) is $+, -$.

Then F is said to be *conditionally more i-dispersed than G*, written as $F \rightarrow^{D(i)} G$.

Note 3.1. (i) We have included (a) in Definition 3.1 for convenience. If the conditional means differ, the cdfs would be translated so that the means coincide. (See Shaked [13] concerning centering.)

(ii) Sometimes for notational ease, if $\mathbf{X} \sim F$ and $\mathbf{Y} \sim G$, we write $\mathbf{X} \rightarrow^{D(i)} \mathbf{Y}$ instead of $F \rightarrow^{D(i)} G$.

(iii) Conditions (a) and (c) imply (see Shaked [13]) for all convex h that,

$$\int h(x_i)\, dF(x_i \mid \mathbf{X}^{(i)} = \mathbf{t}) \geqslant \int h(x_i)\, dG(x_i \mid \mathbf{X}^{(i)} = \mathbf{t}) \qquad \text{for all } \mathbf{t}. \tag{3.1}$$

The condition given by (3.1) can be interpreted as saying that for all $\mathbf{t}$ the conditional distribution $F(x_i \mid \mathbf{X}^{(i)} = \mathbf{t})$ is more dilated (e.g., Marshall and Olkin [10, p. 312]) than $G(x_i \mid \mathbf{X}^{(i)} = \mathbf{t})$.

We next give a one-dimensional concept of convexity for a p-dimensional function. It says simply that the function is convex in the one relevant component for all other values of the remaining component.

DEFINITION 3.2. A function $h(x_1, ..., x_p)$ is i-convex if $h(x_1, ..., x_i, ..., x_p)$ is convex in x_i for all possible fixed values of $\mathbf{x}^{(i)}$.

The main result of this section follows.

THEOREM 3.2. *Let $F(x_1, ..., x_p)$ and $G(x_1, ..., x_p)$ be cdfs and fix $1 \leqslant i \leqslant p$. Suppose $h(x_1, ..., x_p)$ is i-convex. If $F \rightarrow^{D(i)} G$, then $E_F h(\mathbf{X}) \geqslant E_G h(\mathbf{X})$, provided the expectations exist.*

Proof. Observe that for all $\mathbf{t}$, it follows from Shaked [13] and Definition 3.3 (a), (c) that

$$\int h(x_i;\mathbf{t})\,dF(x_i\,|\,\mathbf{X}^{(i)}=\mathbf{t})\geqslant\int h(x_i;\mathbf{t})\,dG(x_i\,|\,\mathbf{X}^{(i)}=\mathbf{t}).$$

By Definition 4.1(b), integration with respect to $F_{R(i)}(\mathbf{t})=G_{R(i)}(\mathbf{t})$ completes thc proof.

COROLLARY 3.1. *Suppose $\partial^2h(\mathbf{x})/\partial x_i^2$ exists for all x and is nonnegative. Then $F\to^{D(i)}G$ implies $E_Fh(\mathbf{X})\geqslant E_Gh(\mathbf{X})$.*

Proof. Obvious.

Conditional dispersiveness derives from comparison of variances for multivariate normal distributions. We state some of those results as lemmas to demonstrate this connection and then give the more general results.

LEMMA 3.1. *Let $\mathbf{Y}\sim N(\mathbf{0},\mathbf{\Sigma})$ and $\mathbf{X}\sim N(\mathbf{0},\mathbf{\Sigma}+\delta\mathbf{I}_{ii})$, and fix $1\leqslant i\leqslant p$. Assume $\delta>0$. Then $\mathbf{X}\to^{D(i)}\mathbf{Y}$.*

Proof. Without loss of generality, assume $i=1$. Then $F(x_1\,|\,\mathbf{X}^{(1)}=\mathbf{t})$ is $N(\mathbf{\Sigma}_{12}\mathbf{\Sigma}_{22}^{-}\mathbf{t},\ \sigma_{11}+\delta-\mathbf{\Sigma}_{12}\mathbf{\Sigma}_{22}^{-}\mathbf{\Sigma}_{21})$ and $G(y_1\,|\,\mathbf{Y}^{(1)}=\mathbf{t})$ is $N(\mathbf{\Sigma}_{12}\mathbf{\Sigma}_{22}^{-}\mathbf{t},\ \sigma_{11}-\mathbf{\Sigma}_{12}\mathbf{\Sigma}_{22}^{-}\mathbf{\Sigma}_{21})$, where $\mathbf{\Sigma}$ is appropriately partitioned. Definitions 3.1(a) and (c) follow because the means are the same and $\mathrm{Var}(X_1\,|\,\mathbf{X}^{(1)}=\mathbf{t})=\mathrm{Var}(Y_1\,|\,\mathbf{Y}^{(1)}=\mathbf{t})+\delta$. Part (b) is obvious.

COROLLARY 3.2. *Let $\mathbf{X}\sim N(\mathbf{0},\mathbf{\Sigma})$ and $h(\mathbf{x})$ be i-convex in each argument. Then $E_{\mathbf{\Sigma}}(h(\mathbf{x}))$ is increasing in the diagonal elements of $\mathbf{\Sigma}$ provided that $E_{\mathbf{\Sigma}}(h(\mathbf{X}))$ exists.*

Proof. Apply Theorem 3.2 and the previous lemma.

COROLLARY 3.3. *Let $\mathbf{X}\sim N(\mathbf{0},\mathbf{\Sigma})$ and $h(\mathbf{x})$ be a function such that $\partial^2h(\mathbf{x})/\partial x_i^2$ exists and is nonnegative for all $\mathbf{x}$, for $i=1,\ldots,p$. Then $E_{\mathbf{\Sigma}}(h(\mathbf{X}))$ is increasing in the diagonal elements of $\mathbf{\Sigma}$, provided that $E_{\mathbf{\Sigma}}(h(\mathbf{X}))$ exists.*

Proof. This is immediate from Corollary 3.2.

Note 3.2. Corollary 3.3 contains part of Proposition 1 of Joag-dev *et al.* [6] (the $i=j$ case), but under weaker moment conditions.

As in Section 2 we use the notation of Cambanis *et al* [2].

LEMMA 3.2. *$\mathbf{Y}\sim EC_p(\mathbf{0},\mathbf{\Sigma},\phi)$ and $\mathbf{X}\sim EC_p(\mathbf{0},\mathbf{\Sigma}+\delta\mathbf{I}_{ii})$, and fix $1\leqslant i\leqslant p$. Assume $\delta>0$. Then $\mathbf{X}\to^{D(i)}\mathbf{Y}$.*

Proof. Without loss of generality assume $i=1$ and partition $\boldsymbol{\Sigma}$ accordingly into dimensions 1 and $p-1$. Denote the cdfs of $\mathbf{X}$ and $\mathbf{Y}$ by F and G, respectively. Suppose $\mathbf{t} \in \mathscr{L}(\boldsymbol{\Sigma}_{22})$. Then by Cambanis *et al.* [2, Corollary 5], $F(x_1 \mid \mathbf{X}^{(1)} = \mathbf{t})$ corresponds to $EC_1(\boldsymbol{\Sigma}_{12}\boldsymbol{\Sigma}_{22}^{-}\mathbf{t}, \boldsymbol{\Sigma}_{1.2} + \delta, \phi_{q(\mathbf{t})})$, and $G(x_1 \mid \mathbf{X}^{(1)} = \mathbf{t})$ corresponds to $EC_1(\boldsymbol{\Sigma}_{12}\boldsymbol{\Sigma}_{22}^{-}\mathbf{t}, \boldsymbol{\Sigma}_{1.2}, \phi_{q(\mathbf{t})})$, where $\boldsymbol{\Sigma}_{1.2} = \sigma_{11} - \boldsymbol{\Sigma}_{12}\boldsymbol{\Sigma}_{22}^{-}\boldsymbol{\Sigma}_{21}$ and $\phi_{q(\mathbf{t})}$ is determined by ϕ and $q(\mathbf{t}) = \mathbf{t}'\boldsymbol{\Sigma}_{22}^{-}\mathbf{t}$. Parts (a) and (b) of Definition 3.1 are obvious and part (c) follows from the fact that for every t, $EC_1(\boldsymbol{\Sigma}_{12}\boldsymbol{\Sigma}_{22}^{-}\mathbf{t}, \boldsymbol{\Sigma}_{1.2} + \delta, \phi_{q(\mathbf{t})})$ and $EC_1(\boldsymbol{\Sigma}_{12}\boldsymbol{\Sigma}_{22}^{-}\mathbf{t}, \boldsymbol{\Sigma}_{1.2}, \phi_{q(\mathbf{t})})$ are univariate cdfs differing only by a scale parameter. For $\mathbf{t} \notin \mathscr{L}_2(\boldsymbol{\Sigma}_{22})$ both conditional distributions are degenerate at 0 and so (a), (b), and (c) are trivially satisfied.

Note. Corollaries similar to Corollary 3.2 and 3.3 follow immediately for elliptically contoured distributions. These provide somewhat more generalized results than Joag-dev *et al.* [6].

4. Conditional Positive and Negative Dependence

In this section, we consider other distributions which are conditionally more (i, j)-PQD ordered. We primarily focus on techniques for constructing such orderings, with particular attention paid to upper and lower bounds, and to comparisons with certain forms of independence.

The following definition formalizes a concept that has appeared in various forms in the literature.

Definition 4.1. A random vector $\mathbf{X}$ with cdf $F(\mathbf{x})$ is conditionally (i, j)-PQD (NQD), $i \neq j$, if

$$F(x_i, x_j \mid \mathbf{X}^{(i,j)} = \mathbf{s}) \geqslant (\leqslant)\ F(x_i, \infty \mid \mathbf{X}^{(i,j)} = \mathbf{s})\, F(\infty, x_j \mid \mathbf{X}^{(i,j)} = \mathbf{s})$$

for all x_i, x_j, $\mathbf{s}$.

Note 4.1. Suppose $F(\mathbf{x})$ is absolutely continuous with pdf $f(\mathbf{x})$. Define $g(\mathbf{x}) = f_{R(j)}(\mathbf{x}^{(j)}) \times f_{R(i)}(\mathbf{x}^{(i)}) / f_{R(i,j)}(\mathbf{x}^{(i,j)})$, when $f_{R(i,j)}(\mathbf{x}^{(i,j)}) > 0$, and 0, otherwise. It is direct to show that (i) g is a pdf, (ii) $g(x_i, x_j \mid \mathbf{X}^{(i,j)} = \mathbf{x}^{(i,j)}) = f_{R(j)}(\mathbf{x}^{(j)}) \times f_{R(i)}(\mathbf{x}^{(i)}) / (f_{R(i,j)}(\mathbf{x}^{(i,j)}))^2$, and (iii) $g_{R(i,j)}(\mathbf{x}^{(i,j)}) = f_{R(i,j)}(\mathbf{x}^{(i,j)})$. Denote by G, the cdf, corresponding to g. Then F is conditionally (i, j)-PQD (NQD) if and only if $F \rightarrow^{P(i,j)} (\leftarrow^{P(i,j)})\ G$.

The next lemma provides a method for constructing multivariate distributions with certain prescribed conditional marginals and, more importantly, having certain conditional positive dependence properties.

Lemma 4.1. *Suppose $F(x, y, \mathbf{z})$, the joint cdf of the random variables X,*

Y, $\mathbf{Z}$, is given. Let $H(u, v)$ be a cdf with marginal distributions that are uniform on $[0, 1]$. *Define*

$$G(x, y, \mathbf{z}) = \int_{-\infty}^{\mathbf{z}} H(F_1(x \mid \mathbf{w}), F_2(y \mid \mathbf{w}))\, dF_{\mathbf{Z}}(\mathbf{w}),$$

where F_1, F_2, and $F_{\mathbf{Z}}$ have the obvious interpretation. Then the following hold:

(a) $G(x, y, \mathbf{z})$ *is a* cdf.

(b) (i) $G_{\mathbf{Z}}(\mathbf{z}) = F_{\mathbf{Z}}(\mathbf{z})$,

(ii) $G_1(x \mid \mathbf{z}) = F_1(x \mid \mathbf{z})$,

(iii) $G_2(y \mid \mathbf{z}) = F_2(y \mid \mathbf{z})$,

(iv) $G(x, y \mid \mathbf{z}) = H(F_1(x \mid \mathbf{z}), F_2(y \mid \mathbf{z}))$.

(c) *If H satisfies any of the following, then $G(x, y \mid \mathbf{z})$ satisfies the same (conditionally):*

(i) *independence,*

(ii) PQD (NQD),

(iii) *upper (lower) Fréchét bound,*

(iv) TP_2 (RR_2).

Proof. (a) This follows directly from the fact that for every $\mathbf{z}$, $H(F_1(x \mid \mathbf{z}), F_2(y \mid \mathbf{z}))$ is a cdf in x, y.

(b) Obvious.

(c) This follows from the result that $G(x, y \mid \mathbf{z}) = H(F_1(x \mid \mathbf{z}), F_2(y \mid \mathbf{z}))$ and requiring for (i) $H(u, v) = uv$, (ii) $H(u, v) \geqslant (\leqslant)\, uv$, (iii) $H(u, v) = \min(u, v)$ $(\max(u + v - 1, 0))$. Result (iv) follows by a standard TP_2 (RR_2) result which gives that increasing functions preserve TP_2- $(\mathrm{RR}_2$-) -ness.

Note 4.2. Suppose H_1 and H_2 are two bivariate distributions with uniform marginals such that H_1 is more PQD than H_2. If corresponding G_1 and G_2 are constructed as in the preceding lemma, then $G_1 \rightarrow^{P(1,2)} G_2$.

Example 4.1. To illustrate the preceding note, consider the family of bivariate uniform cdfs

$$H(x, y; \lambda) = x + y - 1 + (1 - x)^{1-\lambda}(1 - y)^{1-\lambda} \min((1 - x)^{\lambda}, (1 - y)^{\lambda}),$$

where $0 \leqslant x \leqslant 1$, $0 \leqslant y \leqslant 1$, and $0 < \lambda < 1$. This is essentially the Marshall–Olkin bivariate exponential distribution with equal marginals, where the marginals have been transformed. See Kimeldorf and Sampson

[9] for a discussion of this method of transformation. The distribution above is one of those mentioned in Kimeldorf and Sampson [9] but the form given there has an algebraic error. It is direct to show that $H(x, y; \lambda_1) \geqslant H(x, y; \lambda_2)$, whenever $\lambda_1 \leqslant \lambda_2$.

Let $\prod_{ij.R(i,j)} \equiv \Pi(F(x_i, \infty \mid \mathbf{X}^{(i,j)} = \mathbf{x}^{(i,j)}), \quad F(\infty, x_j \mid \mathbf{X}^{(i,j)} = \mathbf{x}^{(i,j)}), F_{R(i,j)}(\mathbf{x}^{(i,j)}))$ denote the class of p-variate cdfs of a r.v. $\mathbf{X}$, where the marginal of $\mathbf{X}^{(i,j)}$ is $F_{R(i,j)}$ and the conditional marginals of X_i and X_j given $\mathbf{X}^{(i,j)}$ are respectively $F(x_i, \infty \mid \mathbf{X}^{(i,j)} = \mathbf{x}^{(i,j)})$ and $F(\infty, x_j \mid X^{(i,j)} = \mathbf{x}^{(i,j)})$. Then if $K(\mathbf{x})$ is in this class,

$$K(\mathbf{x}) \leqslant K^+_{ij.R(i,j)}(\mathbf{x})$$

$$\equiv \int_{-\infty}^{\mathbf{x}^{(i,j)}} \min(F(x_i, \infty \mid \mathbf{X}^{(i,j)} = \mathbf{s}), F(\infty, x_j \mid \mathbf{X}^{(i,j)} = \mathbf{s})\, dF_{\mathbf{R}(i,j)}(\mathbf{s}) \qquad (4.1)$$

and, moreover, the r.h.s. of (4.1) is also in the class. The former statement follows from Dall'Aglio [5] and the latter from Lemma 4.1. Furthermore, for all $K(\mathbf{x}) \in \prod_{ij.R(i,j)}$, $K \leftarrow^{P(i,j)} K^+_{ij.R(i,j)}$.

Thus the preceding corollary states that if h satisfies suitable regularity conditions

$$\max_{K \in \prod_{ij.R(i,j)}} E_K h(\mathbf{X}) = E_{K^+_{ij.R(i,j)}} h(\mathbf{X}).$$

Similarly, the minimum occurs at $E_{K^-_{ij.R(i,j)}} h(\mathbf{X})$, where

$$K^-_{ij.R(i,j)}(\mathbf{x})$$

$$= \int_{-\infty}^{\mathbf{x}^{(i,j)}} \max[F(x_i, \infty \mid \mathbf{X}^{(i,j)} = \mathbf{s}) + F(\infty, x_j \mid \mathbf{X}^{(i,j)} = \mathbf{s}) - 1,\ 0]\, dF_{R(i,j)}(\mathbf{s}).$$

5. FGM Distributions

Johnson and Kotz [7] define the generalized Farlie–Gumbel–Morgenstern distribution as being a cdf $F(\mathbf{x})$ which has representation

$$F(\mathbf{x}) = \prod_{i=1}^{p} F_i(x_i) \left[1 + \sum_{(i_1, ..., i_k) \in I_k} \alpha_{i_1, ..., i_k} \prod_{j=1}^{k} \bar{F}_{i_j}(x_{i_j}) \right], \qquad (5.1)$$

where $I_k = \{(i_1, ..., i_k) | k \geqslant 1,\ 1 \leqslant i_1 < i_2 < \cdots < i_k \leqslant p\}$ and the $\alpha_{i_1, ..., i_k}$ are contained in a multivariate parameter space Θ and where $F_i(x_i) \equiv 1 - \bar{F}_i(x_i)$ is a cdf, $i = 1, ..., p$. In this paper we assume each $F_i(x_i)$ is absolutely continuous so that $F(\mathbf{x})$ has a pdf.

Suppose the FGM family in (5.1) contains the parameter α_{ij}. Fix the remaining parameters at some value $\boldsymbol{\theta}_0$ and denote the cdf. viewed as parametrized by α_{ij} as $F(\mathbf{x}, \alpha_{ij}, \boldsymbol{\theta}_0)$.

THEOREM 5.1. *Let* $\mathbf{Y} \sim F(\mathbf{t}; \alpha_{ij}, \boldsymbol{\theta}_0)$ *and* $\mathbf{X} \sim F(\mathbf{t}; \alpha_{ij} + \delta, \boldsymbol{\theta}_0)$, *where* $\delta > 0$, *F is given by* (5.1), *and* $(\alpha_{ij}, \boldsymbol{\theta}_0), (\alpha_{ij} + \delta, \boldsymbol{\theta}_0) \in \Theta$. *Then* $\mathbf{X} \rightarrow^{P(i,j)} \mathbf{Y}$.

Proof. Without loss of generality, assume $i = 1, j = 2$. It is easy to show that the marginal distributions of $(Y_1, \mathbf{Y}^{(1,2)})$, $(Y_2, \mathbf{Y}^{(1,2)})$, and $\mathbf{Y}^{(1,2)}$ do not depend on α_{12} and, hence, (a) and (b) of Definition 2.1 are satisfied. To show (c), in light of (b), it is sufficient to demonstrate that

$$\frac{\partial^{p-2} F(\mathbf{t}; \alpha_{12} + \delta, \boldsymbol{\theta}_0)}{\partial t_3 \cdots \partial t_p} - \frac{\partial^{p-2} F(\mathbf{t}; \alpha_{12}, \boldsymbol{\theta}_0)}{\partial t_3 \cdots \partial t_p} \geqslant 0 \tag{2}$$

for all $\mathbf{t}$. That (2) holds follows immediately from the assumption $\delta > 0$ and the fact that

$$\frac{\partial^{p-2} F(\mathbf{t}; \alpha_{12}, \boldsymbol{\theta}_0)}{\partial t_3 \cdots \partial t_p} = F_1(t_1) F_2(t_2)(1 + \alpha_{12} \bar{F}_1(t_1) \bar{F}_2(t_2)) \prod_{k=3}^{p} f_k(t_k)$$
$$+ \frac{\partial^{p-2}}{\partial t_3 \cdots \partial t_p} \left[\prod_{k=1}^{p} F_i(t_i) \sum_{\{i_1, \ldots, i_k\} \neq \{1,2\}} \alpha_{i_1, \ldots, i_k} \prod_{j=1}^{k} \bar{F}_{i_j}(t_{i_j}) \right].$$

REFERENCES

[1] BLOCK, H. W., AND SAVITS, T. H. (1980). Multivariate increasing failure rate average distributions. *Ann. Probab.* **8** 793–801.

[2] CAMBANIS, S., HUANG, S., AND SIMONS, G. (1981). On the theory of elliptically contoured distributions. *J. Multivariate Anal.* **11** 368–385.

[3] CAMBANIS, S., AND SIMONS, G. (1982). Probability and expectation inequalities. *Z. Wahrsch. Verw. Gebiete* **59** 6–25.

[4] CAMBANIS, S., SIMONS, G., AND STOUT, W. (1976). Inequalities for $\mathscr{E}k(X, Y)$ when marginals are fixed. *Z. Wahrsch. Verw. Gebiete* **36** 285–294.

[5] DALL'AGLIO, G. (1972). Fréchét classes and compatibility of distribution functions. *Sympos. Math.* **9** 131–150.

[6] JOAG-DEV, K., PERLMAN, M., AND PITT, L. (1983). Association of normal random variables and Slepian's inequality. *Ann. Probab.* **11** 451–455.

[7] JOHNSON, N. L., AND KOTZ, S. (1975). On some generalized Farlie–Gumbel–Morgenstern distributions. *Comm. Statist.* **4** 415–427.

[8] KEMPERMAN, J. H. B. (1977). On the FKG-inequality for measures on a partially ordered space. *Proc. Akad. Wetenschappen Ser. A* **80** 313–331.

[9] KIMELDORF, G., AND SAMPSON, A. R. (1975). Uniform representations of bivariate distributions. *Comm. Statist.* **4** 617–627.

[10] MARSHALL, A. W., AND OLKIN, I. (1979). *Inequalities: Theory of Majorization and its Applications.* Academic Press, New York.

[11] RUSCHENDORF, L. (1980). Inequalities for the expectation of Δ-monotone functions. *Z. Wahrsch. Verw. Gebiete* **54** 341–349.

[12] SHAKED, M. (1980). On mixtures from exponential families. *J. Roy. Statist. Soc. Ser. B* **42** 192–198.

[13] SHAKED, M. (1985). Ordering distributions by dispersion. In *Encyclopedia of Statistical Sciences*, Vol. 6 (Johnson and Kotz, Ed.), pp. 485–490.

[14] SLEPIAN, D. (1962). The one-sided barrier problem for Gaussian noise. *Bell System Tech. J.* **41** 463–501.

[15] TCHEN, A. H. T. (1980). Inequalities for distributions with given marginals. *Ann. Probab.* **8** 814–827.

[16] TONG, Y. L. (1980). *Probability Inequalities in Multivariate Distributions.* Academic Press, New York.

A Discounted Cost Relationship

C. S. CHEN* AND THOMAS H. SAVITS†

University of Pittsburgh

In Savits (1988. *J. Appl. Probab.* **4**, in press) a very general cost mechanism for a maintained system was considered. There he established a relationship between the expected long run cost per unit time for the age and block maintenance policies. In the present paper a similar relationship is obtained for the expected total α-discounted cost.

1. INTRODUCTION

Recently Savits [3] considered a very general cost mechanism for a maintained system. There he established a relatinship between the expected long run cost per unit time for an age replacement policy and that for a block replacement policy.

In this paper we now consider the expected total discounted cost for the same model. Again we show that there is simple cost relationship between the age and block replacement policies.

The basic model is first reviewed in Section 2. In Section 3 we prove our main result. Lastly, some further cost relationships are detailed in Section 4.

2. REVIEW OF THE BASIC MODEL

The model considered in Savits [3] can be described biefly as follows. The basic ingredient consists of a stochastic process $\{R(t); 0 \leqslant t \leqslant \zeta\}$. Here we interpret $R(t)$ as the operational cost of a unit on line during a time interval $[0, t)$. The random variable ζ designates the time of a major

* Partially supported by AFOSR Grant AFOSR-84-0113.
† Supported by ONR Contract N0014-84-K-0084 and AFOSR Grant AFOSR-84-0113.

Multivariate Statistics and Probability
ISBN 0-12-580205-6

Reprinted from *J. Mult. Anal.* **27**(1).

unrepairable breakdown. At this time, we replace the failed item with a new identical unit. Thus we call ζ an unscheduled or unplanned replacement. The cost for such an unplanned replacement is c_1.

The two maintenance policies we consider here are referred to as age replacement and block replacement. In the former case, a scheduled or planned replacement occurs whenever an operating unit reaches age T; in the latter case, a planned replacement occurs at the absolute times T, $2T, \ldots$. In either case, the cost of a planned replacement is c_2.

We assume that items put on line are independent and identical units and that both planned and unplanned replacements take negligible time.

Throughout this paper, we assume (as minimal requirements) that the stochastic process R has right-hand limits on $[0, \zeta)$ and that $R(t+) = \lim_{s \downarrow t} R(s)$ represents the unit operational cost on $[0, t]$. We shall sometimes find it convenient to extend R by setting $R(t) = R(\zeta)$ for $t > \zeta$. In addition, we assume that $R(0+) = R(0) = 0$ and $P\{\zeta > 0\} = 1$.

In order to write down to total operational cost for the maintained system, it is convenient to introduce some further notation. First we consider the age replacement maintenance policy. Let $\{R_i(t); 0 \leqslant t \leqslant \zeta_i\}$, $i = 1, 2, \ldots$, be independent copies of $\{R(t); 0 \leqslant t \leqslant \zeta\}$. Define

$$\eta_i = \min(\zeta_i, T),$$

$$\xi_k = \begin{cases} 0, & \text{if } k = 0 \\ \eta_1 + \cdots + \eta_k, & \text{if } k \geqslant 1, \end{cases} \tag{2.1}$$

and

$$R_i^*(t) = \begin{cases} R_i(t+) & \text{if } 0 \leqslant t < \eta_i \\ R_i(\eta_i) + c_1 I_{\{\zeta_i < T\}} + c_2 I_{\{\zeta_i \geqslant T\}} & \text{if } t \geqslant \eta_i \end{cases}$$

for $i = 1, 2, \ldots$. Here I_A denotes the indicator function of the set A. Then the total operational cost over $[0, t]$ for the age replacement policy, which we denote by $K_A(t)$, is given by

$$K_A(t) = \sum_{i=1}^{k} R_i^*(\eta_i) + R_{k+1}^*(t - \xi_k) \tag{2.2}$$

if $\xi_k \leqslant t < \xi_{k+1}$, $k = 0, 1, \ldots$. We adopt the standard convention that an empty sum is equal to zero.

For the block replacement maintenance policy, we introduce the notation

$$\sigma_k = \begin{cases} 0 & \text{if } k = 0 \\ \zeta_1 + \cdots + \zeta_k & \text{if } k = 1, 2, \ldots \end{cases}$$

and

$$Q(t)=\begin{cases} R_1(t) & \text{if } 0\leqslant t\leqslant \sigma_1 \\ \sum_{i=1}^{k} R_i(\zeta_i)+kc_1+R_{k+1}(t-\sigma_k) & \text{if } \sigma_k<t\leqslant\sigma_{k+1}. \end{cases} \quad (2.3)$$

Next, let $\{Q_i(t);\ 0\leqslant t\}$, $i=1, 2, \ldots,$ be independent copies of $\{Q(t); 0\leqslant t\}$ and set

$$Q_i^*(t)=\begin{cases} Q_i(t+) & \text{if } 0\leqslant t<T \\ Q_i(T)+c_2 & \text{if } t\geqslant T. \end{cases} \quad (2.4)$$

Then the total operational cost over $[0, t]$ for the block replacement policy, denoted by $K_B(t)$, is given by

$$K_B(t)=\sum_{i=1}^{k} Q_i^*(T)+Q_{k+1}^*(t-kT) \quad (2.5)$$

if $kT\leqslant t<(k+1)T$, $k=0, 1, \ldots.$

We also denote the expected total cost over $[0, t]$ by

$$C_A(t)=C_A(t;\ T)=E[K_A(t)]$$

and (2.6)

$$C_B(t)=C_B(t;\ T)=E[K_B(t)],$$

respectively. Consequently, the expected long run cost per unit time is given by the ratio of the average cost per cycle to the average length of a cycle, i.e.,

$$J_A(T)=\lim_{t\to\infty}\frac{C_A(t;\ T)}{t}=\frac{E[R^*(\eta)]}{E[\eta]}$$

and (2.7)

$$J_B(T)=\lim_{t\to\infty}\frac{C_B(t;\ T)}{t}=\frac{E[Q^*(T)]}{T}.$$

The above results follow from the theory of renewal reward process (cf., Ross [2]). We are, of course, making the implicit assumption that $E[|R^*(\eta)|]$ and $E[|Q^*(T)|]$ are finite.

If we denote the corresponding numerators by $A(T)=E[R^*(\eta)]$ and $B(T)=E[Q^*(T)]$, respectively, then it was shown in Savits [3] that

$$B(T)=\int_{[0,\ T)} A(T-x)\, dU(x) \quad (2.8)$$

where $U(x)=\sum_{k=0}^{\infty} P(\sigma_k \leqslant x)$ is the renewal function generated by the independent and identically distributed sequence of random variables $\zeta_1, \zeta_2, \ldots$.

3. Discounted Cost Relationship

In this section we will establish a similar relationship between the discounted costs for the age and block maintenance policies. In order to define the notion of discounting, however, we need to assume that, with probability one, the cost functions $K_A(t)$ and $K_B(t)$ generate a signed measure on $[0, \infty)$. This is indeed the case when the cost parameters c_1 and c_2 are nonnegative and $R(t)$ is a nondecreasing process. *In order to avoid some technical considerations, we shall henceforth only consider the situation described immediately above.*

So let $\alpha > 0$. We then define the α-discounted cost over $[0, t]$ by

$$K_A^{(\alpha)}(t) = \int_{(0,t]} e^{-\alpha u}\, dK_A(u)$$

and (3.1)

$$K_B^{(\alpha)}(t) = \int_{(0,t]} e^{-\alpha u}\, dK_B(u),$$

where K_A and K_B are given by (2.2) and (2.5), respectively. The total α-discounted cost is obtained by replacing $(0, t]$ with $(0, \infty)$.

First we consider the age replacement case. Then

$$\begin{aligned} J_A^{(\alpha)}(T) &= \lim_{t\to\infty} E[K_A^{(\alpha)}(t)] = E\left[\int_{(0,\infty)} e^{-\alpha u}\, dK_A(u)\right] \\ &= \alpha E\left[\int_0^{\infty} e^{-\alpha v}\, K_A(v)\, dv\right] \\ &= \alpha \sum_{k=0}^{\infty} E\left[\int_{\xi_k}^{\xi_{k+1}} e^{-\alpha v} \left\{\sum_{i=1}^{k} R_i^*(\eta_i) + R_{k+1}^*(v-\xi_k)\right\} dv.\right. \end{aligned}$$

In the last step we used the expression (2.2). We now consider each sum separately.

For the second sum, we write

$$\alpha \sum_{k=0}^{\infty} E\left[\int_{\xi_k}^{\xi_{k+1}} e^{-\alpha v} R_{k+1}^*(v-\xi_k)\, dv\right]$$
$$= \alpha \sum_{k=0}^{\infty} E\left[\int_0^{\eta_{k+1}} e^{-\alpha w} e^{-\alpha \xi_k} R_{k+1}^*(w)\, dw\right]$$
$$= \alpha \sum_{k=0}^{\infty} E[e^{-\alpha \xi_k}]\, E\left[\int_0^{\eta_{k+1}} e^{-\alpha w} R_{k+1}^*(w)\, dw\right]$$
$$= \alpha E\left[\int_0^{\eta} e^{-\alpha w} R^*(w)\, dw\right]\left(\sum_{k=0}^{\infty} \{E[e^{-\alpha\eta}]^k\}\right)$$
$$= (1 - E[e^{-\alpha\eta}])^{-1} \alpha E\left[\int_0^{\eta} e^{-\alpha w} R^*(w)\, dw\right].$$

The second and third equalities above follow from independence and the identically distributed assumptions.

Next, we write the first sum as

$$\alpha \sum_{k=0}^{\infty} E\left[\int_{\xi_k}^{\xi_{k+1}} e^{-\alpha v} \sum_{i=1}^{k} R_i^*(\eta_i)\, dv\right]$$
$$= \alpha \sum_{i=1}^{\infty} \sum_{k=i}^{\infty} E\left[R_i^*(\eta_i) \int_{\xi_k}^{\xi_{k+1}} e^{-\alpha v}\, dv\right]$$
$$= \sum_{i=1}^{\infty} E[R_i^*(\eta_i)\, e^{-\alpha \xi_i}] = \sum_{i=1}^{\infty} E[e^{-\alpha \eta_i} R_i^*(\eta_i)]\, E[e^{-\alpha \xi_{i-1}}]$$
$$= (1 - E[e^{-\alpha\eta}])^{-1} E[e^{-\alpha\eta} R^*(\eta)].$$

Consequently,

$$J_A^{(\alpha)}(T) = (1 - E[e^{-\alpha\eta}])^{-1} \left\{E\left[\alpha \int_0^{\eta} e^{-\alpha w} R^*(w)\, dw\right] + E[e^{-\alpha\eta} R^*(\eta)]\right\}$$
$$= \frac{E[\int_{(0,\eta]} e^{-\alpha w}\, dR^*(w)]}{1 - E[e^{-\alpha\eta}]}. \tag{3.2}$$

We shall denote the numerator by $A^{(\alpha)}(T)$, i.e.,

$$A^{(\alpha)}(T) = E\left[\int_{(0,\eta]} e^{-\alpha w}\, dR^*(w)\right]. \tag{3.3}$$

It is the expected α-discounted cost over one cycle. For the denominator, we can also write

$$1 - E[e^{-\alpha\eta}] = \alpha \int_0^T \bar{G}(u)\, e^{-\alpha u}\, du,$$

where $\bar{G}(x) = P\{\zeta > x\}$ is the survival function of ζ. Since

$$E[\eta] = \int_0^T \bar{G}(u)\, du,$$

we note that

$$J_A(T) = \lim_{\alpha \downarrow 0} \alpha J^{(\alpha)}(T). \tag{3.4}$$

Recall that $J_A(T)$ is the expected long run cost per unit time given in Eq. (2.7).

(3.5) *Remark.* One can also derive the result (3.2) from a renewal equation approach. More specifically, if $C_A^{(\alpha)}(t) = E[K_A^{(\alpha)}(t)]$, one can show that $C_A^{(\alpha)}(t)$ satisfies the renewal equation

$$C_A^{(\alpha)}(t) = \left\{\alpha \int_0^t e^{-\alpha v} E[R^*(\eta \wedge v)]\, dv + e^{-\alpha t} E[R^*(\eta \wedge t)]\right\}$$
$$+ \int_{(0,t]} e^{-\alpha x} C_A^{(\alpha)}(t-x)\, dG^*(x),$$

where $G^*(x) = P\{\eta \leqslant x\}$. Since $e^{-\alpha x}\, dG^*(x)$ is a defective probability measure, the result now follows from Feller [1, p. 361].

Next we consider the block replacement policy case. Here

$$J_B^{(\alpha)}(T) = \lim_{t \to \infty} E[K_B^{(\alpha)}(t)] = E\left[\int_{(0,\infty)} e^{-\alpha u}\, dK_B(u)\right].$$

By the same technique as illustrated above, it is easy to derive

$$J_B^{(\alpha)}(T) = \frac{E[\int_{(0,T]} e^{-\alpha w}\, dQ^*(w)]}{1 - e^{-\alpha T}}$$

and (3.6)

$$J_B(T) = \lim_{\alpha \downarrow 0} \alpha J_B^{(\alpha)}(T).$$

In this case we denote the numerator by $B^{(\alpha)}(T)$, i.e.,

$$B^{(\alpha)}(T)=E\left[\int_{(0,T]} e^{-\alpha w}\, dQ^*(w)\right]. \tag{3.7}$$

Our main goal in this section is to relate $A^{(\alpha)}(T)$ and $B^{(\alpha)}(T)$. We proceed as in Savits [3]. Since

$$\int_{(0,t]} e^{-\alpha w}\, dQ^*(w)=\alpha\int_0^t e^{-\alpha w}\, Q^*(v)\, dv+e^{-\alpha t}Q^*(t)-Q^*(0),$$

we can rewrite $\int_{(0,T]} e^{-\alpha w}\, dQ^*(w)$ for $\sigma_k<T\leqslant\sigma_{k+1}$ as

$$\begin{aligned}\int_{(0,T]} e^{-\alpha w}\, dQ^*(w)=&\sum_{j=1}^{k}\,[e^{-\alpha\sigma_{j-1}}R_j^{(\alpha)}(\zeta_j)+e^{-\alpha\sigma_j}c_1]\\ &+e^{-\alpha\sigma_k}R_{k+1}^{(\alpha)}(T-\sigma_k)+e^{-\alpha T}c_2\end{aligned}$$

using Eqs. (2.3) and (2.4). Here we set

$$R_i^{(\alpha)}(t)=\alpha\int_0^t e^{-\alpha v}R_i(v+)\, dv+e^{-\alpha t}R_i(t).$$

It can be thought of as the α- discounted operational cost of the ith unit on line for a time interval $[0, t)$. Consequently,

$$\begin{aligned}B^{(\alpha)}(T)&=E\left[\int_{(0,T]} e^{-\alpha w}\, dQ^*(w)\right]\\ &=\sum_{k=0}^{\infty} E\left[\int_{(0,T]} e^{-\alpha w}\, dQ^*(w);\, \sigma_k<T\leqslant\sigma_{k+1}\right]\\ &=E[R_1^{(\alpha)}(T)+e^{-\alpha T}c_2]+\sum_{k=1}^{\infty} E\left[\sum_{j=1}^{k}\{e^{-\alpha\sigma_{j-1}}R_j^{(\alpha)}(\zeta_j)+e^{-\alpha\sigma_j}c_1\}\right.\\ &\qquad\left.+e^{-\alpha\sigma_k}R_{k+1}^{(\alpha)}(T-\sigma_k)+e^{-\alpha T}c_2;\, \sigma_k<T\leqslant\sigma_{k+1}\right]\\ &=\sum_{k=0}^{\infty} E[e^{-\alpha\sigma_k}R_{k+1}^{(\alpha)}(\zeta_{k+1})+e^{-\alpha\sigma_{k+1}}c_1;\, \sigma_{k+1}<T]\\ &\quad+\sum_{k=0}^{\infty} E[e^{-\alpha\sigma_k}R_{k+1}^{(\alpha)}(T-\sigma_k)+e^{-\alpha T}c_2;\, \sigma_k<T\leqslant\sigma_{k+1}].\end{aligned}$$

We now consider th terms in the first sum in more detail. Since $\sigma_{k+1}=\sigma_k+\zeta_{k+1}$, wc have

$$E[e^{-\alpha\sigma_k} R^{(\alpha)}_{k+1}(\zeta_{k+1}) + e^{-\alpha\sigma_{k+1}} c_1 ; \sigma_{k+1} < T]$$
$$= E[e^{-\alpha\sigma_k} R^{(\alpha)}_{k+1}(\zeta_{k+1}) + e^{-\alpha\sigma_{k+1}} c_1 ; \sigma_k < T, \zeta_{k+1} < T - \sigma_k]$$
$$= E\{e^{-\alpha\sigma_k} E[R^{(\alpha)}_{k+1}(\zeta_{k+1}) + e^{-\alpha\zeta_{k+1}} c_1 ; \zeta_{k+1} < T - x]|_{x=\sigma_k} ; \sigma_k < T\}$$
$$= E\{e^{-\alpha\sigma_k} E[R^{(\alpha)}(\zeta) + e^{-\alpha\zeta} c_1 ; \zeta < T - x]|_{x=\sigma_k} ; \sigma_k < T\}.$$

Hence, the first sum is given by

$$\sum_{k=0}^{\infty} E[e^{-\alpha\sigma_k} R^{(\alpha)}_{k+1}(\zeta_{k+1}) + e^{-\alpha\sigma_{k+1}} c_1 ; \sigma_{k+1} < T]$$
$$= \sum_{k=0}^{\infty} \int_{[0,T)} e^{-\alpha x} E[R^{(\alpha)}(\zeta) + e^{-\alpha\zeta} c_1 ; \zeta < T - x]\, P(\sigma_k \in dx)$$
$$= \int_{[0,T)} e^{-\alpha x} E[R^{(\alpha)}(\zeta) + e^{-\alpha\zeta} c_1 ; \zeta < T - x]\, dU(x)$$

where, as before, $U(x) = \sum_{k=0}^{\infty} P(\sigma_k \leqslant x)$ is the renewal function generated by $\zeta_1, \zeta_2, \ldots$.

Similarly, we can write the terms in the second sum as

$$E[e^{-\alpha\sigma_k} R^{(\alpha)}_{k+1}(T - \sigma_k) + e^{-\alpha T} c_2 ; \sigma_k < T \leqslant \sigma_{k+1}]$$
$$= E\{e^{-\alpha\sigma_k} E[R^{(\alpha)}(T - x) + e^{-\alpha(T-x)} c_2 ; \zeta \geqslant T - x]|_{x=\sigma_k} ; \sigma_k < T\},$$

and so

$$\sum_{k=0}^{\infty} E[e^{-\alpha\sigma_k} R^{(\alpha)}_{k+1}(T - \sigma_k) + e^{-\alpha T} c_2 ; \sigma_k < T \leqslant \sigma_{k+1}]$$
$$= \int_{[0,T)} e^{-\alpha x} E[R^{(\alpha)}(T - x) + e^{-\alpha(T-x)} c_2 ; \zeta \geqslant T - x]\, dU(x).$$

But,

$$A^{(\alpha)}(T) = E\left[\int_{(0,\eta]} e^{-\alpha w}\, dR^*(w)\right]$$
$$= E\left[\alpha \int_0^{\eta} e^{-\alpha v} R^*(v)\, dv + R^*(\eta)\, e^{-\alpha\eta}\right]$$
$$= E\left[\alpha \int_0^{\zeta} e^{-\alpha v} R(v+)\, dv + e^{-\alpha\zeta}\{R(\zeta) + c_1\} ; \zeta < T\right]$$
$$+ E\left[\alpha \int_0^{T} e^{-\alpha v} R(v+)\, dv + e^{-\alpha T}\{R(T) + c_2\} ; \zeta \geqslant T\right]$$
$$= E[R^{(\alpha)}(\zeta) + e^{-\alpha\zeta} c_1 ; \zeta < T] + E[R^{(\alpha)}(T) + e^{-\alpha T} c_2 ; \zeta \geqslant T].$$

Consequently, we obtain the result

$$B^{(\alpha)}(T) = \int_{[0,T)} e^{-\alpha x} A^{(\alpha)}(T-x)\, dU(x). \tag{3.8}$$

We summarize the results of this section in the following theorem.

(3.9) THEOREM. *Under the model of Section 2 with cost parameters c_1 and c_2 nonnegative and $R(t)$ a nondecreasing process, the expected total α-discounted cost for the age and block replacement policies are given by*

$$J_A^{(\alpha)}(T) = E\left[\int_{(0,\infty)} e^{-\alpha u}\, dK_A(u)\right] = \frac{A^{(\alpha)}(T)}{1 - E[e^{-\alpha\eta}]}$$

and

$$J_B^{(\alpha)}(T) = E\left[\int_{(0,\infty)} e^{-\alpha u}\, dK_B(u)\right] = \frac{B^{(\alpha)}(T)}{1 - e^{-\alpha T}},$$

respectively, where $A^{(\alpha)}(T) = E[\int_{(0,\eta]} e^{-\alpha w}\, dR^*(w)]$ *and* $B^{(\alpha)}(T) = E[\int_{(0,T]} e^{-\alpha w}\, dQ^*(w)]$. *Furthermore,*

$$B^{(\alpha)}(T) = \int_{[0,T)} e^{-\alpha x} A^{(\alpha)}(T-x)\, dU(x).$$

(3.10) *Remarks.* (i) It is clear from the proof that the cost parameters c_1 and c_2 need not be constants. Everything remains as above if c_1 and c_2 are random variables. Moreover, we may allow c_1 and c_2 to be different for the two polices of age and block replacement. In this case, the form of (3.8) changes slightly. See Savits [3] for further details.

(ii) One can readily show that if we define a subdistribution function H on $[0,\infty)$ by $H(x) = \int_{[0,x]} e^{-\alpha u}\, dG(u)$, and let W be the associated renewal function generated by H, then $dW(x) = e^{-\alpha x}\, dU(x)$. Thus we many write (3.8) as

$$B^{(\alpha)}(T) = \int_{[0,T)} A^{(\alpha)}(T-x)\, dW(x).$$

Consequently, we can also write

$$A^{(\alpha)}(T) = B^{(\alpha)}(T) - \int_{[0,T)} B^{(\alpha)}(T-x)\, dH(x).$$

4. Other Cost Relationships

Thus far we have established relationships between $A(T)$ and $B(T)$ and also between $A^{(\alpha)}(T)$ and $B^{\alpha()}(T)$. We complete the cycle by considering the relationship between $A(T)$ and $A^{(\alpha)}(T)$ and also between $B(T)$ and $B^{(\alpha)}(T)$. Clearly $A(T)=A^{(0)}(T)$ and $B(T)=B^{(0)}(T)$. It thus remains to express $A^{(\alpha)}(T)$ and $B^{(\alpha)}(T)$ in terms of $A(T)$ and $B(T)$, respectively.

As in Section 3, we shall assume that $R(t)$ is a nondecreasing process and that c_1 and c_2 are nonnegative. In addition, we shall assume that the functions $A(T)$ and $B(T)$ are right-continuous and of bounded variation on compact intervals.

(4.1) Theorem. *Under the above conditions, we have*

(i) $A^{(\alpha)}(T)=\int_{(0,T]} e^{-\alpha x}\,dA(x)+E[c_2 e^{-\alpha(\zeta\wedge T)}]$.

(ii) $B^{(\alpha)}(T)=\int_{(0,T]} e^{-\alpha x}\,dB(x)+e^{-\alpha T}E[c_2]$.

Proof. We will only prove (i) since (ii) is similar. Consider

$$\begin{aligned}
\int_{(0,T]} e^{-\alpha x}\,dA(x) &= \alpha\int_0^T e^{-\alpha v}A(v)\,dv+e^{-\alpha T}A(T)-A(0)\\
&= E\left[\alpha\int_0^\zeta e^{-\alpha v}\{R(v)+c_2\}\,dv;\zeta<T\right]\\
&\quad+E\left[\alpha\int_\zeta^T e^{-\alpha v}\{R(\zeta)+c_1\}\,dv;\zeta<T\right]\\
&\quad+E\left[\alpha\int_0^T e^{-\alpha v}\{R(v)+c_2\}\,dv;\zeta\geqslant T\right]\\
&\quad+e^{-\alpha T}E[R(\zeta)+c_1;\zeta<T]\\
&\quad+e^{-\alpha T}E[R(T)+c_2;\zeta\geqslant T]-E[c_2]\\
&= E\left[\alpha\int_0^\zeta e^{-\alpha v}R(v+)\,dv+e^{-\alpha\zeta}\{R(\zeta)+c_1\};\zeta<T\right]\\
&\quad+E\left[\alpha\int_0^T e^{-\alpha v}R(v+)\,dv+e^{-\alpha T}\{R(T)+c_2\};\zeta\geqslant T\right]\\
&\quad+E[c_2(1-e^{-\alpha\zeta});\zeta<T]\\
&\quad+E[c_2(1-e^{-\alpha T});\zeta\geqslant T]-E[c_2]\\
&= A^{(\alpha)}(T)-E[c_2 e^{-\alpha(\zeta\wedge T)}].
\end{aligned}$$

Thus we have the desired conclusion.

In the above derivation we replaced $R(v)$ with $R(v+)$ in two integrations. This is permissible since an increasing function can have only countably many discontinuities.

References

[1] Feller, W. (1966). *An Introduction to Probability Theory and Its Applications*, Vol. II. Wiley, New York.
[2] Ross, S. M. (1970). *Applied Probability Models with Optimization Applications*. Holden–Day, San Francisco.
[3] Savits, T. H. (1988). A cost relationship between age and block replacement policies. *J. Appl. Probab.* **4**, in press.

Strong Consistency of M-Estimates in Linear Models*

X. R. CHEN AND Y. H. WU

University of Pittsburgh

This article studies the strong consistency of M-estimates in linear regression models directly from the minimization problem

$$\sum_{i=1}^{n} \rho(Y_i - \alpha - \mathbf{X}_i'\boldsymbol{\beta}) := \min,$$

where $\mathbf{X}_1 . \mathbf{X}_2, ...$ can be random observations of a p-dimensional random vector $\mathbf{X}$, or that they are simply known nonrandom p-vectors. It is shown that the solution $(\hat{\alpha}_n, \hat{\boldsymbol{\beta}}_n')$ of this minimization problem converges with probability one to the true parameter $(\alpha_0, \boldsymbol{\beta}_0')$ under very general conditions on the function ρ and the sequence $\{(\mathbf{X}_i', Y_i)\}$.

1. INTRODUCTION

Consider the linear regression model

$$Y_i = \alpha_0 + \mathbf{X}_i'\boldsymbol{\beta}_0 + e_i, \qquad i = 1, 2, ..., \tag{1.1}$$

where $(\alpha_0, \boldsymbol{\beta}_0')$ is the unknown parameter, $e_1, e_2, ...$ are random errors. As for $\{\mathbf{X}_i\}$, two cases will be considered: 1. $\{\mathbf{X}_i\}$ is a sequence of known p-dimensional vectors. 2. $(\mathbf{X}_1', Y_1)$, $(\mathbf{X}_2', Y_2)$, ... are i.i.d. observations of a $(p+1)$-dimensional random vector $(\mathbf{X}', Y)$.

* Research sponsored by the Air Force Office of Scientific Research under Contract F49620-85-C-0008. The United States Government is authorized to reproduce and distribute reprints for governmental purposes not-withstanding any copyright notation hereon.

Multivariate Statistics and Probability
ISBN 0-12-580205-6

Reprinted from *J. Mult. Anal.* **27**(1).

The M-estimate, introduced by Huber [5], takes the solution $(\hat{\alpha}_n, \hat{\boldsymbol{\beta}}'_n)$ of the minimization problem

$$\sum_{i=1}^{n} \rho(Y_i - \alpha - \mathbf{X}'_i \boldsymbol{\beta}) := \min \tag{1.2}$$

as the estimate of $(\alpha_0, \boldsymbol{\beta}'_0)$. This paper seeks the conditions under which $(\hat{\alpha}_n, \hat{\boldsymbol{\beta}}'_n)$ is strongly consistent:

$$\hat{\alpha}_n \to \alpha_0, \qquad \hat{\boldsymbol{\beta}}_n \to \boldsymbol{\beta}_0, \qquad \text{a.s.} \qquad \text{as} \quad n \to \infty. \tag{1.3}$$

In (1.2), ρ is a suitably chosen function on R and $(\alpha, \boldsymbol{\beta}')$ varies over some set $\Theta \subset R^{p+1}$, Θ is the parameter space. Two cases are often considered in the literature: (i) $\Theta = R^{p+1}$, (ii) Θ is a closed subset of R^{p+1} containing the true parameter $(\alpha_0, \boldsymbol{\beta}'_0)$ as an interior point. In the following, unless stated otherwise, we shall only consider the (more general) first case.

An often-made assumption in the literature, for example, [6, 7, 10], is that $\rho'(u) = d\rho(u)/du$ exists everywhere on R. In this case the solution of (1.2) must satisfy the equations

$$\sum_{i=1}^{n} \rho'(Y_i - \alpha - \mathbf{X}'_i \boldsymbol{\beta}) = 0, \qquad \sum_{i=1}^{n} \mathbf{X}_i \rho'(Y_i - \alpha - \mathbf{X}'_i \boldsymbol{\beta}) = 0. \tag{1.4}$$

If, in addition, ρ is convex, then (1.2) and (1.4) are equivalent. However, in many important examples of M-estimates, $\rho'(u)$ does not exist for some u. In such cases, although one may formally write down Eq. (1.4), it may have no solution, or none of its solutions is a solution of (1.2). A well-known example is furnished by $\rho(u) = |u|$ (minimum L_1-norm estimate). Consistency results of the M-estimate in this case were given by [3, 8, 11]. A more sophisticated example, considered in [4], is that $\rho(u) = (1 - \delta)u^2 + \delta |u|$. In the standard form of linear regression $Y_i = \mathbf{X}'_i \boldsymbol{\beta} + e_i$, for this choice of ρ, formally (1.4) reduces to

$$2(1 - \delta) \sum_{i=1}^{n} \mathbf{X}_i (Y_i - \mathbf{X}'_i \boldsymbol{\beta}) + \delta \sum_{i=1}^{n} \mathbf{X}_i \operatorname{sgn}(Y_i - \mathbf{X}'_i \boldsymbol{\beta}) = 0, \tag{1.5}$$

where $\operatorname{sgn}(0) = 0$, $\operatorname{sgn}(u) = u/|u|$ for $u \neq 0$. Although [4] asserts that (1.5) is equivalent to

$$(1 - \delta) \sum_{i=1}^{n} (Y_i - \mathbf{X}'_i \boldsymbol{\beta})^2 + \delta \sum_{i=1}^{n} |Y_i - \mathbf{X}'_i \boldsymbol{\beta}| := \min,$$

this is not true, as has been shown in [1].

We also note that the convexity assumption excludes many functions with practical significance, such as $\rho(u)=\min(|u|, k)$ for some constant $k>0$. Another example is

$$\rho(u)=\begin{cases}|u|, & |u|\leqslant k\\ k/2+|u|/2, & |u|>k.\end{cases} \tag{1.6}$$

So it makes good sense to tackle this estimation problem starting directly from the original formulation (1.2). This we shall do in the following sections.

2. Formulation of Results

First consider the case where $\mathbf{X}_1, \mathbf{X}_2, \ldots$ are i.i.d. random vectors.

Theorem 1. *Suppose that* $(\mathbf{X}_1', Y_1), (\mathbf{X}_2', Y_2), \ldots$ *are i.i.d. observations of a random vector* $(\mathbf{X}', Y)$, *and the following conditions are satisfied*:

(a) *The function* ρ *is continuous everywhere on* R, *nondecreasing on* $[0, \infty)$, *nonincreasing on* $(-\infty, 0]$, *and* $\rho(0)=0$.

(b) *Either* $\rho(\infty)=\rho(-\infty)=\infty$ *and*

$$P(\alpha+\mathbf{X}'\boldsymbol{\beta}=0)<1 \qquad \text{when} \quad (\alpha, \boldsymbol{\beta}')\neq(0, \mathbf{0}') \tag{2.1}$$

or $\rho(\infty)=\rho(-\infty)\in(0, \infty)$ *and*

$$P(\alpha+\mathbf{X}'\boldsymbol{\beta}=0)=0 \qquad \text{when} \quad (\alpha, \boldsymbol{\beta}')\neq(0, \mathbf{0}'). \tag{2.2}$$

(c) *For every* $(\alpha, \boldsymbol{\beta}')\in R^{p+1}$ *we have*

$$Q(\alpha, \boldsymbol{\beta}')\equiv E\rho(Y-\alpha-\mathbf{X}'\boldsymbol{\beta})<\infty \tag{2.3}$$

and Q *attains its minimum uniquely at* $(\alpha_0, \boldsymbol{\beta}_0')$.

Then (1.3) *is true.*

When ρ is a convex function, condition (2.3) can be somewhat weakened.

Theorem 2. *If* ρ *is a convex function, then* (1.3) *is still true when condition* (a) *of Theorem* 1 *is satisfied, condition* (b) *is deleted, and condition* (c) *is replaced by condition* (c′):

(c′) *For every* $(\alpha, \boldsymbol{\beta}')\in R^{p+1}$ *we have*

$$Q^*(\alpha, \boldsymbol{\beta}')\equiv E\{\rho(Y-\alpha-\mathbf{X}'\boldsymbol{\beta})-\rho(Y-\alpha_0-\mathbf{X}'\boldsymbol{\beta}_0)\} \tag{2.4}$$

exists and is finite, and that

$$Q^*(\alpha, \boldsymbol{\beta}') > 0, \qquad \text{for any} \quad (\alpha, \boldsymbol{\beta}') \neq (\alpha_0, \boldsymbol{\beta}_0'). \tag{2.5}$$

The following theorem gives an exponential convergence rate of the estimate $(\hat{\alpha}_n, \hat{\boldsymbol{\beta}}_n')$.

THEOREM 3. *Suppose that the conditions of Theorem 1 are met, and in addition that the moment generating function of $\rho(Y-\alpha-\mathbf{X}'\boldsymbol{\beta})$ exists in some neighbourhood of 0, then for arbitrarily given $\varepsilon > 0$ there exists a constant $c > 0$ independent of n such that*

$$P(|\hat{\alpha}_n - \alpha_0| \geqslant \varepsilon) = O(e^{-cn}), \qquad P(\|\hat{\boldsymbol{\beta}}_n - \boldsymbol{\beta}_0\| \geqslant \varepsilon) = O(e^{-cn}). \tag{2.6}$$

This conclusion remains valid if the conditions of Theorem 2 are met, and the moment generating function of $\rho(Y-\alpha-\mathbf{X}'\boldsymbol{\beta}) - \rho(Y-\alpha_0-\mathbf{X}'\boldsymbol{\beta}_0)$ exists in some neighbourhood of 0.

We next consider the case where $\mathbf{X}_1, \mathbf{X}_2, \ldots$ are nonrandom p-vectors.

THEOREM 4. *Suppose that in model* (1.1) $\mathbf{X}_1, \mathbf{X}_2, \ldots$ *are nonrandom p-vectors and the following conditions are satisfied*:

(a) *Condition* (a) *of Theorem* 1 *is true,* $\rho(\infty) = \rho(-\infty) = \infty$.

(b) $\{\mathbf{X}_i\}$ *is bounded, and if λ_n denotes the smallest eigenvalue of the matrix* $\sum_{i=1}^n (\mathbf{X}_i - \bar{\mathbf{X}}_n)(\mathbf{X}_i - \bar{\mathbf{X}}_n)'$ $(\bar{X}_n = \sum_{i=1}^n \mathbf{X}_i/n)$, *then*

$$\liminf_{n \to \infty} \lambda_n/n > 0. \tag{2.7}$$

(c) $\{e_i\}$ *is a sequence of i.i.d. random errors.*

(d) *For any $t \in R$, $E\rho(e_1 + t) < \infty$, $E\{\rho(e_1+t) - \rho(e_1)\} > 0$ for any $t \neq 0$, and there exists a constant $c_1 > 0$ such that*

$$E\{\rho(e_1+t) - \rho(e_1)\} \geqslant c_1 t^2 \tag{2.8}$$

for $|t|$ sufficiently small.

Then (1.3) *is true. This conclusion remains valid if* (a), (b) *are replaced by*

(a′) *Condition* (a) *of Theorem* 1 *is true,*

$$0 < \rho(\infty) = \rho(-\infty) < \infty. \tag{2.9}$$

(b′) $\lim_{\varepsilon \to 0} \lim_{n \to \infty} \sup \#\{i: 1 \leqslant i \leqslant n, |\alpha + \mathbf{X}_i'\boldsymbol{\beta}| \leqslant \varepsilon\}/n = 0,$

$$(\alpha, \boldsymbol{\beta}') \neq (0, \mathbf{0}'), \tag{2.10}$$

where $\#(B)$ *denotes the number of elements in set B. Note that condition* (2.10) *corresponds to condition* (2.2) *of Theorem* 1.

Also, when ρ is convex, the condition $E\rho(e_1+t)<\infty$ can be weakened to $E|\rho(e_1+t)-\rho(e_1)|<\infty$.

Before proving the theorems, we shall make some comments concerning the conditions assumed:

1. Condition (c) of Theorem 1, which stipulates that Q attains its minimum uniquely at the point $(\alpha_0, \boldsymbol{\beta}_0')$, is closely related to the meaning of the regression. The essence is that the selection of ρ must be compatible with the type of regression considered. For example, when $\alpha_0+\mathbf{x}'\boldsymbol{\beta}_0$ is the conditional median of Y given $\mathbf{X}=\mathbf{x}$ (median regression), we may choose $\rho(u)=|u|$. Likewisely, when $\alpha_0+\mathbf{x}'\boldsymbol{\beta}_0=E(Y|\mathbf{X}=\mathbf{x})$ (the usual mean regression), we may choose $\rho(u)=|u|^2$. An important case is that the conditional distribution of Y given $\mathbf{X}=\mathbf{x}$ is symmetric and unimodal with center $\alpha_0+\mathbf{x}'\boldsymbol{\beta}_0$. In this case, ρ can be chosen as any even function satisfying condition (a), and such that $\rho(t)>0$ when $u\neq 0$. This gives us some freedom in the choice of ρ with the aim of obtaining more robust estimates.

2. Condition (2.8) of Theorem 4 reveals a difference between the two cases of $\{\mathbf{X}_i\}$ mentioned earlier. In the case that $\{\mathbf{X}_i\}$ is a sequence of non-random vectors we can no longer assume only that 0 is the unique minimization point of $E\rho(e_1+u)$, as shown in the counterexample given in [2] for $\rho(u)=|u|$.

Condition (2.8) holds automatically when $\rho(u)=u^2$ and $Ee_1=0$. When $\rho(u)=|u|$, it holds when e_1 has median 0 and a density which is bounded away from 0 in some neighborhood of 0. When ρ is even and e_1 is symmetric and unimodal with center 0, (2.8) holds if one of the following two conditions is satisfied: (i) $\inf\{(\rho(u_2)-\rho(u_1))/(u_2-u_1)\colon\ \varepsilon\leqslant u_1<u_2<\infty\}>0$ for any $\varepsilon>0$, (ii) there exist positive constants $a<b$ and c, such that

$$(\rho(u_2)-\rho(u_1))/(u_2-u_1)\geqslant c, \qquad |f(u_2)-f(u_1)|/(u_2-u_1)\geqslant c$$

for any $a\leqslant u_1<u_2\leqslant b$, where f is the density of e_1.

3. Proof of Theorems 1–3

Our main task is to prove Theorem 1. The proof of Theorem 1 can be easily modified to prove Theorems 2 and 3. For any constant $l>0$, define the sets

$$A_l=[-l, l]^{p+1}, \qquad \tilde{A}_l=[-l, l]^p. \tag{3.1}$$

Without loss of generality, we shall assume in the sequel that

$$\alpha_0 = 0, \qquad \boldsymbol{\beta}_0 = \mathbf{0}. \tag{3.2}$$

LEMMA 1. *Suppose that the conditions of Theorem* 1 *are satisfied. Denote by* $(\tilde{\alpha}_n, \tilde{\boldsymbol{\beta}}_n')$ *a Borel measurable solution of the constrained minimization problem*

$$\sum_{i=1}^{n} \rho(Y_i - \alpha - \mathbf{X}_i'\boldsymbol{\beta}) := \min \qquad \textit{over} \quad (\alpha, \boldsymbol{\beta}') \in A_l, \tag{3.3}$$

where $l > 0$ *is a given constant. Then as* $n \to \infty$,

$$\tilde{\alpha}_n \to 0, \qquad \tilde{\boldsymbol{\beta}}_n \to \mathbf{0} \qquad a.s. \tag{3.4}$$

Proof. Denote the $T = 2^{p+1}$ vertices $(\pm l, \pm l, ..., \pm l)$ of A_l by $(a_1, \mathbf{b}_1'), ..., (a_T, \mathbf{b}_T')$. From condition (a) it can be easily shown that

$$0 \leqslant \rho(Y - \alpha - \mathbf{X}'\boldsymbol{\beta}) \leqslant \sum_{j=1}^{T} \rho(Y - a_j - \mathbf{X}'\mathbf{b}_j) \tag{3.5}$$

for any $(X', Y) \in R^{p+1}$ and $(\alpha, \boldsymbol{\beta}') \in A_l$. From this, the continuity of ρ, and the dominated convergence theorem, one sees that the function Q, defined by (2.3), is continuous. Since $(0, \mathbf{0}')$ is the uniquc minimum point of Q, for any $\varepsilon > 0$ we have

$$q \equiv \inf\{Q(\alpha, \boldsymbol{\beta}') - Q(0, \mathbf{0}') : (\alpha, \boldsymbol{\beta}') \in A_l - A_\varepsilon\} > 0. \tag{3.6}$$

Choose $\varepsilon_1 \in (0, q/6)$ and m sufficiently large such that

$$E\{I((\mathbf{X}', Y) \notin A_m)\, \rho(Y - \alpha - \mathbf{X}'\boldsymbol{\beta})\} < \varepsilon_1, \qquad \text{when} \quad (\alpha, \boldsymbol{\beta}') \in A_l. \tag{3.7}$$

The existence of such m follows from (2.3) and (3.5). Write

$$\{(\mathbf{X}_1^{*\prime}, Y_1^*), ..., (\mathbf{X}_{n'}^{*\prime}, Y_{n'}^*)\} = \{(\mathbf{X}_1', Y_1), ..., (\mathbf{X}_n', Y_n)\} \cap A_m. \tag{3.8}$$

Put $g = \sup\{|Y - a - \mathbf{X}'\mathbf{b}| : (\mathbf{X}', Y) \in A_m, (a, \mathbf{b}') \in A_l\}$. Choose $\varepsilon_2 > 0$ such that

$$\sup\{|\rho(u_2) - \rho(u_1)| : |u_1| \leqslant g, |u_2| \leqslant g, |u_2 - u_1| \leqslant \varepsilon_2\} < \varepsilon_1. \tag{3.9}$$

Choose $\varepsilon_3 > 0$ such that

$$\begin{aligned} \sup\{|a + \mathbf{X}'\mathbf{b} - (\tilde{a} + \mathbf{X}'\tilde{\mathbf{b}})| : (a, \mathbf{b}') \in A_l, (\mathbf{a}, \tilde{\mathbf{b}}') \in A_l, \\ |a - \tilde{a}| \leqslant \varepsilon_3, \|\mathbf{b} - \tilde{\mathbf{b}}\| \leqslant \varepsilon_3, \|\mathbf{X}\| \leqslant pm\} < \varepsilon_2. \end{aligned} \tag{3.10}$$

Choose a finite set $G=\{(\alpha_1, \boldsymbol{\beta}_1'), ..., (\alpha_k, \boldsymbol{\beta}_k')\} \subset A_l - A_\varepsilon$, such that for any $(\alpha, \boldsymbol{\beta}') \in A_l - A_\varepsilon$ there exists j satisfying $|\alpha - \alpha_j| \leqslant \varepsilon_3$, $\|\boldsymbol{\beta} - \boldsymbol{\beta}_j\| \leqslant \varepsilon_3$.

In the following we shall repeatedly use the phrase "with probability one for n sufficiently large." For simplicity we shall abbreviate it by "wpln." Also, "strong law of large numbers" will be simply written as "SLLN."

Now by SLLN, (3.6) and (3.7), we have wpln:

$$
\begin{aligned}
n^{-1} \sum_{i=1}^{n'} \rho(Y_i^* - \alpha_j - \mathbf{X}_i^{*\prime} \boldsymbol{\beta}_j) &> E\{I((\mathbf{X}', Y) \in A_m)\, \rho(Y - \alpha_j - \mathbf{X}' \boldsymbol{\beta}_j)\} - \varepsilon_1 \\
&> E\rho(Y - \alpha_j - \mathbf{X}' \boldsymbol{\beta}_j) - 2\varepsilon_1 \\
&> Q(0, \mathbf{0}') + 4\varepsilon_1, \qquad j = 1, ..., k. \qquad (3.11)
\end{aligned}
$$

Fix $(\alpha, \boldsymbol{\beta}') \in A_l - A_\varepsilon$. Find j such that $|\alpha - \alpha_j| \leqslant \varepsilon_3$, $\|\boldsymbol{\beta} - \boldsymbol{\beta}_j\| \leqslant \varepsilon_3$. According to (3.9)–(3.11), we have

$$
\begin{aligned}
\sum_{i=1}^{n} \rho(Y_i - \alpha - \mathbf{X}_i' \boldsymbol{\beta}) &\geqslant \sum_{i=1}^{n'} \rho(Y_i' - \alpha_j - \mathbf{X}_i^{*\prime} \boldsymbol{\beta}_j) \\
&\quad - \sum_{i=1}^{n'} |\rho(Y_i' - \alpha_j - \mathbf{X}_i^{*\prime} \boldsymbol{\beta}_j) - \rho(Y_i - \alpha - \mathbf{X}_i^{*\prime} \boldsymbol{\beta})| \\
&\geqslant n[Q(0, \mathbf{0}') + 4\varepsilon_1] - n'\varepsilon_1 \\
&\geqslant n[Q(0, \mathbf{0}') + 3\varepsilon_1]. \qquad (3.12)
\end{aligned}
$$

This holds simultaneously for all $(\alpha, \boldsymbol{\beta}') \in A_l - A_\varepsilon$, wpln. On the other hand, by SLLN, we have wpln:

$$
\sum_{i=1}^{n} \rho(Y_i) < n[Q(0, \mathbf{0}) + \varepsilon_1]. \qquad (3.13)
$$

From (3.12) and (3.13), it follows that $|\tilde{\alpha}_n| \leqslant \varepsilon$ and $\|\tilde{\boldsymbol{\beta}}_n\| \leqslant \varepsilon$ wpln, so (3.4) is proved.

LEMMA 2. *Suppose that the conditions of Theorem* 1 *are satisfied. Then there exists a constant* $l > 0$ *such that* $(\tilde{\alpha}_n, \tilde{\boldsymbol{\beta}}_n') \in A_l$ wpln, *where* $(\tilde{\alpha}_n, \tilde{\boldsymbol{\beta}}_n')$ *is defined as a solution of* (3.3).

Proof. Write $S = \{(\alpha, \boldsymbol{\beta}') : (\alpha, \boldsymbol{\beta}') \in R^{p+1}, \alpha^2 + \|\boldsymbol{\beta}\|^2 = 1\}$. By (2.1) we can find $\varepsilon > 0$ such that

$$
v \equiv \inf\{P(|\alpha + \mathbf{X}' \boldsymbol{\beta}| > \varepsilon) : (\alpha, \boldsymbol{\beta}') \in S\} > 0. \qquad (3.14)
$$

Choose $m > 0$ such that $P(\mathbf{X} \in \tilde{A}_m) > 1 - v/4$, and put $u = 3^{-1}(1 + pm)^{-1}\varepsilon$.

Choose a finite set $S_1 \subset S$ such that for each $\boldsymbol{\theta} \in S$ there exists $\boldsymbol{\theta}_1 \in S_1$ for which $\|\boldsymbol{\theta} - \boldsymbol{\theta}_1\| < u$. By (3.14), we have wpln,

$$\#\{i: 1 \leqslant i \leqslant n, |\alpha + \mathbf{X}_i'\boldsymbol{\beta}| > \varepsilon\} \geqslant nv/2, \qquad \text{for every} \quad (\alpha, \boldsymbol{\beta}') \in S_1. \tag{3.15}$$

First consider the case $\rho(\infty) = \rho(-\infty) - \infty$. By SLLN, we have wpln,

$$\#\{i: 1 \leqslant i \leqslant n, \mathbf{X}_i \in A_m\} \geqslant n(1 - v/4). \tag{3.16}$$

Choose a constant $K > 8[Q(0, \mathbf{0}') + 1]/v$. Since ρ is continuous and $\rho(\pm\infty) = \infty$, we can find $h > 0$ such that $\rho(x) \geqslant K$ when $|x| \geqslant h$. Choose $l > 0$ large enough such that

$$\varepsilon l \geqslant 4h, \qquad P(|Y| \leqslant \varepsilon l/4) > 1 - v/8. \tag{3.17}$$

By SLLN, we have wpln:

$$\#\{i: 1 \leqslant i \leqslant n, |Y_i| \leqslant \varepsilon l/4\} \geqslant n(1 - v/8). \tag{3.18}$$

Now choose arbitrarily $(\tilde{\alpha}, \tilde{\boldsymbol{\beta}}') \notin A_l$. Then $(\tilde{\alpha}, \tilde{\boldsymbol{\beta}}') = r(\alpha, \boldsymbol{\beta}')$ for some $r > l$ and $(\alpha, \boldsymbol{\beta}') \in S$. If $(\alpha, \boldsymbol{\beta}') \in S_1$, then from (3.15) and (3.18), we have wpln:

$$\#\{i: 1 \leqslant i \leqslant n, |Y_i - \tilde{\alpha} - \mathbf{X}_i'\tilde{\boldsymbol{\beta}}| \geqslant 3l\varepsilon/4\} \geqslant 3nv/8. \tag{3.19}$$

If $(\alpha, \boldsymbol{\beta}') \notin S_1$, then choose $(\alpha^*, \boldsymbol{\beta}^{*\prime}) \in S_1$ such that $|\alpha \quad \alpha^*| < u$, $\|\boldsymbol{\beta} - \boldsymbol{\beta}^*\| < u$. When $|\alpha^* + \mathbf{X}_i'\boldsymbol{\beta}^*| > \varepsilon$ and $\mathbf{X}_i \in \tilde{A}_m$, we have

$$\begin{aligned} |\alpha + \mathbf{X}_i'\boldsymbol{\beta}| &\geqslant \varepsilon - |\alpha^* - \alpha + \mathbf{X}_i'(\boldsymbol{\beta}^* - \boldsymbol{\beta})| \\ &\geqslant \varepsilon - |\alpha^* - \alpha| - \|\mathbf{X}_i\| \, \|\boldsymbol{\beta}^* - \boldsymbol{\beta}\| \\ &\geqslant \varepsilon - u - pmu \geqslant \varepsilon - (1 + pm)u > \varepsilon/2. \end{aligned}$$

(Recall that $u = 3^{-1}(1 + pm)^{-1}\varepsilon$.) Hence $|\tilde{\alpha} + \mathbf{X}_i'\tilde{\boldsymbol{\beta}}| > l\varepsilon/2$. From this, (3.15), (3.16), and (3.18), we have wpln:

$$\#\{i: 1 \leqslant i \leqslant n, |Y_i - \tilde{\alpha} - \mathbf{X}_i'\tilde{\boldsymbol{\beta}}| > l\varepsilon/4\} \geqslant nv/8. \tag{3.20}$$

By (3.19), (3.20), (3.17), and the choice of h, we have wpln,

$$\sum_{i=1}^{n} \rho(Y_i - \tilde{\alpha} - \mathbf{X}_i'\tilde{\boldsymbol{\beta}}) \geqslant vKn/8 \geqslant [Q(0, \mathbf{0}') + 1]n, \tag{3.21}$$

simultaneous for all $(\tilde{\alpha}, \tilde{\boldsymbol{\beta}}') \notin A_l$. Taking $\varepsilon_1 = \frac{1}{2}$ in (3.13), we see that $(\tilde{\alpha}_n, \tilde{\boldsymbol{\beta}}_n) \in A_l$ wpln.

We now consider the case $0 < \rho(\pm\infty) = c < \infty$. First note that

$Q(0, \mathbf{0}') < c$ by condition (a). Further, condition (2.2) ensures the existence of $\varepsilon > 0$ for given $t < 1$ such that

$$\inf\{P(|\alpha + \mathbf{X}'\boldsymbol{\beta}| > \varepsilon): (\alpha, \boldsymbol{\beta}') \in S\} > t. \tag{3.22}$$

Based on (3.22) and modifying the previous argument appropriately, we can choose $l > 0$ such that for given $\varepsilon_1 > 0$, we have wpln:

$$\sum_{i=1}^{n} \rho(Y_i - \alpha - \mathbf{X}_i'\boldsymbol{\beta}) > n(c - \varepsilon_1), \qquad \text{for all} \quad (\alpha, \boldsymbol{\beta}') \notin A_l. \tag{3.23}$$

Choose $\varepsilon_1 = [c - Q(0, \mathbf{0}')]/3$. From (3.13) and (3.23), it follows that $|\tilde{\alpha}_n| \leqslant \varepsilon$, and $\|\boldsymbol{\beta}_n\| < \varepsilon$ wpln, as before. This concludes the proof of Lemma 2.

Proof of Theorem 1. Apply Lemmas 1 and 2.

Proof of Theorem 2. Since ρ is a convex function, we need only prove that the conclusion of Lemma 1 holds under the assumptions of Theorem 2. For this purpose put $\rho^*(Y - \alpha - \mathbf{X}'\boldsymbol{\beta}) = \rho(Y - \alpha - \mathbf{X}'\boldsymbol{\beta}) - \rho(Y)$, and define q^* as

$$q^* = \inf\{Q^*(\alpha, \boldsymbol{\beta}'): (\alpha, \boldsymbol{\beta}') \in A_l - A_\varepsilon\}, \tag{3.24}$$

where Q^* is defined in (2.4).

Now denote by $(a_1, \mathbf{b}_1'), \ldots, (a_T, \mathbf{b}_T')$ the $T = 2^{p+1}$ vertices $(\pm l, \ldots, \pm l)$ of A_l, $(a_{T+1}, \mathbf{b}_{T+1}'), \ldots, (a_{2T}, \mathbf{b}_{2T}')$ the vertices of A_{2l}. We proceed to show that

$$\begin{aligned} \sup\{|\rho(Y - \alpha - \mathbf{X}'\boldsymbol{\beta}) - \rho(Y)| &: (\alpha, \boldsymbol{\beta}') \in A_l\} \\ &\leqslant 2 \max_{1 \leqslant j \leqslant 2T} |\rho(Y - a_j - \mathbf{X}'\mathbf{b}_j) - \rho(Y)| \\ &\equiv K(\mathbf{X}', Y). \end{aligned} \tag{3.25}$$

Indeed, if $\rho(Y - \alpha - \mathbf{X}'\boldsymbol{\beta}) \geqslant \rho(Y)$, then by condition (a) we have $|\rho(Y - \alpha - \mathbf{X}'\boldsymbol{\beta}) - \rho(Y)| \leqslant \max_{1 \leqslant j \leqslant T} |\rho(Y - a_j - \mathbf{X}'\mathbf{b}_j) - \rho(Y)|$. If $\rho(Y) > \rho(Y - \alpha - \mathbf{X}'\boldsymbol{\beta})$, two cases are possible: $\alpha + \mathbf{X}'\boldsymbol{\beta} > 0$ and $\alpha + \mathbf{X}'\boldsymbol{\beta} < 0$. The handling of these cases being similar, we shall consider only the former case. By convexity of ρ, we have

$$\rho(Y + c) - \rho(Y + c - \alpha - \mathbf{X}'\boldsymbol{\beta}) \geqslant \rho(Y) - \rho(Y - \alpha - \mathbf{X}'\boldsymbol{\beta}), \qquad \text{for any} \quad c \geqslant 0. \tag{3.26}$$

Since $(\alpha, \boldsymbol{\beta}') \in A_l$, there exists $j \leqslant T$ such that $\alpha + \mathbf{X}'\boldsymbol{\beta} \geqslant a_j + \mathbf{X}'\mathbf{b}_j$. Write $\tilde{a} = \alpha - a_j$, $\tilde{\mathbf{b}} = \boldsymbol{\beta} - \mathbf{b}_j$, and set $c = a + \mathbf{X}'\mathbf{b}$ in (3.26). We obtain

$$\rho(Y) - \rho(Y - \alpha - \mathbf{X}'\boldsymbol{\beta}) \leqslant \rho(Y + \tilde{a} + \mathbf{X}'\mathbf{b}) - \rho(Y - a_j - \mathbf{X}'\tilde{\mathbf{b}}_j).$$

Obviously, $(\tilde{a}, \tilde{\mathbf{b}}') \in A_{2l}$. Hence by condition (a) there exists $k \leqslant 2T$ such that $\rho(Y+\tilde{a}+\mathbf{X}'\tilde{\mathbf{b}}) \leqslant \rho(Y-a_k-\mathbf{X}'\mathbf{b}_k)$, and we get

$$\begin{aligned}
&|\rho(Y)-\rho(Y-\alpha-\mathbf{X}'\boldsymbol{\beta})| \\
&\quad = \rho(Y)-\rho(Y-\alpha-\mathbf{X}'\boldsymbol{\beta}) \\
&\quad \leqslant \rho(Y-a_k-\mathbf{X}'\mathbf{b}_k)-\rho(Y-a_j-\mathbf{X}'\mathbf{b}_j) \\
&\quad \leqslant |\rho(Y-a_k-\mathbf{X}'\mathbf{b}_k)-\rho(Y)| + |\rho(Y-a_j-\mathbf{X}'\mathbf{b}_j)-\rho(Y)| \\
&\quad \leqslant 2 \max_{1 \leqslant j \leqslant 2T} |\rho(Y-a_j-\mathbf{X}'\mathbf{b}_j)-\rho(Y)|
\end{aligned}$$

and (3.25) is proved. (3.25) and condition (c') together ensure that Q^* is continuous, and therefore $q^*>0$. The rest of the proof is similar to that of Lemma 1.

Applying Theorem 2 to the case $\rho(u)=|u|$, we obtain the following corollary, which was proved in [3] with the additional conditions that $E|Y|<\infty$, $Y-\alpha_0-\mathbf{X}'\boldsymbol{\beta}_0$ and $\mathbf{X}$ are independent, and $P(\alpha+\mathbf{X}'\boldsymbol{\beta}=0)=0$ when $(\alpha, \boldsymbol{\beta}') \neq (0, \mathbf{0}')$.

COROLLARY 1. *Suppose that* $(\mathbf{X}_1', Y_1)$, $(\mathbf{X}_2', Y_2)$, ... *are i.i.d. samples of the random vector* $(\mathbf{X}', Y)$, *which satisfies the conditions*:

1. $E\|\mathbf{X}\|<\infty$.

2. *The conditional distribution of* Y *given* $\mathbf{X}=\mathbf{x}$ *has a unique median* $\alpha_0+\mathbf{x}'\boldsymbol{\beta}_0$.

Denote by $(\hat{\alpha}_n, \hat{\boldsymbol{\beta}}_n)$ *a solution of* (1.2). *Then* (1.3) *holds.*

Proof of Theorem 3. The proof follows from the following two lemmas.

LEMMA 1$'$. *Suppose that the conditions of Theorem* 3 *are satisfied, and* $l>0$ *is a given constant. Then for any* $\varepsilon>0$ *there exists a constant* $c>0$ *independent of n, such that*

$$P(|\tilde{\alpha}_n-\alpha_0| \geqslant \varepsilon)=O(e^{-cn}), \qquad P(\|\boldsymbol{\beta}_n-\boldsymbol{\beta}_0\| \geqslant \varepsilon)=O(e^{-cn}),$$

where $(\tilde{\alpha}_n, \tilde{\boldsymbol{\beta}}_n')$ *is defined as a solution of* (3.3).

LEMMA 2$'$. *Suppose that the conditions of Theorem* 3 *are satisfied. Then there exist constants* $l>0$ *and* $c>0$ *such that*

$$P\{(\hat{\alpha}_n-\alpha_0, \hat{\boldsymbol{\beta}}_n'-\boldsymbol{\beta}_0') \notin A_l\}=O(e^{-cn}).$$

These lemmas can be proved by the same method used in proving Lemma 1 and Lemma 2, together with the following fact (see [9, p. 288]):

Suppose that $\xi_1, \xi_2, \ldots$ are i.i.d. random variables, $E\xi_1 = 0$ and there exists $\delta > 0$ such that $E\exp(t\xi_1) < \infty$ when $|t| < \delta$. Then for any given $\varepsilon > 0$ we can find a constant $c > 0$ such that

$$P\left(\left|\sum_{i=1}^{n} \xi_i \Big/ n\right| \geqslant \varepsilon\right) = O(e^{-cn}).$$

4. Proof of Theorem 4

We give only the proof of Theorem 4 under conditions (a)–(d). It is easy to modify the proof when (a) and (b) are replaced by (a′) and (b′).

Lemma 3. *Suppose that function ρ is defined on R, $\rho(0) = 0$, is nondecreasing on $[0, \infty)$ and nonincreasing on $(-\infty, 0]$. Let $\{Y_i, i = 1, 2, \ldots\}$ be a sequence of i.i.d. variables such that*

$$E\rho(Y_1 + c) < \infty, \qquad \text{for any} \quad c \in R, \tag{4.1}$$

and $\{c_i, i = 1, 2, \ldots\}$ be a sequence of bounded real constants. Then

$$\lim_{n \to \infty} \frac{1}{n} \sum_{i=1}^{n} [\rho(Y_i - c_i) - E\rho(Y_i - c_i)] = 0, \qquad \text{a.s.} \tag{4.2}$$

Proof. Apply a standard truncation argument.

Lemma 4. *Suppose that the conditions of Lemma 3 are satisfied, and that ρ is continuous everywhere on R, $\{\mathbf{X}_i\}$ is a bounded sequence, and B is a bounded set in R^{p+1}. Then, with probability one, the sequence $\{(1/n)(\sum_{i=1}^{n} \rho(Y_i - \alpha - \mathbf{X}_i'\boldsymbol{\beta}) - \sum_{i=1}^{n} E\rho(Y_i - \alpha - \mathbf{X}_i'\boldsymbol{\beta})) : n = 1, 2, \ldots\}$ of functions of $(\alpha, \boldsymbol{\beta}')$ is equicontinuous and uniformly bounded on B.*

Proof. Denote by F the probability distribution of Y_1. Construct the probability space $(R^\infty, \mathscr{B}^\infty, F^\infty)$. Fix integer $m > 0$, find $h > 0$ such that

$$E\{\rho(Y_1 + T)\, I(\rho(Y_1 + T) \geqslant h) + \rho(Y_1 - T)\, I(\rho(Y_1 - T) \geqslant h)\} < 1/(3m), \tag{4.3}$$

where $T = \sup\{|\alpha + \mathbf{X}_i'\boldsymbol{\beta}| : i = 1, 2, \ldots, (\alpha, \boldsymbol{\beta}') \in B\}$ ($< \infty$ by the boundedness of $\{\mathbf{X}_i\}$ and B). From the assumptions on ρ, it follows that there exists $\varepsilon_m' > 0$ such that we have $|\rho(u_1) - \rho(u_2)| \leqslant 1/(3m)$ when $|u_1 - u_2| \leqslant \varepsilon_m'$ and $\min(\rho(u_1), \rho(u_2)) < h$. Find $\varepsilon_{m1} > 0$ such that $|(\alpha + \mathbf{X}_i'\boldsymbol{\beta}) - (\alpha^* + \mathbf{X}_i'\boldsymbol{\beta}^*)| \leqslant \varepsilon_m'$, $i = 1, 2, \ldots$, whenever $(\alpha, \boldsymbol{\beta}') \in B$, $(\alpha^*, \boldsymbol{\beta}^{*\prime}) \in B$ and $\|(\alpha, \boldsymbol{\beta}') - (\alpha^*, \boldsymbol{\beta}^{*\prime})\| \leqslant \varepsilon_{m1}$.

Now by (4.3) and SLLN, we can find a positive integer N_m and a set $D_m \in \mathscr{B}^\infty$ with $F^\infty(D_m) < 2^{-m}$, such that

$$n^{-1} \sum_{i=1}^{n} \rho(Y_i + T)\, I(\rho(Y_i + T) \geqslant h)$$
$$+ n^{-1} \sum_{i=1}^{n} \rho(Y_i - T)\, I(\rho(Y_i - T) \geqslant h) < 1/(3m), \tag{4.4}$$

whenever $n \geqslant N_m$ and $Y^* \equiv (Y_1, Y_2, \ldots) \notin D_m$. Since $\{\mathbf{X}_i\}$ and B are bounded, for any $Y^* \in R^\infty$ we can find $\varepsilon_{m2}(Y^*)$ such that

$$\left| n^{-1} \sum_{i=1}^{n} (\rho(Y_i - \alpha^* - \mathbf{X}_i' \boldsymbol{\beta}^*) - \rho(Y_i - \alpha - \mathbf{X}_i' \boldsymbol{\beta})) \right| < 1/(3m), \tag{4.5}$$

whenever $1 \leqslant n \leqslant N_m$, $(\alpha, \boldsymbol{\beta}') \in B$, $(\alpha^*, \boldsymbol{\beta}^{*\prime}) \in B$, and $\|(\alpha^*, \boldsymbol{\beta}^{*\prime}) - (\alpha, \boldsymbol{\beta}')\| \leqslant \varepsilon_{m2}(Y^*)$. Take $\varepsilon_m(Y^*) = \min(\varepsilon_{m1}, \varepsilon_{m2}(Y^*))$.

Now suppose that $(\alpha, \boldsymbol{\beta}') \in B$, $(\alpha^*, \boldsymbol{\beta}^{*\prime}) \in B$, $\|(\alpha^*, \boldsymbol{\beta}^{*\prime}) - (\alpha, \boldsymbol{\beta}')\| \leqslant \varepsilon_m(Y^*)$, and $Y^* \notin D_m$. Then for $n \leqslant N_m$ we have (4.5). If $n > N_m$, then

$$\left| n^{-1} \sum_{i=1}^{n} (\rho(Y_i - \alpha^* - \mathbf{X}_i' \boldsymbol{\beta}^*) - \rho(Y_i - \alpha - \mathbf{X}_i' \boldsymbol{\beta})) \right|$$
$$\leqslant n^{-1} \sum_{i=1}^{n} \rho(Y_i - \alpha^* - \mathbf{X}_i' \boldsymbol{\beta}^*)\, I(\rho(Y_i - \alpha^* - \mathbf{X}_i' \boldsymbol{\beta}^*) \geqslant h)$$
$$+ n^{-1} \sum_{i=1}^{n} \rho(Y_i - \alpha - \mathbf{X}_i' \boldsymbol{\beta})\, I(\rho(Y_i - \alpha - \mathbf{X}_i' \boldsymbol{\beta}) \geqslant h)$$
$$+ n^{-1} \left| \sum{}' [\rho(Y_i - \alpha^* - \mathbf{X}_i' \boldsymbol{\beta}^*) - \rho(Y_i - \alpha - \mathbf{X}_i' \boldsymbol{\beta})] \right|$$
$$\equiv J_1 + J_2 + J_3, \tag{4.6}$$

where the summation $\sum'$ is over all i such that $1 \leqslant i \leqslant n$ and $\min(\rho(Y_i - \alpha^* - \mathbf{X}_i' \boldsymbol{\beta}^*), \rho(Y_i - \alpha - \mathbf{X}_i' \boldsymbol{\beta})) < h$. From (4.4), the definition of T and the conditions imposed on ρ, we have

$$J_1 \leqslant \text{the left-hand side of (4.4)} < 1/(3m).$$

Likewise, $J_2 \leqslant 1/(3m)$. Finally, by the definition of ε_m', ε_{m1}, and $\varepsilon_m(Y^*)$, for each i belonging to the range of summation $\sum'$, we have $|\rho(Y_i - \alpha^* - \mathbf{X}_i' \boldsymbol{\beta}^*) - \rho(Y_i - \alpha - \mathbf{X}_i' \boldsymbol{\beta})| < 1/(3m)$. Hence $J_3 < 1/(3m)$. Summing up, we find that (4.5) is still true when $1/(3m)$ on the right-hand side of (4.5) is replaced by $1/m$.

Now write $D = \bigcap_{n=1}^{\infty} \bigcup_{m=n}^{\infty} D_m$. Since $F^\infty(D_m) < 2^{-m}$, we have

$F^\infty(D)=0$. From the above discussion we see that for any $Y^*=(Y_1, Y_2, ...)\notin D$ and any positive integer m, we can find $\varepsilon_m(Y^*)>0$ such that

$$\left|\sum_{i=1}^{n}(\rho(Y_i-\alpha^*-\mathbf{X}_i'\boldsymbol{\beta}^*)-\rho(Y_i-\alpha-\mathbf{X}_i'\boldsymbol{\beta}))\right|\Big/n\leqslant 1/m, \qquad \text{for} \quad n=1, 2, ...$$

whenever $(\alpha, \boldsymbol{\beta}')\in B$, $(\alpha^*, \boldsymbol{\beta}^{*\prime})\in B$, and $\|(\alpha^*, \boldsymbol{\beta}^{*\prime})-(\alpha, \boldsymbol{\beta}')\|\leqslant\varepsilon_m(Y^*)$. This proves the equicontinuity of $\{\sum_{i=1}^n \rho(Y_i-\alpha-\mathbf{X}_i'\boldsymbol{\beta})/n\colon n=1, 2, ...\}$ over B, with probability one. The uniform boundedness of this sequence of functions follows from the fact that when $Y^*\notin D$, we have $Y^*\notin D_m$ for some m. Repeating the above argument, we find that

$$\sum_{i=1}^{n}\rho(Y_i-\alpha-\mathbf{X}_i'\boldsymbol{\beta})/n\leqslant h+1/(3m)$$

for $n\geqslant N_m$ and $(\alpha, \boldsymbol{\beta}')\in B$, while for $n<N_m$ we have

$$\sum_{i=1}^{n}\rho(Y_i-\alpha-\mathbf{X}_i'\boldsymbol{\beta})/n\leqslant\sum_{i=1}^{N_m}(\rho(Y_i+T)+\rho(Y_i-T))$$

for any $(\alpha, \boldsymbol{\beta}')\in B$.

Therefore, in order to prove Lemma 4, we have only to establish that $\{\sum_{i=1}^n E\rho(Y_i-\alpha-\mathbf{X}_i'\boldsymbol{\beta})/n\colon n=1, 2, ...\}$ is uniformly bounded and equicontinuous on B. This is simple, since $E\rho(Y_1+c)$ is continuous for each c, $\sup\{|\alpha+\mathbf{X}_i'\boldsymbol{\beta}|: i=1, 2, ..., (\alpha, \boldsymbol{\beta}')\in B\}=T<\infty$, and

$$E\rho(Y_i-\alpha-\mathbf{X}_i'\boldsymbol{\beta})\leqslant E\rho(Y_1+T)+E\rho(Y_1-T), \quad (\alpha, \boldsymbol{\beta}')\in B, \quad i=1, 2, ...$$

Combining Lemma 3 and Lemma 4, we obtain

LEMMA 5. *If the conditions of Lemma* 3 *and Lemma* 4 *are satisfied, then there exists a set* $D\in\mathscr{B}^\infty$ *such that* $F^\infty(D)=0$, *and when* $(Y_1, Y_2, ...)\notin D$ *we have* $\lim_{n\to\infty}\sum_{i=1}^n[\rho(Y_i-\alpha-\mathbf{X}_i'\boldsymbol{\beta})-E\rho(Y_i-\alpha-\mathbf{X}_i'\boldsymbol{\beta})]/n=0$ *uniformly for* $(\alpha, \boldsymbol{\beta}')\in B$, *B is a given bounded set in* R^{p+1}.

In the following we adhere to (3.2), and put

$$S_M=\{(\alpha, \boldsymbol{\beta}'): \alpha^2+\|\boldsymbol{\beta}\|^2\leqslant M\},$$
$$\bar{S}_M=\{(\alpha, \boldsymbol{\beta}'): \alpha^2+\|\boldsymbol{\beta}\|^2=M\}$$

for any $M>0$.

LEMMA 6. *Suppose that the conditions* (a)–(d) *of Theorem* 4 *are satisfied.*

(i) *There exist* $\varepsilon_1 > 0$ *and* $\varepsilon_2 > 0$ *such that*

$$\inf\{\#(i\colon 1 \leqslant i \leqslant n,\ |\alpha + \mathbf{X}_i'\boldsymbol{\beta}| \geqslant \varepsilon_1)/n\colon (\alpha, \boldsymbol{\beta}') \in \bar{S}_1\} \geqslant \varepsilon_2 \tag{4.7}$$

for n sufficiently large.

(ii) *For each* $M > 0$ *there exists a constant* $\varepsilon_M > 0$ *such that*

$$\sum_{i=1}^{n} E[\rho(Y_i - \alpha - \mathbf{X}_i'\boldsymbol{\beta}) - \rho(Y_i)]/n \geqslant \|(\alpha, \boldsymbol{\beta}')\|^2 \varepsilon_M \tag{4.8}$$

for $(\alpha, \boldsymbol{\beta}') \in S_M$ *and n sufficiently large.*

Proof. Consider

$$\sum_{i=1}^{n} (\alpha + \mathbf{X}_i'\boldsymbol{\beta})^2 = n(\alpha - \bar{\mathbf{X}}_n'\boldsymbol{\beta})^2 + \boldsymbol{\beta}' H_n \boldsymbol{\beta},$$

where $\bar{\mathbf{X}}_n = (\mathbf{X}_1 + \cdots + \mathbf{X}_n)/n$, $H_n = \sum_{i=1}^{n} (\mathbf{X}_i - \bar{\mathbf{X}}_n)(\mathbf{X}_i - \bar{\mathbf{X}}_n)'$. Suppose that $(\alpha, \boldsymbol{\beta}') \in \bar{S}_1$. Write $M_0 = \sup\{1, \|\mathbf{X}_i\|\colon i = 1, 2, \ldots\}$. If $\|\boldsymbol{\beta}\| \geqslant (2pM_0)^{-1}$, we have, according to (2.7), $\boldsymbol{\beta}' H_n \boldsymbol{\beta} \geqslant (2pM_0)^{-2} \delta_1 n$ for some constant $\delta_1 > 0$ and n large. If $\|\boldsymbol{\beta}\| < (2pM_0)^{-1}$, then $\|\bar{\mathbf{X}}_n'\boldsymbol{\beta}\| \leqslant (2p)^{-1}$, and $|\alpha| \geqslant \sqrt{1 - (2p)^{-2}} \geqslant \sqrt{3}/2$. Hence $|\alpha - \bar{\mathbf{X}}_n'\boldsymbol{\beta}| \geqslant \sqrt{3}/2 - \frac{1}{2} > \frac{1}{3}$, and so $n(\alpha - \bar{\mathbf{X}}_n'\boldsymbol{\beta})^2 \geqslant n/9$. Summing up the above gives

$$\sum_{i=1}^{n} (\alpha + \mathbf{X}_i'\boldsymbol{\beta})^2 \geqslant \delta n, \qquad \text{for all } (\alpha, \boldsymbol{\beta}') \in \bar{S}_1 \text{ and } n \text{ large,} \tag{4.9}$$

for some $\delta > 0$.

Now suppose that (4.7) is false. Then we can find $n_j \to \infty$, $0 < \varepsilon_{1j} \to 0$, $0 < \varepsilon_{2j} \to 0$, $(\alpha_j, \boldsymbol{\beta}_j') \in \bar{S}_1$, such that

$$\#\{i\colon 1 \leqslant i \leqslant n_j,\ |\alpha_j + \mathbf{X}_i'\boldsymbol{\beta}_j| \geqslant \varepsilon_{1j}\} \leqslant \varepsilon_{2j} n_j, \qquad j = 1, 2, \ldots,$$

which entails that

$$\sum_{i=1}^{n_j} (\alpha_j + \mathbf{X}_i'\boldsymbol{\beta}_j)^2 \Big/ n_j \leqslant \varepsilon_{1j}^2 + \varepsilon_{2j} T^2, \qquad j = 1, 2, \ldots,$$

where $T = M_0 + 1$. This contradicts (4.9), and (4.7) is proved.

For a proof of (4.8), we notice that since $E\rho(Y_1 + t) > E\rho(Y_1)$ when $t \neq 0$, $E\rho(Y_1 + t)$ is continuous in t and $\alpha + \mathbf{X}_i'\boldsymbol{\beta}$ is uniformly bounded for $i = 1, 2, \ldots$ and $(\alpha, \boldsymbol{\beta}') \in S_M$. Hence it follows from (2.8) that there exists a constant $\delta_M > 0$, depending only on M, such that $E[\rho(Y_i - \alpha - \mathbf{X}_i'\boldsymbol{\beta}) - \rho(Y_i)] \geqslant \delta_M |\alpha + \mathbf{X}_i'\boldsymbol{\beta}|^2$ for $(\alpha, \boldsymbol{\beta}') \in S_M$. From this and (4.9), (4.8) follows.

LEMMA 7. *Suppose that the conditions* (a)–(d) *of Theorem* 4 *are satisfied. Given* $l>0$, *denote by* $(\tilde{\alpha}_n, \tilde{\boldsymbol{\beta}}'_n)$ *the solution of the constrained minimization problem* (3.3). *Then* (3.4) *holds. Moreover, the conclusion of Lemma* 2 *holds.*

Proof. Fix $\varepsilon \in (0, R)$. Let D be the set mentioned in Lemma 5. Since $\|(\alpha, \boldsymbol{\beta}')\| \geqslant \varepsilon$ when $(\alpha, \boldsymbol{\beta}') \notin A_\varepsilon$ (see (3.1)), it follows from Lemma 5 and Lemma 6(b) that

$$\inf\left\{\sum_{i=1}^{n} [\rho(Y_i - \alpha - \mathbf{X}_i'\boldsymbol{\beta}) - \rho(Y_i)] \Big/ n : (\alpha, \boldsymbol{\beta}') \in A_l - A_\varepsilon\right\} \leqslant 2^{-1}\varepsilon^2 \varepsilon_{(p+1)l},$$

for all $(Y_1, Y_2, ...) \notin D$.

By Lemma 6(a), we still have (3.15) in a slightly different notation. Moreover, (3.16) remains true by the boundedness assumption of $\{\mathbf{X}_i\}$. Hence the proof of Lemma 2 remains valid in the present setting.

REFERENCES

[1] BAI, Z. D., CHEN, X. R., MIAO, B. W., AND WU, Y. H. (1987). *On Solvability of an Equation Arising in the Theory of M-Estimates.* Tech. Report 87-46, Center for Multivariate Analysis, University of Pittsburgh.

[2] BAI, Z. D., CHEN, X. R., WU, Y. H., AND ZHAO, L. C. (1987). *Asymptotic Normality of Minimum L_1-Norm Estimates in Linear Models.* Tech. Report 87-35, Center for Multivariate Analysis, University of Pittsburgh.

[3] BLOOMFIELD, P., AND STIEGER, W. L. (1983). *Least Absolute Deviations.* Birkhauser, Basel.

[4] DODGE, Y., AND JURECKOVA, J. (1987). Adaptive combination of least squares and least absolute deviations estimates. *Statistical Data Analysis*, pp, 275–284, North-Holland, Amsterdam.

[5] HUBER, P. J. (1964). Robust estimation of a location parameter. *Ann. Math. Statist.* **35** 73–101.

[6] HUBER, P. J. (1973). Robust regression. *Ann. Statist.* **1** 799–821.

[7] MARONNA, R. A., AND YOHAI, V. J. (1981). Asymptotic behavior of general M-estimates for regression and scale with random carriers. *Z. Wahrsch. Verw. Gebiete* **58** 7–20.

[8] OBERHOFER, W. (1982). The consistency of nonlinear regression minimizing the L_1-norm. *Ann. Statist.* **10** 316–319.

[9] PETROV, V. V. (1975). *Sums of Independent Random Variables.* Springer-Verlag, New York.

[10] YOHAI, V. J., AND MARONNA, R. A. (1979). Asymptotic behavior of M-estimators for the linear model. *Ann. Statist.* **7** 258–268.

[11] WU, Y. H. (1987). *Strong Consistency and Exponential Rate of the "Minimum L_1-norm" Estimates in Linear Regression Models.* Tech. Report 87-18, Center for Multivariate Analysis, University of Pittsburgh.

Minimal Complete Classes of Invariant Tests for Equality of Normal Covariance Matrices and Sphericity

ARTHUR COHEN* AND JOHN I. MARDEN†

Rutgers University and University of Illinois

The problem of testing equality of two normal covariance matrices, $\Sigma_1 = \Sigma_2$ is studied. Two alternative hypotheses, $\Sigma_1 \neq \Sigma_2$ and $\Sigma_1 - \Sigma_2 > 0$ are considered. Minimal complete classes among the class of invariant tests are found. The group of transformations leaving the problems invariant is the group of nonsingular matrices. The maximal invariant statistic is the ordered characteristic roots of $S_1 S_2^{-1}$, where S_i, $i = 1, 2$, are the sample covariance matrices. Several tests based on the largest and smallest roots are proven to be inadmissible. Other tests are examined for admissibility in the class of invariant tests. The problem of testing for sphericity of a normal covariance matrix is also studied. Again a minimal complete class of invariant tests is found. The popular tests are again examined for admissibility and inadmissibility in the class of invariant tests.

INTRODUCTION AND SUMMARY

The problems of testing equality of two normal covariance matrices and testing sphericity of a normal covariance matrix are classical problems in multivariate analysis. See, for example, Anderson [1, Chap. 10] and Muirhead [7, Chap. 8]. In this paper we consider the admissibility of invariant tests in these common testing problems. Two problems (two-sided and one-sided cases) are based on S_1 and S_2, independent, where

$$S_1 \sim W_p(n_1, \Sigma_1) \qquad \text{and} \qquad S_2 \sim W_p(n_2, \Sigma_2), \tag{1.1}$$

and $W_p(n, \Sigma)$ is the Wishart distribution on $p \times p$ matrices with n degrees

* Research supported by NSF MCS-84-18416.
† Research supported by NSF MCS-82-01771.

Multivariate Statistics and Probability
ISBN 0-12-580205-6

Reprinted from *J. Mult. Anal.* **27**(1).

of freedom and expectation $n\Sigma$. We assume that $p \geq 2$, $n_1 \geq p$, and $n_2 \geq p$, and that Σ_1 and Σ_2 are positive definite. We consider testing

$$H_0: \Sigma_1 = \Sigma_2 \qquad \text{versus} \qquad H_A: \Sigma_1 \neq \Sigma_2, \tag{1.2}$$

and

$$H_0: \Sigma_1 = \Sigma_2 \qquad \text{versus} \qquad H_A: \Sigma_1 > \Sigma_2, \tag{1.3}$$

where $\Sigma_1 > \Sigma_2$ means that $\Sigma_1 - \Sigma_2$ is positive definite.

The third problem tests for sphericity of a covariance matrix. That is, we have

$$S \sim W_p(n, \Sigma), \tag{1.4}$$

$n \geq p \geq 2$, $\Sigma > 0$, and test

$$H_0: \Sigma = \sigma^2 I \qquad \text{versus} \qquad H_A: \Sigma \neq \sigma^2 I, \tag{1.5}$$

where $\sigma^2 > 0$ is unspecified and I is the $p \times p$ identity matrix.

Problems (1.2) and (1.3) are invariant under the group $Gl(p)$ of $p \times p$ nonsingular matrices, which acts on (S_1, S_2) via

$$A: (S_1, S_2) \to (AS_1A', AS_2A') \tag{1.6}$$

for $A \in Gl(p)$, and on (Σ_1, Σ_2), similarly. A maximal invariant statistic and parameter are respectively

$$z = \text{diag}\{\text{ordered characteristic roots of } S_1 S_2^{-1}\},$$

and

$$\alpha = \text{diag}\{\text{ordered characteristic roots of } \Sigma_1 \Sigma_2^{-1}\}.$$

See Anderson [1, Theorem 10.6.1]. However, to develop our results it is more convenient to work with the maximal invariants x and θ, where $x_i = (z_i - 1)/(z_i + 1)$ and $\theta_i = (1 - \alpha_{p-i+1})/(1 + \alpha_{p-i+1})$. As such,

$$x = \text{diag}\{\text{ordered characteristic roots of } (S_1 - S_2)(S_1 + S_2)^{-1}\}, \tag{1.7}$$

and

$$\theta = \text{diag}\{\text{ordered characteristic roots of } (\Sigma_2 - \Sigma_1)(\Sigma_1 + \Sigma_2)^{-1}\}.$$

Hence, $x \in \mathscr{D}(p)$, the set of $p \times p$ diagonal matrices, and the diagonal elements of x satisfy $1 \geq x_1 \geq x_2 \geq \cdots \geq x_p \geq -1$. The invariance-reduced problem (1.2) then tests

$$H_0: \theta = 0 \qquad \text{versus} \qquad H_A: \theta \in \Theta - \{0\}, \tag{1.8}$$

where

$$\Theta = \{\theta \in \mathscr{D}(p) \mid 1 > \theta_1 \geqslant \theta_2 \geqslant \cdots \geqslant \theta_p > -1\}, \tag{1.9}$$

based on x with sample space

$$\mathscr{X} = \{x \in \mathscr{D}(p) \mid 1 > x_1 > x_2 > \cdots > x_p > -1\}. \tag{1.10}$$

Note that we have eliminated from the sample space the set of measure zero on which the x_i's are not distinct. A popular test for (1.2), in terms of x, is likelihood ratio test (LRT), which rejects H_0 when

$$|I+x|^{-n_1/2}\,|I-x|^{-n_2/2} > c. \tag{1.11}$$

Another test, which arises from our complete class rejects H_0 when

$$\frac{n_1+n_2}{2}(\operatorname{tr} x)^2 + \operatorname{tr} x^2 > c, \qquad 0 < c < \frac{n_1+n_2}{2}p^2 + p. \tag{1.12}$$

(In each case, the constant c is chosen to provide the desired level.) Other tests, including those based on tr x and the extreme characteristic roots, are listed in Muirhead [7, p. 332]. One such rejects H_0 when

$$\operatorname{tr} x < c_1 \qquad \text{or} \qquad \operatorname{tr} x > c_2, \qquad -p < c_1 < c_2 < p. \tag{1.13}$$

Tests based on the extreme roots of $S_1 S_2^{-1}$, which are equivalent to those based on the extreme roots of $(S_1 - S_2)(S_1 + S_2)^{-1}$ include those which reject H_0 when

$$x_1 < c_1 \qquad \text{or} \qquad x_1 > c_2; \tag{1.14}$$

$$x_p < c_1 \qquad \text{or} \qquad x_p > c_2; \tag{1.15}$$

$$x_p < c_1 \qquad \text{and} \qquad x_1 > c_2; \tag{1.16}$$

and

$$x_p < c_1 \qquad \text{or} \qquad x_1 > c_2. \tag{1.17}$$

In each case, $-1 < c_1 < c_2 < 1$.

Maximal invariants for problem (1.3) are x and θ as in (1.7), but now the alternative parameter space is smaller:

$$H_0\colon \theta = 0 \qquad \text{versus} \qquad H_A\colon \theta \in \Theta^+ - \{0\}, \tag{1.18}$$

where

$$\Theta^+ = \{\theta \in \mathscr{D}(p) \mid 0 > \theta_1 > \theta_2 > \cdots > \theta_p > -1\}. \tag{1.19}$$

The LRT for problem (1.3) modifies (1.11) by using the statistic $\bar{x}$ instead of x, where $\bar{x} \in \mathscr{D}(p)$ is defined by

$$\bar{x}_i = \max \left\{ x_i, \frac{n_1 - n_2}{n_1 + n_2} \right\}. \tag{1.20}$$

The test rejects H_0 when

$$|I + \bar{x}|^{-n_1/2} \, |I - \bar{x}|^{-n_2/2} > c, \qquad c > 0. \tag{1.21}$$

The locally best invariant test rejects H_0 when

$$\operatorname{tr} x > c, \tag{1.22}$$

where $-p < c < p$ (see Giri [4]). The extreme root tests have rejection regions

$$x_1 > c \tag{1.23}$$

and

$$x_p > c, \tag{1.24}$$

where $-1 < c < 1$.

The following theorem summarizes our admissibility/inadmissibility results for problems (1.8) and (1.18).

THEOREM 1.1. (a) *The LRT* (1.11) *when* $n_1 > 2(p-1)$ *and* $n_2 > 2(p-1)$, *and the test* (1.12), *are admissible in the invariant problem* (1.8). *The tests* (1.13)–(1.17) *are inadmissible.* (b) *The test* (1.22) *is admissible in the invariant problem* (1.18). *The LRT* (1.21) *and root tests* (1.23) *and* (1.24) *are inadmissible.*

The result for the test (1.22) follows from the essential uniqueness of its local properties, although it is also easy to prove its admissibility by using Theorem 3.1. The admissibility of the LRT (1.11) in problem (1.8) follows from the stronger result of Kiefer and Schwartz [6] which proves the LRT is admissible Bayes for the original problem (1.2).

The inadmissibility results are all based on violation of the following convexity property. (We represent a test as a measurable function $\phi: x \to [0, 1]$, where $\phi(x)$ is the probability of rejecting H_0 when x is observed.)

PROPERTY 1.2. *The test* ϕ *equals* $1 - I_A$, *a.e.* $[v]$, *for some convex set* $A \subseteq X$ *for which no three points of the boundary in* X *are collinear.*

Here, v is the measure on X when $\theta = 0$, which is absolutely continuous with respect to Lebesgue measure on $\mathbb{R}^p$, and I_A is the indicator function of A. We will prove the next proposition in Sections 2 and 3.

PROPOSITION 1.3. (a) *A necessary condition for a test ϕ to be admissible for problem* (1.8) *is that it equal* $1 - I_A$, *a.e.* $[v]$, *where A is either of the form* $\{x \mid \operatorname{tr} x \leqslant a\}$, *or* $\{x \mid \operatorname{tr} x \geqslant b\}$, *or ϕ satisfy Property* 1.2.

(b) *A necessary condition for a test ϕ to be admissible for problem* (1.18) *is that it equal* $1 - I_A$, *a.e.* $[v]$, *where A is of the form* $\{x \mid \operatorname{tr} x \leqslant a\}$, *or ϕ satisfy Property* 1.2.

It is fairly easy to see that tests (1.13)–(1.17), (1.21), (1.23), and (1.24) are not of the form required by Proposition 1.3.

Now turn to problem (1.5). The invariance group for this problem is the direct product $(0, \infty) \times O(p)$, where the operation for $(0, \infty)$ is multiplication and $O(p)$ is the group of $p \times p$ orthogonal matrices. The action is

$$(c, \Gamma): S \rightarrow c\Gamma S\Gamma^t. \tag{1.25}$$

A maximal invariant statistic and parameter are, respectively,

$$y = \operatorname{diag}\{\text{ordered characteristic roots of } S/\operatorname{tr} S\} \tag{1.26}$$

and

$$\lambda = \operatorname{diag}\{\text{ordered characteristic roots of } \Sigma/\operatorname{tr} \Sigma\}. \tag{1.27}$$

We prefer to use the parameter

$$\omega = p\lambda - I, \tag{1.28}$$

so that the hypotheses in (1.5) become

$$H_0: \omega = 0 \qquad \text{versus} \qquad H_A: \omega \in \Omega - \{0\}, \tag{1.29}$$

where

$$\Omega = \{\omega \in \mathscr{D}(p) \mid (p-1) > \omega_1 \geqslant \cdots \geqslant \omega_p > -1 \text{ and } \operatorname{tr} \omega = 0\}. \tag{1.30}$$

The LRT for problem (1.5) rejects H_0 when

$$|y| < c, \qquad 0 < c < 1, \tag{1.31}$$

where $|y|$ is the determinant of y. The locally most powerful invariant test has rejection region

$$S_y^2 \equiv \frac{1}{p} \Sigma(y_i - \bar{y})^2 > d, \tag{1.32}$$

where $\bar{y} = \Sigma y_i/p = 1/p$. See Sugiura [8]. Relevant root tests have rejection regions

$$y_1 > a, \tag{1.33}$$

$$y_p < b, \tag{1.34}$$

$$y_1 > a \quad \text{and} \quad y_p < b, \tag{1.35}$$

and

$$y_1 > a \quad \text{or} \quad y_p < b, \tag{1.36}$$

where $a \in (1/p, 1)$ and $b \in (0, 1/p)$.

THEOREM 1.4. *The LRT* (1.31) *and LMPI* (1.32) *test are admissible for problem* (1.29). *The root tests* (1.32), (1.34), (1.35), *and* (1.36) *are inadmissible if* $p \geqslant 3$. *When* $p = 2$, *the uniformly most powerful invariant test has rejection region* $\{y \mid y_1 > c\}$, $c \in (\frac{1}{2}, 1)$.

Again the admissibility of the LRT is found in Kiefer and Schwartz [6], and that for the LMPI test is due to its uniqueness. See also Theorem 3.1. The inadmissibility results follow from the next proposition.

PROPOSITION 1.5. *A necessary condition for a test* ϕ *to be admissible for problem* (1.29) *when* $p \geqslant 3$ *is that it satisfy Property* 1.2 (*with* $\mathscr{Y}$, *the space of* y, *in place of* $\mathscr{X}$.)

The proof of this proposition and the $p = 2$ result are given in Section 3. Our main results in the paper are Theorems 2.1, 2.2, and 3.1, which contain the minimal complete classes of tests for the reduced problems (1.8), (1.18), and (1.29). The proofs are in Section 4.

2. TESTING $\Sigma_1 = \Sigma_2$

We will use Brown and Marden [2] heavily, so that our first task is to find the likelihood ratio for x. Recall

$$z = \text{diag}\{\text{ordered characteristic roots of } S_1 S_2^{-1}\}, \tag{2.1}$$

and

$$\alpha = \text{diag}\{\text{ordered characteristic roots of } \Sigma_1 \Sigma_2^{-1}\}.$$

Then from James [5, Eqs. (33) and (65)], we have that

$$f_\alpha(z)/f_I(z) = |\alpha|^{-n_1/2}\, |I+z|^\beta \int_{O(p)} |I+z\Gamma\alpha^{-1}\Gamma'|^{-\beta}\, \rho(d\Gamma),$$
$$\beta = (n_1+n_2)/2, \tag{2.2}$$

where $f_\alpha(z)$ is the density of z when α obtains, and ρ is the Haar probability measure on $O(p)$. Now by (1.7) and (2.1)

$$z = (I+x)(I-x)^{-1} \qquad \text{and} \qquad \dot{\alpha} = (I-\theta)(I+\theta)^{-1}, \tag{2.3}$$

where $\dot{\alpha} = \operatorname{diag}(\alpha_p, ..., \alpha_1)$. Thus the ratio (2.2) in terms of (x, θ) is

$$|I+\theta|^{n_1/2}\, |I-\theta|^{n_2/2} \int_{O(p)} |I+x\Gamma\theta\Gamma'|^{-\beta}\, \rho(d\Gamma). \tag{2.4}$$

(To see this, note that α can be replaced by $\dot{\alpha}$ in (2.2),

$$|\alpha| = |I-\theta|\, |I+\theta|^{-1},$$
$$|I+z| = |I+(I+x)(I-x)^{-1}| = |I-x|^{-1}\, |2I| = |I-x|^{-1}\, 2^p,$$

and

$$\begin{aligned}
|I+z\Gamma\alpha^{-1}\Gamma'| &= |I+(I-x)^{-1}\,(I+x)\,\Gamma(I+\theta)(I-\theta)^{-1}\,\Gamma'| \\
&= |I-x|^{-1}\, |\Gamma'(I-x)\,\Gamma+\Gamma'(I+x)\,\Gamma(I+\theta)(I-\theta)^{-1}| \\
&= |I-x|^{-1}\, |I-\theta|^{-1}\, |\Gamma'(I-x)\,\Gamma(I-\theta)+\Gamma'(I+x)\,\Gamma(I+\theta) \\
&= |I-x|^{-1}\, |I-\theta|^{-1}\, |2I+2\Gamma'x\Gamma\theta| \\
&= |I-x|^{-1}\, |I-\theta|^{-1}\, |I+x\Gamma\theta\Gamma'|\, 2^p.)
\end{aligned}$$

Let $a(\theta) = |I+\theta|^{-n_1/2}\, |I-\theta|^{-n_2/2}$, and define $R_\theta(x)$ to be $a(\theta)$ times the quantity in (2.4), so that

$$R_\theta(x) = \int_{O(p)} |I+x\Gamma\theta\Gamma'|^{-\beta}\, \rho(d\Gamma). \tag{2.5}$$

To define the minimal complete classes, we need the derivatives

$$l(x) = (l_1(x), ..., l_p(x))', \qquad \text{where } l_i(x) = \frac{\partial}{\partial\theta_i} R_\theta(x)\,|_{\theta=0}, \tag{2.6}$$

ad

$$V(x) = \{V_{ij}(x)\}_{i,j=1}^p, \qquad \text{where } V_{ij}(x) = \frac{\partial^2}{\partial\theta_i\,\partial\theta_j} R_\theta(x)\,|_{\theta=0}. \tag{2.7}$$

For $\mu \in \mathbb{R}^p$, $M_0 \in \mathscr{S}(p)$ (the set of nonnegative definite symmetric $p \times p$ matrices), $H \in \mathscr{F}(\bar{\Theta} - \{0\})$, where $\mathscr{F}(\Psi)$ is the set of nonnegative measures on Ψ and $\bar{\Theta}$ is the closure of Θ in $\mathscr{D}(p)$, and $c \in \mathbb{R}$, define

$$\begin{aligned} d(x) &\equiv d(x; \mu, M_0, H, c) \\ &= \mu' l(x) + \tfrac{1}{2} \operatorname{tr} M_0 V(x) + \int_{\bar{\Theta} - \{0\}} \frac{R_\theta(x) - 1 - \boldsymbol{\theta}' l(x)}{\|\theta\|^2} H(d\theta) - c, \end{aligned} \tag{2.8}$$

where $\boldsymbol{\theta}$ is the vector $(\theta_1, ..., \theta_p)$. We have extended the domain of $R_\theta(x)$ to $\bar{\Theta} x \mathscr{X}$ by continuity.

For problem (1.8) define Φ to be the class of all tests of the form

$$\phi(x) = \begin{cases} 1 & \text{if } d(x; \mu, M_0, H, c) > 0 \\ 0 & \text{if } d(x; \mu, M_0, H, c) < 0, \text{ a.e. } [\nu], \end{cases} \tag{2.9}$$

for some

$$(\mu, M_0, H, c) \in C(\Theta) \times \{\gamma J \mid \gamma \geqslant 0\} \times \mathscr{F}_0(\bar{\Theta} - \{0\}) \times \mathbb{R} - \{(0, 0, 0, 0)\}, \tag{2.10}$$

where $C(\Theta)$ is the smallest convex cone containing Θ,

$$C(\Theta) = \{\theta \in \mathscr{D}(p) \mid \theta_1 \geqslant \theta_2 \geqslant \cdots \geqslant \theta_p\}, \tag{2.11}$$

J is the $p \times p$ matrix consisting of all ones, and $\mathscr{F}_0(\bar{\Theta} - \{0\})$ is the set of measures $G \in \mathscr{F}(\bar{\Theta} - \{0\})$ which satisfy

$$\int_{\bar{\Theta} - \{0\}} \frac{\theta_i - \theta_{i+1}}{\|\theta\|^2} G(d\theta) < \infty, \qquad i = 1, ..., p - 1. \tag{2.12}$$

THEOREM 2.1. *The class Φ is minimal complete for problem* (1.8).

The proof will be given in Section 4.

Now we look at the local terms (2.6) and (2.7) more closely. From James [5, Eqs. (13) and (33)], we see that $R_\theta(x)$ in (2.5) is a generalized hypergeometric function of two matrix arguments with zonal polynomial expansion:

$$R_\theta(x) = {}_1F_0(\beta; -\theta, x) = \sum_{k=0}^{\infty} \sum_{\kappa \in \mathscr{P}(k)} \frac{c_\kappa}{k!} \frac{C_\kappa(-\theta)\, C_\kappa(x)}{C_\kappa(I)}. \tag{2.13}$$

Here, $\mathscr{P}(k)$ is the set of partitions of the integer k, and for each partition κ, $C_\kappa(\cdot)$ is the corresponding zonal polynomial and c_κ is a positive constant.

The zonal polynomials for $k \leqslant 6$ are given in the Appendix of James [5]. We need the $k \leqslant 2$ terms,

$$R_\theta(x) = 1 - \frac{\beta}{p} \operatorname{tr} \theta \operatorname{tr} x + \frac{1}{6} \frac{\beta(\beta+1)}{p(p+2)} [(\operatorname{tr} x)^2 + 2 \operatorname{tr} x^2][(\operatorname{tr} \theta)^2 + 2 \operatorname{tr} \theta^2]$$

$$+ \frac{1}{3} \frac{\beta(\beta - 1/2)}{p(p-1)} [(\operatorname{tr} x)^2 - \operatorname{tr} x^2][(\operatorname{tr} \theta)^2 - \operatorname{tr} \theta^2)] + h_\theta(x), \quad (2.14)$$

where

$$h_\theta(x) = \sum_{k=3}^{\infty} \sum_{\kappa \in \mathscr{P}(k)} \frac{c_\kappa}{k!} \frac{C_\kappa(-\theta)\, C_\kappa(x)}{C_\kappa(I)}. \quad (2.15)$$

Since for $\kappa \in \mathscr{P}(k)$ and $A \in \mathscr{D}(p)$, $C_\kappa(A)$ is a symmetric polynomial in $A_1, ..., A_p$ of degree k, and each monomial making up the polynomial has a nonnegative coefficient (see Farrell [3, Problem 13.1.13]), we can derive that

$$|C_\kappa(x)| \leqslant C_\kappa(I) \qquad \text{for } x \in \mathscr{X}, \quad (2.16)$$

since $|x_i| < 1$ for each i, and that for any $\varepsilon \in (0, 1)$, and $k \geqslant 3$,

$$\left| \frac{C_\kappa(-\theta)}{\|\theta\|^2} \right| \leqslant \varepsilon^{k-2} C_\kappa(I)$$

$$\leqslant \varepsilon^{(5/6)k-2} C_\kappa(\varepsilon^{1/6} I) \qquad \text{for } \|\theta\| \leqslant \varepsilon. \quad (2.17)$$

Thus, since $h_\theta(x)$ in (2.15) is a sum of terms with $k \geqslant 3$,

$$\sup_{x \in \mathscr{X}} \sup_{\|\theta\| \leqslant \varepsilon} \left| \frac{h_\theta(x)}{\|\theta\|^2} \right| \leqslant \sum_{k=3}^{\infty} \sum_{\kappa \in \mathscr{P}(k)} \frac{c_\kappa}{k!} C_\kappa(\varepsilon^{1/6} I)\, \varepsilon^{(5/6)k-2}$$

$$\leqslant \varepsilon^{1/2} \sum_{k=0}^{\infty} \sum_{\kappa \in \mathscr{P}(k)} \frac{c_\kappa}{k!} C_\kappa(\varepsilon^{1/6} I)$$

$$= \varepsilon^{1/2} |I - \varepsilon^{1/6} I|^{-\beta}$$

$$= \varepsilon^{1/2} (1 - \varepsilon^{1/6})^{-\beta p}. \quad (2.18)$$

Hence (2.14) and (2.18) make it easy to show that from (2.6) and (2.7),

$$l_i(x) = -\frac{\beta}{p} \operatorname{tr} x, \qquad i = 1, ..., p, \quad (2.19)$$

and

$$V_{ij}(x)=\begin{cases}\dfrac{\beta(\beta+1)}{p(p+2)}\,[(\operatorname{tr}x)^2+2\operatorname{tr}x^2] & \text{if } i=j\\[2ex] \dfrac{1}{3}\dfrac{\beta(\beta+1)}{p(p+2)}\,[(\operatorname{tr}x)^2+2\operatorname{tr}x^2] & \\[2ex] \qquad+\dfrac{2}{3}\dfrac{\beta(\beta-1/2)}{p(p-1)}\,[(\operatorname{tr}x)^2-\operatorname{tr}x^2] & \text{if } i\neq j.\end{cases} \tag{2.20}$$

Hence if we take μ and M_0 as in (2.10),

$$\mu^t l(x)=-\frac{\beta}{p}\,(\Sigma\mu_i)\operatorname{tr}x\equiv\delta\operatorname{tr}x \tag{2.21}$$

and

$$\operatorname{tr}M_0V(x)=\gamma\sum_i\sum_j V_{ij}(x)=\gamma\beta[\beta(\operatorname{tr}x)^2+\operatorname{tr}x^2], \tag{2.22}$$

where $\delta\in\mathbb{R}$ and $\gamma\geqslant 0$. Thus we can alternatively define Φ to consist of all tests of the form

$$\phi(x)=\begin{cases}1 & \text{if } \bar{d}(x;\delta,\gamma,H,c)>0\\ 0 & \text{if } \bar{d}(x;\delta,\gamma,H,c)<0, \text{ a.e. } [\nu],\end{cases} \tag{2.23}$$

for

$$(\delta,\gamma,H,c)\in\mathbb{R}\times[0,\infty)\times\mathscr{F}_0(\bar{\Theta}-\{0\})\times\mathbb{R}-\{(0,0,0,0)\}, \tag{2.24}$$

where

$$\bar{d}(x;\delta,\gamma,H,c)=\delta\operatorname{tr}x+\gamma[\beta\operatorname{tr}x)^2+\operatorname{tr}x^2]$$
$$+\int_{\bar{\Theta}-\{0\}}\frac{(R_\theta(x)-1+(\beta/p)\operatorname{tr}\theta\operatorname{tr}x)}{\|\theta\|^2}\,H(d\theta)-c. \tag{2.25}$$

We turn to Theorem 1.1(a) The test (1.12) is easily seen to be in Φ, hence is admissible for problem (1.8), by taking $(\delta,\gamma,H,c)=(0,1,0,c)$ in (2.23). The remainder of the theorem follows as in the Introduction pending proof of Proposition 1.3(a), which we now give.

Proof of Proposition 1.3.a: We start by showing that $R_\theta(x)$ is strictly convex in x if $\theta\neq 0$. Using the representation of (2.2) obtainable from Wijsman [9], we write

$$\frac{f_\alpha(z)}{f_0(z)}=\frac{|\alpha|^{n_2/2}\int|AA^t|^{\beta-p/2}\,e^{-(1/2)\operatorname{tr}AS_1A^t}e^{-(1/2)\operatorname{tr}\alpha AS_2A^t}\,dA}{\int|AA^t|^{\beta-p/2}\,e^{-(1/2)\operatorname{tr}A(S_1+S_2)A^t}\,dA}, \tag{2.26}$$

where the integrals are over $A \in \mathscr{G}l(p)$. Manipulations familiar in such situations yield the ratio in terms of (x, θ) to be

$$K\,|I+\theta|^{-n_1/2}\,|I-\theta|^{-n_2/2} \int |AA^t|^{\beta-p/2}\, e^{-(1/2)\operatorname{tr} AA^t} e^{-(1/2)\operatorname{tr}\theta A x A^t}\, dA, \qquad (2.27)$$

where K is a positive constant. It is then possible to prove that if $\theta \neq 0$, the expression in (2.27) is strictly convex in x, hence $R_\theta(x)$ is strictly convex in x.

Now consider a test $\phi \in \Phi$ and the corresponding set from (2.23),

$$B \equiv \{x \mid \bar{d}(x; \delta, \gamma, H, c) \leqslant 0\}. \qquad (2.28)$$

Since $R_\theta(x)$ is strictly convex in x if $\theta \neq 0$, $\operatorname{tr} x$ is convex in x, and $\beta(\operatorname{tr} x)^2 + \operatorname{tr} x^2$ is strictly convex in x, we have by (2.25) that

$$\begin{aligned} &\text{(i)} && \bar{d}(x; \delta, \gamma, H, c) \text{ is strictly convex in } x \text{ if } (\gamma, H) \neq (0, 0); && (2.29)\\ &\text{(ii)} && \bar{d}(x; \delta, \gamma, H, c) \equiv -c \quad \text{for } c \neq 0 \text{ if } (\delta, \gamma, H) = (0, 0, 0); &&\\ &\text{(iii)} && \bar{d}(x; \delta, \gamma, H, c) = \delta \operatorname{tr} x - c \quad \text{for } \delta \neq 0 \text{ otherwise}. && \end{aligned}$$

In any of the cases in (2.29), B of (2.28) is convex, and since $\bar{d}$ is continuous in x and ν is absolutely continuous with respect to Lebesgue measure on $\mathbb{R}^p$, the boundary of B in $\mathscr{X}$ equals $\{x \mid \bar{d}(x) = 0\}$ and has ν-measure zero. Hence $\phi = 1 - I_A$, a.e. $[\nu]$.

If case (ii) or (iii) in (2.29) holds, then B is either $\{x \mid \operatorname{tr} x \leqslant \alpha\}$, or $\{x \mid \operatorname{tr} x \geqslant b\}$, where we take a or $b \in [-p, p]$. (In case (ii), B is either empty or $\mathscr{X}$, so we take $a = -p$ or $a = p$, for example.) If case (ii) holds, then since the boundary of B is $\{x \mid \bar{d}(x) = 0\}$, and $\bar{d}$ is strictly convex, no three points on the boundary of B can be collinear, i.e., Property 1.2 holds. Hence Proposition 1.3(a) is proven.

Now turn to the one-sided problem (1.18). Define the class of tests Φ^+, which is a subset of Φ, to consist of all tests of the form

$$\phi(x) = \begin{cases} 1 & \text{if } d^+(x; \delta, H, c) > 0 \\ 0 & \text{if } d^+(x; \delta, H, c) < 0, \text{ a.e. } [\nu], \end{cases} \qquad (2.30)$$

for

$$(\delta, H, c) \in [0, \infty) \times \mathscr{F}(\bar{\Theta}^+ - \{0\}) \times \mathbb{R} - \{(0, 0, 0)\}, \qquad (2.31)$$

where

$$d^+(x; \delta, H, c) = \delta \operatorname{tr} x + \int_{\bar{\Theta}^+ - \{0\}} \frac{(R_\theta(x) - 1)}{\|\theta\|^2}\, H(d\theta) - c. \qquad (2.32)$$

The function $R_\theta(x)$ is given in (2.5). The proof of the next result is in Section 4.

THEOREM 2.2. *The class* Φ^+ *is minimal complete for problem* (1.18).

The proof of Proposition 1.3(b) follows as the proof of part (a) above, where we note that $\delta \geqslant 0$. An additional result is available. Note that, from (2.27),

$$-\operatorname{tr} \theta A x A^t = -\sum_i \sum_j \theta_i x_j a_{ij}^2. \tag{2.33}$$

Since for $\theta \in \Theta^+$, $\theta_i \leqslant 0$ for each i, the expression in (2.33) is nondecreasing in each x_i, hence $R_\theta(x)$ in (2.27) is nondecreasing in each x_i. It is easy to extend the definition of $R_\theta(x)$ to

$$x \in \mathscr{X}^* \equiv \{x \in \mathscr{D}(p) \mid -1 < x_i < 1 \text{ for each } i\}.$$

This new $R_\theta(x)$ and the corresponding $d^+(x)$ are invariant under permutations of the elements of x. See (2.27) which is in terms of ordered x_i's. Together with the convexity of d^+, we have by Proposition 4.C.2d of Marshall and Olkin [10] that d^+ satisfies the weak submajorization monotonicity property, i.e.,

$$\text{If } x, y \in \mathscr{X} \text{ with } x \leqslant y_1, x_1 + x_2 \leqslant y_1 + y_2, \ldots, x_1 + \cdots + x_p \leqslant y_1 + \cdots + y_p, \text{ then } d^+(x) \leqslant d^+(y). \tag{2.34}$$

Thus we have the following:

PROPOSITION 2.3. *A necessary condition for a test* ϕ *to be admissible for problem* (1.18) *is that it equal* $1 - I_B$, *a.e.* $[v]$, *for some set* B *which is monotone nonincreasing in the ordering* (2.34).

3. TESTING SPHERICITY

Let $g_\lambda(y)$ be the density of Y in (1.26) when λ in (1.27) obtains. From Sugiura [8, Eq. (1.3)], we have that

$$\frac{g_\lambda(y)}{g_I(y)} = |\lambda|^{n/2} \int_{O(p)} (\operatorname{tr} y\Gamma\lambda\Gamma^t)^{-\tau} \rho(d\Gamma), \qquad \tau = np/2. \tag{3.1}$$

Recall from Section 2 that ρ is the Haar probability measure on $O(p)$.

Rewriting the ratio (3.1) in terms of ω of (1.28), and multiplying it by $|I+\omega|^{-n/2}$, yields

$$R_{\omega}^{*}(y) \equiv \int_{O(p)} (1 + \operatorname{tr} y\Gamma\omega\Gamma')^{-\tau} \rho(d\Gamma). \tag{3.2}$$

(Recall that $\operatorname{tr} y = 1$.)

We need to find the derivatives corresponding to (2.6) and (2.7). Note that for $|a| \leqslant 1$,

$$(1+a)^{-\tau} = 1 - \tau a + \frac{\tau(\tau+1)}{2} a^2 + o(a^2), \tag{3.3}$$

where $o(a^2)$ is as $a \to 0$, uniformly in $|a| \leqslant \varepsilon$ for any $\varepsilon \in (0, 1)$. Since $y_i \in (0, 1)$ for each i,

$$(\operatorname{tr} y\Gamma\omega\Gamma')^2 \leqslant (\Sigma\, |\omega_i|^2) \leqslant p\, \|\omega\|^2. \tag{3.4}$$

Hence from (3.2) and (3.3) we have

$$R_{\omega}^{*}(y) = 1 - \tau \int_{O(p)} (\operatorname{tr} y\Gamma\omega\Gamma')\, \rho(d\Gamma) + \frac{\tau(\tau+1)}{2} \int_{O(p)} (\operatorname{tr} y\Gamma\omega\Gamma')^2\, \rho(d\Gamma) + o(\|\omega\|^2), \tag{3.5}$$

where $o(\|\omega\|^2)$ is as $\omega \to 0$, uniformly in $y \in \mathscr{Y}$. Using zonal polynomials as in Sugiura [8], or calculating directly, we obtain

$$\int (\operatorname{tr} y\Gamma\omega\Gamma')\, \rho(d\Gamma) = \frac{\operatorname{tr} y \operatorname{tr} \omega}{p} = 0 \qquad (\text{since } \operatorname{tr} \omega = 0) \tag{3.6}$$

and

$$\int (\operatorname{tr} y\Gamma\omega\Gamma')^2\, \rho(d\Gamma) = \frac{2\, \|\theta\|^2}{p(p+2)(p-1)} [p \operatorname{tr} y^2 - 1] = \frac{2\, \|\theta\|^2}{(p+2)(p-1)} S_y^2. \tag{3.7}$$

See (1.32). Thus (3.5), (3.6), and (3.7) show that

$$l_i^{*}(y) \equiv \frac{\partial}{\partial \omega_i} R_{\omega}^{*}(y) \,|_{\omega=0} = 0 \tag{3.8}$$

and

$$V_{ij}^{*}(y) \equiv \frac{\partial^2}{\partial \omega_i\, \partial \omega_j} R_{\omega}^{*}(y) \,|_{\omega=0} = \frac{4}{(p+2)(p-1)} S_y^2 I_{\{i=j\}}. \tag{3.9}$$

Now let Φ^* be the class of tests of the form

$$\phi(y) = \begin{cases} 1 & \text{if } d^+(y; \gamma, H, c) > 0 \\ 0 & \text{if } d^+(y; \gamma, H, c) < 0, \text{ a.e. } [v^*], \end{cases} \tag{3.10}$$

for

$$(\delta, H, c) \in [0, \infty) \times \mathscr{F}(\bar{\Omega} - \{0\}) \times \mathbb{R} - \{(0, 0, 0)\}, \tag{3.11}$$

where

$$d^*(y) \equiv d^*(y; \gamma, H, c) = \gamma S_y^2 + \int_{\bar{\Omega} - \{0\}} \frac{R_\omega^*(y) - 1}{\|\omega\|^2} H(d\omega) - c \tag{3.12}$$

and v^* is the null measure on $\mathscr{Y}$. It is absolutely continuous with respect to Lebesgue measure on $\mathbb{R}^{p-1}$.

THEOREM 3.1. *The class Φ^* in minimal complete for problem* (1.29).

The proof is indicated in Section 4.

Proposition 1.5 is proved as Proposition 1.3, where we note that S_y^2 and $R_\omega^*(y)$ for $\omega \neq 0$ are strictly convex in y. The latter result follows from the facts that $(1+a)^{-\tau}$ is strictly convex in a and $\operatorname{tr} y\Gamma\omega\Gamma'$ is linear in the diagonal elements of y and, with ρ probability one the coefficients multiplying each diagonal element of y are nonzero.

Finally, consider the case $p = 2$ in Theorem 1.4. Extend the definition of $R_\omega^*(y)$ to the set $\{y \in \mathbb{R}^2 \mid y_1 + y_2 = 1,\ y_1 > 0,\ y_2 > 0\}$. Note that $R_\omega^*(y)$ is invariant under the permutation of y_1 and y_2, and S_y^2 and $R_\omega^*(y)$ when $\omega \neq 0$ are strictly convex in y. Thus d^* is also permutation invariant and strictly convex unless $(\delta, H) = (0, 0)$. Thus d^* has a minimum at $(y_1, y_2) = (\frac{1}{2}, \frac{1}{2})$ and is either constant or strictly increasing as y_1 moves away from $\frac{1}{2}$. Thus the only admissible tests are those with acceptance regions essentially of the form $\{y \mid y_1 \leqslant c|,\ c \in [\frac{1}{2}, 1]$.

4. PROOFS OF THEOREMS 2.1, 2.2, AND 3.1

In this section we will refer to Brown and Marden [2] by B – M. We first use B – M Theorem 2.4 to prove the classes Φ, Φ^+, and Φ^* essentially complete for their respective problems (1.8), (1.18), and (1.29). We need to verify B – M Assumptions 2.1, 2.2, and 2.3.

Start with problem (1.8). Assumption 2.1 requires that for each x, $R_\theta(x)$ as a function on $\bar{\Theta}$ satisfies

$$0 < R_\theta(x) < \infty \qquad \text{for } \theta \in \bar{\Theta}. \tag{4.1}$$

By inspection of (2.5), $R_\theta(x)$ is positive. By (2.16) with $-\theta$ instead of x we have that

$$|C_\kappa(-\theta)| \leqslant C_\kappa(I),$$

hence by (2.13) and (2.5)

$$R_\theta(x) \leqslant \prod_{i=1}^{p} (1-|x_i|)^{\beta} < \infty,$$

since each $x_i \in (-1, 1)$. Hence (4.1) holds.

B − M Assumption 2.2 states that the derivatives in (2.7) and (2.8) exist, which we have already shown, and that for sufficiently small $\varepsilon > 0$, for each x there exists $\kappa_x < \infty$ such that

$$\sup_{\|\theta\| \leqslant \varepsilon} \left| \frac{h_\theta(x)}{\|\theta\|^2} \right| \leqslant \kappa_x. \tag{4.2}$$

This result follows from (2.18), where in fact we have the stronger result that

$$\kappa = \sup_{x \in \mathscr{X}} \kappa_x < \infty. \tag{4.3}$$

B − M Assumption 2.3 is trivial in this problem since Θ is bounded. See the remark below Equation (2.5) in B − M. Thus the set $\mathscr{C}$ in B − M consists only of ϕ and $\mathscr{X}$, and hence can be ignored safely.

Now B − M Theorem 2.4 guarantees that an essentially complete class consists of all tests of the form (2.9), where

$$((\mu, M), H, c) \in \Xi - ((0, 0), 0, 0) \tag{4.4}$$

and

$$M_0 = M - \int_{\bar{\Theta} - \{0\}} \frac{\boldsymbol{\theta\theta}'}{\|\theta\|^2} H(d\theta), \tag{4.5}$$

and

$$\Xi = \{((\mu, M), H, c) \mid (\mu, M) \in \Lambda(H), H \in \mathscr{F}(\bar{\Theta} - \{0\}), c \in \mathbb{R}\}. \tag{4.6}$$

(We take α in B − M large enough so that $\theta \in \bar{\Theta} \Rightarrow \|\theta\| < \alpha$.] The set $\Lambda(H)$ is a subset of $\mathbb{R}^p \times \mathscr{S}(p)$ defined in B − M (2.14). We will use B − M Example 4.6 to find $\Lambda(H)$, but first we reparametrize by letting

$$\pi = G\theta \in \mathscr{D}(p), \tag{4.7}$$

where G is the linear transformation from which $\pi_i = \theta_i - \theta_{i+1}$, $i = 1, ..., p-1$, and $\pi_p = \theta_p$. Then the transformed parameter space $G\Theta \equiv \Pi$ is locally one-sided, i.e., for some $\varepsilon > 0$,

$$\Pi \in B_\varepsilon = [[0, \infty)^{p-1} \times \mathbb{R}] \cap B_\varepsilon, \tag{4.8}$$

where B_ε is the ε-ball in $\mathscr{D}(p)$ around 0. From B − M Example 4.6 (with $K_1 = [0, \infty)^{p-1}$ and $q = 1$,) we have that if

$$\int \pi_i GH(d\pi) < \infty, \qquad i = 1, ..., p-1, \tag{4.9}$$

then

$$\Lambda(GH) = \{(\mu^*, M^*) \mid \mu^* \in C(\Pi) \text{ and } M_0^* \in \mathscr{D}(p),$$
$$m_i^* = 0,\ i = 1, ..., p-1,\ m_p^* \geqslant 0\}. \tag{4.10}$$

If (4.9) fails, $\Lambda(GH)$ is empty. Here, GH is the measure induced on Π by G via (4.7). Now it can be seen from the definition of $\Lambda(H)$ in B − M that

$$\Lambda(H) = \{(G^{-1}\mu^*, G^{-1}M^*(G')^{-1}) \mid (\mu^*, M^*) \in \Lambda(GH)]$$
$$= \{(\mu, M) \mid \mu \in C(\Theta), M_0 = \gamma J, \gamma \geqslant 0\}. \tag{4.11}$$

Hence (2.10) is equivalent to (4.4) via (4.11), proving that Φ is essentially complete for problem (1.8).

The verification of B − M Assumptions 2.1, 2.2, and 2.3 for problem (1.18) proceeds as for problem (1.8) since it shares $R_\theta(x)$ and has $\Theta^+ \subseteq \Theta$. Note that Θ^+ is locally pointed as in B − M Example 4.5. That is, there exists $a_0 \in \mathscr{D}(p)$ and $b_0 < 0$ such that for sufficiently small $\varepsilon > 0$,

$$\sup_{\|\theta\| \leqslant \varepsilon} \frac{\mathbf{a}_0'\boldsymbol{\theta}}{\|\theta\|} \leqslant b_0. \tag{4.12}$$

To see this, take $a_0 = I$, and note that by (1.19),

$$\sup_{\theta \in \Theta^+} \frac{\Sigma\theta_i}{\|\theta\|^2} = -1.$$

Thus B − M characterize the complete class as consisting of all tests of the form

$$\phi(x) = \left\{\begin{smallmatrix}1\\0\end{smallmatrix}\right\} \qquad \text{if } \mu' l(x) + \int_{\bar{\Theta}^+ - \{0\}} \frac{R_\theta(x) - 1}{\|\theta\|} H(d\theta) - c \left\{\begin{smallmatrix}>\\<\end{smallmatrix}\right\} 0, \tag{4.13}$$

a.e. $[\nu]$, for some

$$(\mu, H, c) \in C(\Theta^+) \times \mathscr{F}(\bar{\Theta}^+ - \{0\}) \times \mathbb{R} - \{(0, 0, 0)\}. \tag{4.14}$$

But since $\mu' l(x) = -(\beta/p)\, \Sigma\mu_i \operatorname{tr} x$ as in (2.21), and $\mu \in C(\Theta^+)$ implies that $\Sigma\mu_i \leqslant 0$, we see that (4.13) and (4.14) are equivalent to (2.30) and (2.31). Hence Φ^* is essentially complete for problem (1.18).

Now turn to problem (1.29). The B – M Assumptions 2.1, 2.2, and 2.3 are fairly easy to verify by using the approach for the previous two problems, and by noting that

$$\begin{aligned}\inf_{\omega \in \Omega}\ \inf_{\Gamma \in O(p)} (1 + \operatorname{tr} y\Gamma\omega\Gamma') &= p \inf_{\lambda \in \Lambda}\ \inf_{\Gamma \in O(p)} (\operatorname{tr} y\Gamma\lambda\Gamma') \\ &= p \inf_{\lambda \in \Lambda} \Sigma y_i \lambda_{p-i+1} = p y_p > 0,\end{aligned} \tag{4.15}$$

so that $R^*_\omega(y)$ in (3.2) is finite. Since $l^*(x) \equiv 0$ (see (3.8), we can use B – M Remark 2.8 and Example 4.2 to show tha the class of tests of the form

$$\phi(x) = \left\{ {1 \atop 0} \right\} \quad \text{as } \operatorname{tr} M_0 V^+(x) + \int_{\bar{\Omega} - \{0\}} \frac{R^*_\omega(x) - 1}{\|\omega\|^2}\, H(d\omega) - c \left\{ {> \atop <} \right\} 0, \tag{4.16}$$

a.e. $[\nu^*]$, is essentialy complete, where

$$(M, H, c) \in \mathscr{S}(p) \times \mathscr{F}(\bar{\Omega} - \{0\}) \times \mathbb{R} - \{(0, 0, 0)\}. \tag{4.17}$$

Now (3.9) shows that their class is in fact Φ^* of (3.10), (3.11), and (3.12).

To complete the proofs of the theorems, we must show that the classes Φ, Φ^+, and Φ^* are minimal complete. These results follow from B – M Lemma 3.2, which requires verification of B – M Assumption 3.1. We will verify this assumption only for problem (1.8). The verification for the other problems can be dealt with similarly.

Consider problem (1.8). B – M Assumption 3.1 has four parts. Parts (i) and (iii) are trivial since $\mathscr{C} = \{\phi, \mathscr{X}\}$. Part (iv) requires that

$$\nu(\{x \mid d(x; \mu, M_0, H, c) = 0\}) = 0$$

for (μ, M_0, H, c) as in (2.10), which follows from the discussion after (2.29).

Part (ii) requires that for each $\phi \in \Phi$, there exists a sequence $\{J_i\} \subseteq \mathscr{F}(\Theta)$ such that

$$d_i(x) \equiv \int_{\Theta - \{0\}} R_\theta(x)\, J_i(d\theta) - J_i(\{0\}) \xrightarrow{i \to \infty} d(x) \qquad \text{for each } x, \tag{4.18}$$

where $d(x)$ is defined in (2.8), and

$$\lim_{i\to\infty} \int (\phi_i(x)-\phi(x))\, d_i(x)\, \nu(dx)=0, \tag{4.19}$$

where

$$\phi_i(x)=\left\{\begin{smallmatrix}1\\0\end{smallmatrix}\right\} \qquad \text{as } d_i(x)\left\{\begin{smallmatrix}>\\<\end{smallmatrix}\right\} 0. \tag{4.20}$$

Now take $\phi\in\Phi$ and its attendant (μ, M_0, H, c), and define

$$\begin{aligned} \Theta_0&=\{\theta\in\Theta \mid \|\theta\|\leqslant\tfrac{1}{10}\}, &\qquad \Theta_1&=\bar{\Theta}-\Theta_0,\\ H_0(d\theta)&=H(d\theta)\, I_{\Theta_0}, & H_1(d\theta)&=H(d\theta)\, I_{\Theta_1}. \end{aligned} \tag{4.21}$$

Also, for $i\geqslant 1$, let $H_{1i}\in\mathscr{F}(\Theta-\{0\})$ be defined by

$$H_{1i}\left(\frac{i}{i+1}\,A\right)=\left(\frac{i}{i+1}\right)^2 H_1(A) \qquad \text{for } A\subseteq\Theta_1. \tag{4.22}$$

Then using the methods in B − M Lemma 2.5, we can find $\{J_i\}$ such that, from (4.18),

$$d_i(x)=A_i(x)+a_i(x), \tag{4.23}$$

where

$$A_i(x)=\mu_i' l(x)+\tfrac{1}{2}\operatorname{tr} M_i V(x)+\int_{\Theta_0-\{0\}} \frac{h_\theta(x)}{\|\theta\|^2}\, H_{0i}(d\theta)-c_i, \tag{4.24}$$

with

$$\mu_i\to\mu, \qquad M_i\to M_0+\int_{\Theta_0-\{0\}} \frac{\boldsymbol{\theta\theta}'}{\|\theta\|^2}\, H_0(d\theta), \qquad c_i\to c, \tag{4.25}$$

$$\int_{\Theta_0-\{0\}} g(\theta)\, H_{0i}(d\theta)\to\int_{\Theta_0-\{0\}} g(\theta)\, H_0(d\theta) \tag{4.26}$$

for any continuous bounded function g with $g(0)=0$, and

$$a_i(x)=\int_{\Theta_1} \frac{R_\theta(x)-1-\boldsymbol{\theta}' l(x)}{\|\theta\|^2}\, H_{1i}(d\theta). \tag{4.27}$$

It is clear from (4.24) and (4.25) that

$$\begin{aligned} A_i(x)\to A(x)\equiv \mu' l(x)+\tfrac{1}{2}\operatorname{tr}\left(M_0+\int_{\Theta_0-\{0\}} \frac{\boldsymbol{\theta\theta}'}{\|\theta\|^2}\, H_0(d\theta)\right) V(x)\\ +\int \frac{h_\theta(x)}{\|\theta\|^2}\, H_0(d\theta)-c. \end{aligned} \tag{4.28}$$

Now by (2.15) and (2.18) for $h_\theta(x)/\|\theta\|^2$, and by (2.19) and (2.20) for $l(x)$ and $V(x)$, we have that for some $N<\infty$,

$$|A(x)| \leqslant N \qquad \text{and} \qquad |A_i(x)| \leqslant N \qquad \text{for all } i,\ x. \tag{4.29}$$

Also, for a_i in (4.27), since $R_{b\theta}(x) = R_\theta(bx)$, by (4.22),

$$\begin{aligned} a_i(x) &= \int_\Theta \frac{R_\theta(x) - 1 - \boldsymbol{\theta}' l(x)}{\|\theta\|^2} H_{1i}(d\theta) \\ &= \int_{\Theta_1} \frac{R_{(i/i+1)\theta}(x) - 1 - (i/(i+1))\, \boldsymbol{\theta}' l(x)}{\|\theta\|^2} H_1(d\theta) \\ &= \int_{\Theta_1} \frac{R_\theta((i/(i+1))x) - 1 - \boldsymbol{\theta}' l((i/(i+1))x)}{\|\theta\|^2} H_1(d\theta) \\ &= a\left(\frac{i}{i+1} x\right), \end{aligned} \tag{4.30}$$

where

$$a(x) = \int_{\Theta_1} \frac{R_\theta(x) - 1 - \boldsymbol{\theta}' l(x)}{\|\theta\|^2} H_1(d\theta). \tag{4.31}$$

Since the integrand for $a((i/(i+1))x)$ is bounded in i for each fixed x and θ, and continuous in θ, we have that

$$a_i(x) \equiv a\left(\frac{i}{i+1} x\right) \rightarrow a(x). \tag{4.32}$$

Thus (4.23) through (4.27), (4.30), and (4.31) show that (4.18) holds, since $d(x) = A(x) + a(x)$.

Finally, note that $a(0) = 0$, and since $a(x)$ is convex in x (see (2.30)), for $t > 0$,

$$\begin{aligned} a(x) \leqslant t \Rightarrow a\left(\frac{i}{i+1} x\right) \leqslant t \\ \Rightarrow a_i(x) \leqslant t. \end{aligned} \tag{4.33}$$

Turn to (4.19). By (4.20), (4.23), and (4.29), when $a_i(x) > N$, $\phi_i(x) = 1$, and by (4.28), (4.31), and (4.29), when $a(x) > N$, $\phi(x) = 1$. Thus if $d_i(x) > 2N$

then $a_i(x) > N$, hence by (4.33), $a(x) > N$, and $\phi_i(x) = \phi(x) = 1$ (a.e. $[\nu]$). Thus

$$\begin{aligned} \lim_{i\to\infty} \int (\phi_i(x) - \phi(x))\, d_i(x)\, \nu(dx) \\ &= \lim_{i\to\infty} \int_{\{d_i(x) \leqslant 2N\}} (\phi_i(x) - \phi(x))\, d_i(x)\, \nu(dx) \\ &= 0, \end{aligned} \tag{4.34}$$

where the limit and integral can be interchanged by the bounded convergence theorem (the integrand is essentially nonnegative by definition of ϕ_i and d_i in (4.20)), and the limit of the integrand is zero a.e. $[\nu]$ by (2.9), (4.18), and (4.20). Thus (4.34) verifies (4.19), and the proof of Theorem 2.1 is complete.

References

[1] Anderson, T. W. (1984). *An Introduction to Multivariate Statistical Analysis.* Wiley, New York.

[2] Brown, L. D., and Marden, J. I. (1989). Complete class results for hypothesis testing problem with simple null hypotheses. *Ann. Statist.* **17**.

[3] Farrell, R. H. (1976). *Techniques of Multivariate Calculation.* Springer-Verlag, New York/Berlin.

[4] Giri, N. (1968). On tests of equality of two covariance matrices. *Ann. Math. Statist.* **39** 275–277.

[5] James, A. T. (1964). Distributions of matrix variates and latent roots derived from normal samples. *Ann. Math. Statist.* **35** 475–501.

[6] Kiefer, J., and Schwartz, R. (1965). Admissible Bayes character of T^2, R^2, and other fully invariant tests for classical multivariate normal problems. *Ann. Math. Statist.* **36** 747–770.

[7] Muirhead, R. J. (1982). *Aspects of Multivariate Statistical Theory.* Wiley, New York.

[8] Sugiura, N. Locally best invariant test for sphericity and the limiting distributions. *Ann. Math. Statist.* **43** 1312–1316.

[9] Wijsman, R. A. (1967). Cross-sections of orbits and their application to densities of maximal invariants. In *Proceedings, Fifth Berkeley Symp. Math. Statist. and Probab.* 1, pp. 389–400.

[10] Marshall, A. W. and Olkin, I. (1979). *Inequalities: Theory of Majorization and Its Applications.* Academic Press, New York.

Invariance Principles for Changepoint Problems

MIKLÓS CSÖRGŐ*

Carleton University, Ottawa, Canada K1S 5B6

AND

LAJOS HORVÁTH†

Szeged University, Szeged, Hungary

We study the asymptotic behaviour of U-statistics type processes which can be used for detecting a changepoint of a random sequence. Invariance principles are proved for these processes. © 1988 Academic Press, Inc.

1. INTRODUCTION

Let $X_1, ..., X_n$ be independent random variables. Suppose we want to test the null hypothesis

H_0. X_i, $1 \leqslant i \leqslant n$, have the same distribution

versus the alternative hypothesis that there is a changepoint in the sequence $X_1, ..., X_n$, namely that we have

H_1. There is a $\lambda \in (0, 1)$ such that $P\{X_1 \leqslant t\} = P\{X_2 \leqslant t\} = \cdots = P\{X_{[n\lambda]} \leqslant t\}$, $P\{X_{[n\lambda]+1} \leqslant t\} = \cdots = P\{X_n \leqslant t\}$, $-\infty < t < \infty$, and $P\{X_{[n\lambda]} \leqslant t_0\} \neq P\{X_{[n\lambda]+1} \leqslant t_0\}$ for some t_0.

The changepoint problem has been considerably studied in the literature from the parametric as well as the nonparametric point of view. Non-

* Research partially supported by a NSERC Canada grant.

† Research done while at Carleton University, supported by NSERC Canada grants of M. Csörgő, D. A. Dawson, and J. N. K. Rao, and by an EMR Canada grant of M. Csörgő.

Multivariate Statistics and Probability
ISBN 0-12-580205-6

Reprinted from *J. Mult. Anal.* **27**(1).

parametric results are summarized in Wolfe and Schechtman [15]. Recently Csörgő and Horváth [2] proposed statistics based on processes of linear rank statistics with quantile scores. In this paper we study tests for the changepoint problem which are based on processes of U-statistics. They are generalizations of Wilcoxon–Mann–Whitney type statistics.

Let $h(x, y)$ be a symmetric function and consider

$$Z_k = \sum_{1 \leqslant i < k} \sum_{k+1 \leqslant j \leqslant n} h(X_i, X_j), \qquad 1 \leqslant k < n. \tag{1.1}$$

We study Z_k under the null hypothesis in Section 2, and under the alternative hypothesis in Section 3. Typical choices of h are xy, $(x-y)^2/2$ (sample variancie), $|x-y|$ (Gini's mean difference), $\operatorname{sign}(x+y)$ (Wilcoxon's one-sample statistic) (cf. Serfling [13]). The case of $h(x, y) = \operatorname{sign}(x-y)$ has gained special attention in the literature. We cannot apply our results directly in this case, because $\operatorname{sign}(x-y)$ is not a symmetric function. However, $\operatorname{sign}(x-y) = -\operatorname{sign}(y-x)$ $(\operatorname{sign}(0)=0)$, i.e., $\operatorname{sign}(x-y)$ is an antisymmetric kernel. We show in Section 4 that our method can be also used in the case of an antisymmetric kernel.

2. Asymptotics under H_0

In Sections 2 and 3 we assume that h is symmetric, i.e., $h(x, y) = h(y, x)$. Given H_0, $X_1, ..., X_n$ are i.i.d.r.v.'s. We assume

$$Eh^2(X_1, X_2) < \infty \tag{2.1}$$

and let $Eh(X_1, X_2) = \Theta$, $\tilde{h}(t) = E\{h(X_1, t) - \Theta\}$. Condition (2.1) implies that $E\tilde{h}^2(X_1) < \infty$ and we assume

$$0 < \sigma^2 = E\tilde{h}^2(X_1). \tag{2.2}$$

Here we investigate

$$U_k = Z_k - k(n-k)\,\Theta, \qquad 1 \leqslant k < n,$$

which can be expressed as

$$U_k = U_n^{(3)} - \{U_k^{(1)} + U_k^{(2)}\}, \tag{2.3}$$

where

$$U_k^{(1)} = \sum_{1 \leqslant i < j \leqslant k} h(X_i, X_j) - \binom{k}{2}\Theta,$$

$$U_k^{(2)} = \sum_{k+1 \leqslant i < j \leqslant n} h(X_i, X_j) - \binom{n-k}{2}\Theta,$$

and

$$U_n^{(3)} = \sum_{1 \leqslant i < j \leqslant n} h(X_i, X_j) - \binom{n}{2} \Theta.$$

The latter are nondegenerate U-statistics under the conditions (2.1) and (2.2). Thus while U_k itself is not a U-statistic, in (2.3) we concluded that it can be expressed as a linear combination of U-statistics. Hence the basic idea of studying U_k can be based on the projection of a U-statistic on the basic observations (cf. Chap. 5 of Serfling [3]).

In order to state our results we define the Gaussian process Γ by

$$\Gamma(t) = (1-t)\, W(t) + t\{W(1) - W(t)\}, \qquad 0 \leqslant t \leqslant 1, \tag{2.4}$$

where $\{W(t), 0 \leqslant t < \infty\}$ is a Wiener process.

THEOREM 2.1. *We assume that H_0 holds, and (2.1), (2.2) are satisfied. Then we can define a sequence of Gaussian processes $\{\Gamma_n(t), 0 \leqslant t \leqslant 1\}$ such that, as $n \to \infty$,*

$$\sup_{0 \leqslant t < 1} \left| \frac{n^{-3/2}}{\sigma} U_{[(n+1)t]} - \Gamma_n(t) \right| = o_P(1), \tag{2.5}$$

where for each $n \geqslant 1$

$$\{\Gamma_n(t), 0 \leqslant t \leqslant 1\} \overset{\mathscr{D}}{=} \{\Gamma(t), 0 \leqslant t \leqslant 1\}. \tag{2.6}$$

Proof. By Theorem 1 of Hall [6] we have

$$\max_{1 \leqslant k \leqslant n} \left| U_k^{(1)} - k \sum_{i=1}^{k} \tilde{h}(X_i) \right| = O_P(n), \tag{2.7}$$

$$\max_{1 \leqslant k \leqslant n} \left| U_k^{(2)} - (n-k) \sum_{i=k+1}^{n} \tilde{h}(X_i) \right| = O_P(n), \tag{2.8}$$

$$\left| U_n^{(3)} - n \sum_{i=1}^{n} \tilde{h}(X_i) \right| = O_P(n). \tag{2.9}$$

Hence

$$\max_{1 \leqslant k \leqslant n} \left| U_k - \left\{ (n-k) \sum_{i=1}^{k} \tilde{h}(X_i) + k \left(\sum_{i=1}^{n} \tilde{h}(X_i) - \sum_{i=1}^{k} \tilde{h}(X_i) \right) \right\} \right| = O_P(n). \tag{2.10}$$

Thus the result follows from Donsker's theorem (cf. Theorem 2.1.2 and Lemma 4.4.4 in Csörgő and Révész [3]).

One can say more about the weak convergence of U_k if the existence of higher moments is assumed.

THEOREM 2.2. *We assume that H_0 holds,*

$$E\,|h(X_1, X_2)|^{\nu} < \infty \qquad \textit{for some} \quad \nu > 2, \tag{2.11}$$

and (2.2) *is satisfied. Then we can define a sequence of Gaussian processes* $\{\Gamma_n(t), 0 \leqslant t \leqslant 1\}$ *such that* (2.5) *holds,*

$$\sup_{1/(n+1) \leqslant t \leqslant n/(n+1)} \left| \frac{n^{-3/2}}{\sigma} U_{[(n+1)t]} - \Gamma_n(t) \right| \Big/ (t(1-t))^{1/2} = O_P(1), \tag{2.12}$$

and we have (2.6) *for each* $n \geqslant 1$.

Proof. First we note that by (2.11) we have $E\,|\tilde{h}(X_1)|^{\nu} < \infty$. We introduce

$$S_n^{(1)}(x) = \sigma^{-1} \sum_{1 \leqslant i \leqslant x} \tilde{h}(X_i), \qquad 1 \leqslant x \leqslant [n/2],$$

$$S_n^{(2)}(x) = \sigma^{-1} \sum_{n-x < i \leqslant n} \tilde{h}(X_i), \qquad 1 \leqslant x \leqslant n - [n/2],$$

and show that there exist two independent Wiener processes $\{W_n^{(1)}(x), 0 \leqslant x < \infty\}$ and $\{W_n^{(2)}(x), 0 \leqslant x < \infty\}$ such that

$$\sup_{1 \leqslant x \leqslant [n/2]} x^{-1/2}\, |S_n^{(1)}(x) - W_n^{(1)}(x)| = O_P(1), \tag{2.13}$$

$$\sup_{1 \leqslant x \leqslant n-[n/2]} x^{-1/2}\, |S_n^{(2)}(x) - W_n^{(2)}(x)| = O_P(1). \tag{2.14}$$

Using the Skorohod embedding scheme or the Komlós–Major–Tusnády approximation (cf. Theorem 2.2.4 and Theorem 2.6.3 in Csörgő and Révész [3]), we can define a sequence of Wiener processes $\{W_n^{(1)}(x), 0 \leqslant x < \infty\}$ so that

$$\max_{1 \leqslant k \leqslant [n/2]} k^{-1/2}\, |S_n^{(1)}(k) - W_n^{(1)}(k)| = O_P(1). \tag{2.15}$$

By Theorem 1.2.1 of Csörgő and Révész [3] we obtain

$$\sup_{1 \leqslant x \leqslant [n/2]} x^{-1/2}\, |W_n^{(1)}(x) - W_n^{(1)}([x])|$$
$$\leqslant \sup_{1 \leqslant x \leqslant [n/2]} x^{-1/2} \sup_{0 \leqslant s \leqslant 1} |W_n^{(1)}([x]+s) - W_n^{(1)}([x])| = O_P(1). \tag{2.16}$$

Now (2.15) and (2.16) imply (2.13). The proof of (2.14) is similar. Due to the independence of $S_n^{(1)}(x)$ and $S_n^{(2)}(x)$, the Wiener processes $W_n^{(1)}$ and

$W_n^{(2)}$ can be defined independently. Next we define the Wiener process $\{W_n(x), 0 \leqslant x \leqslant n\}$ by

$$W_n(x) = \begin{cases} W_n^{(1)}(x), & 0 \leqslant x \leqslant [n/2], \\ W_n^{(1)}(n) + W_n^{(2)}(n) - W_n^{(2)}(n-x), & [n/2] < x \leqslant n, \end{cases}$$

and conclude from (2.13) and (2.14) that

$$\sup_{1/(n+1) \leqslant t \leqslant n/(n+1)} \Bigg| \left(1 - \frac{[(n+1)\,t]}{n}\right) \sum_{i=1}^{[(n+1)t]} \tilde{h}(X_i) + \frac{[(n+1)t]}{n} \left(\sum_{i=1}^{n} \tilde{h}(X_i) - \sum_{i=1}^{[(n+1)t]} \tilde{h}(X_i) \right) - \sigma\{(1-t)\, W_n((n+1)t) + t(W_n(n+1) - W_n((n+1)t))\} \Bigg| \Big/ (nt(1-t))^{1/2}$$

$$= O_P(1).$$

The latter in turn by (2.10) implies (2.12).

By the construction of the Wiener processes $W_n^{(1)}$ and $W_n^{(2)}$ we obtain

$$\sup_{0<t<1} \Bigg| \left(1 - \frac{[(n+1)t]}{n}\right) \sum_{i=1}^{[(n+1)t]} \tilde{h}(X_i) + \frac{[(n+1)t]}{n} \left(\sum_{i=1}^{n} \tilde{h}(X_i) - \sum_{i=1}^{[(n+1)t]} \tilde{h}(X_i) \right) - \sigma\{(1-t)\, W_n((n+1)t) + t(W_n(n+1) - W_n((n+1)t))\} \Bigg| = o_P(n^{1/\nu}),$$

resulting also in (2.5) via (2.10).

Let Q^* be the class of functions $q\colon (0, 1) \to (0, \infty)$ which are monotone nondecreasing near 0 and monotone nonincreasing near one, and $\inf_{\delta \leqslant t \leqslant 1-\delta} q(t) > 0$ for all $\delta \in (0, 1/2)$. If $q \in Q^*$, we define the integral

$$I(q, c) = \int_0^1 (t(1-t))^{-1} \exp(-cq^2(t)/(t(1-t)))\, dt, \qquad c > 0.$$

This integral appears in the characterization of upper class functions of a Wiener process (cf., e.g., Csörgő *et al.* [1]).

COROLLARY 2.1. *We assume that* H_0 *holds, and* (2.2), (2.11) *are satisfied*:

(a) *If* $q \in Q^*$, *then*

$$\sup_{0<t<1} \left| \frac{n^{-3/2}}{\sigma} U_{[(n+1)t]} - \Gamma_n(t) \right| \Big/ q(t) = o_P(1) \tag{2.17}$$

if and only if $I(q, c) < \infty$ *for all* $c > 0$.

(b) *If* $q \in Q^*$, *then*

$$\frac{n^{-3/2}}{\sigma} \sup_{0<t<1} |U_{[(n+1)t]}|/q(t) \xrightarrow{D} \sup_{0<t<1} |\Gamma(t)|/q(t) \tag{2.18}$$

if and only if $I(q, c) < \infty$ *for some* $c > 0$.

Proof. First we note that $I(q, c) < \infty$ for some $c > 0$ implies (cf. Theorem 3.3 in Csörgő *et al.* [1])

$$\lim_{t \to 0} q(t)/t^{1/2} = \infty. \tag{2.19}$$

We have

$$\sup_{\delta \leqslant t \leqslant 1-\delta} \left| \frac{n^{-3/2}}{\sigma} U_{[(n+1)t]} - \Gamma_n(t) \right| \Big/ q(t) = o_P(1) \tag{2.20}$$

for all $\delta \in (0, \frac{1}{2})$ by Theorem 2.2. Also, by (2.12) and (2.19),

$$\sup_{1/(n+1) \leqslant t \leqslant \delta} \left| \frac{n^{-3/2}}{\sigma} U_{[(n+1)t]} - \Gamma_n(t) \right| \Big/ q(t) = O_P(1) \sup_{0<t\leqslant\delta} \frac{t^{1/2}}{q(t)} \xrightarrow{P} 0 \tag{2.21}$$

as $\delta \to 0$. Next

$$\begin{aligned} \sup_{0<t\leqslant 1/(n+1)} |\Gamma(t)|/q(t) &\leqslant \sup_{0<t\leqslant 1/(n+1)} |W(t)|/q(t) \\ &\quad + \sup_{0<t\leqslant 1/(n+1)} (t/q(t)) \sup_{0\leqslant t\leqslant 1/(n+1)} |W(1)-W(t)| \\ &= o_P(1) \end{aligned}$$

by (2.19) and Theorem 3. of Csörgő *et al.* [1]. One estimates near 1 in a similar way, and the "if" part of (a) is proven.

Assuming now (2.17), we must have

$$\sup_{0<t\leqslant 1/(n+1)} |\Gamma(t)|/q(t) = o_P(1) \tag{2.22}$$

and

$$\sup_{n/(n+1) \leqslant t < 1} |\Gamma(t)|/q(t) = o_P(1). \tag{2.23}$$

It is easy to see that (2.22) and (2.33) imply

$$E\Gamma^2(t)/q^2(t) \to 0 \quad \text{as} \quad t \to 0 \quad \text{or} \quad t \to 1. \tag{2.24}$$

Consequently we havc (2.22) if and only if

$$\sup_{0 < t \leqslant 1/(n+1)} |W(t)|/q(t) = o_P(1). \tag{2.25}$$

Similarly, we have (2.23) if and only if

$$\sup_{n/(n+1) \leqslant t < 1} |W(1) - W(t)|/q(t) = o_P(1), \tag{2.26}$$

which is equivalent to

$$\sup_{0 < t \leqslant 1/(n+1)} |W(t)|/q(1-t) = o_P(1). \tag{2.27}$$

Now Theorem 3.4 of Csörgő *et al.* [1] combined with (2.25) and (2.27) results in the second part of (a).

As to the proof of (b) we first note that (2.19) implies

$$\sup_{1/(n+1) \leqslant t \leqslant n/(n+1)} \left| \frac{n^{-3/2}}{\sigma} U_{[(n+1)t]} - \Gamma_n(t) \right| \Big/ q(t) = o_P(1). \tag{2.28}$$

Hence it suffices to show that

$$\sup_{1/(n+1) \leqslant t \leqslant n/(n+1)} |\Gamma(t)|/q(t) \xrightarrow{\mathscr{D}} \sup_{0 < t < 1} |\Gamma(t)|/q(t),$$

which follows immediately from Theorem 3.3 of Csörgő *et al.* [1]. The proof of the necessary part of (b) is similar to that of (a). Only here we have to use Theorem 3.3 of Csörgő *et al.* [1] instead of their Theorem 3.4.

Remark 2.1. The proof of the necessary part of Corollary 2.1(a) shows that if we have (2.17) with any sequence of Gaussian processes having the same distribution for each $n \geqslant 1$ as that of Γ, then $I(q, c)$ must be finite for all $c > 0$. This means that the necessary part does not depend on our construction.

The desirability of having weight functions q around like in Corollary 2.1 is to make our statistical test more sensitive on the tails. A typical choice of

q in (2.18) is the function $(t(1-t)\log\log(1/t(1-t)))^{1/2}$. The variance of $\Gamma(t)$ is $t(1-t)$, hence another choice of a weight function is $(t(1-t))^{1/2}$. However $I((t(1-t))^{1/2}, c)=\infty$ for every $c>0$, and hence we cannot apply Corollary 2.1. This case is studied in the next theorem. Let $a(y\cdot\log n)=(y+2\log\log n+\frac{1}{2}\log\log\log n-\frac{1}{2}\log\pi)(2\log\log n)^{-1/2}$, $-\infty<y<\infty$.

THEOREM 2.3. *We assume that H_0 holds, and* (2.2), (2.11) *are satisfied. Then*

$$\lim_{n\to\infty} P\left\{\sigma^{-1}\max_{1\leqslant k\leqslant n}\frac{U_k}{(k(n-k+1)n)^{1/2}}\leqslant a(y,\log n)\right\}=\exp(-\exp(-y)), \tag{2.29}$$

and

$$\lim_{n\to\infty} P\left\{\sigma^{-1}\max_{1\leqslant k\leqslant n}\frac{|U_k|}{(k(n-k+1)n)^{1/2}}\leqslant a(y,\log n)\right\}=\exp(-2\exp(-y)). \tag{2.30}$$

We note that it will also follow from the proof of this theorem that the same two limit statements hold for $(n^{-3/2}/\sigma)U_{[(n+1)t]}/(t(1-t))^{1/2}$, $0<t<1$. The proof will be based on the following lemma. Let $b(y,\log n)=(y+2\log\log n+\frac{1}{2}\log\log\log n-\frac{1}{2}\log(4\pi))(2\log\log n)^{-1/2}$, $-\infty<y<\infty$.

LEMMA. *Let $Y_1, Y_2, \ldots$ be i.i.d.r.v.'s with $EY_1=0$, $EY_1^2=1$, and $E|Y_1|^{2+\delta}<\infty$ for some $\delta>0$. Then*

$$\lim_{n\to\infty} P\left\{\max_{1\leqslant k\leqslant n} k^{-1/2}\sum_{i=1}^{k} Y_i\leqslant b(y,\log n)\right\}=\exp(-\exp(-y)) \tag{2.31}$$

and

$$\lim_{n\to\infty} P\left\{\max_{1\leqslant k\leqslant n} k^{-1/2}\left|\sum_{i=1}^{k} Y_i\right|\leqslant b(y,\log n)\right\}=\exp(-2\exp(-y)). \tag{2.32}$$

Also, if $m_n\to\infty$ and $m_n/n\to 0$ $(n\to\infty)$, then

$$\lim_{n\to\infty} P\left\{\max_{m_n\leqslant k\leqslant n} k^{-1/2}\sum_{i=1}^{k} Y_i\leqslant b(y,\log(n/m_n))\right\}=\exp(-\exp(-y)) \tag{2.33}$$

and

$$\lim_{n\to\infty} P\left\{\max_{m_n\leqslant k\leqslant n} k^{-1/2}\left|\sum_{i=1}^{k} Y_i\right| \leqslant b(y, \log(n/m_n))\right\} = \exp(-2\exp(-y)). \tag{2.34}$$

Proof. For the proof of (2.31) and (2.32) we refer to Darling and Erdös [4] and Shorack [14].

Of the two statements (2.33) and (2.34) we verify only (2.34). The proof of (2.33) is similar. First let $1\leqslant m_n\leqslant \log n$. Then by (2.32)

$$(2\log\log n)^{1/2} \max_{1\leqslant k\leqslant m_n} k^{-1/2}\left|\sum_{i=1}^{k} Y_i\right| - \log\log n \xrightarrow{P} -\infty,$$

and

$$\lim_{n\to\infty} P\left\{\max_{m_n\leqslant k\leqslant n} k^{-1/2}\left|\sum_{k=1}^{k} Y_i\right| \leqslant b(y, \log n)\right\} = \exp(-2\exp(-y)).$$

Observing now

$$\left|\left(\log\log\frac{n}{m_n}\right)^{1/2} - (\log\log n)^{1/2}\right| (\log\log n)^{1/2} = o(1)$$

and

$$\left|2\log\log\frac{n}{m_n} + \frac{1}{2}\log\log\lg\frac{n}{m_n} - \left(2\log\log n + \frac{1}{2}\log\log\log n\right)\right| = o(1)$$

we get (2.34). Similarly to the proof of Theorem 2.2, there is a Wiener process W such that

$$\sup_{m_n\leqslant x\leqslant n} x^{-1/2}\left|\sum_{1\leqslant i\leqslant x} Y_i - W(x)\right| = o_P(m_n^{1/(2+\delta)-1/2}) = o_P((\log n)^{-\delta/(2+\delta)}).$$

Let $\{V(t),\ -\infty<t<\infty\}$ be an Ornstein–Uhlenbeck process. Then we have

$$\sup_{m_n\leqslant x\leqslant n} x^{-1/2}\mid W(x) = \sup_{(1/2)\log m_n\leqslant t\leqslant (1/2)\log n} |V(t)| = \sup_{0\leqslant t\leqslant (1/2)\log(n/m_n)} |V(t)|,$$

and consequently by Darling and Erdös [4] we obtain (2.34). For the general m_n sequence of the lemma we consider its subsequence with values in $[0, \log n]$ and that with values in $(\log n, \infty)$.

Proof of Theorem 2.3. Let $k_n^{(1)} = (\log n)^3$ and $k_n^{(2)} = n/(\log n)^2$, and consider

$$\max_{1 \leqslant k \leqslant n} \frac{|U_k|}{(k(n-k+1)n)^{1/2}} = \max_{1 \leqslant k \leqslant k_n^{(1)}} \frac{|U_k|}{(k(n-k+1)n^{1/2}}$$

$$\vee \max_{k_n^{(1)} \leqslant k \leqslant k_n^{(2)}} \frac{|U_k|}{(k(n-k+1)n)^{1/2}}$$

$$\vee \max_{k_n^{(2)} \leqslant k \leqslant n/2} \frac{|U_k|}{(k(n-k+1)n)^{1/2}}$$

$$\vee \max_{n/2 \leqslant k \leqslant n-k_n^{(2)}} \frac{|U_k|}{(k(n-k+1)n)^{1/2}}$$

$$\vee \max_{n-k_n^{(2)} \leqslant k \leqslant n-k_n^{(1)}} \frac{|U_k|}{(k(n-k+1)n)^{1/2}}$$

$$\vee \max_{n-k_n^{(1)} \leqslant k \leqslant n} \frac{|U_k|}{(k(n-k+1)n)^{1/2}}$$

$$= A_n^{(1)} \vee \cdots \vee A_n^{(6)},$$

where $a \vee b = \max(a, b)$. It is easy to see that

$$A_n^{(1)} \leqslant \frac{2}{n} \max_{1 \leqslant k \leqslant k_n^{(1)}} k^{-1/2} \left| \sum_{1 \leqslant i \leqslant k} \sum_{k+1 \leqslant j \leqslant n} \{h(X_i, X_j) - \tilde{h}(X_i)\} \right|$$

$$+ \max_{1 \leqslant k \leqslant k_n^{(1)}} k^{-1/2} \left| \sum_{i=1}^{k} \tilde{h}(X_i) \right|$$

$$= A_n^{(1,1)} + A_n^{(1,2)}. \tag{2.36}$$

First we note that by the definition of $\tilde{h}$ we have

$$E\left(n^{-1} \sum_{1 \leqslant i \leqslant k} \sum_{k+1 \leqslant j \leqslant n} \{h(X_i, X_j) - \tilde{h}(X_i)\}\right)^2 = O(k^2/n),$$

and so

$$P\{A_n^{(1,1)} > 1\} \leqslant \sum_{k=1}^{k_n^{(1)}} P\left\{n^{-1} \left| \sum_{1 \leqslant i \leqslant k} \sum_{k+1 \leqslant j \leqslant n} \{h(X_i, X_j) - \tilde{h}(X_i)\} \right| > k^{1/2}\right\}$$

$$= O(1)\, n^{-1} \sum_{k=1}^{k_n^{(1)}} 1/k = o(1). \tag{2.37}$$

By Lemma we have

$$A_n^{(1,2)} = O_P((\log\log\log n)^{1/2}),$$

and thus by (2.36) and (2.37) we obtain

$$(2\log\log n)^{1/2} A_n^{(1)} - \sigma \log\log n \xrightarrow{P} -\infty. \tag{2.38}$$

By (2.10) we get

$$A_n^{(2)} = \max_{k_n^{(1)} \leqslant k \leqslant k_n^{(2)}} \left| \frac{n-k}{(n(n-k+1))} k^{-1/2} \sum_{i=1}^{k} \tilde{h}(X_i) + \frac{k}{(n(n-k+1)^{1/2}} \sum_{i=k+1}^{n} \tilde{h}(X_i) \right| + O_P(1/\log n). \tag{2.39}$$

It is easy to verify that

$$\begin{aligned}
&\max_{k_n^{(1)} \leqslant k \leqslant k_n^{(2)}} \left(\frac{k}{n}\right)^{1/2} \frac{1}{(n-k+1)^{1/2}} \left| \sum_{i=k+1}^{n} \tilde{h}(X_i) \right| \\
&\quad \overset{\mathscr{D}}{=} \max_{n-k_n^{(2)} \leqslant m \leqslant n-k_n^{(1)}} \left(\frac{n-m}{n}\right)^{1/2} \frac{1}{(m+1)^{1/2}} \left| \sum_{i=1}^{m} \tilde{h}(X_i) \right| \\
&\quad = O(1/\log n) \max_{n-k_n^{(2)} \leqslant m \leqslant n-k_n^{(1)}} \frac{1}{n^{1/2}} \left| \sum_{i=1}^{m} \tilde{h}(X_i) \right| \\
&\quad = O(1/\log n) \max_{1 \leqslant m \leqslant n} \frac{1}{n^{1/2}} \left| \sum_{i=1}^{m} \tilde{h}(X_i) \right| \\
&\quad = O_P(1/\log n).
\end{aligned} \tag{2.40}$$

Using the lemma we have

$$\begin{aligned}
&\max_{k_n^{(1)} \leqslant k \leqslant k_n^{(2)}} \left| \frac{n-k}{(n(n-k+1))^{1/2}} - 1 \right| k^{-1/2} \left| \sum_{i=1}^{k} \tilde{h}(X_i) \right| \\
&\quad = O_P((\log\log n)^{1/2}/(\log n)^2),
\end{aligned}$$

and hence (2.39) and (2.40) yield

$$A_n^{(2)} = \max_{k_n^{(1)} \leqslant k \leqslant k_n^{(2)}} k^{-1/2} \left| \sum_{i=1}^{k} \tilde{h}(X_i) \right| + O_P(1/\log n). \tag{2.41}$$

By the lemma again,

$$(2\log\log k_n^{(2)})^{1/2} \max_{1 \leqslant k \leqslant k_n^{(1)}} k^{-1/2} \left| \sum_{i=1}^{k} \tilde{h}(X_i) \right| - \sigma \log\log k_n^{(2)} \xrightarrow{P} -\infty,$$

and therefore,

$$\lim_{n\to\infty} P\left\{\frac{1}{\sigma}\max_{k_n^{(1)}\leqslant k\leqslant k_n^{(2)}} k^{-1/2}\left|\sum_{i=1}^{k}\tilde{h}(X_i)\right| \leqslant b(y, \log k_n^{(2)})\right\} = \exp(-2\exp(-y)). \tag{2.42}$$

Observing now that

$$|(2\log\log n)^{1/2}-(2\log\log k_n^{(1)})^{1/2}|\,(\log\log k_n^{(1)})^{1/2}=o(1)$$

and

$$|2\log\log n+\tfrac{1}{2}\log\log\log n-(2\log\log k_n^{(1)} + \tfrac{1}{2}\log\log\log k_n^{(1)})|=o(1),$$

(2.41) and (2.42) imply

$$\lim_{n\to\infty} P\left\{\frac{1}{\sigma}A_n^{(2)}\leqslant b(y,\log n)\right\}=\exp(-2\exp(-y)). \tag{2.43}$$

Towards estimating $A_n^{(3)}$, we first note that

$$\max_{k_n^{(2)}\leqslant k\leqslant n/2}\frac{1}{(n-k)^{1/2}}\left|\sum_{i=k+1}^{n}\tilde{h}(X_i)\right| \overset{\mathscr{D}}{=} \max_{n/2\leqslant m\leqslant n-k_n^{(2)}}\frac{1}{m^{1/2}}\left|\sum_{k=1}^{m}\tilde{h}(X_k)\right|=O_P(1).$$

Hence from (2.10) and (2.34) we obtain

$$A_n^{(3)}\leqslant \max_{k_n^{(2)}\leqslant k\leqslant n/2}\frac{n-k}{(k(n-k+1)\,n)^{1/2}}\left|\sum_{i=1}^{k}\tilde{h}(X_i)\right| + \max_{k_n^{(2)}\leqslant k\leqslant n/2}\frac{k}{(n(n-k+1))^{1/2}}\left|\sum_{i=k+1}^{n}\tilde{h}(X_i)\right|+O_P(\log n/n^{1/2})$$
$$=O_P((\log\log\log n)^{1/2}).$$

This in turn implies

$$(2\log\log n)^{1/2}A_n^{(3)}-\sigma^{-1}\log\log n\xrightarrow{P}-\infty. \tag{2.44}$$

The estimation of the r.v.'s $A_n^{(4)}$, $A_n^{(5)}$, and $A_n^{(6)}$ is similar, resulting in the statements

$$(2 \log \log n)^{1/2} A_n^{(i)} - \sigma^{-1} \log \log n \xrightarrow{P} -\infty, \qquad i = 4, 6, \tag{2.45}$$

$$A_n^{(5)} = \max_{n - k_n^{(2)} \leqslant k \leqslant n - k_n^{(1)}} \frac{1}{(n-k)^{1/2}} \left| \sum_{i=k+1}^{n} \tilde{h}(X_i) \right| + O_P(1/\log n), \tag{2.46}$$

and

$$\lim_{n \to \infty} P\left\{ \frac{1}{\sigma} A_n^{(5)} \leqslant b(y, \log n) \right\} = \exp(-2 \exp(-y)). \tag{2.47}$$

The events in (2.43) and (2.47) are asymptotically independent. Therefore the statement follows from (2.35), (2.38), (2.43), and (2.44)–(2.47).

3. Asymptotics under H_1

First we introduce some notations. Let

$$\theta = Eh(X_{[n\lambda]-1}, X_{[n\lambda]}), \qquad \mu = Eh(X_{[n\lambda]+1}, X_{[n\lambda]+2})$$
$$\tau = Eh(X_{[n\lambda]}, X_{[n\lambda]+1}),$$

and we write $\log^+ x = \log(x \vee 1)$.

THEOREM 3.1. *We assume that H_1 holds and*

$$E\,|h(X_{[n\lambda]-1}, X_{[n\lambda]})| < \infty, \qquad E\,|h(X_{[n\lambda]+1}, X_{[n\lambda]+2})| < \infty,$$
$$E\,|h(X_{[n\lambda]}, X_{[n\lambda]+1})|\,|\log^+(|h(X_{[n\lambda]}, X_{[n\lambda]+1})|) < \infty \tag{3.1}$$

are satisfied. Then

$$\lim_{n \to \infty} Z_{[(n+1)t]}/n^2 = \begin{cases} \theta t(\lambda - t) + t\tau(1-\lambda), & 0 < t \leqslant \lambda, \\ \mu(t-\lambda)(1-t) + \tau\lambda(1-t), & \lambda \leqslant t < 1, \end{cases} \tag{3.2}$$

in probability.

Proof. Let $1 \leqslant [(n+1)t] \leqslant [n\lambda]$. Then

$$\begin{aligned} Z_{[(n+1)t]} = & \sum_{1 \leqslant i < j \leqslant [n\lambda]} h(X_i, X_j) + \sum_{1 \leqslant i \leqslant [n\lambda]} \sum_{[n\lambda]+1 \leqslant j \leqslant n} h(X_i, X_j) \\ & - \left\{ \sum_{1 \leqslant i < j \leqslant [(n+1)t]} h(X_i, X_j) \right. \\ & + \sum_{[(n+1)t]+1 \leqslant i < j \leqslant [n\lambda]} h(X_i, X_j) \\ & \left. + \sum_{[(n+1)t]+1 \leqslant i \leqslant [n\lambda]} \sum_{[n\lambda]+1 \leqslant j \leqslant n} h(X_i, X_j) \right\} \\ = & R_n^{(1)} + R_n^{(2)} - \{R_n^{(3)} + \cdots + R_n^{(5)}\}. \end{aligned}$$

By Hoeffding [7] (cf. Theorem A in Section 5.4 of Serfling [13]) we get

$$R_n^{(1)}/n^2 \xrightarrow{\text{a.s.}} \lambda^2\theta/2, \qquad R_n^{(3)}/n^2 \xrightarrow{\text{a.s.}} t^2\theta/2,$$

$$R_n^{(4)}/n^2 \overset{\mathscr{D}}{=} \sum_{1 \leqslant i<j \leqslant [n\lambda]-[(n+1)t]} h(X_i, X_j) \xrightarrow{\text{a.s.}} (t-\lambda)^2\,\theta/2.$$

Now applying Sen [12] and condition (3.1) we obtain

$$R_n^{(2)}/n^2 \xrightarrow{P} \lambda(1-\lambda)\tau, \qquad R_n^{(5)}/n^2 \xrightarrow{P} (\lambda-t)(1-\lambda)\tau.$$

These observations clearly imply the first part of (3.2). The proof of its second part is similar.

Remark 3.1. If we assume the existence of the second moments in Theorem 3.1, then we have an a.s. convergence in (3.2) by the moment inequalities of Grams and Serfling [5].

Theorem 3.1 can be used to study the consistency of tests based on the process $\{U_{[(n+1)t]}, 0 \leqslant t < 1\}$. For example, we conclude that rejecting H_0 vs H_1 when $\sup_{0 \leqslant t<1} (n^{-3/2}/\sigma)\,|U_{[(n+1)t]}|$ is large, then the latter test is consistent except in the case of $\tau=\theta=\mu=0$. The same can be said about the weighted versions of this test.

4. Antisymmetric Kernel

In this section we assume that h is an antisymmetric kernel, i.e.,

$$h(x, y) = -h(y, x). \tag{4.1}$$

In this case $Eh(X_1, X_2)=0$ and similarly to the symmetric case we let $\tilde{h}(t)=Eh(t, X_1)$. We assume

$$Eh^2(X_1, X_2) < \infty \qquad \text{and} \qquad 0<\sigma^2 = E\tilde{h}^2(X_1). \tag{4.2}$$

Accordingly to Section 2 we now have $U_k = Z_k$, where Z_k is defined by (1.1). It is easy to see that (2.3) remains true in the case of an antisymmetric kernel, with Θ taken to be zero, of course.

First we give an analog of Theorem 2.1.

Theorem 4.1. *We assume that H_0 holds, and* (4.1) *and* (4.2) *are satisfied.*

Then we can define a sequence of Brownian bridges $\{B_n(t), 0 \leqslant t \leqslant 1\}$ *such that, as* $n \to \infty$,

$$\sup_{0 \leqslant t \leqslant 1} \left| \frac{n^{-3/2}}{\sigma} U_{[(n+1)t]} - B_n(t) \right| = o_p(1) \tag{4.3}$$

and for each $n \geqslant 0$, $EB_n(t) = 0$, $EB_n(t)\, B_n(s) = \min(t, s) - ts$.

Proof. The proof is similar to that of Theorem 2.1. Instead of Theorem 1 of Hall [6] we use Theorem 2.1 of Janson and Wichura (1983), which gives

$$\max_{1 \leqslant k \leqslant n} \left| U_k^{(1)} - \sum_{i=1}^{k} (k - 2i + 1)\, \tilde{h}(X_i) \right| = O_P(n), \tag{4.4}$$

$$\max_{1 \leqslant k \leqslant n} \left| U_k^{(2)} - \sum_{i=k+1}^{n} (n + k - 2i + 1)\, \tilde{h}(X_i) \right| = O_P(n), \tag{4.5}$$

and

$$\left| U_n^{(3)} - \sum_{i=1}^{n} (n - 2i + 1)\, \tilde{h}(X_i) \right| = O_P(n). \tag{4.6}$$

By (4.4), (4.5), and (4.6) we have

$$\max_{1 \leqslant k \leqslant n} \left| U_k - \left\{ n \sum_{i=1}^{k} \tilde{h}(X_i) - k \sum_{i=1}^{n} \tilde{h}(X_i) \right\} \right| = O_P(n) \tag{4.7}$$

and hence Donsker's theorem implies Theorem 4.1.

Surprisingly, the limiting processes are different in Theorems 2.1 and 4.1. In the special case of $h(x, y) = \operatorname{sign}(x - y)$ (cumulative rank tests) Pettitt [9] (cf. also Pettitt [10]) indicate a proof of Theorem 4.1.

The following Theorem is an analog of Theorem 2.2.

THEOREM 4.2. *We assume that* H_0 *holds,* (4.1) *and* (4.2) *are satisfied, and*

$$E|h(X_1, X_2)|^{\nu} < \infty \qquad \text{for some} \quad \nu > 2. \tag{4.8}$$

Then we can define a sequence of Brownian bridges $\{B_n(t), 0 \leqslant t \leqslant 1\}$ *such that* (4.3) *holds and*

$$\sup_{1/(n+1) \leqslant t \leqslant n/(n+1)} \left| \frac{n^{-3/2}}{\sigma} U_{[(n+1)t]} - B_n(t) \right| \Big/ (t(1-t))^{1/2} = O_P(1). \tag{4.9}$$

Proof. Using (4.4)–(4.6) with the Skorohod embedding scheme (or with the Komlós–Major–Tusnády approximation), the proof goes along the lines of the proof of Theorem 2.2.

The next results are direct consequences of Theorem 4.2. One can give detailed proofs using the methods of the proofs of Corollary 2.1 and Theorem 2.3. Let $\{B(t), 0 \leqslant t \leqslant 1\}$ be a Brownian bridge.

COROLLARY 4.1. *We assume that H_0 holds and* (4.1), (4.2), *and* (4.8) *are satisfied.*

(a) *If $q \in Q^*$, then*

$$\sup_{0<t<1} \left| \frac{n^{-3/2}}{\sigma} U_{[(n+1)t]} - B_n(t) \right| \Big/ q(t) = o_P(1)$$

if and only if $I(q, c) < \infty$ for all $c > 0$.

(b) *If $q \in Q^*$, then*

$$\frac{n^{-3/2}}{\sigma} \sup_{0<t<1} |U_{[(n+1)t]}|/q(t) \xrightarrow{\mathscr{D}} \sup_{0<t<1} |B(t)|/q(t)$$

if and only if $I(q, c) < \infty$ for some $c > 0$.

THEOREM 4.3. *We assume that H_0 holds and* (4.1), (4.2), *and* (4.8) *are satisfied. Then*

$$\lim_{n \to \infty} P\left\{\sigma^{-1} \max_{1 \leqslant k \leqslant n} \frac{U_k}{(k(n-k+1)n)^{1/2}} \leqslant a(y, \log n)\right\} = \exp(-\exp(-y))$$

and

$$\lim_{n \to \infty} P\left\{\sigma^{-1} \max_{1 \leqslant k \leqslant n} \frac{|U_k|}{(k(n-k+1)n)^{1/2}} \leqslant a(y, \log n)\right\} = \exp(-2\exp(-y)).$$

Now we assume that $X_1, ..., X_n$ have a continuous distribution function, and study the case of $h(x, y) = \operatorname{sign}(x-y)$. Under H_0, $E \operatorname{sign}(X_1 - X_2) = 0$ and $\sigma^2 = 1/12$. Then

$$U_k = Z_k = \sum_{1 \leqslant i < k} \sum_{k+1 \leqslant j \leqslant n} \operatorname{sign}(X_i - X_j)$$

is distribution free, and the results of the present section are applicable. By Theorem 4.1, $(12)^{1/2} n^{-3/2} U_{[(n+1)t]}$ converges weakly to a Brownian bridge in $D[0, 1]$. This result was obtained by Pettitt [9] using heuristic arguments.

Sen and Srivastava [11] also mention (without developing any properties) non-parametric tests as analogs to some parametric likelihood ratio procedures. In particular, they suggest rejecting H_0 for large values of

$$D_n = (12)^{1/2} \max_{1 \leqslant k \leqslant n} |U_k|/(k(n-k+1)n)^{1/2}.$$

It follows from Theorem 4.3 that $D_n \to^P \infty$ even under H_0. This is the reason for them finding D_n being superior to other statistics. We can, of course, use D_n for testing H_0 with normalizing factors as given in Theorem 4.3. Naturally then further power studies are also needed in order to conclude any superiority properties.

Acknowledgments

We are very grateful to F. Lombard for his helpful comments and for pointing out an oversight in the first version of this paper. We also wish to thank Vera R. Huse for calling our attention to a slip in this version and for her careful reading of our manuscript.

References

[1] Csörgő, M., Csörgő, S., Horváth, L., and Mason, D. M. (1986). Weighted empirical and quantile processes. *Ann. Probab.* **14** 31–85.

[2] Csörgő, M. and Horváth L. (1987). Nonparametric tests for the changepoint problem. *J. Statist. Plann. Inference* **17** 1–9.

[3] Csörgő, M. and Révész, P. (1981). *Strong Approximations in Probability and Statistics.* Academic Press, New York.

[4] Darling, D. and Erdős, P. (1956). A limit theorem for the maximum of normalized sums of independent random variables. *Duke Math. J.* **23** 143–155.

[5] Grams, W. F. and Serfling, R. J. (1973). Convergence rates for U-statistics and related statistics. *Ann. Statist.* **1** 153–160.

[6] Hall, P. (1979). On the invariance principle for U-statistics. *Stochastic Process. Appl.* **9** 163–174.

[7] Hoeffding, W. (1961). *The Strong Law of Large Numbers for U-Statistics.* University of North Carolina Institute of Statistics Mimeo Series 302.

[8] Janson, S. and Wichura, M. J. (1983). Invariance principles for stochastic area and related stochastic integrals. *Stochastic Process. Appl.* **16** 71–84.

[9] Pettitt, A. N. (1979). A non-parametric approach to the change-point problem. *Appl. Statist.* **28** 126–135.

[10] Pettitt, A. N. (1980). Some results on estimating a change-point using non-parametric type statistics. *J. Statist. Comput. Simul.* **11** 261–272.

[11] Sen, A. and Srivastava M. S. (1975). On tests for detecting changes in mean. *Ann. Statist.* **3** 98–108.

[12] Sen, P. K. (1977). Almost sure convergence of generalized U-statistics. *Ann. Probab.* **5** 287–290.

[13] SERFLING, R. J. (1980). *Approximation Theorems of Mathematical Statistics.* Wiley, New York.

[14] SHORACK, G. R. (1979). Extension of the Darling and Erdős theorem on the maximum of normalized sums. *Ann. Probab.* **7** 1092–1096.

[15] WOLFE, D. A. AND SCHECHTMAN, E. (1984). Nonparametric statistical procedures for the changepoint problem. *J. Statist. Plann. Inference* **9** 389–396.

On the Area of the Circles Covered by a Random Walk

P. ERDÖS

Mathematical Institute, Budapest, Hungary

AND

P. RÉVÉSZ

Mathematical Institute, Budapest, Hungary and Technical University, Vienna, Austria

The area of the largest circle around the origin completely covered by a simple symmetric plane random walk is investigated.

1. INTRODUCTION

Let $X_1, X_2, \ldots$ be a sequence of independent, identically distributed random vectors taking values from R^2 with distribution

$$\mathbb{P}\{X_1=(0,1)\}=\mathbb{P}\{X_1=(0,-1)\}=\mathbb{P}\{X_1=(1,0)\}=\mathbb{P}\{X_1=(-1,0)\}=\tfrac{1}{4}$$

and let

$$S_0=0=(0,0) \qquad \text{and} \qquad S(n)=S_n=X_1+X_2+\cdots+X_n \qquad (n=1,2,\ldots),$$

i.e., $\{S_n\}$ is the simple symmetric random walk on the plane. Further let

$$\xi(x,n)=\#\{k:0<k\leqslant n, S_k=x\}$$

$(n=1,2,\ldots;\ x=(i,j);\ i,j=0,\ \pm 1,\ \pm 2,\ldots)$ be the local time of the random walk. We say that the circle

$$Q(N)=\{x=(i,j): \|x\|=(i^2+j^2)^{1/2}\leqslant N\}$$

Multivariate Statistics and Probability
ISBN 0-12-580205-6

Reprinted from *J. Mult. Anal.* **27**(1).

is covered by the random walk in time n if

$$\xi(x, n) > 0 \qquad \text{for every} \quad x \in Q(N).$$

Let $R(n)$ be the largest integer for which $Q(R(n))$ is covered in n. We are interested in the limit properties of the random variables $R(n)$ as $n \to \infty$. This question was proposed by Erdös and Taylor [5] and they claim "we can show using the methods we have discussed above that" for any $\varepsilon > 0$

$$R(n) \geqslant \exp((\log n)^{1/2 - \varepsilon}) \qquad \text{a.s.}$$

for all but finitely many n "but we have failed to get a satisfactory upper estimate and have no plausible conjecture."

This paper is devoted to the above question and some related problems.

2. A Lower Estimate of $R(n)$

In this section we prove

THEOREM 1. *For any $\varepsilon > 0$ we have*

$$R(n) \geqslant \exp\left(\frac{(\log n)^{1/2}}{(\log_2 n)^{3/4 + \varepsilon}}\right) \qquad a.s.$$

for all but finitely many n where $\log_k$ is the k times iterated logarithm.

Before the proof we present a few notations and lemmas.

Let $\gamma(x, n)$ be the probability that in the first n steps the path does not pass through x i.e.

$$\gamma(x, n) = \mathbb{P}\{\xi(x, n-1) = 0\}.$$

Let $\alpha(r)$ be the probability that the random walk $\{S_n\}$ hits the circle of radius r before returning to the point $0 = (0, 0)$, i.e.,

$$\alpha(r) = \mathbb{P}\{\inf\{n: \|S_n\| \geqslant r\} < \inf\{n: n \geqslant 1, S_n = 0\}\}.$$

Further let $\beta(r, t)$ be the probability that starting from a point of the circle-ring $r \leqslant \|x\| \leqslant r + 1$ the particle hits the point $0 = (0, 0)$ before hiting the circle of radius rt, i.e.,

$$\beta(r, t) = \mathbb{P}\{\inf\{n: S_{n+m} = 0\} < \inf\{n: \|S_{n+m}\| \geqslant rt\} \mid r \leqslant \|S_m\| \leqslant r + 1\}.$$

Finally let

$$\delta(t)=\delta(t,r)=\mathbb{P}\{\max_{k\leqslant tr^2}\|S_k\|<r\}$$

and

$$\mu(x)=\mu(x,n)=\mathbb{P}\{\xi(0,n)<x\log n\}.$$

LEMMA 1. *Let* $\|x\|=\psi^{-1}n^{1/2}$ *with* $20<\psi<n^{1/3}$. *Then*

$$\gamma(x,n)=1-\frac{2\log\psi}{\log n}\left(1+O\left(\frac{\log_2\psi}{\log\psi}\right)\right). \tag{2.1}$$

$$\lim_{n\to\infty}\mu(x,n)=1-\exp(-\pi x) \tag{2.2}$$

for $0<x<(\log n)^{3/4}$ *and the limit is approached uniformly in this range*;

$$\delta(t)=\begin{cases}1-\exp(-O(t^{-1})) & \text{if } t\to 0,\\ \exp(-O(t)) & \text{if } t\to\infty.\end{cases} \tag{2.3}$$

Proof. (2.1) (resp. (2.2)) are proved in Erdös and Taylor [5] cf. (2.18) (resp. Theorem 1). The proof of (2.3) is trivial.

Remark 1. (2.2) implies

$$\mathbb{P}\{\xi(0,n)=0\}\approx\pi/\log n \tag{2.4}$$

(cf. also Dvoretzky and Erdös, [2]).

LEMMA 2. *We have*

$$\lim_{r\to\infty}\alpha(r)\log r=\pi/2. \tag{2.5}$$

Proof. Clearly we have

$$\{\inf\{n:\|S_n\|\geqslant r\}>\inf\{n:n\geqslant 1,S_n=0\}\}$$
$$\subset\{\xi(0,r^2\log r)>0\}\cup\{\max_{0\leqslant k\leqslant r^2\log r}\|S_k\|\leqslant r\}.$$

Since

$$\mathbb{P}\{\xi(0,r^2\log r)=0\}\approx\pi/2\log r \qquad \text{by (2.4)}$$

and

$$\mathbb{P}\{\max_{0\leqslant k\leqslant r^2\log r}\|S_k\|\leqslant r\}=o(1/\log r) \qquad \text{by (2.3)},$$

we have

$$\alpha(r) \geqslant \frac{\pi + o(1)}{2 \log r}.$$

Observe also

$$\alpha(r) \leqslant \mathbb{P}\{\max_{0 \leqslant k \leqslant r^2(\log r)^{-1}} \|S_k\| \geqslant r\} + \mathbb{P}\{\xi(0, r^2(\log r)^{-1}) = 0\}.$$

Applying again (2.3) and (2.4) we obtain (2.5).

LEMMA 3. *For any $\varepsilon > 0$ and r big enough we have*

$$\beta(r, t) \leqslant (1+\varepsilon) \frac{\log_3 r}{\log r} \tag{2.6}$$

provided that $1 < t < o((\log \log r)^{\delta})$ *for any* $\delta > 0$.

Proof. For any $K > 0$ we have

$$\begin{aligned}\beta(r, t) \leqslant\ & \mathbb{P}\{\xi(0, Kr^2 + m) - \xi(0, m) \geqslant 1 \mid r \leqslant \|S_m\| \leqslant r+1\} \\ &+ \mathbb{P}\{\max_{m \leqslant k \leqslant m + Kr^2} \|S_k\| \leqslant rt \mid r \leqslant \|S_m\| \leqslant r+1\} = \mathrm{I} + \mathrm{II}.\end{aligned}$$

By (2.1)

$$\mathrm{I} = 1 - \gamma(x, Kr^2) \approx \frac{2 \log \psi}{\log Kr^2}$$

for any $r \leqslant \|x\| \leqslant r+1$, where $\psi = K^{1/2} r / \|x\|$ and

$$\mathrm{II} \leqslant \mathbb{P}\{\max_{0 \leqslant k \leqslant Kr^2} \|S_k\| \leqslant (t+2) r\} = \delta\left(\frac{K}{(t+2)^2}\right).$$

By choosing $K = (t+2)^2 (\log \log r)^{1+\varepsilon}$ ($\varepsilon > 0$) we obtain

$$\beta(r, t) \leqslant (1+\varepsilon) \frac{\log_3 r}{\log r}$$

for any $\varepsilon > 0$ if r is big enough and $1 < t < o((\log_2 r)^{\varepsilon})$ (for any $\varepsilon > 0$). Hence we have (2.6).

LEMMA 4. *For any $\varepsilon > 0$ and r big enough we have*

$$\beta(r, t) \geqslant 1/\varepsilon \log r \tag{2.7}$$

provided that $t \geqslant (\log \log r)^{1/2 + \delta}$ *for some* $\delta > 0$.

Proof. For any $K>0$ we have

$$\beta(r, t) \geqslant \mathbb{P}\{\xi(0, Kr^2+m)-\xi(0, m) \geqslant 1 \mid r \leqslant \|S_m\| \leqslant r+1\}$$
$$-\mathbb{P}\{\max_{m \leqslant k \leqslant m+Kr^2} \|S_k\| \geqslant rt \mid r \leqslant \|S_m\| \leqslant r+1\} = \mathrm{I}-(1-\mathrm{II}),$$

where

$$\mathrm{I} \approx \log K / \log Kr^2$$

and

$$1-\mathrm{II} \leqslant \mathbb{P}\{\max_{0 \leqslant k \leqslant Kr^2} \|S_k\| \geqslant r(t-1)\} \approx \exp\left(-O\left(\frac{(t-1)^2}{K}\right)\right)$$

provided that $K>400$ is an absolute constant and $t=t(r) \rightarrow \infty$ as $r \rightarrow \infty$. Choosing $t \geqslant (\log_2 r)^{1/2+\delta}$ with some $\delta>0$ we obtain (2.7).

In order to formulate our next lemmas we introduce some further notations. Let

$$\rho_0=0, \qquad \rho_1=\min\{k: k>0, S_k=0\}, \ldots$$
$$\rho_j=\min\{k: k>\rho_{j-1}, S_k=0\} \qquad (j=2, 3, \ldots),$$
$$X_i(r)=\begin{cases} 1 & \text{if} \quad \max_{\rho_{i-1} \leqslant k \leqslant \rho_i} \|S_k\| \geqslant r, \\ 0 & \text{otherwise,} \end{cases}$$
$$Y_n(r)=\sum_{i=1}^n X_i(r),$$
$$Z_n(r)=Y_{\xi(0, n)}(r).$$

Clearly $Y_n(r)$ is the number of those excursions (among the first n) which are going farther than r while $Z_n(r)$ is the same number among the excursions completed before n;

$$\tau_1=\tau_1(r)=\min\{n: \|S_n\| \geqslant r\},$$
$$\tau_2=\tau_2(r, t)=\min\{n: n \geqslant \tau_1, \|S_n\| \geqslant rt\},$$
$$\tau_3=\tau_3(r, t)=\min\{n: n \geqslant \tau_2, \|S_n\| \leqslant r\},$$
$$\cdots\cdots\cdots\cdots\cdots\cdots\cdots\cdots\cdots\cdots$$
$$\tau_{2k}=\tau_{2k}(r, t)=\min\{n: n \geqslant \tau_{2k-1}, \|S_n\| \geqslant rt\},$$
$$\tau_{2k+1}=\tau_{2k+1}(r, t)=\min\{n: n \geqslant \tau_{2k}, \|S_n\| \leqslant r\},$$
$$\Theta_n=\Theta(n; r, t)=\max\{k: \tau_{2k+1} \leqslant n\}.$$

We say that Θ_n is the number of the $r \rightarrow rt$ excursions completed before n.

LEMMA 5. *With probability one for any $\varepsilon > 0$ we have*

$$\frac{\log n}{(\log_2 n)^{1+\varepsilon}} \leqslant \xi(0, n) \leqslant (1+\varepsilon)\, \pi(\log n) \log_3 n$$

for all but finitely many n.

Proof. See Erdös and Taylor [5, Corollary on p. 145 and Theorem 4.C].

LEMMA 6. *Let $r = r_n$ be a sequence of positive numbers with*

$$r_n \nearrow \infty, \qquad \frac{n}{\log r} \geqslant (\log n)^{2+\delta}$$

for some $\delta > 0$. Then for any $\varepsilon > 0$

$$\frac{(1-\varepsilon)\, \pi n}{2 \log r} \leqslant Y_n(r) \leqslant \frac{(1+\varepsilon)\, \pi n}{2 \log r}$$

with probability one for all but finitely many n.

Proof. It is a trivial consequence of Lemma 2.

Lemmas 5 and 6 imply

LEMMA 7. *Let $r = r_n$ be a sequence of positive numbers with*

$$r_n \nearrow \infty, \qquad \frac{\log n}{\log r} > (\log_2 n)^{3+\delta}$$

for some $\delta > 0$. Then for any $\varepsilon > 0$

$$\frac{\log n}{(\log_2 n)^{1+\varepsilon}} \frac{1}{\log r} \leqslant Z_n(r) \leqslant (1+\varepsilon) \frac{\pi^2}{2} \frac{(\log n) \log_3 n}{\log r}$$

with probability one for all but finitely many n.

LEMMA 8. *Let $r = r_n$ be a sequence of positive numbers with*

$$r_n \nearrow \infty, \qquad \frac{\log n}{\log r} > (\log_2 n)^{3+\delta}$$

for some $\delta > 0$. Then for any $\varepsilon > 0$ and for all but finitely many n we have

$$\Theta(n; r, t) \leqslant \varepsilon (\log n) \log_3 n \qquad a.s. \tag{2.8}$$

provided that

$$t \geqslant (\log_2 r)^{1/2+\delta} \qquad \textit{for some} \quad \delta > 0$$

and

$$\Theta(n; r, t) \geqslant \frac{\log n}{(\log_2 n)^{1+\varepsilon}} \frac{1}{\log_3 r} \tag{2.9}$$

provided that

$$t = o((\log_2 r)^{\delta}) \qquad \textit{for all} \quad \delta > 0.$$

Proof. (2.8) follows from Lemmas 4 and 7, (2.9) follows from Lemmas 3 and 7.

Proof of Theorem 1. Let x be an arbitrary point of the circle of radius rt, i.e., $\|x\| \leqslant rt$. Then by (2.1),

$$\begin{aligned} \mathbb{P}\{\xi(x, \tau_{2i-1} + Kr^2t^2) - \xi(x, \tau_{2i-1}) \\ \geqslant 1 \mid S(\tau_{2i-1}(r, t)\} \geqslant \frac{\log K}{\log Kr^2t^2} \qquad \text{a.s.,} \end{aligned} \tag{2.10}$$

provided that $400 \leqslant K \leqslant r^4t^4$. By the law of iterated logarithm one gets that

$$\tau_{(i+1)[(2K\log_2 rt)^{1/2}]}(r, t) - \tau_{i[(2K\log_2 rt)^{1/2}]}(r, t) \geqslant Kr^2t^2. \tag{2.11}$$

Consider the paths

$$\{S_j, \tau_{2i[(2K\log_2 rt)^{1/2}]-1}(r, t) \leqslant j \leqslant \tau_{2i[(2K\log_2 rt)^{1/2}]-1}(r, t) + Kr^2t^2\} \tag{2.12}$$

$$i = 1, 2, 3, ..., \qquad \left[\frac{\log n}{(\log_2 n)^{1+\varepsilon}} \frac{1}{\log_3 r} \frac{1}{(2K\log_2 rt)^{1/2}}\right]$$

and observe that by (2.9) all of these paths are included in the path $\{S_j, 1 \leqslant j \leqslant n\}$. (2.11) implies that the paths (2.12) are disjoint and (2.10) implies that for any x belonging to the circle of radius rt and for any i the probability that the path of (2.12) does not pass through x is less than or equal to

$$1 - \frac{\log K}{\log Kr^2t^2},$$

assuming (2.9) and (2.11).

Consequently assuming again (2.9) and (2.11), the conditional probability that the path $\{S_j,\ 1 \leqslant j \leqslant n\}$ does not pass through x is less than or equal to

$$\left(1-\frac{\log K}{\log Kr^2t^2}\right)^{\log n(\log_2 n)^{-1-\varepsilon}(\log_3 r)^{-1}(2K\log_2 rt)^{-1/2}}$$

$$\leqslant \exp\left(-\frac{\log K \log n}{(\log_2 n)^{1+\varepsilon}\log_3 r(2K\log_2 rt)^{1/2}\log Kr^2t^2}\right)$$

provided that

$$400 \leqslant K \leqslant r^4t^4$$

$$\frac{\log n}{\log r} > (\log_2 n)^{3+\delta} \qquad \text{for some} \quad \delta > 0,$$

$$t = o((\log_2 r)^{\delta}) \qquad \text{for all} \quad \delta > 0.$$

Choosing $K=400$, $t=\log_3 r$, $r=\exp((\log n)^{1/2}\cdot(\log_2 n)^{-(3/4+2\varepsilon)})$, we obtain that the conditional probability that the path does not pass through x is less than or equal to

$$\exp\left(-\frac{(\log n)^{1/2}}{(\log_2 n)^{3/4-\varepsilon}}\right).$$

Consequently the probability that the path does not pass through all points of the circle of radius rt is less than or equal to

$$\exp\left(2\frac{(\log n)^{1/2}}{(\log_2 n)^{3/4+2\varepsilon}}\right)\exp\left(-\frac{(\log n)^{1/2}}{(\log_2 n)^{3/4-\varepsilon}}\right),$$

which easily proves Theorem 1.

3. Circles Covered with Positive Density

Theorem 1 gave a lower estimate of $R(n)$. Unfortunately we do not have any non-trivial upper estimation. The result of Theorem 2 suggests that $R(n)$ can be much bigger. In order to formulate our result, introduce the following notations

$$I(x,n)=\begin{cases}1 & \text{if} \quad \xi(x,n)>0,\\ 0 & \text{if} \quad \xi(x,n)=0,\end{cases} \tag{3.1}$$

$$K(N,n)=(N^2\pi)^{-1}\sum_{x\in Q(N)} I(x,n);$$

i.e., $K(N,n)$ is the density of the points of $Q(N)$ covered by the random walk $\{S_k, 0\leqslant k\leqslant n\}$. We prove

THEOREM 2. *For any* $0<\alpha<1/2$

$$\limsup_{n\to\infty} K(n^{\alpha}, n) \geqslant (1-2\alpha)[1-((1-\alpha)^{-1}-1)^{1/2}] \qquad a.s.$$

The proof is based on the following two lemmas.

LEMMA 9. *Let* $20<\|x\|<n^{1/3}$. *Then*

$$\gamma(x, n)=\frac{2\log\|x\|}{\log n}\left(1+O\left(\frac{\log_3\|x\|}{\log\|x\|}\right)\right). \tag{3.2}$$

Proof. See Erdös and Taylor [5, (2.16)].

LEMMA 10. *We have*

$$\mathbb{E}(I(x, n)\, I(y, n)) \leqslant \frac{(1-\gamma(x-y, n))(1-(\gamma(x, n)+\gamma(y, n))/2)}{1-\gamma(x-y, n)/2}.$$

Proof. For any lattice point z let

$$v_z=\min\{k: k>0, S_k=z\}.$$

Then we have

$$\begin{aligned}
&\mathbb{E}(I(x, n)\, I(y, n))\\
&\quad=\mathbb{P}(I(x, n)=1, I(y, n)=1)\\
&\quad=\sum_{k=0}^{n}\mathbb{P}\{I(x, n)=1, I(y, n)=1 \mid v_x=k<v_y\}\,\mathbb{P}\{v_x=k<v_y\}\\
&\quad\quad+\sum_{k=0}^{n}\mathbb{P}\{I(x, n)=1, I(y, n)=1 \mid v_y=k<v_x\}\,\mathbb{P}\{v_y=k<v_x\}\\
&\quad=\sum_{k=0}^{n}\mathbb{P}\{I(y, n)=1 \mid v_x=k<v_y\}\,\mathbb{P}\{v_x-k<v_y\}\\
&\quad\quad+\sum_{k=0}^{n}\mathbb{P}\{I(x, n)=1 \mid v_y=k<v_x\}\,\mathbb{P}\{v_y-k<v_x\}\\
&\quad=\sum_{k=0}^{n}\mathbb{P}\{I(y-x, n-k)=1\}\,\mathbb{P}\{v_x=k<v_y\}\\
&\quad\quad+\sum_{k=0}^{n}\mathbb{P}\{I(x-y, n-k)=1\}\,\mathbb{P}\{v_y=k<v_x\}\\
&\quad\leqslant\mathbb{P}\{I(x-y, n)=1\}\,\mathbb{P}\left\{\sum_{k=0}^{n}\{\{v_x=k<v_y\}+\{v_y=k<v_x\}\}\right\}\\
&\quad=\mathbb{P}\{I(x-y, n)=1\}\,\mathbb{P}\{I(x, n)=1 \text{ or } I(y, n)=1\}\\
&\quad=\mathbb{P}\{I(x-y, n)=1\}[\mathbb{P}(I(x, n)=1)\\
&\quad\quad+\mathbb{P}(I(y, n)=1)-\mathbb{P}(I(x, n)=1, I(y, n)=1)].
\end{aligned}$$

Hence

$$\mathbb{P}(I(x, n)=1, I(y, n)=1)$$
$$\leqslant \frac{\mathbb{P}(I(x-y, n)=1)[\mathbb{P}(I(x, n)=1)+\mathbb{P}(I(y, n)=1]}{\mathbb{P}(I(x-y, n)=1)+1}$$

and we have the lemma.

Proof of Theorem 2. Apply Lemmas 1 (resp. Lemmas 9 and 10) with

$$\frac{n^\alpha}{\log n} \leqslant \|x\|, \|y\|; \|x-y\| \leqslant n^\alpha \qquad (0<\alpha<\tfrac{1}{2}).$$

We get

$$\mathbb{E}(I(x, n)\, I(y, n)) \leqslant \frac{(1-2\alpha)^2}{(1-\alpha)} \qquad (n \text{ big enough})$$

and

$$\mathbb{E}I(x, n) \approx 1-2\alpha.$$

A simple calculation gives

$$\mathbb{E}(K(n^\alpha, n)-\mathbb{E}K(n^\alpha, n))^2 \leqslant \frac{(1-2\alpha)^2}{1-\alpha}-(1-2\alpha)^2$$

and

$$\mathbb{E}K(n^\alpha, n) \approx 1-2\alpha.$$

Hence by the Chebishev inequality we have

$$\mathbb{P}\{K(n^\alpha, n)>(1-\varepsilon)(1-2\alpha)[1-((1-\alpha)^{-1}-1)^{1/2}]\} \geqslant \delta_\varepsilon>0$$

for any $\varepsilon>0$ if n is big enough. Hence we have Theorem 2.

4. Some Further Problems

In Section 2 we have studied the area of the largest circle around the origin covered by the random walk $\{S_k, k \leqslant n\}$. The analog problem is clearly meaningless since in R^d $(d \geqslant 3)$ the largest covered sphere is finite with probability one. However, one can ask in any dimension about the radius of the largest sphere (not surely around the origin) covered by the random walk in time n. Formally speaking, let

$$Q(N, u) = \{x: \|x-u\| \leqslant N\}$$

and $R^*(n)$ be the largest integer for which there exists a r.v. $u = u(n)$ such that

$$\xi(x, n) \geqslant 1 \qquad \text{if} \quad x \in Q(R^*(n), u).$$

It is trivial to see that in R^d

$$R^*(n) \geqslant \text{Const}(\log n)^{1/d}.$$

However, we do not have any non-trivial estimate.

In case $d=2$ clearly $R^*(n) \geqslant R(n)$. We conjecture that $R^*(n)$ will not be larger than $R(n)$, but cannot settle this question. In fact this question is somewhat related to the problem of favourite values (cf. Bass and Griffin [1], Erdös and Révész [3], (1984), Erdös and Révész [4]).

The analogous question in the case of spheres covered with positive density can be also raised.

We also propose to investigate the area T_n of the smallest convex hull of the path $\{S_k, k \leqslant n\}$. Here we mention only a trivial result,

$$T_n \leqslant 2\pi n \log_2 n \qquad \text{a.s.} \tag{4.1}$$

for all but finitely many n,

$$T_n \geqslant \varepsilon n \log_2 n \qquad \text{a.s.} \qquad \text{i.o.} \tag{4.2}$$

with some suitable $\varepsilon > 0$.

Proof. (4.1) is a trivial consequence of the law of iterated logarithm. Let $S_n = (U_n, V_n)$. Then for any $\varepsilon > 0$

$$\mathbb{P}\{|V_n| \leqslant \varepsilon\sqrt{n}, U_n \geqslant \varepsilon(n \log_2 n)^{1/2}\} = O((\log n)^{-\varepsilon^2/2}).$$

Consider the first crossing of the path after n with the positive y axis assuming that $|V_n| \leqslant \varepsilon\sqrt{n}$, $U_n \geqslant \varepsilon(n \log_2 n)^{1/2}$. Then with a positive probability this crossing point will be farther from the origin than $(\varepsilon/2)(n \log_2 n)^{1/2}$. The time needed to get this point will not be more than n with probability $O((\log n)^{-\varepsilon})$. Hence the path $\{S_k, k \leqslant 2n\}$ meets the points $(\varepsilon(n \log_2 n)^{1/2}, 0)$ and $(0, (\varepsilon/2)(n \log_2 n)^{1/2})$ with probability $O((\log n)^{-2\varepsilon})$. Having this result, (4.2) can be obtained with the usual methods.

Note added in proof. The following result can be obtained trivially:

THEOREM 2*. *For any* $0 < \alpha < 1/2$

$$\limsup_{n \to \infty} K(n^{\alpha}, n) \geqslant 1 - 2\alpha. \qquad a.s.$$

REFERENCES

[1] BASS, R. F., AND GRIFFIN, P. S. (1985). The most visited site of Brownian motion and simple random walk. *Z. Wahrsch. Verw. Gebiete* **70** 417–436.

[2] DVORETZKY, A., AND ERDÖS, P. (1950). Some problems in random walk in space. In *Proceedings, Second Berkeley Symposium*, pp. 353–368.

[3] ERDÖS, P., AND RÉVÉSZ, P. (1984). On the favourite points of a random walk. In Mathematical structures—Computational mathematics— Mathematical modelling Vol. 2. Sofia.

[4] ERDÖS, P., AND RÉVÉSZ, P. (1987). Problems and results of random walk. In *Mathematical Statistics and Probability Theory*, Vol. B (P. Bauer, F. Konecny, and W. Wertz, Eds.), Reidel, Dordrecht.

[5] ERDÖS, P., AND TAYLOR, S. J. (1960). Some problems concerning the structure of random walk paths. *Acta Math. Acad. Sci. Hungar.* **11** 137–162.

Normed Likelihood as Saddlepoint Approximation

D. A. S. FRASER

York University, Toronto. Canada,
University of Toronto, Toronto, Canada, and
University of Waterloo, Waterloo, Canada

Barndorff-Nielsen's formula (normed likelihood with constant-information metric) has been proffered as an approximate conditional distribution for the maximum-likelihood estimate, based on likelihood functions. Asymptotic justifications are available and the formula coincides with the saddlepoint approximation in full exponential models. It is shown that the formula has wider application than is presently indicated, that in local analysis it corresponds to Laplace's method of integration, and that it corresponds more generally to a saddlepoint approximation.

1. INTRODUCTION

The density function for the average $\bar{x}$ of a sample $x_1, ..., x_n$ from a k-variate distribution with known cumulant generating function $K(u)$ can be approximated in terms of simple characteristics of that cumulant generating function. The saddlepoint approximation derived by asymptotic analysis of the cumulant-to-density inversion formula is given by

$$f(\bar{x}) = (2\pi)^{-k/2}[n/|\ddot{K}(\hat{\phi})|]^{1/2} \exp[n(K(\hat{\phi}) - \hat{\phi}'\bar{x})](1 + r_n), \qquad (1.1)$$

where $\hat{\phi} = \hat{\phi}(\bar{x})$, called the saddlepoint, satisfies the saddlepoint equation

$$\dot{K}(\hat{\phi}) = \bar{x}; \qquad (1.2)$$

the cumulant generating function $K(u) = \log M(u)$ is the logarithm of the moment generating function, $\dot{K}(u) = \partial K/\partial u$ is the $k \times 1$ gradient vector and $\ddot{K}(u) = \partial^2 K/\partial u \partial u'$ is the $k \times k$ second derivative matrix; the relative error r_n is $O(n^{-1})$.

Multivariate Statistics and Probability
ISBN 0-12-580205-6

Reprinted from *J. Mult. Anal.* **27**(1).

The univariate version of the saddlepoint was derived by Daniels [5] and the bivariate and multivariate versions by Good [8] and Barndorff-Nielsen and Cox [4]. A comprehensive review of saddlepoint approximations and related statistical inference is given by Reid [10].

The saddlepoint approximation in practice is typically more accurate than the normal approximation or the several-term Edgeworth expansion and often is so accurate as to be indistinguishable from the exact density in a computer plot. It thus seems reasonable to view it as a means to go from an available cumulant generating function to a presumably accurate approximation to the corresponding density. Accordingly we rewrite (1.1) for a variable y with cumulant generating function $H(u)$ (based on the identification $y=\bar{x}$, $H(u)=nK(u/n)$):

$$f(y) \approx (2\pi)^{-k/2} |\ddot{H}(\hat{\phi})|^{-1/2} \exp\{H(\hat{\phi}) - \hat{\phi}' y\}, \tag{1.3}$$

where $\dot{H}(\hat{\phi}) = y$; in effect, this is an $n=1$ version of (1.1). From this present viewpoint we thus treat (1.3) as an empirically based approximation with a good performance record.

We do note as a caution, however, that the asymptotic derivation of the saddlepoint suggests good approximation in normal-like case and perhaps poor approximation far from the normal; thus we would not expect (1.3) to be accurate for a very non-normal distribution such as the uniform (a, b).

The exponential family provides an important extension from the normal; in terms of a natural parameter θ it has density

$$g(x; \theta) = \exp\{\theta' y(x) - \psi(\theta) + h(x)\}, \tag{1.4}$$

where θ and $y(x)$ are k-vectors. The minimal sufficient statistic $y = y(x)$ has cumulant generating function

$$H(u) = \psi(\theta + u) - \psi(\theta). \tag{1.5}$$

The saddlepoint equation for approximating the distribution of y is

$$\dot{\psi}(\theta + \hat{\phi}) = t, \tag{1.6}$$

so that $\hat{\theta} = \theta + \hat{\phi}$ is the maximum likelihood estimate of θ; the saddlepoint approximation is thus

$$f(y) \approx (2\pi)^{-k/2} |\ddot{\psi}(\hat{\theta})|^{-1/2} \exp\{\psi(\hat{\theta}) - \psi(\theta) - (\hat{\theta} - \theta)' y\}. \tag{1.7}$$

As $\ddot{\psi}(\theta) = -\partial^2 \log L(\theta)/\partial\theta\partial\theta' = j(\theta)$ is the observed Fisher information function, we obtain

$$f(y) \approx (2\pi)^{-k/2} |j(\hat{\theta})|^{-1/2} L(\theta)/L(\hat{\theta}), \tag{1.8}$$

where $L(\theta) = L(\theta; y) = f(y; \theta)$, the marginal density of the minimal sufficient statistic y; the approximation uses only a likelihood *ratio* so that

$L(\theta)/L(\hat{\theta}) = f(y;\theta)/f(y;\hat{\theta}) = g(x;\theta)/g(x;\hat{\theta})$ is available from the original density function.

The transformation from y to $\hat{\theta}$ has Jacobian matrix $j(\hat{\theta})$; the density approximation for $\hat{\theta}$ obtained from (1.8) is thus

$$h(\hat{\theta};\theta) \approx (2\pi)^{-k/2}\,|j(\hat{\theta})|^{1/2} L(\theta)/L(\hat{\theta}). \tag{1.9}$$

In the asymptotic context the relative error in (1.9) is $O(n^{-1})$. If the approximation is renormalized

$$h(\hat{\theta};\theta) \approx c\,|j(\hat{\theta})|^{1/2} L(\theta)/L(\hat{\theta}) \tag{1.10}$$

so the right side is a density, the relative error becomes $O(n^{-3/2})$.

The expressions (1.9) and (1.10) involving normed likelihood with respect to the constant-information metric are called Barndorff-Nielsen's formula and were introduced (Barndorff-Nielsen [1]) by an asymptotic argument from which the preceding was derived; the renormalized version (1.10) was also shown to be exact for location and transformation models given the usual conditioning on the Fisher configuration statistic, although for such models the cumulant generating function may not exist.

In Section 2 Barndorff-Nielsen's approximation formula is related to general formulas for exact conditional distributions, and the implicit choice of a Jacobian-type factor in the Barndorff-Nielsen approximation is discussed.

In Section 3 the local form of a density for the maximum-likelihood estimator is examined, and the normed likelihood choice implicit in Barndorff-Nielsen's formula is shown to be in a logical correspondence with the use of Laplace's formula for approximate integration.

In Section 4 a family of saddlepoint approximations for a density function at some point y_0 are discussed. Then in Section 5 a score-based saddlepoint approximation for the density of the maximum likelihood estimator is shown to give Barndorff-Nielsen's formula.

Section 6 contains some concluding remarks; in particular, it is noted that the inversion process from likelihood functions to corresponding density functions is unique, when the statistical model is complete.

2. Barndorff-Nielsen's Formula

Barndorff-Nielsen's [1] formula (1.10) for the distribution of the maximum likelihood estimator $\hat{\theta}$ can be presented as

$$h(\hat{\theta}\,|\,a;\theta)\,d\hat{\theta} \approx c\,\frac{L(\theta;\hat{\theta},a)}{L(\hat{\theta};\hat{\theta},a)}\cdot|j(\hat{\theta},a)|^{1/2}\,d\hat{\theta}, \tag{2.1}$$

where a is some exact or approximate ancillary statistic; in this form it covers the location and transformation model cases which have a standard ancillary statistic a. The choice $c = (2\pi)^{-k/2}$ is indicated by the analysis of the full exponential models as discussed in the Introduction.

The standard context for the formula presupposes a continuous statistical model in which the likelihood function is uniquely determined for each value of the maximum likelihood variable $\hat{\theta}$ under a given value of a. However, in the standard development there is no special guidance for the choice or determination of the conditioning variable a.

The accuracy of (2.1) has been examined asymptotically on the sample space in Barndorff-Nielsen [2, 3] and in terms of cumulants in McCullagh [9].

For the case of a real parameter θ and density $f(y; \theta)$ on an n-dimensional space, an exact formula for the distribution of $\hat{\theta}$ given a *general* $(n-1)$-dimensional statistic a (which determines a curve) is given in Fraser and Reid [6],

$$h(\hat{\theta} \mid a, \theta)\, d\hat{\theta} = c(a, \theta)\, L(\theta; \hat{\theta}, a)\, C(\hat{\theta}, a) \cdot |j(\hat{\theta}, a)|^{1/2}\, d\hat{\theta}, \tag{2.2}$$

where

$$C(\hat{\theta}, a) = \exp\left\{\int^{s} \operatorname{div} v(y)\, ds'\right\} \left|\frac{dS(y; \theta)}{dv(y)}\right| |j(\hat{\theta}, a)|^{1/2}, \tag{2.3}$$

and $c(a, \theta)$ is a normalizing constant, $S(y; \theta)$ is the score function $\partial \log(y; \theta)/\partial\theta$, $v(y)$ is the unit vector tangent to the curve determined by the fixed a at the point y, div $v(y)$ is the divergence $\sum_1^n \partial v_i(y)/\partial y_i$ of the vector field $\{v(y)\}$, $dS(y; \theta)/dv(y)$ is the derivative of $S(y; \theta)$ in the direction $v(y)$, and s designates arc length on the curve for fixed a at the point y. Some current work leads to a generalization of (2.2) for vector θ that uses

$$C(\hat{\theta}, a) = \exp\left\{\int^{s} \mathrm{DIV}\ V(y)\, ds'\right\} |V'(y)\, V(y)|^{-1/2} \left|\frac{\partial S(y; \theta)}{\partial V(y)}\right| |j(\hat{\theta}, a)|^{1/2}, \tag{2.4}$$

where $V(y)$ records k tangent vectors to the $n-k$ dimensional surface $a = \text{constant}$, DIV $V(y)$ is a particular generalization of the divergence, the integral is along a curve from some initial point to the point y on the surface $a = \text{constant}$, and the determinant involves partial derivatives with respect to the vectors in $V(y)$.

Now consider the general formula (2.2) in relation to Barndorff-Nielsen's approximate formula (2.1). If a is ancillary so $c(a, \theta) = c(a)$ then (2.1) involves an implicit choice for the Jacobian-type factor

$$C(\hat{\theta}, a) = 1/L(\hat{\theta}; \hat{\theta}; a). \tag{2.5}$$

This norming of the likelihood $L(\theta; \hat{\theta}, a)$ with respect to its maximum can be interpreted in terms of the approximate density (2.1): as θ varies the maximum of the density function remains constant, where density is examined in the constant information metric. This simple choice for an otherwise difficult Jacobian-type factor has a certain natural appeal, and a clarification of this can be obtained from a local analysis discussed in the next section.

From (2.2) with (2.3) or (2.4) we see that Barndorff-Nielsen's formula provides a valid approximation to the distribution of the maximum likelihood estimate subject only to whatever the support for the approximation (2.5) is. In the next section we present a Laplace integral-approximation justification for (2.5). Higher order calculations can be made which lead to correction terms for the formula (2.1).

In the spirit of the preceding we can comment on the generality of the applicability of the formula (2.1). The formula uses the likelihood function at each value of the variable $\hat{\theta}$. Such a likelihood function can be available, if there is a density function for some initial variable, and a reduction is made to a sufficient statistic, and if then there is an ancillary statistic that complements the maximum likelihood estimate.

For the case of a real parameter θ, differential conditions are discussed in Fraser and Reid [7] for an optimum determination of a conditioning variable a.

3. Maximum Likelihood Estimate: Local Distribution Form

Consider a k-dimensional parameter θ for a statistical model and suppose that the maximum likelihood estimate $\hat{\theta}$ has a continuous distribution and uniquely determines the likelihood function, which we indicate by writing $L(\theta; y) = L(\theta; \hat{\theta})$. In this section we consider how the distribution of $\hat{\theta}$ can be approximated when only a likelihood function $L(\theta; \hat{\theta})$ is available for each value of $\hat{\theta}$. For this we use the general definition of likelihood,

$$L(\theta; y) = L(\theta; y) = c \cdot f(y; \theta), \tag{3.1}$$

which for any given y involves an arbitrary scale factor c; thus only ratios $L(\theta_2; y)/L(\theta_1; y)$ are numerically available.

As discussed in the preceding section this situation can arise if there is a sufficient statistic reduction, or if the maximum likelihood estimate is being examined conditionally given an ancillary, or both; accordingly we omit reference to the ancillary a in the formulas.

From formula (2.2) we have that the probability element for $\hat{\theta}$ has the form

$$h(\hat{\theta}; \theta)\, d\hat{\theta} = cL(\theta; \hat{\theta})\, C(\hat{\theta}) \cdot |j(\hat{\theta})|^{1/2}\, d\hat{\theta}, \tag{3.2}$$

where $L(\theta; \hat{\theta})$ here involves some choice of representative among the θ functions given by (3.1) and the notation is justified by our assumption that the likelihood function is uniquely determined by $\hat{\theta}$. Our concern here is with finding a determination for the factor $C(\hat{\theta})$.

First we make a change of variable in the parameter space so that the observed information determinant is constant. For a real parameter θ let a new parameter η be given by

$$\eta = \int^{\theta} |j(t)|^{1/2}\, dt, \tag{3.3}$$

where the probability integral transformation is used as pattern. In terms of the new parameter η we have constant observed information:

$$j(\hat{\eta}) = j(\hat{\theta}) \left| \frac{d\theta}{d\eta} \right|_{\hat{\eta}}^{2} \equiv 1. \tag{3.4}$$

For a vector parameter θ we seek a new parameter η such that

$$d\hat{\eta} = |j(\hat{\theta})|^{1/2}\, d\hat{\theta}. \tag{3.5}$$

There are many possibilities for this but a simple procedure is to use a modified probability integral transformation radially from some initial point $\theta = \theta_0$, say 0; following Fraser and Reid [7] we define

$$\eta(sv) = v \left\{ k \int_{s0}^{s} |j(tv)|^{1/2} t^{k-1}\, dt \right\}^{1/k}$$

for the value of η at a distance s from $\theta_0 = 0$ in a direction v, where k is the parameter dimension. We then assume that such a reparameterization has been done and use θ now for the new parameter; in terms of this new θ, we have $|j(\hat{\theta})| \equiv 1$.

Second, we investigate the significance of the choice $C(\hat{\theta}) = 1/L(\hat{\theta}; \hat{\theta})$. For this we consider the second-order form of the density function $h(\hat{\theta}; \theta)$ near some $(\hat{\theta}; \theta) = (\theta_0, \theta_0)$, by examining the difference

$$\begin{aligned}
&\log h(\hat{\theta}; \theta) - \log\{C(\hat{\theta})\, L(\hat{\theta}; \hat{\theta})\} - \log c \\
&\quad = \log \left\{ \frac{L(\theta; \hat{\theta})}{L(\hat{\theta}; \hat{\theta})} \right\} \\
&\quad = 0 + l'_{10}(\hat{\theta} - \theta_0) + l'_{01}(\theta - \theta_0) \\
&\qquad + \tfrac{1}{2}(\hat{\theta} - \theta_0)' l_{20}(\hat{\theta} - \theta_0) + (\hat{\theta} - \theta_0)' l_{11}(\theta - \theta_0) \\
&\qquad + \tfrac{1}{2}(\theta - \theta_0)' l_{02}(\theta - \theta_0) + \cdots,
\end{aligned} \tag{3.6}$$

where l_{10}, l_{01} are the $k \times 1$ gradient vectors (with respect to $\hat{\theta}, \theta$) and l_{20}, l_{11}, l_{02} are the $k \times k$ second-derivative matrices of $\log\{L(\theta; \hat{\theta})/L(\hat{\theta}; \hat{\theta})\}$ evaluated at (θ_0, θ_0).

From the definition of $\hat{\theta}$ we have $l_{01}(\hat{\theta}; \hat{\theta}) = 0$ and from the constant maximum of $L(\theta; \hat{\theta})/L(\hat{\theta}; \hat{\theta})$ along $\theta = \hat{\theta}$ we have $l_{10}(\hat{\theta}; \hat{\theta}) + l_{01}(\hat{\theta}; \hat{\theta}) = 0$. If these two properties are used at (θ_0, θ_0) we obtain $l_{10} = l_{01} = 0$. If they are then used at $(\hat{\theta}, \hat{\theta})$ we obtain

$$(\hat{\theta} - \theta_0)' l_{11} + (\hat{\theta} - \theta_0)' l_{02} = 0$$

$$(\hat{\theta} - \theta_0)' l_{20} + (\hat{\theta} - \theta_0)' l_{11} = 0$$

which gives $l_{02} = -l_{11} = l_{20}$. We also have $l_{02} = -j(\theta_0)$. The expression (3.6) can then be rearranged:

$$\log h(\hat{\theta}; \theta) - \log\{C(\hat{\theta})\, L(\hat{\theta}; \hat{\theta})\} = \log c - \tfrac{1}{2}(\hat{\theta} - \theta)' j(\theta_0)(\hat{\theta} - \theta) + \cdots. \quad (3.7)$$

For a similar second-order analysis in a different context, see Fraser and Reid [7].

From (3.7) we now see that the choice $C(\hat{\theta}) = 1/L(\hat{\theta}; \hat{\theta})$ gives the density $h(\hat{\theta}; \theta)$ a location normal form in $(\hat{\theta}, \theta)$ near (θ_0, θ_0):

$$h(\hat{\theta}; \theta) = c \exp\{-\tfrac{1}{2}(\hat{\theta} - \theta)' j(\theta_0)(\hat{\theta} - \theta)\}\{1 + O(|\hat{\theta} - \theta_0|^3, |\theta - \theta_0|^3)\}. \quad (3.8)$$

Thus, to the second order, the density has the $N(\theta; j^{-1}(\theta_0))$ form with inverse variance matrix $j(\theta_0)$ which is constant in that order of expansion. We note that the particular choice of parameterization for θ gives $|j(\theta)| = 1$; thus along the maximum density ridge $\hat{\theta} = \theta$ the "shape" of the inverse variance matrix may change but its determinant remains fixed. The preceding location normal properties are directly linked to the choice $C(\hat{\theta}) = 1/L(\hat{\theta}; \hat{\theta})$.

The density (3.8) based on the choice $C(\hat{\theta}) = 1/L(\hat{\theta}; \hat{\theta})$ has local normal form and the Laplace method of approximate integration based on the second-order approximation gives $c = (2\pi)^{-k/2} |j(\theta_0)|^{1/2} = (2\pi)^{-k/2}$ which is in agreement with the notation c that indicates no θ dependence. In a related way we can see that a different choice for $C(\hat{\theta})$ followed by the Laplace method of integration will give a "constant" c that in fact varies with θ_0: verification by contradiction.

We thus have the interpretation of Barndorff-Nielsen's formula as providing that choice for the Jacobian factor so that the resulting nominal density integrates correctly in accord with the Laplace method for approximate numerical integration.

4. Saddlepoint Approximations

Consider the saddlepoint approximation (1.3) for a density $f(y)$ at some point y_0. In terms of the cumulant generating function $H(u)$ for y we have

$$f(y_0) \approx (2\pi)^{-k/2} |\ddot{H}(\hat{\phi})|^{-1/2} \exp\{H(\hat{\phi}) - \hat{\phi}' y_0\}, \tag{4.1}$$

where $\dot{H}(\hat{\phi}) = y_0$. We can rewrite this in terms of the cumulant generating function $H^0(u) = H(u) - u' y_0$ for the variable $y - y_0$,

$$f(y_0) \approx (2\pi)^{-k/2} |\ddot{H}^0(\hat{\phi})|^{-1/2} \exp\{H^0(\hat{\phi})\}, \tag{4.2}$$

where $\dot{H}^0(\hat{\phi}) = 0$.

One saddlepoint derivation uses an Edgeworth approximation for an exponentially tilted model. If the corresponding exponential family is generated in terms of the variable $y - y_0$ we have

$$f(y;\ \theta) = \exp\{\theta'(y - y_0) - H^0(\theta)\} f(y), \tag{4.3}$$

where the norming constant follows from the cumulant generating property

$$\exp\{H^0(\theta)\} = \int \exp\{\theta'(y - y_0)\} f(y)\, dy; \tag{4.4}$$

the cumulant generating function of $y - y_0$ in this model is $\psi(u) = H^0(\theta + u) - H^0(\theta)$. Let $\hat{\theta}(y)$ be the maximum likelihood estimate in the tilted model $f(y;\theta)$; then $\hat{\theta}(y_0) = \hat{\theta}_0$ is the solution of the score equation

$$\dot{H}(\hat{\theta}) = 0.$$

At $\theta = \hat{\theta}_0$ we have the initial derivatives

$$\psi(0) = 0, \qquad \dot{\psi}(0) = \dot{H}(\hat{\theta}_0), \qquad \ddot{\psi}(0) = \ddot{H}^0(\hat{\theta}_0) \tag{4.5}$$

for the cumulant function of the density of $y - y_0$; it follows that the normal or one-term Edgeworth approximation for the density at $y - y_0 = 0$ is

$$f(y_0; \hat{\theta}_0) \approx (2\pi)^{-k/2} |\ddot{H}^0(\hat{\theta}_0)|^{-1/2} \exp\{0\}, \tag{4.6}$$

which then gives

$$f(y_0) \approx (2\pi)^{-k/2} |\ddot{H}^0(\hat{\theta}_0)|^{-1/2} \exp\{H^0(\hat{\theta}_0)\}, \tag{4.7}$$

where $\dot{H}^0(\hat{\theta}_0) = 0$.

Now suppose we want a saddlepoint approximation for the density $g(x)$

of a variable $x = r(y)$ at the point $x_0 = r(y_0)$. We could proceed directly from the approximation (4.7) for the variable x obtaining

$$g(x_0) \approx (2\pi)^{-k/2} |\ddot{H}^0(\hat{\theta}_0)|^{-1/2} \exp\{H^0(\hat{\theta}_0)\} J(r^{-1}, x_0), \tag{4.8}$$

where $\dot{H}^0(\hat{\theta}_0) = 0$ and

$$J(r^{-1}; x_0) = |\partial r^{-1}(x)/\partial x|_{x_0} \tag{4.9}$$

is the Jacobian of the transformation. Alternatively we could use the cumulant generating function $H_x^0(u)$ for the variable $x - x_0 = r(y) - r(y_0)$,

$$\exp\{H_x^0(u)\} = \int \exp\{u'(r(y) - r(y_0))\} f(y)\, dy, \tag{4.10}$$

and obtain

$$g(x_0) \approx (2\pi)^{-k/2} |\ddot{H}_x^0(\hat{\phi}_0)|^{-1/2} \exp\{H_x^0(\hat{\phi}_0)\}, \tag{4.11}$$

where $\dot{H}_x^0(\hat{\phi}_0) = 0$.

The two methods just described can be combined to produce a saddlepoint approximation to $f(y)$ at y_0 by using the cumulant generating function for $x - x_0 = r(y) - r(y_0)$, for some function $r(y)$:

$$f(y_0) \approx (2\pi)^{-k/2} |\ddot{H}_x^0(\hat{\phi}_0)|^{-1/2} \exp\{H_x^0(\hat{\phi}_0)\} J(r, y_0). \tag{4.12}$$

We can thus have a family of saddlepoint approximations corresponding to a family of alternative transforming variables $r(y)$ that have cumulant generating functions. We examine the choice of a transforming variable in the next section.

5. Normed Likelihood as Saddlepoint Approximation

Consider a variable y that is in one–one correspondence with the maximum likelihood estimate $\hat{\theta}(y)$ of a parameter θ in a statistical model. We suppose, in accord with preceding sections, that the likelihood function $L(\theta, y) = c \cdot f(y;\ \theta)$ is available at each point y, but not the density function itself. This can occur if y is obtained by marginalization under sufficiency, by conditioning under ancillarity, or by both.

For computation we note from the preceding assumptions that the observed information can be written as a function of $\hat{\theta}$:

$$j(\hat{\theta}) = -\partial^2 \ln f(y;\ \theta)/\partial\phi^2|_{\theta = \hat{\theta}(y)}\,.$$

In this section we consider the determination of saddlepoint approximations for the density $f(y; \theta)$ at some point y_0; the available ingredients are taken to be the likelihood function (3.1) at y_0 and the sample space first derivative of the likelihood function at y_0.

First we note that if an approximation is obtained for some parameter value $\theta = \theta_0$ then likelihood modulation extends the approximation to all values for θ:

$$f(y_0; \theta) = \frac{L(\theta; y_0)}{L(\theta_0; y_0)} f(y_0; \theta_0). \tag{5.1}$$

We are thus faced with choosing an appropriate value $\theta = \theta_0$ to use for the initial approximation. Following the implicit rationale for the saddlepoint analysis in Section 1, we choose the maximum likelihood value $\theta_0 = \hat{\theta}(y_0)$.

As indicated in Section 4 a range of possible approximations is available depending on the choice of modified variable $r(y)$ to which the method is applied. Now the derivation of the saddlepoint depends very much on additivity as part of approximating the average (or sum). This argues for using the score function

$$r(y) = S(y; \theta_0) \tag{5.2}$$

in the neighborhood of y_0. We shall make this choice for modified variable, but in fact do so primarily for notational reasons as the method of approximation will be shown to be independent of the choice.

For the change of variable we calculate

$$k(y) = \partial S(y; \theta_0)/\partial y' \tag{5.3}$$

and obtain

$$f(y; \theta) = g(S(y); \theta)\ |k(y)|, \tag{5.4}$$

where $g(S; \theta)$ is the density function for $S(y; \theta_0)$.

We now expand the logarithm of the density $g(S; \theta)$ to the second order in θ at θ_0 and to the first order in $S = S(y; \theta_0)$ at $y = y_0$; in tensor summation notation,

$$\begin{aligned} g(S; \theta) &= g(0; \theta_0) \exp\{a_\alpha S^\alpha + I_{i\alpha}\, \delta^i S^\alpha - \tfrac{1}{2}(j_{ij}\delta^i\delta^j + A_{ij\alpha}\, \delta^i\delta^j S^\alpha) + \cdots\} \\ &= g(0; \theta_0) \exp\{a_\alpha S^\alpha + (I_{i\alpha}\delta^i + \tfrac{1}{2} A_{ij\alpha}\delta^i\delta^j)\, S^\alpha - \tfrac{1}{2} j_{ij}\delta^i\delta^j + \cdots\} \\ &= g(0; \theta_0) \exp\{a'S + \tau'S - \tfrac{1}{2}\tau' j(\theta_0)\,\tau + \cdots\}, \end{aligned} \tag{5.5}$$

where $\delta = \theta - \theta_0$, $I_{i\alpha} = 0$ or 1 according as $i = \alpha$ or $i \neq \alpha$,

$$\begin{aligned} a_\alpha &= \partial \ln g(S; \theta)/\partial S^\alpha |_{0, \theta_0} \\ A_{ij\alpha} &= \partial^3 \ln g(S; \theta)/\partial\theta^i \partial\theta^j \partial S^\alpha |_{0, \theta_0} \end{aligned} \tag{5.6}$$

and

$$\tau^{\alpha} = \delta^{\alpha} + \tfrac{1}{2} A_{ij\alpha} \delta^i \delta^j \tag{5.7}$$

is a quadratic reparameterization in the neighbourhood of $\theta = \theta_0$.

The model (5.5) to the chosen order of expansion coincides with the exponential model

$$c\exp\{a'S + \tau'S - \tfrac{1}{2}\tau' j(\theta_0)\tau + q(\tau)\}, \tag{5.8}$$

where $q(0) = q'(0) = q''(0) = 0$. The saddlepoint approximation for this model at $S = 0$ and $\delta = \tau = 0$ is

$$g(0, \theta_0) \approx (2\pi)^{-k/2} |j(\theta_0)|^{-1/2}. \tag{5.9}$$

It is of interest to note that a range of such exponential models all have the same saddlepoint approximation and one of them is the normal model

$$(2\pi)^{-k/2} |j(\theta_0)|^{-1/2} \exp\{-\tfrac{1}{2}(j^{-1}(\theta_0) S - \tau)' j(\theta_0)(j^{-1}(\theta_0) S - \tau)\} \tag{5.10}$$

for which the approximation (5.9) is obvious.

Now briefly, suppose that some other variable $\tilde{S} = r(S) - r(0)$ is used to examine the exponential models that coincide with the given model to the first order in the variable $\tilde{S}$. Then $dS = Bd\tilde{S}$ at $S = 0$ where B is the Jacobian, and S is replaced by $B\tilde{S}$ in (5.8). The resulting normalization constant in (5.10) is then

$$(2\pi)^{-k/2} |j(\theta_0)|^{-1/2} |B|$$

which is in agreement with the change of probability element

$$\tilde{g}(0; \theta_0)\, d\tilde{S} = g(0; \theta_0)\, |B|\, dS.$$

Thus a change of variable does not affect the effective density approximation implied by (5.9); the use of the score S has the advantages of familiarity.

We can now make the change of variable from $S = S(y; \theta_0)$ to $\hat{\theta}(y)$. The maximum likelihood equation

$$S(y;\ \hat{\theta}(y)) = 0 \tag{5.11}$$

can be differentiated:

$$\frac{\partial S(y, \hat{\theta})}{\partial y}\, dy + \frac{\partial S(y; \hat{\theta})}{\partial \theta}\, d\theta = 0. \tag{5.12}$$

At $y=y_0$ with $\hat{\theta}=\hat{\theta}(y_0)=\theta_0$ we obtain

$$dS(y;\theta_0)-j(\theta_0)\,d\theta=0,$$

giving $dS=|j(\theta_0)|\,d\theta$. Thus the saddlepoint approximation for the density of $\hat{\theta}$ at $\hat{\theta}=\theta_0$ when the parameter $\theta=\theta_0$ is

$$(2\pi)^{-k/2}\,|j(\theta_0)|^{1/2} \tag{5.13}$$

and, for general θ by (5.1), is

$$(2\pi)^{-k/2}\,|j(\theta_0)|^{1/2}\,\frac{L(\theta;y_0)}{L(\theta_0;y_0)}. \tag{5.14}$$

We now rewrite this for an arbitrary point y and obtain the *saddlepoint approximation for the density of* $\hat{\theta}$:

$$h(\hat{\theta};\theta)\approx(2\pi)^{-k/2}\,|j(\hat{\theta}(y))|^{1/2}\,\frac{L(\theta;y)}{L(\hat{\theta}(y);y)}, \tag{5.15}$$

which is Barndorff-Nielsen's formula (1.9).

We can also obtain the *saddlepoint approximation for the original density* $f(y;\theta)$ based on only the likelihood function $L(\theta;y)=cf(y;\theta)$. From (5.9) with (5.4) we obtain

$$f(y;\theta)\approx(2\pi)^{-k/2}\,|j(\hat{\theta}(y))|^{-1/2}\,|k(y)|\,\frac{L(\theta;y)}{L(\hat{\theta}(y);\,y)}. \tag{5.16}$$

6. Remarks

Barndorff-Nielsen's formula (1.9), (1.10) had been proposed as a conditional distribution for a maximum likelihood estimator $\hat{\theta}$ given some approximate ancillary statistic. The conditions under which it can be examined, however, are broader and cover any case where the likelihood function is available marginally or conditionally in unique correspondence with a value of the maximum likelihood statistic.

In this general context the formula can be supported (Section 3) by a local analysis using Laplace's method of approximate integration. It can also be supported as a saddlepoint approximation (Section 5) based on derivatives of the likelihood function. This suggests the use of Barndorff-Nielsen's formula as a likelihood-based alternative to the cumulant-based saddlepoint approximation. A modification of the formula gives an approximate density for a variable y in *one–one correspondence* with the mle $\hat{\theta}$, as determined marginally by sufficiency, conditionally by ancillarity, or by both.

A natural question in relation to Barndorff-Nielsen's formula is whether the availability of the likelihood function at each sample point is enough to determine the statistical model (family of density functions) for the maximum likelihood estimate. The question is whether or not $C(\hat{\theta})$ in (3.2) is uniquely determined by the likelihood functions (3.1) at the various sample points. If $C(\hat{\theta})$ is the factor for the model being examined and $C^*(\hat{\theta}) = C(\hat{\theta})(1 + t(\hat{\theta}))$ is some other factor that produces an alternative statistical model, then $t(\hat{\theta})$ is bounded below and is an unbiased estimate of zero for the statistical model being examined. Thus the factor $C(\hat{\theta})$ is uniquely determined if and only if the statistical model is one-sided boundedly complete; it follows that completeness guarantees a unique $C(\hat{\theta})$. The Barndorff-Nielsen choice can thus be viewed as a first-order determination of this unique $C(\hat{\theta})$, as based on the viewpoints in Sections 3 and 5.

ACKNOWLEDGMENT

The author thanks Professor N. Reid for benefits of joint work on background material and for many fruitful discussions of the present material.

REFERENCES

[1] BARNDORFF-NIELSEN, O. E. (1983). On a formula for the distribution of the maximum likelihood estimator. *Biometrika* **70** 343–365.

[2] BARNDORFF-NIELSEN, O. E. (1986). Inference on full or partial parameters based on the standardized signed log lokelihood ratio. *Biometrika* **73** 307–322.

[3] BARNDORFF-NIELSEN, O. E. (1986). Likelihood and observed geometries. *Ann. Statist.* **14** 856–873.

[4] BARNDORFF-NIELSEN, O. E., AND COX, D. R. (1979). Edgeworth and saddlepoint approximations with statistical applications (with discussion). *J. Roy. Statist. Soc. Ser. B* **41** 279–312.

[5] DANIELS, H. E. (1954). Saddlepoint approximations in statistics. *Ann. Math. Statist.* **25** 631–50.

[6] FRASER, D. A. S., AND REID, N. (1988). On conditional inference for a real parameter: A differential approach on the sample space. *Biometrika* **75**, in press.

[7] FRASER, D. A. S. AND REID, N. (1988). On comparing two methods for approximate conditional inference. *Statist. Papers*, under review.

[8] GOOD, I. J. (1961). The multivariate saddlepoint method and chi-squared for the multinomial. *Ann. Math. Statist.* **32** 535–548.

[9] MCCULLAGH, P. (1984). Local sufficiency, *Biometrika* **71** 233–244.

[10] REID, N. (1988). Saddlepoint methods and statistical inference. *Statist. Sci.* **3**, in press.

Non-uniform Error Bounds for Asymptotic Expansions of Scale Mixtures of Distributions

Y. FUJIKOSHI

Hiroshima University, Hiroshima, Japan

Let $X = \sigma Z$ be the scale mixture of Z with the scale factor $\sigma > 0$. We consider two type expansions $G_{\delta,k}(x)$ and $\Phi_{\delta,k}(x)$ as the approximations to the distribution function $F(x)$ of X. In this paper we derive non-uniform error bounds in approximating $F(x)$ by the asymptotic expansions $G_{\delta,k}(x)$ and $\Phi_{\delta,k}(x)$. The non-uniform bounds are improvements on the uniform bounds in the tail part of the distribution. The results are applied to the asymptotic expansions of t- and F-distributions.

1. INTRODUCTION

Let Z and σ be independent random variables and suppose that $\sigma > 0$ with probability 1. Then $X = \sigma Z$ is said to be a scale mixture of Z with the scale factor σ. The distribution function of X can be expressed as

$$F(x) = E_\sigma\{G(\sigma^{-1}x)\},$$

where $G(x)$ is the distribution function of Z. We are interested in the asymptotic approximations to $F(x)$ in the situation where σ tends to 1. The uniform error bounds in the case when we approximate $F(x)$ by $G(x)$ have been studied by Heyde [7], Heyde and Leslie [8], Hall [5], etc., assuming that Z is distributed as $N(0, 1)$ or the exponential distribution. Recently the following two types of refinements have been considered under the appropriate assumptions on the smoothness of $G(x)$ and the moments of σ:

$$\text{(i)} \qquad G_{\delta,k}(x) = \sum_{j=0}^{k-1} \frac{1}{j!} b_{\delta,j}(x)\, E(\sigma^\delta - 1)^j, \tag{1.1}$$

$$\text{(ii)} \qquad \Phi_{\delta,k}(x) = \sum_{j=0}^{k-1} \frac{1}{j!} a_{\delta,j}(x)\, E(\sigma^{2\delta} - 1)^j, \tag{1.2}$$

Multivariate Statistics and Probability
ISBN 0-12-580205-6

Reprinted from *J. Mult. Anal.* **27**(1).

where $\delta = -1$ or 1. Here it is assumed that the distribution of Z is symmetric about 0 for the second type expansion. If $E(\sigma^{\delta} - 1)^j$ or $E(\sigma^{2\delta} - 1)^j$ is $O(n^{-j})$, the approximation (1.1) or (1.2) is an asymptotic expansion up to the order of $n^{-(k-1)}$. The uniform error bounds for these two types of approximations have been obtained by Fujikoshi [2, 3], Fujikoshi and Shimizu [4], Shimizu [10, 11]. The results have been applied to obtain the error bounds for the asymptotic expansions of t- and F-distributions.

In this paper we refine the uniform error bounds on $|F(x) - G_{\delta,k}(x)|$ or $|F(x) - \Phi_{\delta,k}(x)|$, to reflect dependency on x as well as the moments of σ. In this direction we consider the bounds for

$$\sup_x (1 + |x|^l) |F(x) - G_{\delta,k}(x)| \tag{1.3}$$

and

$$\sup_x (1 + |x|^l) |F(x) - \Phi_{\delta,k}(x)|. \tag{1.4}$$

In general, the non-uniform bounds are improvements on the uniform bounds in the tail part of the distribution of X. It may be noted that the order of (1.3) or (1.4) is known (Bhattacharya and Ranga Rao [1], Hall and Nakata [6], etc.) for asmptotic expansions of the distribution functions of sums of i.i.d. random variables, but its explicit bound is not known. Error bounds for (1.3) and (1.4) are, respectively, given in Sections 3 and 4. In Section 4 we apply our results to the asymptotic expansions of t- and F-distributions.

2. Scale Mixture of a General Distribution

We assume that the support of the distribution of Z is $\Omega = (0, \infty)$ or $(-\infty, \infty)$. The approximation (1.1) with $\delta = -1$ or 1 is based on the following Taylor's expansion of $G(\sigma^{-1}x)$,

$$\begin{aligned} G(\sigma^{-1}x) &= \sum_{j=0}^{k-1} \frac{1}{j!} b_{\delta,j}(x)(\sigma^{\delta} - 1)^j + \Delta_{\delta,k}(x, \sigma) \\ &= G_{\delta,k}(x, \sigma) + \Delta_{\delta,k}(x, \sigma), \end{aligned} \tag{2.1}$$

where

$$b_{\delta,j}(x) = (\partial^j/\partial s^j)\, G(s^{-\delta}x) |_{s=1}, \tag{2.2}$$

$$\Delta_{\delta,k}(x, \sigma) = \frac{1}{k!} (\sigma^{\delta} - 1)^k (\partial^k/\partial s^k)\, G(s^{-\delta}x) |_{s=1+\theta\delta(\sigma^{\delta}-1)} \tag{2.3}$$

and $0<\theta_\delta<1$. In order to obtain the expansion (2.1) and its error estimate, we make the following assumption for some integers $k>0$ and $l \geqslant 0$:

ASSUMPTION 1. *$G(x)$ is k times continuously differentiable on Ω and*

$$\bar{b}_{\delta,k}(l) = \sup_{x \in \Omega} (1+|x|^l)\, |b_{\delta,k}(x)| < \infty. \tag{2.4}$$

The following lemma is fundamental in our error estimates.

LEMMA 2.1. *Letting $\xi_{\delta,k}(x, \sigma, l) = (1+|x|^l)\, \Delta_{\delta,k}(x, \sigma)$, it holds that*

$$\begin{aligned} |\xi_{\delta,k}(x, \sigma, l)| &\leqslant \frac{1}{k!}\, \bar{b}_{\delta,k}(l)(1 \vee \sigma^l)(\sigma \vee \sigma^{-1} - 1)^k \\ &\leqslant \frac{1}{k!}\, \bar{b}_{\delta,k}(l)\{\sigma^l\, |\sigma - 1|^k + |\sigma^{-1} - 1|^k\}, \end{aligned} \tag{2.5}$$

where $\sigma \vee \sigma^{-1} = \mathrm{Max}(\sigma, \sigma^{-1})$.

Proof. Noting that $s^j(\partial^j/\partial s^j)\, G(s^{-\delta}x)$ is a function of $s^{-\delta}x$, we have

$$\begin{aligned} \xi_{\delta,k}(x, \sigma, l) = \frac{1}{k!}\, &[1+|t|^l\, \{1+\theta_\delta(\sigma^\delta - 1)\}^{\delta l}] \\ &\times b_{\delta,k}(t)\{1+\theta_\delta(\sigma^\delta - 1)\}^{-\delta k}(\sigma^\delta - 1)^k, \end{aligned}$$

where $t = \{1+\theta_\delta(\sigma^\delta - 1)\}^{-\delta}x$. It is easy to see that

$$1+\theta_\delta(\sigma^\delta - 1) \leqslant \begin{cases} \sigma^\delta, & \sigma^\delta \geqslant 1. \\ 1, & 0<\sigma^\delta<1, \end{cases}$$

and hence

$$1+|t|^l\, \{1+\theta_\delta(\sigma^\delta - 1)\}^{\delta l} \leqslant (1+|t|^l)(1 \vee \sigma^l).$$

Using these inequalities, we obtain the desired result.

In order to obtain the expansion

$$\begin{aligned} G_{\delta,k}(x) &= E_\sigma[G_{\delta,k}(x, \sigma)] \\ &= \sum_{j=0}^{k-1} \frac{1}{j!}\, b_{\delta,j}(x)\, E(\sigma^\delta - 1)^j \end{aligned} \tag{2.6}$$

and its error estimate, we make the following assumption:

ASSUMPTION 2. $E(\sigma^{l+k})<\infty$, $E(\sigma^{-k})<\infty$.

THEOREM 2.1. *Suppose that $X=\sigma Z$ is a scale mixture of Z. Then, under Assumptions* 1 *and* 2,

$$\begin{aligned}\sup_x\,(1+|x|^l)\,|F(x)-G_{\delta,k}(x)| &\leq \frac{1}{k!}\,\bar{b}_{\delta,k}(l)\,E\{(1\vee\sigma^l)(\sigma\vee\sigma^{-1}-1)^k\}\\ &\leq \frac{1}{k!}\,\bar{b}_{\delta,k}(l)\,E\{\sigma^l\,|\sigma-1|^k+|\sigma^{-1}-1|^k\}.\end{aligned}\tag{2.7}$$

Proof. We can write

$$\begin{aligned}|(1+|x|^l)(F(x)-G_{\delta,k}(x))| &= |E_\sigma\{\xi_{\delta,k}(x,\sigma)\}|\\ &\leq E_\sigma\{|\xi_{\delta,k}(x,\sigma)|\}.\end{aligned}$$

Therefore, using Lemma 2.1 and Assumption 2 we have the desired result.

From (2.7) we have

$$|F(x)-G_{\delta,k}(x)|\leq(1+|x|^l)^{-1}\frac{1}{k!}\,\bar{b}_{\delta,k}(l)\,E\{\sigma^l\,|\sigma-1|^k+|\sigma^{-1}-1|^k\}.\tag{2.8}$$

In a special case of $l=0$,

$$\sup_x|F(x)-G_{\delta,k}(x)|\leq\frac{1}{k!}\,\bar{b}_{\delta,k}E\{|\sigma-1|^k+|\sigma^{-1}-1|^k\},\tag{2.9}$$

where $\bar{b}_{\delta,k}=\frac{1}{2}\bar{b}_{\delta,k}(0)$. This uniform error bounds in the cases of $\delta=-1$ and $\delta=1$ were obtained by Fujikoshi [3] and Fujikoshi and Shimizu [4], respectively. In the comparison with the upper error bounds (2.8) and (2.9), we can say that (2.8) is better than (2.9) if x satisfies

$$|x|^l\geq\frac{\bar{b}_{\delta,k}(l)\,E\{\sigma^l\,|\sigma-1|^k+|\sigma^{-1}-1|^k\}}{\bar{b}_{\delta,k}E\{|\sigma-1|^k+|\sigma^{-1}-1|^k\}}-1.\tag{2.10}$$

So, the error bound (2.8) gives an improvement on (2.9) in the tail part of the distribution of X.

3. Scale Mixtures of a Symmetric Distribution

Suppose that the distribution of Z is symmetric about 0, i.e., $1-G(x)=G(-x)$. It is possible to apply Theorem 2.1 to the distribution of X in this symmetric case. However, the result is not very useful for t-distribution. Here, we consider non-uniform error bounds for the second type of approximation (1.2) that are useful for t-distribution. We can write

$$F(x)=E_\sigma\{\tfrac{1}{2}+\tfrac{1}{2}\,\mathrm{sgn}(x)\,\tilde{G}(\sigma^{-2}x^2)\}, \tag{3.1}$$

where $\mathrm{sgn}(x)=1$ if $x>0$, $=0$ if $x=0$ and $=-1$ if $x<0$, and $\tilde{G}$ is the distribution function of Z^2. Using this relation and considering Taylor's expansions of $\tilde{G}(\sigma^{-2}x^2)$ we have

$$\tfrac{1}{2}+\tfrac{1}{2}\,\mathrm{sgn}(x)\,\tilde{G}(\sigma^{-2}x^2)=\Phi_{\delta,k}(x,\sigma)+\tfrac{1}{2}\,\mathrm{sgn}(x)\,\tilde{\Delta}_{\delta,k}(x^2,\sigma^2), \tag{3.2}$$

where

$$\Phi_{\delta,k}(x,\sigma)=\sum_{j=0}^{k-1}\frac{1}{j!}\,a_{\delta,j}(x)(\sigma^{2\delta}-1)^j \tag{3.3}$$

and

$$a_{\delta,j}(x)=\begin{cases} G(x), & j=0,\\ \tfrac{1}{2}\,\mathrm{sgn}(x)\,\tilde{b}_{\delta,j}(x^2), & j=1,\dots,k.\end{cases} \tag{3.4}$$

Here we use the same notations as the ones used for G in Section 2. So, the expressions $\tilde{b}_{\delta,j}$ and $\tilde{\Delta}_{\delta,k}$ are defined in the same way as the ones for G. In order to obtain the expansion (3.2) with $\delta=-1$ or 1 and its error estimate, we make the following assumption for some integers $k>0$ and $l\geqslant 0$:

Assumption 3. *The distribution function $\tilde{G}$ of Z^2 is k times continuously differentiable on $(-\infty,\infty)$ and*

$$\bar{a}_{\delta,k}(l)=\sup_x\,(1+|x|^l)\,|a_{\delta,k}(x)|<\infty. \tag{3.5}$$

Let

$$\eta_{\delta,k}(x,\sigma,l)=\tfrac{1}{2}\,\mathrm{sgn}(x)(1+|x|^l)\,\tilde{\Delta}_{\delta,k}(x^2,\sigma^2). \tag{3.6}$$

Then, $\tilde{\Delta}_{\delta,k}$ has the same properties as $\Delta_{\delta,k}$. Therefore, we have the following lemma:

LEMMA 3.1. *Under Assumption 3 it holds that*

$$|\eta_{\delta,k}(x, \sigma, l)| \leqslant \frac{1}{k!} \bar{a}_{\delta,k}(l)(1 \vee \sigma^l)(\sigma^2 \vee \sigma^{-2} - 1)^k$$
$$\leqslant \frac{1}{k!} \bar{a}_{\delta,k}(l)\{\sigma^l |\sigma^2 - 1|^k + |\sigma^{-2} - 1|^k\}. \tag{3.7}$$

In order to obtain the expansion

$$\Phi_{\delta,k}(x) = E_\sigma[\Phi_{\delta,k}(x, \sigma)]$$
$$= \sum_{j=0}^{k-1} \frac{1}{j!} a_{\delta,k}(x)\, E(\sigma^{2\delta} - 1)^j \tag{3.8}$$

and its error estimate, we make the following assumption:

ASSUMPTION 4. $E(\sigma^{l+2k}) < \infty$, $E(\sigma^{-2k}) < \infty$.

From (3.1), (3.2), and (3.8) we have

$$(1 + |x|^l)(F(x) - \Phi_{\delta,k}(x)) = E_\sigma[\eta_{\delta,k}(x, \sigma, l)]. \tag{3.9}$$

Therefore, using Lemma 3.1, we have the following theorem:

THEOREM 3.1. *Suppose that $X = \sigma Z$ is a scale mixture of a symmetric random variable Z. Then, under Assumptions 3 and 4, we have*

$$\sup_x (1 + |x|^l)\, |F(x) - \Phi_{\delta,k}(x)|$$
$$\leqslant \frac{1}{k!} \bar{a}_{\delta,k}(l)\, E\{(1 \vee \sigma^l)(\sigma^2 \vee \sigma^{-2} - 1)^k\}$$
$$\leqslant \frac{1}{k!} \bar{a}_{\delta,k}(l)\, E\{\sigma^l |\sigma^2 - 1|^k + |\sigma^{-2} - 1|^k\}. \tag{3.10}$$

Letting $l = 2h$ and $l = 0$ in (3.10), we have

$$|F(x) - \Phi_{\delta,k}(x)| \leqslant (1 + x^2)^{-1} \frac{1}{k!} \bar{a}_{\delta,k}(2h)\, E\{\sigma^{2h} |\sigma^2 - 1|^k + |\sigma^{-2} - 1|^k\} \tag{3.11}$$

and

$$\sup_x |F(x) - \Phi_{\delta,k}(x)| \leqslant \frac{1}{k!} \bar{a}_{\delta,k}\, E\{|\sigma^2 - 1|^k + |\sigma^{-2} - 1|^k\}, \tag{3.12}$$

where $\bar{a}_{\delta,k} = \frac{1}{2}\bar{a}_{\delta,k}(0)$. We can write

$$\bar{a}_{\delta,k}(2h) = \frac{1}{2} \sup_{x>0} (1+x^h)\, \tilde{b}_{\delta,k}(x). \tag{3.13}$$

The uniform error bounds (3.12) in the cases of $\delta = -1$ and $\delta = 1$ were obtained by Fujikoshi [3] and Fujikoshi and Shimizu [4], respectively. The non-uniform error bound (3.11) is better than the uniform error bound (3.12) if x satisfies

$$|x|^{2h} \geqslant \frac{\bar{a}_{\delta,k}(2h)\, E\{\sigma^{2h}\, |\sigma^2 - 1|^k + |\sigma^{-2} - 1|^k\}}{\bar{a}_{\delta,k} E\{|\sigma^2 - 1|^k + |\sigma^{-2} - 1|^k\}} - 1. \tag{3.14}$$

4. Applications

4.1. *t-Distribution*

The t-distribution of n degrees of freedom is defined as the distribution of a scale mixture $T_n = (\chi_n^2/n)^{-1/2} Z$, where Z is the standard normal variable and χ_n^2 is the chi-square variable with n degrees of freedom. Our interest is to find non-uniform error bounds for well-known asymptotic expansions (see, e.g., Johnson and Kotz [9]) of the distribution function $F(x)$ of T_n. Let the pdf and the cdf of the standard normal variable denote by $\phi(x)$ and $\Phi(x)$, respectively. Then it is known (Fujikoshi [3], Fujikoshi and Shimizu [4]) that

$$\begin{aligned} a_{1,j}(x) &= -2^{-j} H_{2j-1}(x)\, \phi(x), \\ a_{-1,j}(x) &= (-1)^{j-1}\, 2^{-j} \left\{ x^{2j-1} + \sum_{i=1}^{j-1} 1 \cdot 3 \cdots (2i-1) \binom{j-1}{i} x^{2j-2i-1} \right\} \phi(x), \end{aligned} \tag{4.1}$$

where $H_j(x)$ is the Hermite polynomial defined by

$$(d^j/dx^j)\, \phi(x) = (-1)^j\, H_j(x)\, \phi(x).$$

For nonnegative integers j and l and $U = \chi_n^2/n$, let

$$\begin{aligned} q_j &= E(U-1)^j, \\ r_j(l) &= E\{U^{-l}(U^{-1} - 1)^j\}, \end{aligned} \tag{4.2}$$

with $r_j = r_j(0)$. The quantities q_j's exist for any j, but the quantities $r_j(l)$ exist for $n - 2l - 2j > 0$. For $j = 1, 2, \ldots, 6$,

$$q_1 = 0, \qquad q_2 = 2/n, \qquad q_3 = 8/n^2, \qquad q_4 = 12(1 + 4n^{-1})/n^2,$$

$$q_5 = 32(5 + 12n^{-1})/n^3, \qquad q_6 = 20(1 + 12n^{-1} + 32n^{-2})/n^3,$$

$$r_1(l) = 2(l+1)\, n^l/N_{l+1},$$

$$r_2(l) = 2\{n + 2(l+1)(l+2)\}\, n^l/N_{l+2},$$

$$r_3(l) = 4\{(3l+7)\, n + 2(l+1)(l+2)(l+3)\}\, n^l/N_{l+3},$$

$$r_4(l) = 4\{3n^2 + 4(3l^2 + 17l + 23)\, n$$
$$+ 4(l+1)(l+2)(l+3)(l+4)\}\, n^l/N_{l+4},$$

$$r_5(l) = 8\{5(3l+11)\, n^2 + 4(5l^3 + 50l^2 + 160l + 163)\, n$$
$$+ 4(l+1)(l+2)(l+3)(l+3)(l+4)(l+5)\}\, n^l/N_{l+5}.$$

$$r_6(l) = 8\{15n^3 + 10(9l^2 + 75l + 152)\, n^2$$
$$+ 4(15l^4 + 230l^3 + 1275l^2 + 3016l + 2556)\, n$$
$$+ 4(l+1)(l+2)(l+3)(l+4)(l+5)(l+6)\}\, n^l/N_{l+6},$$

where $N_j = (n-2)(n-4)\cdots(n-2j)$. Using Theorem 3.1 with the replacement of $l \to 2l$ we have that if $n - 2l - 2k > 0$ and k is even,

$$|F(x) - \Phi_{\delta,k}(x)| \leqslant (1 + x^{2l})^{-1} \frac{1}{k!} \bar{a}_{\delta,k}(2l)\{r_k(l) + q_k\}. \tag{4.3}$$

The first three approximations $\Phi_{\delta,k}(x)$ are given as

$$\Phi_{-1,2}(x) = \Phi(x),$$

$$\Phi_{-1,4}(x) = \Phi_{-1,2}(x) + \phi(x)[-\tfrac{1}{4}n^{-1}(x^3 + x)$$
$$+ \tfrac{1}{6}n^{-2}(x^5 + 2x^3 + 3x)],$$

$$\Phi_{-1,6}(x) = \Phi_{-1,4}(x) + \phi(x)[-\tfrac{1}{32}(1 + 4n^{-1})\, n^{-2}$$
$$\times (x^7 + 3x^5 + 9x^3 + 15x) + \tfrac{1}{120}(5 + 12n^{-1})\, n^{-3}$$
$$\times (x^9 + 4x^7 + 18x^5 + 60x^3 + 105x)],$$

$$\Phi_{1,2}(x) = \Phi(x) - N_1^{-1}\phi(x)\, x,$$

$$\Phi_{1,4}(x) = \Phi_{1,2}(x) - \phi(x)[\tfrac{1}{4}(n+4)\, N_2^{-1}(x^3 - 3x)$$
$$+ \tfrac{1}{12}(7n + 12)\, N_3^{-1}(x^5 - 10x^3 + 15x)].$$

$$\Phi_{1,6}(x) = \Phi_{1,4}(x) - \phi(x)[\tfrac{1}{4}(3n^2 + 92n + 96)\, N_4^{-1}$$
$$\times (x^7 - 21x^5 + 105x^3 - 105x) + \tfrac{1}{4}(55n^2 + 652n + 480)$$
$$\times N_5^{-1}(x^9 - 36x^7 + 378x^5 - 1260x^3 + 945x)].$$

The numerical values of $\bar{a}_{\delta,k}(2l)/k!$ are given for $k=2$, 4, 6 and $l=0$, 1 as follows:

δ	$k=2$		$k=4$		$k=6$	
	$l=0$	$l=1$	$l=0$	$l=1$	$l=0$	$l=1$
-1	0.158	0.339	0.100	0.384	0.076	0.422
1	0.138	0.129	0.074	0.077	0.050	0.049

4.2 *F-Distribution*

Let χ_q^2 and χ_n^2 be mutually independent chi-square variables with q and n degrees of freedom, respectively. Then, the distribution function of $(\chi_n^2/n)^{-1}(\chi_q^2/q)$ can be expressed as

$$E_\sigma\{G(\tfrac{1}{2}\sigma^{-1}xq;\, \tfrac{1}{2}q)\},$$

where $\sigma=(\chi_n^2/n)^{-1}$ and $G(x;\lambda)$ is the cdf of the gamma distribution with the pdf $g(x;\lambda)=x^{\lambda-1}e^{-x}/\Gamma(\lambda)$, if $x>0$ and $=0$, if $x\leqslant 0$. Therefore, we may consider the distribution of $X=\sigma Z$ with $Z\equiv$ the gamma random variable and $\sigma=(\chi_n^2/n)^{-1}$ instead of the F-distribution. Our interest is to find non-uniform error bounds for asymptotic expansions of the distribution function $F(x;\lambda)$ of X when λ is fixed and n is large. It is known (Fujikoshi [3], Fujikoshi and Shimizu [4]) that the expansions (2.6) can be expressed as

$$G_{\delta,k}(x;\lambda)=\sum_{j=0}^{k-1}\frac{1}{j!}\,b_{\delta,j}(x;\lambda)\,E(U^{-\delta}-1)^j, \tag{4.4}$$

where $U=\chi_n^2/n$,

$$\begin{aligned} b_{1,j}(x;\lambda)&=-xL_{j-1}^{(\lambda)}(x)\,g(x;\lambda),\\ b_{-1,j}(x;\lambda)&=(-1)^{j-1}\,x\tilde{L}_{j-1}^{(\lambda)}(x)\,g(x;\lambda). \end{aligned} \tag{4.5}$$

Here $L_p^{(\lambda)}(x)$ is the Laguerre polynomial defined by

$$L_p^{(\lambda)}(x)=(-1)^p\,x^{-\lambda}e^x(d^p/dx^p)(x^{p+\lambda}e^{-x})$$

and

$$\tilde{L}_p^{(\lambda)}(x)=x^p+\sum_{i=1}^{p}(1-\lambda)\cdots(i-\lambda)\binom{p}{i}x^{p-i}.$$

Using Theorem 2.1 we have that if $n-2l-2k>0$ and k is even,

$$|F(x;\lambda)-G_{\delta,k}(x;\lambda)|\leqslant(1+|x|^l)^{-1}\,\bar{b}_{\delta,k}(l;\lambda)\{r_k(l)+q_k\}, \tag{4.6}$$

for any positive x, where

$$\bar{b}_{\delta,k}(l;\lambda) = \sup_{x>0} |(1+x^l)\, b_{\delta,k}(x;\lambda)|. \tag{4.7}$$

We can see that

$$(d/dx)\{1+x^l\}\, b_{\delta,k}(x;\lambda) = (-1)^{(1-\delta)k/2}\, g(x;\lambda)\, D_{\delta,k}(x;\lambda), \tag{4.8}$$

where

$$D_{1,k}(x;\lambda) = (1+x^l)\, L_k^{(\lambda-1)}(x) - lx^l L_{k-1}^{(\lambda)}(x),$$
$$D_{-1,k}(x;\lambda) = (1+x^l)\, \tilde{L}_k^{(\lambda+1)}(x) - lx^l \tilde{L}_{k-1}^{(\lambda)}(x).$$

Since $D_{\delta,k}(x;\lambda)$ are polynomials of degree $k+l$ in x, we can obtain the numerical values of $\bar{b}_{\delta,k}(l;\lambda)$ by computing the values of $|(1+x^l)\, b_{\delta,k}(x;\lambda)|$ on the set of positive roots of $D_{\delta,k}(x;\lambda)=0$. The numerical values of $b_{1,k}(1;\lambda)/k!$ and $b_{-1,k}(1;\lambda)/k!$ for $k=1(1)\,6$ and $\lambda=0.5(0.5)\,10$ are given in Tables I and II.

TABLE I

The Values of $\bar{b}_{1,k}(1;\lambda)/k!$ for $k=1(1)6$ and $\lambda=0.5(0.5)$

k \ λ	1	2	3	4	5	6
0.5	0.415	0.184	0.122	0.096	0.078	0.065
1.0	0.840	0.388	0.330	0.273	0.227	0.192
1.5	1.31	0.658	0.634	0.546	0.465	0.400
2.0	1.81	1.02	1.04	0.934	0.816	0.765
2.5	2.36	1.44	1.57	1.45	1.30	1.32
3.0	2.95	1.94	2.21	2.12	1.95	2.10
3.5	3.57	2.49	2.98	2.96	2.81	3.17
4.0	4.22	3.12	3.89	3.98	3.97	4.58
4.5	4.91	3.80	4.95	5.21	5.42	6.41
5.0	5.63	4.56	6.15	6.65	7.22	8.71
5.5	6.38	5.37	7.51	8.33	9.39	11.56
6.0	7.16	6.25	9.03	10.27	11.99	15.04
6.5	7.97	7.20	10.72	12.49	15.06	19.24
7.0	8.80	8.21	12.58	14.99	18.63	22.25
7.5	9.66	9.28	14.63	17.80	22.77	30.15
8.0	10.55	10.41	16.85	20.94	27.52	37.05
8.5	11.46	11.61	19.27	24.42	32.94	45.07
9.0	12.39	12.87	21.88	28.26	39.07	54.28
9.5	13.35	14.20	24.69	32.49	45.96	64.87
10.0	14.33	15.58	27.71	37.10	53.68	76.83

TABLE II

The Values of $\bar{b}_{-1,k}(1; \lambda)/k!$ for $k = 1(1)6$ and $\lambda = 0.5(0.5)10$

k \ λ	1	2	3	4	5	6
0.5	0.415	0.405	0.415	0.428	0.442	0.456
1.0	0.840	0.907	0.982	1.06	1.13	1.19
1.5	1.31	1.51	1.70	1.89	2.07	2.24
2.0	1.81	2.20	2.58	2.94	3.30	3.65
2.5	2.36	2.99	3.61	4.22	4.83	5.44
3.0	2.95	3.87	4.80	5.74	6.69	7.65
3.5	3.57	4.84	6.14	7.49	8.89	10.33
4.0	4.22	5.89	7.65	9.50	11.46	13.51
4.5	4.91	7.03	9.31	11.78	14.41	17.23
5.0	5.63	8.25	11.14	14.32	17.78	21.53
5.5	6.38	9.55	13.14	17.15	21.58	26.46
6.0	7.16	10.94	15.30	20.26	25.84	32.05
6.5	7.97	12.41	17.63	23.67	30.57	38.34
7.0	8.80	13.96	20.13	27.39	35.79	45.38
7.5	9.66	15.59	22.81	31.43	41.54	53.22
8.0	10.55	17.29	25.66	35.79	47.82	61.89
8.5	11.46	19.08	28.69	40.48	54.66	73.10
9.0	12.39	20.94	31.89	45.51	62.09	87.89
9.5	13.35	22.88	35.28	50.90	70.13	104.7
10.0	14.33	24.90	38.84	56.64	78.79	125.5

Acknowledgments

The author wishes to thank Dr. Sadanori Konishi, The Institute of Statistical Mathematics, for his helpful comments on an earlier version of this paper. Thanks are also due to Misses Yuki Fujima and Chiyo Kamei, Hiroshima University, for their help in numerical computation.

References

[1] Bhattacharya, R. N., and Rao, R. R. (1976). *Normal Approximation and Asymptotic Expansions.* Wiley, New York.

[2] Fujikoshi, Y. (1985). An error bound for an asymptotic expansion of the distribution function of an estimate in a multivariate linear model. *Ann. Statist.* **13** 827–831.

[3] Fujikoshi, Y. (1987). Error bounds for asymptotic expansions of scale mixtures of distributions. *Hiroshima Math. J.* **17** 309–324.

[4] Fujikoshi, Y., and Shimizu, R. (1987). Error bounds for asymptotic expansions of scale mixtures of univariate and multivariate distributions. Technical Report No. 201, Statistical Research Group, Hiroshima Univiversity.

[5] HALL, P. (1979). On measures of the distance of a mixture from its parent distribution. *Stochastic Process. Appl.* **8** 357–365.

[6] HALL, P., AND NAKATA, T. (1986). On non-uniform and global discriptions of the rate of convergence of asymptotic expansions in the central limit theorem. *J. Austral. Math. Soc. Ser. A* **41** 326–335.

[7] HEYDE, C. C. (1975). Kurtosis and departure from normality. In Statistical Distributions in Scientific Work–1 (G. P. Patil *et al.*, Ed.), pp. 193–201, Reidel, Dordrecht.

[8] HEYDE, C. C., AND RESLIE, J. R. (1976). On moment measures of departure from the normal and exponential laws. *Stochastic Process. Appl.* **4** 317–328.

[9] JOHNSON, N. L., AND KOTZ, S. (1970). *Continuous Univariate Distributions–2.* Wiley, New York.

[10] SHIMIZU, R. (1987). Error bounds for asymptotic expansion of the scale mixtures of the normal distribution. *Ann. Inst. Statist. Math.* **39**, Part A, 611–622.

[11] SHIMIZU, R. (1988). Expansion of scale mixtures of the gamma distribution. *J. Statist. Plann. Inference*, in press.

Empirical and Hierarchical Bayes Competitors of Preliminary Test Estimators in Two Sample Problems

MALAY GHOSH*

University of Florida

AND

BIMAL K. SINHA†

University of Maryland Baltimore County and University of Pittsburgh

We consider the problem of estimation of $\boldsymbol{\mu}_1$ when it is suspected that $\boldsymbol{\mu}_1 \approx \boldsymbol{\mu}_2$ based on independent samples from $N_p(\boldsymbol{\mu}_1, \sigma^2 V_1)$ and $N_p(\boldsymbol{\mu}_2, \sigma^2 V_2)$. We assume V_1, V_2 known but σ^2 unknown. First, the EB estimator is derived and its Bayesian and frequentist properties are studied. Second, a modified EB estimator is proposed and shown to dominate a preliminary test estimator. Finally, a hierarchical Bayes approach is proposed as an alternative to EB estimators.

1. INTRODUCTION

Suppose in a laboratory, say Laboratory I, a certain instrument is designed to measure several characteristics and a number of vector-valued measurements is recorded. Our objective is to estimate the unknown population mean. It is known, however, that a similar instrument is used in another laboratory, say Laboratory II for the same purpose, and a number of observations is recorded from the second instrument. It is also suspected that the two population means are equal, in which case, observations recorded in Laboratory II can possibly be used effectively together with those in Laboratory I for estimating the population mean of the first

* Research partially supported by NSF Grant DMS 8600666.
† Research partially supported by Air Force Office of Research (AFOSR) under contract F49620-85-C-0008.

Multivariate Statistics and Probability
ISBN 0-12-580205-6

Reprinted from *J. Mult. Anal.* **27**(1).

instrument. Thus, the question that naturally arises is whether one should use the sample mean from Laboratory I or the pooled mean from the two laboratories.

In problems of this type what is normally sought is a compromise estimator which leans more towards the pooled sample mean when the null hypothesis of the equality of the two population means is accepted, and towards the sample mean from Laboratory I when such a hypothesis is rejected.

A very popular way to achieve this compromise is to use a preliminary test estimator (PTE) which uses the pooled mean when the null hypothesis is accepted at a desired level of significance and uses the sample mean from Laboratory I when the opposite is the case. For an excellent review of PTEs, see Bancroft and Han [1]. It is known, though, in other situations that a PTE is typically not a minimax estimator, and estimators with uniformly smaller mean squared error (MSE) than the PTE can often be produced (see, for example, Sclove *et al.* [7]). Moreover, the degree of evidence for or against the null hypothesis is not reflected in the PTE.

In this paper, we propose instead an empirical Bayes (EB) estimator which achieves the intended compromise. Such an EB estimator is quite often a weighted average of the pooled mean and the first sample mean. The weights are adaptively determined from the data in such a way that the larger the value of the usual F statistic used for testing the equality of the two population means, the smaller is the weight attached to the pooled sample mean. Thus, unlike the PTE, the EB estimator incorporates the degree of evidence for or against the null hypothesis in a very natural way. Also, unlike a subjective Bayes estimator, the EB estimator is quite robust (with respect to its frequentist or Bayesian risk) against a wide class of priors.

Section 2 motivates the EB estimator, and its Bayesian properties are discussed in this section. Among other things, it is shown that the EB estimator has uniformly smaller Bayes risk than the first sample mean. In Section 3, the estimators are compared in terms of their frequentist risks, and sufficient conditions under which an EB estimator dominates the first sample mean are given. Also, in this section, a modified EB estimator is proposed, and sufficient conditions under which it dominates the PTE are given. Finally, in Section 5, a hierarchical Bayes approach is proposed as an alternative to EB estimators. It has recently come to our attention that Saleh and Ahmed [6] have considered estimation of μ_1 under the loss $L(\delta, \mu_1) = (\delta - \mu_1)'V^{-1}(\delta - \mu_1)$, assuming $V_1 = V_2 = V$ unknown, and proposed the shrinkage estimator $\bar{X}_1 + (n_2 c/(n_1 + n_2))(\bar{X}_2 - \bar{X}_1) \cdot n/T_n^2$, where $T_n^2 = (n_1 n_2/(n_1 + n_2))(\bar{X}_2 - \bar{X}_1)'S^{-1}(\bar{X}_2 - \bar{X}_1)$, $nS =$ pooled sum of squares and products matrix, $n = n_1 + n_2 - 2$, and $0 < c < 2(p-2)/(n_1 + n_2 - p + 1)$. A comparison of the risk of the above estimator with those of the PTE as well as $\bar{X}_1$ and $(n_1 X_1 + n_2 \bar{X}_2)/(n_1 + n_2)$ is also undertaken by the above authors.

2. The EB Estimator and Its Bayesian Properties

Let X_{1i} $(i = 1, ..., n_1)$ and X_{2i} $(i = 1, ..., n_2)$ be independent $p(\geqslant 3)$-dimensional random vectors, where X_{1i}'s are i.i.d. $N_p(\mu_1, \sigma^2 V_1)$, while X_{2i}'s are i.i.d. $N_p(\mu_2, \sigma^2 V_2)$. In the above $\mu_1 \in R^p$, $\mu_2 \in R^p$, and $\sigma^2 (>0)$ are unknown, but V_1 and V_2 are known $p \times p$ p.d. matrices. Our goal is to estimate μ_1.

In order to motivate the EB estimator, we need find first a Bayes procedure. It is immediate that the minimal sufficient statistic for (μ_1, μ_2, σ^2) is $(\bar{X}_1, \bar{X}_2, \operatorname{tr}(V_1^{-1} S_1 + V_2^{-1} S_2))$, where $\bar{X}_j = n_j^{-1} \sum_{i=1}^{n_j} X_{ji}$ $(j = 1, 2)$ and $S_j = \sum_{i=1}^{n_j} (X_{ji} - \bar{X}_j)(X_{ji} - \bar{X}_j)^T$, $j = 1, 2$. Note also that $\bar{X}_j \sim N_p(\mu_j, \sigma^2 n_j^{-1} V_j)$ $(j = 1, 2)$, while $\operatorname{tr}(V_1^{-1} S_1 + V_2^{-1} S_2) \sim \sigma^2 \chi^2_{(n_1 + n_2 - 2)p}$.

In a Bayesian framework, the above is treated as a conditional distribution given μ_1 and μ_2. We use the independent $N_p(\nu, \tau^2 n_1^{-1} V_1)$ and $N_p(\nu, \tau^2 n_2^{-1} V_2)$ priors for μ_1 and μ_2; that is, the prior variance–covariance matrix is proportional to the variance–covariance matrix of the corresponding sample mean. The suspicion that μ_1 and μ_2 may be equal is reflected in the choice of a priori common mean ν. For a related prior in the general regression model, see Ghosh *et al.* [3].

In order to find the posterior distribution of $\mu = \binom{\mu_1}{\mu_2}$, first note that conditional on μ_1 and μ_2, $\bar{X}_1$, $\bar{X}_2$, S_1, and S_2 are mutually independent, and the distributions of S_1, S_2 do not depend on μ_1 and μ_2. Hence, we can restrict ourselves to the conditional distributions of $\bar{X}_j$'s given μ_j's. Also, since μ_1 and μ_2 have independent normal priors, standard calculations yield that μ_1 and μ_2 given $\bar{X}_1$ and $\bar{X}_2$ have independent posterior distributions with

$$\mu_j \mid \bar{X}_j = \bar{x}_j \sim N_p((1 - B)\,\bar{x}_j + B\nu, \sigma^2 (1 - B)\, n_j^{-1} V_j), \tag{2.1}$$

$j = 1, 2$, where $B = \sigma^2/(\sigma^2 + \tau^2)$. Now, using the loss

$$L(\mu_1, a) = \sigma^{-2} (a - \mu_1)^T Q (a - \mu_1) \tag{2.2}$$

for estimating μ_1 by a (Q being a known p.d. weight matrix), the Bayes estimator of μ_1 is

$$e_B(\bar{X}_1) = (1 - B)\,\bar{X}_1 + B\nu. \tag{2.3}$$

Note that the Bayes estimator does not depend on the choice of Q. The multiplier σ^{-2} is used in the loss because that makes $\bar{X}_1$ a minimax estimator of μ_1 with the constant risk not depending on any unknown parameter.

In order to find an EB estimator of μ_1, we estimate the unknown parameters B and v in (2.3) from the marginal distributions of $\bar{X}_1, \bar{X}_2$, and $\operatorname{tr}(V_1^{-1} S_1 + V_2^{-1} S_2)$. Note that marginally $\bar{X}_1, \bar{X}_2$, and $\operatorname{tr}(V_1^{-1} S_1 + V_2^{-1} S_2)$ are mutually independent with $\bar{X}_j \sim N_p(v, n_j^{-1}(\sigma^2 + \tau^2) V_j)$ $(j = 1, 2)$, and $\operatorname{tr}(V_1^{-1} S_1 + V_2^{-1} S_2) \sim \sigma^2 \chi^2_{(n_1+n_2-2)p}$. Hence the complete sufficient statistic for (v, τ^2, σ^2) based on this marginal distribution is $(W, Z, \operatorname{tr}(V_1^{-1} S_1 + V_2^{-1} S_2))$, where $W = (n_1 V_1^{-1} + n_2 V_2^{-1})^{-1}(n_1 V_1^{-1} \bar{X}_1 + n_2 V_2^{-1} \bar{X}_2)$ is the pooled sample mean, $Z = Y^T(n_1^{-1} V_1 + n_2^{-1} V_2)^{-1} Y$, and $Y = \bar{X}_1 - \bar{X}_2$. Also, marginally, $W \sim N_p(v, (\sigma^2 + \tau^2)(n_1 V_1^{-1} + n_2 V_2^{-1}))$, $Y \sim N_p(0, (n_1^{-1} V_1 + n_2^{-1} V_2)(\sigma^2 + \tau^2))$, and $\operatorname{tr}(V_1^{-1} S_1 + V_2^{-1} S_2) \sim \sigma^2 \chi^2_{(n_1+n_2-2)p}$. Hence, the UMVUE of v is W, while the UMVUE of $(\sigma^2 + \tau^2)^{-1}$ is $(p-2)/(Y^T(n_1^{-1} V_1 + n_2^{-1} V_2)^{-1} Y)$. The last assertion follows since $Y^T(n_1^{-1} V_1 + n_2^{-1} V_2)^{-1} Y \sim (\sigma^2 + \tau^2) \chi^2_p$. Moreover, since $\operatorname{tr}(V_1^{-1} S_1 + V_2^{-1} S_2) \sim \sigma^2 \chi^2_{(n_1+n_2-2)p}$, the best scale invariant estimator of σ^2 is $((n_1 + n_2 - 2)p + 2)^{-1} \operatorname{tr}(V_1^{-1} S_1 + V_2^{-1} S_2)$. Substituting these estimators for v, $(\sigma^2 + \tau^2)^{-1}$, and σ^2 in (2.3), one gets the EB estimator of μ_1 as

$$e_{\mathrm{EB}}(\bar{X}_1, \bar{X}_2, S_1, S_2) = (1 - \hat{B})\, \bar{X}_1 + \hat{B} W = W + (1 - \hat{B})(\bar{X}_1 - W), \tag{2.4}$$

where

$$\hat{B} = \frac{(p-2) \operatorname{tr}(V_1^{-1} S_1 + V_2^{-1} S_2)}{((n_1 + n_2 - 2)\, p + 2)\, Y^T(n_1^{-1} V_1 + n_2^{-1} V_2)^{-1} Y}. \tag{2.5}$$

Remark 2.1. Note that $0 < B < 1$, while the estimator $\hat{B}$ though positive can take values exceeding one. Accordingly, for practical purposes, one proposes the positive part EB estimator

$$e^+_{\mathrm{EB}}(\bar{X}_1, \bar{X}_2, S_1, S_2) = W + (1 - \hat{B})^+ (\bar{X}_1 - W) \tag{2.6}$$

of μ_1, where $a^+ = \max(a, 0)$. For simplicity of exposition, in the remainder of this section, we shall, however, work with e_{EB} rather than e^+_{EB}.

A question that naturally arises is why this particular method of estimation is used for estimating the prior parameters. We shall answer the question by proving the "optimality" of e_{EB} within the class of estimators

$$\delta_c(\bar{X}_1, \bar{X}_2, S_1, S_2) = W + \left(1 - \frac{c \operatorname{tr}(V_1^{-1} S_1 + V_2^{-1} S_2)}{((n_1 + n_2 - 2)\, p + 2)\, Y^T(n_1^{-1} V_2 + n_2^{-1} V_2)^{-1} Y}\right)(\bar{X}_1 - W), \tag{2.7}$$

where c (>0) is a constant. Note that $e_{EB} = \delta_{p-2}$.

THEOREM 2.1. *The Bayes risk of* δ_c *under the assumed prior (say* ξ*) and the loss (2.2) is given by*

$$r(\xi, \delta_c) = (1-B)\, n_1^{-1} \operatorname{tr}(QV_1) + B \operatorname{tr}(Q(n_1 V_1^{-1} + n_2 V_2^{-1})^{-1})$$
$$+ B \operatorname{tr}(Q\Lambda(n_1^{-1} V_1 + n_2^{-1} V_2)\, \Lambda^T)$$
$$\times \left[\frac{c^2(n_1+n_2-2)}{\{(n_1+n_2-2)\,p+2\}(p-2)} - \frac{2c(n_1+n_2-2)}{(n_1+n_2-2)\,p+2} + 1\right], \tag{2.8}$$

where $\Lambda = (n_1 V_1^{-1} + n_2 V_2^{-1})^{-1} n_2 V_2^{-1}$. *Moreover,* $r(\xi, e_B) \leqslant r(\xi, \delta_c)$.

Proof. The second part of the theorem follows immediately from (2.8). To prove the first part, write

$$r(\xi, \delta_c) = r(\xi, e_B) + \sigma^{-2} E[(e_B - \delta_c)^T Q(e_B - \delta_c)]. \tag{2.9}$$

Note from (2.1) to (2.3) that

$$r(\xi, e_B) = (1-B)\, n_1^{-1} \operatorname{tr}(QV_1). \tag{2.10}$$

Also, writing $\hat{B}_c = c \operatorname{tr}(V_1^{-1} S_1 + V_2^{-1} S_2)/\{((n_1+n_2-2)\,p+2)\;\; Y^T(n_1^{-1} V_1 + n_2^{-1} V_2)^{-1}\, Y\}$, one gets

$$\begin{aligned} e_B - \delta_c &= (1-B)\, \bar{X}_1 + Bv - W - (1-\hat{B}_c)(\bar{X}_1 - W) \\ &= -B(W - v) + (\hat{B}_c - B)(\bar{X}_1 - W) \\ &= -B(W - v) + (\hat{B}_c - B)\, \Lambda Y. \end{aligned} \tag{2.11}$$

Next using the independence of W and $(Y, \operatorname{tr}(V_1^{-1} S_1 + V_2^{-1} S_2))$ and the facts that $E(W) = v$, $\operatorname{Var}(W) = (\sigma^2 + \tau^2)(n_1 V_1^{-1} + n_2 V_2^{-1})^{-1} = \sigma^2 B^{-1}(n_1 V_1^{-1} + n_2 V_2^{-1})^{-1}$, one gets

$$\begin{aligned} &E[(e_B - \delta_c)^T Q(e_B - \delta_c)] \\ &\quad = B^2 E[(W-v)^T Q(W-v)] + E[(\hat{B}_c - B)^2\, Y^T \Lambda^T Q\Lambda Y] \\ &\quad = \sigma^2 B \operatorname{tr}\{Q(n_1 V_1^{-1} + n_2 V_2^{-1})^{-1}\} + E[(\hat{B}_c - B)^2 Y^T \Lambda^T Q\Lambda Y]. \end{aligned} \tag{2.12}$$

Now we find

$$\begin{aligned} &E[(\hat{B}_c - B)^2 Y^T \Lambda^T Q\Lambda Y] \\ &\quad = E\left[\frac{c^2\{\operatorname{tr}(V_1^{-1} S_1 + V_2^{-1} S_2)\}^2}{\{(n_1+n_2-2)\,p+2\}^2\{Y^T(n_1^{-1} V_1 + n_2^{-1} V_2)^{-1} Y\}^2}\,(Y^T \Lambda^T Q\Lambda Y)\right. \\ &\qquad - \frac{2Bc}{\{(n_1+n_2-2)\,p+2\}}\,\frac{\operatorname{tr}(V_1^{-1} S_1 + V_2^{-1} S_2)}{\{Y^T(n_1^{-1} V_1 + n_2^{-1} V_2)^{-1} Y\}}\,(Y^T \Lambda^T Q\Lambda Y) \\ &\qquad \left. + B^2(Y^T \Lambda^T Q\Lambda Y)\right]. \end{aligned} \tag{2.13}$$

Using the independence of Y and $\operatorname{tr}(V_1^{-1} S_1 + V_2^{-1} S_2)$ along with the fact that $\operatorname{tr}(V_1^{-1} S_1 + V_2^{-1} S_2) \sim \sigma^2 \chi^2_{(n_1+n_2-2)p}$, it follows that the right-hand side of (2.13) is

$$E\left[\frac{c^2\sigma^4(n_1+n_2-2)p}{(n_1+n_2-2)p+2} \cdot \frac{Y^T A^T Q A Y}{\{Y^T(n_1^{-1}V_1+n_2^{-1}V_2)^{-1}Y\}^2}\right.$$
$$\left. -\frac{2Bc\sigma^2(n_1+n_2-2)p}{(n_1+n_2-2)p+2} \cdot \frac{Y^T A^T Q A Y}{\{Y^T(n_1^{-1}V_1+n_2^{-1}V_2)^{-1}Y\}} + B^2(Y^T A^T Q A Y)\right]. \tag{2.14}$$

Next observe that $Y^T(n_1^{-1}V_1+n_2^{-1}V_2)^{-1}Y$ is a function of the complete sufficient statistic while $(Y^T A^T Q A Y)/(Y^T(n_1^{-1}V_1+n_2^{-1}V_2)^{-1}Y)$ is ancillary. Now using Basu's theorem (or Lemma 1 of Ghosh *et al.* [3]) along with $E(Y^T A^T Q A Y) = (\sigma^2+\tau^2) \times \operatorname{tr}(QA(n_1^{-1}V_1+n_2^{-1}V_2)A^T)$, $E(Y^T(n_1^{-1}V_1+n_2^{-1}V_2)^{-1}Y) = p(\sigma^2+\tau^2)$, and $E(Y^T(n_1^{-1}V_1+n_2^{-1}V_2)^{-1}Y)^{-1} = (\sigma^2+\tau^2)^{-1}(p-2)^{-1}$, it follows that the right-hand side of (2.14) is

$$\frac{c^2\sigma^2 B(n_1+n_2-2)p\operatorname{tr}(QA(n_1^{-1}V_1+n_2^{-1}V_2)A^T)}{\{(n_1+n_2-2)p+2\}p(p-2)}$$
$$-\frac{2c\sigma^2 B(n_1+n_2-2)p\operatorname{tr}(QA(n_1^{-1}V_1+n_2^{-1}V_2)A^T)}{\{(n_1+n_2-2)p+2\}p}$$
$$+\sigma^2 B\operatorname{tr}(QA(n_1^{-1}V_1+n_2^{-1}V_2)A^T). \tag{2.15}$$

It follows from (2.12)–(2.15) that

$$E[(e_B-\delta_c)^T Q(e_B-\delta_c)]$$
$$=\sigma^2 B\operatorname{tr}(Q(n_1V_1^{-1}+n_2V_2^{-1})^{-1})+\sigma^2 B\operatorname{tr}(QA(n_1^{-1}V_1+n_2^{-1}V_2)A^T)$$
$$\times\left[\frac{c^2(n_1+n_2-2)}{\{(n_1+n_2-2)p+2\}(p-2)}-\frac{2c(n_1+n_2-2)}{(n_1+n_2-2)p+2}+1\right]. \tag{2.16}$$

The proof of the theorem is complete from (2.9), (2.10), and (2.16).

Next we compare the Bayes risks of e_{EB} and $\bar{X}_1$. Note that $\bar{X}_1$ has constant risk, and hence constant Bayes risk (under *any* prior) $\sigma^2 n_1^{-1}\operatorname{tr}(QV_1)$. Rather than comparing the Bayes risks of e_{EB} and $\bar{X}_1$ directly, we find it convenient to introduce the notion of relative savings loss (RSL) as in Efron and Morris [2].

For any estimator e of μ_1, the RSL of e_{EB} with respect to e (under the prior ξ) is defined as

$$\mathrm{RSL}(\xi; e_{\mathrm{EB}}, e) = [r(\xi, e_{\mathrm{EB}}) - r(\xi, e_{\mathrm{B}})]/[r(\xi, e) - r(\xi, e_{\mathrm{B}})]$$
$$= 1 - [r(\xi, e) - r(\xi, e_{\mathrm{EB}})]/[r(\xi, e) - r(\xi, e_{\mathrm{B}})]. \tag{2.17}$$

This is the proportion of the possible Bayes risk improvement over e that is sacrificed by the use of e_{EB} rather than the ideal e_{B} under the prior ξ. From (2.8) with $c = p - 2$ and (2.10), it follows that

$$\begin{aligned} \text{RSL}(\xi; e_{\text{EB}}, \bar{X}_1) = \Bigg[& \operatorname{tr}(Q(n_1 V_1^{-1} + n_2 V_2^{-1})^{-1}) \\ & + \operatorname{tr}(Q\Lambda(n_1^{-1} V_1 + n_2^{-1} V_2) \Lambda^T) \left(\frac{2(n_1 + n_2 - 1)}{(n_1 + n_2 - 2)p + 2} \right) \Bigg] \\ & \times [n_1^{-1} \operatorname{tr}(QV_1)]^{-1}. \end{aligned} \tag{2.18}$$

Note that the above RSL expression does not depend on any unknown parameter. Also, writing

$$\begin{aligned} \Lambda &= (n_1 V_1^{-1} + n_2 V_2^{-1})^{-1} n_2 V_2^{-1} = [V_2^{-1}\{n_1 V_2 + n_2 V_1\} V_1^{-1}]^{-1} n_2 V_2^{-1} \\ &= n_1^{-1} V_1 (n_1^{-1} V_1 + n_2^{-1} V_2)^{-1}, \end{aligned}$$

it follows that

$$\begin{aligned} & (n_1 V_1^{-1} + n_2 V_2^{-1})^{-1} + \Lambda(n_1^{-1} V_1 + n_2^{-1} V_2) \Lambda^T \\ & \quad = n_2^{-1} \Lambda V_2 + n_1^{-1} \Lambda V_1 = \Lambda(n_1^{-1} V_1 + n_2^{-1} V_2) = n_1^{-1} V_1. \end{aligned} \tag{2.19}$$

Now using $2(n_1 + n_2 - 1) < (n_1 + n_2 - 2)p + 2$, it follows from (2.18) that $\text{RSL}(\xi; e_{\text{EB}}, \bar{X}_1) < 1$ which is equivalent to $r(\xi, e_{\text{EB}}) < r(\xi, \bar{X}_1)$. Thus e_{EB} has smaller Bayes risk than $\bar{X}_1$.

Finally, in this section, we compare the Bayes risk of e_{EB} with that of W. Note that W has Bayes risk

$$r(\xi, W) = r(\xi, e_{\text{B}}) + \sigma^{-2} E[(e_{\text{B}} - W)^T Q (e_{\text{B}} - W)]. \tag{2.20}$$

Since $e_{\text{B}} - W = (1 - B)\bar{X}_1 + Bv - W = -B(W - v) + (1 - B)(\bar{X}_1 - W) = -B(W - v) + (1 - B)\Lambda Y$, where Λ is defined following (2.11), using once again the independence of W and Y, it follows that

$$\begin{aligned} & E[(e_B - W)^T Q (e_B - W)] \\ & \quad = \sigma^2 B \operatorname{tr}(Q(n_1 V_1^{-1} + n_2 V_2^{-1})^{-1}) + (1 - B)^2 E(Y^T \Lambda^T Q \Lambda Y) \\ & \quad = \sigma^2 B \operatorname{tr}(Q(n_1 V_1^{-1} + n_2 V_2^{-1})^{-1}) \\ & \qquad + \sigma^2 (1 - B)^2 B^{-1} \operatorname{tr}(Q\Lambda(n_1^{-1} V_1 + n_2^{-1} V_2) \Lambda^T. \end{aligned} \tag{2.21}$$

Thus from (2.10), (2.20), and (2.21),

$$r(\xi, W) = n_1^{-1}(1-B)\,\mathrm{tr}(QV_1) + B\,\mathrm{tr}(Q(n_1 V_1^{-1} + n_2 V_2^{-1})^{-1}) + (1-B)^2 B^{-1}\mathrm{tr}(Q\Lambda(n_1^{-1} V_1 + n_2^{-1} V_2)\Lambda^T). \tag{2.22}$$

Finally, from (2.8) with $c = p-2$, (2.10), and (2.22), it follows that $\mathrm{RSL}(\xi; e_{\mathrm{EB}}, W) = [r(\xi, e_{\mathrm{EB}}) - r(\xi, e_{\mathrm{B}})]/[r(\xi, W) - r(\xi, e_{\mathrm{B}})]$ is

$$\frac{\mathrm{tr}(Q(n_1 V_1^{-1} + n_2 V_2^{-1})^{-1}) + \dfrac{2(n_1+n_2-1)}{(n_1+n2-2)p+2}\mathrm{tr}(Q\Lambda(n_1^{-1} V_1 + n_2^{-1} V_2)\Lambda^T)}{\mathrm{tr}(Q(n_1 V_1^{-1} + n_2 V_2^{-1})^{-1}) + (1-B)^2 B^{-2}\mathrm{tr}(Q\Lambda(n_1^{-1} V_1 + n_2^{-1} V_2)\Lambda^T)} \tag{2.23}$$

which is less than one if and only if

$$\{(1-B)/B\}^2 > 2(n_1+n_2-1)/\{(n_1+n_2-2)p+2\}. \tag{2.24}$$

Remark 2.2. The fact that e_{EB} does not dominate W uniformly is not at all surprising. If, for example, τ^2 is very small and μ_1 is nearly degenerate at v, then W is much closer to v than e_{EB}. Indeed, in this case $B = \sigma^2/(\sigma^2+\tau^2)$ is very close to 1 so that (2.24) cannot hold. However, when $\sigma^2 \leqslant \tau^2$, then $B \leqslant \frac{1}{2} \leftrightarrow (1-B)/B \geqslant 1$ so that (2.24) holds.

3. Minimax Estimation

It is well known that under the loss given in (2.2), $\bar{X}_1$ is a minimax estimator of μ_1 with constant risk $n_1^{-1}\,\mathrm{tr}(QV_1)$. In this section, first we find a class of estimators including e_{EB} as a member which dominates $\bar{X}_1$ under certain conditions, and then investigate whether e_{EB} satisfies these conditions.

With this end, first write

$$F = (Y^T(n_1^{-1} V_1 + n_2^{-1} V_2)^{-1} Y)/\{\mathrm{tr}(V_1^{-1} S_1 + V_2^{-1} S_2)/((n_1+n_2-2)p+2)\} \tag{3.1}$$

and consider the class of estimators

$$\mu_1^{\phi} = \bar{X}_1 - (\phi(F)/F)(\bar{X}_1 - W) \tag{3.2}$$

for estimating μ_1. Note that e_{EB} belongs to this class with $\phi(F) = p-2$. We now compute the frequentist risk of the estimator μ_1^{ϕ} (i.e., without any reference to the prior ξ). Throughout this section, E denotes expectation conditional on μ_1 and μ_2, and we write $V = n_1^{-1} V_1 + n_2^{-1} V_2$.

THEOREM 3.1.

$$E[(\mu_1^\phi-\mu_1)^T Q(\mu_1^\phi-\mu_1)]/\sigma^2$$

$$=n_1^{-1}\operatorname{tr}(QV_1)-2E\left[\frac{\phi(F)}{F}\operatorname{tr}(\Lambda^T Q\Lambda V)+2\left(\phi'(F)-\frac{\phi(F)}{F}\right)\frac{Y^T\Lambda^T Q\Lambda Y}{Y^T V^{-1} Y}\right]$$

$$+\sigma^{-2}E\left[\frac{\phi^2(F)}{F^2}Y^T\Lambda^T Q\Lambda Y\right]. \tag{3.3}$$

Proof. First write

$$\begin{aligned}E[(\mu_1^\phi-\mu_1)^T Q(\mu_1^\phi-\mu_1)]&\\ =E[(\bar{X}_1-\mu_1)^T Q(\bar{X}_1-\mu_1)&\\ -2(\phi(F)/F)\,Y^T\Lambda^T Q(\bar{X}_1-\mu_1)&\\ +(\phi^2(F)/F^2)\,Y^T\Lambda^T Q\Lambda Y],&\end{aligned} \tag{3.4}$$

where we have used the fact that $\bar{X}_1-W=\Lambda Y$. Next writing $\bar{X}_1=W+\Lambda Y$ and correspondingly $\mu_1=\mu_*+\Lambda\mu_0$, where $\mu_*=(n_1V_1^{-1}+n_2V_2^{-1})^{-1}(n_1V_1^{-1}\mu_1+n_2V_2^{-1}\mu_2)$ and $\mu_0=\mu_1-\mu_2$, one gets

$$\begin{aligned}E[(\phi(F)/F)\,Y^T\Lambda^T Q(\bar{X}_1-\mu_1)]&\\ =E[(\phi(F)/F)\,Y^T\Lambda^T Q((W-\mu_*+\Lambda(Y-\mu_0))]&\\ =E[(\phi(F)/F)\,Y^T\Lambda^T Q\Lambda(Y-\mu_0)],&\end{aligned} \tag{3.5}$$

where in the final step of (3.5), one uses the independence of $(Y, \operatorname{tr}(V_1^{-1}S_1+V_2^{-1}S_2))$ with W as well as $E(W)=\mu_*$. Now since V is p.d., there exists a nonsingular D such that $D^{-1}V(D^{-1})^T=I_p$. Write $Z=D^{-1}Y$ and $\eta_0=D^{-1}\mu_0$. Then $Z\sim N_p(\eta_0,\sigma^2 I_p)$. We rewrite

$$Y^T\Lambda^T Q\Lambda(Y-\mu_0)=Z^T U(Z-\eta_0), \tag{3.6}$$

where $U=((u_{ij}))=D^T\Lambda^T Q\Lambda D$. Also, in terms of Z, $F=Z^TZ/\{\operatorname{tr}(V_1^{-1}S_1+V_2^{-1}S_2)/((n_1+n_2-2)p+2)\}$. Now using Stein's identity (cf. Stein [8]), the independence of Z and $\operatorname{tr}(V_1^{-1}S_1+V_2^{-1}S_2)$, and (3.6), we get

$$E[(\phi(F)/F)\,Z^T U(Z-\eta_0)]$$

$$=\sigma^2\sum_{i=1}^p E\left[\frac{\partial}{\partial Z_i}\left\{\frac{\phi(F)}{F}\cdot\sum_{j=1}^p u_{ij}Z_j\right\}\right]$$

$$=\sigma^2\sum_{i=1}^p E\left[\frac{\phi(F)}{F}u_{ii}+\left\{\frac{\phi'(F)}{F}-\frac{\phi(F)}{F^2}\right\}\right.$$

$$\left.\times\frac{2Z_i\sum_{j=1}^p u_{ij}Z_j}{\{\operatorname{tr}(V_1^{-1}S_1+V_2^{-1}S_2)/((n_1+n_2-2)p+2\}}\right]$$

$$= \sigma^2 E\left[\frac{\phi(F)}{F}\cdot \mathrm{tr}(U) + 2\left\{\frac{\phi'(F)}{F} - \frac{\phi(F)}{F^2}\right\}\right.$$

$$\left.\cdot \frac{Z^T U Z}{\{\mathrm{tr}(V_1^{-1} S_1 + V_2^{-1} S_2)/((n_1+n_2-2)p+2)\}}\right]$$

$$= \sigma^2 E\left[\frac{\phi(F)}{F}\,\mathrm{tr}(\Lambda^T Q \Lambda V) + 2\left\{\frac{\phi'(F)}{F} - \frac{\phi(F)}{F^2}\right\}\cdot F \cdot \frac{Y^T \Lambda^T Q \Lambda Y}{Y^T V^{-1} Y}\right]$$

$$= \sigma^2 E\left[\frac{\phi(F)}{F}\,\mathrm{tr}(\Lambda^T Q \Lambda V) + 2\left\{\phi'(F) - \frac{\phi(F)}{F}\right\}\cdot \frac{Y^T \Lambda^T Q \Lambda Y}{Y^T V^{-1} Y}\right]. \tag{3.7}$$

The theorem follows now from (3.3), (3.4), and (3.7).

Next in this section we find an upper bound for $E[(\phi^2(F)/F^2)\, Y^T \Lambda^T Q \Lambda Y]$. We first get the inequality

$$E[(\phi^2(F)/F^2)(Y^T \Lambda^T Q \Lambda Y)]$$

$$= E\left[\frac{\phi^2(F)}{F^2}\cdot F \cdot \frac{Y^T \Lambda^T Q \Lambda Y}{Y^T V^{-1} Y}\cdot \frac{\mathrm{tr}(V_1^{-1} S_1 + V_2^{-1} S_2)}{\{(n_1+n_2-2)p+2\}}\right]$$

$$\leqslant ch_1(\Lambda^T Q \Lambda V)\, E[h^2(F)\, F \cdot \mathrm{tr}(V_1^{-1} S_1 + V_2^{-1} S_1)/((n_1+n_2-2)p+2)], \tag{3.8}$$

where $ch_1(\Lambda^T Q \Lambda V)$ denotes the largest eigen value of $\Lambda^T Q \Lambda V$ and $h(F) = \phi(F)/F$. Next applying (2.18) of Efron and Morris [2], one gets

$$E[h^2(F)\, F\,\mathrm{tr}(V_1^{-1} S_1 + V_2^{-1} S_2)/((n_1+n_2-2)p+2)]$$

$$= E\left[\frac{(n_1+n_2-2)p}{(n_1+n_2-2)p+2}\cdot h^2(F)\, F + \frac{2}{(n_1+n_2-2)p+2}\cdot \frac{\mathrm{tr}(V_1^{-1} S_1 + V_2^{-1} S_2)}{(n_1+n_2-2)p+2}\right.$$

$$\left.\times (2h(F)\, h'(F)\, F + h^2(F))\left(-\frac{F}{\mathrm{tr}(V_1^{-1} S_2 + V_2^{-1} S_2)/((n_1+n_2-2)p+2)}\right)\right]$$

$$= \sigma^2 E\left[\frac{(n_1+n_2-2)p}{(n_1+n_2-2)p+2}\cdot \frac{\phi^2(F)}{F}\right.$$

$$\left. - 2\frac{F}{(n_1+n_2-2)p+2}\left\{2\left(\frac{\phi'(F)}{F} - \frac{\phi(F)}{F^2}\right)\phi(F) + \frac{\phi^2(F)}{F^2}\right\}\right]$$

$$= \sigma^2 E\left[\frac{\phi^2(F)}{F} - \frac{4}{(n_1+n_2-2)p+2}\,\phi(F)\,\phi'(F)\right]. \tag{3.9}$$

From (3.8) and (3.9), one gets

$$E\left[\frac{\phi^2(F)}{F^2}\cdot(Y^T\Lambda^T Q\Lambda Y)\right]$$
$$\leqslant \sigma^2 ch_1(\Lambda^T Q\Lambda V)\, E\left[\frac{\phi^2(F)}{F}-\frac{4}{(n_1+n_2-2)\,p+2}\phi(F)\,\phi'(F)\right]. \quad (3.10)$$

Combining (3.3) and (3.10), one gets

$$\sigma^{-2}E[(\mu_1^\phi-\mu_1)^T Q(\mu_1^\phi-\mu_1)-(\bar{X}_1-\mu_1)^T Q(\bar{X}_1-\mu_1)]$$
$$\leqslant -2E\left[\frac{\phi(F)}{F}\operatorname{tr}(\Lambda^T Q\Lambda V)+2\left(\phi'(F)-\frac{\phi(F)}{F}\right)\frac{Y^T\Lambda^T Q\Lambda Y}{Y^T V^{-1} Y}\right]$$
$$+\, ch_1(\Lambda^T Q\Lambda V)\, E\left[\frac{\phi^2(F)}{F}-\frac{4}{(n_1+n_2-2)\,p+2}\phi(F)\,\phi'(F)\right]. \quad (3.11)$$

The following theorem is now easy to prove from (3.11). Recall that $\Lambda=(n_1V_1^{-1}+n_2V_2^{-1})^{-1}n_2V_2^{-1}$ and $V=n_1^{-1}V_1+n_2^{-1}V_2$.

THEOREM 3.2. *Suppose that*

(i) $\operatorname{tr}(\Lambda^T Q\Lambda V)>2ch_1(\Lambda^T Q\Lambda V)$
(ii) $0<\phi(F)<2[\operatorname{tr}(\Lambda^T Q\Lambda V)/ch_1(\Lambda^T Q\Lambda V)-2]$ *and*
(iii) $\phi(F)\uparrow$ *in* F

hold. Then $\sigma^{-2}E[(\mu_1^\phi-\mu_1)^T Q(\mu_1^\phi-\mu_1)-(\bar{X}_1-\mu_1)^T Q(\bar{X}_1-\mu_1)]<0$ *for all* μ_1 *and* μ_2.

Proof. Using (iii), it follows from (3.11) that

$$\sigma^{-2}E[(\mu_1^\phi-\mu_1)^T Q(\mu_1^\phi-\mu_1)-(\bar{X}_1-\mu_1)^T Q(\bar{X}_1-\mu_1)]$$
$$\leqslant 2E\left[-\frac{\phi(F)}{F}\operatorname{tr}(\Lambda^T Q\Lambda V)+2\frac{\phi(F)}{F}\cdot\frac{Y^T\Lambda^T Q\Lambda Y}{Y^T V^{-1}Y}+\frac{1}{2}\frac{\phi^2(F)}{F}ch_1(\Lambda^T Q\Lambda V)\right]$$
$$\leqslant 2E\left[-\frac{\phi(F)}{F}\operatorname{tr}(\Lambda^T Q\Lambda V)+2\frac{\phi(F)}{F}ch_1(\Lambda^T Q\Lambda V)+\frac{1}{2}\frac{\phi^2(F)}{F}ch_1(\Lambda^T Q\Lambda V)\right]$$
$$=2E\left[-\frac{\phi(F)}{2F}ch_1(\Lambda^T Q\Lambda V)\left\{2\left(\frac{\operatorname{tr}(\Lambda^T Q\Lambda V)}{ch_1(\Lambda^T Q\Lambda V)}-2\right)-\phi(F)\right\}\right]$$
$$<0 \quad (3.12)$$

using conditions (i) and (ii) of the theorem.

Remark 3.1. It is an immediate consequence of the above theorem that if condition (i) of Theorem 3.2 holds, and $0<p-2<2[(\mathrm{tr}(\Lambda^T Q\Lambda V)/ch_1(\Lambda^T Q\Lambda V)-2]$, then the EB estimator e_{EB} dominates $\bar{X}_1$. In particular, if $Q=V_1=V_2=I_p$, then $\mathrm{tr}(\Lambda^T Q\Lambda V)=p\, ch_1(\Lambda^T Q\Lambda V)$, and hence e_{EB} dominates $\bar{X}_1$ for $p\geqslant 3$.

In the remainder of this section we show how a modified EB estimator can dominate the PTE. Once again, an appeal to Theorem 3.1 is made.

A PTE δ_{PTE} of μ_1 is of the form $\delta_{\mathrm{PTE}}=g(F)\,\bar{X}_1+(1-g(F))\,W=\bar{X}_1-(1-g(F))(\bar{X}_1-W)$, where $g(F)=I_{[F>d]}$ for some positive constant d, and I denotes the usual indicator function. The choice of d is governed by the level of significance that is used for testing $H_0\colon \mu_1=\mu_2$. We propose the rival estimator

$$\delta_{\mathrm{MEB}}=\bar{X}_1-\left(1-\left(1-\frac{c}{F}\right)g(F)\right)(\bar{X}_1-W)$$

$$=W+\left(1-\frac{c}{F}\right)g(F)(\bar{X}_1-W) \tag{3.13}$$

which is a modified version of e_{EB} with $p-2$ replaced by a general c. Note that $\delta_{\mathrm{MEB}}=W$ when $g(F)=0$, but $\delta_{\mathrm{MEB}}=\delta_{\mathrm{EB}}$ when $g(F)=1$. The following theorem is then obtained.

THEOREM 3.3. *Suppose condition* (i) *of Theorem* 3.2 *holds and* $0<c<2[\mathrm{tr}(\Lambda^T Q\Lambda V)/ch_1(\Lambda^T Q\Lambda V)-2]$. *Then*

$$\sigma^{-2}E[(\delta_{\mathrm{MEB}}-\mu_1)^T Q(\delta_{\mathrm{MEB}}-\mu_1)-(\delta_{\mathrm{PTE}}-\mu_1)^T Q(\delta_{\mathrm{PTE}}-\mu_1(]<0 \tag{3.14}$$

for all μ_1 *and* μ_2.

Proof. Write $\phi_1(F)=F(1-g(F))$ and $\phi_2(F)=F(1-(1-c/F)\,g(F))=\phi_1(F)+cg(F)$. Then $\delta_{\mathrm{PTE}}=\bar{X}_1-(\phi_1(F)/F)(\bar{X}_1-W)$ while $\delta_{\mathrm{MEB}}=\bar{X}_1-(\phi_2(F)/F)(\bar{X}_1-W)$. Note that both $\phi_1(F)$ and $\phi_2(F)$ are differentiable everywhere except at $F=d$. Thus $\phi_1'(F)$ and $\phi_2'(F)$ are defined a.e. (Lebesgue). Moreover, $\phi_1(F)-\phi_2(F)=-cg(F)$, $\phi_1^2(F)-\phi_2^2(F)=-c^2g^2(F)=-c^2g(F)$ and $\phi_1'(F)=\phi_2'(F)=1-g(F)$ a.e. (Lebesgue). Then, applying Theorem 3.1 twice, once with $\phi(F)=\phi_2(F)$, and next with $\phi(F)=\phi_1(F)$, one gets the left-hand side of (3.14) as

$$-2E\left[\frac{cg(F)}{F}\,\mathrm{tr}(\Lambda^T Q\Lambda V)-\frac{2}{F}\,cg(F)\,\frac{Y^T\Lambda^T Q\Lambda Y}{Y^T V^{-1} Y}\right]$$

$$+\sigma^{-2}\,E\left[\frac{c^2g^2(F)}{F^2}\cdot(Y^T\Lambda^T Q\Lambda Y)\right]$$

$$\leqslant -2E\left[\frac{cg(F)}{F}\operatorname{tr}(\Lambda^T Q\Lambda V)-\frac{2}{F}cg(F)\,ch_1(\Lambda^T Q\Lambda V)\right]$$

$$+\sigma^{-2}E\left[\frac{c^2g^2(F)}{F^2}\cdot F\cdot\frac{\operatorname{tr}(V_1^{-1}S_1+V_2^{-1}S_2)}{(n_1+n_2-2)\,p+2}\cdot\frac{Y^T\Lambda^T Q\Lambda Y}{Y^T V^{-1} Y}\right]$$

$$\leqslant -2E\left[\frac{cg(F)}{F}\operatorname{tr}(\Lambda^T Q\Lambda V)-\frac{2cg(F)}{F}\cdot ch_1(\Lambda^T Q\Lambda V)\right]$$

$$+\sigma^{-2}\,ch_1(\Lambda^T Q\Lambda V)\,E\left[\frac{c^2g^2(F)}{F^2}\cdot F\cdot\frac{\operatorname{tr}(V_1^{-1}S_1+V_1^{-1}S_2)}{(n_1+n_2-2)\,p+2}\right]. \qquad (3.15)$$

Applying (2.18) of Efron and Morris [2] again with $\phi(F)=g(F)$ so that $\phi'(F)=0$ a.e. (Lebesgue), one gets

$$E\left[\left(\frac{g^2(F)}{F^2}\right)F\operatorname{tr}(V_1^{-1}S_1+V_1^{-1}S_2)/((n_1+n_2-2)\,p+2)\right]$$

$$=\sigma^2 E[g^2(F)/F]=\sigma^2 E[g(F)/F]. \qquad (3.16)$$

Now from (3.15) and (3.16), the

left-hand side of (3.14) is

$$\leqslant -E\left[\frac{cg(F)}{F}ch_1(\Lambda^T Q\Lambda V)\left\{2\left(\frac{\operatorname{tr}(\Lambda^T Q\Lambda V)}{ch_1(\Lambda^T Q\Lambda V)}-2\right)-c\right\}\right]<0 \qquad (3.17)$$

by using the upper bound of c given in this theorem. The proof of the theorem is complete.

Remark 3.2. Note that when $Q=V_1=V_2=I_p$, the conditions of the theorem hold when $0<c<2(p-2)$, and in particular when $c=p-2$, $p\geqslant 3$.

4. Hierarchical Bayes Estimation

Section 2 is devoted to classical empirical Bayes estimation, i.e., when the unknown prior parameters are estimated by classical methods of estimation such as uniformly minimum variance unbiased estimation, maximum likelihood estimation, best invariant estimation, etc. Instead, one can assign prior distributions (proper or improper) to the hyperparameters, and come up with hierarchical Bayes (HB) estimators of μ_1. Note that in a classical EB approach, the lower stage Bayesian analysis is performed as if the hyperparameters were known a priori. This approach ignores the error associated with the estimation of the hyperparameters. On the other hand, the HB approach models the uncertainty of the hyperparameters by the second stage prior. Accordingly, unlike positive part EB estimators, the HB estimators are smooth, and bear the potentiality of being admissible.

To introduce the HB model, first note that as in Section 2, one may start with the minimal sufficient statistic $(\bar{X}_1, \bar{X}_2, \operatorname{tr}(V_1^{-1} S_1 + V_2^{-1} S_2))$. Write $r^{-1} = \sigma^2$ and $(\rho r)^{-1} = \tau^2$, i.e., $\rho = \sigma^2/\tau^2$. Now conditional on μ_1, μ_2, and r, $\bar{X}_1$, $\bar{X}_2$, and $U = \operatorname{tr}(V_1^{-1} S_1 + V_2^{-1} S_2)$ are mutually independent with $\bar{X}_1 \sim N_p(\mu_1, (n_1 r)^{-1} V_1)$, $\bar{X}_2 \sim N_p(\mu_2, (n_2 r)^{-1} V_2)$, and $U \sim r^{-1}\chi^2_{(n_1+n_2-2)p}$. Next we assume that conditional on ν, ρ, and r, μ_1 and μ_2 are mutually *independent* with $\mu_1 \sim N(\nu, (r\rho)^{-1} n_1^{-1} V_1)$ and $\mu_2 \sim N(\nu, (\rho r)^{-1} n_2^{-1} V_2)$. Also, it is assumed that ν, ρ, and r are mutually *independent* with ν uniform on R^p, ρ has the type II Beta distribution with pdf $h_1(\rho) \propto \rho^{m-1}(1+\rho)^{-(m+1)} I_{[\rho>0]}$, where m (>0) is known, while r has a gamma distribution with pdf $h_2(r) \propto \exp(-\frac{1}{2}\alpha r) r^{\delta-1}$, α (>0) and δ (>0) being known. We shall aim at finding the posterior distribution of $\mu = (\mu_1^T, \mu_2^T)'$ given $\bar{X}_1$, $\bar{X}_2$, and u.

First note that the joint prior distribution of μ_1, μ_2, ν, r, and ρ is given by

$$f(\mu_1, \mu_2, \nu, r, \rho) \propto (\rho r)^p \times \exp\left[-\frac{\rho r}{2}\{n_1(\mu_1-\nu)^T V_1^{-1}(\mu_1-\nu) + n_2(\mu_2-\nu)^T V_2^{-1}(\mu_2-\nu)\right] \times h_1(\rho) h_2(r). \tag{4.1}$$

Next observe that

$$\begin{aligned} n_1(\mu_1-\nu)^T V_1^{-1}(\mu_1-\nu) &+ n_2(\mu_2-\nu)^T V_2^{-1}(\mu_2-\nu) \\ &= [(\nu-\mu_*)^T V_*^{-1}(\nu-\mu_*)] \\ &\quad + n_1\mu_1^T V_1^{-1}\mu_1 + n_2\mu_2^T V_2^{-1}\mu_2 - \mu_*^T V_*^{-1}\mu_*, \end{aligned} \tag{4.2}$$

where one may recall that $\mu_* = (n_1 V_1^{-1} + n_2 V_2^{-1})^{-1}(n_1 V_1^{-1}\mu_1 + n_2 V_2^{-1}\mu_2) = (V_*^{-1})^{-1}(n_1 V_1^{-1}\mu_1 + n_2 V_2^{-1}\mu_2)$ with $V_*^{-1} = n_1 V_1^{-1} + n_2 V_2^{-1}$. Now integrating with respect to ν, one gets the joint *pdf* of μ_1, μ_2, r, and ρ in the form

$$f(\mu_1, \mu_2, r, \rho) \propto (\rho r)^{p/2} \exp\left[-\frac{\rho r}{2}\{n_1\mu_1^T V_1^{-1}\mu_1 + n_2\mu_2^T V_2^{-1}\mu_2 - \mu_*^T V_*^{-1}\mu_*\}\right] \times h_1(\rho) h_2(r). \tag{4.3}$$

The exponent in (4.3) is easily simplified as

$$\begin{aligned} &n_1\mu_1^T V_1^{-1}\mu_1 + n_2\mu_2^T V_2^{-1}\mu_2 - \mu_*^T V_*^{-1}\mu_* \\ &\quad = \mu_1^T\{n_1 V_1^{-1} - n_1 V_1^{-1} V_* n_1 V_1^{-1}\}\mu_1 + \mu_2^T\{n_2 V_2^{-1} - n_2 V_2^{-1} V_* n_2 V_2^{-1}\}\mu_2 \\ &\qquad - \mu_1^T n_1 V_1^{-1} V_* n_2 V_2^{-1}\mu_2 - \mu_2^T n_2 V_2^{-1} V_* n_1 V_1^{-1}\mu_1, \end{aligned} \tag{4.4}$$

where $V_* = (n_1 V_1^{-1} + n_2 V_2^{-1})^{-1}$. Also, the joint pdf of $\bar{X}_1$, $\bar{X}_2$, and U conditional on μ_1, μ_2, and r is given by

$$\begin{aligned} &f(\bar{x}_1, \bar{x}_2, u|\mu_1, \mu_2, r) \\ &\quad \propto r^p \exp[-r/2\{n_1(\bar{x}_1 - \mu_1)^T V_1^{-1}(\bar{x}_1 - \mu_1) + n_2(\bar{x}_2 - \mu_2)^T V_2^{-1}(\bar{x}_2 - \mu_2)\}] \\ &\qquad \times \exp(-ru/2)\, u^{(n_1+n_2-2)p/2-1}\, r^{(n_1+n_2-2)p/2}. \end{aligned} \tag{4.5}$$

Next we calculate

$$\begin{aligned} G &= n_1(\mu_1 - \bar{x}_1)^T V_1^{-1}(\mu_1 - \bar{x}_1) + n_2(\mu_2 - \bar{x}_2)^T V_2^{-1}(\mu_2 - \bar{x}_2) \\ &\quad + \rho\{n_1\mu_1^T V_1^{-1}\mu_1 + n_2\mu_2^T V_2^{-1}\mu_2 - \mu_*^T V_*^{-1}\mu_*\} \end{aligned} \tag{4.6}$$

which is needed to derive the posterior distribution of μ given $\bar{x}_1$, $\bar{x}_2$, and u. Using (4.4) and straightforward algebra, one gets

$$\begin{aligned} G &= \mu_1^T D_{11}\mu_1 + \mu_2^T D_{22}\mu_2 - 2\mu_1^T D_{12}\mu_2 - 2n_1\bar{x}_1^T V_1^{-1}\mu_1 - 2n_2\bar{x}_2^T V_2^{-1}\mu_2 \\ &\quad + n_1\bar{x}_1^T V_1^{-1}\bar{x}_1 + n_2\bar{x}_2^T V_2^{-1}\bar{x}_2, \end{aligned} \tag{4.7}$$

where

$$\begin{aligned} D_{11} &= n_1 V_1^{-1} + \rho\{n_1 V_1^{-1} - n_1 V_1^{-1} V_* n_1 V_1^{-1}\}, \\ D_{22} &= n_2 V_2^{-1} + \rho\{n_2 V_2^{-1} - n_2 V_2^{-1} V_* n_2 V_2^{-1}\}, \\ D_{12} &= \rho n_1 V_1^{-1} V_* n_2 V_2^{-1}. \end{aligned} \tag{4.8}$$

We now write G as $G_1 + G_2$, where

$$\begin{aligned} G_1 = [&(\mu_1 - A_{11}\bar{x}_1 - A_{12}\bar{x}_2)^T D_{11}(\mu_1 - A_{11}\bar{x}_1 - A_{12}\bar{x}_2) \\ &+ (\mu_2 - A_{21}\bar{x}_1 - A_{22}\bar{x}_2)^T D_{22}(\mu_2 - A_{21}\bar{x}_1 - A_{22}\bar{x}_2) \\ &- 2(\mu_1 - A_{11}\bar{x}_1 - A_{12}\bar{x}_2)^T D_{12}(\mu_2 - A_{21}\bar{x}_1 - A_{22}\bar{x}_2)] \end{aligned} \tag{4.9}$$

and

$$\begin{aligned} G_2 = [&n_1\bar{x}_1^T V_1^{-1}\bar{x}_1 + n_2\bar{x}_2^T V_2^{-1}\bar{x}_2 - (A_{11}\bar{x}_1 + A_{12}\bar{x}_2)^T D_{11}(A_{11}\bar{x}_1 + A_{12}\bar{x}_2) \\ &- (A_{21}\bar{x}_1 + A_{22}\bar{x}_2)^T D_{22}(A_{21}\bar{x}_1 + A_{22}\bar{x}_2) \\ &+ 2(A_{11}\bar{x}_1 + A_{12}\bar{x}_2)^T D_{12}(A_{21}\bar{x}_1 + A_{22}\bar{x}_2)]. \end{aligned} \tag{4.10}$$

From (4.7), (4.9), and (4.10), it follows that A_{11}, A_{12}, A_{21}, and A_{22} satisfy

$$\begin{aligned} D_{11}A_{11} - D_{12}A_{21} &= n_1 V_1^{-1} \\ D_{22}A_{22} - D_{12}^T A_{12} &= n_2 V_2^{-1} \\ D_{11}A_{12} &= D_{12}A_{22} \\ D_{22}A_{21} &= D_{12}^T A_{11} \end{aligned} \tag{4.11}$$

which can be rewritten as

$$\begin{aligned} A_{12} = D_{11}^{-1} D_{12} A_{22}, \ A_{21} &= D_{22}^{-1} D_{12}^T A_{11}, \\ (D_{11} - D_{12} D_{22}^{-1} D_{12}^T) A_{11} &= n_1 V_1^{-1}, \\ (D_{22} - D_{12}^T D_{11}^{-1} D_{12}) A_{22} &= n_2 V_2^{-1}. \end{aligned} \tag{4.12}$$

The following lemma whose proof is omitted (see [4] for details) is crucial to further simplification of G_2. Recall that $B = \sigma^2/(\sigma^2 + \tau^2) = \rho/(1+\rho)$ and $W = (n_1 V_1^{-1} + n_2 V_2^{-1})^{-1}(n_1 V_1^{-1}\bar{x}_1 + n_2 V_2^{-1}\bar{x}_2) = V_*(n_1 V_1^{-1}\bar{x}_1 + n_2 V_2^{-1}\bar{x}_2)$.

LEMMA 4.1.

$$A_{11}\bar{x}_1 + A_{12}\bar{x}_2 = (1-B)\,\bar{x}_1 + BW = b_1 \qquad \text{(say)}, \tag{4.13}$$

$$A_{21}\bar{x}_1 + A_{22}\bar{x}_2 = (1-B)\,\bar{x}_2 + BW = b_2 \qquad \text{(say)}. \tag{4.14}$$

From (4.10), (4.13), and (4.14), G_2 can be simplified as

$$\begin{aligned} G_2 &= n_1 \bar{x}_1^T V_1^{-1} \bar{x}_1 + n_2 \bar{x}_2^T V_2^{-1} \bar{x}_2 \\ &\quad - \{(1-B)\,\bar{x}_1^T + BW^T\}\, D_{11}\{(1-B)\,\bar{x}_1 + BW\} \\ &\quad - \{(1-B)\,\bar{x}_2^T + BW^T\}\, D_{22}\{(1-B)\,\bar{x}_2 + BW\} \\ &\quad + 2\{(1-B)\,\bar{x}_1^T + BW^T\}\, D_{12}\{(1-B)\,\bar{x}_2 + BW\} \\ &= \bar{x}_1^T[n_1 V_1^{-1} - \{(1-B)I \\ &\quad + Bn_1 V_1^{-1} V_*\} D_{11}\{(1-B)\,I + Bn_1 V_* V_1^{-1}\} \\ &\quad - (Bn_1 V_1^{-1} V_*)\, D_{22}(Bn_1 V_* V_1^{-1}) \\ &\quad + 2\{(1-B)\,I + Bn_1 V_1^{-1} V_*\}\, D_{12}(Bn_1 V_* V_1^{-1})]\,\bar{x}_1 \\ &\quad + \bar{x}_2^T[n_2 V_2^{-1} - \{(1-B)\,I \\ &\quad + Bn_2 V_2^{-1} V_*\}\, D_{22}\{(1-B)\,I + Bn_2 V_* V_2^{-1}\} \\ &\quad - (Bn_2 V_2^{-1} V_*)\, D_{11}(Bn_2 V_* V_2^{-1}) + 2\{(1-B)\,I \\ &\quad + Bn_2 V_2^{-1} V_*\}\, D_{12}^T(Bn_2 V_* V_2^{-1})]\,\bar{x}_2 \\ &\quad - \bar{x}_1^T[\{(1-B)\,I + Bn_1 V_1^{-1} V_*\}\, D_{11}(Bn_2 V_* V_2^{-1}) \end{aligned}$$

$$+ (Bn_1 V_1^{-1} V_*) D_{22}(1-B) I$$
$$+ Bn_2 V_* V_2^{-1}\} + 2\{(1-B) I + Bn_1 V_1^{-1} V_*\} D_{12}\{(1-B) I$$
$$+ Bn_2 V_* V_2^{-1}\}]\bar{x}_2$$
$$- \bar{x}_2^T[(Bn_2 V_2^{-1} V_*) D_{11}\{(1-B) I$$
$$+ Bn_1 V_* V_1^{-1}\} + \{(1-B) I + Bn_2 V_2^{-1} V_*\} D_{22}$$
$$\times B(n_1 V_* V_1^{-1}) + 2(Bn_2 V_2^{-1} V_*) D_{12}(Bn_1 V_* V_1^{-1})]\bar{x}_1. \quad (4.15)$$

From (4.8), one gets

$$\begin{aligned} D_{11} + D_{22} - 2D_{12} &= (1+\rho)(n_1 V_1^{-1} + n_2 V_2^{-1}) \\ &\quad - \rho(n_1 V_1^{-1} + n_2 V_2^{-1}) V_*(n_1 V_1^{-1} + n_2 V_2^{-1}) \\ &= n_1 V_1^{-1} + n_2 V_2^{-1} \qquad (\text{since } V_*^{-1} = n_1 V_1^{-1} + n_2 V_2^{-1}) \\ &= V_*^{-1}. \end{aligned} \quad (4.16)$$

Using (4.8) and (4.16), it is possible to simplify G_2 considerably. This is done in the following lemma whose proof is again omitted (see [4] for details).

LEMMA 4.2. $G_2 = B[\bar{x}_1^T\{n_1 V_1^{-1} - n_1 V_1^{-1} V_* n_1 V_1^{-1}\}\, \bar{x}_1 + \bar{x}_2^T\{n_2 V_2^{-1} - n_2 V_2^{-1} V_* n_2 V_2^{-1}\}\, \bar{x}_2 - 2\bar{x}_1^T(n_1 V_1^{-1} V_* n_2 V_2^{-1})\, \bar{x}_2]$.

Therefore, from (4.9), Lemma 4.1, and Lemma 4.2, G can be written as

$$\begin{aligned} G = (\mu_1 - b_1)^T D_{11}(\mu_1 - b_1) + (\mu_2 - b_2)^T D_{22}(\mu_2 - b_2) \\ - 2(\mu_1 - b_1)^T D_{12}(\mu_2 - b_2) \\ + B[\bar{x}_1^T D_{11} * \bar{x}_1 + \bar{x}_2^T D_{22} * \bar{x}_2 - 2\bar{x}_1^T D_{12} * \bar{x}_2], \end{aligned} \quad (4.17)$$

where

$$\begin{aligned} D_{11}* &= n_1 V_1^{-1} - n_1 V_1^{-1} V_* n_1 V_1^{-1} \\ D_{22}* &= n_2 V_2^{-1} - n_2 V_2^{-1} V_* n_2 V_2^{-1} \\ D_{12}* &= n_1 V_1^{-1} V_* n_2 V_2^{-1}. \end{aligned} \quad (4.18)$$

Returning to (4.3) and (4.5), the joint pdf of $\bar{X}_1$, $\bar{X}_2$, U, μ_1, μ_2, r, and ρ is given by

$$\begin{aligned} f(\bar{x}_1, \bar{x}_2, u, \mu_1, \mu_2, r, \rho) \\ \propto r^p(\rho r)^{p/2} \cdot \exp\left[-\frac{r}{2} G\right] \cdot \exp[-ru/2] \\ \times u^{(n_1+n_2-2)p/2-1} r^{(n_1+n_2-2)p/2} h_1(\rho) h_2(r). \end{aligned} \quad (4.19)$$

It follows from (4.17) and (4.19) that conditional on $\bar{x}_1, \bar{x}_2, u, r$, and ρ,

$$\begin{pmatrix} \mu_1 \\ \mu_2 \end{pmatrix} \sim N_{2p}\left[\begin{pmatrix} b_1 \\ b_2 \end{pmatrix}, r^{-1}\begin{pmatrix} D_{11} & -D_{12} \\ -D_{12}^T & D_{22} \end{pmatrix}^{-1}\right]. \tag{4.20}$$

Also, integrating out with respect to μ_1 and μ_2, it follows from (4.19) that the joint pdf of $X_1, \bar{X}_2, U, r$, and ρ is given by

$$f(\bar{x}_1, \bar{x}_2, u, r, \rho) \propto (\rho r)^{p/2} \begin{vmatrix} D_{11} & -D_{12} \\ -D_{12}^T & D_{22} \end{vmatrix}^{-1/2} \exp\left[-\frac{r}{2}\{u + BSS_H\}\right]$$
$$\times r^{(n_1+n_2-2)p/2} \cdot u^{(n_1+n_2-2)p/2-1} \cdot \rho^{m-1}(1+\rho)^{-(m+1)} \cdot \exp(-\alpha r/2)\, r^{\delta-1}, \tag{4.21}$$

where

$$SS_H = \bar{x}_1^T D_{11} * \bar{x}_1 + \bar{x}_2^T D_{22} * \bar{x}_2 - 2\bar{x}_1^T D_{12} * \bar{x}_2. \tag{4.22}$$

Now, from (4.8), one gets

$$\begin{vmatrix} D_{11} & -D_{12} \\ -D_{12}^T & D_{22} \end{vmatrix}$$
$$= \left|(1+\rho)\begin{pmatrix} n_1 V_1^{-1} & 0 \\ 0 & n_2 V_2^{-1} \end{pmatrix} - \rho \begin{pmatrix} n_1 V_1^{-1} V_* n_1 V_1^{-1} & n_1 V_1^{-1} V_* n_2 V_2^{-1} \\ n_2 V_2^{-1} V_* n_1 V_1^{-1} & n_2 V_2^{-1} V_* n_2 V_2^{-1} \end{pmatrix}\right|$$
$$= (1+\rho)^{2p}\left|\begin{pmatrix} n_1 V_1^{-1} & 0 \\ 0 & n_2 V_2^{-1} \end{pmatrix} - B\begin{pmatrix} n_1 V_1^{-1} \\ n_2 V_2^{-1} \end{pmatrix} V_*(n_1 V_1^{-1} : n_2 V_2^{-1})\right|$$
$$= (1+\rho)^{2p}\begin{vmatrix} n_1 V_1^{-1} & 0 \\ 0 & n_2 V_2^{-1} \end{vmatrix}$$
$$\times \left| I_{2p} - B\begin{pmatrix} \frac{V_1^{1/2}}{n_1} & 0 \\ 0 & \frac{V_2^{1/2}}{n_2} \end{pmatrix}\begin{pmatrix} n_1 V_1^{-1} \\ n_2 V_2^{-1} \end{pmatrix} V_*(n_1 V_1^{-1} : n_2 V_2^{-1})\begin{pmatrix} \frac{V_1^{1/2}}{n_1} & 0 \\ 0 & \frac{V_1^{1/2}}{n_2} \end{pmatrix}\right|$$
$$= (1+\rho)^{2p}|n_1 V_1^{-1}|\,|n_2 V_2^{-1}|$$
$$\times \left| I_p - B(n_1 V_1^{-1} : n_2 V_2^{-1})\begin{pmatrix} \frac{V_1}{n_1} & 0 \\ 0 & \frac{V_2}{n_2} \end{pmatrix}\begin{pmatrix} n_1 V_1^{-1} \\ n_2 V_2^{-1} \end{pmatrix} V_*\right|$$
$$= (1+\rho)^{2p}\,|n_1 V_1^{-1}|\,|n_2 V_2^{-1}|\,|I_p - BI_p| \qquad (\text{since } V_*^{-1} = n_1 V_1^{-1} + n_2 V_2^{-1})$$
$$= (1+\rho)^{p}\,|n_1 V_1^{-1}|\,|n_2 V_2^{-1}|. \tag{4.23}$$

Hence, from (4.21) and (4.23), one gets

$$f(\bar{x}_1, \bar{x}_2, u, r, \rho) \propto \rho^{p/2} r^{(n_1+n_2-1)p/2}(1+\rho)^{-p/2} \exp\left[-\frac{r}{2}(u+BSS_H+\alpha)\right]$$

$$\times u^{(n_1+n_2-2)p/2-1} \cdot r^{\delta-1} \rho^{m-1}(1+\rho)^{-(m+1)}. \tag{4.24}$$

Integrating out with respect to r, one gets the joint pdf of $\bar{X}_1$, $\bar{X}_2$, U, and ρ as

$$f(\bar{x}_1, \bar{x}_2, u, \rho) \propto \left(\frac{\rho}{1+\rho}\right)^{p/2} u^{(n_1+n_2-2)p/2-1}(u+BSS_H+\alpha)^{-(n_1+n_2-1)p/2-\delta}$$

$$\times \rho^{m-1}(1+\rho)^{-(m+1)}. \tag{4.25}$$

Using the transformation $\rho/(1+\rho)=B$ provides the joint pdf of $\bar{X}_1$, $\bar{X}_2$, U, and B as

$$f(\bar{x}_1, \bar{x}_2, u, B) \propto B^{p/2+m-1} u^{(n_1+n_2-2)p/2-1}(u+BSS_H+\alpha)^{-(n_1+n_2-1)p/2-\delta}. \tag{4.26}$$

Next observe from (4.20) and (4.13) that

$$E(\mu_1 \mid B, \bar{x}_1, \bar{x}_2, u, r) = b_1 = (1-B)\,\bar{x}_1 + BW.$$

Hence the HB estimator of μ_1 is

$$E(\mu_1 \mid \bar{x}_1, \bar{x}_2, u) = \bar{x}_1 - E(B \mid \bar{x}_1, \bar{x}_2, u)(\bar{x}_1 - W). \tag{4.27}$$

But, from (4.26), one gets

$$E(B \mid \bar{x}_1, \bar{x}_2, u) = \frac{\int_0^1 B^{p/2+m}(u+BSS_H+\alpha)^{-(n_1+n_2-1)p/2-\delta}\,dB}{\int_0^1 B^{p/2+m-1}(u+BSS_H+\alpha)^{-(n_1+n_2-1)p/2-\delta}\,dB}. \tag{4.28}$$

Remark 4.1. From simultaneous diagonalization of $n_1 V_1^{-1}$ and $n_2 V_2^{-1}$, it is easy to show from (4.18) that

$$D_{11}* = D_{22}* = D_{12}* = (n_1 V_1^{-1} + n_2 V_2^{-1})^{-1}, \tag{4.29}$$

so that from (4.22) one gets

$$SS_H = (\bar{x}_1 - \bar{x}_2)^T (n_1 V_1^{-1} + n_2 V_2^{-1})^{-1} (\bar{x}_1 - \bar{x}_2) \tag{4.30}$$

which is precisely the numerator of F defined in (3.1).

Remark 4.2. It is sometimes possible to reduce the above HB estimator to an EB estimator of the form $\bar{x}_1 - (\phi(F)/F)(\bar{x}_1 - W)$. Consider for example the situation when $\alpha = 0$, i.e., R has the improper prior $h_2(r) = r^{\delta-1}$.

Now writing $v = SS_H/u$, we note from (4.30) that $F = ((n_1+n_2-2)p+2)v$. Also, for $\alpha = 0$, it follows from (4.28) that

$$E(B \mid \bar{x}_1, \bar{x}_2, u)$$

$$= \int_0^1 B^{p/2+m}(1+Bv)^{-(n_1+n_2-1)p/2-\delta}\, dB \Big/$$

$$\int_0^1 B^{p/2+m-1}(1+Bv)^{-(n_1+n_2-1)p/2-\delta}\, dB$$

$$= v^{-1}\int_0^1 \left(\frac{1}{1+Bv}\right)^{(n_1+n_2-2)p/2+\delta-m-2}\left(\frac{Bv}{1+Bv}\right)^{p/2+m}\frac{v\,dB}{(1+Bv)^2}$$

$$\div \int_0^1 \left(\frac{Bv}{1+Bv}\right)^{p/2+m-1}\left(\frac{1}{1+Bv}\right)^{(n_1+n_2-2)p/2+\delta-m-1}\frac{v\,dB}{(1+Bv)^2}$$

$$= v^{-1}\int_0^{v/(1+v)} u^{p/2+m}(1-u)^{(n_1+n_2-2)p/2+\delta-m-2}\, du$$

$$\div \int_0^{v/(1+v)} u^{p/2+m-1}(1-u)^{(n_1+n_2-2)p/2+\delta-m-1}\, du. \tag{4.31}$$

From (4.31) it follows that $E(B \mid \bar{x}_1, \bar{x}_2, u)$ can be expressed as $\phi^*(v)/v = \phi(F)/F$. Next note that integration by parts gives numerator of (4.31) equals

$$v^{-1}\left\{-\left(\frac{v}{1+v}\right)^{p/2+m}\frac{(1+v)^{-(n_1+n_2-2)p/2-\delta+m+1}}{(n_1+n_2-2)p/2+\delta-m-1}\right\}$$

$$+\frac{(p/2+m)}{v((n_1+n_2-2)p/2+\delta-m-1)}$$

$$\times \int_0^{v/(1+v)} u^{p/2+m-1}(1-u)^{(n_1+n_2+2)p/2+\delta-m-1}\, du$$

$$\leqslant \frac{p+2m}{v\{(n_1+n_2-2)p+2\delta-2m-2\}}$$

$$\times \int_0^{v/(1+v)} u^{p/2+m-1}(1-u)^{(n_1+n_2-2)p/2+\delta-m-1}\, du. \tag{4.32}$$

Hence from (4.31) and (4.32),

$$E(B \mid \bar{x}_1, \bar{x}_2, u) \leqslant \frac{p+2m}{v\{(n_1+n_2-2)p+2\delta-2m-2\}}$$

$$= \frac{(p+2m)((n_1+n_2-2)p+2)}{F\{(n_1+n_2-2)p+2\delta-2m-2\}} \tag{4.33}$$

so that

$$\phi(F) \leqslant \frac{(p+2m)((n_1+n_2-2)p+2)}{(n_1+n_2-2)p+2\delta-2m-2}$$
$$< 2(p-2)$$

if $(p+2m)((n_1+n_2-2)p+2) < 2(p-2)((n_1+n_2-2)p-2m-2)(\because \delta>0)$ $\leftrightarrow p\{2m(n_1+n_2)+6\} < p(p-4)(n_1+n_2-2)+4m+8$ which holds whenever $p \geqslant 5$ and $m < \{(p-4)(n_1+n_2-2)-6\}/2(n_1+n_2)$, assuming $n_1+n_2>8$. Hence, for this choice of m, $\phi(F)$ satisfies condition (ii) of Theorem 3.2 for $Q=V_1=V_2=I_p$. Also, for $Q=V_1=V_2=I_p$, condition (i) of Theorem 3.2 automatically holds when $p \geqslant 3$.

Finally, noting that v is strictly increasing in F, and using the inequality

$$\int_0^{v/(1+v)} u^{p/2+m}(1-u)^{(n_1+n_2-2)p/2+\delta-m-2}\,du$$
$$= \int_0^{v/(1+v)} \left(\frac{u}{1-u}\right) u^{p/2+m-1}(1-u)^{(n_1+n_2-2)p/2+\delta-m-1}\,du$$
$$\leqslant v\int_0^{v/(1+v)} u^{p/2+m-1}(1-u)^{(n_1+n_2-2)p/2+\delta-m-1}\,du, \qquad (4.34)$$

one gets after direct differentiation $\phi^*(v)' \geqslant 0$. Hence $\phi^*(v)$ is $\uparrow$ in v. Hence, condition (iii) of Theorem 3.2 also holds. Therefore, when $\alpha=0$, $Q=V_1=V_2=I_p$, $p \geqslant 5$, and $0<m<\{(p-4)(n_1+n_1-2)-6\}/2(n_1+n_2)$, the HB estimator obtained in (4.27) is minimax.

Remark 4.3. The conclusion given in Remark 4.2 bears strong resemblance to Strawderman [9] in the one sample problem. However, the formulation here is much more general than the one given in Strawderman [9 or 10]. First, the estimator is not shrunk towards zero or a prespecified point, but is shrunk towards the pooled mean. In Strawderman [9], r is assumed to be known, whereas in Strawderman [10], r is assumed to belong to (γ, ∞) for some $\gamma>0$. Our formulation is also more general than the one given in Morris [5] because there r is assumed known and $(\rho r)^{-1}$ is given a uniform prior on $(0, \infty)$.

References

[1] Bancroft, T. A., and Han, C. P. (1981). Inference based on conditionally specified ANOVA models incorporating preliminary testing. In *Handbook of Statistics*, Vol. 1 (P. R. Krishnaiah, Ed.), pp. 407–441. North-Holland, New York.

[2] Efron, B., and Morris, C. (1973). Stein's estimation rule and its competitors—An empirical Bayes approach. *J. Amer. Statist. Assoc.* **68** 117–130.

[3] Ghosh, M., Saleh, A. K. MD. E., and Sen, P. K. (1987). *Empirical Bayes Subset Estimation in Regression Models.* Tech. Report 281. Department of Statistics, University of Florida.

[4] Ghosh, M. and Sinha, B. K. (1987). *Empirical and Hierarchical Bayes Competitors of Preliminary Test Estimators in Two Sample Problems.* Technical Report, Center for Multivariate Analysis, University of Pittsburgh, and University of Maryland Baltimore County.

[5] Morris, C. (1983). Parametric empirical Bayes confidence intervals. *Scientific Inference, Data Analysis and Robustness* (G. E. P. Box, T. Leonard, and C. F. J. Wu, Eds), pp. 25–50. Academic Press, New York.

[6] Saleh, A. K. Md. E., and Ahmed, S. E. (1987). On preliminary-test and shrinkage estimation of the mean vector of a multivariate normal distribution in a two-sample problem. Private communication.

[7] Sclove, S. L., Morris, C., and Radhakrishnan, R. (1972). Nonoptimality of preliminary test estimators for the multinormal mean. *Ann. Math. Statist.* **43** 1481–1490.

[8] Stein, C. (1981). Estimation of the mean of a multivariate normal distribution. *Ann. Statist.* **9** 1135–1151.

[9] Strawderman, W. E. (1971). Proper Bayes minimax estimators of the multivariate normal mean. *Ann. Math. Statist.* **42** 385–388.

[10] Strawderman, W, E, (1973). Proper Bayes minimax estimators of the multivariate normal mean vector for the case of common unknown variances. *Ann. Statist.* **1** 1189–1194.

On Confidence Bands in Nonparametric Density Estimation and Regression

PETER HALL

Australian National University, Canberra, Australia

AND

D. M. TITTERINGTON

University of Glasgow, Glasgow, Scotland

We describe a unified approach to the construction of confidence bands in nonparametric density estimation and regression. Our techniques are based on interpolation formulae in numerical differentiation, and our arguments generate a variety of bands depending on the assumptions one is prepared to make about derivatives of the unknown function. The bands are *simultaneous*, in the sense that they contain the *entire function* with probability at least an amount. The order of magnitude of the minimum width of any confidence band is described, and our bands are shown to achieve that order. Examples illustrate applications of the technique.

1. INTRODUCTION

There is a prolific recent literature on the topic of nonparametric density estimation and regression. In most of the research, however, the methodology stops at the point of constructing a "point estimate" of the underlying density or regression function. Some form of interval estimation is obviously desirable and, ideally, one would wish for simultaneous confidence bands. This would allow graphical answers to questions like:

(i) Is it plausible that the true density is unimodal?

(ii) Is there clear evidence against the hypothesis that the true regression function is linear?

Multivariate Statistics and Probability
ISBN 0-12-580205-6

Reprinted from *J. Mult. Anal.* **27**(1).

In the case of nonparametric density estimation almost no work has been done on the confidence band aspect of the problem, although Hartigan and Hartigan [3] consider a version of the problem based on cumulative distribution functions. There has been more activity in nonparametric regression. Wahba [8] and Silverman [5] use a Bayesian interpretation of the prescription that leads to curve estimation using splines, to construct confidence bands. However, these are not simultaneous bands in the usual sense of the term. Härdle [2] proposes asymptotic simultaneous confidence bands in a regression context.

The present paper develops a unified procedure for dealing with both types of problems. In contradistinction to Wahba [8], Silverman [5], and Härdle [2], our confidence bands are not constructed as lines on either side of a curve estimate, but are derived from first principles as upper or lower bounds to the curve. In the regression case our confidence bands are related to those of Knafl, Sacks, and Ylvisaker [4], in that they are based on linear (in the data) estimates of the regression function at any given point. However, the linear functions used here are much simpler than those employed by Knafl, Sacks, and Ylvisaker [4], and their foundation is such as to make calculation of the widths of the bands very much easier. In spite of this simplicity, the methods are backed up by reassuring properties of "asymptotic optimality."

Section 2 describes the case of nonparametric density estimation, and shows how formulae from the theory of numerical differentiation may be used to develop a succession of confidence bands under a variety of assumptions. The parallel development for nonparametric regression follows in Section 3. Theoretical results about the widths of the bands are given in Section 4, two illustrative examples are described in Section 5, and proofs are given in Section 6.

2. Nonparametric Density Estimation

The problem of determining confidence bands is closely related to that of numerical differentiation. The bands proposed in this section are based on the number of observations which lie within adjacent intervals ("cells") of width h. The means of these numbers equal integrals of the density over the respective intervals. We numerically differentiate the integrals, to obtain approximate fomulae for the integrands—i.e., for the density itself. The errors in these numerical approximations must somehow be incorporated into the confidence band. Now, the errors in numerical differentiation procedures behave in a manner more complicated than the errors in, say, a Taylor expansion. In particular, if numerical differentiation of a function F is conducted by interpolation among a sequence of points

$a_0 < a_1 < \cdots < a_m$, then usually the error can be expressed in terms of a single value of $F^{(m+1)}$ only when the argument lies *outside* the observation interval (a_0, a_m). If the argument lies inside (a_0, a_m) then the size of the error depends on values of several derivatives, or on several differences of one or more derivatives. See, for example, the discussion in Steffensen [6, pp. 64–65]. It would often be unacceptable to use interpolation within (a_0, a_m) to estimate F' at a point outside (a_0, a_m), since this might involve relatively large error terms. On the other hand, a confidence band which requires knowledge about several different derivatives of the density is not a practical proposition. In Subsection 2.1 below, procedures (i) and (ii) illustrate confidence bands obtained by interpolation outside the interval (a_0, a_m) (there $m = 1$), while procedure (iii) is a compromise which sacrifices a certain amount of "exactness" in return for a smoother confidence band.

The following notation will be used throughout this section. Assume that a random sample of size n is drawn from the distribution with density f. Using these data, we wish to construct a confidence band for f over a certain interval. In that region divide the data among k cells, the cell numbered i comprising the interval $((i-1)h, ih)$ and h being the width of each cell. If the true density is f then

$$p_i \equiv \int_{(i-1)h}^{ih} f(x)\, dx, \qquad 1 \leqslant i \leqslant k,$$

is the probability that a given data point falls into cell i. (Our convention that the first cell starts at the origin serves only to simplify notation.) The confidence bands are developed from simultaneous confidence intervals for the multinomial proportions p_i. Thus, we assume intervals $[\hat{p}_{i1}, \hat{p}_{i2}]$, $1 \leqslant i \leqslant k$, are given such that

$$P(\hat{p}_{i1} \leqslant p_i \leqslant \hat{p}_{i2}, 1 \leqslant i \leqslant k) = \alpha.$$

Define the function $\hat{f}_j$ by interpolating among the function values $\hat{f}_j\{(i+\frac{1}{2})h\} \equiv h^{-1}\hat{p}_{i+1,j}$,

$$\hat{f}_j\{(i+y)h\} \equiv (\tfrac{1}{2} - y)h^{-1}\hat{p}_{ij} + (\tfrac{1}{2} + y)h^{-1}\hat{p}_{i+1,j}, \tag{2.1}$$

for $1 \leqslant i \leqslant k-1$, $-\frac{1}{2} < y \leqslant \frac{1}{2}$, and $j = 1, 2$. Notice that $\hat{f}_1$ and $\hat{f}_2$ are continuous. The band between $\hat{f}_1$ and $\hat{f}_2$ forms the basis for several of our procedures.

The next two subsections list several different types of confidence band. These examples serve to illustrate the theoretical properties of general confidence bands based on the confidence intervals $[\hat{p}_{i1}, \hat{p}_{i2}]$. They form the basis for the practical procedures introduced in Subsection 2.4. Subsection 2.3 describes construction of the intervals $[\hat{p}_{i1}, \hat{p}_{i2}]$.

2.1. *Confidence Bands under the Assumption of a Single Derivative*

(i) Given a sequence $\{c_i\}$ with each $c_i \geqslant 0$, define

$$\bar{f}_1\{(i+y)h\} \equiv h^{-1}\hat{p}_{i1} - \tfrac{1}{2}(2y+1)hc_i$$

and

$$\bar{f}_2\{(i+y)h\} \equiv h^{-1}\hat{p}_{i2} + \tfrac{1}{2}(2y+1)hc_i$$

for $1 \leqslant i \leqslant k$ and $0 < y \leqslant 1$. If

$$\sup_{(i-1)h \leqslant u \leqslant (i+1)h} |f'(u)| \leqslant c_i \qquad \text{for} \quad 1 \leqslant i \leqslant k \tag{2.2}$$

then

$$P\{\bar{f}_1(x) \leqslant f(x) \leqslant \bar{f}_2(x) \qquad \text{for} \quad h \leqslant x \leqslant (k+1)h\} \geqslant \alpha. \tag{2.3}$$

(ii) Given $\varepsilon \geqslant 0$, define

$$\bar{f}_{1,\pm}\{(i+y)h\} \equiv h^{-1}\hat{p}_{i1} - \tfrac{1}{2}(2y+1)h[f'\{(i+y)h\} \pm \varepsilon]$$

and

$$\bar{f}_{2,\pm}\{(i+y)h\} \equiv h^{-1}\hat{p}_{i2} - \tfrac{1}{2}(2y+1)h[f'\{(i+y)h\} - (\pm\varepsilon)]$$

for $1 \leqslant i \leqslant k$ and $0 < y \leqslant 1$, where the $+$, $-$ signs are taken respectively. If

$$|f'(u) - f'(v)| \leqslant \varepsilon$$

whenever $0 \leqslant u \leqslant v \leqslant (k+1)h$ and $|u-v| \leqslant 2h$, then

$$P\{\bar{f}_{1,+}(x) \leqslant f(x) \leqslant \bar{f}_{2,+}(x) \qquad \text{for} \quad h \leqslant x \leqslant (k+1)h\} \geqslant \alpha$$

and

$$P\{\bar{f}_{1,-}(x) \leqslant f(x) \leqslant \bar{f}_{2,-}(x) \qquad \text{for} \quad h \leqslant x \leqslant (k+1)h\} \leqslant \alpha.$$

(iii) Given a sequence $\{c_i\}$ with each $c_i \geqslant 0$, define

$$\bar{f}_1\{(i+y)h\} \equiv \hat{f}_1\{(i+y)h\} - \tfrac{1}{2}h(1-3y^2+2\,|y|^3)c_i$$

and

$$\bar{f}_2\{(i+y)h\} \equiv \hat{f}_2\{(i+y)h\} + \tfrac{1}{2}h(1-3y^2+2\,|y|^3)c_i$$

for $1 \leqslant i \leqslant k-1$ and $-\frac{1}{2} < y \leqslant \frac{1}{2}$. If (2.2) holds then

$$P\{\bar{f}_1(x) \leqslant f(x) \leqslant \bar{f}_2(x) \quad \text{for} \quad \tfrac{1}{2}h \leqslant x \leqslant (k-\tfrac{1}{2})h\} \geqslant \alpha.$$

Remarks. (a) Procedure (ii) is introduced only to illustrate the factors which influence coverage probability of a confidence band; it is not suggested as a practical procedure. It demonstrates that the basic confidence intervals $[\hat{p}_{i2}, \hat{p}_{i2}]$ are biased by an amount $\frac{1}{2}(2y+1)\,hf'\{(i+y)h\}$, plus smaller order terms.

(b) By taking $\varepsilon = 0$ in (ii) we deduce that equality holds in confidence statement (2.3) if f is linear on $(0, (k+1)h)$, if each c_i equals the absolute value of the gradient d of f, and if the intervals $[\hat{p}_{i1}, \hat{p}_{i2}]$ are of the form $[0, \hat{p}_{i2}]$ (for $d < 0$) or $[\hat{p}_{i1}, \infty)$ (for $d \geqslant 0$).

(c) No such "exactness" can be claimed for the confidence band described in (iii). However, that band has certain practical advantages over the earlier procedures. First of all, the function $\frac{1}{2}(1-3y^2+2\,|y|^3)$ lies within the interval $[\frac{1}{2}, 1]$ for $-\frac{1}{2} < y \leqslant \frac{1}{2}$, whereas the function $\frac{1}{2}(2y+1)$ takes values as large as $\frac{3}{2}$ for $0 < y \leqslant 1$. Therefore the band in (iii) can have smaller maximum width than that in (ii). Second, if the c_i's are taken to be identical then the functions $\bar{f}_1$ and $\bar{f}_2$ defined in (iii) are continuous, and so the confidence bands have continuous boundaries.

2.2. *Confidence Bands under the Assumption of Two Derivatives*

(i) Given $\varepsilon > 0$, define

$$\bar{f}_{1,\pm}\{(i+y)h\} \equiv \hat{f}_1\{(i+y)h\} + \tfrac{1}{6}(1-3y^2)h^2[f''\{(i+y)h\} - (\pm\varepsilon)]$$

and

$$\bar{f}_{2,\pm}\{(i+y)h\} \equiv \hat{f}_2\{(i+y)h\} + \tfrac{1}{6}(1-3y^2)h^2[f''\{(i+y)h\} \pm \varepsilon]$$

for $1 \leqslant i \leqslant k-1$ and $-\frac{1}{2} < y \leqslant \frac{1}{2}$. If

$$|f''(u) - f''(v)| \leqslant \varepsilon$$

whenever $0 \leqslant u \leqslant v \leqslant (k+1)h$ and $|u-v| \leqslant 2h$, then

$$P\{\bar{f}_{1,+}(x) \leqslant f(x) \leqslant \bar{f}_{2,+}(x) \quad \text{for} \quad \tfrac{1}{2}h \leqslant x \leqslant (k-\tfrac{1}{2})h\} \geqslant \alpha$$

and

$$P\{\bar{f}_{1,-}(x) \leqslant f(x) \leqslant \bar{f}_{2,-}(x) \quad \text{for} \quad \tfrac{1}{2}h \leqslant x \leqslant (k-\tfrac{1}{2})h\} \leqslant \alpha. \tag{2.4}$$

(ii) Given a sequence $\{c_i\}$ with each $c_i \geqslant 0$, define

$$\bar{f}_1\{(i+y)h\} \equiv \hat{f}_1\{(i+y)h\} - \tfrac{1}{6}(1-3y^2)h^2 c_i$$

and

$$\tilde{f}_2\{(i+y)h\} \equiv \hat{f}_2\{(i+y)h\} + \tfrac{1}{6}(1-3y^2)h^2 c_i$$

for $1 \leqslant i \leqslant k-1$ and $-\frac{1}{2} < y \leqslant \frac{1}{2}$. If

$$\sup_{(i-1)h \leqslant x \leqslant (i+1)h} |f''(x)| \leqslant c_i \qquad \text{for} \quad 1 \leqslant i \leqslant k-1,$$

then

$$P\{\tilde{f}_1(x) \leqslant f(x) \leqslant \tilde{f}_2(x) \qquad \text{for} \quad \tfrac{1}{2}h \leqslant x \leqslant (k-\tfrac{1}{2})h\} \geqslant \alpha. \tag{2.5}$$

Remarks. (a) Procedure (i) is introduced to show that the basic confidence band $(\hat{f}_1, \hat{f}_2)$ (see (2.1)) is biased by an amount $\frac{1}{6}(1-3y^2)\, h^2 f''\{(i+y)h\}$, plus smaller order terms. We do not propose it as a practical method.

(b) By taking $\varepsilon = 0$ in (i) we deduce that equality holds in confidence statement (2.5) if $f(x) \equiv a + bx + \frac{1}{2}dx^2$ for arbitrary constants a, b, and d and $0 \leqslant x \leqslant kh$, provided each $c_i = |d|$ and the intervals $[\hat{p}_{i1}, \hat{p}_{i2}]$ are of the form $[0, \hat{p}_{i1}]$ (for $d<0$) or $[\hat{p}_{i1}, \infty)$ (for $d \geqslant 0$).

(c) If the c_i's are identical then the functions $\tilde{f}_1$ and $\tilde{f}_2$ defined in (ii) are continuous and piecewise linear.

2.3. *Simultaneous Confidence Intervals for Multinomial Probabilities*

Suppose we seek confidence bands whose coverage probability is at least β. The argument given in Subsections 2.1 and 2.2 has reduced the problem of constructing confidence bands to one of deriving simultaneous confidence intervals for multinomial proportions, for which there are several techniques. In particular, if $\hat{p}_{i1}$ and $\hat{p}_{i2}$ are chosen such that

$$P(\hat{p}_{i1} \leqslant p_i \leqslant \hat{p}_{i2}) \geqslant 1-(1-\beta)k^{-1}, \qquad 1 \leqslant i \leqslant k,$$

then

$$P(\hat{p}_{i1} \leqslant p_i \leqslant \hat{p}_{i2} \text{ for } 1 \leqslant i \leqslant k) \geqslant \beta.$$

If $\hat{p}_i$ denotes the relative frequency in cell i then the normal approximation to the binomial suggests taking

$$\hat{p}_{i1} = \hat{p}_i - d_k\{\hat{p}_i(1-\hat{p}_i)\,n^{-1}\}^{1/2} \qquad \text{and} \qquad \hat{p}_{i2} = \hat{p}_i + d_k\{\hat{p}_i(1-\hat{p}_i)\,n^{-1}\}^{1/2}, \tag{2.6}$$

where

$$\Phi(d_k) = 1-(1-\beta)(2k)^{-1} \tag{2.7}$$

and Φ is the standard normal distribution function. This is the approach adopted in Example 5.1 in Section 5. Almost identical results (not reported here) were obtained using the Poisson approximation with square-root transformation, where d_k was defined by

$$\Phi(d_k) = \tfrac{1}{2}(1 + \beta^{1/k})$$

instead of by (2.7). The above definitions are tantamount to approximating the $\hat{p}_i$'s by *independent* normal random variables.

2.4. *Discussion*

The methodology developed in Subsections 2.1–2.3 leads to a variety of practical procedures for constructing confidence bands for an unknown density f. The initial band is formed by the pair of functions $(\hat{f}_1, \hat{f}_2)$ defined at (2.1). To compensate for errors arising from numerical differentiation, extra strips are added to this band. If the absolute value of the first derivative of the density does not exceed $c^{(1)}$, then strips of width (i.e., height) $\frac{1}{2}hc^{(1)}$ added to both sides of the confidence band provide more than adequate compensation. (This follows from Subsection 2.1(iii).) If the absolute value of the second derivative does not exceed $c^{(2)}$ then strips of width $\frac{1}{6}h^2c^{(2)}$ are more than adequate. (See Subsection 2.2(ii).) The bounds $c^{(1)}$ or $c^{(2)}$ may be known from previous empirical experience, or they can themselves be estimated by interpolation. Formulae in Subsections 2.1(iii) and 2.2(ii) show that the widths of these strips do not have to be maintained throughout the bands but can be varied slightly over the cells.

The procedure just described is deliberately designed to be conservative. The confidence bands can be thinned a little if we have additional knowledge about f. For example, suppose we are basing the bands on the second derivative of f. If f is convex within a certain region then only one compensating strip is required there—that strip of width $\frac{1}{6}h^2c^{(2)}$ below the lower function $\hat{f}_1$. If f is concave within a certain region, then only the upper strip of width $\frac{1}{6}h^2c^{(2)}$ above $\hat{f}_2$ is required there. Again, the strips may be reduced in places according to the formulae in subsection 2.2(ii).

An alternative approach is to estimate not just a bound to f' or f'', but the entire function. For example, if the procedure is being based on second derivatives and if $\hat{f}''$ is an estimate of f'', then approximate upper and lower confidence limits are given by

$$\tilde{f}_1\{(i+y)h\} \equiv \hat{f}_1\{(i+y)h\} + \tfrac{1}{6}(1-3y^2)\, h^2 \hat{f}''\{(i+y)h\}$$

and

$$\tilde{f}_2\{(i+y)h\} \equiv \hat{f}_2\{(i+y)h\} + \tfrac{1}{6}(1-3y^2)\, h^2 \hat{f}''\{(i+y)h\},$$

respectively, for $1 \leqslant i \leqslant k-1$ and $-\frac{1}{2} < y \leqslant \frac{1}{2}$; see Subsection 2.2(i). While this approach will give narrower confidence bands, it is difficult to be certain about the direction of the error in coverage probability.

3. Nonparametric Regression

The case of nonparametric regression is similar in many respects to that of density estimation, and so we shall give only an outline. The only essential difference between the two cases is that we no longer estimate an integral, but a sum, the arguments of the terms in the sum being design points in the regression. This change introduces a second error term into the procedure, due essentially to approximation of the integral by the sum. The confidence bands have to be adjusted accordingly.

We shall assume that observations are made at equally spaced design points, distant δ apart. See Section 3.4 for discussion of this restriction. Without loss of generality, the design points are the points $j\delta$ for integers j. The model declares that the observations Y_j have the form

$$Y_j = g(j\delta) + e_j,$$

where g is a smooth function and the e_j's are independent normal $N(0, \sigma^2)$. In the region of interest, divide the Y_j's among k cells, the ith cell containing those pairs $(j\delta, Y_j)$ of observations such that $(i-1)h \leqslant j\delta \leqslant ih$, $1 \leqslant i \leqslant k$, where $h = m\delta$ for an integer m. (Thus, the very ends of the cells overlap.) We shall treat two different estimates of the mean in cell i,

$$\bar{Y}_i^{(1)} \equiv m^{-1}(Y_{(i-1)m} + Y_{(i-1)m+1} + \cdots + Y_{im-1})$$

and

$$\bar{Y}_i^{(2)} \equiv m^{-1}(\tfrac{1}{2}Y_{(i-1)m} + Y_{(i-1)m+1} + \cdots + Y_{im-1} + \tfrac{1}{2}Y_{im}),$$

whose respective means are

$$\mu_i^{(1)} \equiv m^{-1}[g\{(i-1)h\} + g\{(i-1)h+\delta\} + \cdots + g\{(i-1)h + (m-1)\delta\}]$$

and

$$\mu_i^{(2)} \equiv m^{-1}[\tfrac{1}{2}g\{(i-1)h\} + g\{(i-1)h+\delta\} + \cdots + g\{(i-1)h + (m-1)\delta\} + \tfrac{1}{2}g(ih)],$$

and whose variances are $m^{-1}\sigma^2$ and $m^{-2}(m-\frac{1}{2})\sigma^2$. Note that the expressions for $\bar{Y}_i^{(1)}$ and $\bar{Y}_i^{(2)}$ are directly related to the Rectangle Rule and Trapezoidal Rule for numerical integration; see, for instance, Abramowitz and Stegun [9, p. 885]. Let $[\hat{\mu}_{i1}^{(j)}, \hat{\mu}_{i2}^{(j)}]$, $1 \leqslant i \leqslant k$, be simultaneous confidence intervals for the $\mu_i^{(j)}$'s, with

$$P(\hat{\mu}_{i1}^{(j)} \leqslant \mu_i^{(j)} \leqslant \hat{\mu}_{i2}^{(j)}, 1 \leqslant i \leqslant k) = \alpha \tag{3.1}$$

for $j = 1$ and 2.

Both $\bar{Y}_i^{(1)}$ and $\bar{Y}_i^{(2)}$ are normally distributed, and the confidence limits $\hat{\mu}_{il}^{(j)}$ would usually be based on this fact; see Subsection 3.3 below, where methods of constructing the intervals $[\hat{\mu}_{i1}^{(j)}, \hat{\mu}_{i2}^{(j)}]$ are described. The variables $\bar{Y}_i^{(1)}$, $1 \leqslant i \leqslant k$, are independent, although the variables $\bar{Y}_i^{(2)}$ are 1-dependent. This makes it a little easier to construct confidence bands based on the $\bar{Y}_i^{(1)}$'s, than on the $\bar{Y}_i^{(2)}$'s. We use the $\bar{Y}_i^{(2)}$'s when constructing confidence bands under the assumption of bounded second derivatives.

Next we define analogs of the functions $\hat{f}_1$ and $\hat{f}_2$ from Section 2. Set

$$\hat{g}_l^{(j)}\{(i+y)h\} \equiv (\tfrac{1}{2} - y)\,\hat{\mu}_{il}^{(j)} + (\tfrac{1}{2} + y)\hat{\mu}_{i+1,l}^{(j)}, \tag{3.2}$$

for $j = 1, 2$, $l = 1, 2$, $1 \leqslant i \leqslant k-1$, and $-\frac{1}{2} < y \leqslant \frac{1}{2}$.

3.1. *Confidence Bands under the Assumption of a Single Derivative*

(i) Given a sequence $\{c_i\}$ with each $c_i \geqslant 0$, define

$$\tilde{g}_1\{(i+y)h\} \equiv \hat{\mu}_{i1}^{(1)} - \tfrac{1}{2}\{(2y+1)h + \delta\}\,c_i$$

and

$$\tilde{g}_2\{(i+y)h\} \equiv \hat{\mu}_{i2}^{(1)} + \tfrac{1}{2}\{(2y+1)h + \delta\}\,c_i$$

for $1 \leqslant i \leqslant k$ and $0 < y \leqslant 1$. If

$$\sup_{(i-1)h \leqslant u \leqslant (i+1)h} |g'(u)| \leqslant c_i \qquad \text{for} \quad 1 \leqslant i \leqslant k \tag{3.3}$$

then

$$P\{\tilde{g}_1(x) \leqslant g(x) \leqslant \tilde{g}_2(x) \qquad \text{for} \quad h \leqslant x \leqslant (k+1)h\} \geqslant \alpha.$$

(ii) Given $\varepsilon \geqslant 0$, define

$$\tilde{g}_{1,\pm}\{(i+y)h\} \equiv \hat{\mu}_{i1}^{(2)} - \tfrac{1}{2}\{(2y+1)h + \delta\}[g'\{(i+y)h\} \pm \varepsilon]$$

and

$$\tilde{g}_{2,\pm}\{(i+y)h\} \equiv \hat{\mu}_{i2}^{(1)} - \tfrac{1}{2}\{(2y+1)h + \delta\}[g'\{(i+y)h\} - (\pm\varepsilon)]$$

for $1 \leqslant i \leqslant k$ and $0 < y \leqslant 1$, where the $+$, $-$ signs are taken respectively. If

$$|g'(u) - g'(v)| \leqslant \varepsilon$$

whenever $0 \leqslant u \leqslant v \leqslant (k+1)h$ and $|u-v| \leqslant 2h$, then

$$P\{\tilde{g}_{1,+}(x) \leqslant g(x) \leqslant \tilde{g}_{2,+}(x) \quad \text{for} \quad h \leqslant x \leqslant (k+1)h\} \geqslant \alpha$$

and

$$P\{\tilde{g}_{1,-}(x) \leqslant g(x) \leqslant \tilde{g}_{2,-}(x) \quad \text{for} \quad h \leqslant x \leqslant (k+1)h\} \leqslant \alpha.$$

(iii) Given a sequence $\{c_i\}$ with each $c_i \geqslant 0$, define

$$\tilde{g}_1\{(i+y)h\} \equiv \hat{g}_1^{(1)}\{(i+y)h\} - \tfrac{1}{2}\{(1-3y^2+2\,|y|^3)\,h+\delta\}c_i$$

and

$$\tilde{g}_2\{(i+y)h\} \equiv \hat{g}_2^{(1)}\{(i+y)h\} + \tfrac{1}{2}\{(1-3y^2+2\,|y|^3)h+\delta\}c_i$$

for $1 \leqslant i \leqslant k-1$ and $-\frac{1}{2} < y \leqslant \frac{1}{2}$. If (3.3) holds then

$$P\{\tilde{g}_1(x) \leqslant g(x) \leqslant \tilde{g}_2(x) \quad \text{for} \quad \tfrac{1}{2}h \leqslant x \leqslant (k-\tfrac{1}{2})h\} \geqslant \alpha.$$

3.2. *Confidence Bands under the Assumption of Two Derivatives*

(i) Given $\varepsilon > 0$, define

$$\tilde{g}_{1,\pm}\{(i+y)h\} \equiv \hat{g}_1^{(2)}\{(i+y)h\} + \tfrac{1}{6}\{(1-3y^2)h^2+\tfrac{1}{2}\delta^2\}[g''\{(i+y)h\}-(\pm\varepsilon)]$$

and

$$\tilde{g}_{2,\pm}\{(i+y)h\} \equiv \hat{g}_2^{(2)}\{(i+y)h\} + \tfrac{1}{6}\{(1-3y^2)h^2+\tfrac{1}{2}\delta^2\}[g''\{(i+y)h\}\pm\varepsilon]$$

for $1 \leqslant i \leqslant k-1$ and $-\frac{1}{2} < y \leqslant \frac{1}{2}$. If

$$|g''(u)-g''(v)| \leqslant \varepsilon$$

whenever $0 \leqslant u \leqslant v \leqslant (k+1)h$ and $|u-v| \leqslant 2h$, then

$$P\{\tilde{g}_{1,+}(x) \leqslant g(x) \leqslant \tilde{g}_{2,+}(x) \quad \text{for} \quad \tfrac{1}{2}h \leqslant x \leqslant (k-\tfrac{1}{2})h\} \geqslant \alpha$$

and

$$P\{\tilde{g}_{1,-}(x) \leqslant g(x) \leqslant \tilde{g}_{2,-}(x) \quad \text{for} \quad \tfrac{1}{2}h \leqslant x \leqslant (k-\tfrac{1}{2})h\} \leqslant \alpha.$$

(ii) Given a sequence $\{c_i\}$ with each $c_i \geqslant 0$, define

$$\tilde{g}_1\{(i+y)h\} \equiv \hat{g}_1^{(2)}\{(i+y)h\} - \tfrac{1}{6}\{(1-3y^2)h+\tfrac{1}{2}\delta^2\}c_i$$

and

$$\tilde{g}_2\{(i+y)h\} \equiv \hat{g}_2^{(2)}\{(i+y)h\} + \tfrac{1}{6}\{(1-3y^2)h^2 + \tfrac{1}{2}\delta^2\}c_i$$

for $1 \leqslant i \leqslant k-1$ and $-\tfrac{1}{2} < y \leqslant \tfrac{1}{2}$. If

$$\sup_{(i-1)h \leqslant x \leqslant (i+1)h} |g''(x)| \leqslant c_i \qquad \text{for} \quad 1 \leqslant i \leqslant k-1,$$

then

$$P\{\tilde{g}_1(x) \leqslant g(x) \leqslant \tilde{g}_2(x) \text{ for } \tfrac{1}{2}h \leqslant x \leqslant (k-\tfrac{1}{2})h\} \geqslant \alpha.$$

Remarks. The confidence bands in Subsections 3.1 and 3.2 compare directly with those in Subsections 2.1 and 2.2. Remarks similar to those earlier may be made about exactness, bias, etc. The terms in δ and δ^2 in the confidence limits compensate for the extra source of error in the regression case.

3.3. *Simultaneous Confidence Intervals for the μ_i's*

We shall concentrate on the case of two-sided confidence bands. Suppose first that the error variance σ^2 is known. Let Φ denote the standard normal distribution function, and z_γ the solution of $2\Phi(z_\gamma) - 1 = \gamma$, where $0 < \gamma < 1$. Define

$$\begin{aligned} \hat{\mu}_{i1}^{(1)} &\equiv \bar{Y}_i^{(1)} - m^{-1/2}\sigma z_\gamma, & \hat{\mu}_{i2}^{(1)} &\equiv \bar{Y}_i^{(1)} + m^{-1/2}\sigma z_\gamma, \\ \hat{\mu}_{i1}^{(2)} &\equiv \bar{Y}_i^{(2)} - m^{-1}(m-\tfrac{1}{2})^{1/2}\sigma z_\gamma, & \hat{\mu}_{i2}^{(2)} &\equiv \bar{Y}_i^{(2)} + m^{-1}(m-\tfrac{1}{2})^{1/2}\sigma z_\gamma. \end{aligned} \tag{3.4}$$

Then

$$P(\hat{\mu}_{i1}^{(j)} \leqslant \mu_i^{(j)} \leqslant \hat{\mu}_{i2}^{(j)}) = \gamma$$

for $j = 1, 2$. Consequently,

$$P(\hat{\mu}_{i1}^{(1)} \leqslant \mu_i^{(1)} \leqslant \hat{\mu}_{i2}^{(1)} \text{ for } 1 \leqslant i \leqslant k) = \gamma^k$$

and

$$P(\hat{\mu}_{i1}^{(2)} \leqslant \mu_i^{(2)} \leqslant \hat{\mu}_{i2}^{(2)} \text{ for } 1 \leqslant i \leqslant k) \simeq \gamma^k.$$

Taking $\gamma \equiv \alpha^{1/k}$ will give simultaneous coverage probability very nearly α in both cases. To construct a strictly conservative procedure in the case of $\mu_i^{(2)}$, suppose for the sake of argument that k is even. Let $\mathscr{E}_i$ denote the event that $\hat{\mu}_{i1}^{(2)} \leqslant \mu_i^{(2)} \leqslant \hat{\mu}_{i2}^{(2)}$ is false. Since the variables $\bar{Y}_i^{(2)}$ are 1-dependent,

$$P(\hat{\mu}_{i1}^{(2)} \leqslant \mu_i^{(2)} \leqslant \hat{\mu}_{i2}^{(2)} \text{ for } 1 \leqslant i \leqslant k) = 1 - P\left(\bigcup_{\text{odd } i} \mathscr{E}_i \cup \bigcup_{\text{even } i} \mathscr{E}_i \right)$$

$$\geqslant 1 - P\left(\bigcup_{\text{odd } i} \mathscr{E}_i \right) - P\left(\bigcup_{\text{even } i} \mathscr{E}_i \right)$$

$$= 2\gamma^{k/2} - 1.$$

If the error variance σ^2 is unknown, we may construct a slight overestimate of it. Let $\mathscr{S}$ be the set of all differences $Y_{2j} - Y_{2j-1}$ such that neither $2j$ not $2j-1$ is of the form im for an integer i. Assume $\mathscr{S}$ has r elements, and let

$$\hat{\sigma}^2 \equiv \{2(r-1)\}^{-1} \left\{ \sum_{s \in \mathscr{S}} s^2 - r^{-1} \left(\sum_{s \in \mathscr{S}} s \right)^2 \right\}.$$

Then $(r-1)\hat{\sigma}^2/\sigma^2$ has the chi-squared distribution with $r-1$ degrees of freedom and a noncentrality parameter and is independent of $\bar{Y}_1^{(j)}, ..., \bar{Y}_k^{(j)}$ for $j = 1, 2$. Let Φ_{r-1} denote the distribution function of Student's t with $r-1$ degrees of freedom, and t_γ the solution of $2\Phi_{r-1}(t_\gamma) - 1 = \gamma$. Define $\hat{\mu}_{il}^{(j)}$ as in (3.4), but replacing σ by $\hat{\sigma}$ and z_γ by t_γ throughout. Then

$$P(\hat{\mu}_{i1}^{(1)} \leqslant \mu_i^{(1)} \leqslant \hat{\mu}_{i2}^{(1)} \text{ for } 1 \leqslant i \leqslant k) \geqslant \gamma^k$$

and

$$P(\hat{\mu}_{i1}^{(2)} \leqslant \mu_i^{(2)} \leqslant \hat{\mu}_{i2}^{(2)} \text{ for } 1 \leqslant i \leqslant k) \simeq \gamma^k;$$

see Johnson and Kotz [10, p. 193].

3.4. *Discussion*

Here we use the results of Subsections 3.1–3.3 to develop practical procedures for setting confidence bands.

The first derivative of g represents the rate of change of that function. In practice an upper bound to this rate can often be set from physical considerations, from previous empirical experience, or by direct estimation. If it is known that $|g'|$ does not exceed $c^{(1)}$ then the confidence band may be taken to be the band formed by the pair of functions $(\hat{g}_1^{(1)}, \hat{g}_2^{(1)})$ (defined at (3.2)), plus an extra strip on either side of width (i.e., height) $\frac{1}{2}(h+\delta)c^{(1)}$. If $|g''|$ does not exceed $c^{(2)}$ then we add strips of width $\frac{1}{6}(h^2 + \frac{1}{2}\delta^2)c^{(2)}$ to either side of the band formed by the pair $(\hat{g}_1^{(2)}, \hat{g}_2^{(2)})$. In both cases the upper strip may be deleted if it is known g is convex, and the lower strip deleted if it is known g is concave. The full width of the strips does not have to be maintained throughout the band; see the formulae in Subsections 3.1(iii) and 3.2(ii). All these procedures are conservative and give coverage probability at least α, where α is the simultaneous coverage probability of the intervals $[\hat{\mu}_{i1}^{(j)}, \hat{\mu}_{i2}^{(j)}]$; see (3.1).

An alternative approach is to estimate g' or g'' directly. For example, if $\hat{g}''$ is an estimate of g'' then

$$\tilde{g}_1\{(i+y)h\} \equiv \hat{g}_1^{(2)}\{(i+y)h\} + \tfrac{1}{6}\{(1-3y^2)h^2 + \tfrac{1}{2}\delta^2\}\,\hat{g}''\{(i+y)h\}$$

and

$$\tilde{g}_2\{(i+y)h\} \equiv \hat{g}_2^{(2)}\{(i+y)h\} + \tfrac{1}{6}\{(1-3y^2)h^2 + \tfrac{1}{2}\delta^2\}\,\hat{g}''\{(i+y)h\},$$

for $1 \leqslant i \leqslant k-1$ and $-\frac{1}{2} < y \leqslant \frac{1}{2}$, are lower and upper confidence bands, respectively, with coverage probability "approximately" α.

Analogous confidence bands may be described without the assumption that design variables be equally spaced. Then formulae based on more complicated weighted averages should be used in place of the simpler bounds described above. In the case of the procedure proposed by Knafl, Sacks and Ylvisaker [4], similar formulae are required to evaluate the bias bound $B(t)$ which appears in the expressions for their bands.

4. Widths of Confidence Bands

We begin by describing widths of the confidence bands developed for densities in Section 2. Assume that $h \to 0$ like n^{-r} for some $0 < r < 1$, and $k \to \infty$ like h^{-1}. Let $\hat{p}_i$ equal the proportion of the sample falling into the ith cell, and suppose f is bounded away from zero and infinity within the region of interest. In view of results for probabilities of large deviation (e.g., Feller [1]), the numbers ε_i defined by either

$$P(\hat{p}_i - \varepsilon_i \leqslant p_i \leqslant \hat{p}_i + \varepsilon_i) = 1 - (1-\beta)k^{-1}$$

or

$$P(\hat{p}_i - \varepsilon_i \leqslant p_i \leqslant \hat{p}_i + \varepsilon_i) = \beta^{1/k},$$

satisfy

$$\varepsilon_i \sim n^{-1/2} p_i^{1/2} (2 \log k)^{1/2} \sim n^{-1/2}\{2f(ih)\,h \log k\}^{1/2}.$$

(Notice that although ε_i depends on β, the dominant term in an asymptotic expansion of ε_i does not depend on β.) If the confidence intervals $[\hat{p}_{i1}, \hat{p}_{i2}]$ are two-sided then $\hat{p}_{i2} - \hat{p}_{i1} \sim 2\varepsilon_i$, and so the width of the band separating $\hat{f}_1$ and $\hat{f}_2$ (see (2.1)) is asymptotically

$$2h^{-1}\varepsilon_i \sim \{8f(ih)\}^{1/2}\{(nh)^{-1} \log k\}^{1/2}. \tag{4.1}$$

The practical procedures suggested in Subsection 2.4 lead to a confidence band whose width equals this amount, plus an extra term of order h or h^2 to allow for the strips added to the band $(\hat{f}_1, \hat{f}_2)$. Let us assume we are working under the assumption of a bounded second derivative, so that the extra term is of order h^2. If $h = \text{const}\, n^{-r}$ then this extra term is insignificant when $r \geqslant \frac{1}{5}$, but dominates when $r < \frac{1}{5}$. Bearing in mind that $\log k \sim \text{const} \log n$, we see that the minimum confidence interval width is obtained by choosing h such that $(nh)^{-1} \log n$ and h^4 are of the same order of magnitude. This gives $h \sim \text{const}(n^{-1} \log n)^{1/5}$ as the "optimum" achievable by our method and results in a confidence band whose width is approximately $(n^{-1} \log n)^{2/5}$. A similar argument in the case of a bounded first derivative gives the "optimal" h to be of order $(n^{-1} \log n)^{1/3}$, and a confidence band of width approximately $(n^{-1} \log n)^{1/3}$.

Let us assume f has t bounded derivatives. The discussion given above shows that if $t = 1$ or 2, and for a given coverage coefficient $\alpha \in (0, 1)$, we may construct a confidence band of fixed width $C(n^{-1} \log n)^{t/(2t+1)}$ which covers f with probability at least α. Here C is a constant not depending on n. It is possible to generate procedures which give confidence bands with this property for any given $t \geqslant 1$. They are based on higher order interpolation formulae but will not be discussed in detail here since they do not seem to be of general practical interest.

In fact, the constant C may be chosen such that the coverage probability is at least α for all f's in a large class of densities. Suppose the density f is to be estimated in the interval $(0, 1)$. Let $0 < a < 1$, $b > 0$, $c > 0$, and $t \geqslant 1$ be an integer, and let $\mathscr{F} = \mathscr{F}(a, b, c, t)$ denote the class of all functions f satisfying

$$a \leqslant |f(x)| \leqslant a^{-1} \quad \text{and} \quad |f^{(t)}(x)| \leqslant b \quad \text{whenever} \quad -c \leqslant x \leqslant 1 + c.$$

We may choose $C = C(a, b, c, t)$ so large that a confidence band B of width $C(n^{-1} \log n)^{t/(2t+1)}$ covers f with probability at least α, uniformly in densities $f \in \mathscr{F}$:

$$\inf_{f \in \mathscr{F}} P_f\{f(x) \in B \text{ for } 0 \leqslant x \leqslant 1\} \geqslant \alpha, \tag{4.2}$$

$n \geqslant 2$. (The cases $t = 1$ and 2 are dealt with in Subsection 2.4.)

The width of order $(n^{-1} \log n)^{t/(2t+1)}$ is "optimal," in the sense that no procedure can produce fixed-with confidence bands whose width is of a smaller order of magnitude. To see this, we first define the notion of a general fixed-width confidence band B. Let $\xi(\cdot)$: $[0, 1] \to \mathbf{R}$ be a random function, and let $w \geqslant 0$ be a random variable. Both ξ and w may depend on the data, but not on f. Hence they are "nonparametric" in character. Let

$$B \equiv \{(x, h) : 0 \leqslant x \leqslant 1 \text{ and } \xi(x) \leqslant y \leqslant \xi(x) + w\}.$$

In a slight abuse of notation, we say that "$f(x) \in B$ for $0 \leqslant x \leqslant 1$" if the ordered pair $(x, f(x))$ is in B for $0 \leqslant x \leqslant 1$; that is, if the function $f(\cdot)$ restricted to $[0, 1]$ lies between the functions $\xi(\cdot)$ and $\xi(\cdot) + w$. We call B a "confidence band of width w and uniform coverage probability at least α for all $f \in \mathscr{F}$", if (4.2) holds.

An extreme case of this type of band has $\xi \equiv 0$ and

$$w = \begin{cases} 0 & \text{with probability } 1 - \alpha \\ \infty & \text{with probability } \alpha. \end{cases}$$

Any statement we make about the size of w must take account of this pathology. In particular, the limit at (4.3) below may equal α, not 1.

THEOREM 4.1. *Suppose the confidence band B_n, of width w_n, satisfies*

$$\inf_{f \in \mathscr{F}(a,b,c,t)} P_f\{f(x) \in B_n \text{ for } 0 \leqslant x \leqslant 1\} \geqslant \alpha, \qquad n \geqslant 1.$$

If $0 < \alpha < 1$ is fixed then for some $\eta > 0$,

$$\liminf_{n \to \infty} \sup_{f \in \mathscr{F}(a,b,c,t)} P_f\{w_n \geqslant \eta (n^{-1} \log n)^{t/(2t+1)}\} \geqslant \alpha. \tag{4.3}$$

If w_n is non-random, as in the examples considered earlier, then this theorem declares that no fixed-width confidence band can be narrower than $\eta(n^{-1} \log n)^{t/(2t+1)}$, for large n, if it is to have uniform coverage probability at least α.

The regression case is very similar, and so we only sketch the details. Assume the regression function is to be estimated in the interval (0, 1), and that the design points are distant $\delta = n^{-1}$ apart. If the error variance σ^2 is known, then the techniques suggested in Subsection 3.4 (and their analogs for $t \geqslant 3$) give confidence bands of width no more than $\text{const}(n^{-1} \log n)^{t/(2t+1)}$ with probability at least α for all $g \in \mathscr{F}$, provided h is taken to be a constant multiple of $(n^{-1} \log n)^{1/(2t+1)}$. If the error variance is unknown then it should be estimated, as outlined in Subsection 3.3. The resulting confidence band width w_n is a random variable, satisfying

$$\inf_{g \in \mathscr{F}(a,b,c,t)} P_g\{w_n \leqslant \text{const}(n^{-1} \log n)^{t/(2t+1)}\} \to 1$$

as $n \to \infty$. Again, a coverage probability of at least α may be achieved for all $g \in \mathscr{F}$.

The theorem below is an analog of Theorem 4.1 in the regression case.

THEOREM 4.2. *Suppose the confidence band B_n, of width w_n, satisfies*

$$\inf_{g \in \mathscr{F}(a,b,c,t)} P_g\{g(x) \in B_n \text{ for } 0 \leqslant x \leqslant 1\} \geqslant \alpha, \qquad n \geqslant 1.$$

If $0 < \alpha < 1$ is fixed then for some $\eta > 0$,

$$\liminf_{n \to \infty} \sup_{g \in \mathscr{F}(a,b,c,t)} P_g\{w_n \geqslant \eta(n^{-1} \log n)^{t/(2t+1)}\} \geqslant \alpha.$$

In theory it is possible to choose h so as to minimise the area of confidence bands. For example, suppose we are constructing a band for the density f under the assumption that $|f''| \leqslant c$. We start with the band separating $\hat{f}_1$ and $\hat{f}_2$ (see (2.1)). The distance between $\hat{f}_1$ and $\hat{f}_2$ at x is asymptotic to

$$\{8f(x)\}^{1/2}\{(nh)^{-1} \log k\}^{1/2};$$

see (4.1). To this we add two strips of width $\frac{1}{6}h^2c$. Therefore the asymptotic total area of the confidence band for f, drawn between x_1 and x_2, is

$$A(h) \equiv \int_{x_1}^{x_2} [\{8f(x)\}^{1/2}\{(nh)^{-1} \log k\}^{1/2} + \tfrac{1}{3}h^2c] \, dx.$$

If we set $h = d(n^{-1} \log n)^{1/5}$, then $k \sim \text{const}\, h^{-1}$ and

$$A(h) \sim \left\{\left(\frac{8}{5}\right)^{1/2} d^{-1/2} \int_{x_1}^{x_2} f^{1/2}(x) \, dx + \frac{1}{3} cd^2(x_2 - x_1)\right\} (n^{-1} \log n)^{2/5},$$

which is minimised by choosing

$$d = \left\{3.10^{-1/2} c^{-1}(x_2 - x_1)^{-1} \int_{x_1}^{x_2} f^{1/2}(x) \, dx\right\}^{2/5}.$$

Although this formula is not of explicit practical use, it does suggest advice concerning choice of the bandwidth h. In particular, larger values of c and smaller values of f both dictate smaller values of h.

5. ILLUSTRATIVE EXAMPLES

In this section we report on applications of the procedures developed earlier to two particular examples.

EXAMPLE 5.1 (nonparametric density estimation). A set of $n = 900$

independent pseudorandom values were generated, using the NAG Fortran subroutine library, from the mixture density

$$f(x) = 0.2Be(x; 1, 2) + 0.8Be(x; 2, 1), \qquad 0 < x < 1,$$

where $Be(x; \alpha, \beta)$ denotes the density of the $Be(\alpha, \beta)$ distribution. Thus $f(x) = 0.4 + 1.2x$, so that

$$\sup |f'(x)| = 1.2.$$

The value of k was chosen initially to be 30 and h was taken to be $1/k = \frac{1}{30}$. The pairs $\{(\hat{p}_{i1}, \hat{p}_{i2}),\ i = 1, ..., k\}$ were chosen using the normal approximation discussed in subsection 2.3. Specifically, they were given by (2.6) and (2.7) with $\beta = 0.95$ (for a 95% confidence interval).

For the sake of realism it was decided to construct confidence bands under the assumption of a single derivative satisfying

$$\sup_{0<x<1} |f'(x)| \leqslant c.$$

Thus, each $c_i = c$. We took $c = 3$, which is of course conservative.

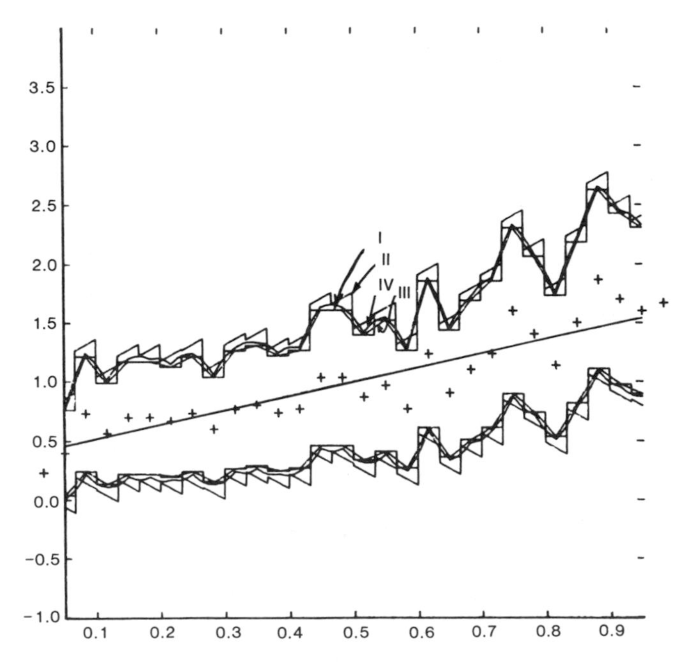

FIG. 1. Bands I through IV in case of nonparametric density estimation, for $k = 30$.

Figure 1 depicts the following functions:

(I) $(\hat{p}_{i1}, \hat{p}_{i2})$, displayed as piecewise constant plots;

(II) $(\hat{f}_1(x), \hat{f}_2(x))$, from (2.1), piecewise linear;

(III) $(\tilde{f}_1(x), \tilde{f}_2(x))$, from Subsection 2.1(i);

(IV) $(\tilde{f}_1(x), \tilde{f}_2(x))$, from Subsection 2.1(iii).

Figure 2 extracts the pair given by (IV). That is the most pleasing of the four pairs in Fig. 1. Of course the results still exhibit a lack of smoothness. Remember, however, that any envelope of a conservative confidence band is also a conservative confidence band, and so one may smooth out the bumps in a variety of ways.

To investigate the effect of changing k, Fig. 3 depicts the results corresponding to Fig. 2 but with $k = 50$. Note that, inevitably, the bands are wider. The appearance would be generally much improved if bounds were placed on $f''(x)$.

EXAMPLE 5.2 (nonparametric regression). The data used here were a subset of larger set of data kindly supplied by Dr. E. M. Scott. The variables are those of radiocarbon age and tree-ring age, both measured in

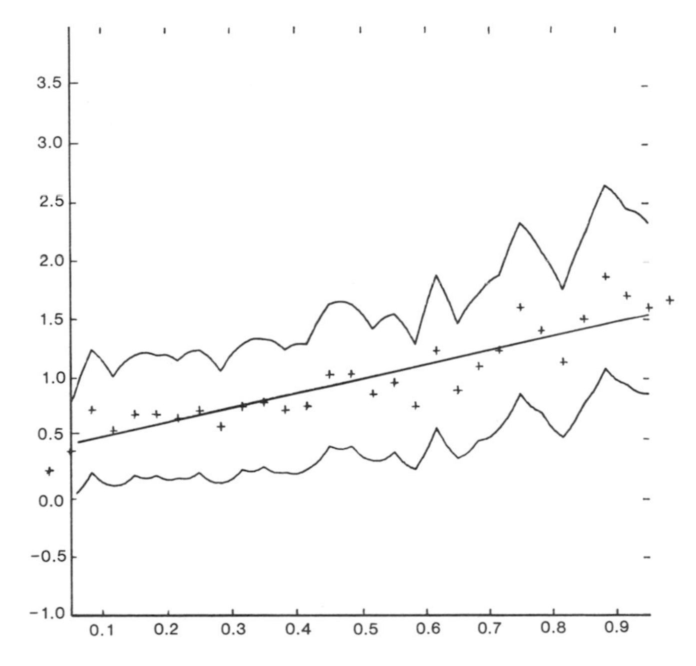

FIG. 2. Band IV in case of nonparametric density estimation, for $k = 30$.

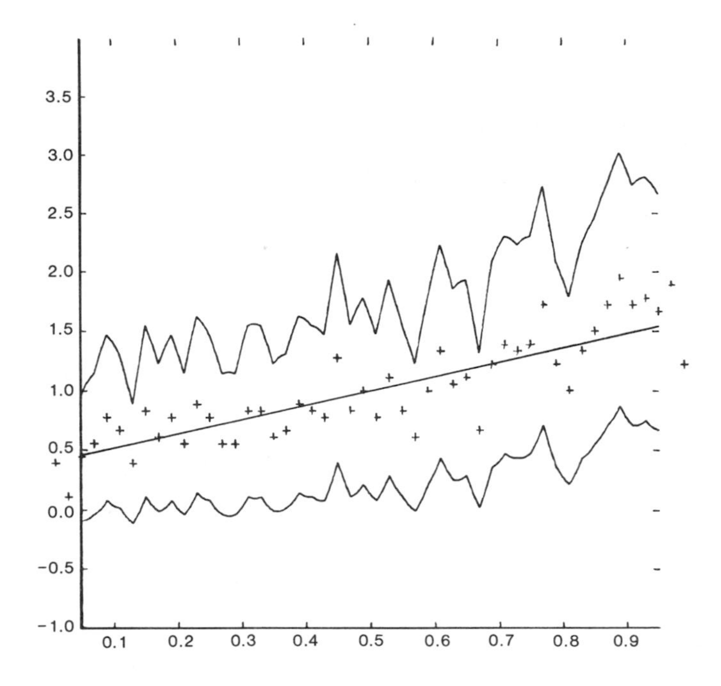

FIG. 3. Band IV in case of nonparametric density estimation, for $k=50$.

years before 1950 A.D. and thinned and rounded so as to achieve equal spacing of the tree-ring ages. Altogether 180 points were included and, initially, we chose $k=30$ so that, in (3.4), $m=6$. For simplicity we used the non-overlapping means $\bar{Y}_i^{(1)}$, and constructed the bands with $\beta=0.95$ and under the assumption of a single derivative, with uniform bound $c=1$ on $|g'(x)|$.

A somewhat different estimator for σ was used than that discussed in Section 3.3. To be specific, we took

$$\hat{\sigma}^2 = \tfrac{2}{3}\gamma^{-1} \sum s_i^2,$$

where s_i is of the form

$$s_i = y_i - (y_{i+1} + y_{i-1})/2$$

and the summation is over all i such that none of $i-1$, i or $i+1$ is of the form im or $(im+1)$ and such that all triples $(i-1, i, i+1)$ are distinct. The symbol r denotes the number of such triples. This estimator is based on the residual of y_i from the straight line based on y_{i-1} and y_{i+1}.

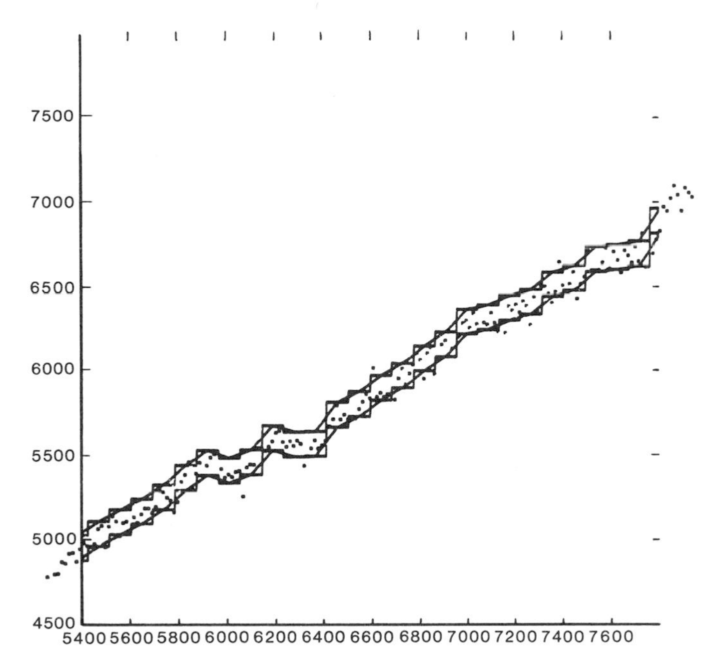

FIG. 4. Bands I through IV in case of nonparametric regression, for $k = 30$.

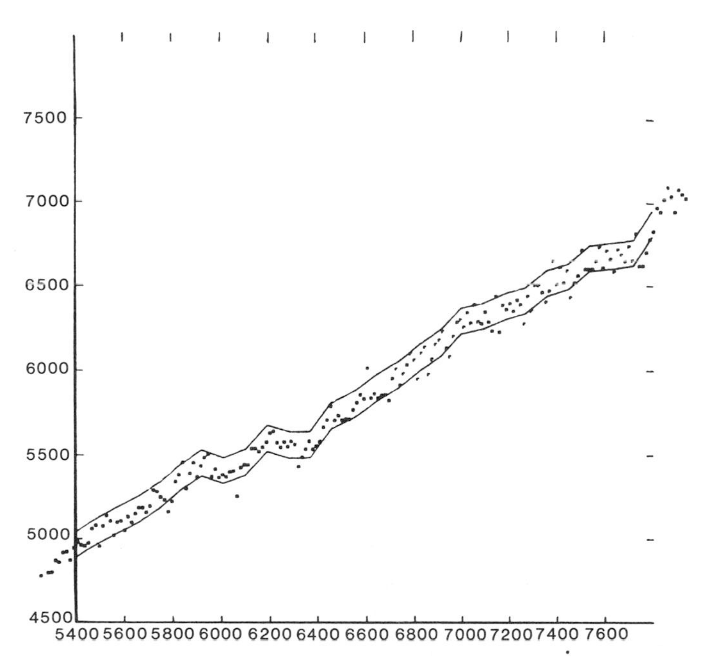

FIG. 5. Band IV in case of nonparametric regression, for $k = 30$.

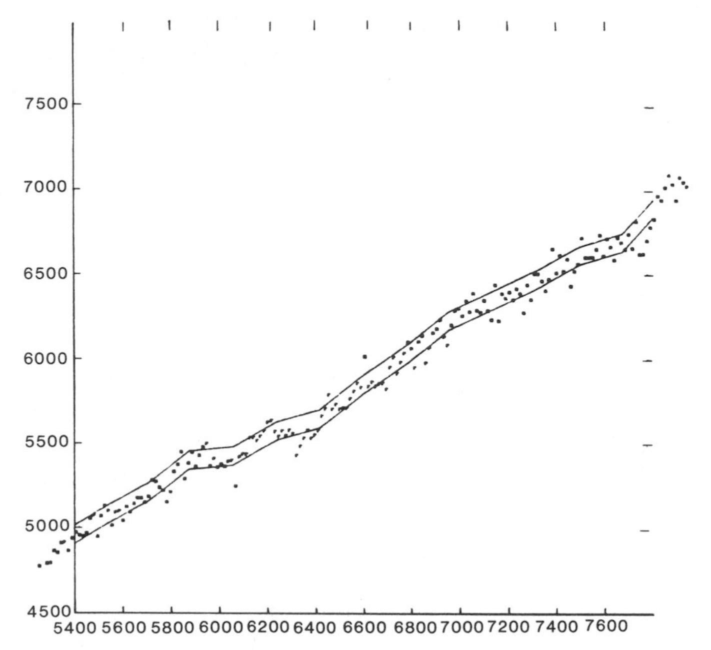

FIG. 6. Band IV in case of nonparametric regression, for $k = 15$.

Figure 4 displays the data points along with the bands:

(I) $(\hat{\mu}_{i1}^{(1)}, \hat{\mu}_{i2}^{(1)})$, displayed as piecewise constant plots;

(II) $(\hat{g}_1^{(1)}(x), \hat{g}_2^{(1)}(x))$, from (3.2), piecewise linear;

(III) $(\tilde{g}_1^{(1)}(x), \tilde{g}_2^{(1)}(x))$ from Subsection 3.1(i);

(IV) $(\tilde{g}_1^{(1)}(x), \tilde{g}_2^{(1)}(x))$ from subsection 3.1(iii).

Figure 5 isolates the bands defined by IV. As in the case of Example 5.1, slight difficulties with the ends of the range of the tree-ring ages led to the bands being drawn only over a restricted range.

The estimate of σ was $\hat{\sigma} = 54.1$, based on $r = 25$. Finally, Fig. 6 gives the version for $k = 15$. In this case $d_{15} = 2.94$, $m = 12$, and $\hat{\sigma} = 54.2$, based on $r = 41$.

6. PROOFS

6.1. *Proofs for Sections 2 and 3*

In the work below, g stands for either f or g.

If g has a continuous first derivative on $((i-1)h, (i+2)h)$, then for $0 \leqslant y \leqslant 1$

$$\int_{(i-1)h}^{ih} g(u)\, du = hg\{(i+y)h\} - \tfrac{1}{2}(2y+1)\, h^2 g'\{(i+\xi)h\}, \tag{6.1}$$

where $-1 \leqslant \xi(i, h, y) \leqslant 1$. (This follows from formula (16), p. 64 of Steffensen [6], on taking "m" = "n" = 1 and "f" equal to an indefinite integral of g.) Results in Subsections 2.1(i) and 2.1(ii) are immediate consequences. (The case where g is not continuous is handled by approximation by a continuous g.) By Taylor expansion,

$$g(a) = \delta^{-1} \int_a^{a+\delta} g(u)\, du - \delta \int_0^1 g'(a + \delta t)(1 - t)\, dt,$$

and so

$$\begin{aligned}
\mu_i^{(1)} &= h^{-1} \int_{(i-1)h}^{ih} g(u)\, du - m^{-1} \delta \int_0^1 \sum_{j=0}^{m-1} g'\{(i-1)h + j\delta + \delta t\}(1-t)\, dt \\
&= g\{(i+y)h\} - \left[\frac{1}{2}(2y+1)\, hg'\{(i+\xi)h\}\right. \\
&\quad \left. + m^{-1} \delta \int_0^1 \sum_{j=0}^{m-1} g'\{(i-1)h + j\delta + \delta t\}(1-t)\, dt\right], \qquad (6.2)
\end{aligned}$$

using (6.1). Results in Subsections 3.1(i) and 3.1(ii) are immediate consequences.

Next we assess the error of piecewise-linear approximants such as $\hat{f}_1$ and $\hat{f}_2$. Observe that the remainder $R_1(y)$ in the formula

$$\left(\frac{1}{2} - y\right) \int_{(i-1)h}^{ih} g(u)\, du + \left(\frac{1}{2} + y\right) \int_{ih}^{(i+1)h} g(u)\, du = hg\{(i+y)h\} + R_1(y), \qquad (6.3)$$

may be written as

$$\begin{aligned}
R_1(y) = h^2 \Bigg[&\left(\frac{1}{2} + y\right)(1-y)^2 \int_0^1 g'\{h(i + y + t(1-y))\}(1-t)\, dt \\
&- 2y^3 \int_0^1 g'\{h(i + y - ty)\}(1-t)\, dt \\
&- \left(\frac{1}{2} - y\right)(1+y)^2 \int_0^1 g'\{h(i + y - t(1+y))\}(1-t)\, dt \Bigg]
\end{aligned}$$

if g is differentiable. (Use the integral formula for the remainder in a Taylor expansion.) Therefore if $|g'| \leqslant c_i$ on $((i-1)h, (i+1)h)$, and $-\frac{1}{2} < y \leqslant \frac{1}{2}$,

$$\begin{aligned}
|R_1(y)| &\leqslant \tfrac{1}{2} h^2 \{(\tfrac{1}{2} + y)(1-y)^2 + 2\,|y|^3 + (\tfrac{1}{2} - y)(1+y)^2\} c_i \\
&= \tfrac{1}{2} h^2 (1 - 3y^2 + 2\,|y|^3) c_i.
\end{aligned}$$

This gives the result in Subsection 2.1(iii). Using the first line of (6.2) we obtain

$$(\tfrac{1}{2}-y)\mu_i^{(1)}+(\tfrac{1}{2}+y)\mu_{i+1}^{(1)}=g\{(i+y)h\}+h^{-1}R_1(y)+R_2(y),$$

where $|R_2(y)|\leqslant\frac{1}{2}\delta c_i$. This gives the result in Subsection 3.1(iii).

If g has two derivatives then the remainder $R_1(y)$ defined by (6.3) may be written as

$$\begin{aligned}R_1(y)=\frac{1}{2}h^3\Bigg[\left(\frac{1}{2}+y\right)&(1-y)^3\int_0^1 g''\{h(i+y+t(1-y))\}(1-t)^2\,dt\\&+2y^4\int_0^1 g''\{h(i+y-ty)\}(1-t)^2\,dt\\&+\left(\frac{1}{2}-y\right)(1+y)^3\int_0^1 g''\{h(i+y-t(1+y))\}(1-t)^2\,dt\Bigg],\end{aligned}$$

again by Taylor expansion. For $-\frac{1}{2}<y\leqslant\frac{1}{2}$ the functions $(\frac{1}{2}+y)(1-y)^3$, $2y^4$ and $(\frac{1}{2}-y)(1+y)^3$ are non-negative and add to $1-3y^2$. Results in Subsections 2.2(i) and 2.2(ii) follow from these properties. In particular to prove (2.4), notice that

$$\begin{aligned}&P\{\tilde{f}_{1,-}(x)\leqslant f(x)\leqslant\tilde{f}_{2,-}(x)\text{ for }\tfrac{1}{2}h\leqslant x\leqslant(k-\tfrac{1}{2})h\}\\&\quad\leqslant P[\tilde{f}_{1,-}\{(i-\tfrac{1}{2})h\}\leqslant f\{(i-\tfrac{1}{2})h\}\leqslant\tilde{f}_{2,-}\{(i-\tfrac{1}{2})h\}\text{ for }1\leqslant i\leqslant k]\\&\quad=P[\tilde{f}_{1,-}\{(i-\tfrac{1}{2})h\}+h^{-1}p_i-f\{(i-\tfrac{1}{2})h\}\leqslant h^{-1}p_i\\&\qquad\leqslant\tilde{f}_{2,-}\{(i-\tfrac{1}{2})h\}+h^{-1}p_i-f\{(i-\tfrac{1}{2})h\}\text{ for }1\leqslant i\leqslant k]\\&\quad\leqslant P[\hat{f}_1\{(i-\tfrac{1}{2})h\}\leqslant h^{-1}p_i\leqslant\hat{f}_2\{(i-\tfrac{1}{2})h\}\text{ for }1\leqslant i\leqslant k]\\&\quad=P(\hat{p}_{i1}\leqslant p_i\leqslant\hat{p}_{i2}\text{ for }1\leqslant i\leqslant k)=\alpha.\end{aligned}$$

By the Euler–Maclaurin expansion,

$$\begin{aligned}&\delta\left[\frac{1}{2}g(a)+g(a+\delta)+\cdots+g\{a+(m-1)\delta\}+\frac{1}{2}g(a+m\delta)\right]\\&\qquad=\int_a^{a+m\delta}g(u)\,du+\frac{1}{2}\delta^3\int_0^1 t(1-t)\left\{\sum_{j=0}^{m-1}g''(a+j\delta+\delta t)\right\}dt,\end{aligned}$$

and so

$$(\tfrac{1}{2}-y)\mu_i^{(2)}+(\tfrac{1}{2}+y)\mu_{i+1}^{(2)}=g\{(i+y)h\}+h^{-1}R_1(y)+R_2(y),$$

where

$$R_2(y) \equiv \frac{1}{2}\delta^2 \int_0^1 t(1-t)\left[\left(\frac{1}{2}-y\right)m^{-1}\sum_{j=0}^{m-1} g''\{(i-1)h+j\delta+\delta t\}\right.$$
$$\left.+\left(\frac{1}{2}+y\right)m^{-1}\sum_{j=0}^{m-1} g''(ih+j\delta+\delta t)\right]dt.$$

Results in Subsections 3.2(i) and 3.2(ii) are immediate consequences.

6.2. *Proofs of Theorems 4.1 and 4.2*

We shall conduct the proofs together. Fix $z_1, z_2 > 0$, let m equal the integer part of $z_1(n^{2t}\log n)^{1/(2t+1)}$, k equal the integer part of $z_2(n/\log n)^{1/(2t+1)}$, and

$$h \equiv mn^{-1} \sim z_1(n^{-1}\log n)^{1/(2t+1)}.$$

Fix $d>0$ and let ψ be a non-degenerate function on $(-\infty, \infty)$ with the properties:

(i) ψ vanishes outside $(0, 1)$; (ii) ψ has at least $t+1$ bounded derivatives on $(-\infty, \infty)$; (iii) $\sup|\psi^{(t)}| \leqslant d^{-1}b$; (iv) $\int\psi = 0$. Given a sequence $\boldsymbol{\theta} = (\theta_0, ..., \theta_{k-1})$ of 0's and 1's, set

$$\phi(x) = \phi(x\,|\,\boldsymbol{\theta}) = d[1+\theta_i h^t \psi\{h^{-1}(x-ih)\}]$$

for $ih < x \leqslant (i+1)h$ and $0 \leqslant i \leqslant k-1$, and $\phi(x) = d$ for $x<0$ and $x>kh$. Then $\int_0^1 \phi\, dx = d$, so ϕ restricted to $[0, 1]$ may be regarded as part of a probability density if $0<d<1$ and n is large. Notice that $\phi \in \mathscr{F}(a, b, c, t)$ if $a<d$ and n is large. We shall take the density f or regression function g to equal $\phi(\cdot\,|\,\boldsymbol{0})$ on $[0, 1]$, for some $\boldsymbol{\theta} \in \Theta = \{0, 1\}^k$.

Let $s \equiv \sup|\psi|$. If the confidence band B_n is of width w_n and $w_n \leqslant \frac{1}{2}ds\,h^t$, we define $\hat\theta_i = 1$ if

$$(x, d[1+h^t\psi\{h^{-1}(x-ih)\}]) \in B_n \qquad \text{for} \quad ih < x \leqslant (i+1)h,$$

and $\hat\theta_i = 0$ otherwise. If $w_n > \frac{1}{2}ds\,h^t$, define $\hat\theta_i$ arbitrarily. Let $\hat{\boldsymbol{\theta}}$ be the k-vector whose ith element is $\hat\theta_i$. If $w_n \leqslant \frac{1}{2}ds\,h^t$ and $f(x) = \phi(x\,|\,\boldsymbol{\theta}) \in B_n$ for $0 \leqslant x \leqslant 1$, then $\theta_i = \hat\theta_i$ for $1 \leqslant i \leqslant k$. Therefore in the density case,

$$P_f(w_n > \tfrac{1}{2}ds\,h^t) \geqslant P_f\{\hat\theta_i \neq \theta_i, \text{ some } i, \text{ and } f(x) \in B_n \text{ for } 0 \leqslant x \leqslant 1\}$$
$$\geqslant \alpha - P(\hat\theta_i = \theta_i,\ 1 \leqslant i \leqslant k).$$

A similar inequality holds in the regression case. Therefore the proof will be complete if we show that for any sequence of estimates $\hat{\theta}_i$ of θ_i, and for z_1 sufficiently small,

$$\limsup_{n\to\infty} \inf_{f\in\mathscr{F}} P_f(\hat{\theta}_i=\theta_i,\ 1\leqslant i\leqslant k)=0. \tag{6.4}$$

(Interpret f as g in the regression case.)

Let $P_{\boldsymbol{\theta}}$ denote the probability measure under the assumption that $\phi(\cdot\,|\,\boldsymbol{\theta})$ is the true density function or true regression function. Define

$$P_*(\mathscr{E})\equiv 2^{-k}\sum_{\boldsymbol{\theta}\in\Theta} P_{\boldsymbol{\theta}}(\mathscr{E})$$

for events $\mathscr{E}$. In the density case, let $X_1, ..., X_n$ be the random n-sample from f, and set

$$\lambda_i\equiv\prod_{\mathscr{C}_i}\,[1+h^t\psi\{h^{-1}(X_j-ih)\}],$$

where $\mathscr{C}_i$ denotes the set of values j such that X_j lies within the interval $\mathscr{I}_i\equiv[ih,(i+1)h)$. In the regression case, let

$$\lambda_i\equiv\exp\left((2\sigma^2)^{-1}\left[2dh^t\sum_{\mathscr{C}_i}(Y_j-d)\,\psi\{h^{-1}(jn^{-1}-ih)\}\right.\right.$$
$$\left.\left.-\,d^2h^{2t}\sum_{\mathscr{C}_i}\psi^2\{h^{-1}(jn^{-1}-ih)\}\right]\right),$$

where $\mathscr{C}_i$ is the set of values j such that $jn^{-1}\in\mathscr{I}_i$. Notice that in both cases, λ_i is a likelihood ratio for $\theta_i=1$ over $\theta_i=0$. Let $\mathscr{S}$ denote the set of all data—either all the sample values X_j in the density case, or all the pairs (jn^{-1}, Y_j) in the regression case. Conditional on $\mathscr{S}$, and under the probability measure P_*, the θ_i's are independent zero-one variables with

$$P_*(\theta_i=1\,|\,\mathscr{S})=\lambda_i/(\lambda_i+1)\equiv p_i,$$

say. Therefore

$$P_*(\hat{\theta}_i=\theta_i,\ 1\leqslant i\leqslant k\,|\,\mathscr{S})=\prod_{i=1}^{k} P_*(\theta_i=\hat{\theta}_i\,|\,\mathscr{S}). \tag{6.5}$$

Conditional on $\mathscr{S}$, the $\hat{\theta}_i$'s are fixed, and so

$$P_*(\theta_i=\hat{\theta}_i\,|\,\mathscr{S})\leqslant\max(p_i, 1-p_i)\equiv q_i, \tag{6.6}$$

say. Let N_i equal the number of subscripts in $\mathscr{C}_i$, and $N. \equiv \sum_{i=1}^{k} N_i$. (In the regression case, the N_i's are fixed.) If

$$P_*(q_i \leqslant 1-\varepsilon \mid N_i, N.) \geqslant \rho_i \tag{6.7}$$

then

$$\begin{aligned} E_*(q_i \mid N_i, N.) &\leqslant (1-\varepsilon)\, P_*(q_i \leqslant 1-\varepsilon \mid N_i, N.) + P_*(q_i > 1-\varepsilon \mid N_i, N.) \\ &\leqslant 1-\varepsilon\rho_i \leqslant \exp(-\varepsilon\rho_i). \end{aligned} \tag{6.8}$$

Conditional on $\mathscr{N} \equiv \{N_1, ..., N_k\}$, the p_i's are independent and the conditional distribution of p_i depends only on N_i and $N.$. Combining this observation with (6.5), (6.6), and (6.8), we conclude that

$$\begin{aligned} P_*(\hat{\theta}_i = \theta_i, 1 \leqslant i \leqslant k \mid \mathscr{N}) &\leqslant \prod_{i=1}^{k} E(q_i \mid N_i, N.) \\ &\leqslant \exp\left(-\varepsilon \sum_{i=1}^{k} \rho_i\right), \end{aligned} \tag{6.9}$$

where ρ_i is any number satisfying (6.7).

Take $\varepsilon \equiv uk^{-1}$, for arbitrary but fixed $u > 0$. For sufficiently large n,

$$\begin{aligned} &P_*(q_i \leqslant 1-\varepsilon \mid N_i, N.) \\ &\quad = P_*\{\varepsilon(1-\varepsilon)^{-1} \leqslant \lambda_i \leqslant (1-\varepsilon)\,\varepsilon^{-1} \mid N_i, N.\} \\ &\quad \geqslant P_* \mid \log \lambda_i \mid \leqslant \tfrac{1}{2} \log \varepsilon^{-1} \mid N_i, N.) \\ &\quad \geqslant \tfrac{1}{2} P_*\{|\log \lambda_i| \leqslant \tfrac{1}{4}(t+1)^{-1} \log n \mid N_i, N., \theta_i = 0\} \\ &\quad \geqslant \tfrac{1}{2}\{1 - 4(t+1)(\log n)^{-1} E(|\log \lambda_i| \mid N_i, N., \theta_i = 0)\}. \end{aligned} \tag{6.10}$$

In the density estimation case, and for large n, it follows from the definition of λ_i that

$$|\log \lambda_i| \leqslant h^t \left| \sum_{\mathscr{C}_i} \psi\{h^{-1}(X_j - ih)\} \right| + h^{2t} \sum_{\mathscr{C}_i} \psi^2\{h^{-1}(X_j - ih)\}.$$

Applying the Cauchy–Schwarz inequality to the first term on the right-hand side, we see that

$$\begin{aligned} E(|\log \lambda_i| \mid N_i, N., \theta_i = 0) &\leqslant h^t(N_i \sup \psi^2)^{1/2} + h^{2t} N_i \sup \psi^2 \\ &\leqslant 2 \max(1, h^{2t} N_i s^2). \end{aligned} \tag{6.11}$$

If each $N_i \leqslant 2dnh$ then by (6.10) and (6.11), and for large n,

$$\begin{aligned} P_*(q_i \leqslant 1-\varepsilon \mid N_i, N.) &\geqslant \tfrac{1}{2}\{1 - 8(t+1)(\log n)^{-1}\, 2dnh^{2t+1} s^2\} \\ &\geqslant \tfrac{1}{2}\{1 - 17(t+1)\, ds^2 z_1^{2t+1}\} \geqslant \tfrac{1}{4} \end{aligned}$$

provided

$$17(t+1)\, ds^2 z_1^{2t+1} \leqslant \tfrac{1}{2}. \tag{6.12}$$

In this case we may take each $\rho_i = \frac{1}{4}$ in (6.7). Therefore by (6.9),

$$P_*(\hat{\theta}_i = \theta_i,\ 1 \leqslant i \leqslant k) \leqslant \exp(-\varepsilon k/4) + \sum_{i=1}^{k} P_*(N_i > 2dnh).$$

Since $E_*(N_i) < dnh$ then it may be proved by Chebychev's inequality that

$$\sum_{i=1}^{k} P_*(N_i > 2dnh) \to 0.$$

In consequence, provided z_1 satisfies (6.12),

$$\limsup_{n \to \infty} \inf_{f \in \mathscr{F}} P_f(\hat{\theta}_i = \theta_i,\ 1 \leqslant i \leqslant k)$$
$$\leqslant \limsup_{n \to \infty} P_*(\hat{\theta}_i = \theta_i,\ 1 \leqslant i \leqslant k) \leqslant \exp(-u/4).$$

(Recall that $\varepsilon = uk^{-1}$.) Since this is true for each $u > 0$, the lim sup on the left-hand side must equal zero. This proves (6.4). The regression case is similar.

Some techniques in this proof are borrowed from Stone [7].

References

[1] Feller, W. (1969). Limit theorems for probabilities of large deviations. *Z. Wahrsch. Verw. Gebiete* **14** 1–20.

[2] Härdle, W. (1986). Asymptotic maximal deviation of *M*-smoothers. *J. Multivariate Anal.*, in press.

[3] Hartigan, J. A., and Hartigan, P. M. (1985). The dip test of unimodality. *Ann. Statist.* **13** 70–84.

[4] Knafl, G., Sacks, J., and Ylvisaker, D. (1985). Confidence bands for regression functions *J. Amer. Statist. Assoc.* **80** 683–691.

[5] Silverman, B. W. (1985). Some aspects of the spline smoothing approach to non-parametric regression curve fitting (with discussion). *J. Roy. Statist. Soc. Ser. B* **47** 1–52.

[6] Steffensen, J. F. (1927). *Interpolation.* Baillière, Tindall & Cox, London.

[7] Stone, C. J. (1982). Optimal global rates of convergence for nonparametric regression. *Ann. Statist.* **10** 1040–1053.

[8] Wahba, G. (1983). Bayesian confidence intervals for the cross-validated smoothing spline. *J. Roy. Statist. Soc. Ser. B* **45** 133–150.

[9] Abramowitz, M., and Stegun, I. A. (1965). *Handbook of Mathematical Functions.* Dover, New York.

[10] Johnson, N. L., and Kotz, S. (1970). *Distributions in Statistics: Continuous Univariate Distributions.* Houghton Mifflin, Boston.

A Note on Generalized Gaussian Random Fields

TAKEYUKI HIDA

Nagoya University,
Nagoya, Japan

Given a generalized Gaussian random field on a domain D in R^d, we are interested in a restriction of the parameter to a lower dimensional submanifold and discuss the variation when the manifold varies.

0. INTRODUCTION

The present work has been motivated by P. Lévy's results [1] and papers [5, 6, 8–10] by others. When we discuss a Gaussian random field, we often meet a conditional expectation or the same as the best linear predictor of its value at a point, under the condition that the values are given on a certain manifold of the parameter space of the random field.

If the manifold changes, we may think of the variation of the conditional expectation which features certain properties of the field. In order to discuss such a property, we have to prepare some basic facts about generalized random fields as well as its restriction to a submanifold of the parameter space. Unlike the one-dimensional parameter case, we have to be careful about how one restricts the random field according to the restriction of the parameter, and we even note that the method is often used in applications, for example, in quantum field theory.

1. WHITE NOISE AND GAUSSIAN RANDOM FIELDS ON D

We start with a white noise on a bounded domain D in the d-dimensional Euclidean space. The boundary ∂D is assumed to be a C^{∞}-manifold. Then the domain D satisfies the cone property (see [2]). Now take the

Multivariate Statistics and Probability
ISBN 0-12-580205-6

Reprinted from *J. Mult. Anal.* **27**(1).

Sobolev space $H^m(D)$ with $m > d/2$, and we wish to establish the imbedding mapping

$$H^m(D) \to L^2(D)$$

which is of the Hilbert–Schmidt type.

Let $C(\xi)$, $\xi \in H^m(D)$, be a characteristic functional given by

$$C(\xi) = \exp\left[-\frac{1}{2}\int_D \xi(t)^2 \, dt\right].$$

Then we obtain a probability measure μ on $H^{-m}(D)$, the dual space of $H^m(D)$, such that

$$C(\xi) = \int_{H^{-m}(D)} \exp[i\langle x, \xi \rangle] \, d\mu(x).$$

The μ thus obtained is called a white noise measure on $H^{-m}(D)$.

Let $\langle x, \xi \rangle$ be the canonical bilinear form connecting $H^{-m}(D)$ and $H^m(D)$. Once ξ is fixed, $\langle x, \xi \rangle = \xi(x)$ is a random variable on the probability space $(H^{-m}(D), \mu)$. The closure, in the Hilbert space $L^2(H^{-m}(D), \mu)$, of the linear space spanned by the $\langle x, \xi \rangle$, $\xi \in H^m(D)$, is denoted by $\mathscr{H}_1(D)$ or simply by $\mathscr{H}_1$.

The $\mathscr{S}$-transform introduced in [7]

$$(\mathscr{S}\varphi)(\xi) = \int_{H^{-m}(D)} \varphi(x + \xi) \, d\mu(x), \qquad \varphi \in \mathscr{H}_1,$$

gives us an isomorphism

$$\mathscr{H}_1 \cong L^2(D)$$

through the correspondence:

$$\varphi \leftrightarrow F \in L^2(D) \qquad \text{(surjection)},$$

where

$$(\mathscr{S}\varphi)(\xi) = \int_D F(u)\xi(u) \, du$$

and $\|\varphi\|_{\mathscr{H}_1} = \|F\|_{L^2(D)}$.

We often meet Gaussian random fields which are expressed as a system of variables in $\mathscr{H}_1$. Such a field is said to be expressed in terms of white noise.

A probability measure ν associated with a generalized Gaussian random field can also be defined in the same manner as a white noise. A generalized Gaussian random field $\mathbb{X} = \{X(\xi), \xi \in E\}$, with a suitable choice of a function space E, is a continuous linear mapping of E to the space of Gaussian random variables. As is well known, the mean $m(\xi) \equiv E((X(\xi))$ and the covariance functional $\Gamma(\xi, \eta) = E\{(X(\xi) - m(\xi))(X(\eta) - m(\eta))\}$ completely determine the probability distribution ν of $\{X(\xi), \xi \in E\}$ on a space of generalized functions. If we are given an ordinary random field denoted by $\{X(t), t \in D\}$, then it is identified with a generalized random field $\{X(\xi), \xi \in E\}$, in such a way that

$$X(\xi) = \int_D X(t)\xi(t)\, dt,$$

where we assume some regularity of $X(t)$ in t so that the mapping

$$\xi \to X(\xi), \qquad \xi \in E,$$

is continuous.

For a generalized Gaussian random field we can define a Hilbert space $\mathscr{H}_1(D)$ as in the case of a white noise, and the space forms a Gaussian system.

2. Restriction of Parameter

Our main topic is concerned with the restriction of the parameter of a generalized Gaussian random field $\mathbb{X}$ to a submanifold of D.

(i) First consider the case where the parameter is restricted to a d-dimensional C^∞-submanifold D' of D. Then, the regular imbedding mapping $D' \to D$ naturally determines the injection

$$\mathscr{H}_1(D') \to \mathscr{H}_1(D). \tag{1}$$

With such a relation, we can proceed to the investigation of various stochastic properties of the field $\mathbb{X}$ (for instance, see [6]).

(ii) We are particularly interested in the case where $\dim(D') < d$. To fix the idea, let $\mathbb{X}$ be a white noise and let D' be a boundary of a d-dimensional C^∞-submanifold D_1 of D: $D' = \partial D_1$. Also, to make the story simpler, the order m of Sobolev space $H^m(D)$ is taken to be $(d+1)/2$. Then, associated with the regular imbedding mapping $D \to D'$, we are given a natural continuous imbedding mapping

$$e\colon H^{m-1/2}(D') \to H^m(D). \tag{2}$$

The white noise measures, denoted by μ and μ_1, are introduced on $H^{-m}(D)$ and $H^{-m+1/2}(D')$, respectively, as was done in Section 1, where it is noted that the injection $H^{m-1/2}(D') \to L^2(D')$ is of the Hilbert–Schmidt type, since $m - 1/2 = d/2 > (d-1)/2$.

There is defined a surjective mapping e^* which is the adjoint of e:

$$e^*: H^{-m}(D) \to H^{-m+1/2}(D'). \tag{3}$$

Summing up what have been discussed, we can prove the following assertion.

PROPOSITION. *Let D and $\partial D = D'$ be C^∞-manifolds in R^d. Set $m = (d+1)/2$. Then, there exist white noise measures μ and μ' on $H^{-m}(D)$ and $H^{-m+1/2}(D')$, respectively, and these two measures are linked in such a way that*

$$(e^*)^{-1} \circ \mu = \mu'.$$

For the proof, we only need to note that the Borel field $\mathscr{B}_1$ generated by subsets of $H^{-m+1/2}(D')$ is equal to the image of Borel field corresponding to $H^{-m}(D)$ under the mapping e^*, and the characteristic functionals of μ and μ_1 are the same in expression.

3. Gaussian Random Fields Depending on a Curve

We use the same notation established in the last section. Consider, in particular, the case $d = 2$, and introduce a class $\mathbb{C}$ of curves given by

$$\mathbb{C} = \{C\colon \text{closed, simple, } C^\infty\text{-curves} \subset D\}.$$

Note that each member of $\mathbb{C}$ is viewed as the boundary of a submanifold of D.

As was discussed in [5], we are interested in a Gaussian system indexed by a domain or a curve. Let $\varphi(x)$ be a $\mathscr{H}_1(D)$-functional. Then the associated U-functional $(\mathscr{S}\varphi)(\xi)$ has the expression

$$U(D, \xi) = \int_D F(u)\xi(u)\, du, \qquad F \in L^2(D).$$

In a similar manner, we have

$$U_1(C, \xi) = \int_C G(u)\xi(u)\, du, \qquad G \in L^2(C),$$

for $\psi_C(x) \in \mathscr{H}_1(C)$. From our discussion in Section 2, $U_1(C, \xi)$ is viewed as a functional obtained from $U(D, \xi)$ by restricting some F to C, or equivalently ψ_C comes from φ by the mapping e^*, if C is a boundary of D. Thus we are able to deal with a family

$$\Psi = \{\psi_C(x); C \in \mathbb{C}\} \tag{4}$$

within a framework of the analysis on $\mathscr{H}_1(D)$.

Under the above setup, we can prove the following theorem (cf. [1]).

THEOREM. *Let Ψ be given by* (4). *Then the variation of $\psi_C(x)$ exists and its U-functional is expressed in the form*

$$\delta U_1(C, \xi) = \int_C \left(\frac{\partial F \cdot \xi}{\partial n}(s) - \kappa(s) F(s)\, \xi(s) \right) \delta n(s)\, ds, \tag{5}$$

where δn denotes the variation δC of C and κ is the curvature.

4. Concluding Remarks

A few remarks are now in order. We have started with a bounded domain, because we wish to use the Sobolev space structure to introduce white noise and to use the trace theorem. However, we may start with the entire space R^d or a half space and still carry out the whole story with slight modification. Hence, there is no difficulty in discussing the variational calculus even when we do not limit our attention to a finite domain.

In Section 3, we have dealt only with functionals of white noise as a prototype of generalized Gaussian random fields. If we choose suitable function spaces like a Sobolev space, we can establish the theory in a similar manner. Also, it is noted that important examples of a Gaussian random field can be realized as functionals expressed in terms of white noise, so that the discussion may be reduced to that of white noise.

The variational calculus of functionals depending on a curve would be generalized to the case where the kernel function F depends on C in addition to s in the expression (5). Important examples are seen in [10]. A general theory will be discussed in a separate paper.

References

[1] LÉVY, P. (1951). *Problèmes concrets d'analyse fonctionnelle.* Gauthier–Villars, Paris.

[2] LÉVY, P. (1948, 1965). *Processus stochastiques et mouvement brownien.* Gauthier–Villars, Paris.

[3] ADAMS, R. A. (1975). *Sobolev Spaces.* Academic Press, New York/London.

[4] HIDA, T. (1975). *Brownian Motion.* Applications of Mathematics, Vol. 11. (Springer-Verlag) New York/Berlin.

[5] HIDA, T., LEE, K.-S., AND SI SI (1987). Multi-dimensional parameter white noise and Gaussian random fields. *In Recent Advances in Communication and Control Theory,* (R. E. Kalman, G. I. Marchuk, A. E. Ruberti, and A. J. Viterbi, Eds.), Optimization Software Inc., New York, pp. 177–183.

[6] KRISHNAIAH, P. R. (Ed). (1979). *Development in Statistics,* Vol. 2. Academic Press, New York/London. In particular, Chap. 5, Stochastic Markovian fields, by Yu. A. Rozanov.

[7] KUBO, I., AND TAKENAKA, S. (1980, 1981, 1982). Calculus on Gaussian white noise, I, II, III, IV, *Proc. Japan Acads. Ser. A Math. Sci.* **56** 376–380, 411–416; **57** 433–437; **58** 186–189.

[8] SI, SI (1987). A note on Lévy's Brownian motion. *Nagoya Math. J.* **108** 121–130.

[9] SI, SI (1987). A note on Lévy's Brownian motion II, to appear.

[10] SI, SI (1987). Gaussian processes and conditional expectations. BiBoS Notes, Nr. 292/87, Universität Bielefeld.

Smoothness Properties of the Conditional Expectation in Finitely Additive White Noise Filtering

H. P. HUCKE

Fern University

G. KALLIANPUR*

University of North Carolina

AND

R. L. KARANDIKAR*

Indian Statistical Institute

It is shown that for a wide class of signal processes and bounded g, the conditional expectation $\pi(g, y)$ in the white noise filtering model is a C^{∞}-functional of the observations in the sense that $\pi(g, y)$ and its Fréchet derivatives (which exist) are random variables on the quasicylindrical probability space on which the observation model is defined.

1. INTRODUCTION

In a reccnt paper, M. Chaleyat-Maurel has shown that the conditional expectations in the nonlinear filtering problem is a C^{∞}-functional in Malliavin's sense [1]. A Malliavin calculus for functionals of finitely additive Gaussian white noise has not yet been developed though, in our view, many of the basic ideas of the former theory carry over naturally to the finitely additive situation.

* Research supported by the Air Force Office of Scientific Research Contract F49620 85 C 0144.

Multivariate Statistics and Probability
ISBN 0-12-580205-6

Reprinted from *J. Mult. Anal.* **27**(1).

In this note, we derive a result close in spirit to Malliavin calculus. We will be concerned with the smoothness properties of the conditional expectation regarded as a functional of the observations. In the same sense as in [1] the result obtained by us may be regarded as a robustness property of the nonlinear filter in the white noise theory. Our result cannot be directly compared with Chaleyat-Maurel's. We are throughout in a Hilbert space setting so that in contrast to the Malliavin theory all directional derivatives are admissible for us. Both the statement and the proof of the main theorems are straightforward. The only thing that sets the proof apart from a standard calculation is the need to show that the various functional derivatives of the filter are also random mappings as defined in [2]. The latter fact is established by relying heavily on properties of lifting maps.

It must be noted that our filtering model assumes signal and noise to be independent whereas in [1] a more general model is considered. However, we are able to prove C^{∞}-smoothness of the filter under less restrictive conditions.

2. Notation and Terminology

For most of the notation, terminology, and definitions used in this paper we refer the reader to [2] since it would take too much space to repeat them here.

H is an infinite dimensional, separable Hilbert space, $\mathscr{C}$ the field of finite dimensional Borel cylinder sets in H, and m the (finitely additive) canonical Gauss measure on H, i.e., the measure with characteristic functional $\exp(-\frac{1}{2}\|h\|^2)$, $(h \in H)$. Let $\mathscr{P}$ denote the class of all orthogonal projections on H with finite dimensional ranges. Let $(\Omega, \mathscr{A}, \Pi)$ be a complete (countably additive) propability space. The triple $(E, \mathscr{E}, \alpha)$ is called a quasicylindrical probability space where $E = \Omega \times H$, $\mathscr{E} = \mathscr{A} \times \mathscr{C}$ and $\alpha = \Pi \odot m$. $\mathscr{E}$ is a field and α is the finitely additive probability on $\mathscr{E}$ such that for any $P \in \mathscr{P}$, the restriction of α to the σ-field $\mathscr{A} \times \mathscr{C}_P$ is the countably additive probability measure $\Pi \times m_P$. Here $\mathscr{C}_P$ is the σ-field of cylinder sets with bases on PH and m_P is the restriction of m to $\mathscr{C}_P$.

Let (L_0, Π_0) be a representation of m with an underlying representation space $(\Omega_0, \mathscr{A}_0, \Pi_0)$ and let $(\tilde{\Omega}, \tilde{\mathscr{A}}, \tilde{\Pi}) = (\Omega, \mathscr{A}, \Pi) \otimes (\Omega_0, \mathscr{A}_0, \Pi_0)$. Writing $\tilde{\omega} = (\omega, \omega_0) \in \tilde{\Omega}$, defining $\rho(\tilde{\omega}) = \omega$ and $L(h)(\tilde{\omega}) = L_0(h)(\omega_0)$ for all $h \in H$, it is seen that $(\rho, L, \tilde{\Pi})$ is a representation of the quasicylindrical probability α on the space $(\tilde{\Omega}, \tilde{\mathscr{A}}, \tilde{\Pi})$. It can, in fact, be shown that $(\rho, L, \tilde{\Pi})$ can be chosen to possess the property that for each $h \in H$, the map $(h, \tilde{\omega}) \rightarrow L(h)(\tilde{\omega})$ is $B(H) \otimes \tilde{\mathscr{A}}$ measurable. It is such a representation that we shall be working with throughout.

Let S be a Polish space, i.e., a complete separable metric space. We shall define classes of S-valued maps on E which form important subclasses of random variables on the finitely additive probability space $(E, \mathscr{E}, \alpha)$.

Let $\mathscr{L}^0(E, \mathscr{E}, \alpha; S)$ be the class of maps f from E to S such that for all $P \in \mathscr{P}$, f_P defined by $f_P(\omega, \eta) = f(\omega, P\eta)$ is $\mathscr{E}_P/B(S)$-measurable and for all sequences $\{P_j\} \subset \mathscr{P}$ converging strongly to the identity $(P_j \to^s I)$, $R_\alpha(f_{P_j})$ is Cauchy in $\tilde{\Pi}$-probability. Elements of $\mathscr{L}^0(E, \mathscr{E}, \alpha; S)$ are called S-valued *accessible* random variables. For $1 \leqslant q < \infty$, define $\mathscr{L}^q(E, \mathscr{E}, \alpha; S)$ as the class of maps f as above with the additional property that

$$\int_{\tilde{\Omega}} |R_\alpha(f_{P_j}) - R_\alpha(f_{P_l})|^q \, d\tilde{\Pi} \to 0.$$

In this case

$$\int_{\tilde{\Omega}} |R_\alpha(f)|^q \, d\tilde{\Pi} < \infty.$$

The notation here is somewhat different from that adopted in [2] where the class $\mathscr{L}^0$ is denoted by $\mathscr{L}^*$ and $\mathscr{L}^1$ by $\mathscr{L}^{1*}$. Wider classes of random variables are also considered in [2]. The symbol S will be suppressed whenever $S = \mathbb{R}^1$.

Let ξ: $\Omega \to H$ be a random variable, i.e., a $B(H)/\mathscr{A}$-measurable map, $B(H)$ being the σ-field of Borel sets in H. The nonlinear filtering model in its abstract form is defined on $(E, \mathscr{E}, \alpha)$ by

$$y = \xi + e, \tag{1}$$

where for $(\omega, \eta) \in E$, $\xi(\omega, \eta) = \xi(\omega)$ and $e(\omega, \eta) = e(\eta) = \eta$. The identity map e on H is called Gaussian white noise, ξ is the signal and y the observation.

Let Q be an arbitrary orthogonal projection on H. If g is a Π-integrable, real random variable on Ω, then the conditional expectation (in the finitely additive theory) $E_\alpha(f \mid Qy)$ exists and is given by the Bayes formula

$$E_\alpha(g \mid Qy) = \frac{\sigma_Q(g, y)}{\sigma_Q(1, y)}, \tag{2}$$

where

$$\sigma_Q(g, y) = \int_\Omega g(\omega) \exp\{(y, Q\xi(\omega)) - \tfrac{1}{2} \|Q\xi(\omega)\|^2\} \, d\Pi(\omega) \tag{3}$$

is called the unnormalized conditional expectation of g. The model (1) covers most of the filtering problems met with in practice including those in which the observation process takes values in a Hilbert space. In applications, the true signal process is denoted by an S-valued process (X_t), $(0 \leqslant t \leqslant T)$ defined on Ω, and (1) takes the form

$$y_t = \xi_t + e_t, \qquad 0 \leqslant t \leqslant T, \tag{4}$$

where

e is K-valued Gaussian white noise. Here $H = L^2([0, T], K)$ and K is a possibly infinite dimensional separable Hilbert space; (5a)

$\xi_t(\omega) = h_t(X_t(\omega))$ where h: $[0, T] \times S \to K$ is measurable and satisfies the condition $\int_0^T \|h_t(X_t(\omega))\|_K^2 \, dt < \infty$ for each ω (or a.a. ω). (5b)

If Q_t is the orthogonal projection on H with range $H_t := \{f \in H: \int_t^T \|f_s\|_K^2 \, ds = 0\}$ then the filter one is interested in is the conditional expectation $E_\alpha(g \mid Q_t y)$ which is given by (2) with $Q = Q_t$. For the sake of notational convenience we shall derive all our results for the abstract model (1) rather than (4).

In what follows we may take, without loss of generality, g to be non-negative and such that $\int g \, d\Pi = 1$. Let $d\Pi_1 = g \, d\Pi$ and $\nu = \Pi_1 \circ (Q\xi)^{-1}$. Then ν is a probability measure on H and

$$\sigma_Q(g, \eta) = \int_H \exp\{(\eta, k) - \tfrac{1}{2} \|k\|^2\} \, d\nu(k), \qquad \eta \in H. \tag{6}$$

Since, throughout this work, g and Q will remain fixed, it is convenient to suppress g and write $\sigma(\eta)$ for $\sigma_Q(g, \eta)$.

For a Banach space B with norm $\|\cdot\|_B$ let $L(H, B)$ denote the class of all bounded linear transformation A: $H \to B$, which is itself a Banach space with operator norm. A mapping f: $H \to B$ is said to be Fréchet differentiable if for every $h \in H$ there exists $f_1(h) \in L(H, B)$ such that

$$\lim_{\|h'\| \to 0} \frac{1}{\|h'\|} \|f(h + h') - f(h) - f_1(h)[h']\|_B = 0,$$

$f_1(h)$ is called the Fréchet derivative of f at h and is written as $(Df)(h)$.

Let $L^0(H) = \mathbb{R}$, $L^1(H) = L(H, \mathbb{R})$, and for $r \geqslant 1$, $L^{r+1}(H) = L(H, L^r(H))$. It is well known that the Banach space $L^r(H)$ can be identified with the class of all linear mappings from the r-fold product $H \times \cdots \times H$ into $\mathbb{R}$. The norm $\|\cdot\|_r$ on $L^r(H)$ under this identification is given by

$$\|f\|_r = \sup\{|f[h_1, ..., h_r]|: h_i \in H, \|h_i\| \leqslant 1\}.$$

A function $f: H \to \mathbb{R}$ is said to be $(r+1)$ times Fréchet-differentiable, if it is r-times Fréchet differentiable, and $D^r f: H \to L^r(H)$ is Fréchet-differentiable and then $D^{r+1} f := D(D^r f)$.

Let $L^r_{(2)}(H)$ be the subclass of $L^r(H)$ consisting of $g \in L^r(H)$ for which

$$\|g\|^2_{r,2} := \sum_{j_1 \cdots j_r} |g[\varphi_{j_1}, ..., \varphi_{j_r}]|^2 < \infty,$$

where $\{\varphi_j\}$ is any CONS in H. It is well known that $\|g\|_{r,2}$ does not depend on the choice of CONS and that $L^{(r)}_{(2)}(H)$ is a Hilbert space with norm $\|\cdot\|_{r,2}$ and that $\|g\|_r \leqslant \|g\|_{r,2}$.

3. Main Results

Lemma 1. *Let the function $\sigma(\eta)$ be defined by* (6). *Then*

(a) *for every $r \geqslant 1$, $\sigma(\eta)$ is r-times Fréchet differentiable and the derivative $D^r\sigma(\eta)$ is given by*

$$D^r\sigma(\eta)[h_1, ..., h_r] = \int [\exp\{(\eta, k) - \tfrac{1}{2}\|k\|^2\}](h_1, k) \cdots (h_r, k)\, d\nu(k). \tag{7}$$

(b) $D^r\sigma(\eta) \in L^r_{(2)}(H)$.

Proof. Denote the right-hand side of (7) by $g_r[h_1, ..., h_r]$. The integral appearing in (7) is finite since

$$|(\eta, k)| \leqslant \tfrac{1}{2}\{\|2\eta\|^2 + \|\tfrac{1}{2}k\|^2\} = 2\|\eta\|^2 + \tfrac{1}{8}\|k\|^2 \tag{8}$$

and

$$|(h, k)| \leqslant \|h\| \cdot \|k\|. \tag{9}$$

Let $\{\varphi_j\}$ be a CONS in H. Note that

$$\begin{aligned}\|g_r\|^2_{r,2} &= \sum_{j_1 \cdots j_r} \left[\int \exp\{(\eta, k) - \tfrac{1}{2}\|k\|^2\} \cdot (\varphi_{j_1}, k) \cdots (\varphi_{j_r}, k)\, d\nu(k)\right]^2 \\ &\leqslant \sum_{j_1 \cdots j_r} \int \exp\{2(\eta, k) - \|k\|^2\} \cdot (\varphi_{j_1}, k)^2 \cdots (\varphi_{j_r}, k)^2\, d\nu(k) \\ &= \int \exp\{2(\eta, k) - \|k\|^2\} \cdot \|k\|^{2r}\, d\nu(k) \\ &< \infty,\end{aligned} \tag{10}$$

in view of (8) and (9). Hence $g_r(\eta) \in L^r_{(2)}(H)$. Let us write $\sigma(\eta) = g_0(\eta)$. To complete the proof, we will show that g_r is Fréchet differentiable for $r \geqslant 0$ and that $Dg_r = g_{r+1}$. This will show that $D^r\sigma(\eta) = g_r(\eta)$. For η, $h \in H$, let

$$v(\eta, h) := \| g_r(\eta + h) - g_r(\eta) - g_{r+1}(\eta)(h) \|_{r,2}.$$

Recall that $g_{r+1}(\eta)(h)[h_1, ..., h_r] = g_r(\eta)[h_1, ..., h_r, h]$ under the identification of $L(H, L^r(H))$ with $L^{r+1}(H)$. Thus

$$\begin{aligned} v^2(\eta, h) &= \sum_{j_1 \cdots j_r} |\{ g_r(\eta + h) - g_r(\eta)\}[\varphi_{j_1}, ..., \varphi_{j_r}] \\ &\quad - g_{r+1}(\eta)[\varphi_{j_1}, ..., \varphi_{j_r}, h]|^2 \\ &= \sum_{j_1 \cdots j_r} \left[\int \exp\{(\eta, k) - \tfrac{1}{2} \|k\|^2\} \right. \\ &\quad \left. \cdot (\varphi_{j_1}, k) \cdots (\varphi_{j_r}, k) \cdot v_1(h, k)\, d\nu(k) \right]^2, \end{aligned} \tag{11}$$

where

$$v_1(h, k) := \exp\{(h, k)\} - 1 - (h, k).$$

Since

$$|v_1(h, k)| \leqslant |(h, k)|^2 \exp((h, k)) \leqslant \|h\|^2 \, \|k\|^2 \exp((h, k)),$$

we get from (1)

$$\begin{aligned} v^2(\eta, h) &\leqslant \int \exp\{2(\eta, k) - \|k\|^2\} \cdot \|k\|^{2r} \\ &\quad \cdot \|h\|^4 \, \|k\|^4 \exp(2(h, k))\, d\nu(k) \\ &\leqslant \|h\|^4 \cdot \exp\{4 \|h\|^2 + 4 \|\eta\|^2\} \\ &\quad \cdot \int \exp\{-\tfrac{1}{2} \|k\|^2\} \, \|k\|^{2r+4}\, d\nu(k) \end{aligned}$$

using (8) and (9). This shows that

$$\lim_{\|h\| \to 0} \frac{1}{\|h\|} \cdot v(\eta, h) = 0$$

and hence that $g_r \colon H \to L^r_{(2)}(H)$ is Fréchet differentiable with $Dg_r = g_{r+1}$. Since $\| \cdot \|_{r,2} \leqslant \| \cdot \|_r$, this implies that $g_r \colon H \to L^r(H)$ is Fréchet differentiable. ∎

Our next step is to show that $D^r\sigma(y)$ is a random variable.

LEMMA 2. *Suppose that* ν *satisfies*

$$\int \|k\|^{2r} \, d\nu(k) < \infty. \tag{12}$$

Then $D^r\sigma(y) \in \mathscr{L}(E, \mathscr{E}, \alpha; L_2^{(r)}(H))$.

Proof. Fix $\{P_i\} \subseteq \mathscr{P}$, $P_i \to^s I$. Let $Z(k, \tilde{\omega})$, $Z_i(k, \tilde{\omega})$ be defined by

$$Z(k, \tilde{\omega}) := \exp\{(k, \xi(\omega)) + L_0(k)(\omega_0) - \tfrac{1}{2}\|k\|^2\}$$

and

$$Z_i(k, \tilde{\omega}) := \exp\{(k, \xi(\omega)) + L_0(P_i k)(\omega_0) - \tfrac{1}{2}\|k\|^2\}, \tilde{\omega} = (\omega, \omega_0).$$

Then from the definition of lifting for cylinder functions (see [2]) it follows that

$$R_\alpha([D^r\sigma \circ y]_{P_i} [\varphi_{j_1} \cdots \varphi_{j_r}](\tilde{\omega}) = \int Z_i(k, \tilde{\omega}) \, f_j(k) \, d\nu(k),$$

where $j = (j_1, ..., j_r)$ and $f_j(k) = (\varphi_{j_1}, k) \cdots (\varphi_{j_r}, k)$. Let $U_{il}(\tilde{\omega}) := \|R_\alpha(\{D^r\sigma \circ y]_{P_i}) - R_\alpha([D^r\sigma \circ y]_{P_l})\|_{r,2}^2 (\tilde{\omega})$. To complete the proof. we will show that $U_{il} \to 0$ in $\tilde{\Pi}$-probability:

$$\begin{aligned}
U_{il}(\tilde{\omega}) &\leqslant \sum_j \left[\int |Z_i(k, \tilde{\omega}) - Z_l(k, \tilde{\omega})| \, |f_j(k)| \, d\nu(k)\right]^2 \\
&\leqslant \sum_j \int |Z_i(k(\tilde{\omega}) - Z_l(k, \tilde{\omega})| \, d\nu(k) \\
&\quad \cdot \int |Z_i(k, \tilde{\omega}) - Z_l(k, \tilde{\omega})| \cdot f_j^2(k) \, d\nu(k) \\
&= \int |Z_i(k, \tilde{\omega}) - Z_l(k, \tilde{\omega})| \, d\nu(k) \\
&\quad \cdot \int |Z_i(k, \omega) - Z_l(k, \tilde{\omega})| \, \|k\|^{2r} \, d\nu(k).
\end{aligned} \tag{13}$$

We have used Hölder's inequality above.

Define a probability measure Π' on $\tilde{\Omega}$ by

$$\begin{aligned}
\frac{d\Pi'}{d\tilde{\Pi}}(\tilde{\omega}) &= C \cdot \exp(-\tfrac{1}{2}\|\xi(\omega)\|^2) \\
&\quad \cdot \exp(-L_0(\xi(\omega))(\omega_0 - \tfrac{1}{2}\|\xi(\omega))\|^2),
\end{aligned}$$

where, $\tilde{\omega} = (\omega, \omega_0) \in \tilde{\Omega}$. The constant C is chosen such that $\Pi'(\tilde{\Omega}) = 1$. It is easy to see that

$$\int \exp(\tfrac{1}{2} \|\xi(\omega)\|^2)\, d\Pi' < \infty.$$

Let μ be any countably additive finite measure on H. Then, it is easy to check that

$$Z_i \to Z \qquad \text{in} \quad \mu \otimes \Pi' \text{ measure}$$

and

$$\iint Z_i(k, \tilde{\omega})\, d\Pi'(\tilde{\omega})\, d\mu(k) \to \iint Z(k, \tilde{\omega})\, d\Pi'(\tilde{\omega})\, d\mu(k),$$

where the integrals appearing above are finite. By arguments similar to the proof of Scheffé's theorem, it follows that

$$\iint |Z_i - Z|\, d\Pi'\, d\mu \to 0$$

and as a consequence

$$\int |Z_i(k, \tilde{\omega}) - Z(k, \tilde{\omega})|\, d\mu(k) \to 0 \tag{14}$$

in Π'-probability. Since $\tilde{\Pi} \ll \Pi'$, (14) also holds in $\tilde{\Pi}$-probability. The assumption (12) implies that ν_1 defined by

$$\frac{d\nu_1}{d\nu}(k) = \|k\|^{2r}$$

is a finite measure. Thus (14) for $\mu = \nu$ and $\mu = \nu_1$ implies that $U_{il} \to 0$ in $\tilde{\Pi}$-probability. ∎

THEOREM 3. *For any integrable function f, for any orthogonal projection Q, $\sigma_Q(g, y)$ is r-times Fréchet differentiable for all $r \geq 1$. Further,* (a) *if*

$$\int \|Q\xi(\omega)\|^{2r} \cdot |g(\omega)|\, d\Pi(\omega) < \infty \tag{15}$$

then

$$D^r \sigma_Q(g, y) \in \mathscr{L}(E, \mathscr{E}, \alpha; L^r_{(2)}(H));$$

(b) *if $\|\xi\|$, g are bounded, then for all $q \geq 1$, $r \geq 1$,*

$$D^r \sigma_Q(g, y) \in \mathscr{L}^q(E, \mathscr{E}, \alpha; L^r_{(2)}(H)).$$

Proof. Part (a) follows from the preceeding lemmas. The proof of (b) is based on the easily verifiable fact that

$$\int [U_{il}]^{q/2}\, d\tilde{\Pi} \to 0$$

as $(i, l) \to \infty$. We are now in a position to prove the main result of this paper.

THEOREM 4. *Suppose ξ and g satisfy* (15). *Then, $\pi_Q(g, y)$ is r-times Fréchet differentiable and*

$$D^r \pi_Q(g, y) \in \mathscr{L}(E, \mathscr{E}, \alpha; L^r_{(2)}(H)). \tag{16}$$

Proof. $\pi_Q(g, y) = f_1(y)/f_2(y)$, where

$$f_1(y) = \sigma_Q(g, y), \qquad f_2(y) = \sigma_Q(1, y).$$

Now, f_1, f_2 are both r-times F-differentiable and $f_2 > 0$. From this it is easy to check that f_1/f_2 is also r-times F-differentiable. It can be shown that $D^r(f_1/f_2)$ can be expressed as

$$D^r(f_1/f_2) = \Lambda_r(f_1, f_2, Df_1, Df_2, ..., D^r f_1, D^r f_2, 1/f_2)$$

where Λ_r is a continuous mapping from

$$\mathbb{R} \times \mathbb{R} \times L^1_{(2)}(H) \times L^1_{(2)}(H) \times \cdots \times L^r_{(2)}(H) \times L^r_{(2)}(H) \times \mathbb{R}$$

into $L^r_{(2)}(H)$. Since f_1, f_2, $1/f_2 \in \mathscr{L}(E, \mathscr{E}, \alpha; \mathbb{R})$, and $D^i f_1$, $D^i f_2 \in \mathscr{L}(E, \mathscr{E}, \alpha; L^i_{(2)}(H))$, $1 \leqslant i \leqslant r$, assertion (16) follows from the continuity of Λ_r. ∎

A functional $f(y)$ will be said to be a C^∞*-functional of the observations* if $D^r f$ exists for all $r \geqslant 1$ and $D^r f(y) \in \mathscr{L}(E, \mathscr{E}, \alpha; L^r_{(2)}(H))$.

We have proved above that if

$$E\,\|\xi\|^r < \infty \qquad \text{for all} \quad r \geqslant 1,$$

then for all g bounded $\pi_Q(g, y)$ is a C^∞-functional of the observations.

A concrete application of this result to the models considered in [2] shows that the conditional expectations in the filtering, prediction, and smoothing problems (for finite or infinite dimensional signals) are C^∞-functionals of the observations.

REFERENCES

[1] CHALEYAT-MAUREL, M. (1986). Robustesse du filtre et calcul des variations stochastique. *J. Funct. Anal.* **18** No. 1 55–71.

[2] KALLIANPUR, G., AND KARANDIKAR, R. L. (1985). White noise calculus and nonlinear filtering theory. *Ann. Probab.* **13** No. 4 1033–1107.

Equivariant Estimation of a Mean Vector μ of $N(\mu, \Sigma)$ with $\mu'\Sigma^{-1}\mu = 1$ or $\Sigma^{-1/2}\mu = c$ or $\Sigma = \sigma^2\mu'\mu I$

TAKEAKI KARIYA

Hitotsubashi University, Japan

N. C. GIRI AND F. PERRON

University of Montreal, Canada

This paper considers the problems of estimating a mean vector μ under constraint $\mu'\Sigma^{-1}\mu = 1$ or $\Sigma^{-1/2}\mu = c$ and derives the best equivariant estimators under the loss $(a-\mu)'\,\Sigma^{-1}(a-\mu)$, which dominate the MLE's uniformly. The results are regarded as multivariate extensions of those with known coefficient of variation in a univariate case. As a particular case for $\mu'\Sigma^{-1}\mu = c$, the case $\Sigma = \sigma^2\mu'\mu I$ is also treated. © 1988 Academic Press, Inc.

1. INTRODUCTION

The problem of estimating the mean μ of a univariate normal population $N_1(\mu, \sigma^2)$ with known coefficient of variation (i.e., $\sigma/\mu = \text{const}$) was originally considered by Fisher a long time ago and recently again focussed upon in the context of a curved model or a model which admits an ancillary statistic (see Efron [6], Cox and Hinkley [4], Hinkley [8], and Amari [1, 2]). The motivation behind the model is based on the empirically observed fact that a standard deviation often becomes large almost proportionally to a corresponding mean so that the coefficient of variation remains constant. This fact is often found also in multivariate (mutually correlated) variates. Though a well-accepted measure for variation between a mean vector μ and a covariance matrix Σ is not

Multivariate Statistics and Probability
ISBN 0-12-580205-6

Reprinted from *J. Mult. Anal.* **27**(1).

available, in this paper we adopt as multivariate versions of the variational coefficient the following measures

$$\lambda = \mu' \Sigma^{-1} \mu \tag{1.1}$$

and

$$\nu = \Sigma^{-1/2} \mu \qquad \text{with} \quad \Sigma^{-1/2} \in \mathrm{GT}(p) \tag{1.2}$$

and consider the problems of estimating μ of a p-variatc normal population $N_p(\mu, \Sigma)$ with either λ or ν known under the quadratic loss function

$$L(a, \mu) = (a - \mu)' \, \Sigma^{-1}(a - \mu), \tag{1.3}$$

where $\mathrm{GT}(p)$ denotes the group of $p \times p$ lower triangular matrices with positive diagonal elements and $\Sigma^{1/2}$ is the unique solution for $\Sigma^{1/2}\Sigma^{1/2\prime} = \Sigma$. The analysis is based on the invariance principle. In the versions of (1.1) and (1.2), the constancy of the measures means that Σ becomes "proportionally" large in the sense of nonnegative definiteness as μ becomes large. Besides these interpretations, some other interpretations are possible for λ and ν. For example, λ is the Mahalanobis distance between $N(0, \Sigma)$ and $N(\mu, \Sigma)$, and ν is a normalized mean vector. As a particular case for which $\mu' \Sigma^{-1} \mu$ becomes constant, the specification $\Sigma = \sigma^2 \mu' \mu I$ with σ^2 known is also considered.

Now let x_i's be a random sample from $N_p(\mu, \Sigma)$ with $\mu \in R^p$ and $\Sigma \in \mathscr{S}(p)$, where $\mathscr{S}(p)$ denotes the set of $p \times p$ positive definite matrices. Then a sufficient statistic is (y, S) with

$$\begin{aligned} y &= \sqrt{n}\, \bar{x} = \sum_{i=1}^{n} x_i / \sqrt{n} \sim N_p(\sqrt{n}\, \mu, \Sigma) \\ S &= \sum_{i=1}^{n} (x_i - \bar{x})(x_i - \bar{x})' \sim W_p(\Sigma, n-1), \end{aligned} \tag{1.4}$$

where $n > p$ and $W_p(\Sigma, m)$ denotes the Wishart distribution with mean $m\Sigma$ and d.f. (degrees of freedom) m. As in the univariate case, when λ (or ν) is known, the model admits an ancillary statistic, that is, a statistic which is a part of a (minimal) sufficient statistic and whose marginal distribution is independent of unknown parameters. Thus an inference on (μ, Σ) may be based on what is called the principle of conditionality. However, in this paper, rather than using the principle directly, we derive a BEE (best equivariant estimator) for each problem under the loss function (1.3). There a conditional argument is inevitably required. The explicit forms of the BEE's are given only for the case of $p = 2$ because the complication of the computation. The MLE's are also derived for comparisons. Since the MLE's are equivariant, which is true in general under a mild condition (see Eaton [5]), the MLE's here are uniformly dominated by the BEE's. In the particular case $\Sigma = \sigma^2 \mu'\mu I$, the BEE and the MLE are also derived (Section 4).

In the literature, not much work has been done on the problems with ancillary statistics from an equivariance viewpoint. Kariya [9] gave a formulation for the equivariant estimation when an ancillary statistic is realized as a maximal invariant. However, he assumed in the formulation that the sample space is homeomorphic to the product of the group leaving the problem invariant and the space of the ancillary statistic. In the first problem with λ known that we treat here, the assumption is not satisfied, though in the second problem with λ known, it is satisfied. A general description of equivariant estimation is found in Ferguson [7], Eaton [5], and Lehmann [11].

2. Problem with λ Known

In this section, we consider the problem of estimating μ of $N_p(\mu, \Sigma)$ with λ in (1.1) known. Without loss of generality, we assume that (μ, Σ) belongs to

$$\textcircled{H} = \{(\mu, \Sigma) \in R^p \times \mathscr{S}(p) \mid \mu'\Sigma^{-1}\mu = 1\}. \tag{2.1}$$

Under the loss function in (1.4), it is easy to see that the problem is left invariant by the group $\mathrm{Gl}(p)$ of $p \times p$ nonsingular matrices acting on (y, S) as

$$(y, S) \longrightarrow (Ay, ASA') \qquad \text{with} \quad A \in \mathrm{Gl}(p), \tag{2.2}$$

which induces the action on (μ, Σ):

$$(\mu, \Sigma) \longrightarrow (A\mu, A\Sigma A') \qquad \text{with} \quad A \in \mathrm{Gl}(p). \tag{2.3}$$

Under the transformation (2.2), the statistic

$$u = y'S^{-1}y \tag{2.4}$$

is a maximal invariant and the distribution of u depends on (μ, Σ) only through the maximal invariant parameter $\lambda = \mu'\Sigma^{-1}\mu$. Therefore by the prior constraint (2.1), u is an ancillary statistic. Further the group $\mathrm{Gl}(p)$ acts transitively on $\textcircled{H}$ in (2.1). This implies that the risk function of an equivariant estimator $\tilde{\mu}$

$$R(\tilde{\mu}, (\mu, \Sigma)) = E_{(\mu, \Sigma)}[(\tilde{\mu} - \mu)' \Sigma^{-1}(\tilde{\mu} - \mu)] \tag{2.5}$$

is constant for all $(\mu, \Sigma) \in \textcircled{H}$ (see Lehmann [11]). Therefore without loss of generality we choose $\mu = e$ and $\Sigma = I$, where $e = (1, 0, ..., 0)' \in R^p$.

Now to find a BEE which minimizes the risk (2.5), we shall characterize an equivariant estimator, that is, an estimator satisfying $\tilde{\mu}(Ay, ASA') = A\tilde{\mu}(y, S)$. Decompose S uniquely as

$$S = WW' \qquad \text{with} \qquad W \in \mathrm{GT}(p) \tag{2.6}$$

and let

$$v = W^{-1}y \qquad \text{and} \qquad q = v/\|v\|, \tag{2.7}$$

where $\|v\|^2 = v'v$. Then $u = \|v\|^2$, where u is given in (2.4).

LEMMA 2.1. *An equivariant estimator $\tilde{\mu}$ is of the form*

$$\tilde{\mu}(y, S) = k(u)\, Wq, \tag{2.8}$$

where k is a measurable function of u.

Proof. Replacing y by $W^{-1}y$, A by W, and S by I in $\tilde{\mu}(Ay, ASA') = A\tilde{\mu}(y, S)$ yields $\tilde{\mu}(y, S) = W\tilde{\mu}(v, I)$. Let Q be an orthogonal matrix with q as the first column. Then $\tilde{\mu}(v, I) = \tilde{\mu}(QQ'v, QQ') = Q\tilde{\mu}(\sqrt{u}\, e, I)$. But since the columns of Q except the first column are arbitrary as far as they are orthogonal to q, it is easy to claim that the elements of $\tilde{\mu}(\sqrt{u}\, e, I)$ except the first element $\tilde{\mu}_1(\sqrt{u}\, e, I)$ are zero. Hence $\tilde{\mu}(v, I) = \tilde{\mu}_1(\sqrt{u}\, e, I)q$, completing the proof.

Consequently the risk function of an equivariant estimator $\tilde{\mu}$ in (2.5) with $\mu = e$ and $\Sigma = I$ is expressed as

$$R(\tilde{\mu}, (e, I)) = E[(k(u)\, Wq - e)'\, (k(u)\, Wq - e)].$$

Hence, using the fact that u is ancillary, a unique BEE is obtained as $\hat{\mu} = k(u)\, Wq$ with $k(u)$ minimizing the conditional risk given u:

$$E[(k(u)\, Wq - e)'\, (k(u)\, Wq - e) \mid u].$$

Therefore we obtain

THEOREM 2.1. *The unique BEE is an estimator $\hat{\mu} = \hat{k}(u)\, Wq$ with*

$$\hat{k}(u) = E[q'W'e \mid u]/E[q'W'Wq \mid u]. \tag{2.9}$$

An explicit evaluation of $\hat{k}(u)$ in (2.9) is rather complicated. Here only the case of $p=2$ is treated. To give a form of $\hat{k}(u)$, let

$$H(a; b; c: u) = \sum_{k=0}^{\infty} \binom{a}{k} B(b, c+k) u^k, \tag{2.10}$$

$$J(\alpha; \beta \mid \gamma; \delta; \varepsilon; u) = \sum_{j=0}^{\infty} \frac{\rho^j}{j!} \frac{\Gamma(\alpha+\beta+j/2)}{\Gamma(\alpha)} \times H\left(\gamma + \frac{j}{2}: \delta + \frac{j}{2}: \varepsilon: u\right), \tag{2.11}$$

and

$$\rho = [2nu/(1+u)]^{1/2}, \tag{2.12}$$

where $B(\alpha, \beta) = \Gamma(\alpha)\,\Gamma(\beta)/\Gamma(\alpha+\beta)$ and $\Gamma(\alpha)$ denotes the gamma function.

THEOREM 2.2. *When $p=2$, the BEE is given by $\hat{\mu} = \hat{k}(u)\, Wq$ with $\hat{k}(u) = \hat{k}_1(u)/\hat{k}_2(u)$, where*

$$\hat{k}_1(u) = \left(\frac{2}{1+u}\right)^{1/2} J\left(\frac{n}{2}; \frac{1}{2} \,\middle|\, \frac{1}{2}; 1: \frac{1}{2}: u\right) \tag{2.13}$$

and

$$\begin{aligned} \hat{k}_2(u) = {} & \frac{2}{1+u} J\left(\frac{n}{2}; 1 \mid 1; \frac{3}{2}; \frac{1}{2}: u\right) + J\left(\frac{n}{2}; 0 \mid -1; \frac{3}{2}; \frac{1}{2}; u\right) \\ & + \frac{2u^2}{1+u} J\left(\frac{n}{2}; 1 \mid -1; \frac{3}{2}; \frac{1}{2}: u\right) \\ & - \frac{4u}{(1+u)^{1/2}} J\left(\frac{n}{2} - \frac{1}{2}; 1 \mid -1; \frac{3}{2}; \frac{3}{2}: u\right) \\ & + (n-1)\, J\left(\frac{n}{2}: 0 \mid -1: \frac{1}{2}: \frac{3}{2}: u\right) \end{aligned} \tag{2.14}$$

The proof is given at the end of this section.

For comparison, we shall derive the MLE, where p is arbitrary here. By using the Lagrange multiplier method, the following theorem is easily obtained.

THEOREM 2.3. *The MLE's of μ and Σ under* (2.1) *are respectively given by*

$$\hat{\mu}_{\mathrm{MLE}} = \frac{u - \sqrt{u(4+5u)}}{2u} \bar{x} \tag{2.15}$$

and

$$\hat{\Sigma}_{\mathrm{MLE}} = \frac{1}{n} S + \frac{u + \sqrt{u(4+5u)}}{2u} \bar{x}\bar{x}'.$$

Proof. Maximizing

$$-\frac{n}{2} \log |\Sigma| - \frac{1}{2} \operatorname{tr} S\Sigma^{-1} - \frac{n}{2} (\bar{x} - \mu)' \Sigma^{-1} (\bar{x} - \mu) - \frac{\gamma}{2} (\mu' \Sigma^{-1} \mu - 1)$$

yields $\hat{\mu} = n\bar{x}/(n+\gamma)$ and $\hat{\Sigma} = (1/n)S + \lambda \bar{x}\bar{x}'/(\lambda + n)$. From $\hat{\mu}'\hat{\Sigma}^{-1}\hat{\mu} = 1$, the result follows.

Clearly $\hat{\mu}_{\mathrm{MLE}}$ is equivariant and hence it is dominated by the BEE in Theorem 2.1 for any p. Also the form of (2.15) is a natural extension of the case $p = 1$, where the MLE is $\frac{1}{2}\bar{x} - [(1/n)S + \frac{5}{4}\bar{x}^2]^{1/2}$. When $p = 1$, some properties on this model associated with the Fisher information are investigated by Hinkley [8]. Amari [1, 2] proposed through a geometric approach what he called the dual MLE, which is also equivariant.

Proof of Theorem 2.2. The joint pdf of (y, S) under $\mu = e$ and $\Sigma = I$ is given by

$$\begin{aligned} &k \exp[-\tfrac{1}{2} \|y - \sqrt{n}\, e\|^2] \\ &\qquad \times \exp(-\tfrac{1}{2} \operatorname{tr} S) \, |\det S|^{(n-p-2)/2} \, dy \, dS. \end{aligned} \tag{2.17}$$

First transforming (y, S) into (v, S) with $v = W^{-1}y$ and $S = WW'$, where $W = W(S) \in \mathrm{GT}(2)$, and next transforming (v, S) into (r, θ, S) with $r = \|v\| = u^{1/2}$ and $q = v/\|v\| = (\cos\theta, \sin\theta)' \equiv (q_1, q_2)$, the joint pdf of (r, θ, S) is given by

$$k \, |I + r^2 qq'|^{-n/2} \exp(\sqrt{n}\, re'Wq) \, g(S \mid r, \theta) \, dS \, d\theta \, dr, \tag{2.18}$$

where $g(S \mid r, \theta)$ is the pdf of $W_p((I + r^2 qq')^{-1}, n)$ and $-\pi \leqslant \theta < \pi$. Noting $|I + r^2 qq'|^{-n/2} = (1 + r^2)^{-n/2}$, the conditional pdf of (θ, S) given $r = u^{1/2}$ is

$$\exp(\sqrt{n}\, re'Wq) \, g(S \mid r, \theta) \, dS \, d\theta / h(r), \tag{2.19}$$

where $h(r)$ is the integral of the numerator over (θ, S). However, in the ratio $k(u)$ in (2.12), $h(r)$'s are cancelled out. Hence in the evaluation of

$k(u)$, $h(u)$ can be ignored. Now to evaluate (2.12), we need the expected value of w_{ij}'s with respect to (2.19). Since $w_{11} = s_{11}^{1/2}$, $w_{21} = s_{11}^{-1/2} s_{12}$, and $w_{22} = s_{22.1}^{1/2} = (s_{22} - s_{21} s_{11}^{-1} s_{12})^{1/2}$ and since it follows from $S = (s_{ij}) \sim W_p(\Delta, n)$,

$$w_{21} \text{ given } s_{11} \sim N(s_{11}^{1/2} \delta_{11}^{-1} \delta_{12}, \delta_{22.1}), \quad s_{11} \sim \delta_{11} \chi^2(n), \quad s_{22.1} \sim \delta_{22.1} \chi^2(n-1) \text{ and } (w_{21}, s_{11}) \text{ and } s_{22.1} \tag{2.20}$$

are independent, the expected values of w_{ij}'s given θ are evaluated by using (2.20), where $\Delta \equiv (\delta_{ij}) = (I + r^2 qq')^{-1}$ and $\delta_{22.1} = \delta_{22} - \delta_{21} \delta_{11}^{-1} \delta_{12}$. Noting $e'Wq = w_{11} q_1$ and

$$\delta_{11} = \frac{1 + r^2 q_2^2}{1 + r^2}, \quad \delta_{12} = -\frac{r^2}{1 + r^2} q_1 q_2, \quad \delta_{22.1} = \frac{1}{1 + r^2 q_2^2}, \tag{2.21}$$

we obtain

LEMMA 2.2. *Let* $\tau = \sqrt{n}\, r/(1 + r^2)^{1/2}$ *and* $m_n(\alpha) = 2^\alpha \Gamma(n/2 + \alpha)/\Gamma(n/2)$:

(1) $E[w_{11}^\alpha \exp(\sqrt{n}\, r q_1 w_{11}) \mid \theta]$

$$= (1 + r^2)^{-a/2} \sum_{j=0}^{\infty} \tau^j q_1^j (1 + r^2 q_2)^{(j+a)/2} m_n((j + a)/2)/j!$$

(2) $E[w_{21} \exp(\sqrt{n}\, r q_1 w_{11}) \mid \theta]$

$$= \sum_{j=0}^{\infty} \tau^j q_1^j (1 + r^2 q_2^2)^{(j-2)/2} m_n(j/2)/j! + r^4 (1 + r^2)^{-1}$$

$$\times \sum_{j=0}^{\infty} \tau^j q_1^{j+2} q_2^2 (1 + r^2 q_2^2)^{(j-2)/2} m_n((j + 2)/2)/j!$$

(3) $E[w_{12} w_{22} \exp(\sqrt{n}\, r q_1 w_{11}) \mid \theta]$

$$= -r^2 (1 + r^2)^{-1/2} \sum_{j=0}^{\infty} \tau^j q_1^{j+1} q_2 (1 + r^2 q_2^2)^{(j-2)/2}$$

$$\times m_n((j + 1)/2)\, m_{n-1}(1/2)/j!$$

(4) $E[w_{22}^2 \exp(\sqrt{n}\, r q_1 w_{11}) \mid \theta]$

$$= \sum_{j=0}^{\infty} \tau^j q_1^j q_2^2 (1 + r^2 q_2^2)^{(j-2)/2} m_n(j/2)\, m_{n-1}(1).$$

Proof. We only prove (2). The other cases are similar. Conditional on s_{11}, $E[w_{21}^2 \mid s_{11}, \theta] = E[(s_{11}^{-1/2} s_{12})^2 \mid s_{11}, \theta] = \delta_{22.1} + s_{11}\, \delta_{11}^{-2}\, \delta_{12}^2$. Using (2.21) and expanding $\exp(\sqrt{n}\, r s_{11}^{1/2} q_1)$, the left side of (2) is evaluated as

$$E\{[(1+r^2 q_2^2)^{-1} + s_{11} q_1^2 q_2^2 r^4 (1+r^2 q_2^2)^{-2}] \exp(\sqrt{n}\, r s_{11}^{1/2} q_1) \mid \theta\}$$

$$= (1+r^2 q_2^2)^{-1} \sum_{j=0}^{\infty} (\sqrt{n}\, r q_1)^j\, \delta_{11}^{j/2} E[(\chi_n^2)^{j/2}]/j!$$

$$+ r^4 (1+r^2 q_2^2)^{-2}\, q_1^2 q_2^2 \sum_{j=0}^{\infty} (\sqrt{n}\, r q_1)^j$$

$$\times \delta_{11}^{(j+2)/2} E[(\chi_n^2)^{(j+2)/2}]/j!$$

gives (2).

Next, using this lemma, we evaluate the numerator of (2.9). Since $e'Wq = w_{11} q_1$, $E[e'Wq] = K \int_{-\pi}^{\pi} q_1 E[w_{11} \exp(\sqrt{n}\, r q_1 w_{11}) \mid \theta]\, d\theta$. Here expanding $(1+r^2 q_2^2)^{\beta}$ as $\Sigma_k \binom{\beta}{k} (r^2 q_2^2)^k$ and using $\int_{-\pi}^{\pi} \cos^a \theta \sin^{2b} \theta\, d\theta = B((a+1)/2, (2b+1)/2)$, we obtain

$$E[e'Wq] = K(1+r^2)^{-1/2}\, 2^{1/2} \sum_{j=0}^{\infty} \left[\rho^j \Gamma\left(\frac{n}{2} + \frac{j}{2} + \frac{1}{2}\right) \Big/ j!\, \Gamma\left(\frac{n}{2}\right) \right]$$

$$\times \Sigma_k \binom{j/2 + 1/2}{k} B\left(\frac{j+2}{2}, \frac{2k+1}{2}\right) r^2 k.$$

This gives the expression (2.13) except the constant $K \equiv K(u)$, which is cancelled out with that of the denominator. Similarly for $q'W'Wq = q_1^2[w_{11}^2 + w_{12}^2] + 2q_1 q_2 w_{21} w_{22} + q_2^2 w_{22}^2$, the expected value of each term is evaluated by using Lemma 2.2. But the details are omitted here.

3. Problem with ν Known

In this section, the problem of estimating μ of $N_p(\mu, \Sigma)$ is considered with the assumption that (μ, Σ) belongs to

$$Ⓗ = \{(\mu, \Sigma) \in R^p \times \mathscr{S}(p) \mid \Sigma^{-1/2} \mu = c\}, \tag{3.1}$$

where $\Sigma^{1/2} \in \mathrm{GT}(p)$ and $c \in R^p$ is known. Assuming the quadratic loss in (1.3), this problem is left invariant under $\mathrm{GT}(p)$ acting on (y, S) by

$$(y, S) \to (Ay, ASA') \qquad \text{with} \quad A \in \mathrm{GT}(p), \tag{3.2}$$

which induces the action on (μ, Σ) as

$$(\mu, \Sigma) \to (A\mu, A\Sigma A') \qquad \text{with} \quad A \in \mathrm{GT}(p). \tag{3.3}$$

Clearly under (3.2) a maximal invariant is

$$v = W^{-1}y, \qquad \text{where} \quad S = WW' \text{ with } W \in \mathrm{GT}(p), \tag{3.4}$$

and since the action of GT(p) on Ⓗ in (3.3) is transitive, the distribution of v does not depend on (μ, Σ) for $(\mu, \Sigma) \in$ Ⓗ. Therefore the risk function of an equivariant estimator defined by

$$R(\tilde{\mu}, (\mu, \Sigma)) = E[(\tilde{\mu} - \mu)' \Sigma^{-1}(\tilde{\mu} - \mu)] \tag{3.5}$$

is constant on $(\mu, \Sigma) \in$ Ⓗ, which implies that without loss of generality we can choose $\mu = c$ and $\Sigma = I$. On the other hand, in a similar manner as in Lemma 2.1, an equivariant estimator is shown to be of the form

$$\tilde{\mu}(y, S) = W\tilde{\mu}(v, I) \equiv W\tilde{\tilde{\mu}}(v). \tag{3.6}$$

Consequently, a BEE is an estimator which minizes the conditioned risk

$$E[(W\tilde{\tilde{\mu}} - c)' (W\tilde{\tilde{\mu}} - c) \mid v] \tag{3.7}$$

with respect to $\tilde{\tilde{\mu}}$, where E denotes the expectation of W given v. Thus we obtain

THEOREM 3.1. *The unique BEE is given by*

$$\hat{\mu} = E[W'W \mid v]^{-1} E[W'c \mid v]. \tag{3.8}$$

Because an explicit of $\hat{\mu}$ in (3.8) is complicated in a general case, the case of $p = 2$ is treated here. In the evaluation, we regard $W = (w_{ij})$ as a function of $S = (s_{ij})$. As in the proof of Theorem 2.2, the joint distribution of (v, S) is given by

$$k\, |I + vv'|^{-n/2} \exp(\sqrt{n}\, c'Wv)\, g(S \mid v)\, dS\, dv, \tag{3.9}$$

where $g(S \mid v)$ is the pdf of $W((I + vv')^{-1}, n)$ (see (2.18)). Noting $c'Wv = c_1 w_{11} v_1 + c_2 w_{21} v_1 + c_2 w_{22} v_2$, $w_{11} = s_{11}^{1/2}$, $w_{21} = s_{11}^{1/2} s_{12}$, and $w_{22} = s_{22.1}^{1/2}$, define the conditional moment generating function of $w_{11} = s_{11}^{1/2}$ and $s_{22.1}^{1/2}$ given v by

$$\delta(t) = E[\exp(ts_{11}^{1/2}) \mid v] = \sum_{j=0}^{\infty} (t\, \delta_{11}^{1/2})^j\, 2^{j/2} \Gamma\left(\frac{n}{2} + \frac{j}{2}\right) \Big/ \Gamma\left(\frac{n}{2}\right) j! \tag{3.10}$$

and

$$\delta(t) = E[\exp(ts_{22.1}^{1/2}) \mid v]$$
$$= \sum_{j=0}^{\infty} (t\, \delta_{22.1}^{1/2})^j\, 2^{j/2} \Gamma\left(\frac{n-1}{2} + \frac{j}{2}\right) \Big/ \Gamma\left(\frac{n-1}{2}\right) j! \tag{3.11}$$

respectively, where $\delta_{11} = (1 + v'vv_2^2)/(1 + v'v)$ and $\delta_{22.1} = 1/(1 + v'vv_2^2)$ with $v = (v_1, v_2)'$ (see (2.20) and (2.21)). And by $\phi^{(i)}(t)$ and $\psi^{(i)}$ denote the ith derivatives of ϕ and ψ, respectively. Further, let

$$d_1 = (\sqrt{n}\, c_1 v_1 + \delta_{11}^{-1}\, \delta_{12} \sqrt{n}\, c_2 v_1)\, \delta_{11}^{1/2} \text{ and } d_2 = \sqrt{n}\, c_2 v_2\, \delta_{22.1}^{1/2}, \tag{3.12}$$

where $\delta_{12} = -v'vv_1^2v_2^2/(1 + v'v)$, and let

$$\begin{aligned} b_1 &= \phi^{(1)}(d_1)\, \psi(d_2), \\ b_2 &= [\delta_{11}^{-1}\, \delta_{12}\phi^{(1)}(d_1) + c_2 v_2 \phi(d_1)]\, \psi(d_2) \\ b_3 &= \phi(d_1)\, \psi^{(1)}(d_2), \\ b_4 &= \phi^{(2)}(d_1)\, \psi(d_2) \\ b_5 &= [(\delta_{11}^{-1}\, \delta_{12})^2\, \phi^{(2)}(d_1) + 2c_2 v_1\, \delta_{11}^{-1}\, \delta_{12}\phi^{(1)}(d_1) \\ &\qquad + c_1^2 v_1^2\, \delta_{22.1}^2\, \phi(d_1)]\, \psi(d_2) \\ b_6 &= [\delta_{11}^{-1}\, \delta_{12}\phi^{(1)}(d_1) + c_2 v_1\, \delta_{22.1}\phi(d_1)]\, \psi^{(1)}(d_1) \\ b_7 &= \phi(d_1)\, \psi^{(2)}(d_2). \end{aligned} \tag{3.13}$$

THEOREM 3.2. *When $p = 2$, the BEE in (3.8) is given by $\hat{\mu} = Wa$, where $a = (a_1, a_2)'$ with*

$$\begin{aligned} a_1 &= [c_1 b_1 (b_4 + b_5) + c_2(b_2 b_7 - b_3 b_6)]/D \\ a_2 &= \{-c_1 b_1 b_6 + c_2[b_3(b_4 + b_5) - b_2 b_6]\}/D. \end{aligned} \tag{3.14}$$

Here $D = (b_4 + b_5) b_7 - b_6$.

Proof. We simply outline the proof since the proof is similar to that of Theorem 2.2. From (3.9), the conditional pdf of S given v is given by $\exp(\sqrt{n}\, c'Wv)\, g(S \mid v)/h(v)$, where $h(v)$ is the normalizing constant. However, it is easy to see that the BEE in (3.8) does not depend on $h(v)$. Hence what we need is the expected values of $w_{11}^{\alpha}\, w_{12}^{\beta} w_{22}^{\gamma} \exp(\sqrt{n}\, c'Wv)$ with respect to $g(S \mid v)$. Then using (2.20) and $E[\exp(\sqrt{n}\, c_2 v_1 w_{21}) \mid s_{11}, v]$ $= \exp(nc_2^2 v_1^2\, \delta_{22.1}/2) \exp(\sqrt{n}\, c_2 v_1\, \delta_{11}^{-1}\, \delta_{12} s_{11}^{1/2})$, we can show that $b_1 = KE[w_{11} Q \mid v]$, $b_2 = KE[w_{12} Q \mid v]$, $b_3 = KE[w_{22} Q \mid v]$, $b_4 = KE[w_{11}^2 Q \mid v]$, $b_5 = KE[w_{21} Q \mid v]$, $b_6 = KE[w_{21} w_{22} Q \mid v]$, and $b_7 = KE[w_{22}^2 Q \mid v]$, where $K = \exp(nc_2^2 v_1^2\, \delta_{22.1}/2)$ and $Q = \exp(\sqrt{n}\, c'Wv)$. From these moments, the result follows.

It is noted that the conditional moment generating functions of ϕ and ψ in (3.10) and (3.11) are Bessel functions and their derivatives can be computed term by term.

On the other hand, the MLE is routinely obtained. First the constrained log-likelihood function is expressed as

$$L=\frac{1}{2}\operatorname{tr}\ \Sigma^{-1}S-\frac{n}{2}\operatorname{tr}\Sigma^{-1}(\bar{x}-\mu)(\bar{x}-\mu)'+\frac{n}{2}\log|\Sigma^{-1}|-n\lambda'(\Sigma^{-1/2}\mu-c)$$

$$=-\frac{1}{2}\operatorname{tr}\tilde{\Phi}\tilde{\Phi}'-\frac{n}{2}\operatorname{tr}\tilde{\Phi}(\tilde{\bar{x}}-\tilde{\mu})(\tilde{\bar{x}}-\tilde{\mu})'\,\tilde{\Phi}+n\log|\tilde{\Phi}|-n\lambda'(\tilde{\Phi}\tilde{\mu}-c),$$

where $\tilde{\Phi}=\Phi W$ with $\Sigma^{-1}=\Phi\Phi'$ and $\Phi\in \mathrm{GT}(p)$, $\tilde{\bar{x}}=W^{-1}\bar{x}$ and $\tilde{\mu}=W^{-1}\mu$. Differentiating L with respect to $\tilde{\mu}$ yields $\tilde{\mu}=\tilde{\bar{x}}-\tilde{\Phi}^{-1}\lambda$ and substituting this $\tilde{\mu}$ into L yields

$$L_1=-\frac{1}{2}\Sigma_{ij}\tilde{\phi}_{ij}^2+n\Sigma\log\tilde{\phi}_{ij}+\frac{3n}{2}\Sigma\lambda_i^2$$

$$-n\Sigma_{i\geq j}\tilde{\phi}_{ij}\tilde{\bar{x}}\lambda_i+n\Sigma\lambda_i c_i.$$

Here differentiating L_1 with respect to $\tilde{\phi}_{ij}$ and λ_i, we obtain

$$\tilde{\phi}_{ii}^2+n\tilde{\phi}_{ii}\tilde{\bar{x}}_i\lambda_i-n=0,\qquad \tilde{\phi}_{ij}=-n\tilde{\bar{x}}_j\lambda_i\quad (i>j)$$

$$\lambda_i=\frac{2}{3}\left[\Sigma_{i\geq j}\tilde{\phi}_{ij}\tilde{\bar{x}}_j-c_i\right].$$

From these equations, $\tilde{\phi}_{ij}$'s are recursively obtained; e.g., since $\lambda_1=\frac{2}{3}(\tilde{\phi}_1\bar{x}_1-c_1)$,

$$\tilde{\phi}_{11}=\left\{\frac{2n\tilde{\bar{x}}_1c_1}{3}+\left[\frac{4}{9}n^2\tilde{\bar{x}}_1c_1^2+4n\left(1+\frac{2}{3}n\tilde{\bar{x}}_1^2\right)\right]^{1/2}\right\}\Big/\left(2+\frac{4}{3}n\tilde{\bar{x}}_1^2\right),$$

etc. Then the MLE of Φ is given by $\hat{\phi}=\tilde{\phi}W^{-1}$ and the MLE of μ is given by $\hat{\mu}=W(\tilde{\bar{x}}-\tilde{\Phi}^{-1}\lambda)$.

4. The Case $\Sigma=\sigma^2\mu'\mu I$

As a particular case for $\mu'\Sigma^{-1}\mu$ constant, in this section we consider the case of $\Sigma=\sigma^2\mu'\mu I$, where σ^2 is known. Then (y, w) is a sufficient statistic where $w=\operatorname{tr} S$ and (y, S) is given in (1.4). Of course, $w/\sigma^2\mu'\mu$ is distributed as $\chi^2_{(n-1)p}$. The loss function in this case becomes

$$L(a,\mu)=(a-\mu)'\,(a-\mu)/\sigma^2\mu'\mu \tag{4.1}$$

and the problem of estimating μ remains invariant under the group $G=R_+\times\mathcal{O}(p)$ which acts on (y, w) by

$$(y,w)\longrightarrow(b\Gamma y, b^2w)\qquad\text{for}\quad (b,\Gamma)\in G, \tag{4.2}$$

where $R_+ = \{b > 0\}$ and $\mathcal{O}(p)$ denotes the group of $p \times p$ orthogonal matrices. The following lemma is similar to Lemma 2.1 and the proof is omitted.

LEMMA 4.1. *An equivariant estimator $\tilde{\mu}(y, w)$ is of the form*

$$\tilde{\mu}(y, w) = h(v)\, y \qquad \text{with} \quad v = y'y/w, \tag{4.3}$$

where $h(\cdot)$ is a measurable function from R^+ into R.

Now to find a BEE which minimizes the risk $R(\tilde{\mu}, \mu) = E_\mu[L(\tilde{\mu}, \mu)]$, note that the action of G on the parameter space is transitive and hence the statistic $v \equiv y'y/w$, which is a maximal invariant, is ancillary. Hence the risk function is constant and so taking $\mu = \mu_0 \equiv (1, 0, ..., 0)'$, the BEE is given by $\hat{\mu}(y, w) = h_0(v)\, y$ with

$$h_0(v) = E_{\mu_0}[y_1 \mid v]/E_{\mu_0}[y'y \mid v], \tag{4.4}$$

where $y = (y_1, ..., y_p)'$. Evaluating $h_0(v)$ yields the following theorem.

THEOREM 4.1. *The BEE is given by $\hat{\mu}(y, w) = h_0(v)\, y$ with*

$$h(v) = \frac{n^{1/2} F(np/2 + 1;\, p/2 + 1 \colon nv/2(1 + v)\sigma^2)}{F(np/2 + 1;\, p/2 \colon nv/2(1 + v)\sigma^2)}, \tag{4.5}$$

where $F(a, b\colon x) = \sum_{j=0}^{\infty} \Gamma(a + i) x^i / \Gamma(b + i) i!$

Proof. In the density of (y, w), transform (y, w) into (y, v) to get the density of (y, v):

$$c \exp\left\{ -\frac{1}{2} \left(\frac{1 + v}{v} \right) y'y + \frac{n}{\sigma^2} \right\} (y'y)^{(n-1)p/2}$$

$$v^{-(n-1)p/2 - 1} \sum_{i=0}^{\infty} (n^{1/2} y_1 \sigma)^i / i!$$

Using this and evaluating the conditional expectations yields the result. The details are left to the readers.

Next we derive the MLE. From the joint density of (y, w), maximizing the log-likelihood equation is equivalent to minimizing

$$\frac{np}{2} \log(\mu'\mu) + \frac{1}{2}[w + y'y - 2n^{1/2} y'\mu]/\sigma^2 \mu'\mu. \tag{4.6}$$

It is then easy to see that the MLE is a solution of

$$np\sigma^2\mu'\mu\mu - n^{1/2}\mu'\mu y - w\mu - y'y\mu + 2n^{1/2}y'\mu\mu = 0. \tag{4.7}$$

We solve this equation as

THEOREM 4.2. *The MLE is given by*

$$\hat{\mu}_M = \left[\frac{(1+4p\sigma^2((1+v)/v)^{1/2} - 1}{2n^{1/2}p\sigma^2}\right] y. \tag{4.8}$$

Proof. First observe that the solutions of (4.5) are of the form $\tilde{\mu} = h(y, w)\, y$. Hence substituting $\mu = cy$, we obtain

$$c[np\sigma^2(y'y)c^2 + (n^{1/2}y'y)c - (y'y + w)] = 0. \tag{4.9}$$

The solutions of this equation are $c_1 = 0$,

$$c_2 = \left\{-1 - \left[1 + 4p\sigma^2\frac{1+v}{v}\right]^{1/2}\right\} \Big/ 2n^{1/2}p\sigma^2,$$

and c_3 where c_3 is [ ] in (4.8). To find the solution which minimizes (4.6), obtain the matrix of the second derivatives of (4.6) and evaluate it at c_i's. Then c_1 is not the solution and for $\mu = c_i y$ with $c_i \neq 0$, the matrix is evaluated as

$$A\left[\frac{1}{b_i} y'yI + yy\right] \quad \text{with} \quad A > 0,$$

where $b_i = 2n^{1/2}p\sigma^2 c_i$. For this to be positive definite, $b_i > 0$ is necessary. Hence c_3 is the only solution, completing the proof.

REFERENCES

[1] AMARI, S. (1982a). Differential geometry of curved exponential families-curvatures and information loss. *Ann. Statist.* **10** 357–385.

[2] AMARI, S. (1982b). Geometrical theory of asymptotic ancillarity and conditional inference. *Biometrika* **69** 1–17.

[3] BARNDORFF-NIELSEN, O. (1980). Conditionality resolution. *Biometrika* **67** 293–310.

[4] COX, D. R., AND HINKLEY, D. V. (1974). *Theoretical Statistics*, Chapman & Hall, London.

[5] EATON, M. L. (1972). *Multivariate Statistical Analysis.* Univ. of Copenhagen, Copenhagen.

[6] EFRON, B. (1978). The geometry of exponential families. *Ann. Statist.* **6** 362–376.

[7] FEARGUSON, T. S. (1967). *Mathematical Statistics—A Decision Theoretic Approach.* Academic Press, New York/London.

[8] Hinkley, D. V. (1977). Conditional inference about a normal mean with known coefficient of variation. *Biometrika* **64** 105–108.

[9] Kariya, T. (1984). *An Invariance Approach to Estimation in a curved Model.* Tech. Rep. 88, Hitotsubashi University, Japan.

[10] Lehmann, E. L. (1981). An interpretation of completeness and Basu's Theorem. *J. Amer. Statist. Assoc.* **76** 335–340.

[11] Lehmann, E. L. (1983). *Theory of Estimation.* Wiley, New York.

A Generalized Cauchy–Binet Formula and Applications to Total Positivity and Majorization*

SAMUEL KARLIN

Stanford University

AND

YOSEF RINOTT

Hebrew University

1. INTRODUCTION

The identification and analysis of multivariate totally positive kernels, log concave densities, Schur-concave functions, and symmetric unimodal functions relies heavily on their conservation under convolution operators. An approach of wide scope incorporating many of the essential composition laws can be based on a generalized Cauchy–Binet formula. The Cauchy–Binet formula for matrix functionals plays an important role in studies of determinants, permanents, and other classes of matrix functions (e.g., Marcus [7], de Oliveira [10]). In this context a generalized matrix function founded on the matrix $\|A(x_i, y_j)\|_1^n$ has the canonical form

$$d_\chi^{\mathscr{H}}(\mathbf{x}, \mathbf{y}, A) = \sum_{\sigma \in \mathscr{H}} \chi(\sigma) \prod_{i=1}^{n} A(x_i, y_{\sigma(i)}), \tag{1.1}$$

where $\mathscr{H}$ is a subgroup of the symmetric group $\mathscr{S}_n$ (permutations on n elements) and $\chi(\sigma)$ is a character on $\mathscr{H}$, i.e., $\chi(e) = 1$, where e is the identity permutation and $\chi(\sigma\tau) = \chi(\sigma)\chi(\tau)$ for $\sigma, \tau \in \mathscr{H}$. The specifications $\mathscr{H} = \mathscr{S}_n$ and $\chi(\sigma) = \text{sign}\,\sigma$ or $\chi(\sigma) \equiv 1$ produces the determinant and permanent functionals, respectively. The classical Cauchy–Binet formula expresses $d_\chi^{\mathscr{H}}(\mathbf{x}, \mathbf{y}, C)$ for the matrix product $C = AB$ in terms of the corresponding

* Supported in part by NIH Grants GM10452 and 1R01-39907-01 and NSF Grant MC582–15131.

Multivariate Statistics and Probability
ISBN 0-12-580205-6

Reprinted from *J. Mult. Anal.* **27**(1).

$d_\chi^{\mathscr{H}}$-functionals of A and B. The Cauchy–Binet formula for continuous matrix multiplication serves abundantly in verifying and generating totally positive (TP) kernels, e.g., Karlin [5].

In this paper we develop an extended Cauchy–Binet formula for multivariate kernels. The setting of matrix functions is generalized to $K(\mathbf{x}, \mathbf{y})$ as a function defined for a direct product domain $\mathbf{R}^{n+n}$ (Euclidean $(2n)$-space) and

$$\begin{aligned} d_\chi^{\mathscr{H}}(\mathbf{x}, \mathbf{y}; K) &= D_K(\mathbf{x}, \mathbf{y}) \\ &= \sum_{\sigma \in \mathscr{H}} \chi(\sigma)\, K(\mathbf{x}, \sigma\mathbf{y}) = \sum_{\sigma \in \mathscr{H}} \chi(\sigma)\, K(x_1, ..., x_n; y_{\sigma(1)}, ..., y_{\sigma(n)}) \end{aligned} \tag{1.2}$$

(the dependence on $\mathscr{H}$ and χ is suppressed where no ambiguity is likely). With $K(\mathbf{x}, \mathbf{y}) = \prod_{i=1}^n A(x_i, y_i)$ based on the matrix kernel $A(x, y)$, we recover (1.1).

The construction (1.2) invites a generalized totally positive (GTP) notion. Thus $K(x, y)$ is said to be GTP with respect to $\mathscr{H} = \mathscr{S}_n$ and $\chi(\sigma) = \operatorname{sign} \sigma$ if $d(\mathbf{x}, \mathbf{y}; K) = \sum_{\sigma \in \mathscr{S}_n} (\operatorname{sign} \sigma)\, K(\mathbf{x}, \sigma\mathbf{y}) \geqslant 0$ for $\mathbf{x} = (x_1, ..., x_n)$ and $\mathbf{y} = (y_1, ..., y_n)$ provided $x_1 < \cdots < x_n$ and $y_1 < \cdots < y_n$. There also occurs the notion of generalized total positivity with respect to subgroups of $\mathscr{S}_n$. In this perspective the property of Schur convexity for $\Phi(\mathbf{z})$, $\mathbf{z} \in \mathbf{R}^n$, is equivalent to the GTP property corresponding to $\mathscr{H}$ consisting of permutation subgroups of two elements operating on the translation kernel $K(\mathbf{x}, \mathbf{y}) = \Phi(\mathbf{x} + \mathbf{y})$, see Theorem 3 below. The fact that the convolution of Schur concave functions remains Schur concave (Marshall and Olkin [8]) is manifest from the Cauchy–Binet formalism. In the same vein the convolution of similarly elliptically contoured unimodal densities is also of the same kind.

The concluding section considers generalizations involving compact groups $\mathscr{G}$ with (1.2) of the form

$$d_\chi^{\mathscr{G}}(\mathbf{x}, \mathbf{y}; K) = \int_{\mathscr{G}} \chi(g)\, K(\mathbf{x}, g\mathbf{y})\, dg, \tag{1.3}$$

where dg refers to the Haar measure of $\mathscr{G}$.

2. A Generalized Cauchy–Binet Formula for the Symmetric Group

Let $d\Gamma(\mathbf{y})$ be an invariant measure with respect to $\mathscr{S}_n$ so that

$d\Gamma(\sigma\mathbf{y}) = d\Gamma(\mathbf{y})$, e.g., when $d\Gamma(\mathbf{y}) = \prod_{i=1}^{n} d\Gamma(y_i)$ is a product measure of identical factors. The following integration formula is elementary

$$\int_{\mathbf{R}^n} \Phi(\mathbf{y})\, d\Gamma(\mathbf{y}) = \sum_{\sigma \in \mathscr{S}_n} \int_A \frac{\Phi(\sigma\mathbf{y})}{t(\mathbf{y})}\, d\Gamma(\mathbf{y}), \tag{2.1}$$

where $\mathbf{y} = (y_1, ..., y_n) \in A$, A is the *increasing orthant* of $\mathbf{R}^n$ (i.e., $\mathbf{y} \in A$ iff $y_1 \leqslant \cdots \leqslant y_n$),

$$t(\mathbf{y}) = \prod m_i! \tag{2.2}$$

and m_i is the number of occurrences of the ith distinct component of $\mathbf{y}$. Equivalently $t(\mathbf{y})$ is the number of $\sigma \in \mathscr{S}_n$ for which $\sigma\mathbf{y} = \mathbf{y}$. Obviously $t(\mathbf{y}) = 1$ when $\mathbf{y} \in A^o$ (the interior of A).

Let $\Delta = \{\tau_1, ..., \tau_r\}$ represent the right coset space $\mathscr{S}_n/\mathscr{H}$ such that $\mathscr{H}\tau_i$ are distinct, $\tau_1 = e$ and $\bigcup_i \mathscr{H}\tau_i = \mathscr{S}_n$. In this case, it is convenient to write the integration formula (2.1) as

$$\int_{\mathbf{R}^n} \Phi(\mathbf{y})\, d\Gamma(\mathbf{y}) = \sum_{\tau \in \Delta} \sum_{\pi \in \mathscr{H}} \int_A \Phi(\pi\tau\mathbf{y}) \frac{1}{t(\mathbf{y})}\, d\Gamma(\mathbf{y}). \tag{2.3}$$

A kernel $K(\mathbf{x}, \mathbf{y})$ defined on a region of $\mathbf{R}^n \times \mathbf{R}^n$ is said to be *invariant* with respect to the group $\mathscr{H} \subset \mathscr{S}_n$ if $K(\pi\mathbf{x}, \pi\mathbf{y}) = K(\mathbf{x}, \mathbf{y})$ for all $\pi \in \mathscr{H}$ and $\mathbf{x}$, $\mathbf{y}$. For a given character χ we define the generalized kernel function

$$D_K(\mathbf{x}, \mathbf{y}) = \sum_{\pi \in \mathscr{H}} \chi(\pi)\, K(\mathbf{x}, \pi\mathbf{y}). \tag{2.4}$$

THEOREM 1 (The Cauchy–Binet formula for generalized kernel functionals). *Let $K(\mathbf{x}, \mathbf{y})$ and $L(\mathbf{y}, \mathbf{z})$ be permutation invariant kernels, square integrable with respect to $d\Gamma(\mathbf{y})$, where Γ is a permutation invariant Borel measure on $\mathbf{R}^n$. Consider the continuous "matrix product"*

$$M(\mathbf{x}, \mathbf{z}) = \int_{\mathbf{R}^n} K(\mathbf{x}, \mathbf{y})\, L(\mathbf{y}, \mathbf{z})\, d\Gamma(\mathbf{y}). \tag{2.5}$$

Then for any subgroup $\mathscr{H} \subset \mathscr{S}_n$ and character χ on $\mathscr{H}$, we have

$$D_M(\mathbf{x}, \mathbf{z}) = \sum_{\tau \in \Delta} \int_A \frac{1}{t(\mathbf{y})} D_K(\mathbf{x}, \tau\mathbf{y})\, D_L(\tau\mathbf{y}, \mathbf{z})\, d\Gamma(\mathbf{y}), \tag{2.6}$$

where $A = \{\mathbf{y} : y_1 \leqslant \cdots \leqslant y_n\}$ is the increasing orthant.

Proof. It is readily verified that the invariance of K and L and the measure Γ implies the invariance of M. Equation (2.5), definition (2.4), and the integration formula produce

$$D_M(\mathbf{x}, \mathbf{z}) = \sum_{\pi \in \mathscr{H}} \chi(\pi)\, M(\mathbf{x}, \pi\mathbf{z}) = \sum_{\pi \in \mathscr{H}} \chi(\pi) \int_{\mathbf{R}^n} K(\mathbf{x}, \mathbf{y})\, L(\mathbf{y}, \pi\mathbf{z})\, d\Gamma(\mathbf{y})$$

$$= \sum_{\pi \in \mathscr{H}} \chi(\pi) \sum_{\tau \in \Delta} \sum_{\varphi \in \mathscr{H}} \int_A \frac{K(\mathbf{x}, \varphi\tau\mathbf{y})\, L(\varphi\tau\mathbf{y}, \pi\mathbf{z})}{t(\mathbf{y})}\, d\Gamma(\mathbf{y})$$

and by virtue of permutation invariance

$$D_M(\mathbf{x}, \mathbf{z}) = \sum_{\tau \in \Delta} \int_A \frac{1}{t(\mathbf{y})} \sum_{\varphi \in \mathscr{H}} \chi(\varphi)\, K(\mathbf{x}, \varphi\tau\mathbf{y}) \sum_{\pi \in \mathscr{H}} \chi(\varphi^{-1}\pi)\, L(\tau\mathbf{y}, \varphi^{-1}\pi\mathbf{z})\, d\Gamma(\mathbf{y})$$

and, since for each $\varphi \in \mathscr{H}$, $\varphi^{-1}\pi$ traverses $\mathscr{H}$ as π traverses $\mathscr{H}$ we achieve

$$D_M(\mathbf{x}, \mathbf{z}) = \sum_{\tau \in \Delta} \int_A \frac{D_K(\mathbf{x}, \tau\mathbf{y})\, D_L(\tau\mathbf{y}, \mathbf{z})}{t(\mathbf{y})}\, d\Gamma(\mathbf{y}) \tag{2.7}$$

as desired.

In some situations the region of integration in (2.7) can be reduced to the subset $B \subset A$ defined as follows. For each $\mathbf{y}$ determine $\mathscr{H}_{\mathbf{y}} = \{\pi \in \mathscr{H} : \pi\mathbf{y} = \mathbf{y}\}$ and let

$$B = \{\mathbf{y} \in A : \chi(\mu) = 1 \text{ for all } \mu \in \mathscr{H}_y\}. \tag{2.8}$$

We claim for $\mathbf{y} \in A - B$ and any $\sigma \in \mathscr{S}_n$ that $D_K(\mathbf{x}, \sigma\mathbf{y}) = 0$. Indeed, let $s = |\mathscr{H}/\mathscr{H}_{\mathbf{y}}|$ and choose left coset representatives $\theta_i \in \mathscr{H}$ such that $\theta_1 \mathscr{H}_{\mathbf{y}}, \ldots, \theta_s \mathscr{H}_{\mathbf{y}}$ are distinct cosets of $\mathscr{H}_{\mathbf{y}}$ in $\mathscr{H}$. Then

$$D_K(\mathbf{x}, \mathbf{y}) = \sum_{\pi \in \mathscr{H}} \chi(\pi)\, K(\mathbf{x}, \pi\mathbf{y}) = \sum_{i=1}^{s} \sum_{\mu \in \mathscr{H}_{\mathbf{y}}} \chi(\theta_i)\, \chi(\mu)\, K(\mathbf{x}, \theta_i \mu\mathbf{y})$$

and since $\mu\mathbf{y} = \mathbf{y}$ for $\mathscr{H}_{\mathbf{y}}$

$$D_K(\mathbf{x}, \mathbf{y}) = \sum_i \chi(\theta_i)\, K(\mathbf{x}, \theta_i\mathbf{y}) \sum_{\mu \in \mathscr{H}_{\mathbf{y}}} \chi(\mu).$$

Because $\mathbf{y} \in A - B$, $\chi(\mu_0) \neq 1$ for some $\mu_0 \in \mathscr{H}_y$ and therefore $a = \sum_{\mu \in \mathscr{H}_{\mathbf{y}}} \chi(\mu) = 0$ (since $\chi(\mu_0)\, a = a$), so that $D_K(\mathbf{x}, \mathbf{y}) = 0$. Replacing $\mathbf{y}$ by $\sigma\mathbf{y}$ and noting that $\mathscr{H}_{\sigma\mathbf{y}} = \sigma \mathscr{H}_{\mathbf{y}} \sigma^{-1}$, we have

$$D_K(\mathbf{x}, \sigma\mathbf{y}) = \sum_{i=1}^{s} \chi(\tilde{\theta}_i)\, K(\mathbf{x}, \tilde{\theta}_i \sigma\mathbf{y}) \sum_{\mu \in H_{\sigma\mathbf{y}}} \chi(\mu),$$

for an appropriate set of $\tilde{\theta}_i$. Also

$$\sum_{\mu \in H_{\sigma y}} \chi(\mu) = \sum_{\mu \in \sigma H_y \sigma^{-1}} \chi(\mu) = \sum_{\mu \in H_y} \chi(\sigma)\, \chi(\mu)\, \chi(\sigma^{-1}) = \sum_{\mu \in H_y} \chi(\mu) = 0.$$

Thus $D_K(\mathbf{x}, \sigma\mathbf{y}) = 0$ for all σ when $\mathbf{y} \in A - B$ and we can replace A by B in the integral of (2.7).

Where the coordinates of $\mathbf{y}$ are all distinct, $\pi\mathbf{y} = \mathbf{y}$ is possible only when $\pi = e =$ the identity permutation so that $H_\mathbf{y} = \{e\}$, and $\mathbf{y} \in B$. Accodingly B contains all points of $A^o = \{\mathbf{y} : y_1 < y_2 < \cdots < y_n\}$. When Γ is a continuous measure the region of integration can obviously be reduced to $A^o =$ interior of A, and $t(\mathbf{y}) = 1$.

Consider $\chi(\sigma) = \operatorname{sign} \sigma$ and $\mathscr{H} = \mathscr{S}_n$. If $\mathbf{y}$ has a pair of coincident coordinates then $\mathscr{H}_\mathbf{y}$ contains the odd permutation σ which only transposes the coincident pair, with $\chi(\sigma) = -1$ so that $\mathbf{y} \notin B$. Hence in this case $B = A^o$. For $\mathscr{H} = \mathscr{S}_n$ we have $\Delta = \{e\}$ so in this case, with $\chi(\sigma) = \operatorname{sign} \sigma$, (2.7) becomes

$$D_M(\mathbf{x}, \mathbf{z}) = \int_{y_1 < \cdots < y_n} \cdots \int D_K(\mathbf{x}, \mathbf{y})\, D_L(\mathbf{y}, \mathbf{z})\, d\Gamma(\mathbf{y}). \tag{2.9}$$

In the example where for some functions Φ_i, $i = 1, 2$,

$$K(\mathbf{x}, \mathbf{y}) = \prod_{i=1}^{n} \Phi_1(x_i, y_i), \qquad L(\mathbf{y}, \mathbf{z}) = \prod_{i=1}^{n} \Phi_2(y_i, z_i), \tag{2.10}$$

and $\chi(\sigma) = \operatorname{sign} \sigma$, the functional D is the classical determinant as mentioned before.

If $\Gamma = \Gamma_1 \times \cdots \times \Gamma_1$, then setting $\Psi(x, z) = \int_R \Phi_1(x, y)\, \Phi_2(y, z)\, d\Gamma_1(y)$ and

$$M(\mathbf{x}, \mathbf{z}) = \int_{\mathbf{R}^n} K(\mathbf{x}, \mathbf{y})\, L(\mathbf{y}, \mathbf{z})\, d\Gamma(\mathbf{y}) = \prod_{i=1}^{n} \Psi(x_i, z_i), \tag{2.11}$$

we obtain the classical determinental Cauchy–Binet formula

$$\Psi\begin{pmatrix} x_1 \cdots x_n \\ z_1 \cdots z_n \end{pmatrix} = \int_{y_1 < \cdots < y_n} \cdots \int \Phi_1\begin{pmatrix} x_1 \cdots x_n \\ y_1 \cdots y_n \end{pmatrix} \Phi_2\begin{pmatrix} y_1 \cdots y_n \\ z_1 \cdots z_n \end{pmatrix} \prod_{i=1}^{n} d\Gamma_1(y_i) \tag{2.12}$$

with the notation $\det(\Phi(x_i, y_j))_{i,j=1}^n = \Phi\binom{x_1 \cdots x_n}{y_1 \cdots y_n}$, which reduces, of course, to the matrix version when Γ_1 is a discrete measure of unit point masses.

With the above notation but $\chi \equiv 1$, the functional D becomes permanent and denoting $\Phi^*\binom{x_1 \cdots x_n}{y_1 \cdots y_n} = \operatorname{per}(\Phi(x_i, y_j))_{i,j=1}^n$, the Cauchy–Binet formula for permanents has the form

$$\Psi^*\begin{pmatrix} x_1 \cdots x_n \\ z_1 \cdots z_n \end{pmatrix}$$
$$= \int \cdots \int_{y_1 \leq \cdots \leq y_n} \frac{1}{t(\mathbf{y})} \Phi_1^* \begin{pmatrix} x_1 \cdots x_n \\ y_1 \cdots y_n \end{pmatrix} \Phi_2^* \begin{pmatrix} y_1 \cdots y_n \\ z_1 \cdots z_n \end{pmatrix} \prod_{i=1}^n d\Gamma_1(y_i), \tag{2.13}$$

with the integration covering all of the increasing orthant A.

3. Generalized Total Positivity

For our immediate purposes, it is useful to describe a complete set of coset representatives for the group $\mathscr{H} = \mathscr{S}_{\{1, 2, ..., k_0\}}$ consisting of all permutations satisfying $\pi(j) = j$ for $j > k_0$, i.e., π permutes only the indices 1, 2, ..., k_0 among themselves. When $\pi_1 \tau = \pi_2 \theta$ with $\pi_1, \pi_2 \in \mathscr{H}$ then $\tau^{-1}(j) = \theta^{-1}(j)$ for all $j > k_0$. Thus with each monotone set $z_1 < \cdots < z_{k_0}$ there are $(n - k_0)!$ permutations τ inducing distinct cosets of $\mathscr{H}$ in the manner that

$$\tau^{-1}(i) = z_i,\ i = 1, 2, ..., k_0, \text{ and } \tau \text{ maps } \{1, ..., n\} - \{z_1, ..., z_{k_0}\}$$
$$\text{onto } \{k_0 + 1, ..., n\}. \tag{3.1}$$

There are $\binom{n}{k_0}$ selections of monotone k_0-element sets from $\{1, ..., n\}$. Since $|\mathscr{H}| = k_0!$ the collection of all τ as described constitutes a complete set of coset representatives for $\mathscr{S}_{\{1, ..., k_0\}}$. In a similar manner we can delineate a complete set of right coset representatives for $\mathscr{H} = \mathscr{S}_{\{i_1, ..., i_{k_0}\}}$ by specifying $z_1 < \cdots < z_{k_0}$ and mapping $\tau^{-1}(i_\nu) = z_\nu$ with a general permutation among the remaining indices.

More generally, the group $\mathscr{S}_{\{\alpha_1, ..., \alpha_{k_1}\}} * \mathscr{S}_{\{\beta_1, ..., \beta_{k_2}\}}$ composed of all permutations that permute the elements $\{\alpha_1, ..., \alpha_{k_1}\}$ among themselves and separately permute the elements $\{\beta_1, ..., \beta_{k_2}\}$ among themselves and leave all other elements fixed, possesses a set of coset representatives $\{\tau\}$ characterized by choosing a set of $k_1 + k_2$ increasing integers and then specifying $z_1 < \cdots < z_{k_1}$ from these, leaving $w_1 < \cdots < w_{k_2}$ and prescribing $\tau^{-1}(\alpha_i) = z_i$, $\tau^{-1}(\beta_i) = w_i$ with an arbitrary permutation otherwise. The extension to $\mathscr{S}_{\{\alpha_1, ..., \alpha_{k_1}\}} * \mathscr{S}_{\{\beta_1, ..., \beta_{k_2}\}} * \mathscr{S}_{\{\gamma_1, ..., \gamma_{k_3}\}}$ etc. is clear.

A permutation invariant kernel $K(\mathbf{x}, \mathbf{y})$ defined on $\mathbf{R}^n \times \mathbf{R}^n$ is said to be χ *generalized sign consistent of order* $p(\chi\text{-GSC}_p)$ if for all pairs of monotone

k-tuples of indices $i_1 < \cdots < i_p$, $j_1 < \cdots < j_p$, and $(\mathbf{x}, \mathbf{y})$ in the corresponding monotone orthant, i.e., $x_{i_1} \leqslant \cdots \leqslant x_{i_p}$ and $y_{j_1} \leqslant \cdots \leqslant y_{j_p}$, we have

$$\sum_{\pi \in \mathscr{S}\{i_1, \ldots, i_p\}} \chi(\pi) K(\mathbf{x}, \pi\mathbf{y}) \geqslant 0. \tag{3.2}$$

Thus the kernel $K(\mathbf{x}, \mathbf{y})$ is GSC_2 (since $\mathscr{S}_{\{i,j\}} = \{e, t\}$, where e is the identity and t specifies the transposition of i to j) if whenever $x_i \leqslant x_j$ and $y_i \leqslant y_j$ holds for some i, j then (for $\chi = \mathrm{sign}$)

$$\begin{aligned} &K(x_1, \ldots, x_{i-1}, x_i, x_{i+1}, \ldots, x_{j-1}, x_j, x_{j+1}, \ldots, x_n;\\ &\qquad y_1, \ldots, y_{i-1}, y_i, y_{i+1}, \ldots, y_{j-1}, y_j, y_{j+1}, \ldots, y_n)\\ &\qquad \geqslant K(x_1, \ldots, x_{i-1}, x_i, x_{i+1}, \ldots, x_{j-1}, x_j, x_{j+1}, \ldots, x_n;\\ &\qquad y_1, \ldots, y_{i-1}, y_j, y_{i+1}, \ldots, y_{j-1}, y_i, y_{j+1}, \ldots, y_n). \end{aligned} \tag{3.3}$$

Note that (3.3) expresses inequality between two values of K, where in the first argument (x_i, x_j), (y_i, y_j) are similarly ordered, whereas in the second they are oppositely ordered. This property was called DT = decreasing in transposition by Hollander, Proschan, and Sethuraman [4], and AI = arrangement increasing by Marshall and Olkin [9]. In our terminology this will be called GSC_2, and if, in addition, $K \geqslant 0$ then it is GTP_2.

A kernel $K(\mathbf{x}, \mathbf{y})$ is said to be *χ-generalized totally positive of order* $p(\chi - \mathrm{GTP}_p)$ if K is χ-GSC_q for all q, $1 \leqslant q \leqslant p$.

EXAMPLES. A kernel of the form $K(\mathbf{x}, \mathbf{y}) = \sum_{i=1}^n \Phi(x_i, y_i)$ is $(\mathrm{sign}\, \sigma)$-GTP_2 if and only if $\Phi \geqslant 0$ and for any $x_1 \leqslant x_2$, $y_1 \leqslant y_2$,

$$\Phi(x_1, y_1) + \Phi(x_2, y_2) - \Phi(x_1, y_2) - \Phi(x_2, y_1) \geqslant 0,$$

i.e., Φ is a positive set function or, equivvalently, $\Psi(x, y) = \exp[\Phi(x, y)]$ is TP_2. $K(\mathbf{x}, \mathbf{y}) = \sum_{i=1}^n \Phi(x_i, y_i)$ is maximized when $\mathbf{x}$ and $\mathbf{y}$ are similarly ordered (Lorentz [6], Rinott [11]).

The following example was stimulated by Boland, Proschan, and Tong [2]. Let $\mathbf{X} = (X_1, \ldots, X_n)$ be a vector of exchangeable random variables. Set

$$H(\mathbf{a}, \mathbf{b}) = \Pr\{\mathbf{a} \leqslant \mathbf{X} \leqslant \mathbf{b}\}, \qquad \mathbf{a}, \mathbf{b} \in \mathbf{R}^n.$$

Then $H(\mathbf{a}, \mathbf{b})$ is GTP_2.

Proof. Let $A = \{(x_3, \ldots, x_n)\colon a_i \leqslant x_i \leqslant b_i,\ i = 3, \ldots, n\}$ and let I_A denote its indicator function. Because

$$H(\mathbf{a},\mathbf{b}) = E[I_A(X_3, \ldots, X_n) \Pr\{(a_1, a_2) \leqslant (X_1, X_2) \leqslant (b_1, b_2) | X_3, \ldots, X_n\}],$$

we see that it suffices to prove the result for

$$H((a_1, a_2), (b_1, b_2)) = \Pr\{(a_1, a_2) \leqslant (X_1, X_2) \leqslant (b_1, b_2) | X_3, ..., X_n\};$$

that is, it suffices to consider the case $n = 2$. Thus the desired result reduces to showing for $a_1 \leqslant a_2 \leqslant b_1 \leqslant b_2$,

$$\begin{aligned} H((a_1, a_2), (b_1, b_2)) &= \Pr\{a_1 \leqslant X_1 \leqslant b_1, a_2 \leqslant X_2 \leqslant b_2\} \\ &\geqslant \Pr\{a_1 \leqslant X_1 \leqslant b_2, a_2 \leqslant X_2 \leqslant b_1\} = H((a_1, a_2), (b_2, b_1)). \end{aligned}$$

Now

$$\begin{aligned} H((a_1, a_2), (b_2, b_1)) &= H((a_1, a_2), (b_1, b_1)) + H((b_1, a_2), (b_2, b_1)) \\ &= H((a_1, a_2), (b_1, b_1)) + H((a_2, b_1), (b_1, b_2)) \end{aligned}$$

(since by exchangeability $H(\mathbf{a}\pi, \mathbf{b}\pi) = H(\mathbf{a}, \mathbf{b})$, where π denotes a permutation)

$$\begin{aligned} &\leqslant H((a_1, a_2), (b_1, b_1)) + H((a_1, b_1), (b_1, b_2)) \\ &= H((a_1, a_2), (b_1, b_2)) \end{aligned}$$

and the proof is complete.

If the underlying kernel function $\Phi(x, y)$ is TP_p in the standard sense then $K(\mathbf{x}, \mathbf{y}) = \prod_{i=1}^{n} \Phi(x_i, y_i)$ is χ-GTP_p for the character $\chi(\sigma) = \operatorname{sign} \sigma$.

THEOREM 2. *If K and L are $\chi - \mathrm{GSC}_p$ then so is their convolution*

$$M(\mathbf{x}, \mathbf{z}) = \int_{\mathbf{R}^n} K(\mathbf{x}, \mathbf{y})\, L(\mathbf{y}, \mathbf{z})\, d\Gamma(\mathbf{y}) \tag{3.4}$$

provided Γ is permutation invariant and the integral exists.

Proof. With a given set of indices $i_1 < \cdots < i_p$ consider the group $\mathscr{H} = \mathscr{S}_{\{i_1, ..., i_p\}}$ and use the coset representatives τ described in (3.1). Then for $\mathbf{x}$ and $\mathbf{z}$ with $x_{i_1} \leqslant \cdots \leqslant x_{i_p}$ and $z_{i_1} \leqslant \cdots \leqslant z_{i_p}$ we have

$$D_M(\mathbf{x}, \mathbf{z}) = \sum_{\tau \in \Delta} \int_A \frac{D_K(\mathbf{x}, \tau\mathbf{y})\, D_L(\tau\mathbf{y}, \mathbf{z})}{t(\mathbf{y})}\, d\Gamma(\mathbf{y}). \tag{3.5}$$

The choice of the coset representatives entails for $\tau \in \Delta$ spanning $\mathscr{S}/\mathscr{S}_{\{i_1, ..., i_p\}}$ that for appropriate $a_1 < \cdots < a_p$, $\tau(a_\nu) = i_\nu$; see the discussion of (3.1). But each $\mathbf{y} \in A$ belongs to the monotone orthant and, in particular,

$$y_{\tau(a_1)} \leqslant \cdots \leqslant y_{\tau(a_p)} \qquad \text{(i.e., } y_{i_1} \leqslant \cdots \leqslant y_{i_p}.)$$

Since K is χ-GSC_p we know for $x_{i_1} \leqslant \cdots \leqslant x_{i_p}$ and $a_{i_1} \leqslant \cdots \leqslant a_{i_p}$ that with $\tau \in \Delta$, $D_K(\mathbf{x}, \tau\mathbf{y}) \geqslant 0$ and similarly $D_L(\tau\mathbf{y}, \mathbf{z}) \geqslant 0$. Since $d\Gamma(\mathbf{y}) \geqslant 0$, it follows that $D_M(\mathbf{x}, \mathbf{z}) \geqslant 0$. Accordingly, Theorem 2 is proved.

There are obvious extensions of the notion of GSC_p relative to groups of the kind $\mathscr{S}_{\{\alpha_1, \ldots, \alpha_r\}} * \mathscr{S}_{\{\beta_1, \ldots, \beta_s\}}$ etc. which lead to tensor products of determinants and permanents.

The following theorem highlights Schur convexity as a special case of χ-GSC_2 ($\chi(\sigma) = \text{sign } \sigma$).

THEOREM 3 (Hollander, Proschan, and Sethuraman [4]). *Let Φ be a real valued function defined on $\mathbf{R}^n$ and define a kernel K by*

$$K(\mathbf{x}, \mathbf{y}) = \Phi(\mathbf{x} + \mathbf{y}).$$

K is GSC_2 if and only if Φ is Schur convex. The kernel L defined by

$$L(\mathbf{x}, \mathbf{y}) = \Phi(\mathbf{x} - \mathbf{y})$$

is GSC_2 if and only if Φ is Schur concave. If $\Phi \geqslant 0$ we can replace GSC_2 by GTP_2.

Proof. It suffices to consider $n = 2$. The notation $\succ$ refers to the majorization ordering, that is $(a, b) \succ (c, d)$ iff $a \geqslant b$, $a \geqslant c$, $a \geqslant d$, and $a + b = c + d$ hold. For $x_1 \geqslant x_2, y_1 \geqslant y_2$, obviously $(x_1 + y_1, x_2 + y_2) \succ (x_1 + y_2, x_2 + y_1)$. On the other hand, if $(a, b) \succ (c, d)$ with $a \geqslant b$, $c \geqslant d$ the choice $x_2 = b, y_2 = 0$, $x_1 = c$, $y_1 = a - c = d - b$ yields $(a, b) = (x_1 + y_1, x_2 + y_2)$, $(c, d) = (x_1 + y_2, x_2 + y_1)$. Hence Φ is Schur convex if and only if

$$\begin{aligned} K((x_1, x_2), (y_1, y_2)) &= \Phi(x_1 + y_1, x_2 + y_2) \geqslant \Phi(x_1 + y_2, x_2 + y_1) \\ &= K((x_1, x_2), (y_2, y_1)). \end{aligned}$$

A similar comparison leads to the conclusion about L.

COROLLARY 1 (Marshall and Olkin [8]). *If f and g are Schur concave on $\mathbf{R}^n$ then so is their convolution h defined by*

$$h(\mathbf{x}) = \int_{\mathbf{R}^n} f(\mathbf{x} - \mathbf{y})\, g(\mathbf{y})\, dy,$$

provided the integral converges absolutely.

Proof. By the translation invariance of Lebesgue measure

$$M(\mathbf{x}, \mathbf{z}) = h(\mathbf{x} - \mathbf{z}) = \int f(\mathbf{x} - \mathbf{y})\, g(\mathbf{y} - \mathbf{z})\, dy = \int K(\mathbf{x}, \mathbf{y})\, L(\mathbf{y}, \mathbf{z})\, dy.$$

The corollary follows from Theorem 2 in view of Theorem 3.

There are many applications of this corollary in the theory of majorization. For example, If $\mathbf{X}$ is a random vector in $\mathbf{R}^n$ having a Schur concave probability density, then $F(x_1, ..., x_n) = \Pr\{X_1 \leqslant x_1, ..., X_n \leqslant x_n\} = \int f(\mathbf{x} - \xi)\, I_+(\xi)\, d\xi$, where $I_+(\xi)$ is the characteristic function of the positive orthant, is Schur concave.

4. Semigroup of Generalized Totally Positive Kernels

In this section we assume that $\Gamma = \mu \times \cdots \times \mu$, where μ is Lebesgue measure on $\mathbf{R}$ or the counting measure assigning unit mass to each integer. We shall exploit the fact that for any Borel set A in $\mathbf{R}$, $\mu(A + x) = \mu(A)$ for all $x \in R$ in the case of Lebesgue measure, $x \in Z$ (=integer) in the case of the counting measure.

In the following definition, $\Lambda \subseteq \mathbf{R}$ will denote a semigroup (under addition) such as $\mathbf{R}$, $\mathbf{R}_+ = [0, \infty)$, $\mathbf{Z} = \{..., -1, 0, 1, ...\}$ or $\mathbf{Z}_+ = \{0, 1, ...\}$. The product Λ^m is again a semigroup in $\mathbf{R}^m$.

Let Λ^m denote a semigroup in $\mathbf{R}^m$. A kernel $K(\boldsymbol{\alpha}, \mathbf{x})$, $(\boldsymbol{\alpha}, \mathbf{x}) \in \Lambda^m \times \mathbf{R}^n$ is said to possess the semigroup property if the identity

$$K(\boldsymbol{\alpha} + \boldsymbol{\beta}, \mathbf{x}) = \int_{\mathbf{R}^n} K(\boldsymbol{\alpha}, \mathbf{x} - \mathbf{y})\, K(\boldsymbol{\beta}, \mathbf{y})\, d\mu(y_1) \cdots d\mu(y_n) \tag{4.1}$$

holds for all $\boldsymbol{\alpha}, \boldsymbol{\beta} \in \Lambda^m$, $\mathbf{x} \in \mathbf{R}^n$.

If $K(\alpha, x)$, $\alpha \in \Lambda$, $x \in \mathbf{R}$ satisfies the semigroup property with respect to μ, then $K(\boldsymbol{\alpha}, \mathbf{x}) = \prod_{i=1}^n K(\alpha_i, x_i)$ satisfies the semi-group property with respect to $\Gamma = \mu \times \cdots \times \mu$ for $\boldsymbol{\alpha} \in \Lambda^n$, $\mathbf{x} \in \mathbf{R}^n$. Any infinitely divisible density $f(\mathbf{x})$ is embedded in a semigroup family of densities in continuous time $f_t(\mathbf{x})$. Sums of i.i.d. variables generate a discrete semigroup family.

The basic result in this section extends a result of (Hollander, Proschan, and Sethuraman [4]), see also [5, Chap. 3].

Theorem 4. *Let $K(\boldsymbol{\lambda}, \mathbf{x})$, $\boldsymbol{\lambda} \in \Lambda^n$, $\mathbf{x} \in \mathbf{R}^n$ be a GSC_p kernel having the semigroup property, and suppose the real valued function Ψ on $\mathbf{R}^n$ is such that the kernel L defined by $L(\mathbf{x}, \mathbf{y}) = \Psi(\mathbf{x} + \mathbf{y})$ is GSC_p. Define*

$$\Phi(\boldsymbol{\lambda}) = \int_{\mathbf{R}^n} \Psi(\mathbf{x})\, K(\boldsymbol{\lambda}, \mathbf{x})\, d\mu(x_1) \cdots d\mu(x_n) \tag{4.3}$$

and $M(\boldsymbol{\alpha}, \boldsymbol{\beta}) = \Phi(\boldsymbol{\alpha} + \boldsymbol{\beta})$. Then M is GSC_p in $\boldsymbol{\alpha}, \boldsymbol{\beta}$.

Proof. Setting $d\Gamma(\mathbf{x}) = d\mu(x_1) \cdots d\mu(x_n)$, we have

$$\begin{aligned}\Phi(\boldsymbol{\alpha}+\boldsymbol{\beta}) &= \int_{\mathbf{R}^n} \Psi(\mathbf{x})\, K(\boldsymbol{\alpha}+\boldsymbol{\beta}, \mathbf{x})\, d\Gamma(\mathbf{x}) \\ &= \int_{\mathbf{R}^n}\int_{\mathbf{R}^n} \Psi(\mathbf{x})\, K(\boldsymbol{\alpha}, \mathbf{x}-\mathbf{y})\, K(\boldsymbol{\beta}, \mathbf{y})\, d\Gamma(\mathbf{y})\, d\Gamma(\mathbf{x}) \\ &= \int_{\mathbf{R}^n} K(\boldsymbol{\beta}, \mathbf{y}) \left\{ \int_{\mathbf{R}^n} K(\boldsymbol{\alpha}, \mathbf{z})\, \Psi(\mathbf{z}+\mathbf{y})\, d\Gamma(\mathbf{z}) \right\} d\Gamma(\mathbf{y}) \\ &= \int K(\boldsymbol{\beta}, \mathbf{y})\, L(\boldsymbol{\alpha}, \mathbf{y})\, d\Gamma(\mathbf{y}).\end{aligned}$$

Applying Theorem 2, we deduce that the inner integral whose resulting kernel is labeled $L(\boldsymbol{\alpha}, \mathbf{y})$ is GSC_p in $\mathbf{y}$, $\boldsymbol{\alpha}$. A second application yields GSC_p in $\boldsymbol{\alpha}$, $\boldsymbol{\beta}$.

COROLLARY 1. *Under the assumptions of the theorem, let*

$$\Phi(\boldsymbol{\lambda}) = \int_{\mathbf{R}^n} \Psi(\mathbf{x})\, C\left(\sum_{i=1}^n \lambda_i, \sum_{i=1}^n x_i\right) K(\boldsymbol{\lambda}, \mathbf{x})\, d\mu(x_1) \cdots d\mu(x_n), \tag{4.4}$$

where C is any non-negative function on $\Lambda \times \mathbf{R}$. Then $M(\boldsymbol{\alpha}, \boldsymbol{\beta}) = \Phi(\boldsymbol{\alpha}+\boldsymbol{\beta})$ is GSC_p.

Proof. For the fixed vectors $\boldsymbol{\lambda}$, $\boldsymbol{\gamma}$ set

$$\Psi^*(\mathbf{x}) = \Psi(\mathbf{x})\, C\left(\sum_{i=1}^n \lambda_i, \sum_{i=1}^n x_i\right)$$

and observe that

$$\sum_{i=1}^n (\lambda_i + \gamma_{\pi(i)}) = \sum_{i=1}^n (\lambda_i + \gamma_i)$$

independent of π. It follows that $\Psi^*(\mathbf{x}+\mathbf{y})$ is GSC_p. The result follows by Theorem 4.

COROLLARY 2. *Suppose $K(\alpha, x)$, $\alpha \in \Lambda \subseteq R$, $x \in \mathbf{R}$ satisfies the semigroup property with respect to μ, and $K(\alpha, x)$ is TP_p. Then $K(\boldsymbol{\alpha}, \mathbf{x}) = \prod_{i=1}^n K(\alpha_i, x_i)$, $\boldsymbol{\alpha} \in \Lambda^n$, $\mathbf{x} \in \mathbf{R}^n$ is GTP_p and satisfies the semigroup property. Hence $\Phi(\boldsymbol{\lambda})$ defined by* (4.4) *has $\Phi(\boldsymbol{\alpha}+\boldsymbol{\beta})$ GTP_p provided $\Psi(\mathbf{x}+\mathbf{y})$ is GTP_p.*

5. Complements

The generalized Cauchy–Binet formula based on the symmetric group (permutations on n elements) is also amenable to representations involving a general compact group acting on $\mathbf{R}^n$. This perspective will be briefly reviewed.

Let $\mathscr{X}$ be a space with G a finite group acting on $\mathscr{X}$ to itself. Suppose μ is a measure on $\mathscr{X}$ invariant under G, that is, $\mu(gE)=\mu(E)$ for any measurable $E\subseteq\mathscr{X}$. Assume there is a fundamental region $A\subseteq\mathscr{X}$ such that

$$\int_{\mathscr{X}} f(\mathbf{x})\,d\mu(\mathbf{x}) = \sum_{g\in G}\int_A f(g\mathbf{x})\,d\mu(\mathbf{x}) \tag{5.1}$$

akin to the integration formula (2.1). Let $K(\mathbf{x},\xi)$ be a kernel invariant under G, that is, $K(g\mathbf{x},g\xi)=K(\mathbf{x},\xi)$ for all $\mathbf{x},\xi\in\mathscr{X}$ and $g\in G$. For $\chi(g)$ a character on G we define

$$D_K(\mathbf{x},\mathbf{z}) = \sum_{g\in G}\chi(g)\,K(\mathbf{x},g\mathbf{z}) \qquad \text{for} \quad \mathbf{x},\mathbf{z}\in\mathscr{X}.$$

The method of Theorem 1 yields: Let

$$h(\mathbf{x},\mathbf{z}) = \int f(\mathbf{x},\mathbf{y})\,g(\mathbf{y},\mathbf{z})\,\mu(d\mathbf{y}). \tag{5.2}$$

Then

$$D_h(\mathbf{x},\mathbf{z}) = \int_A D_f(\mathbf{x},\mathbf{y})\,D_g(\mathbf{y},\mathbf{z})\,\mu(d\mathbf{y}). \tag{5.3}$$

In particular, *if* $D_f(\mathbf{x},\mathbf{y})\geqslant 0$ *and* $D_g(\mathbf{y},\mathbf{z})\geqslant 0$ *for* $\mathbf{x},\mathbf{y},\mathbf{z}\in A$, *then also* $D_h(\mathbf{x},\mathbf{z})\geqslant 0$.

Example 1. Let $\mathscr{X}=\mathbf{R}^n$ and identify G with the reflection group of elements $g=(g_1,\ldots,g_n)$ each $g_i=\pm 1$ and $g\mathbf{x}=(g_1x_1,\ldots,g_nx_n)$. Define $\chi(g)=\prod_{i=1}^n(-1)^{(1-g_i)/2}$. Any positive density of the form $\rho(\mathbf{x})=\rho(|x_1|,\ldots,|x_n|)$ satisfies $\rho(g\mathbf{x})=\rho(\mathbf{x})$ and the measure induced by ρ is invariant with respect to G. In this case (5.1) holds with A the positive orthant.

Consider $n=2$, $\varphi(\mathbf{x})=\varphi(|x_1|+|x_2|)$ and $\psi(\mathbf{x})=\psi(|x_1|+|x_2|)$ with φ and ψ convex (not necessarily decreasing) on the positive axis. Then

$$\sum_{g\in G}\chi(g)\,\varphi(\mathbf{x}-g\mathbf{y})\geqslant 0 \qquad \text{for} \quad \mathbf{x}>0 \quad \text{and} \quad \mathbf{y}>0 \tag{5.4}$$

and similarly for ψ. By virtue of (5.3), the composition (the integral is assumed to exist)

$$\theta(\mathbf{x}, \mathbf{z}) = \int \varphi(\mathbf{x}-\mathbf{y})\, \psi(\mathbf{y}-\mathbf{z})\, \rho(\mathbf{y})\, d\mathbf{y}$$

with $\rho(\mathbf{y}) = \rho(|\mathbf{y}|)$ satisfies $\sum_{g \in G} \chi(g)\, \theta(\mathbf{x}, g\mathbf{z}) \geqslant 0$ for $\mathbf{x}$ and $\mathbf{z} \geqslant \mathbf{0}$.

The result can be extended to the case of n coordinates involving higher order convexity requirements on φ and ψ.

EXAMPLE 2. Let G_i be the group of two elements $\{e, \gamma_i\}$ where $e\mathbf{x} = \mathbf{x}$, $\gamma_i \mathbf{x} = (x_1, ..., x_{i-1}, -x_i, x_{i+1}, ..., x_n)$. For $\varphi(\mathbf{x}) = \varphi(|x_1|, ..., |x_n|)$ and $\chi(g)$ as before then

$$\sum_{g \in G_i} \chi(g)\, \varphi(\mathbf{x} - g\mathbf{y}) \geqslant 0 \qquad \text{for} \quad x_i, y_i > 0 \tag{5.5}$$

iff φ is decreasing in the ith coordinate.

The following composition inequality holds. Let $\varphi(\mathbf{x})$ and $\psi(\mathbf{x})$ satisfy (5.5) for each G_i; i.e., $\varphi(|x_1|, ..., |x_n|)$ and $\psi(|x_1|, ..., |x_n|)$ are decreasing in each coordinate, then the convolution $\theta(\mathbf{x}) = \int \varphi(\mathbf{x} - \xi)\, \psi(\xi)\, d\xi$ is also decreasing in each coordinate.

EXAMPLE 3. The property that a radial function $f(\|\mathbf{x}\|^2)$ in $\mathbf{R}^n$ is decreasing away from the origin (RD) is preserved under convolution, where $h(\mathbf{x}) = \int f(\mathbf{x} - \xi)\, g(\xi)\, d\xi$, follows readily from the identity

$$h(\mathbf{y}-\mathbf{z}) - h(\mathbf{y}+\mathbf{z}) = \int_{\langle \mathbf{y}, \xi \rangle \geqslant 0} [f(\mathbf{y}-\xi) - f(\mathbf{y}+\xi)][g(\xi-\mathbf{z}) - g(\xi+\mathbf{z})]\, d\xi, \tag{5.6}$$

where $\langle \mathbf{y}, \xi \rangle$ denotes the inner product of the vectors $\mathbf{y}$ and ξ. Note, if $\mathbf{z} = \lambda \mathbf{y}$, $0 < \lambda < 1$, then

$$\begin{aligned} &h((1-\lambda)\,\mathbf{y}) - h((1+\lambda)\,\mathbf{y}) \\ &\quad = \int_{\langle \mathbf{y}, \xi \rangle \geqslant 0} [f(\mathbf{y}-\xi) - f(\mathbf{y}+\xi)][g(\xi - \lambda\mathbf{y}) - g(\xi + \lambda\mathbf{y})]\, d\xi \end{aligned}$$

and both factors in the integrand are non-negative, since f and g are RD. The formula (5.6) can be construed as an elementary version of (5.3).

Similar results ensue for convolutions of unimodal elliptically contoured functions of the same tyype. The foregoing are special cases of the Anderson theorem [1] on symmetric unimodal functions.

We close by describing a general group theoretic version of the Cauchy–Binet formula. Consider a locally compact space $\mathscr{X}$ and a compact group $\mathscr{G}$ acting on $\mathscr{X}$. Let $\mathscr{P} = \mathscr{X}/\mathscr{G}$ denote the factor space. Let dg be the unique left and right Haar measure of $\mathscr{G}$ and assume also that $d(g\mathbf{x}) = d\mathbf{x}$; that is, the measure $d\mathbf{x}$ on $\mathscr{X}$ is invariant under the group operation. We postulate the existence of a Fubini type integration formula (analog of (2.1)) of the form

$$\int_{\mathscr{X}} f(\mathbf{x})\, d\mathbf{x} = \int_{\mathscr{P}} u(\mathbf{p})\, d\mathbf{p} \int_{\mathscr{G}} f(g\mathbf{x})\, dg, \tag{5.7}$$

where $u(\mathbf{p})\, d\mathbf{p}$ is an invariant measure on $\mathscr{P}$. Examples will be given below.

A generalized Cauchy–Binet formula based on (5.7) is accessible. Consider the bivariate kernels $K(\mathbf{x}, \mathbf{y})$ and $L(\mathbf{y}, \mathbf{z})$ both invariant with respect to $\mathscr{G}$ (i.e., $K(g\mathbf{x}, g\mathbf{y}) = K(\mathbf{x}, \mathbf{y})$, $L(g\mathbf{x}, g\mathbf{y}) = L(\mathbf{x}, \mathbf{y})$) and form the composed kernel

$$M(\mathbf{x}, \mathbf{z}) = \int K(\mathbf{x}, \mathbf{y})\, L(\mathbf{y}, \mathbf{z})\ \ d\mathbf{y}.$$

It is easy to check that M is invariant since $d(g\mathbf{y}) = d\mathbf{y}$. Analogous to (2.4) we construct the generalized functional

$$D_K(\mathbf{x}, \mathbf{y}) = \int_{\mathscr{G}} \chi(g)\, K(\mathbf{x}, g\mathbf{y})\, dg, \tag{5.8}$$

where $\chi(g)$ is a character defined on $\mathscr{G}$. Paralleling the derivation of (2.6) relying on the integration formula (5.7), we obtain

$$D_M(\mathbf{x}, \mathbf{z}) = \int_{\mathscr{P}} D_K(\mathbf{x}, \mathbf{y})\, D_L(\mathbf{y}, \mathbf{z})\, u(\mathbf{p})\, d\mathbf{p}, \tag{5.9}$$

where the product function $D_K(\mathbf{x}, \mathbf{y})\, D_L(\mathbf{y}, \mathbf{z})$ with respect to $\mathbf{y}$ is actually a function on the coset space $\mathscr{Y}/\mathscr{G}$. In fact, for $\varphi \in \mathscr{G}$,

$$\begin{aligned}
&D_K(\mathbf{x}, \varphi\mathbf{y})\, D_L(\varphi\mathbf{y}, \mathbf{z}) \\
&\quad = \left(\int_{\mathscr{G}} \chi(g)\, K(\mathbf{x}, g\varphi\mathbf{y})\, dg\right)\left(\int_{\mathscr{G}} \chi(h)\, L(\mathbf{y}, \varphi^{-1}h\mathbf{z})\, dh\right) \\
&\quad = \chi(\varphi^{-1})\left(\int_{\mathscr{G}} \chi(g)\, K(\mathbf{x}, g\mathbf{y})\, dg\right) \chi(\varphi) \left(\int_{\mathscr{G}} \chi(g)\, L(\mathbf{y}, g\mathbf{z})\, dg\right) \\
&\quad = D_K(\mathbf{x}, \mathbf{y})\, D_L(\mathbf{y}, \mathbf{z}).
\end{aligned}$$

In the special case $\chi(g) \equiv 1$, the formula (5.9) entails only coset variables such that

$$D_M(\mathbf{px}, \mathbf{pz}) = \int_{\mathscr{P}} u(\mathbf{p})[D_K(\mathbf{px}, \mathbf{py})\, D_L(\mathbf{py}, \mathbf{pz})]\, d\mathbf{p}, \tag{5.10}$$

where **px**, **py**, and **pz** designate the coset representatives of **x**, **y**, and **z**, respectively.

Examples of (5.7) include

1. $\mathscr{X} = \mathbf{R}^n$, $\mathscr{G}$ the orthogonal group on $\mathbf{R}^n$ and $\mathscr{P}$ is identified with the radial value of **x**.

2. Let $\mathscr{X}$ consist of all $r \times n$ matrices **x**, $\mathscr{G}$ again the orthogonal group of $\mathbf{R}^n$ acting on $\mathscr{X}$ by left multiplication $\mathbf{x}g$. The coset space $\mathscr{P}$ is recognized as the collection of all $r \times r$ positive semi-definite matrices. The integration formula (5.7) in this case is

$$\int f(\mathbf{x}) \frac{d\mathbf{x}}{|\mathbf{xx}'|^{n/2}} = \alpha_{n,r} \int_{\mathscr{P}} \frac{d\mathbf{p}}{|\mathbf{p}|^{(r+1)/2}} \int_{\mathscr{G}} f(\mathbf{x}g)\, dg. \tag{5.11}$$

($|\mathbf{p}|$ denotes the determinant of **p**, $\mathbf{x}'$ the transpose matrix to **x**, $\alpha_{n,r}$ is an appropriate constant.) The formula (5.11) underlies the development of the central and noncentral Wishart distribution.

3. Another important example of a Fubini type formula is

$$\int f(\mathbf{x}, \mathbf{w}) \frac{d\mathbf{x}\, d\mathbf{w}}{|\mathbf{w}|^{p+q}} = \alpha(p, q) \int_{u} d\mathbf{u} \int_{\mathscr{C}} f(\mathbf{cx}, \mathbf{cy}) \frac{d\mathbf{c}}{|\mathbf{c}|^{p}}, \tag{5.12}$$

where the integral on the left covers all matrix pairs (**x**, **w**), **x** is $p \times q$, **w** is $p \times p$ nonsingular, **u** is $p \times q$, $d\mathbf{u}$ is ordinary Lesbegue measure on pq space, **c** an arbitrary $p \times p$ nonsingular matrix. The formula (5.12) can be used to obtain the distribution of $\mathbf{u} = \mathbf{w}^{-1}\mathbf{x}$ which is a type of multivariate t-statistic.

Similar integration formulas are available for the generalized Hotelling statistics, canonical correlations, etc. (see [3, Chap. 5]).

The generalized Cauchy–Binet formula of the form (5.10) affords a construction of a compound kernel defined on the coset space of $\mathscr{P} = \mathscr{X}/\mathscr{G}$ resembling compound permanent functions.

References

[1] Anderson, T. W. (1955). The integral of a symmetric unimodal function over a symmetric convex set and some probability inequalities, *Proc. Amer. Math. Soc.* **6** 170–176.

[2] BOLAND, P. J., PROSCHAN, F. AND TONG, Y. L. (1988). Moment and geometric probability inequalities arising from arrangement increasing functions, *Ann. Probab.* **18** 407–413.

[3] FARRELL, R. H. (1985). *Multivariate Calculation: Use of Continuous Groups*, Springer-Verlag, New York.

[4] HOLLANDER, M., PROSCHAN, F., AND SETHURAMAN, J. (1977). Functions decreasing in transposition and their applications in ranking problems, *Ann. Statist.* **5** 722–733.

[5] KARLIN, S. (1968). *Total Positivity*, Stanford Univ. Press, Stanford, CA.

[6] LORENTZ, G. G. (1953). An inequality for rearrangements, *Amer. Math. Monthly* **60** 176–179.

[7] MARCUS, M. (1973–1975). *Finite Dimensional Multilinear Algebra*, Dekker, New York.

[8] MARSHALL, A. W., AND OLKIN, I. (1974). Majorization in multivariate distributions, *Ann. Statist.* **2** 1189–1200.

[9] MARSHALL, A. W., AND OLKIN, I. (1979). *Inequalities: Theory of Majorization and Its Applications*, Academic Press, New York.

[10] DE OLIVEIRA, G. N. (1973). *Generalized Matrix Functions*, Instituto Gulbenkian de Ciência, Centro de Cálculo Cientifico, Oeiras, Portugal.

[11] RINOTT, Y. (1973). Multivariate majorization and rearrangement inequalities with some applications to probability and statistics, *Israel J. Math.* **15** 60–77.

Isotonic M-Estimation of Location: Union-Intersection Principle and Preliminary Test Versions

AZZA R. KARMOUS

Zagazig University, Zagazig, Egypt

AND

PRANAB K. SEN*

University of North Carolina, Chapel Hill, North Carolina

In a k ($\geqslant 2$) sample model, isotonic estimators of locations $\theta_1, \ldots, \theta_k$ take into consideration the prior restraint that $\theta_1 \leqslant \cdots \leqslant \theta_k$. Though these estimators are appealling, they are generally biased. The union–intersection (UI-) principle and the theory of M-estimation of location are incorporated in the formulation of some robust, preliminary test, isotonic (M-) estimators of locations. Associated distribution theory of the test statistic and estimator is studied in a systematic manner.

1. INTRODUCTION

Let X_{ij}, $j = 1, \ldots, n_i$ be n_i independent and identically distributed random variables (i.i.d.r.v.) with a distribution function (d.f.) F_i, defined on the real line R, for $i = 1, \ldots, k$ ($\geqslant 2$); all these k samples are assumed to be independent. Consider the usual location model

$$F_i(x) = F(x - \theta_i), \qquad i = 1, \ldots, k, \tag{1.1}$$

* Work of this author was partially supported by the Office of Naval Research, Contract N00014-83-K-0387.

Multivariate Statistics and Probability
ISBN 0-12-580205-6

Reprinted from *J. Mult. Anal.* **27**(1).

where the θ_i are the location parameters and F is a continuous d.f., assumed to be symmetric about 0. It is desired to develop suitable M-estimators of the vector of location parameters $\boldsymbol{\theta} = (\theta_1, ..., \theta_k)'$ following a preliminary test of

$$H_0: \theta_1 = \cdots = \theta_k \quad \text{against} \quad H_1: \theta_1 \leqslant \cdots \leqslant \theta_k, \tag{1.2}$$

where at least one of the inequalities is strict. The preliminary M-test is an extension of union–intersection (UI-) tests considered by De [5], Chinchilli and Sen [3, 4], and Boyd and Sen [2], while the preliminary test estimator (PTE) is formulated along the lines of Sen and Saleh [10], but for restricted alternatives.

Section 2 deals (succinctly) with the classical M-estimators of location in this multi-sample context; the corresponding UI-M-test for H_0 against H_1 is considered in Section 3. The results of Section 2 and 3 are incorporated in the formulation of isotonic M-estimators and their PTE versions in Sections 4 and 5, respectively. Asymptotic properties of these estimators are studied under a sequence of local alternatives containing H_0 as a special case. The concluding section deals with this relative picture through some simulation studies.

2. M-Estimators of Location and Regularity Conditions

We introduce first a *score function* $\psi: R \to R$, defined by

$$\psi(x) = \psi_1(x) + \psi_2(x), \qquad x \in R = (-\infty, \infty), \tag{2.1}$$

where both ψ_1 and ψ_2 are nondecreasing and skew-symmetric functions with ψ_1 absolutely continuous on any bounded interval in R and ψ_2 a step function having finitely many jumps. We denote these jump-points by $-\infty = a_0 < a_1 < \cdots < a_p < a_{p+1} = \infty$ and assume that there exist real numbers $\alpha_0 < \cdots < \alpha_p$, such that $\psi_2(x) = \alpha_j$, for $x \in (a_j, a_{j+1})$, $j = 0, 1, ..., p$, and, conventionally, we let $\psi_2(a_{j+1}) = (\alpha_j + \alpha_{j+1})/2$, for $j = 0, ..., p-1$. We assume that

$$0 < \sigma_\psi^2 = \int_R \psi^2(x)\, dF(x) < \infty, \tag{2.2}$$

and

$$\int_R \{\psi_1'(x)\}^2\, dF(x) < \infty, \qquad \text{where} \quad \psi_1'(x) = (d/dx)\,\psi_1(x), \quad x \in R. \tag{2.3}$$

Concerning the d.f. F, we assume that it has an absolutely continuous density function f, such that $f'(x) = (d/dx) f(x)$ exists almost everywhere (a.e.), and that

$$\mathscr{I}(f) = \int_R \{f'(x)/f(x)\}^2 \, dF(x) < \infty \quad \text{(i.e., finite Fisher information).} \tag{2.4}$$

These regularity conditions are all adapted from Jurečková [7].

Now, for each i $(=1, ..., k)$ and every real t, we define

$$M_{i,n_i}(t) = \sum_{j=1}^{n_i} \psi(X_{ij} - t), \qquad t \in R, \tag{2.5}$$

and note that by definition $M_{i,n_i}(t)$ is $\searrow$ in $t \in R$. Let then

$$\hat{\theta}^{(1)}_{i,n_i} = \sup\{t\colon M_{i,n_i}(t) > 0\}, \qquad \hat{\theta}^{(2)}_{i,n_i} = \inf\{t\colon M_{i,n_i}(t) < 0\}; \tag{2.6}$$

$$\hat{\theta}_{i,n_i} = (\hat{\theta}^{(1)}_{i,n_i} + \hat{\theta}^{(2)}_{i,n_i})/2, \quad i = 1, ..., k; \qquad \hat{\boldsymbol{\theta}}_n = (\hat{\theta}_{1,n_1}, ..., \hat{\theta}_{k,n_k})'. \tag{2.7}$$

Then, $\hat{\boldsymbol{\theta}}_n$ is the vector of M-estimators of location parameters based on the common score function ψ. In this context, recall that the assumed symmetry of F and the skew-symmetry of ψ (around 0) imply that $\bar{\psi} = \int_R \psi(x) \, dF(x) = 0$, and this motivates the normal equations in (2.6) – (2.7) for the solution of the M-estimators. These M-estimators are translation-equivariant, and depending on the choice of the score function ψ, they are robust too. For later use, we present the following asymptotic results.

First, it follows from Jurečková [7] and Singer and Sen [11] that for any (fixed) $T\colon 0 < T < \infty$, for each i $(=1, ..., k)$, as $n_i \to \infty$,

$$\sup\{n_i^{-1/2} \mid M_{i,n_i}(\theta_i + n_i^{-1/2} t) - M_{i,n_i}(\theta_i) + n_i^{1/2} \gamma t| \colon |t| \leqslant T\} \xrightarrow{P} 0, \tag{2.8}$$

where

$$\gamma = \int_R \psi(x)\{-f'(x)/f(x)\} \, dF(x) \text{ is finite and positive.} \tag{2.9}$$

We let $n = n_1 + \cdots + n_k$ and assume that as n increases,

$$n_i/n \to \lambda_i \colon 0 < \lambda_i < 1, \qquad \text{for each } i \ (=1, ..., k); \quad \sum_{i=1}^{k} \lambda_i = 1. \tag{2.10}$$

A direct consequence of (2.8), (2.9), (2.10), and the asymptotic normality of the M-statistics (studied in detail in Jurečková [7]) is the following: As $n \to \infty$,

$$n^{1/2}(\hat{\boldsymbol{\theta}}_n - \boldsymbol{\theta}) \xrightarrow{\mathscr{D}} \mathscr{N}_k(\mathbf{0}, \gamma^{-2} \sigma^2_\psi \, \boldsymbol{\Lambda}^{-1}); \qquad \boldsymbol{\Lambda} = \text{Diag}(\lambda_1, ..., \lambda_k). \tag{2.11}$$

Finally, (2.11) ensures that

$$n^{1/2} \|\hat{\boldsymbol{\theta}}_n - \boldsymbol{\theta}\| = O_p(1) \qquad \text{(i.e., bounded in probability).} \tag{2.12}$$

3. THE UI-PRELIMINARY M-TEST

Making use of (2.11), we shall incorporate the UI-principle and extend the classical test of Barlow *et al.* [1] to general M-statistics. Let

$$\omega = \{\boldsymbol{\theta}: \theta_1 = \cdots = \theta_k = \theta \in R\} \qquad \text{and} \qquad \omega^* = \{\boldsymbol{\theta}: \theta_1 \leqslant \cdots \leqslant \theta_k\}. \tag{3.1}$$

The (approximate) likelihood function of $\hat{\boldsymbol{\theta}}_n$ is given by

$$L_n(\boldsymbol{\theta}) = \prod_{i=1}^{k} \{\gamma^2 n_i / 2\pi\sigma_\psi^2)^{1/2} \exp[-n_i(\hat{\theta}_{i,n_i} - \theta_i)^2 \gamma^2 / 2\sigma_\psi^2]\}. \tag{3.2}$$

Therefore, we have

$$\sup\{L_n(\boldsymbol{\theta}): \boldsymbol{\theta} \in \omega\} = \text{const} \left\{\exp[-(\gamma^2/2\sigma_\psi^2) \sum_{i=1}^{k} n_i(\hat{\theta}_{i,n_i} - \tilde{\theta}_n)^2]\right\}, \tag{3.3}$$

where

$$\tilde{\theta}_n = \sum_{i=1}^{k} (n_i/n)\, \hat{\theta}_{i,n_i}. \tag{3.4}$$

In passing, we may remark that under ω, a natural estimator of θ might have been obtained by equating $\sum_{i=1}^{k} M_{i,n_i}(t)$ to 0 (in the same fashion as in (2.6)–(2.7)). In view of (2.8), this natural estimator of θ would be square-root n equivalent (in probability) to $\tilde{\theta}_n$. From the computational point of view, given the individual sample M-estimators in (2.6) (2.7), (3.4) involves no extra computation, while the computation of the natural estimator is certainly more involved (although a few iterations should give the estimator up to any desired degree of accuracy). From the point of view of robustness, for small or moderate sample sizes, the natural estimator has some advantage, although in the aymptotic case, there is hardly any difference. Next, we note that

$$\omega^* = \bigcup_{\mathbf{a} \in A} \omega(\mathbf{a}); \qquad \omega(\mathbf{a}) = \{\boldsymbol{\theta}: \theta_i = \theta + \delta a_i,\ 1 \leqslant i \leqslant k,\ a_1 \leqslant \cdots \leqslant a_k\}, \tag{3.5}$$

where δ is a positive scalar constant, and $\mathbf{a}$ belongs to a positively homogeneous cone $\mathscr{A}$. Without any loss of generality, we may set $\bar{a} = n^{-1} \sum_{i=1}^{k} n_i a_i = 0$ and $\sum_{i=1}^{k} n_i a_i^2 = n$.

Under $\omega(\mathbf{a})$, based on (3.2), the MLE of δ and θ are given by

$$\delta_n^*(\mathbf{a}) = \sum_{i=1}^{k} \lambda_i a_i \hat{\theta}_{i,n_i} I\left(\sum_{i=1}^{k} n_i a_i \hat{\theta}_{i,n_i} \geqslant 0\right),$$
$$\theta_n^* = \sum_{i=1}^{k} (n_i/n)\, \hat{\theta}_{i,n_i} = \tilde{\theta}_n, \tag{3.6}$$

where $I(A)$ stands for the indicator function of the set A. Substituting (3.6) in (3.2) and using (3.3) and (3.4), we obtain that

$$\begin{aligned} L_n(\mathbf{a}) &= -2\log\{[\sup\{L_n(\boldsymbol{\theta}): \boldsymbol{\theta} \in \omega\}]/[\sup\{L_n(\boldsymbol{\theta}): \boldsymbol{\theta} \in \omega(\mathbf{a})\}]\} \\ &= n(\gamma^2/\sigma_\psi^2)\left\{\sum_{i=1}^{k} \lambda_i a_i(\hat{\theta}_{i,n_i} - \tilde{\theta}_n)\right\}^2 \cdot I\left(\sum_{i=1}^{k} n_i a_i \hat{\theta}_{i,n_i} \geqslant 0\right). \end{aligned} \tag{3.7}$$

We reject the null hypothesis H_0 in favor of $\omega(\mathbf{a})$ for large values of $L_n(\mathbf{a})$. To obtain an overall test for the entire alternative $\omega = \bigcup_{\mathbf{a} \in A} \omega(\mathbf{a})$, we incorporate the Roy UI-principle, so that on letting

$$\mathscr{A} = \left\{\mathbf{a}: a_1 \leqslant \cdots \leqslant a_k,\ \bar{a} = 0 \text{ and } \sum_{i=1}^{k} \lambda_i a_i^2 = 1\right\}, \tag{3.8}$$

we set the UI-test statistic as

$$L_n^* = \sup\{L_n(\mathbf{a}): \mathbf{a} \in \mathscr{A}\}. \tag{3.9}$$

Our main task is to derive a simple asymptotic expression for L_n^* and to study its distribution theory (under the null as well as local alternatives). Towards this venture, we make use of the basic results in (2.8) through (2.12) along with the Kuhn–Tucker–Lagrange (KTL-) point formula theorem in the nonlinear programming theory. We consider a sequence $\{H_n\}$ of local alternatives

$$H_n: \boldsymbol{\theta} = \boldsymbol{\theta}_{(n)} = \theta\mathbf{1} + n^{-1/2}\boldsymbol{\xi}, \qquad \boldsymbol{\xi} \in \mathscr{A}, \qquad \text{so that } \boldsymbol{\lambda}'\boldsymbol{\xi} = 0, \tag{3.10}$$

where $\boldsymbol{\lambda} = (\lambda_1, ..., \lambda_k)'$ and $\boldsymbol{\xi}$ is an arbitrary (fixed) vector in $\mathscr{A}$. By an appeal to (2.12) and (3.4), we obtain that under (3.10), $n^{1/2}\,|\tilde{\theta}_n - \theta| = O_p(1)$, and hence, by (2.8), we obtain that for each $i\ (=1, ..., k)$, as $n \to \infty$,

$$n^{-1/2}[M_{i,n_i}(\tilde{\theta}_n) - M_{i,n_i}(\hat{\theta}_{i,n_i})] = n^{1/2}\gamma\lambda_i(\hat{\theta}_{i,n_i} - \tilde{\theta}_n) + o_p(1), \tag{3.11}$$

$$n^{-1/2}M_{i,n_i}(\hat{\theta}_{i,n_i}) = o_p(1), \qquad \text{by} \quad (2.5)\text{–}(2.8). \tag{3.12}$$

Consequently, writing $\hat{M}_{i,n_i} = M_{i,n_i}(\tilde{\theta}_n)$ (the residual M-statistic), $i = 1, \ldots, k$; $\hat{\mathbf{M}}_n = (\hat{M}_{1,n_1}, \ldots, \hat{M}_{k,n_k})'$, we obtain from (2.6), (2.11), and the above relations that under $\{H_n\}$ (as well as H_0),

$$n^{1/2}\gamma \sum_{i=1}^{k} \lambda_i a_i(\hat{\theta}_{i,n_i} - \tilde{\theta}_n) = n^{-1/2}\mathbf{a}'\hat{\mathbf{M}}_n + o_p(1); \tag{3.13}$$

$$n^{-1/2}\hat{\mathbf{M}}_n \xrightarrow{\mathscr{D}} \mathscr{N}_k(\gamma\boldsymbol{\Lambda}\boldsymbol{\xi}, \sigma_\psi^2(\boldsymbol{\Lambda} - \boldsymbol{\lambda}\boldsymbol{\lambda}')). \tag{3.14}$$

The use of these residual M-statistics eliminates the need to estimate the unknown parameter γ (for the construction of a suitable test statistic) and also introduces other simplifications to follow. To construct L_n^* in (3.9), we introduce another reparameterization whereby we reduce the problem to an orthant alternative problem, for which the KTL-point formula works out neatly. Let

$$\boldsymbol{\beta} = \mathbf{D}\boldsymbol{\theta}, \qquad \text{where} \quad \underset{(k-1)\times k}{\mathbf{D}} = \begin{pmatrix} -1 & 1 & 0 & \cdots & 0 \\ 0 & -1 & 1 & \cdots & 0 \\ \cdots & \cdots & \cdots & \cdots & \cdots \\ 0 & 0 & \cdots & -1 & 1 \end{pmatrix} \text{is of rank } k-1. \tag{3.15}$$

Then (3.5) can equivalently be written as

$$\omega^* = \bigcup_{\mathbf{b}\in B} \omega^0(\mathbf{b}); \qquad \omega^0(\mathbf{b}) = \{\mathbf{b}: b_j \geqslant 0, j = 2, \ldots, k\}, \tag{3.16}$$

and B is the $(k-1)$-dimensional positive orthant. Let

$$\tilde{M}_{n,i} = \sum_{j=i}^{k} \hat{M}_{j,n_j}, \quad i = 2, \ldots, k; \qquad \tilde{\mathbf{M}}_n = (\tilde{M}_{n,2}, \ldots, \tilde{M}_{n,k})'. \tag{3.17}$$

Then,

$$n^{-1/2}\mathbf{a}'\hat{\mathbf{M}}_n = n^{-1/2}\mathbf{b}'\tilde{\mathbf{M}}_n \qquad \text{whenever} \quad a_i = a_{i-1} + b_i, \quad i = 2, \ldots, k; \tag{3.18}$$

$$n^{-1/2}\tilde{\mathbf{M}}_n \xrightarrow{\mathscr{D}} \mathscr{N}_{k-1}(\gamma\mathbf{U}\boldsymbol{\Lambda}\boldsymbol{\xi}, \sigma_\psi^2\boldsymbol{\Lambda}^*); \quad \mathbf{U} = ((u_{jj'})), \quad \boldsymbol{\Lambda}^* = ((\lambda_{jj'}^*)); \tag{3.19}$$

where

$$u_{jj'} = 0 \quad \text{if} \quad j' < j \quad \text{and} \quad 1 \quad \text{if} \quad j' \geqslant j \qquad (2 \leqslant j \leqslant k, 1 \leqslant j' \leqslant k), \tag{3.20}$$

$$\lambda_{jj'}^* = \sum_{i=j\vee j'}^{k} \lambda_i - \left(\sum_{i=j}^{k} \lambda_i\right)\left(\sum_{i=j'}^{k} \lambda_i\right), \qquad \text{for} \quad j, j' = 2, \ldots, k. \tag{3.21}$$

Thus, the maximization problem in (3.9) reduces (asymptotically) to that of maximizing $n^{-1/2}\mathbf{b}'\tilde{\mathbf{M}}_n I(\mathbf{b}'\tilde{\mathbf{M}}_n > 0)$ over the nonnegative orthant $\{\mathbf{b} \geqslant \mathbf{0}\}$, where we may set without any loss of generality that $\mathbf{b}'\mathbf{\Lambda}^*\mathbf{b} = 1$. For this maximization problem, the KTL-point formula may be adapted. Avoiding the details of this formulation (by cross reference to Chinchilli and Sen [3, 4], where the rank procedures have been considered in detail), we may formulate the ultimate solution as follows.

Let $\mathscr{J}$ be the set of 2^{k-1} possible subsets of $\{2, ..., k\}$ and let J be a typical element of $\mathscr{J}$, and J' be the complementary element. For each J, partition (and rearrange) $\tilde{\mathbf{M}}_n$ and $\mathbf{\Lambda}^*$ as

$$\tilde{\mathbf{M}}_n = (\tilde{\mathbf{M}}'_{n(J)}, \tilde{\mathbf{M}}'_{n(J')})' \qquad \text{and} \qquad \mathbf{\Lambda}^* = \begin{pmatrix} \mathbf{\Lambda}^*_{(JJ)} & \mathbf{\Lambda}^*_{(JJ')} \\ \mathbf{\Lambda}^*_{(J'J)} & \mathbf{\Lambda}^*_{(J'J')} \end{pmatrix}. \tag{3.22}$$

Also, let k_J be the number of elements in the set J. For each J: $\varnothing \subseteq J \subseteq \mathscr{J}$, let

$$\tilde{\mathbf{M}}_{n(J:J')} = \tilde{\mathbf{M}}_{n(J)} - \mathbf{\Lambda}^*_{(JJ')}\mathbf{\Lambda}^{*-1}_{(J'J')}\tilde{\mathbf{M}}_{n(J')}, \tag{3.23}$$

$$\mathbf{\Lambda}^*_{(JJ:J')} = \mathbf{\Lambda}^*_{(JJ)} - \mathbf{\Lambda}^*_{(JJ')}\mathbf{\Lambda}^{*-1}_{(J'J')}\mathbf{\Lambda}^*_{(J'J)}. \tag{3.24}$$

Then, for the orthant problem in (3.16), he UI-statistic based on the $\tilde{M}_n$ is given by

$$\mathscr{L}^*_n = (n\hat{\sigma}^2_\psi)^{-1} \sum_{\phi \subseteq J \subseteq \mathscr{J}} \{\tilde{\mathbf{M}}'_{n(J:J')}\mathbf{\Lambda}^{*-1}_{(JJ:J')}\tilde{\mathbf{M}}_{n(J:J')}\}$$
$$\times I(\tilde{\mathbf{M}}_{n(J:J')} \geqslant \mathbf{0})\, I(\mathbf{\Lambda}^{*-1}_{(J'J')}\tilde{\mathbf{M}}_{n(J')} \leqslant \mathbf{0}), \tag{3.25}$$

where

$$\hat{\sigma}^2_\psi = n^{-1} \sum_{i=1}^{k} \sum_{j=1}^{n_i} \psi^2(X_{ij} - \hat{\theta}_{i,n_i}). \tag{3.26}$$

Following the arguments in Chinchilli and Sen [4], it follows that under H_0, the asymptotic distribution of $\mathscr{L}^*_n$ is the so-called chi-squared bar distribution; i.e.,

$$P\{\mathscr{L}^*_n \leqslant c \mid H_0\} \to \sum_{r=0}^{k-1} w_r P\{\chi^2_r \leqslant c\}, \qquad \forall c \in R^+, \tag{3.27}$$

where the w_r are nonnegative weights adding upto 1, χ^2_r has the central chi-square distribution with r degrees of freedom (DF), and specifically,

$$w_r = \sum_{J:k_J = r} \lim_{n\to\infty} P\{\tilde{\mathbf{M}}_{n(J:J')} \geqslant \mathbf{0}, \mathbf{\Lambda}^{*-1}_{(J'J')}\tilde{\mathbf{M}}_{n(J')} \leqslant \mathbf{0} \mid H_0\}$$
$$= \sum_{J:k_J = r} \lim_{n\to\infty} P\{\tilde{\mathbf{M}}_{n(J:J')} \geqslant \mathbf{0} \mid H_0\}\, P\{\mathbf{\Lambda}^{*-1}_{(J'J')}\tilde{\mathbf{M}}_{n(J')} \leqslant \mathbf{0} \mid H_0\} \tag{3.28}$$

for $r=0, \ldots, k-1$. These orthant probabilities may be computed by reference to the asymptotic normality result in (3.19) (where under H_0, $\xi=0$) and the tables for the multinormal orthant probabilities considered by Gupta [6] and others. Once these w_r are computed, the critical level c for which (3.27) corresponds to $1-\alpha$, for some prespecified level of significance α $(0<\alpha<1)$, can easily be obtained from the tables for the central chi-square distributions, available extensively in the literature. We denote this critical level by c_α^*. Then, the UI-preliminary test for H_0 vs. H_1, based on the residual M-statistics, may be carried out as follows:

$$\text{Reject or accept } H_0 \text{ according as } \mathscr{L}_n^* \text{ is} \geqslant \text{or} < c_\alpha^*. \tag{3.29}$$

A key factor in the simplification of this asymptotic null distribution of the proposed UI-test statistic is the (asymptotic) independence (for each J: $\phi \subseteq J \subseteq \mathscr{J}$) of the quadratic form and the two indicator functions in the right-hand side of (3.25). Unfortunately the non-null distribution (even for local alternatives) is not expressible in terms of averages of appropriate non-central chi-squared distributions. This problem arises mainly due to the fact that when the null hypothesis is not true, though $\mathbf{\Lambda}^{*-1}_{(J'J')}\tilde{\mathbf{M}}_{n(J')}$ in (3.25) is (asymptotically) independent of $\tilde{\mathbf{M}}_{n(J:J')}$ and the quadratic form $\tilde{\mathbf{M}}'_{n(J:J')}\mathbf{\Lambda}^{*-1}_{(JJ:J')}\tilde{\mathbf{M}}_{n(J:J')}$, the later random variable is not independent of $I(\tilde{\mathbf{M}}_{n(J:J')} \geqslant \mathbf{0})$. As such, the best we can do is to express the asymptotic non-null distribution of $\mathscr{L}_n^*$, under $\{H_n\}$, in the form

$$\begin{aligned} P\{\mathscr{L}_n^* \leqslant c \mid H_n\} \sim \sum_{\phi \subseteq J \subseteq \mathscr{J}} & P\{\tilde{\mathbf{M}}_{n(J:J')}\mathbf{\Lambda}^{*-1}_{(JJ:J')}\tilde{\mathbf{M}}_{n(J:J')} \\ & \leqslant nc\sigma_\psi^2,\ \tilde{\mathbf{M}}_{n(J:J')} \geqslant \mathbf{0} \mid H_n\} \\ & \times P\{\mathbf{\Lambda}^{*-1}_{(J'J')}\tilde{\mathbf{M}}_{n(J')} \leqslant \mathbf{0} \mid H_n\}. \end{aligned} \tag{3.30}$$

For the right-hand side, the second factor can be evaluated using the normal orthant probability tables, but for appropriate shifts, while evaluation of the first factor may be quite involved. Though the non-central chi-square (bar) distribution may not generally hold for (3.30), there are alternative forms involving central chi-square distributions with mixing coefficients depending on the alternative hypothesis which have been worked out by some authors (viz., Tsai and Sen [12]), and these may be used (to a limited extent) to study the asymptotic power properties of the UI-test. Equation (3.30) is quite amenable for simulation studies of the asymptotic power function, and for some numerical results, we may refer to Karmous [8].

4. Isotonic M-Estimation of Location

We may refer to Barlow *et al.* [1] for an excellent account of iotonic estimation of the multi-sample normal mean problem. Borrowing their general line of attack and the basic philosophy of M-estimation theory, we may present isotonic M-estimators of the location vector $\boldsymbol{\theta}$ as the solution of

$$\sum_{i=1}^{k} \left| \sum_{j=1}^{n_i} \psi(X_{ij} - \theta_i) \right| \quad \left(\text{or} \sum_{i=1}^{k} \left[\sum_{j=1}^{n_i} \psi(X_{ij} - \theta_i) \right]^2 \right) = \text{minimum},$$
$$\text{subject to the restraint that } \theta_1 \leqslant \cdots \leqslant \theta_k. \tag{4.1}$$

However, in view of the fact that (unlike the normal mean case), the ψ-function is not generally linear (though it could be piece-wise linear as in the Huber case), the computational algorithm (such as the "pool adjacent violators") discussed in Barlow *et al.* [1] may not be totally adaptable here. Leurgans [9] has addressed the basic issues underlying the use of the "partitionng algorithms" in the case of isotonic M-estimation and stressed the lack of robustness aspects. Although in our case, we have a well-defined replicated design (ensuring robustness), her study reveals the general weakness of the usual "partitioning algorithms" in robust isotonic estimation problems. On the other hand, by virtue of the Jurečková [7] linearity of M-statistics (with related first-order asymptotic expansions for M-estimators) and the asmptotic normality results discussed in the last two sections, it is possible to formulate a simple algorithm directly along the lines of Barlow *et al.* [1]. We shall follow this approach here.

We start with the approximate likelihood function in (3.2), and based on this reduced data set (i.e., $\hat{\boldsymbol{\theta}}_n$ and $n_1, ..., n_k$), we construct isotonic M-estimators of $\theta_1, ..., \theta_k$. The isotonized M-estimator of $\boldsymbol{\theta}$, denoted by $\boldsymbol{\theta}_n^*$, is obtained by minimizing (with respect to $\boldsymbol{\theta}$)

$$\sum_{i=1}^{k} n_i [\hat{\theta}_{i,n_i} - \theta_i]^2 \qquad \text{subject to } \theta_1 \leqslant \cdots \leqslant \theta_k. \tag{4.2}$$

The algorithm for the computation of $\boldsymbol{\theta}_n^*$ is the same one as for the ordered mean problem considered in detail in Section 1.2 of Barlow *et al.* [1]. In particular, there exist a positive integer l: $1 \leqslant l \leqslant k$ and l positive integers $k_1 < \cdots < k_l = k$, such that on letting

$$n_j^* = \sum_{i=k_{j-1}+1}^{k_j} n_i \qquad \text{and} \qquad \theta_{n,j}^{**} = \sum_{i=k_{j-1}+1}^{k_j} n_i \hat{\theta}_{i,n_i} / n_j^*, \qquad j = 1, ..., l, \tag{4.3}$$

we have

$$\theta^*_{i,n} = \theta^{**}_{n,j}, \qquad \text{for} \quad i = k_{j-1}+1, ..., k_j, \quad j = 1, ..., l; \quad k_0 = 0. \tag{4.4}$$

Note that l, k_j, and n_j^* are all stochastic in nature and they depend on the relative ordering of the basic M-estimators $\hat{\theta}_{1,n_1}, ..., \hat{\theta}_{k,n_k}$. However, there are only finitely many possible realizations for these stochastic elements. Further, note that the $\theta^{**}_{n,j}$ for a monotone (nondecreasing) sequence while within each of the l buckets, the individual $\hat{\theta}_{i,n_i}$ violate this monotone principle. Finally, note that the isotonic M-estimators are weighted linear combinations of the basic M-estimators, although the weights are themselves stochastic elements and depend on the relative ordering of the initial k estimators. Thus we can conceive of a finite set Π of partitions $\{\pi\}$ such that $\Pi = \bigcup \{\pi\}$ and R^k, the sample space of $\hat{\boldsymbol{\theta}}_n$, is the set theoretic union of disjoint sub-spaces R_π, $\pi \in \Pi$. For each $\pi \in \Pi$, there exists a matrix $\mathbf{D}_\pi$, such that

$$\boldsymbol{\theta}^*_n = \mathbf{D}_\pi \hat{\boldsymbol{\theta}}_n \qquad \text{for} \quad \hat{\boldsymbol{\theta}}_n \in R_\pi, \quad \forall \pi \in \Pi, \tag{4.5}$$

where the $\mathbf{D}_\pi$ depend on $n_1, ..., n_k$ through l and $n_1^*, ..., n_l^*$ which are held fixed for the individual partitionings. A a result, we may write in a compact form

$$\boldsymbol{\theta}^*_n = \sum_{\pi \in \Pi} I(\hat{\boldsymbol{\theta}}_n \in R_\pi)\, \mathbf{D}_\pi \hat{\boldsymbol{\theta}}_n. \tag{4.6}$$

Incorporating (4.6), we have for every $\mathbf{x} \in R^k$,

$$P\{n^{1/2}(\boldsymbol{\theta}^*_n - \boldsymbol{\theta}) \leqslant \mathbf{x} \mid \boldsymbol{\theta}\} = \sum_{\pi \in \Pi} P\{n^{1/2}(\mathbf{D}_\pi \hat{\boldsymbol{\theta}}_n - \boldsymbol{\theta}) \leqslant \mathbf{x}, \hat{\boldsymbol{\theta}}_n \in R_\pi \mid \boldsymbol{\theta}\}, \tag{4.7}$$

and this form is quite amenable for further analysis. The asymptotic normality results on the classical M-estimators studied in earlier sections can thus be used to study the asymptotic distribution theory of isotonic M-estimators.

5. The Preliminary Test Isotonic M-Estimator (PTIME)

It is quite clear from (4.6) and the partitionings R_π, $\pi \in \Pi$, that the isotonic M-estimator $\boldsymbol{\theta}^*_n$ may not be unbiased unless the individual θ_i are quite apart from each other in the domain $\theta_1 < \cdots < \theta_k$. Particularly, for $\boldsymbol{\theta}$ close to the line $\theta_1 = \cdots = \theta_k$, the isotonic M-estimator may be considerably biased. For this reason, it may be quite conceivable to incorporate the preliminary test in Section 3 for constructing a PTE which should behave more properly for small departure of $\boldsymbol{\theta}$ from the line $\theta_1 = \cdots = \theta_k$ and which for large departures should behave closely to the isotonic estimator $\boldsymbol{\theta}_n^*$. With this objective, we propose the following PTIME.

Corresponding to a preassigned level of significance α $(0 < \alpha < 1)$, as in (3.29), let c_α^* be the critical level of the test statistic $\mathscr{L}_n^*$ in (3.25). Also, let $\tilde{\boldsymbol{\theta}}_n = \tilde{\theta}_n \mathbf{1}$ and $\boldsymbol{\theta}_n^*$ be defined as in Sections 3 and 4. Define then

$$\hat{\boldsymbol{\theta}}_n^{PT} = \tilde{\boldsymbol{\theta}}_n I(\mathscr{L}_n^* < c_\alpha^*) + \boldsymbol{\theta}_n^* I(\mathscr{L}_n^* \geqslant c_\alpha^*). \tag{5.1}$$

Thus the PTIME is a convex combination of the classical and isotonic M-estimators of $\boldsymbol{\theta}$ where the mixing coefficient is data based and rests on the preliminary test for the homogeneity of the θ_i against isotonic alternatives. As is generally the case with the PTE, this PTIME is not unbiased for $\boldsymbol{\theta}$, even when $\boldsymbol{\theta}$ deviates from the line $\theta_1 = \cdots = \theta_k$. However, the relative bias of the PTIME and the isotonic M-estimator generally signals a clear cut preference for the PTIME. A similar picture can be obtained with respect to the risk of the two estimators with suitable quadratic error loss functions. A study of the risk of the PTIME and the isotonic ME (IME) demands the knowledge of the exact distribution theory of these estimators. Unfortunately, the distribution of the PTIME or IME is not very simple, even in the asymptotic case. Moreover, in the finite sample case, the distribution may depend on the underlying density function in a rather involved manner. For the IME or the PTIME, the main complication arises due to the distribution theory of $\boldsymbol{\theta}_n^*$ and its close relation with the preliminary test statistic $\mathscr{L}_n^*$ To obtain some meaningul results in this direction we consider some relevant asymptotic theory and use the asymptotic distributional risk measure to compare these estimates.

In the asymptotic setup of Sections 2 and 3, we assume that (2.10) holds and n is large. Next, we note that if H_0 in (1.2) does not hold and H_1 holds, the test based on $\mathscr{L}_n^*$ is consistent (against any fixed alternative within the class depicted by H_1), and as such, by (5.1), $\hat{\boldsymbol{\theta}}_n^{PT}$ and $\boldsymbol{\theta}_n^*$ will be asymptotically equivalent, in probability. However, under H_0 or for local alternatives, this asymptotic stochastic equivalence may not hold, and hence, the relative picture becomes an important issue for closer study. For this reason, we carry out our investigation in two phases:

Phase I. Relative picture of the PTIME and IME for local alternatives and under H_0.

Phase II. Asymptotic properties of the IME for fixed alternatives.

To frame the local alternatives, we conceive of a fixed vector $\boldsymbol{\tau} = (\tau_1, ..., \tau_k)$, such that $\tau_1 \leqslant \cdots \leqslant \tau_k$, and set

$$H_{1(n)}: \boldsymbol{\theta} = \boldsymbol{\theta}_{(n)} = \theta \mathbf{1} + n^{-1/2} \boldsymbol{\tau}, \qquad \theta \text{ arbitrary}; \tag{5.2}$$

by virtue of the translation equivariance of the M-estimators of location, we may set without any loss of generality that $\theta = 0$. The null hypothesis H_0 relates to $\boldsymbol{\tau} = \mathbf{0}$. The asymptotic distribution of the unrestricted M-estimator (UME) $\hat{\boldsymbol{\theta}}_n$, given in (2.11), remains intact irrespective of any alternative (with appropriate change for $\boldsymbol{\theta}$), but the other versions of the M-estimators would have different forms. For the restricted M-estimator (RME) $\tilde{\theta}_n$ in (3.4), (2.11) and (5.2) can readily be used to show that under $\{H_{1(n)}\}$,

$$n^{1/2}(\tilde{\theta}_n - \theta) \xrightarrow{\mathscr{D}} \mathscr{N}(\boldsymbol{\lambda}'\boldsymbol{\tau}, \gamma^{-2}\sigma_{\psi}^2). \tag{5.3}$$

For the IME and PTIME, the asymptotic distributions are of much more complicated forms. First, we consider the case of the IME, and denote by

$$\mathbf{D}_{\pi}^{0} = \mathbf{I} - \mathbf{D}_{\pi}, \qquad \boldsymbol{\tau}_{\pi}^{0} = \mathbf{D}_{\pi}^{0}\boldsymbol{\tau}, \qquad \text{for} \quad \pi \in \Pi. \tag{5.4}$$

Then, by virtue of (4.7), we have under (5.2),

$$P\{n^{1/2}(\boldsymbol{\theta}_n^* - \boldsymbol{\theta}_{(n)}) \leqslant \mathbf{x}\}$$

$$= \sum_{\pi \in \Pi} P\{n^{1/2}\mathbf{D}_{\pi}(\hat{\boldsymbol{\theta}}_n - \boldsymbol{\theta}_{(n)}) \leqslant \mathbf{x} + \boldsymbol{\tau}_{\pi}^{0}, \hat{\boldsymbol{\theta}}_n \in R_{\pi} \mid \boldsymbol{\theta}_{(n)}\}. \tag{5.5}$$

At this stage, we may note that for each $\pi \in \Pi$,

$$n^{1/2}[\mathbf{D}_{\pi}(\hat{\boldsymbol{\theta}}_n - \boldsymbol{\theta}_{(n)})] \xrightarrow{\mathscr{D}} \mathscr{N}(\mathbf{0}, \gamma^{-2}\sigma_{\psi}^2 \mathbf{D}_{\pi}'\boldsymbol{\Lambda}^{-1}\mathbf{D}_{\pi}). \tag{5.6}$$

However, $n^{1/2}\mathbf{D}_{\pi}(\hat{\boldsymbol{\theta}}_n - \boldsymbol{\theta}_{(n)})$ and $n^{1/2}(\hat{\boldsymbol{\theta}}_n - \boldsymbol{\theta}_{(n)})$ are not asymptotically independent (even under H_0), for every $\pi \in \Pi$. Thus, the right-hand side of (5.5) may not be factorized into two terms involving the marginal normal probabilities. Nor is R_{π} a linear subspace of R^k (typically, R_{π} is a cone), and hence, $n^{1/2}(\hat{\boldsymbol{\theta}}_n - \boldsymbol{\theta}_{(n)})$ may not belong to a linearly transformed form of R_{π}. On the other hand, the individual terms on the right-hand side of (5.5) can be expressed in terms of the multi-normal probability integrals (for large values of n) over specific sub-spaces in R^k, and (2.11) provides the access for this asymptotic simplification. Unfortunately, for such inequality-restrained sub-spaces in R^k, for $k \geqslant 3$, compact forms for the probability contents based on multi-normal distributions are not available, and numerical integration seems to be a feasible way. On the other hand, by (4.6),

$$E(\boldsymbol{\theta}_n^*) = \sum_{\pi \in \Pi} \mathbf{D}_{\pi}\{E[\hat{\boldsymbol{\theta}}_n I(\hat{\boldsymbol{\theta}}_n \in R_{\pi})]\}, \tag{5.7}$$

so that using the fact that the components of $\hat{\boldsymbol{\theta}}_n$ are independent, this expectation may often be computed relatively easily. A similar simplification also holds for the second-order moments.

Let us proceed to the case of the PTIME. First, using the asymptotic linearity results in (2.8), it follows from (3.22) through (3.26) that under $\{H_{1(n)}\}$ (as well as H_0), $\mathscr{L}_n^*$ in (3.25) is equivalent in probability to

$$\mathscr{L}_n^0 = \gamma^2 \sigma_\psi^{-2} \sum_{i=1}^{k} n_i (\theta_{i,n}^* - \tilde{\theta})^2, \tag{5.8}$$

where $\tilde{\theta}_n$ is defined by (3.4) and $\boldsymbol{\theta}_n^*$ is the IME of $\boldsymbol{\theta}$, defined by (4.2)–(4.4). As such, using (4.6) and (5.8), we have

$$\mathscr{L}_n^0 = \sum_{\pi \in \Pi} I(\hat{\boldsymbol{\theta}}_n \in R_\pi)\, n\hat{\boldsymbol{\theta}}_n' \mathbf{A}_\pi \hat{\boldsymbol{\theta}}_n, \tag{5.9}$$

where

$$\mathbf{A}_\pi = (\mathbf{D}_\pi' \boldsymbol{\Lambda} \mathbf{D}_\pi - \boldsymbol{\lambda}\boldsymbol{\lambda}')\, \gamma_2 \sigma_\psi^{-2}, \qquad \text{for} \quad \pi \in \Pi. \tag{5.10}$$

Using (5.1), (5.8), (5.9), and (5.10), we may consider the following asymptotically equivalent (in probability) version of the PTIME:

$$\begin{aligned} \hat{\boldsymbol{\theta}}_n^{PT} &= \sum_{\pi \in \Pi} I(\hat{\boldsymbol{\theta}}_n \in R_\pi) \{\tilde{\theta}_n \mathbf{1} I(n\hat{\boldsymbol{\theta}}_n' \mathbf{A}_\pi \hat{\boldsymbol{\theta}}_n \leqslant c_\alpha^*) + \mathbf{D}_\pi \hat{\boldsymbol{\theta}}_n I(n\hat{\boldsymbol{\theta}}_n' \mathbf{A}_\pi \hat{\boldsymbol{\theta}}_n > c_\alpha^*)\} \\ &= \sum_{\pi \in \Pi} \{I(\hat{\boldsymbol{\theta}}_n \in R_{\pi n}^{(1)})(\mathbf{1}\boldsymbol{\lambda}')\, \hat{\boldsymbol{\theta}}_n + I(\hat{\boldsymbol{\theta}}_n \in R_{\pi n}^{(2)})\, \mathbf{D}_\pi \hat{\boldsymbol{\theta}}_n\}, \end{aligned} \tag{5.11}$$

where

$$\begin{aligned} R_{\pi n}^{(1)} &= \{\hat{\boldsymbol{\theta}}_n : \hat{\boldsymbol{\theta}}_n \in R_\pi \text{ and } n\hat{\boldsymbol{\theta}}_n' \mathbf{A}_\pi \hat{\boldsymbol{\theta}}_n \leqslant c_\alpha^*\}, \\ R_{\pi n}^{(2)} &= R_\pi \setminus R_{\pi n}^{(1)}, \qquad \pi \in \Pi. \end{aligned} \tag{5.12}$$

Thus, $\{R_{\pi n}^{(j)},\ j = 1, 2,\ \pi \in \Pi\}$ is a finer partitioning of R^k, and we may rewrite the right-hand side of (5.11) as $\sum_{\pi \in \Pi} \sum_{j=1}^{2} I(\hat{\boldsymbol{\theta}}_n \in R_{\pi n}^{(j)})\, \mathbf{D}_\pi^{(j)} \hat{\boldsymbol{\theta}}_n$, where $\mathbf{D}_\pi^{(1)} = \mathbf{1}\boldsymbol{\lambda}'$ and $\mathbf{D}_\pi^{(2)} = \mathbf{D}_\pi$, $\pi \in \Pi$. As such, parallel to (5.5), we have under (5.2)

$$\begin{aligned} &P\{n^{1/2}(\hat{\boldsymbol{\theta}}_n^{PT} - \boldsymbol{\theta}_{(n)}) \leqslant \mathbf{x}\} \\ &\qquad \simeq \sum_{\pi \in \Pi} \sum_{j=1}^{2} P\{n^{1/2} \mathbf{D}_\pi^{(j)} (\hat{\boldsymbol{\theta}}_n - \boldsymbol{\theta}_{(n)}) \leqslant \mathbf{x} + \boldsymbol{\tau}_{\pi j}^0,\ \hat{\boldsymbol{\theta}}_n \in R_{\pi n}^{(j)} \mid \boldsymbol{\theta}_{(n)}\}, \end{aligned} \tag{5.13}$$

and (2.11) can then be used to express (5.13) in terms of an appropriate multi-normal distribution over specific sectors of R^k; in this definition,

$$\boldsymbol{\tau}_{\pi 1}^0 = (\mathbf{I} - \mathbf{D}_\pi^{(1)})\, \boldsymbol{\tau} \qquad \text{and} \qquad \boldsymbol{\tau}_{\pi 2}^0 = \boldsymbol{\tau}^0 = \mathbf{D}_\pi \boldsymbol{\tau}, \qquad \text{for} \quad \pi \in \Pi. \tag{5.14}$$

Equation (5.7) also extends in a natural way to the case of the PTIME.

Let us next consider the asymptotic distribution theory of IME in a relatively more general setup. Recall that the restricted alternatives we have in mind relate to ω^*, defined in (3.1). This is a positively homogeneous cone in R_k, and the asymptotic distribution theory of the IME depends on whether $\boldsymbol{\theta}$ belongs to the interior of this cone or near any of its edges. Consider an l-dimensional subspace of ω^*, where for l positive integers $k_1 < \cdots < k_l = k$, we have

$$\theta_{k_{j-1}+1} = \cdots = \theta_{k_j} < \theta_{k_{j+1}}, \qquad \text{for} \quad j=1, ..., l; k_0 = 0; \qquad \text{and} \qquad \theta_{k_{l+1}} = \infty. \tag{5.15}$$

Here, l is a positive integer less than or equal to k. It is easy to verify that when $l = k$, i.e., the θ_i are all distinct and ordered, as n increases, the IME and classical M-estimator (based on the common score function) become equivalent, in probability. On the other hand, for every l: $1 \leqslant l \leqslant k-1$, the IME and ME are not equivalent in probability, and they have different asymptotic distributions. Keeping this in mind, we would like to study the asymptotic distribution theory of the IME when $\boldsymbol{\theta}$ belongs to (or lies on the boundary of) such a lower dimensional subspace of ω^*. We may, however, note that for $\theta_k > \theta_1$, the preliminary M-test considered in Section 3 is consistent, and hence, the PTIME and IME would have the same asymptotic behaviour for every l: $2 \leqslant l \leqslant k$. For $l = 1$, the picture has already been drawn earlier. Thus, there is no need to bring the PTIME into this asymptotic study.

Consider a partitioning of $\{1, ..., k\}$ into l subsets $[k_{j-1}+1, k_j]$, $j = 1, ..., l$, where the k_j are defined by (5.15), and $2 \leqslant l \leqslant k-1$. We denote the centroids of the θ-values within these subsets as $\theta^*_{(1)}, ..., \theta^*_{(l)}$, respectively. Consider then a sequence $\{H^*_{1(n)}\}$ of local alternatives:

$$H^*_{1(n)}: \theta_{k_{j-1}+1+r} = \theta^*_{(j)} + n^{-1/2}\tau^*_r,$$
$$\text{for} \quad r - 0, ..., k_j - k_{j-1} - 1, \quad j = 1, ..., l; \tag{5.16}$$

where the τ^*_r are all fixed numbers, and within each bucket, the τ^*_r are ordered. Note that by definition $\theta^*_{(1)} < \cdots < \theta^*_{(l)}$. We shall show that the asymptotic distribution of the normalized form of the IME exists and is different from that of the classical ME, for each of these local alternatives.

We denote by Π_l the subset of R^k for which

$$\max_{r \leqslant k_{j-1}} \hat{\theta}_{r,n_r} < \min_{k_{j-1} < r \leqslant k_j} \hat{\theta}_{r,n_r} \leqslant \max_{k_{j-1} < r \leqslant k_j} \hat{\theta}_{r,n_r} < \min_{r > k_j} \hat{\theta}_{r,n_r}, \qquad 1 \leqslant j \leqslant l, \tag{5.17}$$

where the $\hat{\theta}_{r,n_r}$ are the classical M-estimators of the θ_r. This subspace Π_l may then be partitioned into further subsets π_l: $\pi_l \in \Pi_l$, and these are defined as in after (4.4), but restricted to Π_l. We then refer to (5.5) where

$\boldsymbol{\theta}_{(n)}$ now belongs to the lower dimensional space in (5.16) (actually the boundary of an l-dimensional subset of ω^*). It is easy to show that under (5.16), $P\{\hat{\boldsymbol{\theta}}_n \in R_\pi \mid \boldsymbol{\theta}_{(n)}\}$ converges to 0 as n increases, for every π not belonging to Π_l. On the other hand, for π belonging to Π_l, $P\{n\mathbf{D}_\pi(\hat{\boldsymbol{\theta}}_n - \boldsymbol{\theta}_{(n)}) \leqslant \mathbf{x} + \boldsymbol{\tau}^{*0}, \hat{\boldsymbol{\theta}}_n \in R_\pi \mid \boldsymbol{\theta}_{(n)}\}$ has a nondegenerate limit, where the τ^{*0} are defined as in (5.4) with the τ_r being replaced by the τ_r^*. Thus, under (5.16), the asymptotic distribution function of the IME is given by

$$\lim_{n \to \infty} P\{n^{1/2}(\boldsymbol{\theta}_n^* - \boldsymbol{\theta}_{(n)}) \leqslant \mathbf{x} \mid (5.16)\}$$
$$= \lim_{n \to \infty} \left[\sum_{\pi \in \Pi_l} P\{n^{1/2}\mathbf{D}_\pi(\hat{\boldsymbol{\theta}}_n - \boldsymbol{\theta}_{(n)}) \leqslant \mathbf{x} + \boldsymbol{\tau}^{*0}, \hat{\boldsymbol{\theta}}_n \in R_\pi \mid (5.16)\} \right]. \tag{5.18}$$

It may be noted that for $l=1$, $\Pi_l = \Pi$ and (5.18) reduces to (5.5), while for $l \geqslant 2$, (5.18) involves a subset of the terms appearing in (5.5), and hence, the two forms are not isomorphic. In passing, we may remark that if (5.17) holds for $l=k$ then within each of the k buckets, there is only one element, and hence, Π_k consists of the cone $\hat{\theta}_{1,n_1} \leqslant \cdots \leqslant \hat{\theta}_{k,n_k}$. As such, (4.5) holds with $\mathbf{D}_\pi = \mathbf{I}$ with probability converging to 1 as $n \to \infty$. Thus, in this case, the classical M-estimator and the IME based on the same score function becomes asymptotically equivalent, in probability. Thus, (2.11) applies to the IME as well.

It is quite clear that the computation of the exact bias and mean product matrix of the IME and PTIME is highly involved; even the asymptotic case is not that simple to handle. For small values of k (viz., $k=3, 4$, etc.), term by term evaluation of (5.5) or (5.18) is possible, although the task becomes prohibitively laborious as k increases. For this reason, we take recourse to simulation studies of the relative bias and efficiency of the PTIME and IME. In this context, we interpret the relative efficiency (e^*) of the PTIME with respect to the IME in the usual way as the inverse ratio of the generalized variance of their respective asymptotic distributions.

6. Some Simulation Studies

We consider specifically the case of three samples (i.e., $k=3$) and for the M-estimators of location, we choose the Huber score function with $K=1.5$, i.e., we take

$$\psi(x) = \begin{cases} x, & |x| \leqslant K = 1.5, \\ K \operatorname{sign} x, & |x| > K. \end{cases} \tag{6.1}$$

All the samples are generated by random normal deviates with appropriate shifts in the location parameters. Since the M-estimators are translation-

TABLE I

Asymptotic Bias and Asymptotic Relative Efficiency of the PTE and IME under H_0

	Bias						
	PTIME Component			IME Component			Relative efficiency
n	1	2	3	1	2	3	e^*
10	0.0061	0.0136	0.0329	−0.1099	0.0058	0.1318	4.3311
15	0.0069	0.0141	0.0251	−0.0931	0.0074	0.1071	4.9139
20	0.0070	0.0130	0.0250	−0.0795	0.0068	0.0955	4.2977
25	0.0039	0.0094	0.0220	−0.0748	0.0021	0.0848	4.7852
30	0.0057	0.0095	0.0195	−0.0685	0.0043	0.0776	5.4195
35	0.0063	0.0103	0.0209	−0.0601	0.0053	0.0727	5.0517
40	0.0063	0.0101	0.0180	−0.0566	0.0062	0.0692	4.7341

equivariant, we have taken the location parameter of the first distribution as 0. All the three samples are taken to be of equal size (n) and various combinations of n and possibly uneven spacings of the location parameters. Tables I–VII pertain to the simulation results on the *bias* and *relative efficiency* (e^*) of the PTE and IME.

Recall that here n stands for the (equal) individual sample sizes, so that the combined sample size is $3n$. It is clear from Tables I and II that under the null hypothesis H_0 or for small departures from H_0, the PTIME performs better than the IME both in terms of the bias and mean product

TABLE II

Same Entries for $\boldsymbol{\theta} = (0, 0.1, 0.2)$ (i.e., Equally Spaced Means)

	Bias						
	PTIME Component			IME Component			Relative efficiency
n	1	2	3	1	2	3	e^*
10	0.0894	0.0111	−0.0428	−0.0704	0.0052	0.0928	2.0132
15	0.0822	0.0138	−0.0449	−0.0542	0.0068	0.0689	1.7756
20	0.0799	0.0119	−0.0421	−0.0425	0.0064	0.0588	1.5676
25	0.0749	0.0066	−0.0420	−0.0382	0.0046	0.0486	1.4826
30	0.0750	0.0064	−0.0439	−0.0338	0.0039	0.0433	1.1027
35	0.0740	0.0078	−0.0425	−0.0270	0.0052	0.0396	1.3054
40	0.0717	0.0077	−0.0438	−0.0237	0.0054	0.0371	1.2109

TABLE III

Average Bias and Relative Efficiency for $\boldsymbol{\theta} = (0, 0.05, 0.15)$ (Uneven Spacing)

	Bias						
	PTIME Component			IME Component			Relative efficiency e^*
n	1	2	3	1	2	3	
10	0.0636	0.0287	−0.0305	−0.0844	0.0158	0.0963	2.5935
15	0.0615	0.0308	−0.0376	−0.0678	0.0170	0.0723	2.3186
20	0.0614	0.0278	−0.0376	−0.0554	0.0165	0.0616	2.1086
25	0.0558	0.0235	−0.0345	−0.0506	0.0113	0.0513	2.0008
30	0.0580	0.0230	−0.0376	−0.0456	0.0132	0.0459	1.9718
35	0.0579	0.0237	−0.0365	−0.0381	0.0142	0.0418	1.8193
40	0.0565	0.0236	−0.0380	−0.0347	0.0144	0.0391	1.6909

TABLE IV

Same Entries for $\boldsymbol{\theta} = (0.\ 0.2, 0.5)$ (i.e., Uneven Spacings)

	Bias						
	PTIME Component			IME Component			Relative efficiency e^*
n	1	2	3	1	2	3	
10	0.1672	0.0325	−0.1174	−0.0395	0.0186	0.0485	1.2090
15	0.1446	0.0333	−0.1081	−0.0261	0.0182	0.0295	1.0649
20	0.1265	0.0293	−0.0906	−0.0188	0.0160	0.0256	0.9652
25	0.1103	0.0208	−0.0791	−0.0163	0.0093	0.0191	0.9274
30	0.0905	0.0223	−0.0641	−0.0153	0.0113	0.0174	0.7065
35	0.0671	0.0186	−0.0419	−0.0076	0.0107	0.0156	0.8390

TABLE V

Same Entries for $\boldsymbol{\theta} = (0, 0.5, 1.0)$ (i.e., Large Equal Spacing)

	Bias						
	PTIME Component			IME Component			Relative efficiency e^*
n	1	2	3	1	2	3	
10	0.1315	0.0084	−0.1045	−0.0019	0.0015	0.0280	0.7553
15	0.0756	0.0072	−0.0572	0.0031	0.0041	0.0143	0.7620
20	0.0387	0.0052	−0.0194	0.0045	0.0043	0.0139	0.7978
25	0.0211	0.0002	−0.0071	0.0013	−0.0012	0.0119	0.8278
30	0.0087	0.0039	0.0033	−0.0001	0.0023	0.0111	0.8743
35	0.0091	0.0037	0.0071	0.0037	0.0030	0.0112	0.9082
40	0.0063	0.0048	0.0095	0.0028	0.0043	0.0116	0.9499

TABLE VI

Average Bias and Relative Efficiency for $\boldsymbol{\theta} = (0, 0.2, 0,8)$
(Uneven Large Spacings)

	Bias						
	PTIME Component			IME Component			Relative efficiency
n	1	2	3	1	2	3	e^*
10	0.1379	0.0766	−0.1431	−0.0366	0.0404	0.0238	0.9315
15	0.0933	0.0597	−0.0978	−0.0241	0.0340	0.0117	0.8735
20	0.0546	0.0446	−0.0538	−0.0177	0.0283	0.0121	0.8454
25	0.0314	0.0302	−0.0311	−0.0159	0.0174	0.0106	0.8580
30	0.0163	0.0259	−0.0137	−0.0149	0.0179	0.0104	0.8732
35	0.0118	0.0216	−0.0044	−0.0089	0.0161	0.0107	0.8947
40	0.0066	0.0194	−0.0008	−0.0075	0.0149	0.0113	0.8940

matrix-risk. Also, the bias of the PTIME and IME are not in concordance with each other. A somewhat diferent picture emerges in the uneven spacing case and for alternatives not so close to the null one. The last three tables indicate the superiority of the IME to PTIME. This is not surprising: We have both uneven spacings and moderate deviations from the null hypothesis. Thus, for alternatives close to the null hypothesis (of the homogeneity of the θ_i), the PTIME performs better than the IME, while the opposite picture hols when $\boldsymbol{\theta}$ moves away from the line of homogeneity. In any case, if $\boldsymbol{\theta}$ is too far away from this line, the PTIME and IME both

TABLE VII

Same Entries for $\boldsymbol{\theta} = (0, 0.5, 1.5)$ (i.e., Large Uneven Spacings)

	Bias						
	PTIME Component			IME Component			Relative efficiency
n	1	2	3	1	2	3	e^*
10	0.0349	0.0187	−0.0182	−0.0018	0.0128	0.0166	0.8411
15	0.0126	0.0120	0.0010	0.0031	0.0099	0.0085	0.9319
20	0.0077	0.0086	0.0082	0.0045	0.0073	0.0110	0.9615
25	0.0026	0.0019	0.0094	0.0013	0.0011	0.0096	0.9727
30	0.0006	0.0050	0.0103	−0.0001	0.0035	0.0100	0.9629
35	0.0043	0.0045	0.0111	0.0037	0.0036	0.0105	0.9717
40	0.0038	0.0054	0.0113	0.0028	0.0046	0.0113	0.9596

perform very similarly. Moreover, the PTIME is never too inefficient relative to the IME, although it can be considerably more efficient (see Table I). Thus, the PTIME can be posed as an efficiency-robust competitor of the usual IME. For some further numerical studies, we refer to Karmous [8].

Acknowledgments

We are grateful to Professor Vernon M. Chinchilli for his most critical reading of the manuscript leading to various improvements.

References

[1] Barlow, R. E., Bartholomew, D. J., Bremner, J. M., and Brunk, H. D. (1972). *Statistical Inference under Order Restrictions.* Wiley, New York.

[2] Boyd, M. N., and Sen, P. K. (1983). Union–intersection rank tests for ordered alternatives in some simple linear models. *Comm. Statist.—Theory Methods* **12** 1737–1758.

[3] Chinchilli, V. M., and Sen, P. K. (1981) Multivariate linear rank statistics and the union–intersection principle for hypothesis testing under restricted alternatives. *Sankhyā Ser. B* **43** 135–151.

[4] Chinchilli, V. M., and Sen, P. K. (1981). Multivariate linear rank statistics and the union–intersection principle for the orthant restriction problem. *Sankhyā Ser. B* **43** 152–171.

[5] De, N. (1975). Rank tests for randomized blocks against ordered alternatives. *Calcutta Statist. Assoc. Bull.* **25** 1–27.

[6] Gupta, S. S. (1963). Probability integrals of multivariate normal and multivariate *t*. *Ann. Math. Statist.* **34** 792–838.

[7] Jurečková, J. (1977). Asymptotic relations of *M*-estimates and *R*-estimates in linear regression models. *Ann. Statist.* **5** 464–472.

[8] Karmous, A. R. (1986). *Robust, Isotonic and Preliminary Test Estimators in Some Linear Models under Rrestraints.* Doctoral dissertation. University North Carolina, Chapel Hill, NC.

[9] Leurgans, S. (1986). Isotonic *M*-estimation. In *Advances in Order Restricted Statistical Inference* (R. Dykstra, T. Robertson, and F. T. Wright, Eds.), pp. 48–68, Springer-Verlag, New York.

[10] Sen, P. K., and Saleh, A. K. M. E. (1987). On preliminary test and shrinkage *M*-estimation in linear models. *Ann. Statist.* **15** 1580–1592.

[11] Singer, J. M., and Sen, P. K. (1985). *M*-methods in multivariate linear models. *J. Multivariate Anal.* **17** 168–184.

[12] Tsai, M.-T. M., and Sen, P. K. (1987). Asymptotic distribution of UI-LMPR tests for restricted alternatives. Submitted for publication.

Some Asymptotic Inferential Problems Connected with Elliptical Distributions

C. G. KHATRI*

Gujarat University, Ahmedabad, India

Asymptotic confidence bounds on the location parameters of the linear growth curve, asymptotic distribution of the canonical correlations and asymptotic confidence bounds on the discriminatory value for the linear discriminant function are established when a set of independent observations are taken from an elliptical distribution (or from a distribution possessing some properties on the moments).

1. INTRODUCTION

Exact confidence bounds on the location parameters of the linear growth curve model,

$$X = B\xi A' + \varepsilon; \qquad \text{column vectors of } \varepsilon \text{ being } \mathrm{IN}(\mathbf{0}, \Sigma),$$

were given by Khatri [2]. What will happen to the confidence bounds when the column vectors of ε are independent and have a common elliptical distribution instead of normal distribution? This question is answered using the well-known asymptotic theory based on central limit theorem or the convergence theorem. For this problem, we require the asymptotic joint distribution of

$$(Z - B\xi)\sqrt{n} \qquad \text{and} \qquad \sqrt{n}\left(S - \frac{n-m}{n} b_1 \Sigma\right)\Big/ b_1, \tag{1.1}$$

where $S = (XX' - ZA'AZ')/n$ and $Z = XA(A'A)^{-1}$. Here, A and B are assumed to be of full rank matrices (i.e., $A'A$ and $B'B$ are nonsingular), $m = \text{Rank } A$, and b_1 is a constant depending on the structure of the

* Deceased on March 31, 1989.

Multivariate Statistics and Probability
ISBN 0-12-580205-6

Reprinted from *J. Mult. Anal.* **27**(2).

(elliptical) distribution. Further, the following assumptions on the $n \times m$ matrix A are made for large n:

(i) elements of A are finite so that the elements of AA' are finite and (1.2a)

(ii) the limit of $(A'A/n)$ for large n tends to a nonsingular matrix C. (1.2b)

The above two conditions are essential for the application of the central limit theorem. The asymptotic normality results are established in Section 2 and Section 3 justifies the asymptotic confidence bounds on ξ similar to those mentioned by Khatri [2] based on $\hat{\xi} = (B'S^{-1}B)^{-1} B'S^{-1}Z$.

Since the sample canonical correlations between the two sets of variables depend on the elements of S, we consider the problem of establishing the asmptotic distribution of canonical correlations similar to normal variates. This was first established by Krishnaiah *et al.* [5] for the elliptical variates. We reestablish this for a wider class of distribution in Section 4.

In a particular case, the matrix Z and S have been utilized by Khatri *et al.* [4] in the study of performance of linear discriminant function for the normal variates and developed the asmptotic results concerning the confidence bounds on the discriminatory values in different situations when $B = I$ and $m = 2$. If $\xi = (\boldsymbol{\mu}_1, \boldsymbol{\mu}_2)$ and $Z = (\bar{\mathbf{x}}_1, \bar{\mathbf{x}}_2)$, then the discriminatory value of linear Fisher's discriminant function $\mathbf{w}'\mathbf{x}$ (or $\mathbf{w}'\mathbf{x} + c$) for the future observation $\mathbf{x}$ is

$$D_w = [E(\mathbf{w}'\mathbf{x} \mid \pi_1) - E(\mathbf{w}'\mathbf{x} \mid \pi_2)]/\mathrm{Var}(\mathbf{w}'\mathbf{x}))^{1/2},$$

where π_i is the population having the mean $\boldsymbol{\mu}_i$ and the covariance matrix Σ, so that

$$D_w = w'(\boldsymbol{\mu}_1 - \boldsymbol{\mu}_2)/(\mathbf{w}'\Sigma\mathbf{w})^{1/2}$$

which is a function of unkown parameters. The three situations considered for Khatri *et al.* [4] are based on the following situations:

(i) $\boldsymbol{\mu}_1 - \boldsymbol{\mu}_2$ is known, Σ is unknown, and $\mathbf{w} = S^{-1}(\boldsymbol{\mu}_1 - \boldsymbol{\mu}_2)$,

(ii) $\boldsymbol{\mu}_1 - \boldsymbol{\mu}_2$ is unknown, Σ is known, and $\mathbf{w} = \Sigma^{-1}(\bar{\mathbf{x}}_1 - \bar{\mathbf{x}}_2)$, and

(iii) $\boldsymbol{\mu}_1, \boldsymbol{\mu}_2$, and Σ are unknown and $\mathbf{w} = S^{-1}(\bar{\mathbf{x}}_1 - \bar{\mathbf{x}}_2)$,

giving rise to the three functions D'_a, D''_a, and D_a (for D_w), respectively. Asymptotic confidence bounds on these values similar to those for normal variates are established for elliptical variates in Section 5.

Thus, it appears that in the problems where Z and S are utilized, one can develop the asymptotic results similar to those developed for the above three types of problems. Here, we mention that similar results for the complex elliptical distributions are available but will be presented in a later communication.

2. Asymptotic Distribution of Z and S

Let $\mathbf{y}$ be a random vector such that

$$Ey = \mathbf{0}, \qquad E(\mathbf{y}\mathbf{y}') = b_1 I_p, \qquad E(y_i y_j y_k) = 0 \qquad \text{for all } i, j, k,$$
$$Ey_i^4 = 3b_2, \qquad E(y_i^2 y_j^2) = b_2 \qquad \text{for} \quad i \neq j,$$
$$\text{and all other } E(y_i y_j y_k y_l) = 0, \tag{2.0}$$

where y_i denotes the ith component of $\mathbf{y}$. We observe that if $\mathbf{y}$ has spherical distribution or its characteristic function (c.f.) is $\psi(\sum_{i=1}^{p} t_i^2)$ and the first four moments exist, then the moment relations (2.0) hold with $b_1 = -2\psi'(0)$ and $b_2 = 4\psi''(0)$. It may be noted that the moment relations (2.0) may be true for the wider class of distributions including spherical ones. Suppose $\mathbf{x}$ is a random vector such that $E\mathbf{x} = \boldsymbol{\mu}$ and $\operatorname{Var} \mathbf{x} = E(\mathbf{x} - \boldsymbol{\mu})(\mathbf{x} - \boldsymbol{\mu})' = \Sigma$ is positive definite and $\mathbf{y} = \Sigma_1^{-1}(\mathbf{x} - \boldsymbol{\mu})$ satisfies the moment relations given in (2.0). Here $\Sigma = \Sigma_1 \Sigma_1'$ and Σ_1 is nonsingular. These conditions are satisfied for the elliptical distribution whose c.f. is

$$\exp(\sqrt{1}\, \mathbf{t}'\boldsymbol{\mu})\, \psi(\mathbf{t}'\Sigma\mathbf{t}) \qquad \text{for all } \mathbf{t} \in \mathscr{R}^p,$$

and this is denoted by $\mathbf{x} \sim E_p(\boldsymbol{\mu}, \Sigma; \psi)$, an elliptical distribution.

Let there be n independent observations on $\mathbf{y}$ whose distribution function $G((\mathbf{y})$ satisfies (2.0) and let us define

$$W = \sum_{i=1}^{n} (\mathbf{y}_i \mathbf{y}_i' - b_1 I_p)/\sqrt{n}\, b_1 \qquad \text{with} \quad n > p. \tag{2.1}$$

Let $W = (w_{ij})$, $w_1' = (w_{11}, w_{22}, ..., w_{pp})$, $w_2' = (w_{12}, w_{13}, ..., w_{1p}, w_{23}, ..., w_{2p}, ..., w_{p-1,p})$ and $\mathbf{w}'(\mathbf{w}_1', \mathbf{w}_2')$. Let vec W be defined as the column vector obtained by putting vectors one by one; (i.e., if $W = (\mathbf{v}_1, \mathbf{v}_2, ..., \mathbf{v}_p)$, then $(\operatorname{vec} W)' = (\mathbf{v}_1', \mathbf{v}_2', ..., \mathbf{v}_p')$). Notice that from (2.1), we have

$$\operatorname{vec} W = \bar{\mathbf{z}}\sqrt{n} \qquad \text{with} \quad \mathbf{z}_i = (\mathbf{y}_i \otimes \mathbf{y}_i - b_1 \operatorname{vec} I_p)/b_1 \quad \text{and} \quad \bar{\mathbf{z}} = \sum_{i=1}^{n} \mathbf{z}_i/n,$$

where $A \otimes B$ denotes the Kronecker product of A with B and is defined by $(a_{ij}B)$ if $A = (a_{ij})$. Using the central limit theorem for independent and identically distributed random variables, z's (see, for example, Cramér [1, pp. 213–217]), we see that

$$\sqrt{n}\,\bar{\mathbf{z}} \qquad \text{is asymptotically normal}$$

which is equivalent to the statement that $\mathbf{w}_1$ and $\mathbf{w}_2$ are asymptotically independent normal variates and

$$\mathbf{w}_2 \sim N(\mathbf{0}, (\kappa+1)\, I_{p(p-1)/2}) \qquad \text{and} \qquad \mathbf{w}_1 \sim N(\mathbf{0}, \Sigma_0), \tag{2.2}$$

where $\kappa+1=b_2/b_1^2$, and $\Sigma_0=2(\kappa+1)\, I_p+\kappa \mathbf{1}_p \mathbf{1}_p'$ with $\mathbf{1}_p$ being a p-vector of unit elements.

Assume that the column vectors of ε in (1.1) are independently distributed such that if $\Sigma=\Sigma_1\Sigma_1'$, Σ_1 is nonsingular and

$$(\mathbf{y}_1, ..., \mathbf{y}_n)=Y=\Sigma_1^{-1}\varepsilon=\Sigma_1^{-1}(X-B\xi A) \tag{2.3}$$

then $\mathbf{y}_i$ $(i=1, 2, ..., n)$ are identical and independent and satisfy the moment conditions (2.0).

If $Z_1=YA(A'A)^{-1}=YAC_{1n}^{-1}/n=\sum_{i=1}^n \mathbf{y}_i\mathbf{d}_i'/n$, where $C_{1n}=A'A/n$ and $C_{1n}^{-1}A'=(\mathbf{d}_1, \mathbf{d}_2, ..., \mathbf{d}_n)$, then

$$\operatorname{vec} Z_1=\sum_{i=1}^{n} (\mathbf{d}_i \otimes \mathbf{y}_i)/n, \qquad E \operatorname{vec} Z_1=0, \tag{2.4}$$

$$\begin{aligned} En(\operatorname{vec} Z_1)(\operatorname{vec} Z_1)' &= b_1 \sum_{i=1}^{n} (\mathbf{d}_i\mathbf{d}_i' \otimes I_p)/n \\ &= b_1(C_{1n}^{-1} \otimes I_p) \to b_1(C^{-1} \otimes I_p) \end{aligned} \tag{2.4a}$$

as $n \to \infty$, using assumption (ii) of (1.2).

In order to use, the Lyapunov's theorem for independent random variables (see, for example, Cramér [1, p. 215–217]), we observe that

$$\sum_{i=1}^{n} E\,|\mathbf{d}_i \otimes \mathbf{y}_i|^3/n^{3/2} \to 0 \qquad \text{as} \quad n \to \infty \tag{2.5}$$

because

$$E\,|\mathbf{d}_i \otimes \mathbf{y}_i|^3=E(\mathbf{y}'\mathbf{y})^{3/2}(\mathbf{d}_i'\mathbf{d}_i)^{3/2}$$

$$\mathbf{d}_i=C_{1n}^{-1}\mathbf{f}_i, \qquad A'=(\mathbf{f}_1, \mathbf{f}_2, ..., \mathbf{f}_n),$$

$$\mathbf{d}_i'\mathbf{d}_i=\mathbf{f}_i' C_{1n}^{-2}\mathbf{f}_i \leqslant \lambda_{1n}^2 \mathbf{f}_i'\mathbf{f}_i \leqslant M\lambda_{1n}^2$$

with $\lambda_{1n}=$ maximum eigen value of (C_{1n}^{-1}), and

$$\sum_{i=1}^{n} (\mathbf{d}_i'\mathbf{d}_i)^{3/2}/n^{3/2} \leqslant \lambda_{1n}^3 M^{3/2}/n^{1/2} \to 0 \qquad \text{as} \quad n \to \infty,$$

where $M = \max_i(\mathbf{f}_i'\mathbf{f}_i)$ is finite by assumption (i) of (1.2), and $lt_{n\to\infty} \lambda_{1n}$ = maximum eigen value of C^{-1} (by assumption (ii)). Hence

$$\sqrt{n}\,\text{vec}\, Z_1 = \sqrt{n}\,\text{vec}(YA(A'A)^{-1}) \overset{\text{asy}}{\simeq} N(\mathbf{0}, b_1(C^{-1} \otimes I_p)). \tag{2.5}$$

Further, using (2.1) and (2.4), we have

$$\begin{aligned} &\text{Cov}(\text{vec}\, W, \sqrt{n}\,\text{vec}\, Z_1) \\ &= \sum_{i=1}^{n} \text{Cov}(\mathbf{y}_i \otimes \mathbf{y}_i - b_1 \,\text{vec}\, I_p, \mathbf{d}_i \otimes \mathbf{y}_i)/nb_1 = 0. \end{aligned} \tag{2.6}$$

Hence W and $\sqrt{n}\,\text{vec}\, Z_1$ are stochastically independent normal variates. Now since

$$\begin{aligned} nS &= XX' - XA(A'A)^{-1} A'X' \\ &= \Sigma_1 YY'\Sigma_1' - \Sigma_1 YA(A'A)^{-1} A'Y'\Sigma_1', \end{aligned}$$

we get

$$\begin{aligned} &\sqrt{n}\left(\Sigma_1^{-1} S \Sigma_1'^{-1} - \frac{n-m}{n} b_1 I_p\right)\Big/ b_1 \\ &= W - b_1^{-1}[\sqrt{n}\, Z_1(A'A/n)\, Z_1' \sqrt{n} - mb_1 I_p]/\sqrt{n}. \end{aligned} \tag{2.7}$$

We observe that

$$\underset{n\to\infty}{\text{Plim}}\, [(\sqrt{n}\, Z_1)(A'A/n)(\sqrt{n}\, Z_1)' - mb_1 I_p] = \underset{n\to\infty}{\text{Plim}}\, (T_n) = 0 \quad \text{(say)}. \tag{2.8}$$

Hence, from (2.6), (2.7), and (2.8), we get

THEOREM 1. *Let* $\mathbf{x}_i \sim IE_p(\boldsymbol{\mu}_i, \Sigma; \psi)$ $(i = 1, 2, ..., n)$. *Then,* $(Z - B\xi)\sqrt{n}$ *and* $\sqrt{n}\,(S - (n-m)n^{-1} b_1 \Sigma)/b_1$ *are asymptotic independent, and are normally distributed, under the assumptions* (i) *and* (ii) *of* (1.2).

Further,

$$\sqrt{n}\,\text{vec}(Z - B\xi) \overset{\text{asy}}{\simeq} N(0, b_1(C^{-1} \otimes \Sigma))$$

and

$$\sqrt{n}\,\text{vec}\left(S - \frac{n-m}{n} b_1 \Sigma\right)\Big/ b_1 \overset{\text{asy}}{\simeq} N(0, \Sigma_2)$$

with

$$\Sigma_2 = \text{Var}[(\mathbf{x} \otimes \mathbf{x})\, b_1^{-1} - \text{vec}\, \Sigma] \qquad \textit{and} \qquad \mathbf{x} \sim E_p(\mathbf{0}, \Sigma; \psi).$$

Here $Z = XA(A'A)^{-1}$, $S = (XX' - ZA'AZ')/n$, *and* $X = (\mathbf{x}_1, \mathbf{x}_2, ..., \mathbf{x}_n)$. *From* (2.7)

$$(\Sigma_1^{-1} S \Sigma_1'^{-1})^{-1} = \left(\frac{b_1 W}{\sqrt{n}} + \frac{n-m}{n} b_1 I - \frac{1}{n} T_n\right)^{-1}$$

$$= \left(\frac{n b_1^{-1}}{n-m}\right)\left[I + \frac{1}{\sqrt{n}} W + \frac{1}{n} T_{1n}\right]^{-1}$$

where $T_{1n} = -T_n[n/b_1(n-m)] + (n/(n-m))(W/\sqrt{n})$. *Then with* $B_1 = \Sigma_1^{-1} B$, $\hat{\xi} - \xi = (B'S^{-1}B)^{-1} B'S^{-1}(Z - B\xi) = [B_1'(I + (1/\sqrt{n}) W + (1/n) T_{1n})^{-1} B_1]^{-1} B_1'(I + (1/\sqrt{n}) W + (1/n) T_{1n})^{-1} Z_1$, *or*

$$(\hat{\xi} - \xi) = (B_1' B_1)^{-1} B_1' Z_1 + O(1/n) \tag{2.9}$$

and for $\hat{\Sigma} = S + (I - B(B'S^{-1}B)^{-1} B'S^{-1}) Z C_{1n} Z'(I - S^{-1}B(B'S^{-1}B)^{-1} B')$,

$$\sqrt{n}\,(\Sigma_1^{-1} \hat{\Sigma} \Sigma_1'^{-1} - b_1 I_p)/b_1 = W + O(1/\sqrt{n}). \tag{2.10}$$

From (2.9) *and* (2.10), *we have*

THEOREM 2. *With the notations of Theorem* 1, $\sqrt{n}\,(\hat{\xi} - \xi)$ *and* $\sqrt{n}\,(\hat{\Sigma} - b_1 \Sigma)/b_1$ *are asymptotic independent,*

$$\sqrt{n}\,\mathrm{vec}(\hat{\xi} - \xi) \overset{\text{asy}}{\simeq} N(\mathbf{0}, b_1 C^{-1} \otimes (B'\Sigma^{-1}B)^{-1})$$

and

$$\sqrt{n}\, b_1^{-1}\,\mathrm{vec}(\hat{\Sigma} - b_1 \Sigma) \overset{\text{asy}}{\simeq} N(\mathbf{0}, \Sigma_2),$$

where

$$\Sigma_2 = \mathrm{Var}[(\mathbf{x} \otimes \mathbf{x})\, b_1^{-1} - \mathrm{vec}\,\Sigma] \qquad \text{and} \qquad \mathbf{x} \sim E_p(0, \Sigma; \psi).$$

3. ASYMPTOTIC CONFIDENCE BOUNDS ON ξ

Let us consider the nonzero eigen values $l_1 > l_2 > \cdots > l_t > 0$ of

$$n(B'\hat{\Sigma}^{-1}B)(\hat{\xi} - \xi)(A'A/n)(\hat{\xi} - \xi)'/b_1 = T_n \qquad \text{(say)}, \tag{3.1}$$

where $t = \min(q, m)$. We observe that

$$\operatorname*{Plim}_{n \to \infty} (B'\hat{\Sigma}^{-1}B) = (B'\Sigma^{-1}B) \qquad \text{by Theorem 2}$$

and by assumption (ii), $\lim_{n \to \infty}(A'A/n) = C$. Then, the asymptotic distribution of $l_1, l_2, ..., l_t$ is the same as the eigenvalues of the $t \times t$ Wishart matrix V distributed as $W_t(u, I_t)$, where $u = \max(q, m)$. For this, one can obtain the asymptotic distribution of l_1, or the asymptotic distribution of $\sum_{i=1}^{t} l_i = \operatorname{tr} T_n$. Suppose,

$$P(l_1 \leqslant c_\alpha) = 1 - \alpha.$$

Then for all non-null vectors $\mathbf{a} \in \mathscr{R}^r$ and $\mathbf{b} \in \mathscr{R}^m$,

$$|\mathbf{a}'(\hat{\xi} - \xi)\,\mathbf{b}|^2/\{(\mathbf{b}'(A'A)^{-1}\,\mathbf{b})(\mathbf{a}'(B'\hat{\Sigma}^{-1}B)^{-1}\,\mathbf{a})\} \leqslant c_\alpha b_1.$$

or the simultaneous confidence bounds for $\mathbf{a}'\xi\mathbf{b}$ for all $\mathbf{a} \in \mathscr{R}^r$ and $\mathbf{b} \in \mathscr{R}^m$ are

$$\mathbf{a}'\hat{\xi}\mathbf{b} \pm \{b_1 c_\alpha(\mathbf{b}'(A'A)^{-1}\mathbf{b})(\mathbf{a}'(B'\hat{\Sigma}^{-1}B)^{-1}\mathbf{a})\}^{1/2}. \tag{3.2}$$

We can use $\operatorname{tr} T_n \leqslant c_{1\alpha}$ to find the confidence bounds on ξ.

4. Asymptotic Distribution of Canonical Correlations

In this section, we shall consider without loss of generality,

$$\Sigma = \begin{pmatrix} I_{p_1} & D_\rho & 0 \\ D_\rho & I_{p_1} & 0 \\ 0 & 0 & I_{p_2 - p_1} \end{pmatrix}, \qquad D_\rho = \operatorname{diag}(\rho_1 I_{g_1}, ..., \rho_{k-1} I_{g_{k-1}}, \rho_k I_{g_k})$$

with $\sum_{i=1}^{k} g_i = p_1$, $p_2 > p_1$, and $\rho_1 > \rho_2 > \cdots > \rho_{k-1} > \rho_k = 0$, its estimate $\hat{\Sigma}$ and the asymptotic distribution of $\hat{\Sigma}$ as given in Theorem 2. Let us write

$$\Sigma_1 = \left(\begin{array}{c|c|c} D_1 & D_\rho & 0 \\ \hline 0 & I_{p_1} & 0 \\ \hline 0 & 0 & I_{p_2 - p_1} \end{array}\right) \qquad \text{with} \quad D_1 = \operatorname{diag}(\sqrt{1 - \rho_1^2}\, I_{g_1}, ..., \sqrt{1 - \rho_k^2}\, I_{g_k})$$

and

$$W = \sqrt{n}\,(\Sigma_1^{-1}\hat{\Sigma}\Sigma_1'^{-1} - b_1 I_p)/b_1 \qquad \text{or} \qquad b_1[n^{-1/2}\Sigma_1 W \Sigma_1' + \Sigma] = \hat{\Sigma}.$$

Let us partition $\hat{\Sigma}$ and W as

$$\hat{\Sigma} = \begin{pmatrix} \hat{\Sigma}_2 & \hat{\Sigma}_3 \\ \hat{\Sigma}_3' & \hat{\Sigma}_4 \end{pmatrix} \begin{matrix} p_1 \\ p_2 \end{matrix} \qquad \text{and} \qquad W = \begin{pmatrix} W_1 & W_2 \\ W_2' & W_3 \end{pmatrix} \begin{matrix} p_1 \\ p_2. \end{matrix}$$
$$\quad p_1 \quad p_2 \qquad\qquad\qquad\qquad p_1 \quad p_2$$

Then

$$\hat{\Sigma}_2 = b_1 \left[\frac{D_1 W_1 D_1 + (D_\rho, 0) W_2' D_1 + D_1 W_2 \begin{pmatrix} D_\rho \\ 0 \end{pmatrix} + (D_\rho, 0) W_3 \begin{pmatrix} D_\rho \\ 0 \end{pmatrix}}{\sqrt{n}} + I_{p_1} \right],$$

$$\hat{\Sigma}_3 = b_1 \left[\frac{D_1 W_2 + (D_\rho, 0) W_3}{\sqrt{n}} + (D_\rho, 0) \right] \quad \text{and} \quad \hat{\Sigma}_4 = b_1 \left[\frac{1}{\sqrt{n}} W_3 + I_{p_2} \right].$$

Then

$$\begin{aligned} b_1^{-1} P(r) &= (r^2 \hat{\Sigma}_2 - \hat{\Sigma}_3 \hat{\Sigma}_4^{-1} \hat{\Sigma}_3')/b_1 \\ &= (r^2 I_{p_1} - D_\rho^2) + \frac{1}{\sqrt{n}} \left[r^2 D_1 W_1 D_1 + (D_\rho, 0) W_2' D_1 (r^2 - 1) \right. \\ &\quad \left. + (r^2 - 1) D_1 W_2 \begin{pmatrix} D_\rho \\ 0 \end{pmatrix} + (r^2 - 1)(D_\rho, 0) W_3 \begin{pmatrix} D_\rho \\ 0 \end{pmatrix} \right] \\ &\quad - \frac{1}{n} (W_2 W_2') + O(n^{-3/2}). \end{aligned} \tag{4.1}$$

Let us denote

$$w_\alpha = \sqrt{n}\,(r - \rho_\alpha)/(1 - \rho_\alpha^2) \qquad \text{for} \quad \alpha = 1, 2, \ldots, k-1 \tag{4.2}$$

$$\text{and} \qquad w_k = \sqrt{n}\, r. \tag{4.2a}$$

If $P(r) = (P_{\alpha\alpha'}$ for $\alpha, \alpha' = 1, 2, \ldots, k)$ and $P_{\alpha\alpha'}$ is a $g_\alpha \times g_{\alpha'}$ sub-matrix of $P(r)$, then

$$\begin{aligned} \sqrt{n}\, P_{\alpha\alpha}/b_1 &= 2\rho_\alpha (1 - \rho_\alpha^2)\, w_\alpha I_{g_\alpha} + \rho_\alpha (1 - \rho_\alpha^2)[\rho_\alpha (W_{1,\alpha\alpha} - W_{3,\alpha\alpha}) \\ &\quad - (W_{2,\alpha\alpha}' + W_{2,\alpha\alpha})(1 - \rho_\alpha^2)^{1/2}] + O(n^{-1/2}) \\ &\qquad \text{for} \quad \alpha = 1(1)\,k-1, \end{aligned} \tag{4.3}$$

$$n P_{kk}/b_1 = w_k I_{g_k} - (W_2 W_2')_{kk} + O(n^{-1/2}), \tag{4.4}$$

$$\begin{aligned} P_{\alpha\alpha'}/b_1 &= O(n^{-1/2}) \qquad \text{for} \qquad \alpha \neq \alpha' (\leqslant k-1) \\ (P_{\alpha k}/b_1 &\text{ or } P_{k\alpha}/b_1) = O(n^{-1}). \end{aligned} \tag{4.5}$$

From (4.3) to (4.5), it is obvious that

$$\begin{aligned} |P(r)| = 0 \Rightarrow & \left\{ \prod_{\alpha=1}^{k-1} |w_\alpha I_{g_\alpha} - B_\alpha + O(n^{-1/2})| \right\} \\ & \times |w_k^2 I_{g_k} - B_k + O(n^{-1/2})| = 0, \end{aligned} \tag{4.6}$$

where $B_k = (W_2 W_2')_{kk}$ is a submatrix of order $g_k \times g_k$ obtained from $W_2 W_2'$ by taking the last g_k rows and g_k columns, and

$$B_\alpha = \tfrac{1}{2}[(1-\rho_\alpha^2)^{1/2}(W'_{2,\alpha\alpha} + W_{2,\alpha\alpha}) - \rho_\alpha(W_{1,\alpha\alpha} - W_{3,\alpha\alpha})] \tag{4.7}$$

for $\alpha = 1, 2, ..., k-1$. Let $B_\alpha = (b_{\alpha,ij}; i, j = 1, 2, ..., g_\alpha)$. Then

$$b_{\alpha,ii} = (1-\rho_\alpha^2)^{1/2}\, w_{2,\alpha\alpha,ii} - \tfrac{1}{2}(w_{1,\alpha\alpha,ii} - w_{3,\alpha\alpha,ii})\,\rho_\alpha$$

and

$$b_{\alpha,ij} = \tfrac{1}{2}(1-\rho_\alpha^2)^{1/2}(w_{2,\alpha\alpha,ij} + w_{2,\alpha\alpha,ji}) - \tfrac{1}{2}\rho_\alpha(w_{1,\alpha\alpha,ij} - w_{3,\alpha\alpha,ij})$$

for $i \neq j$, $i, j = 1, 2, ..., g_\alpha$.

We observe that $w_{1,\alpha\alpha,ii} - w_{3,\alpha\alpha ii}$ and $w_{1,\alpha'\alpha'i'i'} - w_{3,\alpha'\alpha'ii'}$ (for $\alpha \neq \alpha'$ or $i \neq i'$) are asymptotic independent, and hence B_α $(\alpha = 1, 2, ..., k-1)$ and B_k are asymptotic independent, B_α is symmetric, and the elements are independent normals or the joint density of the elements of B_α is

$$2^{-g_\alpha/2}(\pi(\kappa+1))^{-g_\alpha(g_\alpha+1)/2} \exp[-\operatorname{tr} B_\alpha^2/2(\kappa+1)] \tag{4.8}$$

and $B_k \overset{\text{asy}}{\simeq} W_g(p_2, (\kappa+1) I_{g_k})$. Now, if $r_1 > r_2 > \cdots > r_{p_1} > 0$ are the sample canonical correlations (or the square root of the eigen values of $\hat{\Sigma}_{11}^{-1}\hat{\Sigma}_{12}^2\hat{\Sigma}_{22}^{-1}\hat{\Sigma}_{12}'$) and $r_{(\alpha)j} = r_{g_1 + \cdots + g_{\alpha-1}+j}$ for $j = 1, 2, ..., g_\alpha$ and $\alpha = 1, 2, ..., k$ with $g_0 = 0$, we see that

$$w_{\alpha,j} = \sqrt{n}\,(r_{(\alpha)j} - \rho_\alpha)/(1-\rho_\alpha^2), \qquad j = 1, 2, ..., g_\alpha$$

are the eigen values of B_α (for $\alpha = 1, 2, ..., k-1$), while $w_{k,j}^2 = r_{(k)j}^2$ are the eigen values of B_k. These distributions can be easily obtained from (4.8). In particular, if all the population canonical correlations are nonzero and they are of multiplcity one, then

$$\sqrt{n}\,(r_j - \rho_j)/(1-\rho_j^2) \overset{\text{asy}}{\simeq} IN(0, \kappa+1), \qquad j = 1, 2, ..., p_1. \tag{4.9}$$

These results are similar to those of Krishnaiah *et al.* [5] and Khatri [3] but here we have given a simple proof.

5. Asymptotic Confidence Bounds on Discriminatory Values

Let us denote $A = \begin{pmatrix} \mathbf{1}_{n_1} & 0 \\ 0 & \mathbf{1}_{n_2} \end{pmatrix}$, $n = n_1 + n_2$, $m = 2$, $B = I_p$, $XA(A'A)^{-1} = (\bar{\mathbf{x}}_1, \bar{\mathbf{x}}_2)$ and $\hat{\Sigma} = S = (1/n)[\sum_{j=1}^{n_1} (\mathbf{x}_{1j} - \bar{\mathbf{x}}_1)(\mathbf{x}_{1j} - \bar{\mathbf{x}}_1)' + \sum_{j=1}^{n_2} (\mathbf{x}_{2j} - \bar{\mathbf{x}}_2)$

$(\mathbf{x}_{2j}-\bar{\mathbf{x}}_2)']$. We shall assume that n_1 and n_2 are large so that $\lim_{n\to\infty} n_1/n = \kappa_0$ is fixed and constant. Let $\Sigma=\Sigma_1\Sigma_1'$ and $\boldsymbol{\delta}=\Sigma_1^{-1}(\boldsymbol{\mu}_1-\boldsymbol{\mu}_2)$.

Case (i). When $\boldsymbol{\delta}_1=\boldsymbol{\mu}_1-\boldsymbol{\mu}_2$ is known but Σ is unknown, then

$$D_a'=\boldsymbol{\delta}_1'S^{-1}\boldsymbol{\delta}_1/(\boldsymbol{\delta}_1'S^{-1}\Sigma S^{-1}\boldsymbol{\delta}_1)^{1/2} \quad \text{and} \quad D'=(\boldsymbol{\delta}_1'S^{-1}\boldsymbol{\delta}_1)^{1/2}. \tag{5.1}$$

We know from Section 2 that if

$$W=\sqrt{n}\,(\Sigma_1^{-1}S\Sigma_1'^{-1}-b_1I)/b_1=(w_{ij}) \quad \text{or} \quad S/b_1=(\Sigma_1W\Sigma_1'/\sqrt{n})+\Sigma,$$

then $(w_{11},\ldots,w_{pp})$ and w_{ij}'s $(i\neq j)$ are asymptotic independent normals, $w_{ij}\simeq^{\text{asy}} IN(0,\kappa+1)$ and $(w_{11},\ldots,w_{pp})'\simeq^{\text{asy}} N(0,2(\kappa+1)I_p+\kappa\mathbf{1}_p\mathbf{1}_p')$ with $\kappa+1=b_2/b_1^2$. Notice that

$$D_a'=\boldsymbol{\delta}'(I+W/\sqrt{n})^{-1}\,\boldsymbol{\delta}/\{\boldsymbol{\delta}'(I+W/\sqrt{n})^{-2}\,\boldsymbol{\delta}\}^{1/2},$$
$$\sqrt{b_1}\,D'=\{\boldsymbol{\delta}'(I+W/\sqrt{n})^{-1}\,\boldsymbol{\delta}\}^{1/2}.$$

By expanding $(I+W/\sqrt{n})^{-1}$ in powers of n^{-1}, we get

$$D_a'=\Delta\left[1-\frac{1}{2n}(\kappa+1)\,\chi_{p-1}^2\right]+O(n^{-3/2}) \tag{5.2}$$

and

$$D_1'=\sqrt{b_1}\,D'=\Delta\left[1-\frac{1}{2\sqrt{n}}u+\frac{1}{2n}(\kappa+1)\,\chi_{p-1}^2+\frac{3}{8n}u^2\right] + O(n^{-3/2}), \tag{5.3}$$

where $\Delta=(\boldsymbol{\delta}'\boldsymbol{\delta})^{1/2}$ or $\Delta^2=\boldsymbol{\delta}'\boldsymbol{\delta}=\boldsymbol{\delta}_1'\Sigma^{-1}\boldsymbol{\delta}_1$,

$$u=\boldsymbol{\delta}'W\boldsymbol{\delta}/\boldsymbol{\delta}'\boldsymbol{\delta} \quad \text{and} \quad (\kappa+1)\,\chi_{p-1}^2=(\boldsymbol{\delta}'W^2\boldsymbol{\delta}/\boldsymbol{\delta}'\boldsymbol{\delta})-u^2 \tag{5.4}$$

and it can be easily verified that u and χ_{p-1}^2 are asymptotic independent,

$$u\overset{\text{asy}}{\simeq}N(0,3\kappa+2) \quad \text{and} \quad \chi_{p-1}^2\overset{\text{asy}}{\simeq}\text{Chi-square with } (p-1) \tag{5.5}$$

degrees of freedom.

If $(\boldsymbol{\delta}/\sqrt{\boldsymbol{\delta}'\boldsymbol{\delta}},\Gamma_1)=\Gamma$ is an orthogonal matrix and $V=\Gamma'W\Gamma=(v_{ij})$, then $v_{11}=u$, $v_{12},\ldots,v_{1p}$ are asymptotic independent normals and

$$\chi_{p-1}^2(\kappa+1)=\sum_{i=2}^{p}v_{1i}^2. \tag{5.6}$$

Then it is easy to write

$$\left[D'_a - D'_1\left(1 - \frac{\kappa(8p+1)+2(4p-1)}{8n}\right)\right] \Big/ D'_1$$

$$= \frac{1}{2\sqrt{n}} u - \frac{1}{n}(\kappa+1)(\chi^2_{p-1} - (p-1))$$

$$-\frac{1}{8n}(u^2 - 3\kappa - 2) + \frac{3\kappa+2}{4n} + O(n^{-3/2}).$$

Hence,

$$\frac{2\sqrt{n}\left\{D'_a - D'_1\left(1 - \frac{\kappa(8p+1)+2(4p-1)}{8n}\right)\right\}}{D'_1(3\kappa+2)^{1/2}} \overset{\text{asy}}{\simeq} N(0,1)$$

or $2\sqrt{n}(D'_a - D'_1)/D'_1(3\kappa+2)^{1/2} \simeq^{\text{asy}} N(0,1)$ and hence the simultaneous confidence bound on D'_a is

$$D'_1[1 \pm \{d_\alpha(3\kappa+2)^{1/2}/2\sqrt{n}\}],$$

where $\int_{-d_\alpha}^{d_\alpha} \phi(x)\,dx = 1-\alpha$ with ϕ denotes the density of $N(0,1)$.

Case (ii). When $\boldsymbol{\delta}_1 = \boldsymbol{\mu}_1 - \boldsymbol{\mu}_2$ is unknown but Σ is known, then

$$D''_a = (\bar{\mathbf{x}}_1 - \bar{\mathbf{x}}_2)' \, \Sigma^{-1} \boldsymbol{\delta}_1 / D'' \qquad \text{and}$$
$$D'' = \{(\bar{\mathbf{x}}_1 - \bar{\mathbf{x}}_2)' \, \Sigma^{-1} (\bar{\mathbf{x}}_1 - \bar{\mathbf{x}}_2)\}^{1/2}. \tag{5.7}$$

By Section 2,

$$\sqrt{n}\,[\Sigma_1^{-1}(\bar{\mathbf{x}}_1 - \bar{\mathbf{x}}_2) - \boldsymbol{\delta}]/\sqrt{b_{(1)}} = \mathbf{y} \overset{\text{asy}}{\simeq} N(\mathbf{0}, I_p), \tag{5.8}$$

where $b_{(1)} = b_1/\kappa_0(1-\kappa_0)$. Taking $\boldsymbol{\delta}'\boldsymbol{\delta} = \Delta^2$, we can write

$$D''_a = \Delta[1 - b_{(1)}\chi^2_{1,p-1}/2n\Delta^2] + O(n^{-3/2}) \tag{5.9a}$$

and

$$D'' = \Delta\left[1 + \frac{\sqrt{b_{(1)}}}{\sqrt{n}\,\Delta} u_1 + \frac{b_{(1)}}{2n\Delta^2}\chi^2_{1,p-1}\right] + O(n^{-3/2}), \tag{5.9b}$$

where $u_1 = \mathbf{y}'\boldsymbol{\delta}/\Delta$ and $\chi^2_{1,p-1} = \mathbf{y}'\mathbf{y} - u_1^2$. u_1 and $\chi^2_{1,p-1}$ are asymptotic independent, $u_1 \simeq^{\text{asy}} N(0, 1)$ and $\chi^2_{1,p-1}$ is asymptotic Chi-square with $(p-1)$ degrees of freedom. Notice that

$$D''_a - D'' + b_{(1)}(p-1)/nD''$$
$$= -\frac{\sqrt{b_{(1)}}}{\sqrt{n}} u_1 \left(1 + \frac{b_{(1)}(p-1)}{n\Delta^2}\right) - \frac{b_{(1)}}{n\Delta}(\chi^2_{1,p-1} - p + 1) + O(n^{-3/2})$$

or

$$\sqrt{n}(D''_a - D'' + b_{(1)}(p-1)/nD'')/\sqrt{b_{(1)}}\,(1 + 2b_{(1)}(p-1)/nD''^2)^{1/2}$$
$$= -u_1 + O(n^{-1/2})$$

because $(1 + b_{(1)}(p-1)/n\Delta^2)/(1 + 2b_{(1)}(p-1)/nD''^2) = 1 + O(n^{-1/2})$. Therefore,

$$\sqrt{n}\,\{D''_a - D'' + b_{(1)}(p-1)/nD''\}/\sqrt{b_{(1)}}$$
$$\times (1 + 2b_{(1)}(p-1)/nD''^2)^{1/2} \overset{\text{asy}}{\simeq} N(0, 1),$$

and the simultaneous confidence bound on D''_a is

$$\{D'' - b_{(1)}(p-1)(nD'')^{-1}\}$$
$$\pm \{d_\alpha^2 b_{(1)}(1 + 2b_{(1)}(p-1)/nD''^2)/n\}^{1/2},$$

where d_α is defined at the end of Case (i).

Case (iii). When $\boldsymbol{\mu}_1$, $\boldsymbol{\mu}_2$, and Σ are unknown, then

$$D_a = (\bar{\mathbf{x}}_1 - \bar{\mathbf{x}}_2)' S^{-1}\boldsymbol{\delta}_1/\{(\bar{\mathbf{x}}_1 - \bar{\mathbf{x}}_2)' S^{-1}\Sigma S^{-1}(\bar{\mathbf{x}}_1 - \bar{\mathbf{x}}_2)\}^{1/2} \quad (5.10)$$

and

$$D = \{(\bar{\mathbf{x}}_1 - \bar{\mathbf{x}}_2)' S^{-1}(\bar{\mathbf{x}}_1 - \bar{\mathbf{x}}_2)\}^{1/2}. \quad (5.10a)$$

If $\boldsymbol{\delta} = \Sigma_1^{-1}\boldsymbol{\delta}_1$ and $\Sigma = \Sigma_1\Sigma_1'$, then by Theorem 1, $\sqrt{b_{(1)}}\mathbf{y} = \sqrt{n}\,(\Sigma_1^{-1}(\bar{\mathbf{x}}_1 - \bar{\mathbf{x}}_2) - \boldsymbol{\delta})$ and $b_1 W = \sqrt{n}\,(\Sigma_1^{-1}S\Sigma_1'^{-1} - b_1 I)$ are asymptotic independent normals. Let $\delta'\delta = \Delta^2$, $\mathbf{y}'\boldsymbol{\delta}/\Delta = u$, and $\chi^2_{p-1} = \mathbf{y}'\mathbf{y} - u^2$.

Let $\Gamma = ((\boldsymbol{\delta}, \mathbf{y})(\begin{smallmatrix}\Delta & u \\ \cdot & \chi_{p-1}\end{smallmatrix})^{-1}, \Gamma_2)$ be an orthogonal matrix. Then, it is easy to verify that $\Gamma' W \Gamma = V$ and W are identically distributed, and further

$$\boldsymbol{\delta}'\Gamma = \Delta\mathbf{e}_1', (\bar{\mathbf{x}}_1 - \bar{\mathbf{x}}_2)' \Sigma_1'^{-1}\Gamma$$
$$= [(b_{(1)}/n)^{1/2}u + \Delta]\,\mathbf{e}_1' + (b_{(1)}/n)^{1/2}\chi_{p-1}\mathbf{e}_2',$$

where $\mathbf{e}_1$ and $\mathbf{e}_2$ are the first and the second column vectors of I_p. Notice that V, u, and χ^2_{p-1} are asymptotically independent, $u \simeq N(0, 1)$, and χ^2_{p-1} is distributed as Chi-square with $(p-1)$ degrees of freedom. Then

$$b_1(\bar{\mathbf{x}}_1 - \bar{\mathbf{x}}_2)' S^{-1}\boldsymbol{\delta}_1 = \Delta \Bigg[\Delta + (b_{(1)}/n)^{1/2} u - (\Delta v_{11}/\sqrt{n}) - n^{-1}\Big(f\sqrt{b_{(1)}} u v_{11} + \sqrt{b_{(1)}}\,\chi_{p-1} v_{12} - \Delta \sum_{i=1}^{p} v_{1i}^2\Big)\Bigg] + O(n^{-3/2})$$

$$b_1^2(\bar{\mathbf{x}}_1 - \bar{\mathbf{x}}_2)' S^{-1}\Sigma S^{-1}(\bar{\mathbf{x}}_1 - \bar{\mathbf{x}}_2) = \Delta^2 + 2\Delta(\sqrt{b_{(1)}}u - \Delta v_{11})\, n^{-1/2} + n^{-1}\Big(b_{(1)}\chi^2_{p-1} + b_{(1)}u^2 - 4\sqrt{b_{(1)}}\Delta(u v_{11} + v_{12}\chi_{p-1}) + 3\sum_{i=1}^{p} v_{1i}^2 \Delta^2 \Big) + O(n^{-3/2})$$

and

$$b_1(\bar{\mathbf{x}}_1 - \bar{\mathbf{x}}_2)' S^{-1}(\bar{\mathbf{x}}_1 - \bar{\mathbf{x}}_2) = \Delta^2 + \Delta(2\sqrt{b_{(1)}}u - \Delta v_{11})\, n^{-1/2} + n^{-1}\Big(b_{(1)}\chi^2_{p-1} + b_{(1)}u^2 - 2\sqrt{b_{(1)}}\Delta(u v_{11} + v_{12}\chi_{p-1}) + \sum_{i=1}^{p} v_{1i}^2 \Delta^2 \Big) + O(n^{-3/2})$$

Hence,

$$D_a = \Delta + (2n)^{-1}\Big(2\sqrt{b_{(1)}}\,\chi_{p-1} v_{12} - \Delta \sum_{i=2}^{p} v_{1i}^2 - \Delta^{-1} b_{(1)}\chi^2_{p-1} \Big) + O(n^{-3/2}) \tag{5.11}$$

and

$$D_1 = \sqrt{b_1}D = \Delta + n^{-1/2}(\sqrt{b_{(1)}}\,u - \Delta_{11}/2) + (2n)^{-1} \times \Bigg[\Delta \sum_{i=2}^{p} v_{1i}^2 + (3/4)\Delta v_{11}^2 - \sqrt{b_{(1)}}\,(u v_{11} + 2 v_{12}\chi_{p-1}) + \Delta^{-1} b_{(1)}\chi^2_{p-1} \Bigg] + O(n^{-3/2}). \tag{5.12}$$

Therefore,

$$\sqrt{n}\,(D_1 - D_a) = (\sqrt{b_{(1)}}\,u - \Delta v_{11}/2) + n^{-1/2}\left[\Delta \sum_{i=2}^{p} v_{1i}^2 + \tfrac{3}{8}\Delta v_{11}^2 \right.$$

$$\left. + \Delta^{-1} b_{(1)} \chi_{p-1}^2 - 2\sqrt{b_{(1)}}\,(v_{12}\chi_{p-1} + u v_{11}/4)\right] + O(n^{-1}),$$

and if

$$y = \sqrt{n}\,(D_1 - D_a) - n^{-1/2}[(\kappa+1)(p-1)\,D_1 + (3/8)(3\kappa+2)\,D_1 + b_{(1)}(p-1)\,D_1^{-1}], \tag{5.13}$$

then

$$E(y) = O(n^{-1})$$

and

$$Ey^2 = b_{(1)} + \Delta^2(3\kappa+2)/4 + 2n^{-1}[\Delta^2(\kappa+1)^2(p-1) + (\tfrac{3}{8})^2\Delta^2(3\kappa+2)^2 + \Delta^{-2}b_{(1)}^2(p-1) + 2b_{(1)}((\kappa+1)(p-1) + (3\kappa+2)/16)] + O(n^{-2}).$$

Hence if

$$y_2 = (b_{(1)} + D_1^2(3\kappa+2)/4) + (4n)^{-1} \times [D_1^2(\kappa+1)(p-1)(5\kappa+6) + D^2(3\kappa+2)^2(16)^{-1} + 8D_1^{-2}b_{(1)}^2(p-1) + b_{(1)}(p-1)(13\kappa+14)], \tag{5.14}$$

then

$$Ey_2 = Ey^2 + O(n^{-2}) \qquad \text{and} \qquad y/\sqrt{y_2} \overset{\text{asy}}{\simeq} N(0, 1). \tag{5.15}$$

This can be utilized to get an approximate confidence bound on D_a. The first approximate confidence bound on D_a is

$$D_1 \pm \{d_\alpha^2(b_{(1)} + D_1^2(3\kappa+2)/4)/n\}^{1/2}, \tag{5.16}$$

where d_α is defined at the end of Case (i), and from (5.15), we get an approximate confidence bound on D_a as

$$y_1 \pm \{d_\alpha^2\, y_2/n\}^{1/2}, \tag{5.17}$$

where

$$y_1 = D_1 - n^{-1}((\kappa+1)(p-1) D_1 + (3/8)(3\kappa+2) D_1 + b_{(1)}(p-1) D_1^{-1}$$

and y_2 is defined in (5.14).

References

[1] Cramér, H. (1951). *Mathematical Methods of Statistics.* Princeton Univ. Press, Princeton, NJ.

[2] Khatri, C. G. (1966). A note on a MANOVA model applied to problems in growth curve. *Ann. Inst. Statist. Math.* **18** 75–86.

[3] Khatri, C. G. (1986). On elliptical and spherical distributions (unpublished).

[4] Khatri, C. G., Rao, C. R., Schaafma, W., Steerneman, A. G. M., and Van Vark, G. N. (1986). *Inference about the Performance of Fisher's Linear Discriminant Function.* Technical Report 86-39, Center for Multivariate Analysis, University of Pittsburgh.

[5] Krishnaiah, P. R., Lin, J., and Wang, L. (1985). *Inference on the Ranks of the Canonical Correlation Matrices for Elliptically Symmetric Populations.* Technical Report 85–14, Center for Multivariate Analysis, University of Pittsburgh.

Stochastic Integrals of Empirical-Type Processes with Applications to Censored Regression

TZE LEUNG LAI* AND ZHILIANG YING

Stanford University and University of Illinois

Motivated by the analysis of linear rank estimators and the Buckley–James nonparametric EM estimator in censored regression models, we study herein the asymptotic properties of stochastic integrals of certain two-parameter empirical processes. Applications of these results on empirical processes and their stochastic integrals to the asymptotic analysis of censored regression estimators are also given. © 1988 Academic Press, Inc.

1. INTRODUCTION

Consider the linear regression model

$$y_i = \alpha + \beta x_i + \varepsilon_i \qquad (i = 1, 2, ...), \tag{1.1}$$

where the ε_i are i.i.d. random variables with mean 0, and the x_i are either non-random or are independent random variables independent of $\{\varepsilon_i\}$. Suppose that the responses y_i are not completely observable and that the observations are (x_i, z_i, δ_i), where $z_i = \min\{y_i, t_i\}$, $\delta_i = I_{\{y_i \leqslant t_i\}}$, and the t_i are independent random variables, independent of $\{\varepsilon_i\}$. This is often called the "censored regression model" and the t_i are called the "censoring variables."

In 1979, Buckley and James [3] proposed the following method to estimate α and β. They started by replacing y_i by

$$y_i^* = y_i \delta_i + E(y_i \mid y_i > t_i)(1 - \delta_i), \tag{1.2}$$

* Research supported by the National Science Foundation and the Army Research Office.

Multivariate Statistics and Probability
ISBN 0-12-580205-6

Reprinted from *J. Mult. Anal.* **27**(2).

and regressing the y_i^* (instead of the y_i) on the x_i to obtain

$$\hat{\beta} = \left\{ \sum_1^n y_i^*(x_i - \bar{x}_n) \right\} \Big/ \sum_1^n (x_i - \bar{x}_n)^2, \tag{1.3}$$

$$\hat{\alpha} = \bar{y}_n^* - \hat{\beta}\bar{x}_n, \tag{1.4}$$

noting that $Ey_i^* = Ey_i = \alpha + \beta x_i$, where $\bar{x}_n = n^{-1}\sum_1^n x_i$. Since $E(y_i | y_i > t_i)$ in (1.2) is unknown, they replaced (1.3) by an iterative scheme in which $E(y_i | y_i > t_i)$ is substituted by its successive estimates. Specifically, let $e_i(b) = z_i - bx_i$ and order the uncensored $e_i(b)$ as $e_{(1)}(b) \leqslant \cdots e_{(k)}(b)$, assuming that there are k uncensored observations. Let

$$n_i(b) = \#\{j: e_j(b) \geqslant e_{(i)}(b)\}, \tag{1.5}$$

where $\# A$ denotes the number of elements of a set A. Buckley and James first used the Kaplan–Meier estimator

$$\hat{F}_{n,b}(u) = 1 - \prod_{i: e_{(i)}(b) \leqslant u} (n_i(b) - 1)/n_i(b) \tag{1.6}$$

to estimate the common distribution function F of $e_i \triangleq \alpha + \varepsilon_i$. Assuming the x_i to be nonrandom, they then replaced $E(y_i | y_i > t_i) = \beta x_i + E(e_i | e_i > t_i - \beta x_i)$ by

$$z_i(b) = bx_i + \int_{u > t_i - bx_i} u \, d\hat{F}_{n,b}(u) / (1 - \hat{F}_{n,b}(t_i - bx_i)). \tag{1.7}$$

Replacing (1.2) by $y_i^*(b) = y_i \delta_i + z_i(b)(1 - \delta_i)$, they proposed to estimate β by iterative solution of the equation

$$b = \left\{ \sum_{i=1}^n (x_i - \bar{x}_n) \, y_i^*(b) \right\} \Big/ \sum_{i=1}^n (x_i - \bar{x}_n)^2, \tag{1.8}$$

in analogy with (1.3). Note that (1.8) is equivalent to the equation

$$W_n(b) = 0,$$

where

$$W_n(b) = \sum_{i=1}^n \delta_i (x_i - \bar{x}_n)(y_i - bx_i) + \sum_{i=1}^n (1 - \delta_i)(x_i - \bar{x}_n)(z_i(b) - bx_i). \tag{1.9}$$

Once a slope estimator b^* is determined, an estimator of α can be obtained as the mean of $\hat{F}_{b^*}$.

To analyze the asymptotic properties of the Buckley–James estimator, a crucial step is to study the random function $W_n(b)$ as $n \to \infty$. Of particular importance is the behavior of $W_n(b)$ for b near β. Useful tools to study this kind of problems are provided by the concept of metric entropy of empirical-type processes and their stochastic integrals, which are discussed in Sections 2 and 3 below. Applications of these results to the random function $W_n(b)$, or more precisely, to a slight modification thereof, are discussed in Section 5. In this modification, we ignore the factors $1 - n_i^{-1}(b)$ in the Kaplan–Meier estimator (1.6) when $n_i(b)/n$ is too small, causing instability in the estimator. Specifically, we redefine $\hat{F}_{n,b}$ by

$$\hat{F}_{n,b}(u) = 1 - \prod_{i: e_{(i)}(b) < u} \{1 - p_n(n^{-1} n_i(b))/n_i(b)\}, \tag{1.10}$$

where p_n is a smooth weight function on $[0, 1]$ that will be specified in Section 5. In addition, we also use the weight function p_n to modify the definition (1.7) of $z_i(b)$ in Section 5.

In Section 4, we apply the results of Sections 2 and 3 to another class of estimators of β in the censored regression model, introduced in [7] as extensions of the classical rank estimators with complete (uncensored) data. The rank estimators of β in [7] are defined by the equation

$$S_n(b) = 0, \tag{1.11}$$

where

$$S_n(b) = \sum_{i=1}^{k} \psi \cdot p_n(\hat{F}_{n,b}(e_{(i)}(b)))\{x_{(i)} - \bar{x}(i, b)\}\, p_n(n^{-1} n_i(b)), \tag{1.12}$$

$$\bar{x}(i, b) = \left[\sum_{j=1}^{n} x_j I_{\{e_j(b) \geq e_{(i)}(b)\}} \right] \Big/ n_i(b), \tag{1.13}$$

$\hat{F}_{n,b}$ is defined in (1.10), p_n is a smooth function on $[0, 1]$ that will be specified in Section 4, and $\psi \cdot p_n$ denotes the product of p_n and ψ, which is a given "score function" (cf. [7]), i.e., $\psi \cdot p_n(x) = \psi(x)\, p_n(x)$. Since Eq. (1.11) may not have a solution, we define a rank estimator $\tilde{\beta}_n$ of β as a zero-crossing of the step function $S_n(b)$, i.e., the right and left hand limits $S_n(\tilde{\beta}_n +)$ and $S_n(\tilde{\beta}_n -)$ do not have the same sign. This zero-crossing notion of a solution of the equation $W_n(b) = 0$ was also used by James and Smith [5] to give a more precise definition of the Buckley–James estimator.

The functions $W_n(b)$ and $S_n(b)$, defined by (1.9) and (1.12), respectively, appear to be rather intractable analytically. An important step in our analysis of these functions is to express them using stochastic integrals of empirical-type processes. In particular, as shown in [7],

$$S_n(b) = \int_{s=-\infty}^{\infty} \psi \cdot p_n(\hat{F}_{n,b}(s))\, p_n(n^{-1}\#_n(b,s)) \times \left[dY_n(b,s) - \frac{X_n(b,s)}{\#_n(b,s)}\, dL_n(b,s) \right], \tag{1.14}$$

where

$$\#_n(b,s) = \sum_{j=1}^{n} I_{\{e_j \wedge (t_j - \beta x_j) \geqslant s + (b-\beta)x_j\}}, \tag{1.15a}$$

$$X_n(b,s) = \sum_{j=1}^{n} x_j I_{\{e_j \wedge (t_j - \beta x_j) \geqslant s + (b-\beta)x_j\}}, \tag{1.15b}$$

$$L_n(b,s) = \sum_{j=1}^{n} I_{\{e_j \leqslant (t_j - \beta x_j) \wedge (s + (b-\beta)x_j)\}}, \tag{1.15c}$$

$$Y_n(b,s) = \sum_{j=1}^{n} x_j I_{\{e_j \leqslant (t_j - \beta x_j) \wedge (s + (b-\beta)x_j)\}}. \tag{1.15d}$$

Here and in the sequel, $e_j = \alpha + \varepsilon_j$, $x \wedge y$ denotes $\min(x, y)$, and $x \vee y$ denotes $\max(x, y)$. We call the two-parameter processes $\#_n - E\#_n$, $X_n - EX_n$, $L_n - EL_n$, $Y_n - EY_n$ *empirical-type processes* because they are similar to empirical processes and can be analyzed by techniques similar to those recently developed in empirical process theory, as will be shown in Section 2. In particular, these techniques enable us to obtain probability bounds, which are uniform in b and s, in the approximation of the random function $\#_n(b,s) - \#_n(\beta,s)$ (or $L_n(b,s) - L_n(\beta,s)$, etc.) by its mean $E\#_n(b,s) - E\#_n(\beta,s)$. In Section 3, we apply these results to analyze stochastic integrals involving empirical-type processes. Making use of these stochastic integrals, we then study the asymptotic properties of $\hat{F}_{n,b}$, $S_n(b)$, and $W_n(b)$ in Sections 4 and 5.

2. Metric Entropy and Convergence Properties of Empirical-type Processes

In this section we first review some recent results in empirical process theory due to Alexander [1] and then extend these results to the empirical-type processes (1.15). Let $\xi_1, \xi_2, \ldots$, be independent random variables taking values in a measurable space $(S, \mathscr{B})$ and let P_i denote the probability distribution of ξ_i (i.e., $P_i(B) = P\{\xi_i \in B\}$). Consider the empirical measure and process

$$\pi_n = n^{-1} \sum_{i=1}^{n} \delta_{\xi_i}, \qquad \nu_n = n^{1/2}(\pi_n - \bar{P}_n),$$

where $\bar{P}_n = n^{-1} \sum_{i=1}^n P_i$ and δ_x denotes the unit point mass (delta function) at x. Let $\mathscr{F}$ be a class of real-valued measurable functions on S such that $|f| \leqslant A$ for all $f \in \mathscr{F}$ and some $A > 0$. Let

$$v_n(f) = \int f \, dv_n = n^{-1/2} \sum_{i=1}^{n} (f(\xi_i) - Ef(\xi_i)).$$

An important concept in Alexander's [1] analysis of $\sup_{f \in \mathscr{F}} |v_n(f)|$ is the "metric entropy" of $\mathscr{F}$ defined as follows. Given $\varepsilon > 0$, $p > 0$, and a probability measure μ on $(S, \mathscr{B})$, let

$$N_p(\varepsilon, \mathscr{F}, \mu) = \min\{k\text{: There exist } f_1, ..., f_k \in \mathscr{F} \text{ such that } \min_{i \leqslant k} \|f - f_i\|_p < \varepsilon \text{ for all } f \in \mathscr{F}\},$$

$$N_p^B(\varepsilon, \mathscr{F}, \mu) = \min\{k\text{: There exist } f_1^U, f_1^L, ..., f_k^U, f_k^L \in \mathscr{F} \text{ such that } f_i^L \leqslant f \leqslant f_i^U \text{ for some } i \text{ for every } f \in \mathscr{F}, \text{ and } \|f_i^U - f_i^L\|_p < \varepsilon \text{ for all } i\}.$$

The "metric entropy" and "metric entropy with bracketing" of $\mathscr{F}$ in $L^p(\mu)$ are $\log N_p$ and $\log N_p^B$, respectively.

Given a class $\mathscr{F}$ with finite $L^p(\bar{P}_n)$ entropy and $\delta_0 > \delta_1 > \cdots > \delta_K > 0$, there exist $\mathscr{F}_j \subset \mathscr{F}$ $(j \leqslant m)$ such that $|\mathscr{F}_j| = N_p(\delta_j, \mathscr{F}, \bar{P}_n)$ and for each $f \in \mathscr{F}$ there exists $f_j(f) \in \mathscr{F}_j$ with $\|f - f_j(f)\|_p < \delta_j$. A basic idea in Alexander's probability bounds for $\sup_{\mathscr{F}} |v_n(f)|$ is the following "chaining argument" (cf. also [4]). Writing

$$v_n(f) = v_n(f_0(f)) + \sum_{j=0}^{K-1} v_n[f_{j+1}(f) - f_j(f)] + v_n[f - f_K(f)], \quad (2.1)$$

we have

$$\begin{aligned} P^*\{\sup_{\mathscr{F}} |v_n(f)| > M\} &\leqslant |\mathscr{F}_0| \sup_{\mathscr{F}} P\{|v_n(f)| > (1 - \varepsilon/4)M\} \\ &\quad + \sum_{j=0}^{K-1} |\mathscr{F}_j| \, |\mathscr{F}_{j+1}| \\ &\quad \times \sup_{\mathscr{F}} P\{|v_n[f_{j+1}(f) - f_j(f)]| > \eta_j\} \\ &\quad + P^*\{\sup_{\mathscr{F}} |v_n(f_K(f) - f)| > \varepsilon M/8 + \eta_K\} \\ &\triangleq R_1 + R_2 + R_3, \end{aligned} \quad (2.2)$$

where the $\eta_j > 0$ are so chosen that $\sum_{j=0}^{K} \eta_j < \varepsilon M/8$, and P^* denotes outer measure. Bounds for the terms R_1 and R_2 in (2.2) are provided by Bennett's [2] inequality for sums of bounded independent random variables: If $X_1, ..., X_n$ are independent random variables such that $EX_i = 0$ and $|X_i| \leqslant A$, then for $\alpha \geqslant n^{-1} \sum_1^n \mathrm{Var}(X_i)$,

$$P\left\{\left|n^{-1/2} \sum_{i=1}^{n} X_i\right| > M\right\} \leqslant 2 \exp\left\{-\frac{1}{2} M^2 \alpha^{-1} g(AMn^{-1/2}\alpha^{-1})\right\}, \tag{2.3}$$

where

$$g(\lambda) = 2\lambda^{-2}\{(1+\lambda)\log(1+\lambda) - \lambda\}.$$

Making use of (2.2) and (2.3) together with an appropriate choice of the δ_j and η_j, Alexander [1] obtained sharp probability bounds for $\sup_{\mathscr{F}} |v_n(f)|$ under a variety of metric entropy assumptions on $\mathscr{F}$; the method to bound R_3 in (2.2) varies with these assumptions on $\mathscr{F}$. In particular, he showed that for $\varepsilon > 0$, $0 < r < 2$, and $\theta > 0$, there exists $C = C(r, \theta, \varepsilon)$ such that if

$$\log N_\infty(\delta, \mathscr{F}, \bar{P}_n) \leqslant \theta\, \delta^{-r} \qquad \text{for all} \quad 0 < \delta \leqslant 1 \tag{2.4}$$

and if

$$M \geqslant C\{\alpha^{(2-r)/4} \vee n^{(r-2)/2(r+2)}\}, \tag{2.5}$$

then analogous to (2.3),

$$P^*\{\sup_{\mathscr{F}} |v_n(f)| > M\} \leqslant 5 \exp\{-\tfrac{1}{2}(1-\varepsilon) M^2 \alpha^{-1} g(AMn^{-1/2}\alpha^{-1})\}, \tag{2.6}$$

where $\alpha \geqslant \sup_{\mathscr{F}} n^{-1} \sum_{i=1}^{n} \mathrm{Var}\, f(\xi_i)$. The term R_3 in this case is handled by taking $\delta_K = \varepsilon M n^{-1/2}/16$, so that

$$|v_n(f_K(f) - f)| \leqslant 2n^{1/2}\, \|f_K(f) - f\|_\infty \leqslant \varepsilon M/8. \tag{2.7}$$

Let $\mathscr{D}$ be a class of measurable subsets of S and let $\mathscr{F} = \{I_D : D \in \mathscr{D}\}$. Alexander [1] showed that if we replace (2.4) by

$$\log N_2^B(\delta, \mathscr{F}, \bar{P}_n) \leqslant \theta\, \delta^{-r} \qquad \text{for all} \quad 0 < \delta \leqslant 1, \tag{2.4*}$$

then (2.6) still holds for M satisfying both (2.5) and

$$M \leqslant \varepsilon \alpha n^{1/2}/16. \tag{2.8}$$

Note that in this case with $f=I_D$, $\sup_{\mathscr{F}} |v_n(f)| = \sup_{\mathscr{D}} |v_n(D)|$ and $\alpha \geqslant \sup_{\mathscr{D}} n^{-1} \sum_1^n P_i(D)(1-P_i(D))$. The term R_3 in (2.2) is handled by taking $\delta_K^2 = \varepsilon M n^{-1/2}/16$ and using the bound

$$
\begin{aligned}
|v_n[f_K^U(f) - f]| &\leqslant |v_n[f_K^U(f) - f_K^L(f)]| + 2n^{1/2} \, \|f_K^U(f) - f_K^L(f)\|_1 \\
&\leqslant |v_n[f_K^U(f) - f_K^L(f)]| + 2n^{1/2} \, \delta_K^2,
\end{aligned} \tag{2.9}
$$

since $EI_D = EI_D^2 = \|I_D\|_2^2$. Hence

$$R_3 \leqslant |\mathscr{F}_K| \sup_{\mathscr{F}} P\{|v_n[f_K^U(f) - f_K^L(f)]| > \eta_K\},$$

which can then be bounded by using Bennett's inequality (2.3).

As a corollary of (2.6), we obtain the following result on empirical-type processes, which will be used in Section 3. Throughout the sequel, replacing $t_i - \beta x_i$ in (1.15) by t_i, we shall assume without loss of generality that $\beta = 0$. We shall also restrict b in (1.15) to a bounded interval $|b| \leqslant \rho$. For notational simplicity we shall write $\sup_{b,s}$ to denote supremum over the region $|b| \leqslant \rho$ and $-\infty < s < \infty$.

LEMMA 1. *Let (e_i, x_i, t_i), $i = 1, 2, \ldots$, be independent random vectors such that for some nonrandom constant A,*

$$|x_i| \leqslant A \qquad \textit{for all} \quad i. \tag{2.10}$$

Let $Z_n(b, s)$ be any of the four empirical-type processes defined in (1.15) *with $\beta = 0$. Let u_n: $[-\rho, \rho] \times (-\infty, \infty) \to (-\infty, \infty)$ be a nonrandom Borel function such that*

$$
\begin{aligned}
&|u_n(b, s)| \leqslant A, \\
&|u_n(b, s) - u_n(b', s')| \leqslant A\{|b - b'| + |s - s'|\}, \textit{for all } n, b, b', s, s'.
\end{aligned} \tag{2.11}
$$

Then for every $0 \leqslant \gamma < 1$ and $\varepsilon > 0$,

$$
\begin{aligned}
&\sup_{|b - b'| \leqslant n^{-\gamma}} \left| \int_{s=-\infty}^{\infty} [u_n(b, s) - u_n(b', s)] \, d(Z_n(b, s) - EZ_n(b, s)) \right| \\
&\quad = O(n^{(1-\gamma)/2 + \varepsilon}) \qquad \textit{a.s.}
\end{aligned} \tag{2.12}
$$

Proof. We shall only consider the case $Z_n = Y_n$. First note that

$$
\begin{aligned}
&\int_{s=-\infty}^{\infty} [u_n(b, s) - u_n(b', s)] \, dY_n(b, s) \\
&\quad = \sum_{i=1}^{n} x_i [u_n(b, e_i - bx_i) - u_n(b', e_i - bx_i)] \, I_{\{e_i \leqslant t_i\}}.
\end{aligned}
$$

For fixed n, let $\psi_{b,b'}(e_i, x_i, t_i) = x_i[u_n(b, e_i - bx_i) - u_n(b', e_i - bx_i)] I_{\{e_i \leqslant t_i\}}$. Letting $\xi_i = (e_i, x_i, t_i)$, the class $\mathscr{F} = \{\psi_{b,b'}: |b| \leqslant \rho, |b'| \leqslant \rho\}$ clearly satisfies the entropy assumption (2.4) for every $r > 0$, in view of (2.10) and (2.11) (which in fact implies that $\log N_\infty(\delta, \mathscr{F}, \bar{P}_n) = O(\log \delta)$ as $\delta \to 0$). Moreover, by (2.11), there exists A' such that $\operatorname{Var} \psi_{b,b'}(e_i, x_i, t_i) \leqslant A' |b - b'|$ for all i. Hence the desired conclusion (2.12) follows from (2.6) with $M = n^{-\gamma/2 + \varepsilon}$ and the Borel–Cantelli lemma. ∎

We next modify Alexander's arguments sketched above to prove the following result, which will be used repeatedly in the subsequent sections.

THEOREM 1. *Let $e_1, e_2, \ldots$ be i.i.d. random variables whose common distribution function F satisfies the Lipschitz condition $|F(x) - F(y)| \leqslant C|x - y|$ for all x, y and some $C > 0$. Let (x_i, t_i), $i = 1, 2, \ldots$, be independent random vectors that are independent of $\{e_n\}$. Assume that* (2.10) *holds and*

$$\sup_{|b| \leqslant \rho, -\infty < s < \infty} \sum_1^n P\{s \leqslant t_i - bx_i \leqslant s + h\}$$
$$= O(nh) \text{ as } n \to \infty \text{ and } h \to 0 \text{ with } nh \to \infty, \tag{2.13}$$

$$\sup_i E(|e_1 \wedge t_i|^r) < \infty \qquad \text{for some} \quad r > 0. \tag{2.14}$$

Let $Z_n(b, s)$ be any of the four empirical-type processes defined in (1.15) *with $\beta = 0$. For $0 < d \leqslant 1$ let*

$$\alpha_{n,d} = \sup_{|b - b'| + |s - s'| \leqslant d} n^{-1} \operatorname{Var}\{Z_n(b, s) - Z_n(b', s')\}. \tag{2.15}$$

Then for every $0 < \varepsilon < 1$, as $n \to \infty$ and $M = o(n^{1/2}\alpha_{n,d})$ but $M/\{\alpha_{n,d}^{(1-\varepsilon)/2} \vee n^{-(1-\varepsilon)/2}\} \to \infty$,

$$P\{\sup_{|b - b'| + |s - s'| \leqslant d} n^{-1/2} |Z_n(b, s) - EZ_n(b, s) - Z_n(b', s') + EZ_n(b' s')| > M\}$$
$$= O(\exp\{-\tfrac{1}{2}(1 - \varepsilon) M^2 \alpha_{n,d}^{-1}\}). \tag{2.16}$$

Consequently, for every $0 \leqslant \gamma < 1$ and $\theta > 0$,

$$\sup_{|b - b'| + |s - s'| \leqslant n^{-\gamma}} |Z_n(b, s) - EZ_n(b, s) - Z_n(b', s') + EZ_n(b', s')|$$
$$= O(n^{(1-\gamma)/2 + \theta}) \qquad a.s. \tag{2.17}$$

Proof. We shall only consider the case $Z_n = X_n$. To prove (2.16), note that the assumptions on M here satisfy Alexander's conditions (2.8) and (2.5) (with sufficiently small r). Let

$$\Delta_n(b, s; b', s') = n^{-1/2}\{X_n(b, s) - EX_n(b, s) - X_n(b', s') + EX_n(b', s')\}.$$

As in Alexander's argument outlined above, choose $\delta_0 > \cdots > \delta_K$ with $\delta_K \sim C_\varepsilon M n^{-1/2}$, where C_ε is some positive constant depending on ε. For fixed $j = 0, 1, ..., K$, partition the interval $[-\rho, \rho]$ by points $\beta_v^{(j)} < \beta_{v+1}^{(j)}$ such that $\beta_{v+1}^{(j)} - \beta_v^{(j)} \leqslant \delta_j$ ($v = 1, 2, ...$), with equality except possibly for the case $v = 1$ ($\beta_1^{(j)} = -\rho$). Thus, the number N_j of sub-intervals is the smallest integer $\geqslant 2\rho/\delta_j$, so $\log N_j \sim \log \delta_j$ (in analogy with (2.4*)). For $j = 0, ..., K$ and $-\rho \leqslant b < \rho$, define $v(b, j)$ by $\beta_{v(b,j)}^{(j)} \leqslant b < \beta_{v(b,j)+1}^{(j)}$. In view of (2.14),

$$\sup_i P\{|e_i \wedge t_i| \geqslant \delta^{-1/r}\} = O(\delta) \quad \text{as} \quad \delta \to 0. \tag{2.18}$$

For $j = 0, ..., K$, partition the interval $[-\delta_j^{-1/r}, \delta_j^{-1/r}]$ by points $\sigma_m^{(j)} < \sigma_{m+1}^{(j)}$ such that $\sigma_{m+1}^{(j)} - \sigma_m^{(j)} \leqslant \delta_j$ ($m = 1, 2, ..., M_j$) with equality except possibly for the case $m = 1$ ($\sigma_1^{(j)} = -\delta_j^{-1/r}$). Thus, the number M_j of such sub-intervals is the smallest integer $\geqslant 2\,\delta_j^{-1/r-1}$, so $\log M_j \sim \log \delta_j$. Let $\sigma_0^{(j)} = -\infty$, $\sigma_{M_j+2}^{(j)} = \infty$. For any given s, define $m(s, j)$ by $\sigma_{m(s,j)}^{(j)} \leqslant s < \sigma_{m(s,j)+1}^{(j)}$. As in (2.1), note that

$$\begin{aligned}\Delta_n(b, s; b', s') = {} & \Delta_n(\beta_{v(b,0)}^{(0)}, \sigma_{m(s,0)}^{(0)}; \beta_{v(b',0)}^{(0)}, \sigma_{m(s',0)}^{(0)}) \\ & + \sum_{j=0}^{K-1} [\Delta_n(\beta_{v(b,j+1)}^{(j+1)}, \sigma_{m(s,j+1)}^{(j+1)}; \\ & \beta_{v(b',j+1)}^{(j+1)}, \sigma_{m(s',j+1)}^{(j+1)}) - \Delta_n(\beta_{v(b,j)}^{(j)}, \sigma_{m(s,j)}^{(j)}; \\ & \beta_{v(b',j)}^{(j)}, \sigma_{m(s',j)}^{(j)})] \\ & + [\Delta_n(b, s; b', s') - \Delta_n(\beta_{v(b,K)}^{(K)}, \sigma_{m(s,K)}^{(K)}; \\ & \beta_{v(b',K)}^{(K)}, \sigma_{m(s',K)}^{(K)})],\end{aligned} \tag{2.19}$$

and apply the chaining argument (2.2) with v_n replaced by Δ_n. Since $|x_i I_{\{e_i \wedge t_i \geqslant s + bx_i\}}| \leqslant A$ and the (e_i, x_i, t_i) are independent, we can apply Bennett's inequality (2.3) to obtain probability bounds as in Alexander's argument [1], noting that by the Lipschitz continuity of F and the assumption (2.13) on t_i,

$$\begin{aligned}\sup_{|b_1 - b_2| \vee |b'_1 - b'_2| \vee |s_1 - s_2| \vee |s'_1 - s'_2| \leqslant h} & \operatorname{Var}[\Delta_n(b_1, s_1; b'_1, s'_1) \\ & - \Delta_n(b_2, s_2; b'_2, s'_2)] \\ & = O(h) \text{ as } n \to \infty \text{ and } h \to 0 \text{ such that } nh \to \infty.\end{aligned} \tag{2.20}$$

The rest of the proof of (2.16) is similar to that in Alexander [1, proof of Theorem 2.3]. In particular, the last term in (2.19) can be handled by a "bracketing argument" as in (2.9), noting that $n\,\delta_K \sim C_\varepsilon M n^{1/2} \to \infty$ and that $X_n(b, s)$ can be decomposed as monotone functions in b and s:

$$X_n(b, s) = \sum_{j \leqslant n, x_j \geqslant 0} x_j I_{\{e_j \wedge t_j \geqslant s + b x_j\}} - \sum_{j \leqslant n, x_j < 0} |x_j|\, I_{\{e_j \wedge t_j \geqslant s - b|x_j|\}}.$$

Setting $M = n^{-\gamma/2 + \theta}$ in (2.16) and noting that $\alpha_{n, n^{-\gamma}} = O(n^{-\gamma})$ as in (2.20), (2.17) follows from (2.16) and the Borel–Cantelli lemma. ∎

In the preceding proof, the chain $\delta_0 > \cdots > \delta_K$ terminates with $\delta_K \sim C_\varepsilon M n^{-1/2}$, and therefore we can apply condition (2.13) with $h = \delta_j$ (since $\min_{j \leqslant K} n\,\delta_j \to \infty$). Since the chain $\delta_0 > \cdots > \delta_K$ in Alexander's proof of (2.6) under the assumption (2.4) also terminates with $\delta_K \sim \varepsilon M n^{-1/2}/16$, we can introduce the following relaxation of the assumption (2.11) in Lemma 1, which we have shown to be a corollary of (2.6) by setting $M = n^{-\gamma/2 + \varepsilon}$ (and therefore $n(Mn^{-1/2}) \to \infty$).

LEMMA 2. *Suppose that in Lemma* 1 *we replace the assumption* (2.11) *by*

$$\sup_{b,s} |u_n(b, s)| = O(1) \quad \text{and} \quad \sup_{|b-b'| + |s-s'| \leqslant h} |u_n(b, s) - u_n(b', s')| = O(h)$$

$$\text{as } n \to \infty \text{ and } h \to 0 \text{ such that } nh \to \infty. \tag{2.21}$$

Then the conclusion (2.12) *still holds for every* $0 \leqslant \gamma < 1$ *and* $\varepsilon > 0$.

Under the assumptions of Theorem 1 we can further strengthen the conclusion (2.12) of Lemma 1 for our main result in Section 3. This is the content of

LEMMA 3. *With the same notation and assumptions as in Theorem* 1, *let* $u_n\colon [-\rho, \rho] \times (-\infty, \infty) \to (-\infty, \infty)$ *be nonrandom Borel functions satisfying* (2.21). *Then for every* $0 \leqslant \gamma < 1$ *and* $\varepsilon > 0$,

$$\sup_{|b-b'| \leqslant n^{-\gamma}, -\infty < y < \infty} \left| \int_{s=-\infty}^{y} [u_n(b, s) - u_n(b', s)]\, d(Z_n(b, s) - EZ_n(b, s)) \right|$$
$$= O(n^{(1-\gamma)/2 + \varepsilon}) \quad \text{a.s.} \tag{2.22}$$

Proof. We shall only consider the case $L_n(b, s)$. For fixed n, denote $L_n(b, s)$, $EL_n(b, s)$, $u_n(b, s) - u_n(b', s)$ by $L_b(s)$, $\bar{L}_b(s)$, $u_{b,b'}(s)$, respectively, and let $V(b, b', s) = \int_{-\infty}^{s} u_{b,b'}(t)\, d(L_b(t) - \bar{L}_b(t))$. As in the proof of

Theorem 1, choose $\delta_0 > \cdots > \delta_K$, and for $j=0, .., K$, partition the real line by the points $\sigma_0^{(j)} = -\infty < \sigma_1^{(j)} < \cdots < \sigma_{M_j+1}^{(j)} < \infty = \sigma_{M_j+2}^{(j)}$, and the interval $[-\rho, \rho]$ by the points $\beta_1^{(j)} = -\rho < \cdots < \beta_{N_j+1}^{(j)} = \rho$. Analogous to (2.19), we now have

$$\begin{aligned} V(b, b', s) = {} & V(\beta_{v(b,0)}^{(0)}, \beta_{v(b',0)}^{(0)}, \sigma_{m(s,0)}^{(0)}) \\ & + \sum_{j=0}^{K-1} [V(\beta_{v(b,j+1)}^{(j+1)}, \beta_{v(b',j+1)}^{(j+1)}, \sigma_{m(s,j+1)}^{(j+1)}) \\ & - V(\beta_{v(b,j)}^{(j)}, \beta_{v(b',j)}^{(j)}, \sigma_{m(s,j)}^{(j)})] \\ & + [V(b, b', s) - V(\beta_{v(b,K)}^{(K)}, \beta_{v(b',K)}^{(K)}, \sigma_{m(s,K)}^{(K)})]. \end{aligned}$$

Note that for $\sigma \leqslant s$,

$$\begin{aligned} V(b, b', s) - V(a, a', \sigma) = {} & [V(b, b', \sigma) - V(a, a', \sigma)] \\ & + \int_\sigma^s u_{b,b'}(t)\, d(L_b(t) - \bar{L}_b(t)). \end{aligned}$$

The rest of the proof is similar to that of Theorem 1 and Lemma 1. ∎

An argument similar to the proof of Theorem 1 can also be used to prove the following result, which will be used in Sections 4 and 5.

LEMMA 4. *With the same notation and assumptions as in Theorem* 1, *for every* $0 \leqslant \gamma < 1$ *and* $\theta > 0$,

$$\begin{aligned} & \sup_{(b,s):\ \mathrm{Var}\, Z_n(b,s) \leqslant n^{-\gamma}} |Z_n(b, s) - EZ_n(b, s)| \\ & \quad = O(n^{(1-\gamma)/2+\theta}) \qquad a.s. \end{aligned}$$

3. Stochastic Integrals of Empirical-type Processes

In this section we apply the results of Section 2 to study stochastic integrals of the form

$$\int_{s=-\infty}^{y} U_n(b, s)\, dL_n(b, s) \quad \text{or} \quad \int_{s=-\infty}^{y} U_n(b, s)\, dY_n(b, s),$$

where L_n and Y_n are the empirical-type processes defined by (1.15c) and (1.15d), and $U_n(b, s)$ are random variables for which there exist nonrandom Borel functions $u_n(b, s)$ satisfying the following assumptions for some $\xi \geqslant 0$: For every $0 \leqslant \gamma < 1$ and $\varepsilon > 0$,

(A1) $$\sup_{|b-a| \leqslant n^{-\gamma},\, -\infty < s < \infty} |U_n(b, s) - u_n(b, s) - U_n(a, s) + u_n(a, s)| = O(n^{-1/2-\gamma/2+\xi+\varepsilon}) \text{ a.s.}$$

(A2) $$\sup_{b,s} |U_n(b, s) - u_n(b, s)| = O(n^{-1/2+\xi+\varepsilon}) \text{ a.s.}$$

(A3) For fixed $b \in [-\rho, \rho]$, $U_n(b, s)$ has bounded variation in s and

$$\sup_{|b| \leqslant \rho} \int_{s=-\infty}^{\infty} |dU_n(b, s)| = O(n^{\xi}) \qquad \text{a.s.}$$

(A4) $n^{-\xi} u_n$ satisfies condition (2.21).

An example of such stochastic integrals is the linear rank statistic $S_n(b)$ defined in (1.12). In view of (1.14), we can express $S_n(b)$ in the form

$$S_n(b) = \int_{s=-\infty}^{\infty} U_n(b, s)\, dY_n(b, s) - \int_{s=-\infty}^{\infty} \tilde{U}_n(b, s)\, dL_n(b, s),$$

where $U_n(b, s) = \psi \cdot p_n(\hat{F}_{n,b}(s))\, p_n(n^{-1} \#_n(b, s))$ and $\tilde{U}_n = U_n X_n / \#_n$. Another example is given by (1.10), which can be expressed in the form

$$\log(1 - \hat{F}_{n,b}(y)) = \int_{-\infty < s < y} \log\{1 - p_n(n^{-1} \#_n(b, s))/\#_n(b, s)\}\, dL_n(b, s).$$

Theorem 2 below, which will be applied to these two examples in Section 4, shows that under certain conditions we can approximate the stochastic integral $\int_{-\infty}^{y} U_n(b, s)\, dZ_n(b, s)$ by the nonrandom function $\int_{-\infty}^{y} u_n(b, s)\, dEZ_n(b, s)$ with $Z_n = L_n$ or Y_n, and also provides two kinds of error bounds for the approximation. The first kind of results, given in (3.3) below, shows that the difference between the stochastic integral and its nonrandom approximation is of the order $O(n^{1/2+\xi+\varepsilon})$, where $\varepsilon > 0$ can be arbitrarily small. Hence if $\xi < \frac{1}{2}$, the approximation error is of the order $o(n)$. For example, in the case of the linear rank statistic $S_n(b)$ to be studied in Section 4, this implies that $\sup_{|b| \leqslant \rho} n^{-1} |S_n(b) - h_n(b)| \to 0$ a.s., where $h_n(b)$ is a nonrandom function defined in (4.3). This result can be used to establish the consistency of the rank estimator $\tilde{\beta}_n$ (which is a zero-crossing of $S_n(b)$) under certain assumptions on $h_n(b)$. To prove that $n^{1/2}(\tilde{\beta}_n - \beta)$

has a limiting normal distribution, however, the order $O(n^{1/2+\xi+\varepsilon})$ in the approximation of $S_n(b)$ by $h_n(b)$ is obviously too crude, and we need another kind of results, given by (3.2) in Theorem 2 below. Applying (3.2) to $S_n(b)$ yields that with probability 1,

$$S_n(b) = S_n(\beta) + \{h_n(b) - h_n(\beta)\} + O(n^{1/2+(\xi-\gamma/2)+\varepsilon})$$

uniformly in $|b-\beta| \leqslant n^{-\gamma}$. Thus, if $\xi < \gamma/2$, we can approximate $S_n(b) - S_n(\beta)$ by $h_n(b) - h_n(\beta)$ with an error of the order $o(n^{1/2})$ for $|b-\beta| \leqslant n^{-\gamma}$. This result is important for establishing the asymptotic normality of $\tilde{\beta}_n$, as will be discussed further in Section 4. Hence, (3.2) enables us to dampen the factor n^{ξ} in the assumptions (A1)–(A4) on U_n by using the proximity of b to β, and its usefulness will be illustrated by the applications in Sections 4 and 5.

THEOREM 2. *Let $e_1, e_2, \ldots$ be i.i.d. random variables having a continuously differentiable density function f such that*

$$\int_{-\infty}^{\infty} \Big(\sup_{s \leqslant t \leqslant s+d} |f'(t)| \Big)\, ds < \infty \qquad \textit{for some} \quad d > 0. \tag{3.1}$$

Let (x_i, t_i), $i = 1, 2, \ldots$, be independent random vectors that are independent of $\{e_n\}$ and such that conditions (2.10), (2.13), *and* (2.14) *are satisfied. Define $L_n(b, s)$ and $Y_n(b, s)$ by* (1.15c) *and* (1.15d) *with $\beta = 0$. Let $U_n(b, s)$, $u_n(b, s)$ be the same as above (satisfying* (A1)–(A4) *for some $\xi \geqslant 0$). Then for every $0 \leqslant \gamma < 1$ and $\varepsilon > 0$,*

$$\begin{aligned}
\sup_{|b-a| \leqslant n^{-\gamma},\, -\infty < y < \infty} \Bigg| &\int_{s=-\infty}^{y} U_n(b, s)\, dL_n(b, s) \\
&- \int_{s=-\infty}^{y} u_n(b, s)\, dEL_n(b, s) - \int_{s=-\infty}^{y} U_n(a, s)\, dL_n(a, s) \\
&+ \int_{s=-\infty}^{y} u_n(a, s)\, dEL_n(a, s) \Bigg| \\
&= O(n^{(1-\gamma)/2+\xi+\varepsilon}) \quad \textit{a.s.}
\end{aligned} \tag{3.2}$$

$$\begin{aligned}
\sup_{|b| \leqslant \rho,\, -\infty < y < \infty} \Bigg| &\int_{s=-\infty}^{y} U_n(b, s)\, dL_n(b, s) \\
&- \int_{s=-\infty}^{y} u_n(b, s)\, dEL_n(b, s) \Bigg| = O(n^{1/2+\xi+\varepsilon}) \ \textit{a.s.}
\end{aligned} \tag{3.3}$$

Moreover, (3.2) *and* (3.3) *still hold if L_n is replaced by Y_n.*

Proof. For fixed n, denote $U_n(b, s)$, $u_n(b, s)$, $L_n(b, s)$, $EL_n(b, s)$ by $U_b(s)$, $u_b(s)$, $L_b(s)$, and $\bar{L}_b(s)$, respectively, to simplify the notation. Note that

$$\int_{-\infty}^{y} U_b\, dL_b - \int_{-\infty}^{y} u_b\, d\bar{L}_b - \int_{-\infty}^{y} U_a\, dL_a + \int_{-\infty}^{y} u_a\, d\bar{L}_a$$

$$= \int_{-\infty}^{y} (U_b - u_b - U_a + u_a)\, dL_b + \int_{-\infty}^{y} U_a\, d(L_b - \bar{L}_b - L_a + \bar{L}_a)$$

$$+ \int_{-\infty}^{y} (u_b - u_a)\, d(L_b - \bar{L}_b) + \int_{-\infty}^{y} (U_a - u_a)\, d(\bar{L}_b - \bar{L}_a).$$

Since $\sup_{n \geqslant 1, |b| \leqslant \rho} n^{-1} \int_{-\infty}^{\infty} dL_b \leqslant 1$, it then follows from (A1) that

$$\sup_{|b-a| \leqslant n^{-\gamma}} \int_{-\infty}^{\infty} |U_b - u_b - U_a + u_a|\, dL_b$$
$$= O(n^{(1-\gamma)/2 + \xi + \varepsilon}) \quad \text{a.s.}$$

Likewise, by (A3) and Theorem 1,

$$\sup_{|b-a| \leqslant n^{-\gamma}} \int_{-\infty}^{\infty} |L_b - \bar{L}_b - L_a + \bar{L}_a|\, |dU_a|$$
$$= O(n^{(1-\gamma)/2 + \xi + \varepsilon}) \quad \text{a.s.}$$

By (A4) and Lemma 3,

$$\sup_{|b-a| \leqslant n^{-\gamma}, -\infty < y < \infty} \left| \int_{-\infty}^{y} n^{-\xi}(u_b - u_a)\, d(L_b - \bar{L}_b) \right|$$
$$= O(n^{(1-\gamma)/2 + \varepsilon}) \quad \text{a.s.}$$

We shall show that

$$\sup_{|b-a| \leqslant n^{-\gamma}, -\infty < y < \infty} \left| \int_{-\infty}^{y} (U_a - u_a)\, d(\bar{L}_b - \bar{L}_a) \right|$$
$$= O(n^{1/2 - \gamma + \xi + \varepsilon}) \quad \text{a.s.} \tag{3.4}$$

Hence the desired conclusion (3.2) follows.

To prove (3.4), first note that

$$d\bar{L}_b(s) - d\bar{L}_a(s) = \sum_{j=1}^{n} E[f(s + bx_j)\, I_{\{t_j \geqslant s + bx_j\}}$$
$$- f(s + ax_j)\, I_{\{t_j \geqslant s + ax_j\}}]\, ds. \tag{3.5}$$

By (2.10) and (2.13),

$$\sup_{|b-a| \leqslant n^{-\gamma}} E\left| \sum_{j=1}^{n} [f(s+bx_j) - f(s+ax_j)] I_{\{t_j \geqslant s+bx_j\}} \right.$$

$$\left. + \sum_{j=1}^{n} f(s+ax_j)(I_{\{t_j \geqslant s+bx_j\}} - I_{\{t_j \geqslant s+ax_j\}}) \right|$$

$$\leqslant \sup_{s-A\rho \leqslant z \leqslant s+A\rho} \left[An^{1-\gamma} |f'(z)| \right.$$

$$\left. + f(z) \sup_{|b| \leqslant \rho} \sum_{1}^{n} P\{s - An^{-\gamma} \leqslant t_j - bx_j \leqslant s + An^{-\gamma}\} \right].$$

Since $\sup_{s-A\rho \leqslant z \leqslant s+A\rho} f(z) \leqslant f(s) + A\rho \sup_{s-A\rho \leqslant z \leqslant s+A\rho} |f'(z)|$, (3.4) follows from (3.1), (3.5), and (A2).

To prove (3.3), apply (A2)–(A4) and Lemma 3 together with the bounds

$$\left| \int_{-\infty}^{y} U_b \, dL_b - \int_{-\infty}^{y} u_b \, d\bar{L}_b \right|$$

$$\leqslant \int_{-\infty}^{y} |U_b - u_b| \, d\bar{L}_b + \left| \int_{-\infty}^{y} U_b \, d(L_b - \bar{L}_b) \right|$$

$$\leqslant \int_{-\infty}^{\infty} |U_b - u_b| \, d\bar{L}_b + \int_{-\infty}^{\infty} |L_b - \bar{L}_b| \, |dU_b|$$

$$+ (|U_b(y) - u_b(y)| + |u_b(y)|) \, |L_b(y) - \bar{L}_b(y)|. \quad \blacksquare$$

4. Applications to Censored Rank Estimators

In this section we apply Theorems 1 and 2 to study the properties of the linear rank estimator $\tilde{\beta}_n$ of the slope β in the censored regression model described in Section 1. Since $\tilde{\beta}_n$ is defined as a zero crossing of the function $S_n(b)$ defined in (1.12), it is important to study the function $S_n(b)$ first. The function $S_n(b)$, however, is not a smooth function in b and therefore one cannot apply standard techniques (based on Taylor's expansion of the random function defining the estimator in a neighborhood of the true parameter) that are commonly used to prove asymptotic normality of maximum likelihood estimators, M-estimators, etc. Moreover, $S_n(b)$ is not a monotone function in b, so one cannot make use of the monotonicity and contiguity arguments (cf. [6]) that have been applied to prove asymptotic normality of rank estimators of β in the regression model (1.1) based on complete (uncensored) data (x_i, y_i). Without loss of generality, we shall assume that $\beta=0$. Theorems 1 and 2 enable us to approximate $S_n(b)$, in a neighborhood of $\beta(=0)$, by $S_n(\beta)+\{h_n(b)-h_n(\beta)\}$, where h_n is a nonrandom function which is much more tractable than $S_n(b)$. This is the content of

THEOREM 3. *With the same notation and assumptions as in Theorem 2, define $\hat{F}_{n,b}$ by (1.10) and $S_n(b)$ by (1.14), where ψ is a twice continuously differentiable function on $(0, 1)$ such that for some $\theta \geq 0$ and $i=0, 1, 2$,*

$$|\psi^{(i)}(u)| = O(u^{-\theta-i} \vee (1-u)^{-\theta-i}) \quad \text{as} \quad u(1-u) \to 0, \tag{4.1}$$

and the weight function p_n is of the form

$$p_n(x) = p(n^{\lambda}(x - cn^{-\lambda})), \qquad 0 \leq x \leq 1, \tag{4.2a}$$

with $c>0$, $0<\lambda<1$, and p being a twice continuously differentiable function on the real line such that

$$p(y)=0 \text{ for } y \leq 0, \qquad p(y)=1 \text{ for } y \geq 1. \tag{4.2b}$$

Define

$$\Lambda_{n,b}(y) = -\int_{-\infty<s<y} [p_n(n^{-1}E\#_n(b,s))/E\#_n(b,s)]\, dEL_n(b,s),$$

$$h_n(b) = \int_{-\infty}^{\infty} \psi \cdot p_n(1-e^{\Lambda_{n,b}(s)})\, p_n(n^{-1}E\#_n(b,s)) \tag{4.3}$$

$$\times \left[dEY_n(b,s) - \frac{EX_n(b,s)}{E\#_n(b,s)}\, dEL_n(b,s) \right].$$

Then for every $0 \leq \gamma < 1$ and $\varepsilon > 0$,

$$\sup_{|b-a| \leq n^{-\gamma},\, -\infty<s<\infty} |\log(1-\hat{F}_{n,b}(s)) - \Lambda_{n,b}(s) - \log(1-\hat{F}_{n,a}(s)) + \Lambda_{n,a}(s)|$$

$$= O(n^{-1/2-\gamma/2+3\lambda+\varepsilon}) \quad \text{a.s.}, \tag{4.4}$$

$$\sup_{b,s} |\log(1-\hat{F}_{n,b}(s)) - \Lambda_{n,b}(s)| = O(n^{-1/2+3\lambda+\varepsilon}) \quad \text{a.s.}, \tag{4.5}$$

$$\sup_{|b-a| \leq n^{-\gamma}} |S_n(b) - h_n(b) - S_n(a) + h_n(a)|$$

$$= O(n^{(1-\gamma)/2+(3+\theta)\lambda+\varepsilon}) \quad \text{a.s.} \tag{4.6}$$

Proof. To apply Theorem 2 we shall make use of the following inequality: For any twice continuously differentiable function g on $(0, 1)$,

$$
\begin{aligned}
|g(x_1)-g(x_2)-g(y_1)+g(y_2)| \leqslant (\sup_t |g'(t)|)\, |x_1-x_2-y_1+y_2| \\
+ (\sup_t |g''(t)|)\, |y_1-y_2| \\
\times \{|x_1-x_2|+|y_1-y_2|+|x_2-y_2|\}.
\end{aligned} \tag{4.7}
$$

Since

$$
p_n(n^{-1}\,\#(b,s))=0 \qquad \text{if} \quad \#_n(b,s)\leqslant cn^{1-\lambda}, \tag{4.8}
$$

it follows from (1.10) that

$$
\begin{aligned}
\log(1-\hat{F}_{n,b}(u)) = -\int_{-\infty<s<u} \{p_n(n^{-1}\,\#_n(b,s))/\#_n(b,s) \\
+O(\#_n^{-2}(b,s))\}\, dL_n(b,s).
\end{aligned} \tag{4.9}
$$

Let $g_n(x)=n^{-3\lambda}p(n^{\lambda}(x-cn^{-\lambda}))/x$ for $0<x\leqslant 1$. Then $\sup_{0<x\leqslant 1}(|g_n'(x)|+|g_n''(x)|)=O(1)$. By (2.13) and the continuity of f, as $n\to\infty$ and $h\to 0$ such that $nh\to\infty$,

$$
\sup_{|b-b'|+|s-s'|\leqslant h} |n^{-1}E\,\#_n(b,s)-n^{-1}E\,\#_n(b',s')| = O(h). \tag{4.10}
$$

Hence it follows from Theorem 1, Lemma 4, and (4.7) that for every $0\leqslant\gamma<1$ and $\varepsilon>0$,

$$
\begin{aligned}
\sup_{|b-a|\leqslant n^{-\gamma},\,-\infty<s<\infty} &|g_n(n^{-1}\,\#_n(b,s))-g_n(n^{-1}E\,\#_n(b,s)) \\
&-g_n(n^{-1}\,\#_n(a,s))+g_n(n^{-1}E\,\#_n(a,s))| \\
&= O(n^{-1/2-\gamma/2+\varepsilon}) \quad \text{a.s.},
\end{aligned}
$$

$$
\sup_{b,s} |g_n(n^{-1}\,\#_n(b,s))-g_n(n^{-1}E\,\#_n(b,s))| = O(n^{-1/2+\varepsilon}) \quad \text{a.s.}
$$

Moreover, $\int_{s=-\infty}^{\infty} |dg_n(n^{-1}\,\#_n(b,s))| \leqslant \sup_t |g_n'(t)|$. Noting that

$$
\begin{aligned}
\int_{-\infty<s<u} &[p_n(n^{-1}\,\#_n(b,s))/\#_n(b,s)]\, dL_n(b,s) \\
&= n^{3\lambda-1}\int_{-\infty<s<u} g_n(n^{-1}\,\#_n(b,s))\, dL_n(b,s),
\end{aligned}
$$

conclusions (4.4) and (4.5) follow from Theorem 2 (with $\xi=0$).

To prove (4.6), let $\phi_n(x) = \psi \cdot p_n(1 - e^{-x})$ for $x \geqslant 0$, so that $\psi \cdot p_n(\hat{F}_{n,b}(s)) = \phi_n(-\log(1 - \hat{F}_{n,b}(s)))$. Using (4.8), (4.9), and $dL_n \leqslant |d \#_n|$, it can be shown that there exists $K > 0$ such that

$$\sup_{b,s} |\log(1 - \hat{F}_{n,b}(s))| \leqslant \log(Kn^{\lambda}) \quad \text{for all large } n. \tag{4.11}$$

In view of (4.1) and (4.2), $\sup_{2 \leqslant e^x \leqslant Kn^{\lambda}} n^{-\theta\lambda}(|\phi_n(x)| + |\phi_n'(x)| + |\phi_n''(x)|) = O(1)$; moreover, $\sup_{1/2 \leqslant e^{-x} \leqslant 1} n^{-(2+\theta)\lambda}(|\phi_n(x)| + |\phi_n'(x)| + |\phi_n''(x)|) = O(1)$. Hence using a similar argument as before, we obtain the desired conclusion (4.6) for (1.14) by applying Theorem 2 to the cases $U_n(b, s) = n^{-(3+\theta)\lambda} \phi_n(-\log(1 - \hat{F}_{n,b}(s)))\, p_n(n^{-1} \#_n(b, s))$ and $U_n(b, s) = n^{-(3+\theta)\lambda} \times \phi_n(-\log(1 - \hat{F}_{n,b}(s))) \times n^{-1}X_n(b, s) \times p_n(n^{-1} \#_n(b, s))/[n^{-1} \#_n(b, s)]$, respectively, making use of (4.4), (4.5), and Theorem 1 in this connection. ∎

Suppose that λ in the weight function (4.2) is so chosen that $6(3 + \theta)\lambda < 1$. Then by (4.6), with probability 1,

$$S_n(b) - S_n(a) = h_n(b) - h_n(a) + o(n^{1/2}) \quad \text{uniformly in } a, b \in [-\rho, \rho] \text{ with } |b - a| \leqslant n^{-1/3}, \tag{4.12}$$

$$|S_n(b) - S_n(a) - h_n(b) + h_n(a)| = o(n^{2/3}) = o(n\,|b - a|) \quad \text{uniformly in } a, b \in [-\rho, \rho] \text{ with } |b - a| \geqslant n^{-1/3}. \tag{4.13}$$

Since $n^{-1}\,|S_n(b) - h_n(b)| \to 0$ a.s. for every fixed b, it follows from (4.12) and (4.13) that

$$\sup_{|b| \leqslant \rho} n^{-1}\,|S_n(b) - h_n(b)| \to 0 \quad \text{a.s.} \tag{4.14}$$

Under certain assumptions on the nonrandom function h_n, it can be shown by making use of (4.12)–(4.14) that the rank estimator $\tilde{\beta}_n$, which is a zero-crossing of $S_n(b)$, is strongly consistent and asymptotically normal. The details are given in [7]. In particular, the following steps are used in [7] to prove the asymptotic normality of $\tilde{\beta}_n$ after establishing its consistency. First, by (4.12) and (4.13) with $a = \beta$, we have with probability 1,

$$S_n(b) = S_n(\beta) + \{h_n(b) - h_n(\beta)\} + o(n^{1/2} \vee n\,|b - \beta|) \quad \text{uniformly in } |b| \leqslant \rho. \tag{4.15}$$

Next, an asymptotic analysis of the nonrandom function $h_n(b)$ (defined in (4.3)) shows that under certain conditions,

$$h_n(b) - h_n(\beta) \sim Cn(b - \beta) \qquad \text{as} \quad n \to \infty \text{ and } b \to \beta, \tag{4.16}$$

for some nonrandom $C \neq 0$. The third step uses a martingale central limit theorem which can be used to show, under certain assumptions, that as $n \to \infty$,

$$n^{-1/2} S_n(\beta) \text{ has a limiting normal } N(0, \tau) \text{ distribution,} \tag{4.17}$$

for some constant τ. After showing that $\tilde{\beta}_n$ converges to β a.s. and recalling that $\tilde{\beta}_n$ is a zero crossing of $S_n(b)$, we then obtain from (4.15)–(4.17) that $n^{1/2}(\tilde{\beta}_n - \beta)$ has a limiting $N(0, \tau/C^2)$ distribution. In view of (4.14), a sufficient condition for the consistency of $\tilde{\beta}_n$ is

$$\varliminf_{n \to \infty} \inf_{|b-\beta| \geq \delta} n^{-1} |h_n(b)| > 0 \qquad \text{for every} \quad \delta > 0. \tag{4.18}$$

5. Applications to the Buckley–James Estimator

In this section we consider the Buckley–James estimator, which is a zero-crossing of the function $W_n(b)$ defined in (1.9). Instead of the Kaplan–Meier-type estimator (1.6) originally used by Buckley and James, we use here the modified version (1.10), involving a weight function p_n as in Section 4, for the $\hat{F}_{n,b}$ in $z_i(b)$. In addition, we change the definition (1.7) of $z_i(b)$ as follows. Noting that

$$\begin{aligned} E(e_i \mid e_i > z) &= \int_{s>z} s \, dF(s)/(1 - F(z)) \\ &= z + \int_{s>z} (1 - F(s)) \, ds/(1 - F(z)), \end{aligned}$$

we replace (1.7) by

$$\begin{aligned} z_i(b) = t_i + \bigg\{ \int_{s > t_i - bx_i} &(1 - \hat{F}_{n,b}(s)) \\ &\times p_n(n^{-1} \#_n(b, s)) \, ds \bigg\} \bigg/ (1 - \hat{F}_{n,b}(t_i - bx_i)). \end{aligned}$$

Using this definition of $z_i(b)$ in (1.9), we obtain that

$$\begin{aligned} W_n(b) - W_n(\beta) = (\beta - b) \sum_1^n (x_i - \bar{x}_n)^2 &+ \sum_1^n (1 - \delta_i)(x_i - \bar{x}_n) \\ &\times [U_n(b, t_i - bx_i) - U_n(\beta, t_i - \beta x_i)], \end{aligned} \tag{5.1}$$

where

$$U_n(b, z) = \bigg\{ \int_{s>z} (1 - \hat{F}_{n,b}(s)) \, p_n(n^{-1} \#_n(b, s)) \, ds \bigg\} \bigg/ (1 - \hat{F}_{n,b}(z)). \tag{5.2}$$

Our analysis of $W_n(b)$ depends on the following theorem on the approximation of $U_n(b, z)$ by the nonrandom function

$$u_n(b, z) = \int_{s>z} p_n(n^{-1}E \#_n(b, s)) \exp\{\Lambda_{n,b}(s) - \Lambda_{n,b}(z)\} \, ds, \qquad (5.3)$$

where $\Lambda_{n,b}$ is defined in (4.3). Without loss of generality we shall again assume that $\beta = 0$.

THEOREM 4. *With the same notation and assumptions as in Theorem 2, define $\hat{F}_{n,b}$ by (1.10) and $U_n(b, z)$, $u_n(b, z)$ by (5.2), and (5.3), where the weight function p_n is of the form (4.2a) with $c > 0$, $0 < \lambda < \frac{1}{2}$, and p being a twice continuously differentiable function satisfying (4.2b). Assume furthermore that*

$$M \triangleq \inf\{a: P[e_1 \leqslant a] = 1\} < \infty, \ f(M) > 0, \text{ and}$$

$$\liminf_{n \to \infty} n^{-1} \sum_1^n P\{t_i \geqslant M\} > 0. \qquad (5.4)$$

Then for every $0 \leqslant \gamma < 1$, $\theta \geqslant 0$, and $\varepsilon > 0$,

$$\sup_{|b| \leqslant n^{-\gamma}, z \geqslant -n^{\theta}} |U_n(b, z) - u_n(b, z)| = O(n^{-1/2 + [(\lambda - \gamma)^+ \vee \theta] + \varepsilon}) \quad a.s. \qquad (5.5)$$

Moreover, if $\gamma > \lambda$ and $\theta < \gamma/2$, then

$$\sup_{|b| \vee |\tilde{b}| \vee |a| \leqslant n^{-\gamma}, z \geqslant n^{-\theta}} |U_n(b, z) - u_n(b, z) - U_n(\tilde{b}, z + a) + u_n(\tilde{b}, z + a)| = o(n^{-1/2}) \quad a.s. \qquad (5.6)$$

Proof. From (4.8) and Lemma 4, it follows that

$$\begin{aligned} p_n(n^{-1} \#_n(b, s)) > 0 &\Rightarrow \#_n(b, s) > cn^{1-\lambda} \quad \text{and} \quad \#_n(b, s) \sim E \#_n(b, s), \\ p_n(n^{-1}E \#_n(b, s)) > 0 &\Rightarrow E \#_n(b, s) > cn^{1-\lambda} \quad \text{and} \quad \#_n(b, s) \sim E \#_n(b, s). \end{aligned} \qquad (5.7)$$

Since $p'_n(x) = 0$ if $x \leqslant cn^{-\lambda}$ or $x \geqslant (c+1)n^{-\lambda}$ and since $p'_n(x) = O(n^{\lambda}) = O(x^{-1})$ for $cn^{-\lambda} < x < (c+1)n^{-\lambda}$, it then follows that there exists $K > 0$ such that

$$|p_n(x)/x - p_n(y)/y| \leqslant K |x - y|/x^2 \text{ if } \tfrac{1}{2} \leqslant x/y \leqslant \tfrac{3}{2} \ (x, y \in (0, 1)). \qquad (5.8)$$

From (5.7) and (5.8) together with Lemma 4, we obtain that with probability 1,

$$\int_{z \leqslant s < y} \left| \frac{p_n(n^{-1} \#_n(b, s))}{n^{-1} \#_n(b, s)} - \frac{p_n(n^{-1}E \#_n(b, s))}{n^{-1}E \#_n(b, s)} \right| d[n^{-1}EL_n(b, s)]$$

$$= O\left(n^{-1/2+\varepsilon} \left\{ \int_{z \leqslant s < y} (n^{-1}E \#_n(b, s))^2 \, d[n^{-1}EL_n(b, s)] \right\}\right)$$

$$= O(n^{-1/2+\varepsilon}/n^{-1}E \#_n(b, y)), \tag{5.9}$$

uniformly in $z < y$ with $E \#_n(b, y) \geqslant \frac{1}{2} cn^{1-\lambda}$. Here and in the sequel, ε is chosen to be an arbitrarily small positive number. Moreover, using integration by parts and Lemma 4, it can be shown that with probability 1,

$$\left| \int_{z \leqslant s < y} \frac{p_n(n^{-1} \#_n(b, s))}{n^{-1} \#_n(b, s)} d[n^{-1}L_n(b, s) - n^{-1}EL_n(b, s)] \right|$$

$$= O(n^{-1/2+\varepsilon}/n^{-1} \#_n(b, y)), \tag{5.10}$$

uniformly in $z < y$ with $\#_n(b, y) \geqslant \frac{1}{2} cn^{1-\lambda}$, noting that by (5.8),

$$|d[p_n(n^{-1} \#_n(b, s))/n^{-1} \#_n(b, s)]|$$

$$= O((n^{-1} \#_n(b, s))^{-2} \, d(n^{-1} \#_n(b, s))).$$

We now apply (5.9) and (5.10) to prove (5.5). Let $\hat{G}_{n,b} = 1 - \hat{F}_{n,b}$, $G_{n,b} = \exp(\Lambda_{n,b})$. It follows from (4.3) and (4.9) that

$$\frac{\hat{G}_{n,b}(y)}{\hat{G}_{n,b}(z)} - \frac{G_{n,b}(y)}{G_{n,b}(z)}$$

$$= \frac{G_{n,b}(y)}{G_{n,b}(z)} \left(\exp\left\{ -\int_{z \leqslant s < y} \frac{p_n(n^{-1} \#_n(b, s))}{\#_n(b, s)} dL_n(b, s) \right.\right.$$

$$\left.\left. + \int_{z \leqslant s < y} \frac{p_n(n^{-1}E \#_n(b, s))}{E \#_n(b, s)} dEL_n(b, s) + O(n^{\lambda-1}) \right\} - 1 \right). \tag{5.11}$$

First consider the case $\gamma = 0$. From (5.7), (5.9), and (5.10), it follows that

$$\sup_{|b| \leqslant \rho} \int_z^{M+A\rho} |\hat{G}_{n,b}(s) \, p_n(n^{-1} \#_n(b, s))/\hat{G}_{n,b}(z) - G_{n,b}(s)$$

$$\times p_n(n^{-1}E \#_n(b, s))/G_{n,b}(z)| \, ds \left(= \int_{(M-1) \vee z}^{M+A\rho} + \int_z^{(M-1) \vee z} \right)$$

$$= O(n^{-1/2+\varepsilon+\lambda} + n^{-1/2+\varepsilon} |z|) \quad \text{a.s.}$$

and therefore (5.5) follows. Here and in the sequel we use the convention $\int_v^u = 0$ if $v \geqslant u$. Note in this connection that, by (1.10), $\hat{G}_{n,b}(z)$ remains constant for all $z \geqslant \inf\{s\colon \#_n(b, s) \leqslant n^{1-\lambda}\}$ and that $G_{n,b}(z)$ remains constant for all $z \geqslant \inf\{s\colon E\,\#_n(b, s) \leqslant n^{1-\lambda}\}$, by (4.3). Moreover, since $|bx_i| \leqslant A\rho$ and $e_i \leqslant M$ a.s., the range of integration in (5.2) or (5.3) can be restricted to be $\leqslant M + A\rho$.

We next consider the case $\gamma > 0$. Then by (5.4), with probability 1, as $n \to \infty$ and $s \to M$ such that $M - s \geqslant n^{-\gamma+\varepsilon}$,

$$\begin{aligned} n^{-1}\,\#_n(b, s) &\sim n^{-1}E\,\#_n(b, s) \\ &\sim f(M)(M-s)\,n^{-1}\sum_{i=1}^{n} P\{t_i \geqslant s + bx_i\} \\ &\qquad \text{uniformly in } |b| \leqslant n^{-\gamma}, \end{aligned} \tag{5.12}$$

since $|bx_i| \leqslant An^{-\gamma} = o(M-s)$. Moreover, by (4.3) and (5.4), as $n \to \infty$ and $y \to M$ such that $M - y \geqslant n^{-\gamma+\varepsilon}$,

$$\begin{aligned} G_{n,b}(y) &= \exp(\Lambda_{n,b}(y)) \\ &= (M-y)^{1+o(1)} \quad \text{uniformly in } |b| \leqslant n^{-\gamma}. \end{aligned} \tag{5.13}$$

To prove (5.5), it suffices to assume that $\gamma \leqslant \lambda$. From (5.9)–(5.13), it then follows that with probability 1,

$$\begin{aligned} &\int_z^{M-n^{-\gamma+\varepsilon}} |\hat{G}_{n,b}(s)\,p_n(n^{-1}\,\#_n(b, s))/\hat{G}_{n,b}(z) - G_{n,b}(s) \\ &\qquad \times p_n(n^{-1}E\,\#_n(b, s))/G_{n,b}(z)|\,ds \left(= \int_{(M-1)\vee z}^{M-n^{-\gamma+\varepsilon}} + \int_z^{(M-1)\vee z} \right) \\ &= O(n^{-1/2+2\varepsilon} + n^{-1/2+\varepsilon}\,|z|) \text{ uniformly in } z \text{ and in } |b| \leqslant n^{-\gamma}, \end{aligned} \tag{5.14}$$

noting in view of (5.12) and (5.4) that $p_n(n^{-1}E\,\#_n(b, s)) = 1$ for $s \leqslant M - n^{-\gamma+\varepsilon}$ and large n, since $\gamma \leqslant \lambda$. For $M - n^{-\gamma+\varepsilon} \leqslant s \leqslant M + An^{-\gamma}$, we use the bounds $G_{n,b}(s)/G_{n,b}(z) \leqslant 1$ if $s \geqslant z$, and

$$\begin{aligned} &|\hat{G}_{n,b}(s)\,p_n(n^{-1}\,\#_n(b, s))/\hat{G}_{n,b}(z) - G_{n,b}(s) \\ &\qquad \times p_n(n^{-1}E\,\#_n(b, s))/G_{n,b}(z)| \\ &\leqslant |\hat{G}_{n,b}(s)/\hat{G}_{n,b}(z) - G_{n,b}(s)/G_{n,b}(z)|\;p_n(n^{-1}\,\#_n(b, s)) \\ &\qquad + [G_{n,b}(s)/G_{n,b}(z)]|\;n^{-1}\,\#_n(b, s) \\ &\qquad - n^{-1}E\,\#_n(b, s)\;|n^{\lambda} \sup_x |p'(x)|. \end{aligned} \tag{5.15}$$

From (5.9)–(5.11) and (5.15) together with Lemma 4, it follows that with probability 1,

$$\int_{(M-n^{-\gamma+\varepsilon})\vee z}^{M+An^{-\gamma}} |\hat{G}_{n,b}(s)\, p_n(n^{-1}\,\#_n(b,s))/\hat{G}_{n,b}(z)$$
$$-G_{n,b}(s)\, p_n(n^{-1}E\,\#_n(b,s))/G_{n,b}(z)|\, ds$$
$$=O(n^{-1/2+\lambda+2\varepsilon-\gamma}) \text{ uniformly in } z \text{ and in } |b|\leqslant n^{-\gamma}. \qquad (5.16)$$

From (5.14) and (5.16), we obtain (5.5) (with ε replaced by $\tilde{\varepsilon}=2\varepsilon$, which can be arbitrarily small).

We now assume that $\gamma>\lambda$ and $\theta<\gamma/2$ to prove (5.6). First note that for $|b|\leqslant n^{-\gamma}$, $\sup_i |bx_i|\leqslant An^{-\gamma}=o(n^{-\lambda})$. Hence analogous to (5.12), we now have for $|b|\leqslant n^{-\gamma}$,

$$\#_n(b,s)\geqslant cn^{1-\lambda} \quad \text{and} \quad s\to M\Rightarrow \#_n(b,s)$$
$$\sim E\,\#_n(b,s)\sim f(M)(M-s)\sum_1^n P\{t_i\geqslant s+bx_i\}. \qquad (5.17)$$

Moreover, analogous to (5.13), we now have for $|b|\leqslant n^{-\gamma}$,

$$E\,\#_n(b,s)\geqslant cn^{1-\lambda} \quad \text{and} \quad s\to M\Rightarrow G_{n,b}(s)=(M-s)^{1+o(1)}. \qquad (5.18)$$

Since $E\,\#_n(b,s)\sim f(M)(M-s)\sum_1^n P\{t_i\geqslant s+bx_i\}=O(n^{1-\xi})$ uniformly in $|b|\leqslant n^{-\gamma}$ and $s\geqslant M-n^{-\xi}$, we obtain from Lemma 4 together with (5.7) and (5.8) the following refinement of (5.9) and (5.10): With probability 1,

$$\left|\int_{z\vee(M-n^{-\xi})\leqslant s<y}\left\{\frac{p_n(n^{-1}\,\#_n(b,s))}{\#_n(b,s)}\,dL_n(b,s)\right.\right.$$
$$\left.\left.-\frac{p_n(n^{-1}E\,\#_n(b,s))}{E\,\#_n(b,s)}\,dEL_n(b,s)\right\}\right|$$
$$=O(n^{-1/2-\xi/2+\varepsilon}/n^{-1}E\,\#_n(b,y)). \qquad (5.19)$$

From (5.11), (5.17), (5.18), and (5.19), it follows that with probability 1,

$$\sup_{|b|\leqslant n^{-\gamma}}\int_{z\vee(M-n^{-\xi})}^{M+An^{-\gamma}} |\hat{G}_{n,b}(s)\, p_n(n^{-1}\,\#_n(b,s))/\hat{G}_{n,b}(z)-G_{n,b}(s)$$
$$\times p_n(n^{-1}E\,\#_n(b,s))/G_{n,b}(z)|\, ds$$
$$=O(n^{-1/2+2\varepsilon-\xi/2}), \quad \text{uniformly in } z, \qquad (5.20)$$

where $\xi>0$ and $\varepsilon>0$ are so chosen that

$$\lambda>\xi>4\varepsilon, \quad 3\xi+\theta+\varepsilon<\gamma/2, \quad 6\xi+2\varepsilon+\theta<\tfrac{1}{2}. \qquad (5.21)$$

Since $\xi < \lambda$, $p_n(n^{-1}E\,\#_n(b, s)) = 1$ and $n^{-1}E\,\#_n(b, s) \geqslant \text{constant} \times n^{-\xi}$ for $s \leqslant M - n^{-\xi}$ and large n. Hence the same argument used to prove (4.4) and (4.5) of Theorem 3 can be used to show that

$$\sup_{|b| \vee |\bar{b}| \vee |a| \leqslant n^{-\gamma}, s \leqslant M - n^{-\xi}} |\log \hat{G}_{n,b}(s) - \Lambda_{n,b}(s)$$
$$- \log \hat{G}_{n,\bar{b}}(s+a) + \Lambda_{n,\bar{b}}(s+a)|$$
$$= O(n^{-1/2 - \gamma/2 + 3\xi + \varepsilon}) \quad \text{a.s.} \tag{5.22}$$

$$\sup_{|b| \leqslant \rho, s \leqslant M - n^{-\xi}} |\log \hat{G}_{n,b}(s) - \Lambda_{n,b}(s)|$$
$$= O(n^{-1/2 + 3\xi + \varepsilon}) \quad \text{a.s.} \tag{5.23}$$

From (5.22) and (5.23) together with the inequality (4.7) applied to $g(x) = e^x$ with $x \leqslant 1$, it follows that

$$\sup_{|b| \vee |\bar{b}| \vee |a| \leqslant n^{-\gamma}, -n^{\theta} \leqslant z \leqslant M - n^{-\xi}} \left| \int_z^{M - n^{-\xi}} \left(\frac{\hat{G}_{n,b}(s)}{\hat{G}_{n,b}(z)} - \frac{G_{n,b}(s)}{G_{n,b}(z)} \right) ds \right.$$
$$\left. - \int_{z+a}^{M - n^{-\xi}} \left(\frac{\hat{G}_{n,\bar{b}}(s)}{\hat{G}_{n,\bar{b}}(z+a)} - \frac{G_{n,\bar{b}}(s)}{G_{n,\bar{b}}(z+a)} \right) ds \right|$$
$$= O(n^{-1/2 - \gamma/2 + 3\xi + \varepsilon + \theta} + n^{-1 + 6\xi + 2\varepsilon + \theta}) \text{ a.s.} \tag{5.24}$$

From (5.20), (5.21), and (5.24), (5.6) follows. ∎

Suppose that λ in the weight function p_n above is so chosen that $\frac{1}{4} < \lambda < \frac{1}{2}$. Then making use of (5.1) and Theorem 4 and following the steps similar to those outlined at the end of Section 4 for the rank estimator $\tilde{\beta}_n$, we can prove the consistency and asymptotic normality of the Buckley–James estimator under certain regularity conditions. The details are given in [8].

References

[1] Alexander, K. (1984). Probability inequalities for empirical processes and a law of the iterated logarithm. *Ann. Probab.* **12** 1041–1067. Correction in *Ann. Probab.* **15** (1987), 428–430.

[2] Bennett, G. (1962). Probability inequalities for the sums of independent random variables. *J. Amer. Statist. Assoc.* **57** 33–45.

[3] Buckley, J., and James, I. (1979). Linear regression with censored data. *Biometrika* **66** 429–436.

[4] Dudley, R. M. (1978). Central limit theorems for empirical measures. *Ann. Probab.* **6** 899–929.

[5] James, I. R., and Smith, P. J. (1984). Consistency results for linear regression with censored data. *Ann. Statist.* **12** 590–600.

[6] JUREČKOVÁ, J. (1969). Asymptotic linearity of a rank statistic in regression parameter. *Ann. Math. Statist.* **40** 1889–1900.
[7] LAI, T. L., AND YING, Z. (1987). *Linear Rank Statistics in Regression Analysis of Censored or Truncated Data.* Stanford University Technical Report.
[8] LAI, T. L., AND YING, Z. (1988). *Consistency and Asymptotic Normality of Buckley-James-Type Statistics in Regression Analysis of Censored or Truncated Data.* Stanford University Technical Report.

Nonminimum Phase Non-Gaussian Deconvolution

KEH-SHIN LII

Department of Statistics, University of California, Riverside, Riverside, California 92521

AND

MURRAY ROSENBLATT

Department of Mathematics, University of California, San Diego, La Jolla, California 92093

A procedure for deconvolution of nonminimum phase non-Guassian time series based on the estimation of higher order (greater than two) spectra is given. This can be applied to the analysis of seismograms. The procedure allows estimation of the wavelet. Knowledge of cumulant spectra of order greater than two allows estimation of the phase of the wavelet. In this way one has access to information not available in the ordinary second-order deconvolution procedures. Computational details of the method for estimating the phase of the wavelet are given. There are simulated illustrative examples. One of the examples is based on an actual reflectivity series from a sonic well log. The method is effective asymptotically in the nonminimum phase non-Gaussian context where the Wiener–Levinson procedure does not apply.

INTRODUCTION

We shall make use of a model that has been used often in deconvolution. It is that of a linear process

$$x_t = \sum \alpha_k \xi_{t-k},$$

where $\{\alpha_k\}$ is the wavelet sequence, $\{x_t\}$ the seismogram, and $\{\xi_t\}$ the reflectivity sequence which is here assumed to be a sequence of independent, identically distributed non-Gaussian random variables. It has been claimed that many seismograms are non-Gaussian [1, 12] and we shall indicate how a non-Gaussian character (as contrasted with a

Multivariate Statistics and Probability
ISBN 0-12-580205-6

Reprinted from *J. Mult. Anal.* **27**(2).

Gaussian character) allows us to resolve most of the phase information. We shall just deal with this model and not consider many real difficulties like multiple reflection and multipaths. We shall assume that the seismogram sequence $\{x_t\}$ is observed but that the wavelet and reflectivity are unknown. The object is to estimate as much as one can about the wavelet and to deconvolve x_t so as to estimate the reflectivity series ξ_t. This will be accomplished by making use of higher order moment (cumulant) or spectral estimates. A discussion of the method has been given elsewhere [3, 9] but our object here is to give an exposition in a geophysical context. The method described has the positive feature that for a non-Gaussian nonminimum phase stationary sequence, it will yield estimates that converge to the wavelet with probability one as the sample size increases and correspondingly will also effect deconvolution with probability 1 (see [3]). Such a result has not been established in Donoho [1], Matsuoka and Ulrych [7], and Wiggins [12], where computational aspects of related procedures are described. Wiener–Levinson deconvolution will not converge to a nonminimum phase wavelet asymptotically and thus will not deconvolve in such a context. In the spirit of exposition of what appears to us a fruitful procedure which does not solve by any means many of the real difficulties but does represent an advance relative to an important aspect of deconvolution, we try to describe relevant features. One of our examples has some attempted aspect of a geophysical context. We should mention that the method discussed is only effective in the non-Gaussian case and is suggested for nonminimum phase series. We shall presently give a more detailed discussion of the model. In the next section we shall describe the computational procedures associated with the method. In the third section, a number of illustrations will be given. One example will involve a wavelet with three nonzero values and an exponentially distributed reflectivity series. Other examples will have spikey data with trinomial reflectivity series. The wavelet then has 20 nonzero values. In the last example, using a well-log reflectivity series provided by Henkart and a wavelet that is a recorded water gun signature, we will generate by convolution a possible seismogram x_t. By using our method, we shall estimate the wavelet and deconvolve. This will be compared with a Wiener–Levinson deconvolution (see [8]). It should be noted that the wavelet is not strictly minimum phase. Of course, the reflectivity series we give is obtained by a sonic measuring device and there is consequently a distortion of the real reflectivity that we shall discuss later.

Assume that $\{\xi_t\}$ is a non-Gaussian sequence with mean zero and kth order cumulant $\gamma_k \neq 0$ for some $k > 2$. Further let the α_k's be real with

$$\sum_k |\alpha_k|^2 < \infty.$$

Actually stronger assumptions will be made later on. Then the spectral density of the x_t sequence is

$$\frac{1}{2\pi}|\alpha(e^{-i\lambda})|^2\sigma^2,$$

where

$$\alpha(z)=\sum \alpha_k z^k$$

and $\sigma^2>0$ is the ξ variance. The kth order cumulant of random variables $Y_1, ..., Y_k$ is given in terms of moments by the relation

$$\operatorname{cum}(Y_1, ..., Y_k)=\sum(-1)^{p-1}(p-1)!\\ \times E\left(\prod_{j\in v_1} Y_j\right)\cdots E\left(\prod_{j\in v_p} Y_j\right),$$

where $v_1, ..., v_p$ is a partition of $(1, 2, ..., k)$ and the sum is over all such partitions. We write out these relations in the case $k=2, 3, 4$ when $EY_j=0$, $j=1, ..., k$. Notice that then the cumulants of order 2 and 3 are the same as the corresponding moments

$$\operatorname{cum}(Y_1, Y_2)=E(Y_1 Y_2)$$
$$\operatorname{cum}(Y_1, Y_2, Y_3)=E(Y_1 Y_2 Y_3)$$

but the cumulant of order 4 differs, as is the case with higher order cumulants,

$$\begin{aligned}\operatorname{cum}(Y_1, Y_2, Y_3, Y_4)=&E(Y_1 Y_2 Y_3 Y_4)\\ &-E(Y_1 Y_2)E(Y_3 Y_4)\\ &-E(Y_1 Y_3)E(Y_2 Y_4)\\ &-E(Y_1 Y_4)E(Y_2 Y_3).\end{aligned}$$

If $Y_1=Y_2=Y_3=Y_4$ the corresponding 4th cumulant is sometimes called the coefficient of kurtosis. It is more appropriate to consider Fourier transforms (series) in higher order cumulants rather than the corresponding higher order moments. Further, the kth order cumulant for the process x_t is

$$\operatorname{cum}(x_{t_0}, x_{t_1}, ..., x_{t_{k-1}})=\sum_s \alpha_s\alpha_{s+t_1-t_0}\cdots\alpha_{s+t_{k-1}-t_0}\gamma_k$$

and so the kth order cumulant spectral density [10] of the process x_t is

$$b_k(\lambda_1, ..., \lambda_{k-1}) = \frac{\gamma_k}{(2\pi)^{k-1}} \sum_{j_1, ..., j_{k-1}} \operatorname{cum}(x_0, x_{j_1}, ..., x_{j_{k-1}}) \times \exp\left(-\sum_{s=1}^{k-1} i j_s \lambda_s\right)$$

$$= \frac{\gamma_k}{(2\pi)^{k-1}} \alpha(e^{-i\lambda_1}) \cdots \alpha(e^{-i\lambda_{k-1}}) \, \alpha(e^{i(\lambda_1 + \cdots + \lambda_{k-1})}). \quad (1)$$

Introduce the function

$$h(\lambda) = \arg\left\{\alpha(e^{-i\lambda}) \frac{\alpha(1)}{|\alpha(1)|}\right\},$$

assuming that $\alpha(1) \neq 0$. Relation (1) implies that

$$h(\lambda_1) + \cdots + h(\lambda_{k-1}) - h(\lambda_1 + \cdots \lambda_{k-1}) = \arg\left[\left\{\frac{\alpha(1)}{|\alpha(1)|}\right\}^k \alpha_k^{-1} b_k(\lambda_1, ..., \lambda_{k-1})\right],$$

since $h(-\lambda) = -h(\lambda)$. This relation clearly implies that knowledge of the kth order cumulant spectral density $b_k(\lambda_1, ..., \lambda_{k-1})$ gives one information about $h(\lambda)$.

We shall actually require that

$$\sum |k\alpha_k| < \infty \quad (2)$$

because we want to have continuous differentiability of $\alpha(e^{-i\lambda})$. One can show that there is an integer linear indeterminacy in the phase of $\alpha(e^{-i\lambda})$ for these stochastic models under the conditions we specify [3]. The linear indeterminacy in the phase corresponds to an indeterminacy in the time indexing of the ξ_t process. For convenience we shall actually assume more than (2), specifically that $\alpha(z)$ is analytic in an annulus containing the unit circle. Then, of course, $\alpha(e^{-i\lambda})$ can have zeros but they are at most finite in number.

To effect deconvolution in the non-Gaussian case one must estimate the argument of $\alpha(e^{-i\lambda})$ or $h(\lambda)$. Information of this character requires knowledge about higher order moments or cumulants. It cannot be obtained from information on the covariances alone. The deconvolution is carried out by estimating $\alpha(e^{-i\lambda})^{-1}$. Information on the absolute value of $\alpha(e^{-i\lambda})$ (or its inverse) can be obtained from the second-order spectral density. But information on the argument or phase of $\alpha(e^{-i\lambda})$ can only be obtained from data on kth order cumulant spectra with $k > 2$.

COMPUTATION

We shall now consider computational questions. For convenience, the case $k=4$ will be discussed in some detail but the case $k=3$ can be considered in quite an analogous manner. We focus on $k=4$ assuming $\gamma_4 \neq 0$ because the skewness of data encountered often seems to be small [2, p. 2110; 13, p. 2723]. Since $k=4$, we shall be dealing with fourth-order cumulant spectral estimates. Initially we will assume that

$$\alpha(e^{-i\lambda}) \neq 0 \tag{3}$$

for any λ and later see how to remove this assumption. Because of (2) and (3) h is continuously differentiable and

$$h(\lambda) = \int_0^\lambda \{h'(u) - h'(0)\}\, du + c\lambda = h_1(\lambda) + c\lambda, \qquad c = h'(0). \tag{4}$$

Now $h(\pi)$ has to be an integral multiple of π because the α_j's are real. One therefore can rewrite (4) as

$$h(\lambda) = h_1(\lambda) - \frac{h_1(\pi)}{\pi}\lambda + a\lambda$$

with "a" an indeterminate integer. Thus

$$h'(0) - h'(\lambda) = \lim_{\Delta \to 0} \frac{1}{2\Delta} \{h(\lambda) + 2h(\Delta) - h(\lambda + 2\Delta)\}$$

up to an indeterminancy in sign. Let us set $\Delta = \Delta(n)$, $k\Delta = \lambda$, and consider $\Delta = \Delta(n) \to o$ as $n \to \infty$. Now $b(0, 0, 0)$ is positive if $\gamma_4 > 0$. For the sake of simplicity, assume $\gamma_4 > 0$.

Notice that

$$\begin{aligned}\sum_{j=1}^{k-1} \arg b(j\Delta, \Delta, \Delta) &= \sum_{j=1}^{k-1} \{h(j\Delta) + 2h(\Delta) - h(j\Delta + 2\Delta)\} \\ &= 2[kh(\Delta) - h(k\Delta)] + B\end{aligned}$$

with

$$B = h(2\Delta) - h(\Delta) + h(k\Delta) - h((k+1)\Delta)$$

and so if $\lambda = k\Delta$,

$$h_1(\lambda) = h(\lambda) - h'(0)\lambda \cong -\frac{1}{2}\sum_{j=1}^{k-1} \arg b(j\Delta, \Delta, \Delta) - \frac{1}{2}B.$$

We start with phase zero at frequency zero and then proceed by proximity or continuity. If Δ is small, B would also be expected to be small. A plausible estimate of $h_1(\lambda)$ would then be given by

$$G_n(\lambda) = -\frac{1}{2} \sum_{j=1}^{k-1} \arg {}_n b(j\Delta, \Delta, \Delta),$$

where ${}_n b(j\Delta, \Delta, \Delta)$ is an estimate of the fourth-order cumulant spectral density $b(j\Delta, \Delta, \Delta)$ based on a sample of size n.

Of course estimates ${}_n b(j\Delta, \Delta, \Delta)$ can be computed in terms of the fast Fourier transform of the data. A more detailed discussion of this procedure using FFT can be found in [6]. If there are 1000 data points and there are at most ten nonzero contiguous α_k's, this method based on FFT appears to lead to reasonable results. However, if one still has 1000 data points and the number of nonzero contiguous α_k's is as long as 50 (as often is the case with real data) methods based on FFT do not appear to give reasonable estimates. This might be due to the fact that a third- (fourth-) order periodogram using a FFT based on data of length m has a variance of the order m^2 (m^3) (see [6] 1976) and reduction of the size of this variance is accomplished in part by smoothing over disjoint sections in frequency domain. It is perhaps startling that better estimates (in terms of resolution) than those obtained by FFT are obtained by making use of classical Fourier analysis in our experience. One estimates cumulants and then Fourier transforms them with appropriate weights. Of course, the weights have to be appropriately chosen. Our computations, for the most part in this paper, will be based on this classical Fourier transform procedure.

We shall briefly describe such a computation. Our estimates of the moments

$$E(x_0 x_j x_k x_l), \qquad |j|, |k|, |l| \leqslant M \ll n$$

on the basis of a sample $x_0, ..., x_n$ are

$$\frac{1}{n-2M+1} \sum_{t=M}^{n-M} x_t x_{t+j} x_{t+k} x_{t+l}.$$

Here we assume $Ex_t \equiv 0$. The second moments

$$E(x_0, x_u), \qquad |u| \leqslant M,$$

are estimated by

$$\frac{1}{n-2M+1} \sum_{t=M}^{n-M} x_t x_{t+u}.$$

The natural estimates of the cumulants

$$\begin{aligned} c_{j,k,l} &= \mathrm{cum}(x_0, x_j, x_k, x_l) \\ &= E(x_0 x_j x_k x_l) - E(x_0 x_j)\,E(x_k x_l) \\ &\quad - E(x_0 x_k)\,E(x_j x_l) - E(x_0 x_l)\,E(x_j x_k) \end{aligned}$$

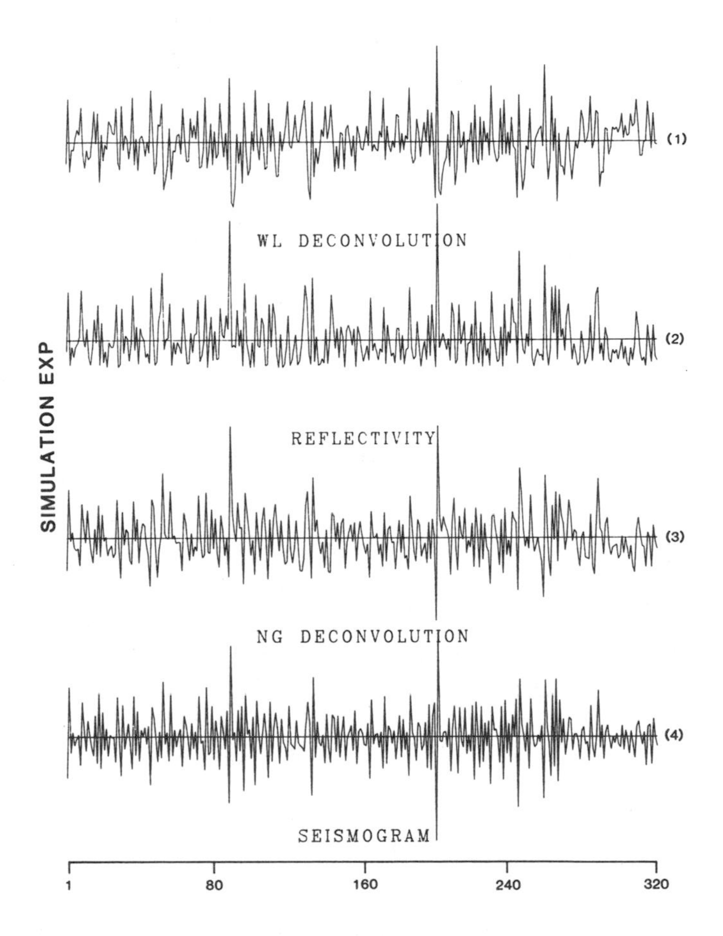

FIG. 1. Deconvolution of a second-order moving average $x_t = \varepsilon_t - 5\varepsilon_{t-1} + 6\varepsilon_{t-2}$ which has roots $\frac{1}{2}$ and $\frac{1}{3}$. The reflectivity ε_t's are generated by independent identically distributed exponential random variables with parameter 1. (2) is the reflectivity ε_t which generates the seismogram x_t, (4). (1) and (3) are deconvolution of x_t to estimate ε_t by the Wiener–Levinson method and the non-Gaussian method, respectively. The horizontal scale is from 1 to 320 units. Vertical scale is normalized to mean 0 and variance 1.

are given in terms of the moment estimates. The estimate

$$ {}_n b(\lambda_1, \lambda_2, \lambda_3) = \frac{1}{(2\pi)^3} \sum_{|j|,|k|,|l| \leq m} c_{j,k,l} \times w_{j,k,l}^{(n)} \exp\{-i(j\lambda_1 + k\lambda_2 + l\lambda_3)\}, $$

with $w_{j,k,l}^{(n)}$ an appropriately chosen set of weights. In our case we often chose

$$ w_{j,k,l} = \left(1 - \frac{|j|}{M}\right)\left(1 - \frac{|k|}{M}\right)\left(1 - \frac{|l|}{M}\right). $$

If one appears to have zeros of $a(e^{-i\lambda})$ it is appropriate to add a small

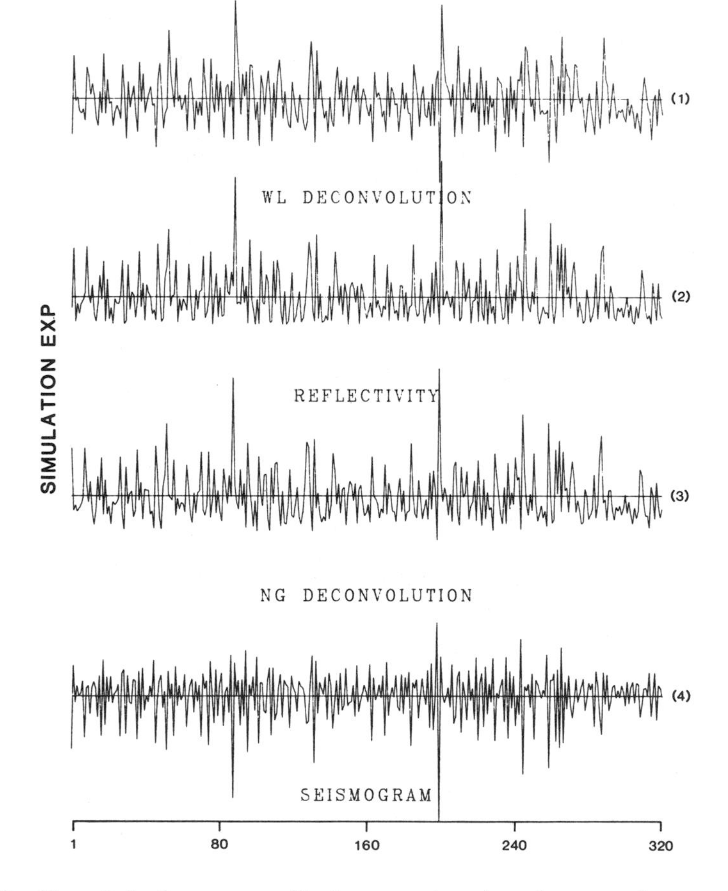

FIG. 2. Figure 2 is the same as Fig. 1 except that the seismogram is generated by $x_t = \varepsilon_t - 2.33\varepsilon_{t-1} + 0.867\varepsilon_{t-2}$ which has the roots $\frac{1}{2}$ and 3.

amount of Gaussian white noise to the data and then deconvolve as suggested above. This type of procedure has been suggested in Treitel and Wang [11]. A more formal justification can be found in Lii and Rosenblatt [5]. These procedures appear to be robust relative to the addition of a mild amount of Gaussian noise (see [4]).

ILLUSTRATIONS AND EXAMPLES. The first and second examples are the moving averages

$$x_t = \varepsilon_t - 5\varepsilon_{t-1} + 6\varepsilon_{t-2}$$

and

$$x_t = \varepsilon_t - 2.33\varepsilon_{t-1} + 0.667\varepsilon_{t-2},$$

respectively with the ε_t's independent, identically distributed exponential

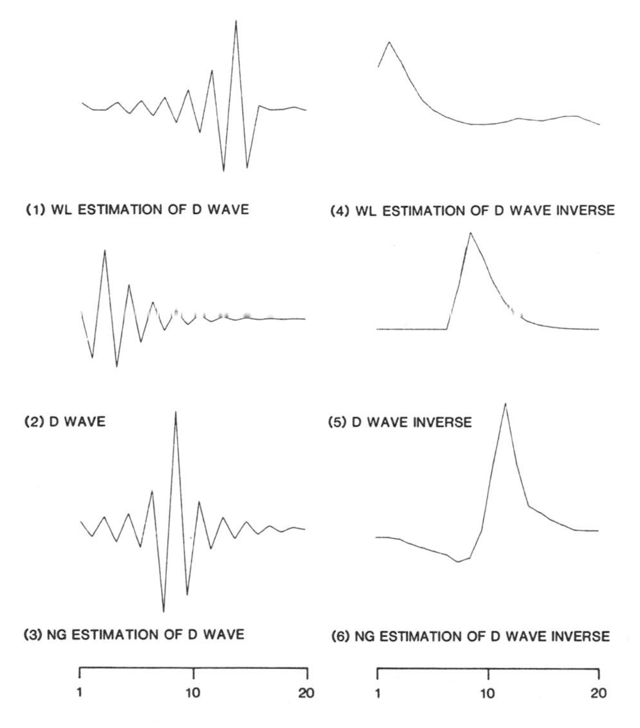

FIG. 3. The D wavelet is generated by expanding $(1 - 5e^{-i\lambda} + 6e^{-2i\lambda})/(1 - 0.667e^{-i\lambda})$ and then truncating at $e^{-i20\lambda}$. This wavelet has two roots inside of the unit circle ($\frac{1}{2}$ and $\frac{1}{3}$) and seventeen roots outside of the unit circle. The D wavelet is given in (2). (1) and (3) are estimates of (2) by the Wiener–Levinson method and the non-Gaussian method, respectively. (4), (5), and (6) are the inverses of (1), (2), and (3), respectively. The horizontal scale is 1 through 20 with arbitrary time shift. The vertical scale is arbitrary.

random variables with parameter 1. The first and second figures have graphs of the reflectivity series (the ε_t's), the seismogram (x_t series) generated, as well as the results of our non-Gaussian deconvolution and the Wiener–Levinson deconvolution for these two examples. In all cases, the sample size is 1280 points and a fourth-order cumulant spectrum is used in the deconvolution. It is apparent in both these cases that the non-Gaussian deconvolution does a better job of reproducing the reflectivity than the Wiener–Levinson deconvolution. Of course, both these examples are nonminimum phase and non-Gaussian.

The second and third examples have as their wavelets the D and F wavelets as given in Figs. 3 and 4, respectively. The reflectivity series for these examples are generated from a sequence of independent, identically distributed trinomial variables with the instantaneous distribution

$$\varepsilon_t = \begin{cases} 1 & \text{with probability } 0.05 \\ -1 & \text{with probability } 0.05 \\ 0 & \text{with probability } 0.90. \end{cases}$$

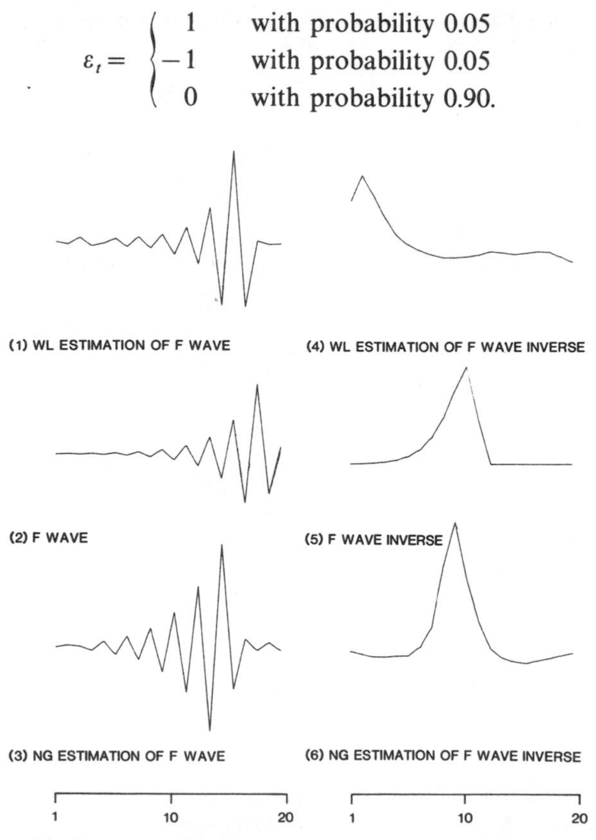

FIG. 4. Figure 4 is the same as Fig. 3 except that the wavelet F is obtained by the reversing of the time direction in wavelet D. Roots of the z-transform of wavelet D are the inverse of the roots of the z-transform of the F wavelet. Notice that (1) and (4) are the same as (1) and (4) in Fig. 3.

The D wavelet is obtained by expanding

$$\frac{1-5e^{-i\lambda}+6e^{-2i\lambda}}{1-0.7e^{-i\lambda}} \tag{5}$$

and truncation at $e^{-i20\lambda}$. The F wavelet is obtained by replacing $e^{-i\lambda}$ in (5) by $e^{i\lambda}$, expanding, and truncating at $e^{i20\lambda}$. A graph is given of the zero locations of the z-transforms of the D and F wavelets in Fig. 5. Notice that the roots in the case of the D z-transform are the inverses of the roots of the F z-transform. In Figs. 6 and 7 graphs are given of the reflectivity series, the seismogram generated, the results of our non-Gaussian deconvolution and the Wiener–Levinson deconvolution. The version of Wiener–Levinson we have used is based on the computation of the one step prediction error. In these two examples the non-Gaussian deconvolution does give a closer estimate of the reflectivity series than the Wiener–Levinson deconvolution. The object in the case of the D and F wavelets was to generate simulated

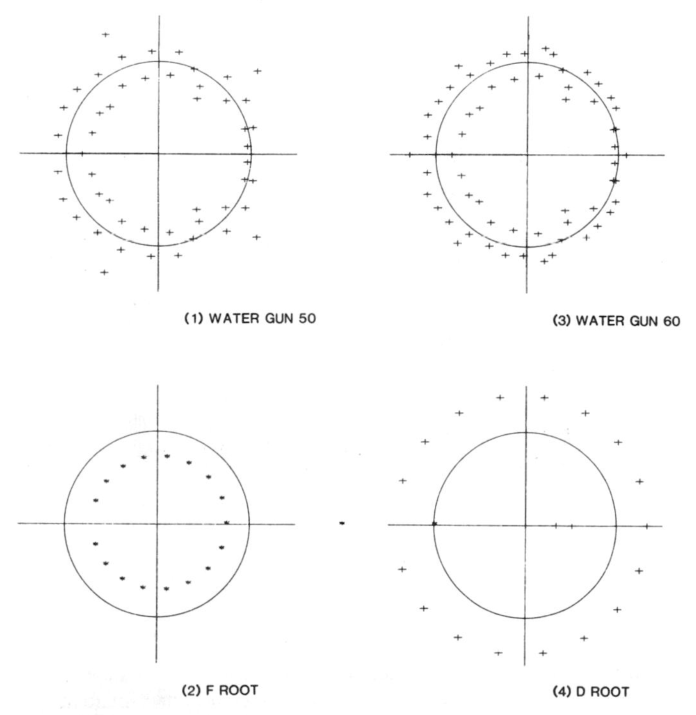

FIG. 5. (2) and (4) give locations of the roots of the z-transform of wavelets F and D, respectively. Locations are given relative to the unit circle on the complex plane. (1) and (3) give locations of roots of the z-transform of a water-gun signature truncated at 50 and 60 lags, respectively.

series with a larger number of lags, relative to the data sample size, than in the first two examples. Notice that in Figs. 3 and 4 both the wavelet and the Fourier inverse of the wavelet are graphed. Then the Wiener–Levinson and non-Gaussian estimates of the wavelet and the inverse are also given.

The last example concerns an actual set of well-log reflectivity readings

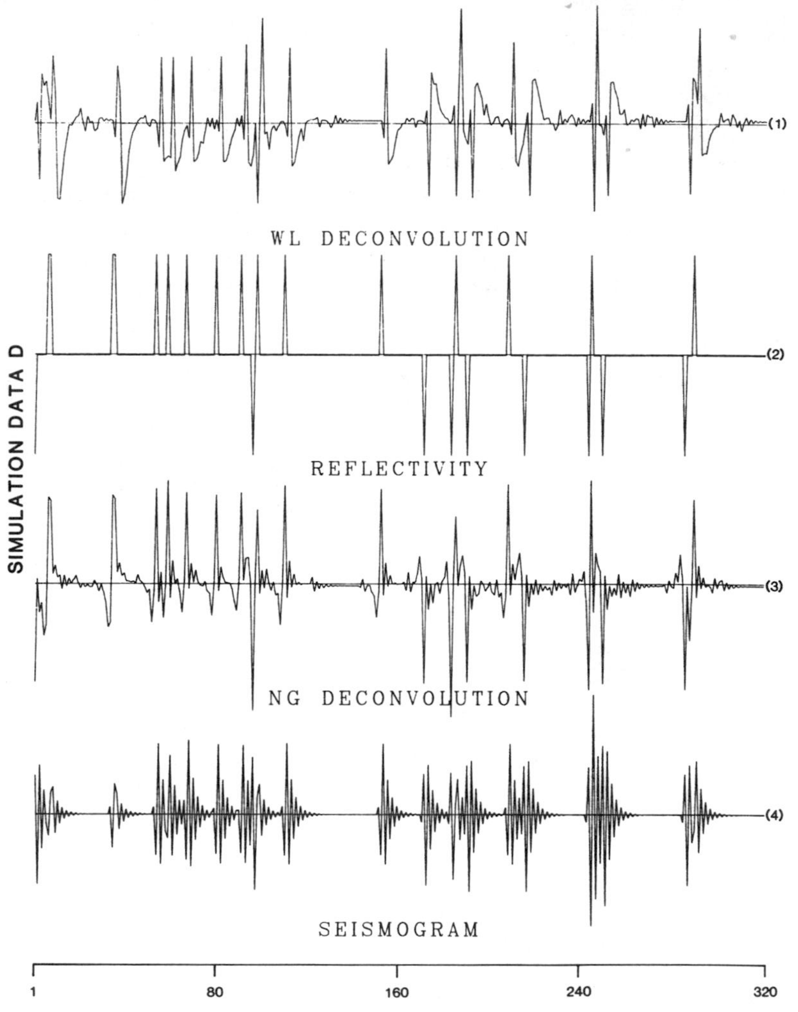

FIG. 6. Reflectivity in (2) is generated from a sequence of independent, identically distributed trinomial random variables which take values 0 with probability 0.9 and take values 1 and -1 with probability 0.05 each. Seismogram (4) is generated by the convolution of the D wavelet with (2). Deconvolution results by the Wiener–Levinson method and the non-Gaussian method are given in (1) and (3), respectively. The horizontal scale is from 1 to 320; the vertical scale is normalized to mean zero and variance one.

obtained by an oil company and supplied to us by Paul Henkart. This reflectivity series was passed through the filter corresponding to the water gun wavelet shown in Fig. 8. The result was the simulated seismogram pictured in Fig. 9. This was then deconvolved by the non-Gaussian and the Wiener–Levinson deconvolution procedures. We note that the effective length of the water gun signature is about 50 to 60 lags. The non-Gaussian deconvolution does appear to give a series closer to the reflectivity than does the Wiener–Levinson deconvolution. The contrast of the non-Gaussian deconvolution with the Wiener–Levinson deconvolution in this

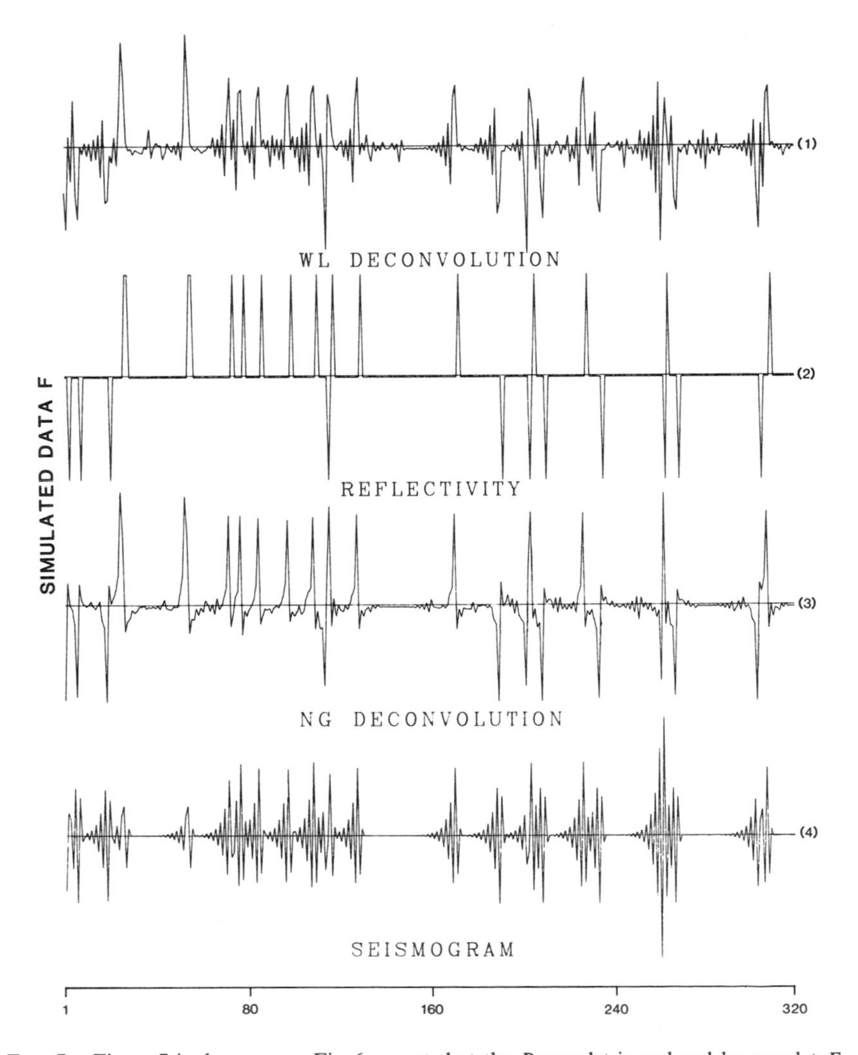

FIG. 7. Figure 7 is the same as Fig. 6 except that the D wavelet is replaced by wavelet F.

case is not as pronounced as previous simulated cases and is perhaps due to the relative length of the wavelet with respect to the length of the data (seismogram). Asymptotic theory tells us that the longer the length of the seismogram relative to the wavelet length the better the filter estimates and the deconvolution. However, in comparing the deconvolution with the reflectivity we should note that the assumption of independence of reflectivity readings in our model is certainly not satisfied by the actual reflectivity readings. These readings are made by a sonic device from overlapping sections in the descent. For this reason it might be better to model the reflectivity readings as a moving average.

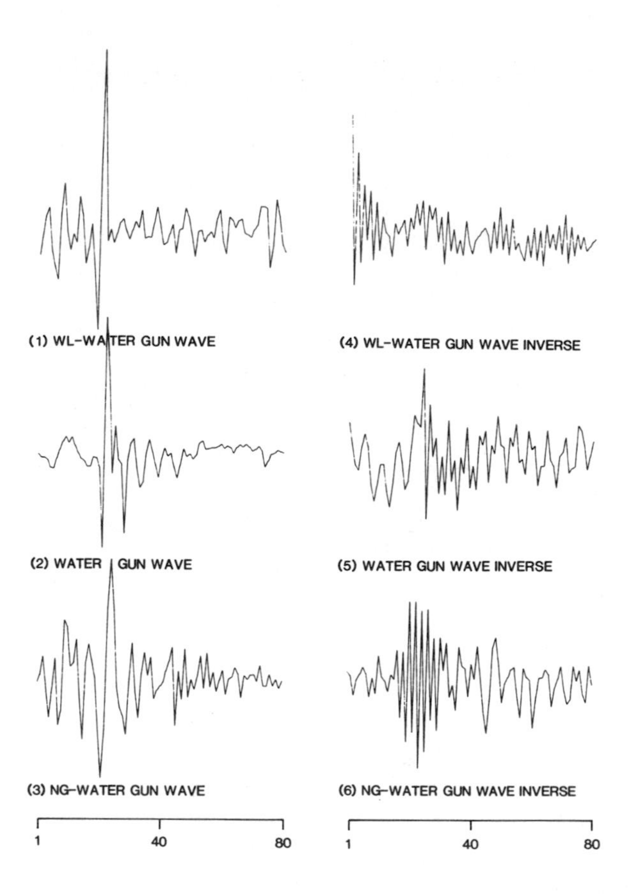

FIG. 8. A water-gun wavelet is plotted in (2). (1) and (3) give estimates of (2) by the Wiener–Levinson method and non-Gaussian method, respectively. (4), (5), and (6) give the inverses of (1), (2), and (3), respectively. The horizontal scale is 1 through 80. The vertical scale is arbitrary.

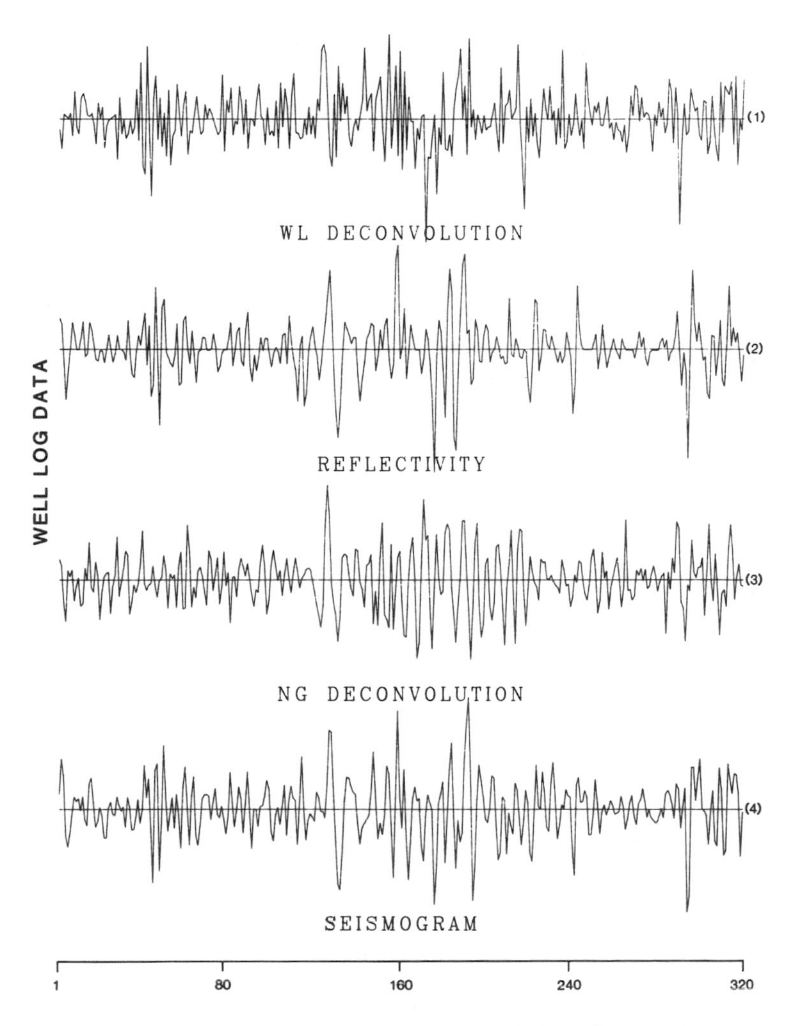

FIG. 9. Reflectivity (2) is from well-log data. Convolution of (2) with the watergun signature is given in (4). (1) and (3) are deconvolutions of (4) using the Wiener–Levinson method and the non-Gaussian method, respectively. The horizontal scale is from 1 to 320. The vertical scale is normalized to mean zero and variance one.

CONCLUSIONS

In this paper we describe and illustrate a procedure for deconvolution that allows us to estimate the phase of the transfer function in the non-Gaussian case without making use of the ad hoc minimum phase assumption. This method converges asymptotically as the sample size increases relative to the effective length of the wavelet. This is not true of the Wiener–Levenson procedure in the nonminimum phase context. Questions relating to multiple reflections, multipath data, and heavy noise are not addressed.

Acknowledgments

We are indebted to Paul Henkart for the well-log data, the water gun wavelet, and for some stimulating discussions. We also wish to acknowledge the great assistance given us by the National Science Foundation in providing access to the Boeing Cray computer. This allowed us to carry out the analysis of the well-log data. This research was also supported by the Office of Naval Research.

References

[1] Donoho, D. (1981). On minimum entropy deconvolution. In *Applied Time Series Analysis II* (D. F. Findley, Ed.), pp. 565–608, New York/London.

[2] Jurkevics, A., and Wiggins, R. (1984). A critique of seismic deconvolution methods, *Geophysics* **49** 2109–2116.

[3] Lii, K. S., and Rosenblatt, M. (1982). Deconvolution and estimation of transfer function phase and coefficients for nonGaussian linear processes, *Ann. Statist.* **10** 1195–1208.

[4] Lii, K. S., and Rosenblatt, M. (1984). Remarks on nonGaussian linear processes with additive Gaussian noise, Lecture Notes in Statistics, Vol. 26, pp. 185–197.

[5] Lii, K. S., and Rosenblatt, M. (1986). Deconvolution of non-Gaussian linear processes with vanishing spectral values, *Proc. Nat'l. Acad. Sci. USA* **86** 199–200.

[6] Lii, K. S., Rosenblatt, M., and Van Atta, C. (1976). Bispectral measurements in turbulence, *J. Fluid Mech.* **77** 45–62.

[7] Matsuoka, T., and Ulrych, T. (1984). Phase estimation using the bispectrum, *Proc. IEEE* **72** 1403–1411.

[8] Peacock, K. L., and Treitel, S. (1969). Predictive deconvolution: Theory and practice, *Geophysics* **34** 155–169.

[9] Rosenblatt, M. (1980). Linear processes and bispectra, *J. Appl. Probab.* **17** 265–270.

[10] Rosenblatt, M. (1985). "Stationary Sequences and Random Fields," Birkhaüser, Basel.

[11] Treitel, S., and Wang, R. J. (1976). The determination of digital Wiener filters from an ill-conditioned system of normal equations, *Geophys. Prospecting* **24** 317–327.

[12] Wiggins, R. (1978). Minimum entropy deconvolution, *Geoexploration* **16** 21–35.

[13] Wiggins, R. (1985). Entropy guided deconvolution, *Geophysics* **50** 2720–2726.

Inference in a Model with at Most One Slope-Change Point

B. Q. MIAO

University of Pittsburgh

In this paper the problem of slope-change point in linear regression model is discussed with the help of the theory of Gaussian process. The distribution of the estimators of the change point proposed in this paper can be approximated by the first type of extremal distribution. Based on this fact, the detection and interval estimation of a change-point in various situations are discussed. © 1988 Academic Press, Inc.

1. INTRODUCTION

Consider the model

$$x(t) = f(t) + \varepsilon_t, \qquad 0 < t \leqslant 1, \tag{1.1}$$

where $f(t)$ is a nonrandom function with thc form

$$f(t) = \begin{cases} \mu + \beta_1(t - t_0), & 0 < t \leqslant t_0 \\ \mu + \beta_2(t - t_0), & t_0 < t \leqslant 1. \end{cases} \tag{1.2}$$

t_0 is called the slope change point (of $f(t)$, or the model (1.1)), ε_t is the random error of the model, while μ, β_1, β_2, and t_0 are unknown parameters.

For given integer n we take observations of $x(t)$ at $t = i/n$, $i = 1, ..., n$. For simplicity of writing, $x(i/n)$ and $\varepsilon(i/n)$ will be abbreviated to x_i and ε_i, respectively.

* Research sponsored by the Air Force Office of Scientific Research under Contract F49620-58-C-0008. The U.S. Government's right to retain a nonexclusive royalty-free license in and to the copyright covering this paper, for governmental purposes, is acknowledged.

Multivariate Statistics and Probability
ISBN 0-12-580205-6

Reprinted from *J. Mult. Anal.* **27**(2).

The problem of making statistical inference in this model is important in practical applications and of much theoretical interest. Many authors have contributed to it. To name a few among others, Hudson [6], Hinkley [4, 5], Feder [3], Krishnaiah and Miao [10], and Csörgö and Horváth [2].

In this paper we shall propose a method of dealing with this problem. Our method possesses a desirable feature in that the asymptotic distribution of the proposed statistic is very simple, which allows us to derive simple procedures for various inference problems in this model. The basic idea of the method is motivated by recent works of Yin [12] and Chen [1].

In Section 2 we treat the case where $\varepsilon_1, ..., \varepsilon_n$ are normal with zero mean and known variance σ^2. In Section 3 we consider the normal case with unknown σ^2. Section 4 considers the nonnormal case. Finally, in Section 5 we discuss the estimation of the slope change $\beta_1 - \beta_2$ under some mild conditions.

2. Normal Error with Known Variance

In this section we suppose that $\varepsilon_1, ..., \varepsilon_n$ are i.i.d. with mean zero and known variance σ^2. Our method is based on the following theorem:

THEOREM 1. *Suppose that*

$$x_k = a + \frac{k}{n}\beta + \varepsilon_k, \qquad k = 1, ..., n, \tag{2.1}$$

where $\varepsilon_1, ..., \varepsilon_n$ *are i.i.d.,* $\varepsilon_1 \sim N(0, \sigma^2)$. *Let* $m = m_n$ *be a positive integer such that*

$$n \gg m \gg n^{2/3} \log^{2/3} n. \tag{2.2}$$

Here and in the sequel, $u_n \gg v_n > 0$ *means* $\lim_{n\to\infty}(u_n/v_n) = \infty$. *Set*

$$Y_k = \frac{1}{2\sqrt{m}} [(x_{k-4m+1} + \cdots + x_{k-3m}) - (x_{k-3m+1} + \cdots + x_{k-2m}) - (x_{k-2m+1} + \cdots + x_{k-m}) + (x_{k-m+1} + \cdots + x_k)], \qquad k = 4m, 4m+1, ..., n, \tag{2.3}$$

$$\xi_n = \max_{4m \leqslant k \leqslant n} |Y_k|,$$

and

$$A_n(x)=\left(2\log\left(\frac{5n}{4m}-5\right)\right)^{-1/2}$$
$$\times\left(x+2\log\left(\frac{5n}{4m}-5\right)+\frac{1}{2}\log\log\left(\frac{5n}{4m}-5\right)-\frac{1}{2}\log\pi\right). \quad (2.4)$$

Then

$$\lim_{n\to\infty} P\left(\frac{\xi_n}{\sigma}\leqslant A_n(x)\right)=\exp\{-2e^{-x}\}, \qquad -\infty<x<\infty. \quad (2.5)$$

Proof. Construct a standard Brownian Motion $\{W(t): t\geqslant 0\}$, such that

$$W\left(\frac{5k}{4m}\right)=\sqrt{\frac{5}{4m}}\left(x_1+\cdots+x_k-ka-\frac{k(k+1)}{2n}\beta\right)\Big/\sigma, \qquad k=4m, ..., n. \quad (2.6)$$

Define a Gaussian process $Z(t)$ by

$$Z(t)=\frac{1}{\sqrt{5}}\left[W(t+5)-2W\left(t+\frac{15}{4}\right)+2W\left(t+\frac{5}{4}\right)-W(t)\right], \qquad t\geqslant 0. \quad (2.7)$$

It is easy to see that

$$Y_k=\sigma Z\left(\frac{5k}{4m}-5\right), \qquad k=4m, ..., n, \quad (2.8)$$

and the covariance function $\rho(\tau)$ of $Z(t)$ is

$$\rho(\tau)=\begin{cases}1-|\tau| & |\tau|\leqslant\frac{5}{4}\\ -\frac{1}{5}|\tau| & \frac{5}{4}\leqslant|\tau|\leqslant\frac{5}{2}\\ \frac{3}{5}|\tau|-2 & \frac{5}{2}\leqslant|\tau|\leqslant\frac{15}{4}.\\ 1-\frac{1}{5}|\tau| & \frac{15}{4}\leqslant|\tau|\leqslant 5\\ 0 & |\tau|>5\end{cases} \quad (2.9)$$

Set

$$\tilde{\xi}_n=\sup\left\{|Z(t)|: 0\leqslant t\leqslant\frac{5n}{4m}-5\right\},$$

$$\eta_n=\tilde{\xi}_n-\sigma^{-1}\xi_n.$$

Similarly to Chen [1] it can be shown that

$$\lim_{n \to \infty} \eta_n \sqrt{\log n} = 0, \quad \text{a.s.} \tag{2.10}$$

For the Gaussian process $Z(t)$ with covariance function $\rho(\tau)$, the conditions of the theorem of Qualls and Watanable [11] are satisfied, and we get

$$\lim_{n \to \infty} P(\tilde{\xi}_n \leqslant A_n(x)) = \exp\{-2e^{-x}\}. \tag{2.11}$$

Since $A_n(x)$ is linear in x, for n large we have

$$\begin{aligned} &P(\tilde{\xi}_n \leqslant A_n(x - |\Delta x|)) - P(\eta_n \geqslant |\Delta x| / \sqrt{2 \log n}) \\ &\quad \leqslant P(\xi_n / \sigma \leqslant A_n(x)) \\ &\quad \leqslant P(\tilde{\xi}_n \leqslant A_n(x + |\Delta x|)) + P(\eta_n \geqslant |\Delta x| / \sqrt{2 \log n}). \end{aligned} \tag{2.12}$$

From (2.10) to (2.12), letting $n \to \infty$, then $\Delta x \to 0$, we obtain (2.5).

This theorem suggests a way to test the null hypothesis that no change points exists, i.e.,

$$H_0 \colon \theta \equiv \beta_2 - \beta_1 = 0. \tag{2.13}$$

For this purpose, we have only to solve the equation $\exp(-2e^{-x}) = 1 - \alpha$ for a chosen level $\alpha \in (0, 1)$. The solution is

$$x(\alpha) = -\log(-\tfrac{1}{2} \log(1 - \alpha)).$$

Set

$$d = \frac{4m}{n}, \qquad C_n(\alpha, d) = A_n(x(\alpha)). \tag{2.14}$$

The null hypothesis (2.13) is rejected when and only when

$$\xi_n > \sigma C_n(\alpha, d). \tag{2.15}$$

From Theorem 1 it is seen that this test has an asymptotic level α as the sample size n tends to infinity.

We can give an approximate power $\beta_n = \beta_n(\beta_1, \beta_2, \sigma)$ of this test. Let r be the integer such that

$$\frac{r}{n} \leqslant t < \frac{r+1}{n}.$$

Then

$$Y_{r+2m} \sim N\left(\frac{m^{3/2}}{2n}(\beta_2-\beta_1), \sigma^2\right).$$

Hence,

$$\begin{aligned}\beta_n(\beta_1, \beta_2, \sigma) &\geqslant P(|Y_{r+2n}| > \sigma C_n(\alpha, d)) \\ &> \Phi\left(\frac{m^{3/2}}{2n\sigma}|\beta_2-\beta_1| - C_n(\alpha, d)\right),\end{aligned} \tag{2.16}$$

where

$$\Phi(x) = \int_{-\infty}^{x} \frac{1}{\sqrt{2\pi}} e^{-t^2/2}\, dt.$$

Next consider the interval estimate of the slope change point t_0. The existence of t_0 may be a fact known in advance, but usually it is evidenced by the rejection of the null hypothesis.

RULE. *Find an integer k such that* $|Y_k| = \xi_n$. *Take* $[(k-4m)/n, k/n]$ *as the confidence interval of* t_0.

The length of this interval is $4m/n$. Hence, the smaller the value of m, the more accurate is the estimate. m cannot be taken too small, for from (2.16) it can be seen that the risk of false acceptance of the hypothesis (2.13) will increase. We can give an approximate value of the confidence coefficient γ of this rule:

$$\begin{aligned}\gamma &= P\left(\frac{k-4m}{n} \leqslant t_0 \leqslant \frac{k}{n}\right) \\ &\geqslant P\Big(\Big\{\sup_{k \notin [r, r+4m]} |Y_k| \leqslant \sigma C_n(\alpha, d)\Big\} \cap \{|Y_{r+2m}| > \sigma C_n(\alpha, d)\}\Big).\end{aligned}$$

Set

$$A = \Big\{\sup_{4m \leqslant k < r} |Y_k| \leqslant \sigma C_n(\alpha, d)\Big\},$$

$$B = \Big\{\sup_{r+4m < k \leqslant n} |Y_k| \leqslant \sigma C_n(\alpha, d)\Big\},$$

$$B_1 = \Big\{\sup_{r+6m < k \leqslant n} |Y_k| \leqslant \sigma C_n(\alpha, d)\Big\},$$

and

$$C = \{|Y_{r+2m}| > \sigma C_n(\alpha, d)\}.$$

Notice that the event B_1 is independent of both A and C, and $B \subset B_1$, we have

$$\gamma \geqslant P((A \cup B)C) \geqslant P(C) - (P(B_1) - P(B)) - P(\bar{A})\, P(\bar{B}_1),$$

where $\bar{D}$ denotes the complementary event of D. Again, using Theorem 1, we get

$$\begin{aligned}\gamma \geqslant \Phi&\left(\frac{m^{3/2}\,|\beta_2 - \beta_1|}{2n\sigma} - C_n(\alpha, d)\right)\\ &- (\exp\{-2e^{-x_3(\alpha)}\} - \exp\{-2e^{-x_2(\alpha)}\})\\ &- (1 - \exp\{-2e^{-x_1(\alpha)}\})(1 - \exp\{-2e^{-x_3(\alpha)}\}),\end{aligned} \tag{2.17}$$

where

$$\begin{aligned}x_1 = x_1(\alpha) &= C_n(\alpha, d)\left(2\log\left(\frac{5r}{4m} - 5\right)\right)^{1/2}\\ &-\left(2\log\left(\frac{5r}{4m} - 5\right) + \frac{1}{2}\log\log\left(\frac{5r}{4m} - 5\right) - \frac{1}{2}\log \pi\right),\end{aligned} \tag{2.18}$$

$$\begin{aligned}x_2 = x_2(\alpha) &= C_n(\alpha, d)\left(2\log\left(\frac{5(n-r)}{4m} - 5\right)\right)^{1/2}\\ &-\left(2\log\left(\frac{5(n-r)}{4m} - 5\right) + \frac{1}{2}\log\log\left(\frac{5(n-r)}{4m} - 5\right) - \frac{1}{2}\log \pi\right),\end{aligned} \tag{2.19}$$

and

$$\begin{aligned}x_3 = x_3(\alpha) &= C_n(\alpha, d)\left(2\log\left(\frac{5(n-r)}{4m} - 7.5\right)\right)^{1/2}\\ &-\left(2\log\left(\frac{5(n-r)}{4m} - 7.5\right)\right.\\ &\left.+\frac{1}{2}\log\log\left(\frac{5(n-r)}{4m} - 7.5\right) - \frac{1}{2}\log \pi\right).\end{aligned} \tag{2.20}$$

Since

$$\begin{aligned}P\Big(\sup_{k \notin [r, r+4m]} |Y_k| \leqslant \sigma C_n(\alpha, d)\Big) &\geqslant P\Big(\sup_{4m \leqslant k \leqslant n} |Y_k| \leqslant \sigma C_n(\alpha, d)\Big)\\ &\approx 1 - \alpha.\end{aligned} \tag{2.21}$$

We get

$$\gamma > \Phi\left(\frac{m^{3/2}\,|\beta_2 - \beta_1|}{2n\sigma} - C_n(\alpha, d)\right) - \alpha. \tag{2.22}$$

By the above inequalities, we see that γ increases with $(2n\sigma)^{-1}m^{3/2}\,|\beta_2-\beta_1|$. But the length of the confidence interval is $4m/n$. So in the choice of m we must strike a balance between these two considerations. Usually the slope-change point is of practical importance only when $|\beta_2-\beta_1|$ is reasonably large as coompared with σ, say $|\beta_2-\beta_1|/2\sigma \geqslant M$, where M is a constant decided by practical considerations.

In practical applications we often have to give an answer to the following important question: How can we choose suitable integers m and n so that the confidence interval of t_0 formed above has a length not greater than d_0 and confidence coefficient not smaller than $1-\alpha_0$? For this purpose, take $\alpha=\alpha_0/2$ in (2.22). Solve the equations

$$\begin{aligned} \Phi\left(M\frac{m^{3/2}}{n}-C_n(\alpha_0/2,\, d_0)\right)-\alpha_0/2 &= 1-\alpha_0, \\ d_0 &= 4m/n; \end{aligned} \tag{2.23}$$

we obtain

$$m=(4/d_0M)^2(C_n(\alpha_0/2,\, d_0)+u_{\alpha_0/2})^2, \qquad n=4m/d_0, \tag{2.24}$$

where $u_{\alpha_0/2}$ is the upper percentile $(\alpha_0/2)$-point of $N(0, 1)$.

If we know in advance that $a \leqslant t_0 \leqslant b$, for some known constants a, b, $0<a<b<1$, then $an \leqslant r \leqslant bn$. From (2.18)–(2.20) we can calculate the minimum value $\tilde{x}_1(\alpha_0)$ of $x_1(\alpha_0)$ and the maximum values $\tilde{x}_2(\alpha_0)$, $\tilde{x}_3(\alpha_0)$ of $x_2(\alpha_0)$, $x_3(\alpha_0)$, all under the restriction that $an \leqslant r \leqslant bn$. (2.17) suggests that in this case we should choose m as the solution of the equation

$$\begin{aligned} &\Phi\left(\frac{M}{2}\sqrt{m}-C_n(\alpha_0,\, d_0)\right)-(\exp\{-2e^{-\tilde{x}_3(\alpha_0)}\}-\exp\{-2e^{-\tilde{x}_2(\alpha_0)}\}) \\ &\qquad -(1-\exp\{-2e^{\;x_1(\alpha_0)}\})(1-\exp\{-2e^{-x_3(\alpha_0)}\})=1-\alpha_0, \end{aligned} \tag{2.25}$$

and $n=4m/d_0$, as before.

From this we see that if some prior information about t_0 is available, then it can be utilized to construct a confidence interval with greater confidence coefficient. Also, the related test will have a smaller critical value.

3. Normal Error with Unknown Variance

When σ^2 is unknown, we form an estimate, say $\hat{\sigma}_n^2$. Substitute $\hat{\sigma}_n$ for σ in (2.15) to perform the test. Following Chen [1], we can prove the following theorem.

THEOREM 2. *Under the conditions of Theorem* 1, *if* $\hat{\sigma}_n^2$ *is an estimator of* σ^2 *satisfying*

$$\lim_{n \to \infty} |\hat{\sigma}_n^2 - \sigma^2| \log n = 0, \qquad \textit{in probability.} \tag{3.1}$$

Then

$$\lim_{n \to \infty} P(\xi_n/\hat{\sigma}_n - A_n(x)) = \exp\{-2e^{-x}\}.$$

Our problem is to find an estimator satisfying (3.1). We propose to use the MLE of σ^2 given below in (3.5). It will be shown that this estimator satisfies (3.1).

Suppose $x_1, ..., x_n$ are observations from the model (1.1) and (1.2) such that

$$x_i = \begin{cases} \mu_1 + \dfrac{i-n_1}{n}\beta_1 + \varepsilon_i, & i = 1, ..., n_1 \\[2ex] \mu_2 + \dfrac{i-n_1}{n}\beta_2 + \varepsilon_i, & i = n_1 + 1, ..., n. \end{cases} \tag{3.2}$$

$\varepsilon_1, ..., \varepsilon_n$ are random errors. We assume that the slope-change point t_0 falls into $[n_1/n, (n_1+1)/n)$. By (1.2), we have

$$|\mu_1 - \mu_2| < \frac{1}{n}|\beta_2 - \beta_1|. \tag{3.3}$$

Let

$$\bar{x}_{1c} = \frac{1}{c}\sum_{i=1}^{c} x_i, \qquad \bar{x}_{2c} = \frac{1}{n-c}\sum_{i=c+1}^{n} x_i,$$

$$\Sigma_{Lc} = \frac{2}{c(c-1)}\sum_{i=1}^{c}(c-i)\,x_i, \qquad \Sigma_{Rc} = \frac{2}{(n-c)(n-c+1)}\sum_{i=c+1}^{n}(i-c)\,x_i.$$

THEOREM 3. *Suppose that* $\varepsilon_1, ..., \varepsilon_n$ *are i.i.d., and* $\varepsilon_1 \sim N(0, \sigma^2)$. *Set*

$$S_{nc}^2 = \sum_{i=1}^{c}(x_i - \bar{x}_{1c})^2 + \sum_{i=c+1}^{n}(x_i - \bar{x}_{2c})^2 - \frac{3c(c-1)}{c+1}(\Sigma_{Lc} - \bar{x}_{1c})^2$$
$$- \frac{3(n-c)(n-c+1)}{n-c-1}(\Sigma_{Rc} - \bar{x}_{2c})^2; \tag{3.4}$$

$$\hat{\sigma}_{nc}^2 = \frac{1}{n} S_{nc}^2, \qquad c = m+1, ..., n-m. \tag{3.5}$$

Then

$$\min_{m \leqslant c \leqslant n-m} |\hat{\sigma}^2_{nc} - \sigma^2| \log n \xrightarrow{p} 0. \tag{3.6}$$

Proof. Write

$$F_c = \begin{pmatrix} e_c & -\frac{1}{n} f_c & 0 & 0 \\ 0 & 0 & e_{n-c} & \frac{1}{n} g_{n-c} \end{pmatrix} \tag{3.7}$$

$$\begin{aligned} e'_j &= (1, ..., 1)'_{1\times j}, & f_j &= (j-1, j-2, ..., 1, 0)'_{1\times j} \\ g'_j &= (1, ..., j)'_{1\times j}, & \beta &= (\mu_1, \beta_1, \mu_2, \beta_2)' \\ x &= (x_1, ..., x_n)', & \varepsilon &= (\varepsilon_1, ..., \varepsilon_n)'. \end{aligned} \tag{3.8}$$

Then

$$x = F_{n_1}\beta + \varepsilon \tag{3.9}$$

and

$$\begin{aligned} S^2_{nc} &= x'(I - F_c(F'_cF_c)^{-1}F'_c)x; \\ F_c(F'_cF_c)^{-1}F'_c &= \begin{pmatrix} f_{11} & 0 \\ 0 & f_u \end{pmatrix}, \end{aligned} \tag{3.10}$$

where

$$\begin{aligned} f_{11} &= a_{1c}e_ce'_c - a_{2c}f_ce'_c - a_{2c}e_cf'_c + n^{-1}a_{3c}f_cf'_c, \\ f_{22} &= b_{1c}e_{n-c}e'_{n-c} - b_{2c}g_{n-c}e'_{n-c} - b_{2c}e_{n-c}g'_{n-c} + n^{-1}b_{3c}g_{n-c}g'_{n-c}; \end{aligned} \tag{3.11}$$

$$\begin{aligned} a_{1c} &= \frac{2(2c-1)}{c(c-1)}, & a_{2c} &= \frac{6}{c(c+1)}, & a_{3c} &= \frac{12n}{c(c^2-1)} \\ b_{1c} &= \frac{2(2n-2c+1)}{(n-c)(n-c-1)}, & b_{2c} &= \frac{6}{(n-c)(n-c-1)}, & b_{3c} &= \frac{12n}{(n-c)(n-c+1)(n-c-1)}. \end{aligned} \tag{3.12}$$

Without loss of generality, we assume that $n > c > n_1$. Set $k \cong c - n_1$:

$$F_{c-n_1} \cong F_c - F_{n_1}. \tag{3.13}$$

We have

$(F_c - F_{n_1})' F_c (F_c' F_c)^{-1} F_c' =$

$$\left[\begin{array}{ll} \frac{k}{c^3}(4c^2 - 9kc + 6k^2)\, e'_{n_1} - \frac{6k(c-k)}{c^3} f'_{n_1} \\ -\frac{k}{nc^3}((c^3 - 2kc^2 + 4k^2c - 2k^3)\, e'_{n_1} + k(3c-2k) f'_{n_1}) \\ -\frac{k}{c^3}((4c^2 - 9kc + 6k^2)\, e'_{n_1} - 6(c-k) f'_{n_1}) \\ -\frac{k^2}{nc^3}((2(c-k)^2\, e'_{n_1} - (3c-2k) f'_{n_1}) \end{array}\right.$$

$$\left.\begin{array}{ll} \frac{k}{c^3}(c(4c-3k)\, e'_{c-n_1} - 6(c-k) f'_{c-n_1}) & 0 \\ -\frac{k}{nc^3}(c(c-k)^2\, e'_{c-n_1} + k(3c-2k) f'_{c-n_1}) & 0 \\ -\frac{k}{c^3}(c(4c-3k)\, e'_{c-n_1} - 6(c-k) f'_{c-n_1}) & 0 \\ -\frac{k^2}{n^3}(c(2c-k)\, e'_{c-n_1} - (3c-2k) f'_{c-n_1}) & -\frac{k}{c} e'_{n-c} \end{array}\right] + O\left(\frac{1}{n}\right). \tag{3.14}$$

Set $G = F_c(F_c'F_c)^{-1}F_c' - F_{n_1}(F_{n_1}'F_{n_1})^{-1}F_{n_1}' = (g_{ij})_{n\times m}$. By a tedious calculation, we can get

$$E\left|\sum_{i=1}^{n} g_{ii}\varepsilon_1^2\right| \leqslant E\sum_{i=1}^{n} \{\mathrm{tr}(F_c(F_c'F_c)^{-1}F_c') + \mathrm{tr}(F_{n_1}(F_{n_1}'F_{n_1})^{-1}F_{n_1}')\}\, \varepsilon_i^2 = 8\sigma^2 + O\left(\frac{1}{n}\right), \tag{3.15}$$

$$E\left|\sum_{i\neq j} g_{ij}\varepsilon_i\varepsilon_j\right|^2 = \sum g_{ij}^2\sigma^4 \leqslant 280\sigma^4 + O\left(\frac{1}{n}\right). \tag{3.16}$$

Write $\gamma' = \beta' F'_{c-n_1}(I - F_c(F_c'F_c)^{-1}F_c')$. From (3.14) and (3.3), we get

$$\mathrm{Var}(\gamma'\varepsilon) = \sigma^2\,\mathrm{tr}(\gamma\gamma') = \sigma^2\gamma'\gamma \tag{3.17}$$

$$\frac{k^4 n_1^3}{4n^2c^4}(\beta_2 - \beta_1)^2 \leqslant \gamma'\gamma \leqslant \frac{3k^4 n_1^3}{n^2c^4}(\beta_2 - \beta_1)^2 + \frac{100k^2(n-c)}{n^2}\beta_2^2. \tag{3.18}$$

By (3.8), (3.9), and (3.5),

$$\hat{\sigma}_{nc}^2 - \hat{\sigma}_{nn_1}^2 = -2\gamma'\varepsilon + \varepsilon' G\varepsilon + \gamma'\gamma. \tag{3.19}$$

Now consider $(\hat{\sigma}_{nc}^2 - \hat{\sigma}_{nn_1}^2)$.

Case 1. $\beta_1 \neq \beta_2$ and $k = |c - n_1| \geqslant n/\log^2 n$. We have, by (3.15)–(3.18),

$$\begin{aligned}
&P(\hat{\sigma}_{nc}^2 - \hat{\sigma}_{nn_1}^2 \leqslant 0)\\
&\quad = P(-2\gamma'\varepsilon + \varepsilon' G\varepsilon \leqslant -\gamma'\gamma)\\
&\quad \leqslant P(|\gamma'\varepsilon| \geqslant \gamma'\gamma/4) + P\left(|\varepsilon' G\varepsilon| \geqslant \frac{\gamma'\gamma}{2}\right)\\
&\quad \leqslant \frac{16}{(\gamma'\gamma)^2}\operatorname{Var}(\gamma'\varepsilon) + P\left(\left|\sum_{i=1}^{n} g_{ii}\varepsilon_i^2\right| \geqslant \frac{\gamma'\gamma}{4}\right)\\
&\qquad + P\left(\left|\sum_{i \neq j} g_{ij}\varepsilon_i\varepsilon_j\right| \geqslant \frac{\gamma'\gamma}{4}\right)\\
&\quad \leqslant \frac{64\sigma^2}{\gamma'\gamma} + \frac{32}{\gamma'\gamma}\sigma^2 + \frac{64 \times 280\sigma^4}{(\gamma'\gamma)^2}\\
&\quad < 130\sigma^2(\beta_2 - \beta_1)^{-2}(\log n)^{-2} \to 0.
\end{aligned} \tag{3.20}$$

This shows that the minimization point h of $\{\hat{\sigma}_{nc}^2\}$ satisfies

$$|h - n_1| < n/\log^2 n$$

with probability approaching 1 as $n \to \infty$.

Case 2. $\beta_1 \neq \beta_2$, $k = |c - n_1| < n/\log^2 n$. It follows that for any $u > 0$,

$$\begin{aligned}
&P\left(|\hat{\sigma}_{nc}^2 - \hat{\sigma}_{nn_1}| \geqslant \frac{u}{\log n}\right)\\
&\quad \leqslant P\left(|-2\gamma'\varepsilon + \varepsilon' G\varepsilon| \geqslant \frac{un}{2\log n}\right)\\
&\quad \leqslant P\left(|2\gamma'\varepsilon| \geqslant \frac{un}{4\log n}\right) + P\left(|\varepsilon' G\varepsilon| \geqslant \frac{un}{4\log n}\right)\\
&\quad \leqslant \frac{64\log^2 n}{u^2 n^2}\cdot \gamma'\gamma\sigma^2 + \frac{4\log n}{un}\cdot 8\sigma^2 + \frac{280\sigma^4}{\tau^2\log^2 n} \to 0,
\end{aligned} \tag{3.21}$$

by (3.15)–(3.18).

Now note that $\sum_{i=1}^{n}(\varepsilon_i^2-\sigma^2)$ is a martingale and $A_{n_1}(A'_{n_1}A_{n_1})^{-1}A'_{n_1}\geqslant 0$. Hence by Marcinkiewicz–Zygmund–Burkholder's martingale inequality, we have, for any τ, δ, and $u: 0<\tau<\delta/(1+\delta)$, $u>0$:

$$
\begin{aligned}
&P(|\hat{\sigma}^2_{nn_1}-\sigma^2|\geqslant un^{-\tau})\\
&\quad\leqslant P\left(|\varepsilon'\varepsilon-n\sigma^2|\geqslant\frac{un^{1-\tau}}{2}\right)+P\left(\varepsilon' A_{n_1}(A'_{n_1}A_{n_1})^{-1}A'_{n_1}\varepsilon\geqslant\frac{un^{1-\tau}}{2}\right)\\
&\quad\leqslant c_{\delta,u}E|\varepsilon_1|^{2+\delta_n-(1+\delta)(1-\tau)}\cdot n\\
&\qquad+P\left(\operatorname{tr}(A_{n_1}(A'_{n_1}A_{n_1})^{-1}A'_{n_1})\,\varepsilon'\varepsilon\geqslant\frac{un^{1-\tau}}{2}\right)\\
&\quad\leqslant c_{\delta,u}E|\varepsilon_1|^{2+\delta_n-(\delta-(1+\delta)\tau)}+\frac{n^{\tau}\cdot n\sigma^2}{\mu(n_1+1)(n-n_1+1)}\to 0. \qquad (3.22)
\end{aligned}
$$

From Cases 1 and 2, the theorem is true if $\beta_1\neq\beta_2$. When $\beta_1=\beta_2$, a similar argument gives

$$|\hat{\sigma}^2_{nc}-\hat{\sigma}^2_{n0}|\log n\to 0$$

and

$$\hat{\sigma}^2_{n0}\to\sigma^2$$

in probability. Thus we complete the proof.

4. Nonnormal Error

When the distribution of random error $\varepsilon(t)$ is nonnormal, we can use the theory of strong approximation of partial sums of i.i.d. variables by Brownian Motion process to extend Theorem 1 to such cases.

THEOREM 4. *Let $\varepsilon_1, \varepsilon_2, ...$ be i.i.d. random errors, and the moment generating function of e_1 exists in some neighborhood or zero, i.e.,*

$$E\exp(t\varepsilon_1)<\infty \qquad \textit{for} \quad |t| \textit{ small enough}, \qquad (4.1)$$

then the conclusion of Theorem 1 remains valid.

Proof. Put

$$S_k\cong S_{nk}=\sum_{i=1}^{k}\left(x_i-a-\frac{i}{n}\beta\right)\Big/\sigma, \qquad k=1,2,...,n,$$

then, by Komlós–Major–Tusnády [7, 8], there exists a Brownian Motion process $\{W(t), t \geqslant 0\}$ such that

$$\lim_{n \to \infty} \sup\{\sup_{k \leqslant n} |S_k - W(k)|/\log n\} < \infty, \qquad \text{a.s.} \tag{4.2}$$

Since

$$\frac{Y_k}{\sigma} = \frac{1}{2\sqrt{m}} (S_k - 2S_{k-m} + 2S_{k-3m} - S_{k-4m}),$$

we have for $4m \leqslant k \leqslant n$,

$$\left| \frac{Y_k}{\sigma} - \frac{1}{2\sqrt{m}} (W(k) - 2W(k-m) + 2W(k-3m) - W(k-4m)) \right|$$
$$\leqslant \frac{6}{2\sqrt{m}} \sup_{4m \leqslant k \leqslant n} |S_k - W(k)|. \tag{4.3}$$

By (4.2), and noticing that $\log n/\sqrt{m} \to 0$ as $n \to \infty$, we get

$$\lim_{n \to \infty} \left(\max_{4m \leqslant k \leqslant n} \left| \frac{Y_k}{\sigma} - \frac{1}{2\sqrt{m}} (W(k) - 2W(k-m) \right.\right.$$
$$\left.\left. + 2W(k-3m) - W(k-4m)) \right) \right| = 0, \quad \text{a.s.} \tag{4.4}$$

From Theorem 1, we get

$$\lim_{n \to \infty} P \left\{ \sup_{4m \leqslant k \leqslant n} \left| \frac{1}{2\sqrt{m}} (W(k) - 2W(k-m) + 2W(k-3m) \right.\right.$$
$$\left.\left. - W(k-4m)) \right| \leqslant A_n(x) \right\} = \exp\{-2e^{-x}\}, \tag{4.5}$$

where $A_n(x)$ is defined by (2.5). Thus, (2.6) is also true in view of (4.3)–(4.5). Theorem 4 is proved.

A close inspection of the proof of Theorem 3 convinces us that this theorem is still true under assumption (4.1). Therefore, the method of the previous two sections can be applied.

Further, using a result of Major [9], the following theorem can be established.

THEOREM 5. *Let* $\varepsilon_1, \varepsilon_2, \ldots$ *be i.i.d. random errors with finite* $(2+\delta)$th *moment, where* $\delta > 0$, *and* $n \gg m \gg n^{2/(2+\delta)}$. *Then* (2.6) *remains true.*

Also, the conclusion of Theorem 3 remains valid under the conditions of

Theorem 5. So the previous methods still apply. We note, however, that the requirement on m is more stringent in this case.

5. Estimation of the Slope Change $\beta_1 - \beta_2$

In order to form a point estimate of the slope change $\theta = \beta_1 - \beta_2$, we first find c such that $|Y_c| = \xi_n = \max_{4m \leqslant k \leqslant n} |Y_k|$, and compute

$$\begin{aligned}\hat{\theta} = \hat{\beta}_1 - \hat{\beta}_2 &= \frac{12n}{c(c^2-1)} \sum_{i=1}^{c} \left(i - \frac{c+1}{2}\right) x_i \\ &\quad - \frac{12n}{(n-c)((n-c)^2-1)} \sum_{i=c+1}^{n} \left(i - \frac{n+c+1}{2}\right) x_i \\ &= (F_c' F_c)^{-1} F_c' x, \end{aligned} \tag{5.1}$$

which is taken as an estimator of θ. Generally, if c is near $4m$ or n, then the slope change point t_0 is near 0 or 1, and the samples at our disposal are perhaps not enough to give a reasonable estimate. For an interval estimate of θ, we prove the following asymptotic theorem for $\hat{\theta}$.

Theorem 6. *Suppose that t_0 is the slope change point and $E|e_1|^{2+\delta} < \infty$ for some $\delta > \frac{2}{3}$, and $m \ll n^{3/4}$. Then, as $n \to \infty$,*

$$\sqrt{\frac{n}{12\sigma^2} (t_0^{-3} + (i - t_0)^{-3})^{-1}} \, (\hat{\theta} - \theta) \xrightarrow{L} N(0, 1), \tag{5.2}$$

where $\xrightarrow{L}$ means "converges in law."

Proof. Without loss of generality, we assume $q = 1$. Choose c such that $|Y_c| = \max_{4m \leqslant j \leqslant n} |Y_j|$. Then, for any $0 < \alpha < 1$ and $\alpha > 0$,

$$\begin{aligned} &P(nt_0 \leqslant c \leqslant nt_0 + 4m) \\ &\quad = P\left(t_0 \leqslant \frac{c}{n} \leqslant t_0 + \frac{4m}{n}\right) \\ &\quad \geqslant P(\{\sup_{j/n \notin [t_0, t_0 + 4m/n]} |Y_j| \leqslant c_n(\alpha, d)\} \cap \{|Y_c| > c_n(\alpha, d)\}) \\ &\quad = P(\sup_{j/n \notin [t_0, t_0 + (4m/n)]} |Y_j| \leqslant c_n(\alpha, d)) \, P(|Y_c| > c_n(\alpha, d)). \end{aligned} \tag{5.3}$$

Using Theorem 5 and slightly modifying the argument of Section 2, we can easily prove that

$$\lim_{n \to \infty} P(nt_0 \leqslant c \leqslant nt_0 + 4m) = 1. \tag{5.4}$$

Denote $n_1 = \min\{l: l/n \geqslant t_0,\ 4m \leqslant l \leqslant n - 4m\}$. Without loss of generality assume $n_1 \leqslant c \leqslant n - 4m$. By (3.7) and (3.8), $\hat{\theta}$ can be rewritten as

$$\begin{aligned}\hat{\beta}_1 - \hat{\beta}_2 &= (0, 1, 0, -1)(F_c'F_c)^{-1}F_c'x \\ &= (0, 1, 0, -1)(F_c'F_c)^{-1}F_c'(F_{n_1}\beta + \varepsilon).\end{aligned} \tag{5.5}$$

So it follows that

$$(\hat{\beta}_1 - \hat{\beta}_2) - (\beta_1 - \beta_2) = (0, 1, 0, -1)(F_c'F_c)^{-1}F_c'(-F_{c-n_1}\beta + \varepsilon), \tag{5.6}$$

where F_{c-n_1} is defined as (3.11). We can easily calculate that

$$(F_c'F_c)^{-1}F_c' = \begin{pmatrix} (a_{1c} - ka_{2c})\, e_m' - a_{2c} f_m' & a_{1c} e_k' - a_{2c} f_k' & 0 \\ (na_{2c} - a_{4c}h)\, e_m' - a_{4c} f_m' & na_{2c} e_k' - a_{4f} f_k' & 0 \\ 0 & 0 & b_{1c} e_{n-c}' - b_{2c}\, g_{n-c}' \\ 0 & 0 & -nb_{2c} e_{n-c}' + b_{4c}\, g_{n-c}' \end{pmatrix}, \tag{5.7}$$

where a_{jc}, b_{jc}, $j = 1, 2, 4$, e_m, f_m, etc., are defined in (3.8) and (3.12), and $k = c - n_1$. According to (3.3) and (3.13), on replacing $pn - qn_1 \pm 1$ by $pn - qn_1$, where p, q are some integers, we get

$$\begin{aligned}|E\hat{\theta} - \theta| &= |-(0, 1, 0, -1)(F_c'F_c)^{-1}F_c'F_{c-n_1}\beta| \\ &= \frac{6nkn_1}{c^3}(\mu_2 - \mu_1) + \frac{k^2(c + 2n_1)}{c^3}(\beta_2 - \beta_1)| \\ &\leqslant \left|\left(\frac{6kn_1}{c^3} + \frac{3k^2c}{c^3}\right)(\beta_2 - \beta_1)\right| \leqslant \frac{4k^2}{c^2}|\beta_2 - \beta_1|,\end{aligned} \tag{5.8}$$

and

$$\begin{aligned}&\mathrm{Var}\{(\hat{\beta}_1 - \hat{\beta}_2) - (\beta_1 - \beta_2)\} \\ &= (0, 1, 0, -1)(F_c'F_c)^{-1}(F_c'F_c)^{-1}(0, 1, 0, -1)' \\ &= (0, 1, 0, -1)(F_c'F_c)^{-1}(0, 1, 0, -1)' \\ &= 12n^2(c^{-1}(c^2 - 1)^{-1} + (n - c)^{-1}((n - c)^2 - 1)^{-1}).\end{aligned} \tag{5.9}$$

To justify the use of the standard CLT, we note the following three easily verified facts:

1. From the expressions (5.1) and (5.6), we have

$$\begin{aligned}
&\operatorname{Var}\{(\hat{\beta}_1-\hat{\beta}_2)-(\beta_1-\beta_2)\}^{-(2+\delta)/2}\\
&\quad\times\left\{\sum_{i=1}^{c}\left(\frac{12n}{c(c^2-1)}\right)^{2+\delta}\left|i-\frac{c+1}{2}\right|^{2+\delta}E|e_i|^{2+\delta}\right.\\
&\quad\left.+\sum_{i=c+1}^{n}\left(\frac{12n}{(n-c)[(n-c)^2-1]}\right)^{2+\delta}\left|i-\frac{n+c+1}{2}\right|^{2+\delta}E|e_i|^{2+\delta}\right\}\\
&\quad\leqslant KE|e_1|^{2+\delta}\cdot\frac{n^{2+\delta}(c^{-3(2+\delta)+(3+\delta)}+(n-c)^{-3(2+\delta)+(3+\delta)})}{n^{2+\delta}(c^{-3(2+\delta)/2}+(n-c)^{-3(2+\delta)/2})}\\
&\quad\leqslant 2K(\max(c,n-c))^{-\delta/2}\leqslant 2Kc^{-\delta/2}\leqslant 2Kt_0^{-\delta/2}n^{-\delta/2}\to 0,
\end{aligned}\tag{5.10}$$

where K is a constant.

2. Since $n^{3/4}\gg k$, we get

$$\begin{aligned}
&\lim_{n\to\infty}\frac{|E\{(\hat{\beta}_1-\hat{\beta}_2)-(\beta_1-\beta_2)\}|}{\sqrt{\operatorname{Var}\{(\hat{\beta}_1-\hat{\beta}_2)-(\beta_1-\beta_2)\}}}\\
&\qquad\leqslant\lim_{n\to\infty}\frac{4k^2}{c^2}|\beta_2-\beta_1|\cdot(12n^2c^{-3})^{-1/2}\\
&\qquad\leqslant\lim_{n\to\infty}\frac{2k^2}{\sqrt{3}\,t_0n^{3/2}}=0.
\end{aligned}\tag{5.11}$$

3. It is easy to see that

$$12n^3(c^{-1}(c^2-1)^{-1}+(n-c)^{-1}((n-c)^2-1)^{-1})\to 12(t_0^{-3}+(1-t_0)^{-3}).\tag{5.12}$$

Theorem 6 is proved.

Notice that $\hat{t}_0-(c-2m)/n$ is a consistent estimator of t_0. (Of course, only when $\theta\neq 0$, hence t_0 is well defined.) In Section 3 we introduced a consistent estimator $\hat{\sigma}_n$ of σ. Substituting $\hat{t}_0$ to t_0 and $\hat{\sigma}_n$ for σ, we have the following result.

THEOREM 7. *Suppose that the conditions of Theorem 6 are satisfied. We then have*

$$\left\{\frac{n}{12\hat{\sigma}_n^2}(\hat{t}_0^{-3}+(t-\hat{t}_0)^{-3})^{-1}\right\}^{1/2}\{\hat{\theta}-\theta\}\xrightarrow{L}N(0,1),\tag{5.13}$$

as $n\to\infty$.

When $\beta_1 = \beta_2$, though t_0 does not exist, the statistic $\hat{t}_0$ is still well defined. Since it is not known whether or not (5.13) is true for $\beta_1 = \beta_2$, so (5.13) cannot be used to give a test for the hyperthesis $\beta_1 = \beta_2$. However, (5.13) can be utilized to form a confidence interval of $(\beta_1 - \beta_2)$ if we know $\beta_1 \neq \beta_2$ a priori, when the null hypothesis (2.13) is rejected.

REFERENCES

[1] CHEN, X. R. (1987). *Testing and Interval Estimation in a Change-Point Model Allowing at Most One Change.* Technical Report No. 87-25. Center for Multivariate Analysis, University of Pittsburgh. *Sci. Sinica, Ser. A*, in press.

[2] CSÖRGÖ, M., AND HORVÁTH, L. (1986). Nonparametric methods for change-point problems. In *Handbook of Statistics*, Vol. 7. North-Holland, Amsterdam.

[3] FEDER, P. I. (1975). On asymptotic distribution theory in segmented regression problems—Identified case. *Ann. Statist.* **3**, 49–83.

[4] HINKLEY, D. V. (1970). Inference about the change point in a sequence of random variables. *Biometrika* **57**, 1–17.

[5] HINKLEY, D. V. (1971). Inference in two-phase regression. *J. Amer. Statist. Assoc.* **66**, 736–743.

[6] HUDSON, D. J. (1966). Fitting segmented curves whose join points have to be estimated. *J. Amer. Statist. Assoc.* **61**, 1097–1129.

[7] KOMLÓS, J., MAJOR, P., AND TUSNÁDY, G. (1975). An approximation of partial sums of independent R.V.'s, and the sample DF. Part I. *Z. Wahrsch. Verw. Gebiete* **32**, 111–131.

[8] KOMLÓS, J., MAJOR, P., AND TUSNÁDY, G. (1976). An approximation of partial sums of independent R.V.'s and the sample DF. Part II. *Z. Wahrsch. Verw. Gebiete* **34**, 33–58.

[9] MAJOR, P. (1976). The approximation of partial sums of indepdent R.V.'s *Z. Wahrsch. Verw. Gebiete* **35**, 213–220.

[10] KRISNAIAH, P. R., AND MIAO, B. Q. (1987). *Review about Estimation of Change-Point.* Technical Report No. 87-48. Center for Multivariate Analysis, University of Pittsburgh. In *Handbook of Statistics*. Vol. 7. North-Holland, Amsterdam.

[11] QUALLS, C., AND WATANABE, H. (1972). Asymptotic properties of Gaussian processes. *Ann. Math. Statist.* **43** 580–596.

[12] YIN, Y. Q. (1986). *Detection of the Number, Locations and Magnitudes of Jumps.* Technical Report. University of Arizona.

Maximum Likelihood Principle and Model Selection when the True Model Is Unspecified

R. NISHII

University of Pittsburgh and Hiroshima University, Hiroshima, Japan

Suppose that independent observations come from an unspecified unknown distribution. Then we consider the maximum likelihood based on a specified parametric family which provides a good approximation of the true distribution. We examine the asymptotic properties of the maximum likelihood estimate and of the maximum likelihood. These results will be applied to the model selection problem.

1. INTRODUCTION

The maximum likelihood principle is a basic and useful technique in statistics. It has a long history and there is quite a bit of literature treating its asymptotic properties, e.g., Wald [16] and LeCam [10]. These classical results are based on the assumption that the unknown density function lies in a specified parametric family. However, if this assumption is not true, do similar results remain valid? Cox [4, 5] first considered such a problem in testing separated families (see also Berk [2, 3]). Huber [8] pointed out that this problem is connected with robust estimation. White [17] reviewed this problem and showed the consistency and the asymptotic normality under the assumptions corresponding to the regularity conditions in the classical theory. Additional related references are Akaike [1] and Foutz and Srivastava [6].

In Section 2 we give the consistency order of the maximum likelihood estimate and of the maximum likelihood under the usual conditions and the additional assumptions on higher order derivatives of the specified densities. Further we treat the testing problem of two families. Section 3 is concerned with model selection. We prove the strong consistency of BIC

Multivariate Statistics and Probability
ISBN 0-12-580205-6

Reprinted from *J. Mult. Anal.* **27**(2).

type criteria in a very general setting. The inconsistency of AIC will also be shown. However we reconsider the consistency in model selection in Section 4.

2. Observations and a Family of Densities

Let n observations (which may be multivariate) $x_1, ..., x_n$ $(\in \mathbb{R}^d)$ be independently and identically distributed as a probability density function (pdf) g with respect to a fixed measure ν on $\mathbb{R}^d$. Note that ν may be discrete. Suppose that $\int |\log g(x)|\, g(x)\, d\nu(x) < \infty$. Next consider the family of pdf's:

$$\mathcal{M} = \{f(x \mid \theta) \mid \theta \in \Theta\}, \tag{2.1}$$

where Θ is a convex set in $\mathbb{R}^p$. Define the quasi log-likelihood and the quasi maximum likelihood estimate (QMLE) based on n observations as

$$L_n(\theta) = \sum_{i=1}^{n} \log f(x_i \mid \theta) \qquad \text{and} \qquad L_n(\hat{\theta}) = \max_{\theta \in \Theta} L_n(\theta). \tag{2.2}$$

Recall the Kullback–Leibler information:

$$I(g; f, \theta) = \int g(x) \log\{g(x)/f(x \mid \theta)\}\, d\nu \geqslant 0 \tag{2.3}$$

provides some closeness from g to $f(\cdot \mid \theta)$. Define the expected log-likelihood $e(g; f, \theta)$ and the quasi true parameter θ_g as

$$e(g; f, \theta) = \int g(x) \log f(x \mid \theta)\, d\nu \quad \text{and} \quad e(g; f, \theta_g) = \max_{\theta \in \Theta} e(g; f, \theta). \tag{2.4}$$

Obviously $I(g; f, \theta)$ is minimized at $\theta = \theta_g$. We call the density $f(\cdot \mid \theta_g)$ the quasi true pdf. If $g(x)$ is exactly specified by $\mathcal{M}$, i.e., $g(x) = f(x \mid \theta_0)$ for $\theta_0 \in \Theta$, then the quasi true parameter θ_g is given by θ_0.

Example 1. Let $x_1, ..., x_n$ be random samples from a pdf $g(x) = \{\phi(x-1-\delta) + \phi(x+1)\}/2$, where δ is a constant and $\phi(x)$ is the standard normal density function. When we approximate $g(x)$ by a set of normal densities

$$\mathcal{M} = \{\sqrt{\theta_2}^{-1} \phi((x-\theta_1)/\sqrt{\theta_2}) \mid \theta = (\theta_1, \theta_2) \in \Theta\}, \qquad \Theta = \mathbb{R} \times (0, \infty),$$

the QMLE of θ is given by $\hat{\theta}=(\bar{x}, n^{-1}\sum_{i=1}^{n}(x_i-\bar{x})^2)$, where $\bar{x}$ is the sample mean. The quasi true density in $\mathscr{M}$ is given by the normal density with mean $\delta/2$ and variance $2+\delta(\delta+4)/4$, i.e., $\theta_g=(\delta/2,\ 2+\delta(\delta+4)/4)$ since $E_g(x_1)=\delta/2$ and $E_g(x_1-\delta/2)^2=2+\delta(\delta+4)/4$, where E_g denotes the expectation with respect to the true density g. Also the maximized expected log-likelihood is given by $-\frac{1}{2}\log[2\pi e\{2+\delta(\delta+4)/4\}]$.

Now we make assumptions on $(g, \mathscr{M})$ which will enable us to study the asymptotic behavior of the maximum likelihood principle.

ASSUMPTION A1. *There exists the quasi true parameter θ_g uniquely, and θ_g is an interior point of Θ.*

ASSUMPTION A2. (a) *The derivatives $l_\alpha(x\mid\theta)=\partial l(x\mid\theta)/\partial\theta_\alpha$ and $l_{\alpha\beta}(x\mid\theta)=\partial^2 l(x\mid\theta)/\partial\theta_\alpha\,\partial\theta_\beta$ $(\alpha, \beta=1, ..., p)$ of $l(x\mid\theta)=\log f(x\mid\theta)$ are measurable with respect to $x\in\mathbb{R}^d$ for each $\theta\in\Theta$ and continuous with respect to θ for each x.* (b) *$|l(x\mid\theta)|$, $|l_\alpha(x\mid\theta)|$, $|l_{\alpha\beta}(x\mid\theta)|$, $|l_\alpha(x\mid\theta)\,l_\beta(x\mid\theta)|$ are dominated by integrable functions with respect to $g(x)$, which do not depend on θ.*

ASSUMPTION A3. *Define $p\times p$ matrices $V(\theta)$ and $W(\theta)$ by*

$$V(\theta)=E_g\left[\frac{\partial}{\partial\theta}l(X\mid\theta)\frac{\partial}{\partial\theta^{\mathrm{T}}}l(X\mid\theta)\right] \quad \text{and} \quad W(\theta)=-E_g\left[\frac{\partial^2}{\partial\theta\,\partial\theta^{\mathrm{T}}}l(X\mid\theta)\right],$$

where E_g denotes the expectation with respect to the true density g, a random variable X has the true pdf $g(x)$ $(X\sim g(x))$, and $l(x\mid\theta)=\log f(x\mid\theta)$. Then $V(\theta_g)$ and $W(\theta_g)$ are positive definite, where θ_g is the quasi true parameter.

ASSUMPTION A4. *There exists the quasi maximum likelihood estimate $\hat{\theta}=\hat{\theta}_n$ which tends to θ_g with probability* 1.

ASSUMPTION A5. (a) *$l_{\alpha\beta\gamma}(x\mid\theta)=\partial^3 l(x\mid\theta)/\partial\theta_\alpha\,\partial\theta_\beta\,\partial\theta_\gamma$ $(\alpha, \beta, \gamma=1, ..., p)$ are measurable with respect to x for each θ.* (b) *$|l_\alpha(x\mid\theta)|^2$, $|l_{\alpha\beta}(x\mid\theta)|^2$, $|l_{\alpha\beta\gamma}(x\mid\theta)|$ are dominated by integrable functions with respect to g, which do not depend on θ.*

Remark on A4. (i) When $g(x)=f(x\mid\theta_0)$: Several sufficient conditions ensuring the assumption A4 are known, e.g., Wald [16], Huber [8], and 5e.2 of Rao [11]. (ii) When $g(x)$ is not a member of $\mathscr{M}$: White [17] showed that A1–A3 with

ASSUMPTION A4′. *The parameter space Θ is a compact set of $\mathbb{R}^p$, ensure* A4. *Conditions by Huber, derived without assuming that g lies in $\mathscr{M}$, suffice* *for* A4. *Also Wald's assumptions can be modified to this situation by substituting* $df(x, \theta_0)$ *for* $g(x)\,dv$ *and* θ_0 *for* θ_g, *which meet* A4.

If the true density does not lie in $\mathscr{M}$ and is completely unknown, any of our conditions is not checked. However, if $\mathscr{M}$ gives a good approximation to g and $\mathscr{M}$ meets conditions A1–A5 for $g(x) = f(x \mid \theta_0)$, then $(g, \mathscr{M})$ will satisfy A1–A5.

The assumptions A1–A4 correspond to the regularity conditions in the classical theory. They ensure the strong consistency of $\hat{\theta}_n$ on $L_n(\hat{\theta})$. Further, the asymptotic normality of $\hat{\theta}_n$ can be shown, e.g., White [17] and Foutz and Srivastava [6]. If we assume A5 additionality, the consistency order may be evaluated as in the following theorem which will play a key role in studying model selection criteria.

THEOREM 1. *Let n independent observations $x_1, ..., x_n$ come from the distribution with the density g and let $\mathscr{M}$ be a parametric family as* (2.1). *If* $(g, \mathscr{M})$ *meets* A1–A5, *the stochastic orders relating to the QMLE $\hat{\theta}_n$ and the quasi log-likelihood are*:

(i) $\hat{\theta}_n = \theta_g + O(\sqrt{(n^{-1} \log\log n)})$ *a.s.*,

(ii) $L_n(\hat{\theta}_n) = L_n(\theta_g) + O(\log\log n)$ *a.s.*,

(iii) $(1/n)\, L_n(\hat{\theta}_n) = e(g; f, \theta_g) + O(\sqrt{(n^{-1} \log\log n)})$ *a.s.*,

where θ_g is the quasi true parameter, $L_n(\theta)$ is the quasi log-likelihood of (2.2) *and $e(g; f, \theta)$ is the expected log-likelihood of* (2.4).

Proof. From A1 and A4, $\hat{\theta}_n = \hat{\theta}$ exists and is an interior point of Θ for large n. Employing Taylor's expansion we get

$$\mathbf{0} = \frac{1}{n}\, \partial L_n(\hat{\theta})/\partial\theta = \frac{1}{n}\, \partial L_n(\theta_g)/\partial\theta - W_n(\theta_g)(\hat{\theta} - \theta_g) + \mathbf{r}_n, \tag{2.5}$$

where

$$W_n(\theta) = -\frac{1}{n}\, \partial^2 L_n(\theta)/\partial\theta\, \partial\theta^{\mathrm{T}}: p \times p, \qquad \mathbf{r}_n = (r_{1n}, ..., r_{pn})^{\mathrm{T}},$$

$$r_{\alpha n} = (\hat{\theta} - \theta_g)^{\mathrm{T}} \frac{1}{n}\left[\partial^2 \left\{\frac{\partial}{\partial\theta_\alpha} L_n(\bar{\theta})\right\} \Big/ \partial\theta\, \partial\theta^{\mathrm{T}}\right](\hat{\theta} - \theta_g) \qquad (\alpha = 1, ..., p), \tag{2.6}$$

$$\bar{\theta} = \theta_g + \varepsilon(\hat{\theta} - \theta_g), \qquad 0 < \varepsilon < 1.$$

The expected log-likelihood $e(g; f, \theta)$ of (2.4) is maximized at $\theta = \theta_g$, which yields that $\partial e(g; f, \theta_g)/\partial\theta = 0$. Hence $E_g[\partial l(X \mid \theta_g)/\partial\theta] = \partial E_g[l(X \mid \theta_g)]/\partial\theta = \partial e(g; f, \theta_g)/\partial\theta = 0$ (by A2), where $X \sim g(x)$ and $l(x \mid \theta) = \log f(x \mid \theta)$. Hence $\partial l(x_i \mid \theta_g)/\partial\theta$ $(i = 1, ..., n)$ are i.i.d. with mean zero vector and finite variance-covariance matrix (by A3). Therefore by the law of iterated logarithm, we have

$$\partial L_n(\theta_g)/\partial\theta = O(\sqrt{(n \log\log n)}) \qquad \text{a.s.} \tag{2.7}$$

Similarly by the law of the iterated logarithm and A2 and A4,

$$W_n(\theta_g) = W(\theta_g) + O(\sqrt{(n^{-1} \log\log n)}) \qquad \text{a.s.} \tag{2.8}$$

From A3 $W(\theta_g)$ is positive definite, and so is $W_n(\theta_g)$ when n is large. Solving (2.5) with respect to $\hat{\theta} - \theta_g$, we get

$$\hat{\theta} - \theta_g = W_n(\theta_g)^{-1} \left\{ \frac{1}{n} \partial L_n(\theta_g)/\partial\theta + \mathbf{r}_n \right\}. \tag{2.9}$$

By A5 there exist an integrable function H with respect to $g(x)$ and a constant $K > 0$ such that for any $\alpha, \beta, \gamma = 1, ..., p$,

$$\left| \frac{1}{n} \partial^3 L_n(\bar{\theta})/\partial\theta_\alpha\, \partial\theta_\beta\, \partial\theta_\gamma \right| \leqslant \frac{1}{n} \sum_{i=1}^{n} H(x_i) < K. \tag{2.10}$$

Consequently by (2.6) we know that $r_{\alpha n} = (\hat{\theta} - \theta_g)^{\mathrm{T}} O(1)(\hat{\theta} - \theta_g) = O(1)(\hat{\theta} - \theta_g)$ a.s. and that $\mathbf{r}_n = O(1)(\hat{\theta} - \theta_g)$ a.s. since $\hat{\theta} - \theta_g = o(1)$ a.s. (by A4), where $O(1)$ denotes a random vector or a random matrix whose all elements are $O(1)$, and $o(1)$ is similarly defined. Thus by (2.9)

$$\hat{\theta} - \theta_g = O(\sqrt{(n^{-1} \log\log n)}) \qquad \text{a.s.} \tag{2.11}$$

Again by the law of the iterated logarithm we know

$$\frac{1}{n} L_n(\theta_g) = e(g; f, \theta_g) + O(\sqrt{(n^{-1} \log\log n)}) \qquad \text{a.s.} \tag{2.12}$$

Using Taylor's expansion we get

$$L_n(\theta_g) - L_n(\hat{\theta}) = (\hat{\theta} - \theta_g)^{\mathrm{T}} \partial L_n(\theta_g)/\partial\theta + \tfrac{1}{2}(\hat{\theta} - \theta_g)^{\mathrm{T}} [\partial^2 L_n(\bar{\theta})/\partial\theta\, \partial\theta^{\mathrm{T}}](\hat{\theta} - \theta_g),$$

and by the relations (2.11), (2.7), and (2.10),

$$L_n(\hat{\theta}) = L_n(\theta_g) + O(\log\log n) \qquad \text{a.s.} \tag{2.13}$$

Hence, by (2.12) and (2.13)

$$\begin{aligned} \frac{1}{n} L_n(\hat{\theta}) &= \frac{1}{n} L_n(\theta_g) + \frac{1}{n} \{L_n(\hat{\theta}) - L_n(\theta_g)\} \\ &= e(g; f, \theta_g) + O(\sqrt{(n^{-1} \log\log n)}) \qquad \text{a.s.} \end{aligned} \tag{2.14}$$

This completes the proof.

Note that Theorem 1 is new even if g is exactly specified by $\mathscr{M}$. Under non-regular case the consistency order of $\hat{\theta}_n$ may be higher than $O(\sqrt{(n^{-1} \log \log n)})$. However, (ii) still remains valid because the order of (ii) is based on the law of iterated logarithm for $L_n(\theta) = \sum_{i=1}^{n} \log f(x_i \mid \theta)$.

Cox [4, 5] introduced the problem: Which family specifies the true density? He proposed the corrected likelihood ratio test. Our problem is: Which family is closer to the true density? We take a simple likelihood ratio approach. Let

$$\mathscr{M}_i = \{f_i(x \mid \theta_i) \mid \theta_i \in \Theta_i\} \qquad (i = 1, 2)$$

be families of densities (which may not be separated). Assume both $(g, \mathscr{M}_i)$ satisfy A1–A5. Let θ_{ig} be the quasi true parameter in Θ_i associated with the true density $g(x)$, and put

$$\varepsilon_i = \int g(x) \log f_i(x \mid \theta_{ig})\, dv(x) \qquad (i = 1, 2)$$

which is the maximized expected log-likelihood in $\mathscr{M}_i$. Then test the hypothesis

$$H_0: \varepsilon_1 = \varepsilon_2 \qquad \text{versus} \qquad H_1: \varepsilon_1 > \varepsilon_2. \tag{2.15}$$

If H_1 is true, from (iii) of Theorem 1 the likelihood ratio

$$\lambda_n = \sum_{i=1}^{n} \log\{f_1(x_j \mid \hat{\theta}_1)/f_2(x_j \mid \hat{\theta}_2)\} \tag{2.16}$$

tends to infinity since $n^{-1}\lambda_n \to \varepsilon_1 - \varepsilon_2 > 0$ a.s., which implies the likelihood ratio can asymptotically find the family closer to the unknown true density $g(x)$. To make more detailed discussion, we get:

THEOREM 2. *Consider the testing hypothesis* (2.15) *under the conditions* A1–A5. *Then the likelihood ratio test is consistent.*

Proof. The asymptotic normality of the likelihood ratio λ_n of (2.16) is known by Foutz and Srivastava [6] as

$$\sqrt{n^{-1}}\{\lambda_n - n(\varepsilon_1 - \varepsilon_2)\} \xrightarrow{L} N(0, \sigma^2) \qquad \text{as} \quad n \to \infty,$$

where $\sigma^2 = E_g[\log\{f_1(X \mid \theta_{1g})/f_2(X \mid \theta_{2g})\}]^2$, θ_{ig} $(i = 1, 2)$ are the quasi true parameters and $X \sim g(x)$. Define a estimator of σ^2 as

$$\hat{\sigma}_n^2 = n^{-1} \sum_{i=1}^{n} [\log\{f_1(x_i \mid \hat{\theta}_1)/f_2(x_1 \mid \hat{\theta}_2)\}]^2.$$

Using Theorem 1, we can show that $\hat{\sigma}_n^2$ is a consistent estimator of σ^2. Now we make the rejection region of H_0 by

$$R_n^{(\eta)} = \{\lambda_n > \sqrt{n}\xi_\eta \hat{\sigma}_n\},$$

where ξ_η is the upper 100η-percent point of the standard normal distribution. The significance level of this test procedure is asymptotically η because $\lambda_n/\hat{\sigma}_n \to N(0, 1)$ under H_0. On the other hand, under H_1, $\varepsilon_1 - \varepsilon_3$ ($=\mu$, say) is positive. Hence

$$P[R_n^{(\eta)} \mid H_1] = P\left[\frac{1}{\sqrt{n}}(\lambda_n - n\mu) \geqslant \hat{\sigma}_n \xi_\eta - \sqrt{n}\,\mu \mid H_1\right]$$
$$\to 1 \qquad (n \to \infty),$$

because $\sqrt{n}^{-1}(\lambda_n - n\mu) \to^L N(0.\ \sigma^2)$ and $\hat{\sigma}_n \xi_\eta - \sqrt{n}\,\mu \to^P -\infty$. This completes the proof.

Let σ^2 be the asymptotic variance of $\sqrt{n}^{-1}\lambda_n$. Then if $d \equiv |\varepsilon_1 - \varepsilon_2|/\sigma$ is large, we can discriminate the families by using small data. However, when d is small we need a large data. Hence in such a case it would be preferable to develop similar discussion as the corrected likelihood ratio proposed by Cox. See also Kent [9].

3. Model Selection

We have shown that the likelihood ratio test is useful when two models are under consideration. When we have more than two models which are candidates for the true density g, a multiple divergence criterion are proposed, e.g., see Sawyer [13]. Alternatively we take the model selection procedures. Consider k models $\mathcal{M}_i = \{f_i(x \mid \theta_i) \mid \theta_i \in \Theta_i\}$ $(i = 1, ..., k)$. We treat here the information criteria (IC) given by the form

$$\text{IC}(i) = -2L_n^{(i)}(\hat{\theta}_i) + c_n p_i \qquad (i = 1,, ..., k), \tag{3.1}$$

where $\hat{\theta}_i$, $L_n^{(i)}(\theta_i)$, and p_i are respectively the QMLE, the quasi log-likelihood, and the number of parameters under the model $\mathcal{M}_i$. The model minimizing (3.1) will be regarded as the best model. This procedure is a sort of maximum likelihood principle. Akaike [1] proposed to take $c_n \equiv 2$ (AIC), Schwarz [14] and Rissanen [12] proposed $c_n \equiv \log n$ (BIC), and Hannan and Quinn [7] proposed $c_n = K \log\log n$ $(K > 0)$. Suppose the expected log-likelihood of $\mathcal{M}_1$ is largest among those of k families. By Theorem 2, IC(i) $(i = 1, ..., k)$ will take almost surely its minimum value at IC(1) for large n if $\lim_{n \to \infty} n^{-1}c_n = 0$. Every criterion above satisfies this condition. Hence we can find asymptotically the model which is closest to g. Further we treat the case that the closest model $\mathcal{M}_1$ ($\mathcal{M}$; say) is divided into several subfamilies (nested case).

Let $\theta^{(0)} = (\theta_1^{(0)}, \ldots, \theta_p^{(0)})$ be a fixed and given interior point of Θ. Then define a subfamily $\mathcal{M}(\{1, 2\})$ of $\mathcal{M}$ by

$$\mathcal{M}(\{1, 2\}) = \{f(x \mid \theta) \in \mathcal{M} \mid \theta = (\theta_1, \theta_2, \theta_3^{(0)}, \ldots, \theta_p^{(0)}) \in \Theta\}.$$

This subfamily has two free parameters θ_1 and θ_2 and the set $\{1, 2\}$ specifies such indices of parameters. For simplicity we call $\mathcal{M}(\{1, 2\})$ a model $\{1, 2\}$. In general let $J = \{j_1, \ldots, j_l\}$ be a subset of the set of all indices $J_p = \{1, \ldots, p\}$. Then the submodel of $\mathcal{M}$ specified by J, say $\mathcal{M}(J)$, is defined by $\{f(x \mid \theta(J)) \mid \theta \in \Theta\}$, where $\theta(J)$ is a $p \times 1$ vector whose j_tth ($t = 1, \ldots, l$) elements are given by θ_{j_t} and remaining elements are given by those of $\theta^{(0)}$. For simplicity we call $\mathcal{M}(J)$ a model J and call $\mathcal{M} = \mathcal{M}(J_p)$ the full model.

Now suppose the unknown quasi true density lies in the model $\mathcal{M}(\{1, \ldots, q\})$, $1 \leqslant q \leqslant p$, i.e., the quasi true parameter vector θ_g can be written as

$$\theta_g = (\theta_1^*, \ldots, \theta_q^*, \theta_{q+1}^{(0)}, \ldots, \theta_p^{(0)}), \quad \theta_1^* \neq \theta_1^{(0)}, \ldots, \theta_q^* \neq \theta_q^{(0)}.$$

This assumption implies that the parameters $\theta_{q+1}, \ldots, \theta_p$ are redundant. We denote $\{1, \ldots, q\}$ by J^* and call it the quasi true model.

Example 2 (continued). Let $\theta^{(0)} = (\theta_1^{(0)}, \theta_2^{(0)}) = (0, 2)$ and the full model $J_2 = \{1, 2\}$. Then the submodels of $\mathcal{M}$ are given by

full model: $\mathcal{M}(\{1, 2\}) = \{\sqrt{\theta_2}^{-1} \phi((x - \theta_1)/\sqrt{\theta_2}) \mid (\theta_1, \theta_2) \in \mathbb{R} \times (0, \infty)\}$,
model $\{1\}$: $\mathcal{M}(\{1\}) = \{\sqrt{2}^{-1} \phi((x - \theta_1)/\sqrt{2}) \mid \theta_1 \in \mathbb{R}\}$,
model $\{2\}$: $\mathcal{M}(\{2\}) = \{\sqrt{\theta_2}^{-1} \phi(x/\sqrt{\theta_2}) \mid \theta_2 \in (0, \infty)\}$,
model $\{\ \}$: $\mathcal{M}(\{\ \}) = \{\sqrt{2}^{-1} \phi(x/\sqrt{2})\}$.

Recall that the true parameter is $\theta_g = (\delta/2, 2 + \delta(\delta + 4)/4)$. Hence the quasi true model $J^* = \{\ \}$ if $\delta = 0$; $= \{1\}$ if $\delta = -4$; $= \{1, 2\} = J_2$ if $\delta \neq 0, -4$.

Suppose $(g, \mathcal{M}(J))$ meet the assumptions A1–A5 for every submodel $\mathcal{M}(J)$, $J \subset J_p$, and write the quasi true parameter and the QMLE in the model J by θ_{Jg} and $\hat{\theta}_J$, respectively. Hence the relation between the expected log-likelihoods of a model J and of the full model is $e(g; f, \theta_{Jg}) = e(g; f, \theta_g)$ if the model J is bigger than or equal to the quasi true model J^*, and $e(g; f, \theta_{Jg}) < e(g; f, \theta_g)$ if the model J does not include the quasi true model J^*.

Theorem 3. *Let λ_n be the likelihood ratio $L_n(\hat{\theta}_J) - L_n(\hat{\theta}_{J^*})$ associated with the models J and J^*. Then if J is bigger than or equal to the quasi true* *model* J^*, λ_n *is nonnegative and has almost surely the order* $O(\log\log n)$. *On the contrary if* J *does not include the quasi true model* J^*, $n^{-1}\lambda_n$ *tends almost surely to* $e(g; f, \theta_{Jg}) - e(g; f, \theta_g) < 0$ (*which yields that* λ_n *tends to minus infinity*).

Proof. If the model J is bigger than the quasi true model J^*, $\lambda_n = L_n(\hat{\theta}_J) - L_n(\hat{\theta}_{J^*}) \geqslant 0$, and by (ii) of Theorem 1, we get $L_n(\hat{\theta}_J) = L_n(\theta_{Jg}) + O(\log\log n)$ and $L_n(\hat{\theta}_{J^*}) = L_n(\theta_{J^*g}) + O(\log\log n)$, where θ_{Jg} and θ_{J^*g} are quasi true parameters in the model $\mathscr{M}(J)$ and $\mathscr{M}(J^*)$, respectively. By the definition of the quasi true model and $J \supset J^*$, we know that $\theta_{Jg} = \theta_{J^*g} = \theta_g$. Hence $\lambda_n = O(\log\log n)$. If the model J does not include J^*, by (iii) of Theorem 1

$$\frac{1}{n} L_n(\hat{\theta}_J) = e(g; f, \theta_J) + O(\sqrt{(n^{-1} \log\log n)})$$

and

$$\frac{1}{n} L_n(\hat{\theta}_{J^*}) = e(g; f, \theta_{J^*}) + O(\sqrt{(n^{-1} \log\log n)}).$$

Hence

$$\frac{1}{n} \lambda_n = e(g; f, \theta_{Jg}) - e(g; f, \theta_{J^*g}) + O(\sqrt{(n^{-1} \log\log n)})$$

$$\to e(g; f, \theta_{Jg}) - e(g; f, \theta_{J^*g}) < 0.$$

THEOREM 4. *Let* $\hat{J}_n$ *be a selected model by the information criterion* (3.1), *i.e.*, J_n *minimizes*

$$\mathrm{IC}(J) = -2L_n(\hat{\theta}_J) + c_n {}^{\#}J$$

based on n samples with respect to submodels $J = \{j_1, ..., j_l\}$, *where* ${}^{\#}J = l$ *denotes a number of free parameters. If* c_n *satisfies both conditions*

$$\lim_{n \to \infty} \frac{1}{n} c_n = 0 \quad \textit{and} \quad \lim_{n \to \infty} \frac{c_n}{\log\log n} = +\infty, \tag{3.2}$$

then $\hat{J}_n$ *is a strongly consistent estimator of the quasi true model* J^*, *i.e.*, $\lim_{n \to \infty} \hat{J}_n = J^*$ *a.s.*

Proof. When the quasi true model J^* is a proper subset of a model J, then by Theorem 3,

$$\begin{aligned}\mathrm{IC}(J)-\mathrm{IC}(J^*) &= ({}^{\#}J-q)c_n - 2\{L_n(\hat{\theta}_J)-L_n(\hat{\theta}_{J^*})\}\\ &= ({}^{\#}J-q)c_n - O(\log\log n)\\ &= \log\log n\{({}^{\#}J-q)c_n/\log\log n - O(1)\}\\ &\to +\infty \qquad \text{a.s.,}\end{aligned}$$

since ${}^{\#}J-q>0$ and $\lim_{n\to\infty} c_n/\log\log n=+\infty$. This implies for large n, $\mathrm{IC}(J)>\mathrm{IC}(J^*)$ a.s. Now we are finding the model which minimizes the information criterion function IC, henceforth for large n, the selected model $\hat{J}_n$ will not be bigger than the true model J^*.

When a model J does not include the true model J^*,

$$\begin{aligned}\mathrm{IC}(J)-\mathrm{IC}(J^*) &= 2n\left\{\frac{1}{n}L_n(\hat{\theta}_{J^*})-\frac{1}{n}L_n(\hat{\theta}_J)-({}^{\#}J-q)c_n/(2n)\right\}\\ &\to \infty \qquad \text{a.s.,}\end{aligned}$$

since $(1/n)\,L_n(\theta_{J^*}) - (1/n)\,L_n(\theta_J) \to e(g; f, \theta_g) - e(g; f, \theta_{Jg}) > 0$ and $n^{-1}c_n \to 0$. Thus $\mathrm{IC}(J)>\mathrm{IC}(J^*)$ for large n. Therefore the information criterion prefers J^* to J. Combining two cases, $\hat{J}_n=J^*$ for $n\geqslant N$, where N depends on the sequence of $x_1, x_2, ..., x_n$.

Note that if we relax the conditions of (3.2) as

$$\lim_{n\to\infty}\frac{1}{n}c_n=0 \qquad \text{and} \qquad \lim_{n\to\infty} c_n=+\infty, \tag{3.3}$$

then $\hat{J}_n$, obtained by such an information criterion, is a weakly consistent estimator of J^*, i.e., $\lim_{n\to\infty} P[\hat{J}_n=J^*]=1$.

However, we need extensive calculation to obtain $\hat{J}_n$ when p is large because there are 2^p-1 possible models. Our alternate procedure, due to Zhao, Krishnaiah, and Bai [18], saves computation. Let $J_{-j}=\{1, ..., j-1, j+1, ..., p\}$ be a $p-1$ set omitted j from J_p for $j=1, ..., p$. Define

$$\tilde{J}_n=\{1\leqslant j\leqslant p \mid \mathrm{IC}(J_{-j})\geqslant \mathrm{IC}(J_p)\}.$$

This set consists of the indices j of the parameters which is important in the sense that the information criterion prefers the full model to the model omitted the jth parameters. This model is obtained by calculating $\mathrm{IC}(J_{-1}), ..., \mathrm{IC}(J_{-p})$ and $\mathrm{IC}(J_p)$ only. However, by the similar lines of the proof of Theorem 4, we get

THEOREM 5. *If c_n satisfies* (3.2) *or* (3.3), *then $\tilde{J}_n$ is also a strongly or weakly consistent estimator of J^*.*

AIC is not consistent because $c_n \equiv 2$ does not meet (3.2) nor (3.3). It will overestimate the quasi true model. The probability $\lim_{n \to \infty} P[\hat{J}_{n,\mathrm{AIC}} = J] > 0$, for $J \supset J^*$ will be expressed using positive linear combinations of independent chi-square variates, however, its formula is hard to evaluate in a simple form.

4. DISCUSSION

We may again note that the condition A5 is not assumed in the usual regularity conditions. Under strong regularity conditions A1–A5, we can evaluate the stochastic orders relating to the QMLE and the quasi log-likelihood by Theorem 1, which are useful to show the strong consistency of the information criteria satisfying (3.2). Our results are based on the i.i.d. assumption. However, Theorems 1–5 still remain valid even if n observations have weak dependency which ensure the central limit theorem and the law of the iterated logarithm. Hence our results are quite general.

Next we try to reconsider the consistency in the model selection problem. From the point of view that the model is an approximation with finite parameters to the true density with infinite parameters (see Shibata [15]), the quasi true model under $\mathscr{M}$ becomes the full model in many cases. Then AIC also becomes consistent since it does not underestimate the quasi true model. Unfortunately our observations do not provide the difference of AIC and BIC in this case.

The purpose of the model selection may be to find the model by which we can get some good prediction for future observation, not the model which provides a good fitting for given observations. Recall AIC is proposed as an estimator of the predictive density. The consistency is one criterion for classifying the model selection procedures, and this criterion may not always lead a suitable conclusion in practical situation.

ACKNOWLEDGMENTS

The author would like to express his thanks to Professor C. R. Rao of University of Pittsburgh and to Professor M. B. Rao of North Dakota State University for their helpful comments. This work was supported by Center for Multivariate Analysis, University of Pittsburgh.

References

[1] Akaike, H. (1973). Information theory and an extension of the maximum likelihood principle. In *Proceedings, 2nd Internat. Symp. on Information Theory* (B. N. Petrov *et al.*, Eds.), pp. 267–281, Akademiai Kiado, Budapest.

[2] Berk, R. H. (1966). Limiting behavior of posterior distributions when the model is incorrect. *Ann. Math. Statist.* **37** 51–58.

[3] Berk, R. H. (1970). Consistency a posteriori. *Ann. Math. Statist.* **41** 894–906.

[4] Cox, D. R. (1961). Tests of separate families of hypotheses. In *Proceedings, 4th Berkeley Symp.*, Vol. 1, pp. 105–123.

[5] Cox, D. R. (1962). Further results on tests of separate families of hypothesis. *J. Roy. Statist. Soc. B* **24** 406–424.

[6] Foutz, R. V., and Srivastava, R. C. (1977). The performance of the likelihood ratio test when the model is incorrect. *Ann. Statist.* **5** 1183–1194.

[7] Hannan, E. J., and Quinn, B. G. (1979). The determination of the order of an autoregression. *J. Roy. Statist. Soc. B* **41** 190–195.

[8] Huber, P. (1976). The behavior of maximum likelihood estimates under non-standard conditions. In *Proceedings, 5th Berkeley Symp.*, Vol. 1, pp. 221–233.

[9] Kent, J. T. (1986). The underlying structure of nonnested hypothesis tests. *Biometrika* **73** 333–343.

[10] LeCam, L. (1953). On some asymptotic properties of maximum likelihood estimates and related Bayes estimates. *Univ. Calif. Publ. Statist.* **1** 227–330.

[11] Rao, C. R. (1973). *Linear Statistical Inference and Its Applications.* Wiley, New York.

[12] Rissanen, J. (1978). Modeling by shortest data description. *Automatica* **14** 465–471.

[13] Sawyer, K. R. (1982). A multiple divergence criterion for testing between separate hypotheses. *Statist. Probab. Lett.* **1** 26–30.

[14] Schwarz, G. (1978). Estimating the dimension of a model. *Ann. Statist.* **6** 461–464.

[15] Shibata, R. (1980). Asymptotically efficient selection of the order of the model for estimating parameters of a linear process. *Ann. Statist.* **8** 147–164.

[16] Wald, A. (1949). Note on the consistency of the maximum likelihood estimate. *Ann. Math. Statist.* **60** 595–601.

[17] White, H. (1982). Maximum likelihood estimation of misspecified models. *Econometrica* **50** 1–25.

[18] Zhao, L. C., Krishnaiah, P. R., and Bai, Z. D. (1986). On detection of the number of signals in presence of white noise. *J. Multivariate Anal.* **20** 1–25.

An Asymptotic Minimax Theorem of Order $n^{-1/2}$

J. PFANZAGL

University of Cologne, Cologne, West Germany

The asymptotic minimax theorem of LeCam and Hájek is refined by inclusion of terms of order $n^{-1/2}$. This renders more precise informations about the local properties of superefficient estimator-sequences. © 1988 Academic Press, Inc.

1. THE RESULTS

Let $(X, \mathscr{A})$ be a measurable space, and $P_\vartheta | \mathscr{A}$, $\vartheta \in \Theta$, a family of probability measures with parameter set $\Theta \subset \mathbb{R}$. Assume that P_ϑ has density $p(\cdot, \vartheta)$ with respect to some dominating measure, say μ. In regular cases $\sigma(\vartheta) := (\int((\partial/\partial\vartheta) \log p(x, \vartheta))^2 P_\vartheta(dx))^{-1/2}$ exist. The socalled asymptotic minimax theorem, specialized to the loss function $1 - 1_{(-u, u)}$, implies the following.

For any sequence of estimators $\vartheta^{(n)}: X^n \to \mathbb{R}$, $n \in \mathbb{N}$, and any $u > 0$,

$$\lim_{a \uparrow \infty} \overline{\lim_{n \to \infty}} \inf_{|t| \leqslant a} P^n_{\vartheta + n^{-1/2} t}\{n^{1/2} | \vartheta^{(n)} - (\vartheta + n^{-1/2} t)| < u\}$$
$$\leqslant N(-u/\sigma(\vartheta), u/\sigma(\vartheta)), \tag{1.1}$$

where N denotes the standard normal distribution.

Relation (1.1) implies in particular that for any sequence $a_n \uparrow \infty$

$$\overline{\lim_{n \to \infty}} \inf_{|t| \leqslant a_n} P^n_{\vartheta + n^{-1/2} t}\{n^{1/2} | \vartheta^{(n)} - (\vartheta + n^{-1/2} t)| < u\}$$
$$\leqslant N(-u/\sigma(\vartheta), u/\sigma(\vartheta)). \tag{1.2}$$

It does, however, not exclude the possibility that

$$\underline{\lim_{n \to \infty}} \inf_{|t| \leqslant a} P^n_{\vartheta + n^{-1/2} t}\{n^{1/2} | \vartheta^{(n)} - (\vartheta + n^{-1/2} t)| < u\} > N(-u/\sigma(\vartheta), u/\sigma(\vartheta))$$

Multivariate Statistics and Probability
ISBN 0-12-580205-6

Reprinted from *J. Mult. Anal.* **27**(2).

for all $a>0$, i.e., that superefficiency holds uniformly on all neighborhoods of ϑ which are of the order $n^{-1/2}$.

The purpose of this paper is to "quantify" the possible amount of superefficiency. We shall show that superefficiency of order $O(n^{-1/2})$ is impossible on all neighborhoods of ϑ which are of the order $n^{-1/4}$.

To formulate this result appropriately, we have to take into account that the normal approximation $N(-u/\sigma(\tau), u/\sigma(\tau))$ to $P_\tau^n\{n^{1/2}|\vartheta^{(n)}-\tau|<u\}$ deviates from $N(-u/\sigma(\vartheta), u/\sigma(\vartheta))$ by an amount of order $|\tau-\vartheta|$, hence by an amount of order $n^{-1/2}t$ if $\tau=\vartheta+n^{-1/2}t$.

To seize on differences of order $n^{-1/2}$ in an appropriate way, we have, therefore, to replace the normal approximation $N(-u/\sigma(\vartheta), u/\sigma(\vartheta))$ by $N(-u/\sigma(\vartheta+n^{-1/2}t), u/\sigma(\vartheta+n^{-1/2}t))$.

Let

$$\begin{aligned} \Delta_n(t) := & P^n_{\vartheta+n^{-1/2}t}\{n^{1/2}|\vartheta^{(n)}-(\vartheta+n^{-1/2}t)|<u\} \\ & -N(-u/\sigma(\vartheta+n^{-1/2}t), u/\sigma(\vartheta+n^{-1/2}t)). \end{aligned} \tag{1.3}$$

(Since ϑ and u remain fixed throughout the following considerations, they are omitted in the symbol $\Delta_n(t)$.)

With this notation, relation (1.1) may be rewritten as

$$\lim_{a\uparrow\infty}\ \overline{\lim_{n\to\infty}}\ \inf_{|t|\leqslant a} \Delta_n(t)\leqslant 0$$

(presuming that σ is continuous at ϑ).

Our paper is concerned with the asymptotic behavior for $n\uparrow\infty$ of

$$\inf_{|t|\leqslant an^{1/4}} n^{1/2}\Delta_n(t).$$

In regular cases, $N(-u/\sigma(\tau), u/\sigma(\tau))$ is certainly an appropriate standard for the asymptotic evaluation of $P_\tau^n\{n^{1/2}|\vartheta^{(n)}-\tau|<u\}$. This follows from the fact that the bound, implicit in the interpretation of (1.1), is attained by certain estimator-sequences. The same argument justifies the use of $N(-u/\sigma(\tau), u/\sigma(\tau))$ as a reference for an evaluation of $P_\tau^n\{n^{1/2}|\vartheta^{(n)}-\tau|<u\}$ taking into consideration also terms of order $O(n^{-1/2})$. Is it plausible that no terms of order $n^{-1/2}$ are needed for the "standard"? The answer is "yes," because the $n^{-1/2}$-term of the Edgeworth-approximation to the distributions of estimator-sequences which are maximally concentrated up to $o(n^{-1/2})$ is *odd* and cancels out in approximations for *symmetric* intervals: In regular cases, $N(-u/\sigma(\tau), u/\sigma(\tau))$ is a bound of order $o(n^{-1/2})$ (and not just $o(n^0)$) for the concentration of estimator-sequences (see [6, p. 35/6] for the parametric case, and [7, Theorem 9.2.7, p. 295] for a "nonparametric" version).

THEOREM. *Assume that the family P_ϑ, $\vartheta\in\Theta$, is regular in the sense specified in Section 4 by (i)–(v). Assume that there exists $a_0>0$ and a subsequence $\mathbb{N}_0\subset\mathbb{N}$ such that*

$$\lim_{n\in\mathbb{N}_0}\ \inf_{|t|\leqslant a_0 n^{1/4}} n^{1/2}\Delta_n(t)>0. \tag{1.4}$$

Then there exists $a_1 > a_0$ such that

$$\overline{\lim_{n \in \mathbb{N}_0}} \inf_{|t| \leqslant a_1 n^{1/4}} n^{1/2} \Delta_n(t) < 0. \tag{1.5}$$

To obtain another equivalent formulation of the theorem, we provide the following lemma which refers to *arbitrary* sequences of nonincreasing functions $D_n : \mathbb{R}^+ \to \mathbb{R}$. The theorem asserts that property A is fulfilled for

$$D_n(a) = \inf_{|t| \leqslant a n^{1/4}} n^{1/2} \Delta_n(t). \tag{1.6}$$

LEMMA. *For any sequence of nonincreasing functions $D_n : \mathbb{R}^+ \to \mathbb{R}$, the following two properties are equivalent:*

A. *For every subsequence $\mathbb{N}_0 \subset \mathbb{N}$,*

$$\lim_{n \in \mathbb{N}_0} D_n(a_0) > 0 \qquad \textit{for some } a_0 \in \mathbb{R}^+$$

implies

$$\overline{\lim_{n \in \mathbb{N}_0}} D_n(a_1) < 0 \qquad \textit{for some } a_1 > a_0.$$

B. *For every subsequence $\mathbb{N}_0 \subset \mathbb{N}$,*

$$\underline{\lim}_{n \in \mathbb{N}_0} D_n(a) \geqslant 0 \qquad \textit{for every } a \in \mathbb{R}_+$$

implies

$$\lim_{n \in \mathbb{N}_0} D_n(a) = 0 \qquad \textit{for every } a \in \mathbb{R}_+.$$

ADDENDUM. A *or* B *imply*

$$\overline{\lim_{n \in \mathbb{N}}} D_n(a_n) \leqslant 0 \qquad \textit{for every sequence } a_n \uparrow \infty. \tag{1.7}$$

The idea to describe the local properties of superefficient estimator-sequences by an as. minimax theorem (of order n^0) goes back to LeCam [3].

Deviations of higher order in the as. minimax theorem are thoroughly investigated in Bickel, Götze, and van Zwet [1]. Using Bayes-type arguments, these authors arrive at results of order $O(n^{-1/2})$ and $O(n^{-1})$ for symmetric bowl-shaped loss functions. Specialized to the loss function $1-1_{(-u,u)}$ their $O(n^{-1/2})$-result (see [1, Theorem 1a, p. 753]) leads to

$$\overline{\lim_{n\in\mathbb{N}}}\, D_n(an^{1/2})\leqslant 0,$$

a result weaker than (1.7).

A detailed study of second-order differences in the asymptotic minimax theorem for estimators of the means of normal distributions is due to Levit (see [5] and the references cited there).

Proof. A *implies* B. Let $\mathbb{N}_0\subset\mathbb{N}$ be an arbitrary subsequence. If $\lim_{n\in\mathbb{N}_0} D_n(a)\geqslant 0$ for every $a\in\mathbb{R}_+$, and $\overline{\lim}_{n\in\mathbb{N}_0} D_n(a_0)>0$ for some $a_0\in\mathbb{R}_+$, choose a subsequence $\mathbb{N}_1\subset\mathbb{N}_0$ such that $\lim_{n\in\mathbb{N}_1} D_n(a_0)>0$. By A there exists $a_1>a_0$ such that $\overline{\lim}_{n\in\mathbb{N}_1} D_n(a_1)<0$, in contradiction to $\lim_{n\in\mathbb{N}_0} D_n(a)\geqslant 0$ for every $a\in\mathbb{R}_+$.

B *implies* A. Let $\mathbb{N}_0\subset\mathbb{N}$ be an arbitrary subsequence. Assume that $\lim_{n\in\mathbb{N}_0} D_n(a_0)>0$ and $\overline{\lim}_{n\in\mathbb{N}_0} D_n(a)\geqslant 0$ for all $a>a_0$, hence for all $a\in\mathbb{R}_+$. Choose a subsequence $\mathbb{N}_1\subset\mathbb{N}_0$ and a sequence $a_n\uparrow\infty$ such that $\lim_{n\in\mathbb{N}_1} D_n(a_n)\geqslant 0$. This implies $\underline{\lim}_{n\in\mathbb{N}_1} D_n(a)\geqslant 0$ for all $a\in\mathbb{R}_+$, hence, by B, $\lim_{n\in\mathbb{N}_1} D_n(a)=0$ for all $a\in\mathbb{R}^+$, in contradiction to $\lim_{n\in\mathbb{N}_0} D_n(a_0)>0$.

Proof of the Addendum. If $\overline{\lim}_{n\in\mathbb{N}} D_n(a_n)>0$ for some sequence $a_n\uparrow\infty$, there exists a subsequence $\mathbb{N}_0\subset\mathbb{N}$ for which $\lim_{n\in\mathbb{N}_0} D_n(a_n)>0$. Hence $\underline{\lim}_{n\in\mathbb{N}_0} D_n(a)>0$ for every $a\in\mathbb{R}_+$, which is impossible by B.

The question poses itself whether the theorem can be improved, for instance, by showing that (1.5) follows from a weaker version of (1.4), say one in which the infimum over $|t|\leqslant a_0 n^{1/4}$ is replaced by an infimum over $|t|\leqslant a_n n^{1/4}$ with $a_n\downarrow 0$. The following example shows that improvements of this kind are impossible, in general: The order $n^{-1/4}$ is a sort of threshold for the region of superefficiency of order $n^{-1/2}$.

EXAMPLE. For the location parameter family of normal distributions, $\{N_{(\vartheta,1)}: \vartheta\in\mathbb{R}\}$, the following holds true:

(a) For every $a>0$ there exists an estimator-sequence such that

$$\underline{\lim_{n\in\mathbb{N}}}\ \inf_{|t|\leqslant an^{1/4}} n^{1/2}\Delta_n(t)>0.$$

(b) For every sequence $a_n\downarrow 0$ there exists an estimator-sequence such that

$$\lim_{n\in\mathbb{N}}\ \inf_{|t|\leqslant a_n n^{1/4}} n^{1/2}\Delta_n(t)>0$$

and

$$\liminf_{n \in \mathbb{N}} \inf_{t \in \mathbb{R}} n^{1/2} \Delta_n(t) \geq 0.$$

Remark. The theorem is stated for one-parameter families to keep the regularity conditions transparent. It holds, in fact, for an arbitrary family $\mathscr{P}$, and any twice differentiable functional $\kappa: \mathscr{P} \to \mathbb{R}$. A precise statement for this general case requires, however, an unrestricted use of concepts like tangent space, canonical gradient, etc. To obtain a proof of the general version replace $s \to P_{\vartheta+s}$ by a twice differentiable path $s \to P_s$ and let

$$\Delta_n(t) := P^n_{n^{-1/2}t}\{n^{1/2}|\kappa^{(n)} - \kappa(P_{n^{-1/2}t})| < u\}$$
$$- N(-u/\sigma(P_{n^{-1/2}t}), u/\sigma(P_{n^{-1/2}t})),$$

with $\sigma(P) = (\int \kappa^*(x, P)^2\, P(dx))^{1/2}$.

The proof goes through with $Q_{n,k} = P^n_{n^{-1/2}t_{n,k}}$, where $t_{n,k}$ is defined inductively by $t_{n,k+1} = t_{n,k} + 2u/\sigma(Q_{n,k})^2 + n^{-1/2} r_{n,k}$, with $r_{n,k}$ chosen such that $\kappa(Q_{n,k+1}) - \kappa(Q_{n,k}) > 2un^{-1/2}$.

Instead of Lemma 1 use [7, relation (4.5.6), Theorem 6.6.3, pp. 194–195, in particular (6.6.4) and (6.6.5)], instead of Lemma 2 use [7, relation (4.5.6), p. 125. See also 9.2.1(ii), pp. 291–292.].

The literature now has plenty of nonparametric minimax theorems of order n^0. The idea of such nonparametric versions goes back to Levit [4] (who takes suprema over non-shrinking neighborhoods of P).

2. Proof of the Theorem

Throughout the proof, r_n, $n \in \mathbb{N}$, denotes a generic null-sequence, and n_* a generic element of $\mathbb{N}$, with "$n > n_*$" indicating that a certain statement holds for all sufficiently large $n \in \mathbb{N}$.

(i) We use the following notations:

$$\vartheta_{n,k} := \vartheta + 2ukn^{-1/2},$$
$$u_{n,k} := u/\sigma(\vartheta_{n,k})$$
$$Q_{n,k} := P^n_{\vartheta_{n,k}}$$
$$\alpha^-_{n,k} := Q_{n,k}\{n^{1/2}(\vartheta^{(n)} - \vartheta_{n,k}) \leq -u\}$$
$$\alpha^+_{n,k} := Q_{n,k}\{n^{1/2}(\vartheta^{(n)} - \vartheta_{n,k}) \geq u\}.$$

(ii) If the assertion is wrong, we have

$$\lim_{n \in \mathbb{N}_0} D_n(a_0) > 0, \qquad \text{and} \qquad \overline{\lim_{n \in \mathbb{N}_0}} D_n(a) \geqslant 0 \qquad \text{for every } a > 0.$$

W.l.o.g. we assume $D_n(a_0) \geqslant A > 0$ for $n \in \mathbb{N}_0$. Moreover, there exists a sequence $a_n \uparrow \infty$, $n \in \mathbb{N}$, such that $\overline{\lim}_{n \in \mathbb{N}_0} D_n(a_n) \geqslant 0$. We may assume that the convergence of a_n, $n \in \mathbb{N}$, to infinity is sufficiently slow so that $\lim_{n \in \mathbb{N}_0} a_n n^{-1/4} = 0$. Hence we obtain the following relations:

(a) There exists a sequence $c_n \uparrow \infty$, $n \in \mathbb{N}$, with $\lim_{n \in \mathbb{N}_0} c_n n^{-1/4} = 0$ such that for all integers $k \in [0, c_n n^{1/4}]$ and $n \in \mathbb{N}_0$

$$Q_{n,k}\{n^{1/2} | \vartheta^{(n)} - \vartheta_{n,k}| < u\} \geqslant N(-u_{n,k}, u_{n,k}) + n^{-1/2} r_n; \tag{2.1}$$

equivalently

$$\alpha_{n,k}^- + \alpha_{n,k}^+ \leqslant 2\Phi(-u_{n,k}) + n^{-1/2} r_n. \tag{2.1$'$}$$

(b) There exists $c_0 > 0$ such that for all integers $k \in [0, c_0 n^{1/4}]$ and $n \in \mathbb{N}_0$

$$Q_{n,k}\{n^{1/2} | \vartheta^{(n)} - \vartheta_{n,k}| < u\} \geqslant N(-u_{n,k}, u_{n,k}) + n^{-1/2} A; \tag{2.2}$$

equivalently

$$\alpha_{n,k}^- + \alpha_{n,k}^+ \leqslant 2\Phi(-u_{n,k}) - n^{-1/2} A. \tag{2.2$'$}$$

In the following we replace the somewhat clumsy expression "for all integers $k \in [0, m_n]$" by "for $k \leqslant m_n$."

Notice that $\alpha_{n,k}^+ \in (0, 1)$. This can be seen as follows: $\alpha_{n,k}^+ = Q_{n,k}\{n^{1/2}(\vartheta^{(n)} - \vartheta_{n,k}) \geqslant u\} = 0$, implies $Q_{n,k+1}\{n^{1/2}(\vartheta^{(n)} - \vartheta_{n,k}) \geqslant u\} = 0$. By definition of $\vartheta_{n,k}$ we have

$$\{n^{1/2}(\vartheta^{(n)} - \vartheta_{n,k}) \geqslant u\} = \{n^{1/2}(\vartheta^{(n)} - \vartheta_{n,k+1}) \geqslant -u\}; \tag{2.3}$$

hence $\alpha_{n,k+1}^- = 1$, in contradiction to (2.1$'$).

From (2.2$'$), applied for $k = 0$, we obtain that at least one of the following inequalities holds for infinitely many $n \in \mathbb{N}_0$:

$$\alpha_{n,0}^- \leqslant \Phi(-u_{n,0}) - n^{-1/2} A/2, \tag{2.4$'$}$$

$$\alpha_{n,0}^+ \leqslant \Phi(-u_{n,0}) - n^{-1/2} A/2. \tag{2.4$''$}$$

W.l.o.g. we may assume that this is the case with (2.4$''$).

Let $\mathbb{N}_1 \subset \mathbb{N}_0$ denote the infinite subsequence for which (2.4$''$) holds true. For $k \leqslant c_n n^{1/4}$, $n \in \mathbb{N}_1$, we define numbers $\eta_{n,k}$ by

$$\alpha_{n,k}^+ = \Phi(-u_{n,k} - \eta_{n,k}). \tag{2.5}$$

From (2.4″) we have for $n \in \mathbb{N}_1$

$$\Phi(-u_{n,0}-\eta_{n,0}) \leqslant \Phi(-u_{n,0}) - n^{-1/2} A/2.$$

Hence there exists $A_0 > 0$ such that

$$\eta_{n,0} > A_0 n^{-1/2} \qquad \text{for} \quad n \in \mathbb{N}_1. \tag{2.6}$$

(iii) Considering $\{n^{1/2}(\vartheta^{(n)} - \vartheta_{n,k}) > u\}$ as a critical region for testing the hypothesis $Q_{n,k}$ at level $\alpha_{n,k}^+$ against the alternative $Q_{n,k+1}$ we obtain from Lemma 1 and relation (2.5) for $k \leqslant c_n n^{1/4}$, $n \in \mathbb{N}_1$,

$$\begin{aligned}
&Q_{n,k+1}\{n^{1/2}(\vartheta^{(n)} - \vartheta_{n,k}) \geqslant u\} \\
&\quad \leqslant \Phi(\Phi^{-1}(\alpha_{n,k}^+) + 2u_{n,k} + n^{-1/2}\tfrac{1}{3}u_{n,k}\sigma^3(\vartheta_{n,k}) \\
&\qquad \times [2u_{n,k}(3a(\vartheta_{n,k}) + b(\vartheta_{n,k})) - \Phi^{-1}(\alpha_{n,k}^+)\, b(\vartheta_{n,k})]) + n^{-1/2} r_n \\
&\quad = \Phi(u_{n,k} + n^{-1/2}u_{n,k}^2\,\sigma_{n,k}^3(2a(\vartheta_{n,k}) + b(\vartheta_{n,k})) \\
&\qquad - \eta_{n,k}(1 - n^{-1/2}\tfrac{1}{3}u_{n,k}\sigma^3(\vartheta_{n,k})\, b(\vartheta_{n,k})) + n^{-1/2} r_n.
\end{aligned} \tag{2.7}$$

By a Taylor expansion of $s \to \sigma(\delta + n^{-1/2}s)$ up to $o(n^{-1/2})$ which holds uniformly for δ and s varying in bounded sets, we obtain that

$$\sigma(\vartheta_{n,k+1})^{-1} = \sigma(\vartheta_{n,k})^{-1} + n^{-1/2}u\sigma(\vartheta_{n,k})[2a(\vartheta_{n,k}) + b(\vartheta_{n,k})] + n^{-1/2} r_n,$$

for $k \leqslant c_n n^{1/2}$ and $n \in \mathbb{N}_1$; hence

$$u_{n,k+1} = u_{n,k} + n^{-1/2}u_{n,k}^2\,\sigma(\vartheta_{n,k})^3[2a(\vartheta_{n,k}) + b(\vartheta_{n,k})] + n^{-1/2} r_n. \tag{2.8}$$

For $k \leqslant c_n n^{1/4}$ we have $|\vartheta_{n,k} - \vartheta| \leqslant 2uc_n n^{-1/4} = o(n^0)$. Since σ and b are continuous at ϑ, $B_{n,k} := \frac{1}{3}u_{n,k}\sigma^3(\vartheta_{n,k})\, b(\vartheta_{n,k})$ is uniformly bounded for $k \leqslant c_n n^{1/4}$, $n \in \mathbb{N}_1$.

From (2.7) and (2.8) we obtain for $k \leqslant c_n n^{1/4}$, $n \in \mathbb{N}_1$,

$$\begin{aligned}
&Q_{n,k+1}\{n^{1/2}(\vartheta^{(n)} - \vartheta_{n,k}) \geqslant u\} \\
&\quad \leqslant \Phi(u_{n,k+1} - \eta_{n,k}(1 - n^{-1/2} B_{n,k})) + n^{-1/2} r_n.
\end{aligned} \tag{2.9}$$

From (2.3) and (2.9) we obtain

$$\begin{aligned}
&Q_{n,k+1}\{n^{1/2}(\vartheta^{(n)} - \vartheta_{n,k+1}) \geqslant -u\} \\
&\quad \leqslant \Phi(u_{n,k+1} - \eta_{n,k}(1 - n^{-1/2} B_{n,k})) + n^{-1/2} r_n,
\end{aligned}$$

whence

$$\Phi(-u_{n,k+1} + \eta_{n,k}(1 - n^{-1/2} B_{n,k})) \leqslant \alpha_{n,k+1}^- + n^{-1/2} r_n. \tag{2.10}$$

Relations (2.1′) and (2.2′), applied with k replaced by $k+1$ read as follows: For $k \leqslant c_n n^{1/4}$, $n \in \mathbb{N}_1$,

$$\alpha^-_{n,k+1} + \alpha^+_{n,k+1} \leqslant 2\Phi(-u_{n,k+1}) + n^{-1/2} r_n. \tag{2.11}$$

Uniformly for $k \leqslant c_0 n^{1/4}$, $n \in \mathbb{N}_0$,

$$\alpha^-_{n,k+1} + \alpha^+_{n,k+1} \leqslant 2\Phi(-u_{n,k+1}) - n^{-1/2} A. \tag{2.12}$$

By definition of $\eta_{n,k+1}$ (see (2.5) with k replaced by $k+1$), we obtain from (2.10)

$$\begin{aligned}\Phi(-u_{n,k+1} + \eta_{n,k}(1 - n^{-1/2} B_{n,k})) &+ \Phi(-u_{n,k+1} - \eta_{n,k+1}) \\ &\leqslant \alpha^-_{n,k+1} + \alpha^+_{n,k+1} + n^{-1/2} r_n. \end{aligned}\tag{2.13}$$

From (2.11)–(2.13) we obtain for $k \leqslant c_n n^{1/4}$, $n \in \mathbb{N}_1$,

$$\begin{aligned}\Phi(-u_{n,k+1} + \eta_{n,k}(1 - n^{-1/2} B_{n,k})) &+ \Phi(-u_{n,k+1} - \eta_{n,k+1}) \\ &\leqslant 2\Phi(-u_{n,k+1}) - n^{-1/2} A\, 1_{[0, c_0 n^{1/4}]}(k) + n^{-1/2} r_n. \end{aligned}\tag{2.14}$$

The proof will be concluded by showing that (2.14) is contradictory. To prepare this proof, we apply Lemma 2 to (2.14) and obtain

$$\eta_{n,k}(1 - n^{-1/2} B_{n,k})\, \varphi(u_{n,k}) \leqslant \Phi(-u_{n,k}) + n^{-1/2} r_n \tag{2.15′}$$

and

$$\eta_{n,k}(1 - n^{-1/2} B_{n,k}) \leqslant \eta_{n,k+1} + n^{-1/2} r_n. \tag{2.15″}$$

Hence $\{\eta_{n,k} : k \leqslant c_n n^{1/4},\ n \in \mathbb{N}_1\}$ is bounded, and positive because of (2.6).

(iv) Let $u_0 := u/\sigma(\vartheta)$. In this section we shall prove the existence of $n_* \in \mathbb{N}$ such that

$$\eta_{n,k} \leqslant \frac{3}{2} u_0 \qquad \text{for} \quad k \leqslant \frac{c_n}{2} n^{1/4},\ n \in \mathbb{N}_1,\ n \geqslant n_*. \tag{2.16}$$

Assume that, on the contrary, there exists an infinite subsequence $\mathbb{N}_2 \subset \mathbb{N}_1$ and, for each $n \in \mathbb{N}_2$, an integer $k_n \leqslant (c_n/2)\, n^{1/4}$ such that

$$\eta_{n,k_n} > \tfrac{3}{2} u_0, \qquad n \in \mathbb{N}_2. \tag{2.17}$$

Let

$$c(v) := (\tfrac{3}{2} - 2\Phi(-v) - \Phi(2v))/\varphi(0). \tag{2.18}$$

Notice that $c(v) > 0$ for $v > 0$.

W.l.o.g. we may assume that the elements of $\mathbb{N}_2$ are large enough so that the following relations hold for $n \in \mathbb{N}_2$, $k \leqslant (c_n/2)\, n^{1/4}$:

$$|u_{n,k} - u_0| < u_0/8 \tag{2.19'}$$

$$c(u_{n,k}) > 3c(u_0)/4 \tag{2.19''}$$

$$b_n^k \geqslant \tfrac{3}{4}, \tag{2.19'''}$$

where $b_n := 1 - n^{-1/2} B$, with $B := \sup\{B_{n,k} : k \leqslant (c_n/2)\, n^{1/4},\ n \in \mathbb{N}_1\}$.

Let $k_n \leqslant (c_n/2)\, n^{1/4}$, $n \in \mathbb{N}_2$, be a sequence fulfilling (2.17). We shall show that for $n \in \mathbb{N}_2$, $v \leqslant (c_n/2)\, n^{1/4}$,

$$\eta_{n,k_n+v} \geqslant (\eta_{n,k_n} + v\Delta)\, b_n^v, \qquad \text{with} \quad \Delta = c(u_0)/2. \tag{2.20}$$

From (2.14) we obtain for $k \leqslant c_n n^{1/4}$, $n \in \mathbb{N}_2$,

$$\begin{aligned} &\Phi(-u_{n,k+1} + \eta_{n,k} b_n) + \Phi(-u_{n,k+1} - \eta_{n,k+1}) \\ &\quad \leqslant 2\Phi(-u_{n,k+1}) + n^{-1/2} r_n. \end{aligned} \tag{2.21}$$

Relation (2.20) is trivial for $v = 0$. Assume now that (2.20) is true for $v-1$. From (2.21), applied for $k = k_n + v - 1$ we obtain

$$\begin{aligned} &1 - \Phi(u_{n,k_n+v} + \eta_{n,k_n+v}) + \Phi(-u_{n,k_n+v} + \eta_{n,k_n+v-1} b_n) \\ &\quad \leqslant 2\Phi(-u_{n,k_n+v}) + n^{-1/2} r_n. \end{aligned} \tag{2.22}$$

If (2.20) holds true with v replaced by $v-1$, we obtain from (2.17), (2.19″), and (2.19‴)

$$\eta_{n,k_n+v-1} b_n \geqslant \eta_{n,k_n} b_n^v \geqslant \tfrac{9}{8} u_0 > u_{n,k_n+v}. \tag{2.23}$$

Let $\Delta_{n,v}$ be defined by

$$\eta_{n,k_n+v} = \eta_{n,k_n+v-1} b_n + \Delta_{n,v}. \tag{2.24}$$

From (2.22), (2.24), and (2.23),

$$\begin{aligned} &1 - 2\Phi(-u_{n,k_n+v}) \\ &\quad \leqslant \Phi(u_{n,k_n+v} + \eta_{n,k_n+v-1} b_n + \Delta_{n,v}) \\ &\qquad - \Phi(-u_{n,k_n+v} + \eta_{n,k_n+v-1} b_n) + n^{-1/2} r_n \\ &\quad \leqslant \Phi(u_{n,k_n+v} + \eta_{n,k_n+v-1} b_n) - \Phi(-u_{n,k_n+v} + \eta_{n,k_n+v-1} b_n) \\ &\qquad + \varphi(0)\, \Delta_{n,v}\, 1_{(0,\infty)}(\Delta_{n,v}) + n^{-1/2} r_n \\ &\quad \leqslant \Phi(2u_{n,k_n+v}) - \tfrac{1}{2} + \varphi(0)\, \Delta_{n,v}\, 1_{(0,\infty)}(\Delta_{n,v}) + n^{-1/2} r_n. \end{aligned}$$

For the last inequality, use $\eta_{n,k_n+v-1} b_n > u_{n,k_n+v}$ (see (2.23)).

Using (2.18) and (2.19″) we obtain for $n \geqslant n_*$

$$\Delta_{n,\nu}\, 1_{(0,\infty)}(\Delta_{n,\nu}) \geqslant c(u_{n,k_n+\nu}) - n^{-1/2} r_n/\varphi(0)$$
$$\geqslant c(u_0)/2 = \Delta > 0.$$

This implies $\Delta_{n,\nu} \geqslant \Delta$ for $n \geqslant n_*$.

From (2.24) and (2.20) with ν replaced by $\nu-1$ we obtain

$$\eta_{n,k_n+\nu} \geqslant \eta_{n,k_n+\nu-1}\, b_n + \Delta \geqslant (\eta_{n,k_n} + (\nu-1)\,\Delta)\, b_n^{\nu} + \Delta$$
$$\geqslant (\eta_{n,k_n} + \nu\Delta)\, b_n^{\nu}.$$

This concludes the proof of (2.20).

Let $\bar{k}_n := [(c_n/2)\, n^{1/4}]$. From (2.20), applied with $\nu = \bar{k}_n$, and (2.19‴) we obtain for $n \in \mathbb{N}_2$

$$\eta_{n,k_n+\bar{k}_n} \geqslant \bar{k}_n \Delta b_n^{\bar{k}_n} \geqslant \tfrac{3}{4}\Delta \bar{k}_n;$$

i.e., the sequence $\eta_{n,k_n+\bar{k}_n}$, $n \in \mathbb{N}_2$, tends to infinity. Since $k_n + \bar{k}_n \leqslant c_n n^{1/4}$, this contradicts (2.21). This concludes the proof of (2.16).

(v) From (2.16) we obtain the existence of $n_* \in \mathbb{N}$ such that

$$\eta_{n,k} \leqslant \frac{3}{2} u_0 < 2u_{n,k} \qquad \text{for} \quad k \leqslant \frac{c_n}{2} n^{1/4},\ n \in \mathbb{N}_1,\ n \geqslant n_*.$$

By Lemma 2 there exists $\Delta > 0$ such that for $k \leqslant (c_n/2)\, n^{1/4}$, $n \in \mathbb{N}_1$, $n \geqslant n_*$,

$$\Phi(-u_{n,k+1} + \eta_{n,k} b_n)$$
$$\geqslant \Phi(-u_{n,k+1}) + \eta_{n,k}\, b_n\, \varphi(u_{n,k+1}) + \eta_{n,k}^2\, b_n^2 \Delta \tag{2.25'}$$

and

$$\Phi(-u_{n,k+1} - \eta_{n,k+1})$$
$$\geqslant \Phi(-u_{n,k+1}) - \eta_{n,k+1}\ \varphi(u_{n,k+1}) + \eta_{n,k+1}^2\, \Delta. \tag{2.25''}$$

Together with (2.14) this implies for $k \leqslant (c_n/2)\, n^{1/4}$, $n \in \mathbb{N}_1$, $n \geqslant n_*$,

$$(\eta_{n,k}\, b_n - \eta_{n,k+1})\, \varphi(u_{n,k+1}) + (\eta_{n,k}^2\, b_n^2 + \eta_{n,k+1}^2)\, \Delta$$
$$\leqslant -n^{-1/2} A 1_{[0,c_0 n^{1/4}]}(k) + n^{-1/2} r_n. \tag{2.26}$$

With $0 < c \leqslant \varphi(u_{n,k+1})$ for $n \in \mathbb{N}_1$, $n \geqslant n_*$, $k \leqslant (c_n/2)\, n^{1/4}$, we obtain

$$(\eta_{n,k}\, b_n - \eta_{n,k+1}) + (\eta_{n,k}^2 b_n^2 + \eta_{n,k+1}^2)\, \Delta/c$$
$$\leqslant -n^{-1/2} \frac{A}{c} 1_{[0,\, c_0 n^{1/4}]}(k) + n^{-1/2} r_n.$$

With $\xi_{n,k} := \eta_{n,k}\,\Delta/c$, $A_1 := A\Delta/2c^2$ we obtain for $k \leqslant (c_n/2)\,n^{1/4}$, $n \in \mathbb{N}_1$, $n \geqslant n_*$,

$$(\xi_{n,k} b_n - \xi_{n,k+1}) + (\xi_{n,k}^2 b_n^2 + \xi_{n,k+1}^2) \leqslant -n^{-1/2}\,2A_1\,1_{[0,\,c_0 n^{1/4}]}(k) + n^{-1/2} r_n. \tag{2.27}$$

(vi) Relation (2.27) implies in particular for $k \leqslant c_0\,n^{1/4}$, $n \in \mathbb{N}_1$, $n \geqslant n_*$,

$$\xi_{n,k+1} \geqslant \xi_{n,k}\,b_n + n^{-1/2}\,A_1. \tag{2.28}$$

We shall show that for $k \leqslant c_0 n^{1/4}$, $n \in \mathbb{N}_1$, $n \geqslant n_*$,

$$\xi_{n,k} \geqslant n^{-1/2}\,A_1\,\frac{1-b_n^k}{1-b_n}. \tag{2.29}$$

For $k=0$ we have $\xi_{n,0} > 0$ (since $\eta_{n,0} > 0$ by (2.6)), hence (2.29) is trivially true. Relation (2.29) now follows from (2.28) by induction.

With $k_n := [c_0 n^{1/4}]$ we obtain

$$1 - b_n^{k_n} = 1 - (1 - n^{-1/2}\,B)^{k_n} \geqslant 1 - (1 - \tfrac{1}{2} n^{-1/2}\,B k_n) \geqslant \tfrac{1}{2} B c_0 n^{-1/4}. \tag{2.30}$$

Therefore, (2.29), applied for $k = k_n$, yields for $n \in \mathbb{N}_1$, $n \geqslant n_*$,

$$\xi_{n,k_n} \geqslant A_2 n^{-1/4} \qquad \text{with} \quad A_2 > 0. \tag{2.31}$$

Let now

$$\omega_{n,v} := \xi_{n,k_n+v}. \tag{2.32}$$

From (2.27), applied for $k = k_n + v$, we obtain for $v \leqslant (c_n/3)\,n^{1/4}$, $n \in \mathbb{N}_1$, $n \geqslant n_*$,

$$\omega_{n,v}\,b_n - \omega_{n,v+1} + \omega_{n,v}^2 b_n^2 + \omega_{n,v+1}^2 \leqslant n^{-1/2}\,\bar{r}_n. \tag{2.33}$$

We write $\bar{r}_n$ rather than r_n, because from now on $\bar{r}_n$, $n \in \mathbb{N}$, is a fixed (rather than generic) null-sequence.

From (2.31),

$$\omega_{n,0} \geqslant A_2 n^{-1/4} > 0. \tag{2.34}$$

Moreover,

$$0 < \omega_{n,v} \leqslant \frac{1}{4} \qquad \text{for} \quad v \leqslant \frac{c_n}{3}\,n^{1/4},\ n \in \mathbb{N}_1,\ n \geqslant n_*. \tag{2.35}$$

The first inequality follows immediately from (2.33) and (2.34). To establish the second inequality, observe that (2.33) is equivalent to

$$(\omega_{n,v} b_n + \tfrac{1}{2})^2 + (\omega_{n,v+1} - \tfrac{1}{2})^2 \leqslant \tfrac{1}{2} + n^{-1/2} \bar{r}_n.$$

Therefore,

$$(\omega_{n,v} b_n + \tfrac{1}{2})^2 \leqslant \tfrac{1}{2} + n^{-1/2} \bar{r}_n,$$

from which the second inequality follows easily.

(vii) The proof will be concluded by showing that (2.33) and (2.35) are contradictory. For this purpose we derive from (2.33) the following weaker inequality. For $v \leqslant (c_n/3)\, n^{1/4}$, $n \in \mathbb{N}_1, n \geqslant n_*$,

$$\omega_{n,v} b_n \leqslant \omega_{n,v+1} - \omega_{n,v+1}^2 + n^{-1/2} \bar{r}_n. \tag{2.36}$$

Let

$$m_n := [2/\omega_{n,0}] - 4. \tag{2.37}$$

By this choice of m_n we achieve that

$$1 - \tfrac{1}{2} m_n \omega_{n,0} \geqslant 2\omega_{n,0}, \tag{2.37$'$}$$

a relation needed later on. Because of (2.34), we have $m_n \leqslant (c_n/3) n^{1/4}$ for $n \in \mathbb{N}_1$, $n \geqslant n_*$, so that (2.36) holds, in particular, for all $v \leqslant m_n$.

Let

$$a_{n,v} := \omega_{n,0}\, b_n^v (1 - \tfrac{1}{2} v \omega_{n,0})^{-1}. \tag{2.38}$$

For later use we remark that

$$a_{n,v} < \tfrac{1}{2} \qquad \text{for} \quad v \leqslant m_n \tag{2.39$'$}$$

$$a_{n,m_n} > \tfrac{1}{4} \qquad \text{for} \quad n \geqslant n_*. \tag{2.39$''$}$$

(For (2.39″) observe that $b_n^{m_n} \to 1$, so that $a_{n,m_n} > \omega_{n,0}\, \tfrac{3}{4}(1 - \tfrac{1}{2} m_n \omega_{n,0})^{-1}$.)

We shall show that $v \leqslant m_n$, $n \in \mathbb{N}_1$, $n \geqslant n_*$, implies

$$a_{n,v-1} b_n \geqslant a_{n,v} - a_{n,v}^2 + n^{-1/2} \bar{r}_n. \tag{2.40}$$

An elementary computation shows that (2.40) is equivalent to

$$\frac{\omega_{n,0}^2}{4} \geqslant \omega_{n,0} \left(1 - \frac{v-1}{2} \omega_{n,0}\right) \left[\frac{1}{2} - b_n^v + n^{-1/2} \bar{r}_n \omega_{n,0}^{-2} b_n^{-v} \left(1 - \frac{v}{2} \omega_{n,0}\right)^2\right]. \tag{2.41}$$

Since $1-\frac{1}{2}\nu\omega_{n,0}>0$ for $\nu\leqslant m_n$, relation (2.41) follows from

$$\frac{1}{2}-b_n^\nu+n^{-1/2}\bar{r}_n\,\omega_{n,0}^{-2}b_n^{-\nu}\left(1-\frac{\nu}{2}\omega_{n,0}\right)^2<0. \tag{2.42}$$

For $\nu\leqslant m_n$ we have $b_n^\nu>\frac{3}{4}$ and $|1-(\nu/2)\,\omega_{n,0}|\leqslant 1$ by (2.34) and (2.37′). Together with (2.34) this implies that the left-hand side of (2.42) is smaller than

$$\tfrac{1}{2}-\tfrac{3}{4}+n^{-1/2}\bar{r}_n(A_2n^{-1/4})^{-2}\tfrac{4}{3}=-\tfrac{1}{4}+\bar{r}_n\tfrac{4}{3}A_2^{-2}<0 \qquad \text{for} \quad n\geqslant n_*.$$

This concludes the proof of (2.40).

(viii) Now we shall show that

$$\omega_{n,\nu}\geqslant a_{n,\nu} \qquad \text{for} \quad \nu\leqslant m_n,\, n\in\mathbb{N}_1,\, n\geqslant n_*. \tag{2.43}$$

For $\nu=0$ this follows immediately from (2.38). Assume now that (2.43) is true for $\nu-1$. From (2.33), (2.40), and the inductive assumption, we obtain

$$\begin{aligned}\omega_{n,\nu}-\omega_{n,\nu}^2&\geqslant\omega_{n,\nu-1}\,b_n-n^{-1/2}\,\bar{r}_n\\&\geqslant a_{n,\nu}b_n-n^{-1/2}\,\bar{r}_n\geqslant a_{n,\nu}-a_{n,\nu}^2.\end{aligned} \tag{2.44}$$

From (2.35) and (2.39′) we have $\omega_{n,\nu}<\frac{1}{2}$ and $a_{n,\nu}<\frac{1}{2}$. Since $v\to v-v^2$ is increasing for $v\in[0,\frac{1}{2}]$, relation (2.44) implies $\omega_{n,\nu}\geqslant a_{n,\nu}$. This concludes the proof of (2.43).

From (2.43) and (2.35) we obtain $a_{n,\nu}\leqslant\frac{1}{4}$ for $\nu\leqslant m_n,\, n\in\mathbb{N}_1,\, n\geqslant n_*$, which contradicts (2.39″).

3. Construction of the Estimator–Sequence

(i) To prepare the construction, let g be an arbitrary odd and increasing function with the following properties: $|g|\leqslant 1$, g' is nonincreasing on $[0,\infty)$, and $0\leqslant g'\leqslant\frac{1}{2}$. Then the following relations hold for $v, w\in\mathbb{R}$ and $\varepsilon\in[0,1]$:

$$w<v+\varepsilon g(v)-\varepsilon^2 \qquad \text{implies} \quad w-\varepsilon g(w)<v \tag{3.1}$$

$$v+\varepsilon g(v)+\varepsilon^2<w \qquad \text{implies} \quad v<w-\varepsilon g(w). \tag{3.2}$$

We prove (3.1). Since $w\to w-\varepsilon g(w)$ is increasing, $w<v+\varepsilon g(v)-\varepsilon^2$ implies $w-\varepsilon g(w)<(v+\varepsilon g(v)-\varepsilon^2)-\varepsilon g(v+\varepsilon g(v)-\varepsilon^2)\leqslant v$, since

$$g(v)-g(v+\varepsilon g(v)-\varepsilon^2)\leqslant\varepsilon.$$

(If $v \geqslant 0$, this is trivial. If $v < 0$ we have $v + \varepsilon g(v) - \varepsilon^2 < v$, hence $g(v) - g(v + \varepsilon g(v) - \varepsilon^2) < g'(v)(\varepsilon |g(v)| + \varepsilon^2) < \varepsilon$.)

From (3.1), applied with $w = t + y$, $v = t + u$, and (3.2), applied with $w = t + y$, $v = t - u$, we obtain for arbitrary $t \in \mathbb{R}$, $u > 0$, $\varepsilon \in [0, 1]$,

$$(-u + \varepsilon g(t-u) + \varepsilon^2, u + \varepsilon g(t+u) - \varepsilon^2) \subset \{y \in \mathbb{R}: |y - \varepsilon g(y+t)| < u\}. \quad (3.3)$$

For $u, \Delta \subset \mathbb{R}$,

$$|\Phi(u + \Delta) - \Phi(u) - \Delta \varphi(u)| \leqslant \Delta^2/4.$$

Hence we obtain from (3.3) for arbitrary $t \in \mathbb{R}$, $u > 0$, $\varepsilon \in [0, 1]$,

$$\begin{aligned} &N\{y \in \mathbb{R}: |y - \varepsilon g(y+t)| < u\} \\ &\qquad \geqslant N(-u, u) + \varepsilon \varphi(u)[g(t+u) - g(t-u)] - 3\varepsilon^2. \end{aligned} \quad (3.4)$$

(ii) For $\alpha > 0$ let

$$g_\alpha(v) := v/(\alpha + |v|), \qquad v \in \mathbb{R}.$$

Observe that g_α fulfills the assumptions imposed in (i) on g, provided $\alpha \geqslant 2$. We shall show that $\alpha \geqslant u$ implies

$$\inf\{g_\alpha(t+u) - g_\alpha(t-u): |t| \leqslant \alpha\} \geqslant u/2\alpha. \quad (3.5)$$

If $t \in [u, \alpha]$, we have

$$g_\alpha(t+u) - g_\alpha(t-u) = \frac{2u\alpha}{(\alpha + t + u)(\alpha + t - u)} \geqslant \frac{2u\alpha}{(\alpha + t)^2} \geqslant \frac{u}{2\alpha}.$$

If $t \in [0, u]$, we have

$$\begin{aligned} g_\alpha(t+u) - g_\alpha(t-u) &= g_\alpha(t+u) + g_\alpha(u-t) \\ &= \frac{2u\alpha + 2(u^2 - t^2)}{(\alpha + t + u)(\alpha + u - t)} \geqslant \frac{2u\alpha}{(\alpha + u)^2} \geqslant \frac{u}{2\alpha}. \end{aligned}$$

Hence

$$g_\alpha(t+u) - g_\alpha(t-u) \geqslant u/2\alpha \qquad \text{for} \quad t \in [0, \alpha].$$

Since $g_\alpha(-t+u) - g_\alpha(-t-u) = g_\alpha(t+u) - g_\alpha(t-u)$, the same inequality holds for $t \in [-\alpha, 0]$.

Inequalities (3.4) and (3.5) together imply for $\varepsilon \in [0, 1]$, $\alpha \geqslant \max\{u, 2\}$,

$$\begin{aligned} &\inf_{|t| \leqslant \alpha} (N\{y \in \mathbb{R}: |y - \varepsilon g_\alpha(y+t)| < u\} - N(-u, u)) \\ &\qquad \geqslant u\varphi(u)\, \varepsilon/2\alpha - 3\varepsilon^2. \end{aligned} \quad (3.6)$$

(iii) Given sequences $\varepsilon_n \downarrow 0$ and $\alpha_n \uparrow \infty$, we define the estimators $\vartheta^{(n)}$ by

$$\vartheta^{(n)}(x_1, ..., x_n) = \bar{x}_n - n^{-1/2} \varepsilon_n g_{\alpha_n}(n^{1/2} \bar{x}_n) \qquad \text{with} \quad \bar{x}_n = n^{-1} \sum_1^n x_\nu. \quad (3.7)$$

Let N_t denote the normal distribution with mean t and variance 1. (As above, we write N for N_0.) Since the distribution of $n^{1/2} \bar{x}_n$ under $N^n_{n^{-1/2}t}$ is N_t, we obtain

$$N^n_{n^{-1/2}t}\{n^{1/2} |\vartheta^{(n)} - n^{-1/2} t| < u\} = N\{y \in \mathbb{R}: |y - \varepsilon_n g_{\alpha_n}(y + t)| < u\}. \quad (3.8)$$

With

$$\Delta_n(t) := N^n_{n^{-1/2}t}\{n^{1/2} |\vartheta^{(n)} - n^{-1/2} t| < u\} - N(-u, u),$$

we obtain from (3.6) and (3.8)

$$\inf_{|t| \leqslant \alpha_n} n^{1/2} \Delta_n(t) \geqslant u\varphi(u)\, n^{1/2} \varepsilon_n / 2\alpha_n - 3n^{1/2} \varepsilon_n^2. \quad (3.9)$$

(iv) Given $a > 0$, we choose $\varepsilon_n = (u\varphi(u)/8a)\, n^{-1/4}$ and $\alpha_n = an^{1/4}$. (3.9) implies for all $n \in \mathbb{N}$

$$\inf_{|t| \leqslant an^{1/4}} n^{1/2} \Delta_n(t) \geqslant u^2 \varphi(u)^2 / 64a^2 > 0.$$

This proves part (a) of the example.

(v) Given a sequence $a_n \downarrow 0$, let $\bar{a}_n := \max\{a_n, (2+u)\, n^{-1/4}\}$ and $\varepsilon_n = \bar{a}_n n^{-1/4}$, $\alpha_n = \bar{a}_n n^{1/4}$. (3.9) implies for all sufficiently large $n \in \mathbb{N}$,

$$\inf_{|t| \leqslant a_n n^{1/4}} n^{1/2} \Delta_n(t) \geqslant \inf_{|t| \leqslant \bar{a}_n n^{1/4}} n^{1/2} \Delta_n(t)$$
$$\geqslant u\varphi(u)/2 - 3\bar{a}_n^2 > 0.$$

Since g_α is increasing, we obtain from (3.4) and (3.8)

$$\inf_{t \in \mathbb{R}} n^{1/2} \Delta_n(t) \geqslant \inf_{t \in \mathbb{R}} (n^{1/2} \varepsilon_n \varphi(u)[g_{\alpha_n}(t+u) - g_{\alpha_n}(t-u)] - 3n^{1/2} \varepsilon_n^2)$$
$$\geqslant -3n^{1/2} \varepsilon_n^2 = -3\bar{a}_n^2 = o(n^0).$$

This proves part (b) of the example.

4. Lemmas

In the proof of the theorem, we need an asymptotic expansion of order $o(n^{-1/2})$ for a power function. Such an expansion holds true under appropriate regularity conditions on the densities $p(\cdot, \vartheta)$, $\vartheta \in \Theta$. Various sets of sufficient conditions are available. The result of Götze [3, Theorem 1.4, p. 262] seems particularly useful for our purpose because it asserts the validity of this expansion without a Cramér-type condition. Strictly speaking, we need slightly more than Götze's theorem asserts, namely uniformity over ϑ in bounded sets. Lemma 1 below is the specialization of such a uniform version to families with one real parameter.

Let $l(x, \vartheta) := \log p(x, \vartheta)$. Let $l^{(k)}(x, \vartheta)$ denote the partial derivative of $\vartheta \to l(x, \vartheta)$ of order k.

Regularity Conditions

(i) The probability measures P_ϑ, $\vartheta \in \Theta$, are mutually absolutely continuous.

(ii) The functions $l^{(k)}(\cdot, \vartheta)$, $k = 1, 2$, are not linearly dependent μ-a.e.

(iii) $\int l^{(1)}(x, \vartheta)\, P_\vartheta(dx) = 0$, $\int (l^{(1)}(x, \vartheta)^2 + l^{(2)}(x, \vartheta))\, P_\vartheta(dx) = 0$,

$$\int (l^{(1)}(x, \vartheta)^3 + 3l^{(1)}(x, \vartheta)\, l^{(2)}(x, \vartheta) + l^{(3)}(x, \vartheta))\, P_\vartheta(dx) = 0.$$

(iv) For every $\vartheta \in \Theta$ there exists an open neighborhood U_ϑ of ϑ such that

$$\sup_{\delta \in U_\vartheta} \int l^{(k)}(x, \vartheta)^4\, P_\delta(dx) < \infty \qquad \text{for} \quad k = 1, 2, 3.$$

(v) $l^{(3)}$ fulfills a local Lipschitz condition: For every $\vartheta \in \Theta$ there exists an open neighborhood U_ϑ of ϑ and a function $m(\cdot, \vartheta)\colon X \to \mathbb{R}$ with $\sup_{\delta \in U_\vartheta} \int m(x, \vartheta)^4\, P_\delta(dx) < \infty$ such that for all $\delta', \delta'' \in U_\vartheta$,

$$|l^{(3)}(x, \delta') - l^{(3)}(x, \delta'')| \leqslant |\delta' - \delta''|\, m(x, \vartheta).$$

Let

$$\sigma(\vartheta) := \left(\int l^{(2)}(x, \vartheta)\, P_\vartheta(dx)\right)^{-1/2}$$

$$a(\vartheta) := \int l^{(1)}(x, \vartheta)\, l^{(2)}(x, \vartheta)\, P_\vartheta(dx) \tag{4.4}$$

$$b(\vartheta) := \int l^{(1)}(x, \vartheta)^3\, P_\vartheta(dx).$$

LEMMA 1. *Assume regularity conditions* (i)–(v). *Given a sequence of critical functions* φ_n, $n \in \mathbb{N}$, let $\alpha_n(\delta) := \int \varphi_n(\mathbf{x})\, P^n_\delta(d\mathbf{x})$. *Assume there exists a neighborhood* U_ϑ *of* ϑ *such that* $\{\alpha_n(\delta) : \delta \in U_\vartheta, n \in \mathbb{N}\}$ *is bounded away from* 0 *and* 1. *Then uniformly for* $\delta \in U_\vartheta$, $t \in \mathbb{R}$,

$$\int \varphi_n(\mathbf{x})\, P^n_{\delta + n^{-1/2}t}(d\mathbf{x})$$

$$\leqslant \Phi(\Phi^{-1}(\alpha_n(\delta)) + t\sigma(\delta)^{-1} + n^{-1/2}\, \tfrac{1}{6}\, t\sigma(\delta)[t(3a(\delta) + b(\delta))$$
$$- \Phi^{-1}(\alpha_n(\delta))\, \sigma(\delta)\, b(\delta)]) + o(n^{-1/2}).$$

LEMMA 2. *Given* $0 < u' < u''$ *and* $0 < v'$ *there exists* $\Delta > 0$ *such that*

$$\Phi(-u+v) \geqslant \Phi(-u) + v\varphi(u) + v^2\Delta$$

for $-v' \leqslant v \leqslant 2u$, $u' \leqslant u \leqslant u''$.

Proof. Let $\Psi(u, v) := \Phi(-u+v) - \Phi(-u) - v\varphi(u)$. We have

$$\frac{\partial}{\partial v}\,\Psi(u, v) \begin{cases} > 0 & \text{for} \quad 0 < v < 2u \\ < 0 & \text{for} \quad v < 0. \end{cases}$$

Since $\Psi(u, 0) = 0$ for $u \in \mathbb{R}$, we have $\Psi(u, v) > 0$ for $v \leqslant 2u$, $v \neq 0$. The function

$$w(u, v) := \begin{cases} v^{-2}\,\Psi(u, v), & v \neq 0 \\ u\varphi(u)/2, & v = 0 \end{cases}$$

is continuous on $W := \{(u, v) \in \mathbb{R}^2 : -v' \leqslant v \leqslant 2u,\ u' \leqslant u \leqslant u''\}$ and positive. Hence $\Delta := \inf\{w(u, v) : (u, v) \in W\} > 0$.

ACKNOWLEDGMENTS

The author thanks H. Strasser for helpful discussions about the literature, and H. Plesske and L. Schröder for checking the proofs.

REFERENCES

[1] BICKEL, P. J., GÖTZE, F., AND VAN ZWET, W. R. (1985). A simple analysis of third-order efficiency of estimates. In *Proceedings, Berkeley Conference in Honor of Jerzy Neyman and Jack Kiefer* (L. M. LeCam and R. A. Olshen, Eds.), Vol. II, pp. 749–768. Statistics/Probability Series, Wadsworth, Monterey, CA.

[2] Götze, F. (1981). Second-order optimality of randomized estimation and test procedures. *J. Multivariate Anal.* **11** 260–272.

[3] LeCam, L. (1953). On some asymptotic properties of maximum likelihood estimates and related Bayes' estimates. *Univ. Calif. Publ. Statist.* **1** 277–330.

[4] Levit, B. Ya. (1974). On optimality of some statistical estimates. In *Proceedings, Prague Symposium on Asymptotic Statistics*, Vol. 2 (J. Hájek, Ed.), pp. 215–238. Charles University, Prague.

[5] Levit, B. Ya. (1980). On asymptotic minimax estimates of the second order. *Theor. Probab. Appl.* **25** 552–568.

[6] Pfanzagl, J. (1980). Asymptotic expansions in parametric statistical theory. In *Developments in Statistics*, Vol. 3. (P. R. Krishnaiah, Ed.) pp. 1–97, Academic Press, New York.

[7] Pfanzagl, J., and Wefelmeyer, W. (1985). *Asymptotic Expansions for General Statistical Models.* Lecture Notes in Statistics Vol. 31, Springer-Verlag, Berlin/Heidelberg/New York/Tokyo.

An Improved Estimation Method for Univariate Autoregressive Models

TARMO M. PUKKILA

University of Tampere, Tampere, Finland

Autoregressive models are important in describing the behaviour of the observed time series. One of the reasons is that a covariance stationary process can be approximated by an autoregressive model. Thus, e.g., the spectrum of a covariance stationary time series can be approximated by the spectrum of an autoregressive process. The estimation of the autoregressive parameters is therefore of special importance in time series analysis. Several methods have been introduced to estimate autoregressive models. The most popular method has been the Yule–Walker method. The Yule–Walker estimates for the autoregressive parameters are known to have poor statistical properties in certain cases. On the other hand, the Burg estimates have better statistical properties. For example the Burg estimates are less biased than the Yule–Walker estimates. In this paper an alternative to the Burg estimates will be introduced. In the proposed method the true correlation matrix of the lagged variables is calculated for the lags 1, 2, ... From each correlation matrix the corresponding partial autocorrelation can be calculated. These, on the other hand, will lead to autocorrelation estimates with improved statistical properties. From the autocorrelation estimates the autoregressive parameters can be estimated by solving the Yule–Walker equations. The statistical properties of the new estimates are studied by simulations.

1. INTRODUCTION

Assume that the observed time series $X_1, X_2, ..., X_n$ is generated by a univariate autoregressive process of order p, i.e.,

$$X_t - \mu = \sum_{k=1}^{p} \phi_k(X_{t-k} - \mu) = a_t, \tag{1}$$

where $\{a_t\}$ is a normal white noise process with mean zero and variance σ^2. Besides σ^2 the model (1) also contains the parameters $\phi_1, \phi_2, ..., \phi_p$ and

Multivariate Statistics and Probability
ISBN 0-12-580205-6

Reprinted from *J. Mult. Anal.* **27**(2).

μ to be estimated on the basis of the observations. It is assumed that (1) represents a stationary model. This requirement is satisfied if the roots of the equation

$$1 - \sum_{k=1}^{p} \phi_k B^k = 0 \tag{2}$$

lie outside the unit circle. In a stationary case $\mu = E\{X_t\}$, i.e., the mean of $\{X_t\}$.

Especially in the past solving the Yule–Walker equations has been a popular means of estimating autoregressive models. The resulting Yule–Walker estimates $\hat{\phi}_1, \hat{\phi}_2, ..., \hat{\phi}_p$ possess some nice properties. First, they are obtained by solving a system of linear equations. Second, the Yule–Walker estimates lead to stationary models, i.e.,

$$1 - \sum_{k=1}^{p} \hat{\phi}_k B^k \neq 0 \qquad \text{for} \quad |B| \leqq 1$$

(see Anderson and Mentz [2]). Third, the Yule–Walker estimates can be calculated iteratively for $p = 1, 2, ...$

In this paper we will introduce a new method of estimating univariate autoregressive models. The first step in the new method is to estimate partial autocorrelations which will lead to autocorrelation estimates with improved statistical properties compared with the estimates calculated in ordinary fashion. Finally the autocorrelation estimates are used to solve the Yule–Walker equations to produce the estimates for the autoregressive parameters. Also in the Burg method to estimate autoregressive models the first step is to estimate partial autocorrelations. Here we, however, use a different method to estimate partial autocorrelations, or at least we will give a different interpretation to the estimates of partial autocorrelation estimates.

Tjostheim and Paulsen [10] study the bias of Yule–Walker and least squares estimates for univariate and multivariate autoregressive processes. They also give explicit formulae for the large sample bias of Yule–Walker estimates in the scalar first- and second-order processes and for least squares estimates in the general case. Lysne and Tjostheim [7] show that autoregressive spectral analysis depends on the method used for estimating the autoregressive parameters. Because of the large bias in the Yule–Walker estimates Lysne and Tjostheim [7] state that least squares estimates should be preferred to the Yule–Walker estimates.

The paper is organized as follows. In Section 2 we will demonstrate the statistical properties of the Yule–Walker estimates for the parameters of the autoregressive parameters using simulated time series from an AR(4) model as an example. In Section 3 we will discuss the Yule–Walker and Burg methods of estimating autoregressive models. In Section 4 we will introduce improved methods to estimate partial autocorrelations, autocorrelation, and autoregressive parameters. We will also discuss the relation of the method to the method of Burg. We will also describe the performance of the method by using simulations. Finally in Section 5, we will offer some concluding remarks.

2. Simulation Results

In practice, the above properties of the Yule–Walker estimates are, of course, important. Besides these, even more important, however, is that the statistical properties of the autoregressive estimates should be good. In spite of the fact that the Yule–Walker estimates are asymptotically equivalent with the maximum likelihood estimates, in finite samples the performance of the Yule–Walker estimates can be really poor. This can be seen, for example, using the univariate AR(4) model

$$X_t = 2.7607X_{t-1} - 3.8106X_{t-2} + 2.6535X_{t-3} - 0.9238X_{t-4} + a_t, \quad (3)$$

where $\{a_t\}$ is a normal white noise process with mean zero and σ^2 as its variance.

The model (3) was considered by Beamish and Priestley [4], Priestley [9, p. 609], as well as Newton and Pagano [8] to illustrate the biasedness of the univariate Yule–Walker estimates.

In order to see how poor the statistical properties the Yule–Walker estimates can have in finite samples, we generated 1000 time series of length 50, 100, and 200 from (3). For each sample size we calculated the means and standard deviations of the estimates over 1000 realizations. For comparative purposes we calculated the same statistics also for the Burg's estimates (see Burg [6], Ulrych and Bishop [11], Anderson [1] and Newton and Pagano [8]. The results are given in Table I.

In Table I we see that the Yule–Walker estimates are extremely biased. We can see that the bias of these estimates is reduced only marginally as n increases from 50 to 200. A striking feature is that the variances of the Yule–Walker estimates become larger as n increase from 50 to 200. In this study we did not, however, go beyond the sample size 200 to see how long time series would be needed in order that the observed variances of the Yule–Walker estimates would begin to decrease. On the other hand, the Burg estimates behave as would be expected on the basis of the asymptotic theory for the maximum likelihood estimates.

TABLE I

Means and Standard Deviations of the Yule–Walker (YW) and Burg Estimates over 1000 Realizations of Length 50, 100, and 200 from the AR(4) Model (3)

	Means		Standard deviations	
Par	YW	Burg	YW	Burg
$n=50$				
2.7607	1.3164	2.7278	0.3183	0.0836
−3.8106	−0.9206	−3.7008	0.4681	0.1809
2.6535	0.0538	2.5359	0.3537	0.1757
−0.9238	−0.0662	−0.8646	0.0995	0.0806
$n=100$				
2.7607	1.5041	2.7424	0.3417	0.0518
−3.8106	−1.1888	−3.7521	0.5938	0.1206
2.6535	0.2328	2.5927	0.4967	0.1169
−0.9238	−0.0803	−0.8940	0.1461	0.0533
$n=200$				
2.7607	1.7179	2.7474	0.3573	0.0345
−3.8106	−1.5508	−3.7702	0.6909	0.0781
2.6535	0.5196	2.6116	0.6161	0.0741
−0.9238	−0.1441	−0.9040	0.1996	0.0311

3. The Yule–Walker and Burg Methods

In the univariate case the partial autocorrelation ϕ_{kk} at lag k, $k=1, 2, \ldots$, is defined as an ordinary partial correlation between the variables X_t and X_{t+k} given $X_{t+1}, \ldots, X_{t+k-1}$. The partial autocorrelations can be obtained by solving the Yule–Walker equations

$$\begin{bmatrix} \gamma(0) & \gamma(1) & \cdots & \gamma(k-1) \\ \gamma(1) & \gamma(0) & \cdots & \gamma(k-2) \\ \vdots & \vdots & & \vdots \\ \gamma(k-1) & \gamma(k-2) & \cdots & \gamma(0) \end{bmatrix} \begin{bmatrix} \phi_{k1} \\ \phi_{k2} \\ \vdots \\ \phi_{kk} \end{bmatrix} = \begin{bmatrix} \gamma(1) \\ \gamma(2) \\ \vdots \\ \gamma(k) \end{bmatrix} \tag{4}$$

with respect to ϕ_{kk}, $k=1, 2, \ldots$. In (4) we have written

$$\gamma(k) = E\{(X_t - \mu)(X_{t+k} - \mu)\},$$

for the autocovariance of $\{X_t\}$ at lag k. In univariate case we can replace the autocovariances $\gamma(k)$ in (4) by the corresponding autocorrelations

$$\rho(k) = \frac{\gamma(k)}{\gamma(0)},$$

$k = 0, 1, \ldots$. Therefore if the autocovariances $\gamma(k)$ are known, by solving (4) for $k = 1, 2, \ldots, p$ the partial autocorrelations ϕ_{kk} can be obtained. On the other hand, if the partial correlations ϕ_{kk} are known, we can calculate $\rho(k)$.

Using the autocorrelations the solutions of Eq. (4) can be expressed as the ratio of two determinants as

$$\phi_{kk} = \frac{|P_{kk}|}{|P_k|} \tag{5}$$

for $k = 1, 2, \ldots$, where

$$P_k = \begin{bmatrix} \rho(0) & \rho(1) & \cdots & \rho(k-1) \\ \rho(1) & \rho(0) & \cdots & \rho(k-2) \\ \vdots & \vdots & & \vdots \\ \rho(k-1) & \rho(k-2) & \cdots & \rho(0) \end{bmatrix}.$$

The matrix P_{kk} is obtained from P_k by replacing the last column of P_k by the vector ρ_k where $\rho_k^{\mathrm{T}} = (\rho(1), \rho(1), \ldots, \rho(k))$. Here the superscript T refers to the transpose of a matrix. Therefore, for an example we have

$$\phi_{11} = \rho(1), \qquad \phi_{22} = \frac{\rho(2) - \rho^2(1)}{1 - \rho^2(1)}.$$

On the other hand,

$$\rho(1) = \phi_{11}, \qquad \rho(2) = \rho^2(1) + \phi_{22}(1 - \rho^2(1)).$$

For example, from (5) it is easy to see that an autocorrelation $\rho(j)$ can be calculated from ϕ_{jj} and $\rho(1), \ldots, \rho(j-1)$; i.e., the autocorrelations can be calculated recursively from the partial autocorrelations. On the other hand, the partial autocorrelations can also be calculated recursively from the autocorrelations (see, e.g., Box and Jenkins [5, pp. 82–84]).

In practice the autocovariances $\gamma(h)$ and autocorrelations $\rho(h)$ are usually estimated by the quantities

$$c(h) = \frac{1}{n} \sum_{t=1}^{n} (X_t - \bar{x})(X_{t+h} - \bar{x}), \quad r(h) = \frac{c(h)}{c(0)}, \tag{6}$$

$h=0, 1, 2, \ldots$. Using the definition $c(-h)=c(h)$, $h=1, 2, \ldots$, the autocovariances and autocorrelation can be estimated also at negative lags.

If we replace $\gamma(h)$ by $c(h)$ or $r(h)$ in (4) we can obtain the estimates $\hat{\phi}_{kk}$, $k=1, 2, \ldots$ for the partial autocorrelations. The solutions $\hat{\phi}_{k1}, \ldots, \hat{\phi}_{kk}$ of (4) are then called the Yule–Walker estimates for the parameters of an autoregressive model of order k. As we have seen above, in finite samples the statistical properties of the Yule–Walker estimates can be really poor. It would be surprising if the Yule–Walker estimates would not suffer from the corresponding weaknesses in the multivariate case.

The method of Burg provides us with an alternative approach to autoregressive estimation. In the estimation method developed by Burg, partial autocorrelations are first obtained. These are then transformed into autoregressive parameter estimates. It can be seen that the Burg estimates are calculated by applying the definition of partial correlations. This means that in order to obtain an estimate for the partial autocorrelation at lag h, both the forward autoregression

$$X_t = \alpha_{h1} X_{t-1} + \cdots + \alpha_{h,h-1} X_{t-h+1} + \varepsilon_t, \tag{7}$$

and the backward autoregression

$$X_{t-h} = \beta_{h1} X_{t-1} + \cdots + \beta_{h,h-1} X_{t-h+1} + \delta_t \tag{8}$$

are estimated and the corresponding residual series $\hat{\varepsilon}_t$ and $\hat{\delta}_t$ are calculated. By definition, the correlation between $\hat{\varepsilon}_t$ and $\hat{\delta}_t$ is then the partial autocorrelation estimate at lag h.

In the univariate case the coefficients for forward and backward are, however, theoretically the same, i.e., $\alpha_{hj} = \beta_{hj}$, $j=1, \ldots, h-1$. Therefore, only one-way autoregressions need to be estimated.

Suppose that the estimates $\hat{\alpha}_{h1}, \ldots, \hat{\alpha}_{h,h-1}$ are available. Then we can calculate the forward residuals

$$\hat{\varepsilon}_t = X_t - \hat{\alpha}_{h1} X_{t-1} - \cdots - \hat{\alpha}_{h,h-1} X_{t-h} \tag{9}$$

and the bachward residuals

$$\hat{\delta}_t = X_{t-h} - \hat{\alpha}_{h1} X_{t-1} - \cdots - \hat{\alpha}_{h,h-1} X_{t-h}. \tag{10}$$

The correlation estimate calculated from $\hat{\varepsilon}_t$ and $\hat{\delta}_t$ then gives an estimate $\hat{\phi}_{hh}$ for ϕ_{hh}. If we originally have the observations $X_1, \ldots, X_n$, we can calculate the forward residuals $\hat{\varepsilon}_t$ for $t=h+1, \ldots, n$ and the backward residuals $\hat{\delta}_t$ for $t=1, \ldots, n-h$. For this reason only $n-2h$ pairs $(\hat{\varepsilon}_t, \hat{\delta}_t)$, $t=h+1, \ldots, n-h$, are available for the estimation of the correlation coefficient.

In the method of Burg $\hat{\phi}_{hh}$ is obtained by applying the formula

$$\hat{\phi}_{hh} = \frac{2 \sum \hat{\varepsilon}_t \hat{\delta}_t}{\sum \hat{\varepsilon}_t^2 + \sum \hat{\delta}_t^2}. \tag{11}$$

In (11) the sums are formed over those t's for which both the forward and backward residuals are available, i.e., for $t = h+1, ..., n-h$. If $n-2h$ is large, then we have approximately

$$\sum \hat{\varepsilon}_t^2 = \sum \hat{\delta}_t^2,$$

which implies that $\hat{\phi}_{hh}$ defined by (11) is approximately the ordinary Pearson's product moment correlation between $\hat{\varepsilon}_t$ and $\hat{\delta}_t$. In the method of Burg the partial correlation estimates can be calculated recursively for $h = 1, 2, ...$

4. Improved Estimation of Autoregressions

In the following we will consider the estimation method of an autoregressive model which is similar to the method of Burg in the sense that at the first stage the partial autocorrelations are estimated. The second stage then produces the autocorrelation and autoregressive estimates. As, e.g., Newton and Pagano [8] demonstrates, the poor statistical properties of the Yule–Walker estimates are caused by the way the end effects are treated in the estimation of the autocovariances. In what follows we will provide an alternative method to handle the problem. The method can also be applied in the estimation of multivariate autoregressions.

The first step in the proposed method is to estimate the partial correlations ϕ_{hh}. To calculate $\hat{\phi}_{hh}$, $h = 1, 2, ...,$ we form the ordinary correlation matrix for the variables $X_t, X_{t-1}, ..., X_{t-h}$. Because the variable X_{t-h} has defined observed values for $t > h$, we can calculate the correlations using the the observations for $t = h+1, ..., n$. Let it be mentioned that the resulting $h+1 \times h+1$ matrix is not a Toeplitz matrix. Of course, the theoretical correlation matrix of the variables $X_t, X_{t-1}, ..., X_{t-h}$ has the Toeplitz property. Let us denote the estimated correlation matrix by R_h. It can be written in the form

$$R_h = \begin{bmatrix} r(0,0) & r(0,1) & \cdots & r(0,h) \\ r(1,0) & r(1,1) & \cdots & r(1,h) \\ \vdots & \vdots & & \vdots \\ r(h,0) & r(h,1) & \cdots & r(h,h) \end{bmatrix}.$$

The correlations $r(i, j)$ are the ordinary correlation coefficients calculated from the formula

$$r(i, j) = \frac{c(i, j)}{\sqrt{c(i, i)\, c(j, j)}}, \tag{13}$$

where

$$c(i, j) = \frac{1}{n-h} \sum_{t=1}^{n-h} (X_{t+h-i} - \bar{x}_{(i)})(X_{t+h-j} - \bar{x}_{(j)}) \tag{14}$$

and

$$\bar{x}_{(i)} = \frac{1}{n-h} \sum_{t=1}^{n-h} X_{t+h-i}. \tag{15}$$

From the correlation matrix R_h we then calculate the ordinary partial autocorrelation $\hat{\phi}_{hh}$ between the variables X_t and X_{t-h} given $X_{t-1}, ..., X_{t-h+1}$ (see, e.g., Anderson [3, pp. 125–130]. These partial correlations are denoted here by $\hat{\phi}_{hh}$. They are estimates for the true partial autocorrelations of the process $\{X_t\}$. Therefore ϕ_{hh} are also called estimated partial autocorrelations at lag h. In this way we can calculate the partial autocorrelation estimates $\hat{\phi}_{hh}$, $h = 1, 2, ..., p$.

It is clear that the autocovariance estimators $c(h)$ and $c(i, j)$ defined correspondingly by (6) and (14), have the same asymptotic distributions. Therefore the estimated partial autocorrelations $\hat{\phi}_{hh}$ as defined here have the same asymptotic distributions as the partial autocorrelations considered by Box and Jenkins (see [5, p. 65]).

Using the estimated partial autocorrelations we can calculate the corresponding autocorrelation estimates $r(h)$ using the relation (5) such that we replace ϕ_{hh} in (5) by $\hat{\phi}_{hh}$ and solve the resulting equation with respect to $\rho(h)$, $h = 1, 2, ...$ The solution will provide us with an alternative autocorrelation estimate $r(h)$. As indicated above, the partial autocorrelations $\hat{\phi}_{11}$, $\hat{\phi}_{22}, ...,$ $\hat{\phi}_{pp}$ can also be transformed into the autoregressive parameter estimates $\hat{\phi}_1$, $\hat{\phi}_2, ..., \hat{\phi}_p$. These are the final autoregressive parameter estimates. In the following this estimation method is called the first modified Yule–Walker method (FMYW).

The proposed autoregressive estimation method is based on the true covariances $c(i, j)$ defined in (14). These covariances are calculated from the centered data such that for centering actual means $\bar{x}_{(i)}$ are used. An alternative and natural way to center the data is to use the mean $\bar{x}$ of all the observations X_1, $X_2, ..., X_n$, instead of $\bar{x}_{(i)}$. In this way we can also obtain autoregressive estimates which, in small samples, can slightly differ from FMYW estimates. The latter autoregressive estimates will be called the second modified Yule–Walker estimates, shortened as SMYW estimates in the following.

It is worth mentioning that as a biproduct of our method we also obtain an alternative autocorrelation function estimate $r(h)$, $h=1, 2, \ldots$. From $r(h)$ we can calculate the corresponding autocovariance estimators $c(h) = c(0)\, r(h)$, where

$$c(0) = \frac{1}{n} \sum_{t=1}^{n} (X_t - \bar{x})^2$$

is the usual formula for the variance of the observed time series.

It is clear that the proposed method, similarily to the method of Burg, leads to the estimated stationary models. This is equivalent to the property that the estimated autocorrelation and autocovariance sequences $\{r(h)\}$ and $\{c(h)\}$ are positive semidefinite.

To illustrate the performance of the proposed two estimation methods we generated 1000 time series of length 50, 100, and 200 from the model (3). For each time series an AR(4) model was estimated using both of the proposed methods. Similarly as for Table II we can calculated the means

TABLE II

Means and Standard Deviations of the FMYW and SMYW Estimates over 1000 Realizations of Length 50, 100, and 200 from the AR(4) Model (3)

	Means		Standard deviations	
Par	FMYW	SMYW	FMYW	SMYW
$n=50$				
2.7607	2.7176	2.7278	0.0808	0.0823
−3.8106	−3.6971	−3.6995	0.1814	0.1809
2.6535	2.5413	2.5333	0.1778	0.1776
−0.9238	−0.8756	−0.8629	0.0812	0.0830
$n=100$				
2.7607	2.7412	2.7427	0.0503	0.0504
−3.8106	−3.7527	−3.7527	0.1180	0.1178
2.6535	2.5946	2.5930	0.1147	0.1146
−0.9238	−0.8962	−0.8940	0.0525	0.0526
$n=200$				
2.7607	2.7470	2.7474	0.0344	0.0343
−3.8106	−3.7701	−3.7702	0.0778	0.0778
2.6535	2.6120	2.6117	0.0736	0.0737
−0.9238	−0.9045	−0.9041	0.0307	0.0307

TABLE III

The Estimation Results for Four Realizations of Length 50 Generated from the AR(4) Model (3)

Par	Realization 1	2	3	4
FMYW				
2.7607	2.7465	2.6927	2.6709	2.7058
−3.8106	−3.8058	−3.6734	−3.5385	−3.7162
2.6535	2.6760	2.5354	2.3406	2.5735
−0.9238	−0.9693	−0.8906	−0.7711	−0.9012
SMYW				
2.7607	2.7434	2.6962	2.6919	2.7034
−3.8106	−3.8323	−3.6689	−3.5303	−3.7107
2.6535	2.6973	2.5248	2.3059	2.5679
−0.9238	−0.9651	−0.8807	−0.7323	−0.8988
Burg				
2.7607	2.7250	2.7024	2.6206	2.8264
−3.8106	−3.8058	−3.6865	−3.4015	−3.8763
2.6535	2.6750	2.5429	2.1966	2.6808
−0.9238	−0.9621	−0.8887	−0.6986	−0.8959

and standard deviations of the parameter estimates FMYW and SMYW over 1000 replications for each sample size. The statistics given in both tables were calculated using the same time series for each sample size.

As we can see, the means and the standard errors for the two estimation methods are practically the same. Furthermore, when we compare the numbers in Table I and Table II, we observe that the Burg method, FMYW and SMYW produce estimates whose means and standard deviations are practically the same. Therefore, and because the SMYW estimates are easier to calculate than the FMYW estimates, we recommend the usage of the SMYW method.

In order to illustrate further the three estimation methods, in Table III we give the estimation results for 5 realizations of length 50, generated from the model (3). Also these results show that all of the methods produce similar estimates.

5. Concluding Remarks

In this paper we have introduced a new method to estimate univariate autoregressive models. As the first step of the method, partial autocorrelations are estimated. The partial autocorrelations lead to improved autocorrelation estimates. These can be used to obtain autoregressive parameter estimates by solving the Yule–Walker equations. Simulation results show that the proposed methods leads to autoregressive estimates which have similar statistical properties as the Burg estimates of the autoregressive parameters.

One of the striking features observed in the simulations carried out for this paper was that the variance of the Yule–Walker estimates for the autoregressive parameters increased as the number of observations increased from 50 to 200 in the case of an AR(4) model considered also by Beamish and Priestley [4]. Of course, the consistency of the Yule–Walker estimates implies that the variances of the estimates finally approach zero, but for the model considered it was observed to be the exception rather than the rule. In applications one has often to rely on asymptotic results. For the model studied, asymptotics do not, however, work, in spite of the fact that the number of observations is as high as 200.

How can we explain the increase of variances of the Yule–Walker estimates in the case of the model studied in this paper? An explanation might be due to the large bias of the estimates. When the number of observations is increased, estimates closer to the true parameters are obtained more often. This causes increased variability in the parameters and this increase is faster than the bias reduction in the parameter estimates. Of course, these considerations are only valid for the AR(4) model considered in the paper and for the number of observations varrying between 50 and 200.

The next step of our study in the future will be the generalization of the method to cover also multivariate time series. To estimate the multivariate partial autocorrelation matrices will be a straightforward generalization of the ideas presented in the paper.

Acknowledgments

This study was initiated while the author visited the Center for Multivariate Analysis during the Academic Year 1985–1986. The visit to the Center and the opportunity to work with Professor Krishnaiah was a privilege.

References

[1] Anderson, N. (1974). On the calculation of filter coefficients for maximum entropy spectral analysis. *Geophysics* **39** 69–72.

[2] Anderson, T. W., and Mentz, R. P. (1982). Notes on the estimation of parameters in vector autoregressive models. In *A Festschrift for Erich Lehman* (P. J. Bickel, K. A. Docksum, and J. L. Hodges, Eds.), pp. 1–13, Wadsworth, Belmont, CA.

[3] ANDERSON, T. W. (1984). *An Introduction to Multivariate Statistical Analysis*, 2nd ed. Wiley, New York.

[4] BEAMISH, N., AND PRIESTLEY, M. B. (1981). A study of AR and window spectral estimation. *Appl. Statist.* **30** 41–58.

[5] BOX, G. E. P., AND JENKINS, G. M. (1970). *Time Series Analysis, Forecasting and Control.* Holden-Day, San Francisco.

[6] BURG, J. P. (1967). Maximum entropy spectral analysis. In *37th Annual International S.E.G. Meeting, Oklahoma City, OK.*

[7] LYSNE, D., AND TJOSTHEIM, D. (1987). Loss of spectral peaks in autoregressive spectral estimation. *Biometrika* **74** 200–206.

[8] NEWTON, H. J., AND PAGANO, M. (1983). Computing for autoregressions. In *Computer Science and Statistics: The Interface* (J. E. Gentle, Ed.), pp. 113–118, North-Holland, New York/Amsterdam.

[9] PRIESTLEY, M. B. (1981). *Spectral Analysis and Time Series.* Vol. 1, Univariate Series. Academic Press, New York.

[10] TJOSTHEIM, D., AND PAULSEN, D. (1983). Bias of some commonly-used time series estimates. *Biometrika* **70** 389–399. Correction (1984), **71** 656.

[11] ULRYCH AND BISHOP (1975). Maximum entropy spectral analysis and autoregressive decomposition. *Rev. Geophys. Space Phys.* **13** 183–200.

Paradoxes in Conditional Probability

M. M. RAO

University of California, Riverside

It is shown that paradoxes arise in conditional probability calculations, due to incomplete specification of the problem at hand. This is illustrated with the Borel and the Kac–Slepian type paradoxes. These are significant in applications including Bayesian inference. Also Rényi's axiomatic setup does not resolve them. An open problem on calculation of conditional probabilities in the continuous case is noted.

1. INTRODUCTION

In presenting his famous twenty-three problems in 1900, Hilbert [4] begins his sixth problem as: "the investigations on the foundation of geometry suggest the problem: *To treat in the same manner, by means of axioms, those physical sciences in which mathematics plays an important part; in the first rank are the theory of probabilities and mechanics.*" At that time, Hilbert was influenced by a published lecture given for high school teachers by Bohlmann, containing a brief account of the axioms of probability which clearly were not satisfactory. In presenting a solution of this sixth problem as it concerns probability theory, Kolmogorov went further in 1933 and included a general definition of conditional probability [6]. The latter concept was, until then, used only for discrete random variables and probability spaces. However, no systematic method of calculating these general conditional probabilities was given in [6]. In some of its practical applications, ad hoc methods of calculation usually resulted in different answers for the same problem, giving rise to paradoxes. These difficulties have not been adequately addressed in the literature and are skipped often by indicating heuristic advice.

The purpose of this article is to discuss these troubles in some detail by using an analog of the Borel and the Kac–Slepian paradoxes, and a

Multivariate Statistics and Probability
ISBN 0-12-580205-6

Reprinted from *J. Mult. Anal.* **27**(2).

"strange" identity for the (conditional) expectations. It will be shown that, except in the elementary case of discrete probability spaces, the problem of finding the conditional probability or expectation given a condition or hypothesis on a set of negligible probability is not well posed for the traditional calculations using the L'Hôpital type approximation procedure, and to make it unique additional restricts that are inherent in the Kolmogorov model should be specified. Thus after presenting a precise framework (to avoid ambiguities) for Kolmogorov's general definition in the next section, integration relative to the conditional probability measure and a resulting difficulty will be sketched in Section 3. The paradoxes mentioned above are analyzed in Section 4 and the final section contains some complements on a related problem regarding a computational method to obtain conditional probabilities unambiguously. *Thus although known examples are used to illustrate the problems, the main focus of this paper is to point out the difficulty, to present a solution, and to bring the just-noted (unavailable) nontrivial constructive mathematical procedure to the user's attention.*

2. The Framework

To state the questions precisely, let (Ω, Σ, P) be a probability space. Thus Ω is a point set representing all possible outcomes of an experiment, Σ is a σ-algebra containing all the events of interest to the experimenter, and P is a probability function on Σ describing the experiment. Then a random variable (r.v.) is a mapping $f: \Omega \to \mathbb{R}$ such that $f^{-1}(I) \in \Sigma$ for each interval $I \subset \mathbb{R}$. The expectation of f, denoted $E(f)$, is

$$E(f) = \int_\Omega f\,dP, \tag{1}$$

and $E(f)$ is a Lebesgue integral so that $E(f)$ exists iff $E(|f|) < \infty$. For any event A (i.e., $A \in \Sigma$), $P(A) > 0$, the *conditional probability* of an event B given A, denoted $P(B \mid A)$, is defined as

$$P_A(B) = P(B \mid A) = P(B \cap A)/P(A). \tag{2}$$

Clearly $P(\cdot \mid A): \Sigma \to \mathbb{R}^+$ is a probability, and then the *conditional expectation of f* given A becomes

$$E_A(f) = \int_A f\,dP_A = \frac{1}{P(A)} \int_A f\,dP, \tag{3}$$

whenever $E(f)$ exists. Two events C, D are *independent* if $P(C \cap D) = P(C)\,P(D)$, so that in general $P_A(\cdot)$ and $E_A(\cdot)$ vary with A.

Both (2) and (3) are easily extended to countable partitions $\mathscr{P} = \{A_i, i \geqslant 1\}$ of events of Ω, i.e., if $P(A_i) > 0$, $\Omega = \bigcup_{i=1}^{\infty} A_i$, $A_i \cap A_j = \varnothing$, $i \neq j$. Indeed for each r.v. f with $|E(f)| < \infty$, the conditional expectation relative to $\mathscr{P}$ is

$$E^{\mathscr{P}}(f) = \sum_{i=1}^{\infty} E_{A_i}(f) \cdot \chi_{A_i} \tag{4}$$

and then the conditional probability is given by

$$P^{\mathscr{P}}(B) = E^{\mathscr{P}}(\chi_B) = \sum_{n=1}^{\infty} P_{A_n}(B) \cdot \chi_{A_n}, \qquad B \in \Sigma. \tag{5}$$

In applications, frequently one has to apply these formulas to events of the form: $A = \{\omega\colon g(\omega) = y\}$, $B = \{\omega\colon f(\omega) < x\}$ for r.v.'s f, g. If $A \notin \mathscr{P}$, then it is necessary to extend (5). For this, it is useful to express (4) and (5) alternately. If $E(f)$ exists, for any $A \in \mathrm{alg}(\mathscr{P})$, one has, on noting that the event $A \subset \bigcup_{i \in J} A_i$, $J \subset \mathbf{N}$ (natural numbers),

$$\begin{aligned}\int_A E^{\mathscr{P}}(f)\,dP &= \int_A \sum_{i=1}^{\infty} E_{A_i}(f) \cdot \chi_{A_i}\,dP, \qquad \text{by (4)},\\ &= \sum_{i \in J} \int_{A \cap A_i} f\,dP = \int_A f\,dP. \end{aligned} \tag{6}$$

Taking $f = \chi_B$ one gets a similar set of equations for $P^{\mathscr{P}}$:

$$\int_A P^{\mathscr{P}}(B)\,dP = \int_A \chi_B\,dP = P(A \cap B), \qquad B \in \Sigma, A \in \mathscr{P}. \tag{7}$$

If $P(A) = 0$, then $P_A(\cdot)$ in (2) is undefined. Moreover, if $\mathscr{B}$ is the smallest σ-algebra containing such a $\mathscr{P}$, then (4) and (5) easily extend. But if $\mathscr{B} \subset \Sigma$ is a more general σ-algebra, this constructuve procedure fails. However, (6) and (7) show how $E^{\mathscr{B}}$, $P^{\mathscr{B}}$ can still be defined, but with a sophisticated idea. If $\nu_f\colon A \mapsto \int_A f\,dP$, $A \in \mathscr{B}$, then the P-integrability of f implies ν_f is σ-additive on $\mathscr{B}$ and is absolutely continuous relative to $P_{\mathscr{B}}$, the restriction of P to $\mathscr{B}$, still a probability. Hence by the Radon–Nikodým theorem there is a $P_{\mathscr{B}}$-unique function $\tilde{f}$, measurable relative to $\mathscr{B}$, such that

$$\nu_f(A) = \int_A \tilde{f}\,dP_{\mathscr{B}}, \qquad A \in \mathscr{B}. \tag{8}$$

Then the mapping $E^{\mathscr{B}}: f \mapsto \tilde{f}$ is well defined on $L^1(\Omega, \Sigma, P)$, is linear, and has range $L^1(\Omega, \mathscr{B}, P_{\mathscr{B}})$. $E^{\mathscr{B}}$ and $P^{\mathscr{B}}$ coincide with $E^{\mathscr{P}}$ and $P^{\mathscr{P}}$ of (4) and (5) on $\mathscr{P}$, and $P^{\mathscr{B}}(B)$ is $E^{\mathscr{B}}(\chi_B)$, $B \in \Sigma$. These are called the (abstract) *conditional expectation* and *probability*, respectively, following [6]. Since they are only $P_{\mathscr{B}}$-unique, one chooses a member of the equivalence class and calls it a *version.* Note that in contrast to (2) and (3), the general theory with (8) first yields the conditional expectation *from which the conditional probability* is obtained. Further the constructions of $E^{\mathscr{B}}(f)$ and $P^{\mathscr{B}}(B)$, given by (8), are not easy. Ad hoc methods to obtain them lead to paradoxes, as illustrated below. Also it is seen that (1), (6), and (8) imply the identity

$$E(E^{\mathscr{B}}(f)) = E(f), \qquad f \in L^1(\Omega, \Sigma, P). \tag{9}$$

Several properties of the operator $E^{\mathscr{B}}$ may be found in [7, 10], and an extended analysis of $E^{\mathscr{B}}$ and $P^{\mathscr{B}}$ is in [8].

3. Conditional Probability as an Integrator

Here the standard practice of integrating relative to conditional probability (and their "densities") will be discussed and some "side effects" analyzed. Thus let X, Y be a pair of r.v.'s on (Ω, Σ, P), with an absolutely continuous distribution F. Let its density be $f_{X,Y}$, so that

$$F_{X,Y}(x, y) = P(\{\omega: X(\omega) < x, Y(\omega) < y\}), \qquad (x, y) \in \mathbb{R} \times \mathbb{R}, \tag{10}$$

and $f_{X,Y}(x, y) = (\partial^2 F_{X,Y}/\partial x\, \partial y)(x, y)$. The marginal distributions are then given by $F_X(x) = \lim_{y \to \infty} F_{X,Y}(x, y)$, $F_Y(y) = \lim_{y \to \infty} F_{X,Y}(x, y)$, which have densities f_X, f_Y (say). A common problem in applications, with such r.v.'s, is to find explicitly $P(B \mid A)$, where $B = \{\omega: X(\omega) < x\}$ and $A = \{\omega: Y(\omega) = y\}$. Since $P(A) = 0$, formula (2) is not applicable. To simplify matters, let $\Omega = \mathbb{R}^2$, $\Sigma =$ the smallest σ-algebra containing all rectangles of $\mathbb{R}^2$, $X, Y: \mathbb{R}^2 \to \mathbb{R}$ be functions such that $X(x, y) = x$, $Y(x, y) = y$, $(x, y) \in \Omega$, and $P(E) = \iint_E f(x, y)\, dx\, dy$, where f is a probability density. It is verfied that X, Y are coordinate functions, $F_{X,Y}(x, y) = \int_{-\infty}^{x} \int_{-\infty}^{y} f(u, v)\, dv\, du$, defines $F_{X,Y}$ to be the distribution of (X, Y) in (10), with $f_{X,Y} = f$, $f_X: x \mapsto \int_{\mathbb{R}} f(x, y)\, dy$, $f_Y: y \mapsto \int_{\mathbb{R}} f(x, y)\, dx$. Let us also define

$$f_{X \mid Y}(x \mid y) = \begin{cases} f_{X,Y}(x, y)/f_Y(y), & \text{if } f_Y(y) \neq 0, \\ \delta \geq 0, & \text{if } f_Y(y) = 0. \end{cases} \tag{11}$$

Then $\int_{\mathbb{R}} f_{X|Y}(x \mid y)\,dx = 1$, $f_{X|Y}(\cdot \mid y)$ is termed a conditional density of X given $Y = y$. For definiteness take $\delta = 0$ hereafter. It is not obvious that this new "definition" giving the conditional probability $P(X < x \mid Y = y) = \int_{-\infty}^{x} f_{X|Y}(u \mid y)\,du$, satisfies (7). It must be shown that this does imply (7) so that there is no conflict between the definition of $P(X < x \mid Y = y)$ using (11) and the general Kolmogorov concept.

For this verification, one takes $\mathscr{B}$ as the σ-algebra generated by (= smallest σ-algebra containing) the "cylinders" or strips $\mathbb{R} \times I$, $I \subset \mathbb{R}$ being an interval. Let $\mathscr{B}_2$ be the σ-algebra generated by the intervals of $\mathbb{R}$, and $\pi_i \colon \mathbb{R}^2 \to \mathbb{R}$ be the ith $(i = 1, 2)$ coordinate projection. Then it follows that $\mathscr{B} = \pi_2^{-1}(\mathscr{B}_2) \subset \Sigma$, and $\mathscr{B}$ is also the σ-algebra generated by Y, i.e., by $\{Y^{-1}(I) \colon I \subset \mathbb{R} \text{ intervals}\}$. Observe that $\mathscr{B}$ or $\mathscr{B}_2$ is not generated by countable partitions of $\mathbb{R}^2$ or $\mathbb{R}$. Now define $P^{\mathscr{B}}$ by the equation:

$$P^{\mathscr{B}}(E)(\omega) = \int_{I_1} f_{X|Y}(u \mid y)\,du = \int_{\pi_1(E)} f_{X|Y}(u \mid \pi_2(\omega))\,du, \tag{12}$$

for all $\omega = (x, y) \in \Omega$, $(u, y) \in E = I_1 \times I_2$, a rectangle of $\mathbb{R}^2$. Standard results in real analysis show that $P^{\mathscr{B}}(\cdot)(\omega)$ is σ-additive on the algebra of all such rectangles, $P^{\mathscr{B}}(\mathbb{R}^2)(\omega) = 1$, and has a unique extension to be a probability on Σ, for each $\omega \in \Omega = \mathbb{R}^2$. It is also measurable relative to $\mathscr{B}$, and a computation (using Tonelli's theorem) shows that for any $A \in \mathscr{B}$,

$$\int_A P^{\mathscr{B}}(E)(\omega)\,dP(\omega) = P(\pi_1(A) \times \pi_2(E)) = P(A \cap E). \tag{13}$$

(The omitted detail can be found, e.g., in [10, p. 118].) Thus $P^{\mathscr{B}}$ satisfies (7). Consequently by the essential uniqueness, $P^{\mathscr{B}}$ is a version of the (image) conditional probability, thereby showing that the concrete definition provided by (11) and the abstract version given by Kolmogorov agree on their image space. Note that this verification, usually omitted, is not entirely trivial; but it becomes necessary in order to use the abstract theory.

Since $P^{\mathscr{B}}(E)(\omega) = E^{\mathscr{B}}(\chi_E)(\omega)$, Eqs. (11)–(13) imply, first for simple and then for general r.v.'s $X \geqslant 0$, the representation,

$$E^{\mathscr{B}}(X)(\omega) = \int_{\Omega} X(\omega')\,P^{\mathscr{B}}(d\omega')(\omega) = \int_{\mathbb{R}} x f_{X|Y}(x \mid \pi_2(\omega))\,dx, \tag{14}$$

for all $\omega \in \Omega$, with $\omega' = (x, y) \in \Omega$, $X(\omega') = x$. This equation is usually expressed symbolically as

$$E(X \mid Y)(y) = E(X \mid Y = y) = \int_{\mathbb{R}} x f_{X|Y}(x \mid y)\,dx. \tag{15}$$

In this form, the integral is defined for all random variables for which (15) is meaninful. On the other hand, it is natural to ask whether an expression $E(X \mid Y)(\cdot)$ of (15) always represents a conditional expectation of X given Y. A negative answer is provided by the following:

EXAMPLE. Let (Ω, Σ, P) be as defined for (11), and let $f_{X,Y}$ be

$$f_{X,Y}(x, y) = \begin{cases} \frac{1}{\pi} \exp\{-y(1+x^2)\}, & -\infty < x < \infty, 0 < y < \infty, \\ 0, & \text{otherwise}. \end{cases} \tag{16}$$

Then $f_Y(y) = (\pi y)^{-1/2} e^{-y}$, $y > 0$, and $= 0$ for $y \leqslant 0$. It follows from (11) that $f_{X \mid Y}(x \mid y) = (y/\pi)^{-1/2} \exp(-x^2 y)$, for $-\infty < x < \infty$, $y > 0$ and $= 0$ elsewhere. Hence (12) holds and $f_{X \mid Y}$ is a conditional density of X given $Y = y$. It results from (15) that $E(X^n \mid Y)(y) = 0$ for all $n = 2m - 1$, $m \geqslant 1$, and all $y > 0$. If $E(X^n \mid Y)$ is the conditional expectation of X^n given Y, then $E(E(X^n \mid Y)) = E(0) = 0$, where $\mathscr{B} = \sigma$-algebra generated by Y. However, by (9) this must also equal $E(X^n)$ which does not exist for any $n \geqslant 1$, since $E(X^n) = \int_\Omega x^n f_{X \mid Y}(x, y)\, dx\, dy = (1/\pi) \int_{\mathbb{R}} (x^n/(1+x^2))\, dx$. Thus (9) is not valid! This example is essentially given in [3]. (Here m, n are integers.)

What has gone wrong here? A direct calculation shows that $E(X^n \mid Y)$ exists for all $n \geqslant 1$, while for no $n \geqslant 1$, $E(X^n)$ exists on (Ω, Σ, P). Here the set function $\nu_{X^n}(\cdot)$ in (8) is *not* σ-additive for $n = 2m - 1$, $m \geqslant 1$, and the Radon–Nikodým theorem is not applicable. Since the latter is the basis for Kolmogorov's generalization from which the identity (9) is deduced, it is not valid in this case. Note that if $n = 2m$, $m \geqslant 1$, then $\nu_{X^n}(\cdot)$ is σ-additive and nonnegative for which (8) is well defined and (9) holds with both sides becoming $+\infty$. It follows that (9) is true for all r.v.'s f for which the positive or negative part of f is integrable.

At this point another remark is in order. In the special case considered for (11), $P^{\mathscr{B}}(\cdot)(\cdot)$ defined by (12) and verified by (13) has the following two properties: (i) $P^{\mathscr{B}}(\cdot)(\omega)$ is an honest probability measure, $\omega \in \Omega$, and (ii) $P^{\mathscr{B}}(E)(\cdot)$ is $\mathscr{B}$-measurable for each $E \in \Sigma$. These two properties (especially (i)) need not hold for $P^{\mathscr{B}}$, given by (8) abstractly. If they hold, $P^{\mathscr{B}}(\cdot)(\cdot)$ is termed *regular*. Since by definition $P^{\mathscr{B}}(A) = E^{\mathscr{B}}(\chi_A)$, $A \in \Sigma$, one can extend this by linearity of $E^{\mathscr{B}}$ to express

$$E^{\mathscr{B}}(f) = \int_\Omega f(\omega)\, P^{\mathscr{B}}(d\omega), \tag{17}$$

first for step functions and then for all bounded measurable (for Σ) functions using a standard argument. The appropriate procedure here turns out to be the Dunford–Schwartz integral. This coincides with the Lebesgue integral iff $P^{\mathscr{B}}$ is regular (cf., e.g., [13, Theorem 2.3.11). It follows that the conditional expectation cannot always be evaluated by an elementary procedure such as that implied by (4)–(6). Further formula (2) when $P(A)=0$, using some form of the L'Hôpital rule to calculate $P(\cdot \mid A)$, leads to paradoxes, as is shown by the examples in the next section.

It should be noted, however, that there are several important applications in which $P^{\mathscr{B}}$ is regular. If, for instance, X, Y are random variables (or vectors) which are representable as coordinate functions (extending the case of the above example of (16)) and $\mathscr{B}$ is the σ-algebra generated by Y, then $P^{\mathscr{B}}(\cdot)$ is regular. A general discussion of this nontrivial problem is given in [7, p. 360ff] and in more detail in [10, p. 119ff].

4. Two Types of Paradoxes

If X, Y are a pair of r.v.'s on a nonatomic (or diffuse) probability space (Ω, Σ, P) with an absolutely continuous distribution, having a density $f_{X,Y}$, then the work in (11)–(15) shows that one can calculate the following conditional probability:

$$P[X<x \mid A]=\int_{-\infty}^{x} f(u \mid y)\, du, \qquad A=[Y=y]. \tag{18}$$

Also writing the left side as $P(B \mid y)$, $B=[X<x]$, it represents a regular conditional probability and satisfies the system of Eqs. (13). However, $P(A)=0$ now and $P(B \mid y)$ is not directly obtainable from formula (2). It will now be shown, by two types of examples, that $P(B \mid y)$ is not uniquely determined with computations often used in applications, and the underlying difficulties will be exposed.

(a) *The Borel-type paradox.* The problem here is analogous to that considered in [6, p. 51]. A simple but vivid case is detailed for computational clarity. Let X, Y be independent r.v.'s having a common distribution:

$$P[X<x]=P[Y<x]=\begin{cases} 1-e^{-x}, & x>0, \\ 0, & x \leqslant 0. \end{cases} \tag{19}$$

For any $a>0$, let $Z=(X-a)/Y$, so that $-\infty<Z<+\infty$. If $\alpha \in \mathbb{R}$, and $A=[Z=\alpha]$, then $P(A)=0$. The problem is to calculate $P[Y<y \mid A]$. If $f_{Y,Z}$ and f_Z are the density functions of (Y, Z) and Z, then using (19) and an elementary change of variables technique one finds

$$f_{Y,Z}(y, z)=\begin{cases} y \exp[-(yz+a)-y], & y>0 \text{ and } yz>-a, \\ 0, & \text{otherwise}, \end{cases} \tag{20}$$

and $f_{Y|Z}(y|\alpha) = f_{Y,Z}(y, \alpha)/f_Z(z)$ becomes

$$f_{Y|Z}(y|\alpha) = \begin{cases} y(1+\alpha)^{-2} \exp[-y(1+\alpha)], & y>0, \alpha \geqslant 0, \\ 0, & \text{otherwise,} \end{cases} \tag{21}$$

and somewhat more complicated expression for $\alpha < 0$. It is not needed here.

Since clearly the event $[Z = \alpha]$ is the same, in this example, as the event $[X - Y\alpha - a]$, the corresponding conditional density $f_{Y|U}$ of (Y, U) is obtained from a similar computation, after setting $U = X - Y\alpha$, as

$$f_{Y,U}(y, u) = \begin{cases} \exp[-y(1+\alpha) - u], & y>0, \alpha y + u > 0, \\ 0, & \text{otherwise,} \end{cases}$$

and

$$f_{Y|U}(y|a) = (1+\alpha)\exp[-y(1+\alpha)], \qquad y \geqslant 0. \tag{22}$$

It is now evident that $f_{Y|Z}$ and $f_{Y|U}$ agree for almost no values of y ($\alpha \geqslant 0$ being fixed). Consequently the conditional probabilities calculated with (18) using the densities (21) and (22) will be different. Thus a paradox has resulted!

In [6, p. 51] discussing an analogous problem originally raised by Borel [1], Kolmogorov makes a brief statement: "the concept of a conditional probability with regard to an isolated given hypothesis whose probability equals zero is inadmissible." Since the above type calculations frequently occur in many probabilistic and statistical practices, with (11) playing a key role, a deeper reason should be found. Indeed, this paradox can be satisfactorily explained with the general theory as follows.

The problem involved is the calculation of $P(B|A_\alpha)$ $(= E(\chi_B|Z = \alpha))$, $A_\alpha = [Z = \alpha]$ with $P(A_\alpha) = 0$. The desired value should be the same as $E(\chi_B|Z)(\alpha)$ of the general theory, by (15) and (18). Now for any integrable r.v. Y, $E(Y|Z) = g(Z)$ by the Doob–Dynkin lemma, where $g: \mathbb{R} \to \mathbb{R}$ is a (Borel) measurable function. This is essentially a standard fact (cf., e.g., [7, p. 343; or 10, Proposition 4, p. 102]). Hence $P(B|Z = \alpha) = g(\alpha)$ if $Y = \chi_B$. Here *the function g is uniquely defined by the conditioning* σ-**algebra** $\mathscr{B}_Z$ *of the r.v. Z, and hence by Z*. For (21) and (22) two *different* σ-algebras $\mathscr{B}_Z$ and $\mathscr{B}_U$ are at work and $A_\alpha \in \mathscr{B}_U \cap \mathscr{B}_Z$. Consequently $P(B|Z = \alpha)$ and $P(B|U = \alpha)$ are different. Thus in lieu of a paradox, the meaning of Kolmogorov's statement should be understood as follows. The problem of *calculating* $P[B|A_\alpha]$ *with* $P(A_\alpha) = 0$ *is not completely specified* and so a unique solution is not possible; in other words, *the problem is not well posed.* Here the analogy with the classical Bertrand paradox *is* appropriate. On the latter, with an accompanying discussion regarding its incomplete specification, see [11, Section 3].

(b) *The Kac–Slepian paradox.* Instead of evaluating $P(B \mid A)$ by (18) when $P(A)=0$, one can use the formula (2) in which A is replaced by a sequence of events $A_n \downarrow A$ with $P(A_n) > 0$ for each n. Consequently, with a type of L'Hôpital's rule, it is reasonable to define

$$P(B \mid A) = \lim_n P(B \mid A_n) = \lim_n \frac{P(B \cap A_n)}{P(A_n)}, \qquad B \in \Sigma, \tag{23}$$

whenever this limit exists. It is not obvious, however, that this definition is not in conflict with the earlier accepted concept from [6]. The fact that $P(\cdot \mid A)$ is σ-additive and hence is a probability is also nontrivial, but this follows from the classical Vitali–Hahn–Saks theorem [12, p. 176], and a more elementary proof is in [15, p. 190]. Since $P(\cdot \mid A)$ is thus a probability, for each bounded random variable X, let $\tilde{E}_A(X) = \int_\Omega X(\omega)\, P(d\omega \mid A)$. This is well defined. To see that it satisfies the Kolmogorov definition in the sense that it is a version of a conditional expectation of X given A, let $\sigma(\{A_n, n \geqslant 1\}) = \mathscr{B}$, the σ-algebra generated by the sets shown. Then $\mathscr{B} \subset \Sigma$, $A \in \mathscr{B}$, $E_{A_n}(X)$ is given by (3), and for each A_{n_0} of the generators, with $P_{A_n}(\cdot)$ for $P(\cdot \mid A_n)$,

$$\begin{aligned} \int_{A_{n_0}} X\, dP = \int_{A_{n_0}} E_{A_n}(X)\, dP &= \int_{A_{n_0}} \left(\int_\Omega X\, dP_{A_n} \right) dP, \qquad n \geqslant 1, \\ &\to \int_{A_{n_0}} \left(\int_\Omega X\, dP_A \right) dP \qquad \text{as} \quad n \to \infty, \\ &= \int_{A_{n_0}} \tilde{E}_A(X)\, dP, \end{aligned} \tag{24}$$

where the preceding fact that $P_A(\cdot)$ is a probability and the Helly–Bray theorem are used (or one can reduce this to the Lebesgue bounded convergence through the Skorokhod mapping theorem, cf. [10, pp. 336 and 218]). Since A_{n_0} is a generator of $\mathscr{B}$, (24) implies that $\tilde{E}_A(X)$ is a version of $E^{\mathscr{B}}(X)$ as asserted.

In this argument, it is evident that such an A may be determined by several sequences $\{A_n, n \geqslant 1\}$. Then the corresponding $\mathscr{B}$ families are different. To illustrate this, consider a stationary ergodic Gaussian process $\{X(t), t \geqslant 0\}$ with mean 0 and covariance function $r(\cdot)$. Suppose that the pointwise derivative $X'(t)$ of $X(t)$ exists so that it is the slope of the continuous curve $X(\cdot)$ at t. The existence of such a process follows from the general theory. The problem is to find the conditional probability (or density) of $X'(0)$ given that $(X(0) = a)$ for any fixed a. Since X' is obtained by a linear operation, it follows that $X'(0)$ is also normally distributed with mean 0, and variance σ^2 (>0, say). The event $A = [X(0) = a]$ has probability 0, and let us use one of the approximations indicated above. Thus if $\delta > 0$, and $m \in \mathbb{R}$, consider

$$A_\delta^m = \{\omega: X_t(\omega) \text{ passes through the line } y = a + mt \text{ of length } \delta \text{ for some } t \leqslant \delta(1+m^2)^{-1/2}\}$$
$$= \{\omega: X_t(\omega) = a + mt \text{ for some } 0 \leqslant t \leqslant \delta(1+m^2)^{-1/2}\}.$$

Clearly $A_\delta^m \downarrow A$ for each m, as $\delta \downarrow 0$ through a sequence. By (23), one has

$$P[X'(0) < x \mid A] = \lim_{\delta \downarrow 0} P[X'(0) < x \mid A_\delta^m], \qquad 0 \leqslant m \leqslant \infty. \tag{25}$$

A standard but nontrivial argument shows (for a detailed computation, see, e.g., [10, p. 128]) that (25) becomes

$$\lim_{\delta \downarrow 0} P[X'(0) < x \mid A_\delta^m] = \int_{-\infty}^{x} \frac{|v - m|\, e^{-v^2/2\sigma^2}\, dv}{2\sigma^2 e^{-m^2/2\sigma^2} + \int_{-n}^{m} e^{-v^2/2\sigma^2}\, dv}. \tag{26}$$

From (25) and (26) one sees that $P(X'(0) < x \mid A]$ is different for each value of $0 \leqslant m \leqslant \infty$, and hence there are uncountably many answers to the problem at hand so that one has a "bad" paradox. There is no single correct answer here. This example is extracted from [5].

Letting $m \to \infty$ in (26), one gets the limit through the vertical line, called a "vertical window" (v.w.) solution, and letting $m \to 0$, one has a "horizontal window" (h.w.) solution given respectively by

$$P[X'(0) < x \mid A]_{\text{v.w.}} = \int_{-\infty}^{x} e^{-v^2/2\sigma^2}(2\pi\sigma^2)^{-1/2}\, dv, \tag{27}$$

$$P[X'(0) < x \mid A]_{\text{h.w}} = \int_{-\infty}^{x} |v|\, e^{-v^2/2\sigma^2}(2\sigma^2)^{-1}\, dv. \tag{28}$$

Here (27) corresponds to the fact that $X'(0)$ and $X(0)$ are independent, and this explanation ignores part of the information that $X'(0)$ is obtained as a limit of the quotients $(X(t) - X(0))/t$ as $t \downarrow 0$. On the other hand, the h.w. solution (28) seems to have some special relation to the "mean recurrence time" studied in statistical mechanics as noted in [5]. Considering other approximations of A (e.g., through circles with center $(a, 0)$ and radius δ) still different values for the left side of (25) can be obtained. Thus the problem is again *not well posed* as in the last subsection.

To understand the problem, consider the abstract theory. Since $A_\delta^m \supset A_{\delta'}^m$ for $\delta > \delta'$, let $\mathscr{B}^m$ be the σ-algebra generated by $\{A_\delta^m, \delta > 0\}$. Then $A \in \bigcap_m \mathscr{B}^m$ and $P(A) = 0$. The above computation merely shows that $P^{\mathscr{B}^m}(X'(0) < x)(a)$ gives different values for different m, since the $\mathscr{B}^m$ vary with m, and there is no paradox and for a unique solution the *conditioning σ-algebra should be specified* (but a lattice will not be sufficient as classical measure theory shows, cf., e.g. [12, p. 459]).

There is no universal recipe to calculate $P^{\mathscr{B}}(\cdot)$ for a given $\mathscr{B}$, in contrast with the elementary case. The work here naturally leads to differentiation theory and is relatively involved. For some discussion on the problem, see [10, p. 130]).

5. Another Approach and Complements

An alternative method to the above difficulties is an axiomatic approach to conditional probability concept itself. This was proposed by Rényi [14] and his axioms may be stated as follows. If (Ω, Σ) is a measurable space, $\mathscr{B}_0 \subset \Sigma$ is a nonempty class (*not* a ring), let $P(\cdot \mid \cdot): \Sigma \times \mathscr{B}_0 \to \mathbb{R}^+$ be a mapping which satisfies the axioms:

I. $A \in \Sigma$, $B \in \mathscr{B}_0 \Rightarrow 0 \leqslant P(A \mid B) \leqslant 1$, $P(B \mid B) = 1$,

II. $P(\cdot \mid B)$ is σ-additive (i.e., a measure) for each $B \in \mathscr{B}_0$,

III. (a) $A \in \Sigma$, $B \in \mathscr{B} \Rightarrow P(A \mid B) = P(A \cap B \mid B)$, and

(b) $A \in \Sigma$, $\{B, C\} \subset \mathscr{B}$, $A \subset B \subset C \Rightarrow P(A \mid B)\, P(B \mid C) = P(A \mid C)$.

The class $\{\Omega, \Sigma, \mathscr{B}_0, P(\cdot \mid \cdot)\}$ is then termed a *conditional probability space* (in the sense of Rényi). From axioms I and II, it follows that $\phi \notin \mathscr{B}_0$. Also I and III imply a disintegration formula, i.e., $\{B_n; n \geqslant 1\} \subset \mathscr{B}_0$, disjoint, $B = \bigcup_n B_n$, then for any $C \in \mathscr{B}_0$, $C \subset B$ and for each $C \in \mathscr{B}_0$, $C \subset B$, with $C \cap B_n \in \mathscr{B}_0$ one has

$$P(A \mid C) = \sum_{k=1}^{\infty} P(A \mid B_k)\, P(B_k \mid C), \qquad A \in \Sigma. \tag{29}$$

It is clear that P_A of (2) satisfies this system for each $A \in \Sigma$ with $P(A) > 0$. Also Rényi [15, p. 40], and later Császár [2, p. 351] in somewhat more generality, proved that if $\Omega \in \mathscr{B}_0$, then $P(A \mid B) = \tilde{P}(A \cap B)/\tilde{P}(B)$ for a probability $\tilde{P}(\cdot) = P(\cdot \mid \Omega)$. A number of properties including a treatment of the Borel paradox for a class called "Cavalieri spaces" are in [14]. But the solution obtained in [14] differs from the earlier work and, as expected, depends on the method used. The problems of Kac–Slepian type seem harder to fit in this system. An enlargement of $\mathscr{B}_0$ to treat the latter type introduces the same difficulties as in the previous case. A further analysis with examples of this, and an elaboration of the preceding section, appears in Chapters III and IV of a monograph [13].

A consequence of this analysis in current practice should be recorded. Conditional probability theory is basic in such areas as Markov processes and Bayesian inference. In the discrete case (i.e., for Markov chains) the original model (2) suffices. In the general case, one *assumes* that $P^{\mathscr{B}}(\cdot)(\cdot)$ is regular and develops the subject. The theory is valid with any fixed version. But again computational difficulties appear in the general case. The problem in the Bayesian case has the following structure.

Let $X_1, ..., X_n$ be random variables with a joint distribution $F_n(x_1, ..., x_n \mid \theta)$, depending on a parameter θ. Suppose that F_n is either absolutely continuous, or discrete, with density $f_n(x_1, ..., x_n \mid \theta)$ relative to μ_n, a Lebesgue or a counting measure, respectively. In the Bayesian analysis, θ is a value of a random variable Θ. If the latter takes values in $T \subset \mathbb{R}^k$ with density $\xi(\theta)$, then

$$h_n(t_1, ..., t_n, \theta) = f_n(t_1, ..., t_n \mid \theta)\, \xi(\theta)$$

is the joint density of the vector $(X_1, ..., X_n, \Theta)$ in $\mathbb{R}^n \times T$. Thus the conditional density of Θ, given $X_1 = x_1, ..., X_n = x_n$, called the *posterior* density, is $\xi_n(\Theta \mid x_1, ..., x_n)$ as in (11). Hence the posterior probability of Θ given the X_i-values is obtained as usual by the equation

$$P(\Theta \in A \mid X_1 = x_1, ..., X_n = x_n) = \int_A \xi_n(\theta \mid x_1, ..., x_n)\, d\theta. \qquad (30)$$

If $\mathscr{B}_n = \sigma(X_1, ..., X_n)$ and $\tilde{P}^{\mathscr{B}_n}(\tilde{A})(x_1, ..., x_n)$ is calculated with the Kolmogorov definition, where $\tilde{A} = \mathbb{R}^n \times A$, then our examples and analysis of the last sections show that this and the value given by (30) need not agree. The situation becomes more pronounced for stochastic processes. Since one accepts the Kolmogorov model in the current practice of these subjects, the correct value is $\tilde{P}^{\mathscr{B}_n}(\cdot)(\cdot)$, and not necessarily that given by (30). There are several conditions on the basic probability model, derived from the classical differentiation theory, to calculate $\tilde{P}^{\mathscr{B}_n}$. Unfortunately, an efficient and implementable procedure to actually use in practical problems is still not available. The methods leading to (30), and the only other place [17, Chap. 9]; cf., also [16], where such a problem is discussed prescribing a similar procedure, do not give a recipe for calculating the correct value. The L'Hôpital type ratio approximations are necessarily not well posed, yielding essentially always nonunique solutions. A rigorous analysis of this note and the exact reasons for the difficulties with the traditional calculations seem to be missing in the literature for too long. Further detail, discussion, and applications are included in [13], cited above.

ACKNOWLEDGMENTS

This is an expanded and revised version of the main part of an invited talk presented at the "Conference on Teaching of Probability and Statistics," May 22, 1983 at California State University, Fullerton. The work is prepared with the partial support of ONR Contract N00014-84-K-0356.

References

[1] Borel, E. (1925). *Traité des probabilités*, Gauthier-Villars, Paris.

[2] Császár, A. (1955), Sur la structure des espaces de probabilité conditionnelle, *Acta Math. Acad. Sci. Hungar.* **6**, 337–361.

[3] Ennis, P. (1973). On the equation $E(E(X|Y)) = E(X)$, *Biometrika* **60**, 432–433.

[4] Hilbert, D. (1902). Mathematical problems (ICM lecture, Paris, 1900), English translation, *Bull. Amer. Math. Soc.* **8** 437–479.

[5] Kac, M., and Slepian, D. (1959). Large excursions of Gaussian processes. *Ann. Math. Statist.* **30** 1215–1228.

[6] Kolmogorov, A. N. (1956). *Foundations of the Theory of Probability*, 2nd ed., Chelsea, New York, (English translation)

[7] Loève, M. (1963). *Probability Theory*, 3rd ed., Van Nostrand, Princeton, NJ.

[8] Rao, M. M. (1975). Conditional measures and operators, *J. Multivariate Anal.* **4** 330–413.

[9] Rao, M. M. (1981). *Foundations of Stochastic Analysis*, Academic Press, New York.

[10] Rao, M. M. (1984). *Probability Theory with Applications*. Academic Press, New York.

[11] Rao, M. M. (1987). Probability. *Encyclopedia Phys. Sci. Technol.* **11** 289–309.

[12] Rao, M. M. (1987). *Measure Theory and Integration*. Wiley-Interscience, New York.

[13] Rao, M. M. (1988). *Conditional Measures and Expectations*. Monograph in preparation.

[14] Rényi, A. (1955). On a new axiomatic theory of probability, *Acta Math. Acad. Sci. Hungar.* **6** 285–333.

[15] Rényi, A. (1970). *Foundations of Probability*. Holden-Day, San Francisco.

[16] Tjur, T. (1974). *Conditional Probability Distributions*. Lecture Notes No. 2. Inst. of Statistics, Univ. of Copenhagen.

[17] Tjur, T. (1980). *Probability Based on Radon Measures*. Wiley, New York.

Inference Properties of a One-Parameter Curved Exponential Family of Distributions with Given Marginals

CARMEN RUIZ-RIVAS

Universidad Autónoma de Madrid, Madrid, Spain

AND

CARLES M. CUADRAS

Universitat de Barcelona, Barcelona, Spain

This paper introduces a one-parameter bivariate family of distributions whose marginals are arbitrary and which include Fréchet bounds as well as the distribution corresponding to independent variables. Some geometrical and statistical properties on the stochastic dependence parameter are studied, considering this family as a member of Efron's curved exponential families of distributions.

1. INTRODUCTION

Let X, Y be two random variables with continuous distribution functions $F(x)$, $G(y)$. Let us consider the class $\mathscr{F}$ of all possible joint cdf's H for (X, Y).

Hoeffding [11] and Fréchet [10] stated that the following extremal cdf's

$$H^{+}(x, y) = \min\{F(x), G(y)\}$$
$$H^{-}(x, y) = \max\{F(x) + G(y) - 1, 0\}$$

define two elements of $\mathscr{F}$ with associated extreme correlations, i.e.,

Multivariate Statistics and Probability
ISBN 0-12-580205-6

Reprinted from *J. Mult. Anal.* **27**(2).

$\rho^- \leqslant \rho \leqslant \rho^+$, where ρ^-, ρ, ρ^+ are the correlation coefficients for H^-, H, H^+, respectively. H^-, H^+ are called the Fréchet bounds. It is verified that

$$H^-(x, y) \leqslant H(x, y) \leqslant H^+(x, y) \qquad \forall (x, y) \in \mathbb{R}^2.$$

Furthermore, if $H = H^-$ then

$$F(X) + G(Y) = 1 \qquad \text{(a.s.)},$$

and if $H = H^+$ then

$$F(X) = G(Y) \qquad \text{(a.s.)}.$$

Many authors have been interested in constructing parametric families of cdf's with given marginals F and G. Fréchet states that every family should include H^- and H^+. Kimeldorf and Sampson [12], proposed five desirable conditions that should be satisfied by any one-parameter family $\{H_\theta : -1 \leqslant \theta \leqslant +1\}$ of cdf's with absolutely continuous marginals F and G. These conditions are:

(a) $H_1(x, y) = H^+(x, y)$;

(b) $H_0(x, y) = F(x)\, G(y)$;

(c) $H_{-1}(x, y) = H^-(x, y)$ (i.e., the family contains the Fréchet bounds as well as the stochastic independence case);

(d) H_θ is continuous in $\theta \in [-1, 1]$;

(e) H_θ is absolutely continuous for fixed $\theta \in (-1, 1)$.

The uniform representation (Kimeldorf and Sampson [13]) and the notion of copula (Schweizer and Sklar [18]) provide the natural framework in which to study certain dependence properties of bivariate distributions and non-parametric measures of correlation. The uniform representatin or copula of H_θ is

$$U_H(u, v) = H(F^{-1}(u), G^{-1}(v)) \qquad (u, v) \in [0, 1]^2,$$

the marginal distributions of U_H then being uniform on [0, 1].

Fréchet, Farlie, Gumbel, Morgenstern, Plackett, Mardia., Kimeldorf, Sampson, Ruiz-Rivas, Cuadras, Auge, Algarra, Nelsen, and others, have proposed one-parameter families. One of these families (see Section 2.2) is studied here.

Some applications deal with:

(a) Variance reduction in statistical simulation (Fishman [9], Whitt [20]).

(b) The construction of non-negative quantum-mechanical distribution functions, given the marginal distribution of position and moment (Cohen and Zaparovanny [4], O'Connell and Wigner [15], Cohen [3]).

(c) The construction of upper and lower bounds of the cdf's when the marginal are given, under the additional condition that $X \leqslant Y$ with probability one (Smith [19]).

2. One-Parameter System

2.1. *Definition*

Cuadras and Auge [6] defined the cdf on R^2,

$$H_\theta(x, y) = F(x)^{1-\theta}\, G(y) \qquad \text{if} \quad F(x) \geqslant G(y),$$
$$H_\theta(x, y) = F(x)\, G(y)^{1-\theta} \qquad \text{if} \quad F(x) < G(y),$$

θ being a parameter satisfying $0 \leqslant \theta \leqslant 1$. The general definition, including the negative parameter case, is:

$$\begin{aligned} H_\theta(x, y) &= [\min\{F(x), G(y)\}]^\theta \cdot [F(x)\, G(y)]^{1-\theta} \qquad \text{for} \quad 0 \leqslant \theta \leqslant 1, \\ H_\theta(x, y) &= F(x) - [\min\{F(x), 1 - G(y)\}]^{-\theta} \cdot [F(x)(1 - G(y))]^{1+\theta} \\ &\qquad \text{for} \quad -1 \leqslant \theta < 0. \end{aligned} \tag{1}$$

2.2 *General properties*

The one-parameter system H_θ of cdf's has some interesting properties:

(1) If (X, Y) is distributed as $H_\theta(x, y)$, $0 \leqslant \theta \leqslant 1$, and Z verifies $G(Z) = 1 - G(Y)$ (a.s.) then (X, Z) is distributed as $H_{-\theta}(x, y)$.

(2) $H_1 = H^+$, $H_0 = FG$, $H_{-1} = H^-$, and H_θ is continuous in θ.

(3) H_θ is not absolutely continuous for $\theta \neq 0$, but can be decomposed as

$$H_\theta = H_\theta^{(1)} + H_\theta^{(2)}, \tag{2}$$

$H_\theta^{(1)}$ being its absolutely continuous part with density function (for $\theta \in [0, 1]$)

$$h_\theta(x, y) = (1 - \theta) f(x)\, g(y) \max\{F(x), G(y)\}^{-\theta} \qquad \forall (x, y) \in \mathbb{R}^2, \tag{3}$$

provided that F, G are absolutely continuous with densities f, g, and $H_\theta^{(2)}$ being its singular part corresponding to a positive mass over the curve

$F(x)=G(y)$. (The negative case $\theta\in[-1,0)$ is straightforward considering property (1).) In fact,

$$\begin{aligned}H_\theta^{(1)}(x,y)&=\int_{-\infty}^{x}\int_{-\infty}^{y}h_\theta(u,v)\,du\,dv\\&=-\frac{\theta}{2-\theta}\min\{F(x),G(y)\}^{2-\theta}\\&\quad+\min\{F(x),G(y)\}[\max\{F(x),G(y)\}]^{1-\theta}\\&=-H_\theta^{(2)}(x,y)+H_\theta(x,y)\qquad\forall(x,y)\in\mathbb{R}^2.\end{aligned}$$

(4) Let $P_\theta=P_\theta^{(1)}+P_\theta^{(2)}$ be the probability measure related to H_θ. The family $\{P_\theta\colon\theta\in[0,1]\}$ is dominated by a σ-finite measure μ, and

$$f_\theta(x,y)=h_\theta(x,y)\,I_{\bar C}(x,y)+\tilde h_\theta(x)\,I_C(x,y)\qquad\forall(x,y)\in\mathbb{R}^2,\quad\theta\in[0,1]\tag{4}$$

are the corresponding Radon–Nykodim derivatives, where $C=\{(x,y)\mid F(x)=G(y)\}$, I is the indicator function, and

$$\tilde h_\theta(x)=\theta f(x)\,F(x)^{1-\theta}.$$

Proof. Let λ^2, λ be the Lebesgue measures in $\mathbb{R}^2$ and $\mathbb{R}$, respectively. For any Borel set B in $\mathbb{R}^2$ let us define

$$\mu_0(B)=\lambda\{x\in\mathbb{R}\mid(x,G^{-1}F(x))\in B\},$$
$$\mu=\lambda^2+\mu_0;$$

μ_0 can be characterized as a product measure on $(\mathbb{R}^2,\beta^2)$

$$\mu_0(A\times B)=\int_A\tilde\mu(x,B)\,d\lambda(x),\qquad A,B\in\beta,$$

where β is the Borel σ-algebra of $\mathbb{R}$, and

$$\begin{aligned}\tilde\mu(x,B)&=1\qquad\text{if}\quad G^{-1}F(x)\in B,\\&=0\qquad\text{if}\quad G^{-1}F(x)\notin B.\end{aligned}$$

Applying Fubini's theorem it is easily shown that

$$H_\theta^{(2)}(x,y)=\int_{-\infty}^{x}\int_{-8}^{y}\tilde h_\theta(u)\,d\mu_0(u,v)$$

so that $\tilde h_\theta=dP_\theta^{(2)}/d\mu_0$.

Noting that $\lambda^2(C)=0$, $\mu_0(\bar{C})=0$, and $h_\theta = dP_\theta^{(1)}/d\lambda^2$, the result (4) follows.

(5) $P_\theta(F(X)>G(Y))+P_\theta(F(X)<G(Y))=2(1-\theta)/(2-\theta)$,

$$P_\theta(F(X)=G(Y))=\theta/(2-\theta).$$

(6) The relations among θ and the Pearson's ρ, Kendall's τ, and Spearman's ρ_s correlations are

$$\rho=\frac{3\theta}{4-|\theta|} \qquad \text{(for uniform marginals)},$$

$$\tau=\frac{\theta}{2-|\theta|}, \qquad \rho_s=\frac{3\theta}{4-|\theta|}$$

(Cuadras and Auge [6]; Cuadras [5]).

(7) If (X_1, Y_1), (X_2, Y_2) are i.i.d. as H_θ, then

$$\theta=2-[P_\theta((X_1-X_2)(Y_1-Y_2)>0)]^{-1}.$$

Hence θ is invariant under monotone transformations of X and Y (Cuadras [5]).

3. Some Statistical Properties

3.1. *One-Parameter Curved Exponential Family*

Let $(X_1, Y_1), \ldots, (X_n, Y_n)$ be a bivariate random sample from H_θ, $\theta \in [0, 1]$ (the study of the negative case $\theta \in [-1, 0)$ is straightforward using suitable modifications).

Let $\alpha \subset \{1, 2, \ldots, n\}$ be the set of indexes of points in the sample lying on the curve C (i.e., $i \in \alpha$ iff $(x_i, y_i) \in C$).

Using the density function (4) with respect to the measure μ, the joint density function of the sample can be expressed as

$$\begin{aligned} f_\theta(\{x_i, y_i\}) &= \left[\prod_{i\notin\alpha} h_\theta(x_i, y_i)\right]\left[\prod_{i\in\alpha} \tilde{h}_\theta(x_i)\right] \\ &= (1-\theta)^n J(\{x_i, y_i\}) \exp\{\theta T + n_c \log(\theta/(1-\theta))\}, \qquad \theta\in[0,1] \end{aligned} \tag{5}$$

where

$$J(\{x_i, y_i\}) = \left[\prod_{i=1}^{n} f(x_i)\right]\left[\prod_{i\notin\alpha} g(y_i)\right] \exp\left\{\sum_{i\in\alpha} \log F(x_i)\right\}$$

does not depend on θ,

$$n_c = \# \alpha, \qquad T = -\sum_{i=1}^{n} \log \max\{F(x_i), G(y_i)\}. \tag{6}$$

The family of densities (5) constitutes a curved exponential family as named by Efron [7], where its curvature is the geometric curvature of $\mathscr{L} = \{(\theta, \log(\theta/(1-\theta)) \colon \theta \in [0, 1]\}$ with respect to the inner product Σ_θ, being Σ_θ the covariance matrix of (T, n_c).

It immediately follows that (T, n_c) is a minimal sufficient statistic for θ.

3.2. *Joint and Marginal Distribution of* (T, n_c)

(1) As n_c is the number of points in the sample lying on the curve C and $P_\theta(F(X) = G(Y)) = \theta/(2-\theta)$, n_c is a Binomial random variable $B(n, \theta/(2-\theta))$.

(2) T is a gamma random variable $G(2-\theta, n)$.

Proof. Let $Z = \max\{F(X), G(Y)\}$. Then $P_\theta(Z \leqslant z) = H_\theta(F^{-1}(z), G^{-1}(z)) = z^{2-\theta}$, $0 \leqslant z \leqslant 1 \Rightarrow P_\theta(-\log Z > u) = e^{-u(2-\theta)}$, $u > 0$. Thus $-\log \max\{F(X), G(Y)\} \sim G(2-\theta, 1)$ and hence $T \sim G(2-\theta, n)$.

(3) T and n_c are independent random variables.

Proof. $n_c = \sum_{i=1}^n U_i$, where $U_i = 1$ if $F(x_i) = G(y_i)$, and $U_i = 0$ if $F(x_i) \neq G(y_i)$. Then $U_i \sim B(1, \theta/(2-\theta))$, $i = 1, ..., n$, are all independent. $T = \sum_{i=1}^n V_i$ being $V_i = -\log \max\{F(X_i), G(Y_i)\} \sim G(2-\theta, 1)$. It is obvious that U_i is independent of V_j for $i \neq j$. In the case $i = j$, let $Z_1 = F(X_i)$, $Z_2 = G(Y_i)$. Then

$$P(U_i = 1, V_i > v) = P(Z_1 = Z_2, Z_1 < e^{-v}, Z_2 < e^{-v})$$

$$= \begin{cases} \theta/(2-\theta) & \text{if} \quad v < 0, \\ \displaystyle\int_0^{e^{-v}} \theta x^{1-\theta}\, dx = \frac{\theta}{2-\theta} e^{-v(2-\theta)} & \text{if} \quad v \geqslant 0, \end{cases}$$

and thus

$$P(U_i = 1, V_i > v) = P(U_i = 1) \cdot P(V_i > v).$$

Analogously

$$P(U_i = 0, V_i > v) = P(U_i = 0) \cdot P(V_i > v).$$

Let us remark that (T, n_c) is not a complet statistic. For instance, from (1) and (2), we have

$$E_\theta(2T - n_c) = n \qquad \forall \theta \in [0, 1].$$

3.3 *Curvature and Fisher Information Measure*

Let us denote $\eta_\theta = (\theta, \log(\theta/(1-\theta))'$, Σ_θ the covariance matrix of (T, n_c) and

$$M_\theta = \begin{pmatrix} \dot{\eta}'_\theta \Sigma_\theta \dot{\eta}_\theta & \dot{\eta}'_\theta \Sigma_\theta \ddot{\eta}_\theta \\ \ddot{\eta}'_\theta \Sigma_\theta \dot{\eta}_\theta & \ddot{\eta}' \Sigma_\theta \ddot{\eta}_\theta \end{pmatrix}$$

the point meaning componentwise derivatives with respect to θ. If $i_\theta(X)$ represents the Fisher information measure obtained for the r.v. X, we have (Efron [7])

$$\dot{\eta}'_\theta \Sigma_\theta \dot{\eta}_\theta = \frac{n(\theta(1-\theta)+2)}{(2-\theta)^2\, \theta(1-\theta)} = i_\theta(T) + i_\theta(n_c) = i_\theta(T, n_c).$$

The curvature being

$$\gamma_\theta = \left(\frac{|M_\theta|}{i^3_\theta(T, n_c)}\right)^{1/2} = \frac{(2\theta-1)(2-\theta)}{(\theta(1-\theta)+2)^{3/2}} \sqrt{\frac{2}{n}}, \qquad \theta \in (0, 1).$$

These properties may be used to study second-order efficiency and to construct confidence intervals for the estimation of θ (Efron [8], Moolgavkar and Venzon [14]).

3.4. *Rao Distance*

Let $\psi = \psi(\theta)$ be an admissible transformation of the parameter θ. The Fisher information measure on ψ contained in (T, n_c) satisfies

$$i_\psi(T, n_c) = \left(\frac{d\psi}{d\theta}\right)^2 i_\theta(T, n_c).$$

Thus, $i_\theta(T, n_c)$ can be considered as a covariant tensor of the second order for all $\theta \in (0, 1)$ and we can obtain the Rao distance [17] for the family H_θ (see Burbea and Rao [2]; Burbea [1]; Oller and Cuadras [16]). The Rao distance between θ_1 and θ_2 is given by

$$S(\theta_1, \theta_2) = \int_{\theta_1}^{\theta_2} \frac{\sqrt{n[\theta(1-\theta)+2]}}{(2-\theta)\sqrt{\theta(1-\theta)}}\, d\theta.$$

This distance is invariant under any admissible transformation of the parameter θ and the random vector (X, Y).

Using the function

$$\Phi(\varphi)=\sum_{i=0}^{\infty}\left(-\frac{1}{2}\right)^{i}\beta_{2i}(\varphi)\sum_{j=0}^{i}\binom{-\frac{1}{2}}{j}(9/8)^{j}$$

where

$$\beta_{2i}(\varphi)=\frac{(1;2;i)}{2^{i}i!}\varphi-\frac{1}{2}\sin\varphi\cos\varphi\sum_{k=1}^{i}\frac{(2i-1;-2;k-1)}{2^{k-1}(i;-1;k)}\sin^{2(i-k)}(\varphi)$$

with

$$(a;b;c)=a(a+b)(a+2b)\cdots(a+(c-1)b)$$

for real numbers a, b, and integer number c, we obtain

$$S(\theta_1,\theta_2)=\sqrt{n}\left[\Phi\left(\sin^{-1}\sqrt{\frac{2\theta_2}{1+\theta_2}}\right)-\Phi\left(\sin^{-1}\sqrt{\frac{2\theta_1}{1+\theta_1}}\right)\right].$$

If $\theta_1\simeq\theta_2$, it is easy to check that

$$\frac{1}{\sqrt{n}}S(\theta_1,\theta_2)=\frac{\sqrt{2+\theta_1(1-\theta_1)}}{(2-\theta_1)}\left[\sin^{-1}(2\theta_2-1)-\sin^{-1}(2\theta_1-1)\right]$$
$$+O((\theta_2-\theta_1)(\sqrt{\theta_2}-\sqrt{\theta_1})),$$

which provides a useful approximation for $S(\theta_1,\theta_2)$.

4. Maximum Likelihood Estimation of θ

From expression (5) we obtain the log-likelihood function

$$\log L(\{x_i,y_i\};\theta)=(n-n_c)\log(1-\theta)+n_c\log\theta+\theta T$$

and by solving the equation

$$\frac{\partial}{\partial\theta}\log L(\{x_i,y_i\};\theta)=0$$

we get the maximum likelihood estimation of θ

$$\hat{\theta}=\frac{T-n+\sqrt{(n-T)^2+4n_cT}}{2T}.$$

Let $a = \sqrt{(n-T)^2 + 4n_c T}$. Since $|n-T| \leqslant a \leqslant n+T$, we see that

$$0 \leqslant \frac{T-n+|n-T|}{2T} \leqslant \frac{T-n+a}{2T} \leqslant \frac{n+T+T-n}{2T} = 1.$$

Thus, we check that $0 \leqslant \hat{\theta} \leqslant 1$.

Let $(x_1, y_1), ..., (x_n, y_n)$ be a bivariate random sample from H_θ, $-1 \leqslant \theta < 0$. Let Z_i be such that $G(Z_i) = 1 - F(X_i)$ (a.s.), $i = 1, ..., n$, so $(x_1, z_1), ..., (x_n, z_n)$ is a sample from $H_{-\theta}$ and we obtain the maximum likelihood estimate for θ,

$$\hat{\theta} = \frac{n - T - \sqrt{(n-T)^2 + 4n_c T}}{2T},$$

where now n_c is the number of pairs (x_i, y_i) satisfying $F(x_i) + G(y_i) = 1$ and $T = -\sum_{i=1}^{n} \log \max\{F(x_i), 1 - G(y_i)\}$.

References

[1] Burbea, J. (1986). Informative geometry of probability spaces. *Exposition Math.* **4**, 347–378.

[2] Burbea, J., and Rao, C. R. (1982). Entropy differential metric, distance and divergence measures in probability spaces: A unified approach. *J. Multivariate Anal.* **12** 575–596.

[3] Cohen, L. (1984). Probability distributions with given marginals *J. Math. Phys.* **25** 2402–2403.

[4] Cohen, L., and Zaparovanny, Y. I. (1980). Positive quantum joint distributions. *J. Math. Phys.* **21** 794–796.

[5] Cuadras, C. M. (1985). Sobre medidas de dependencia estocástica invariantes. *Homenatge a F. d'A. Sales. Contribucions Cientifiques*, pp. 28–47, Faculty of Mathematics, Univ. of Barcelona.

[6] Cuadras, C. M., and Auge, J. (1981). A continuous general multivariate distribution and its properties. *Comm. Statist. A—Theory Methods* **10** 339–353.

[7] Efron, B. (1975). Defining the curvature of a statistical problem with applications to second order efficiency (with discussion). *Ann. Statist.* **3** 1189–1242.

[8] Efron, B. (1978). The geometry of exponential families. *Ann. Statist.* **6** 362–376.

[9] Fishman, G. S. (1972). Variance reduction in simulation studies. *J. Statist. Comput. Simulation* **1** 173–182.

[10] Fréchet, M. (1951). Sur les tableaux de correlations dont les marges sont données. *Ann. Univ. Lyon Sect. A Ser. 3* **14** 53–77.

[11] Hoeffding, W. (1940). Masstabinvariante Korrelations theorie. *Schriftenreihe Math. Inst. Angew. Math. Univ. Berlin* **5** 179–233.

[12] Kimeldorf, G., and Sampson, A. (1975). One-parameter families of bivariate distributions with fixed marginals. *Comm. Statist.* **4** 293–301.

[13] Kimeldorf, G., and Sampson, A. (1975). Uniform representations of bivariate distributions. *Comm. Statist.* **4** 617–627.

[14] Moolgavkar, S. H., and Venzon, D. J. (1987). Confidence regions in curved exponential families: Application to matched case-control and survival studies with general relative risk function. *Ann. Statist.* **15** 346–359.

[15] O'CONNELL, R. F., AND WIGNER, E. P. (1981). Some properties of a non-negative quantum-mechanical distribution function. *Phys. Lett. A* **85** 121–126.

[16] OLLER,, J. M., AND CUADRAS, C. M. (1985). Rao's distance for negative multinomial distributions. *Sankhyā A* **47** 75–83.

[17] RAO, C. R. (1945). Information and accuracy attainable in the estimation of statistical parameters. *Bull. Calcutta Math. Soc.* **37** 81–91.

[18] SCHWEIZER, R., AND SKLAR, E. F. (1981). *Probabilistic Metric Spaces.* Elsevier Science, New York.

[19] SMITH, W. (1983). Inequalities for bivariate distributions with $X \leqslant Y$ and marginals given. *Comm. Statist. A—Theory Methods* **12** 1371–1379.

[20] WHITT, W. (1976). Bivariate distributions with given marginals. *Ann. Statist.* **4** 1280–1289.

Asymptotically Precise Estimate of the Accuracy of Gaussian Approximation in Hilbert Space

V. V. SAZONOV, V. V. ULYANOV, AND B. A. ZALESSKII

V. A. Steklov Mathematical Institute, Moscow, U.S.S.R.

1. INTRODUCTION

In [1] employing F. Götze's ideas (see [2]) V. V. Yurinskii obtained the following result.

Let $X_1, X_2, \ldots$ be independent random variables with the same distribution P on a separable Hilbert space H. Assume that $EX_1 = 0$, $\beta = E|X_1|^3 < \infty$. Denote by V the covariance operator of $\sigma^{-1}X_1$, where $\sigma^2 = E|X_1|^2$, and let Y be a $(0, V)$ Gaussian H-valued random variable. Put $S_n = n^{-1/2}\sigma^{-1}\sum_1^n X_i$. Then for all $a \in H$, $r \geqslant 0$,

$$\begin{aligned} \Delta_n(a, r) &= |P(|S_n - a| < r) - P(|Y - a| < r)| \\ &\leqslant c(V)\, \beta\sigma^{-3}(1 + |a|^3)\, n^{-1/2}, \end{aligned} \tag{1}$$

where $c(V)$ depends only on the eigenvalues $\sigma_1^2 \geqslant \sigma_2^2 \geqslant \cdots$ of V.

From V. V. Yurinskii's proof it follows that $c(V)$ in (1) depends on no more than the first 13 eigenvalues of V. Later S. V. Nagaev proved (see [3]) that $c(V)$ may be taken to be $c((\prod_1^7 \sigma_i)^{-6/7} + (\sigma_1\sigma_2\sigma_7^2)^{-1})$, where c is an absolute constant (see also [4, 5]).

On the other hand, from [6, 7] one can deduce that for any $c_0 > 0$ and any $1 \geqslant \tau_1^2 \geqslant \cdots \geqslant \tau_6^2$ there exist $a \in H$, $|a| > c_0$ and a probability distribution P on H such that if $X_1, X_2, \ldots$ are independent random variables with distribution P, then they satisfy the above-mentioned conditions, $\sigma_i^2 = \tau_i^2$, $i = \overline{1, 6}$, and

$$\liminf_{n \to \infty} n^{1/2} \sup_r \Delta_n(a, r) \geqslant c\left(\prod_1^6 \sigma_i^{-1}\right) \beta\sigma^{-3}|a|^3, \tag{2}$$

where c is an absolute constant. This implies that, in general, it is

Multivariate Statistics and Probability
ISBN 0-12-580205-6

Reprinted from *J. Mult. Anal.* **28**(2).

impossible to construct an estimate of type (1) with $c(V)$ depending on less than the first six eigenvalues of V.

In this paper an asymptotically precise estimate of type (1) with $c(V)$ depending on the first six eigenvalues of V will be obtained. Our proof uses F. Götze's approach (see [2]) to the estimation of the characteristic function of $|\sum_1^n X_i|^2$ as well as some ideas due to V. V. Yurinskii (see the proof of his Theorem 1, p. 82 in [8]). Note that in the special case when, in certain basis the first six coordinates of X_1 are independent of the others, an estimate with $c(V)$ depending only on the first six eigenvalues of V was also constructed by V. V. Senatov [5]. Before V. V. Senatov in an even more special case (when all coordinates of X_1 are independent), the first steps in this direction were made by S. V. Nagaev and V. I. Chebotarev [9].

In what follows χ_A denotes the indicator function of a set A, i.e., $\chi_A(x)=1$ or 0 according to $x\in A$ or $x\notin A$; $B_r(a)$ is the open ball of radius r with center at a, $B_r=B_r(0)$; if P is a measure then P^n is the n-fold convolution of P with itself.

2. The Main Result

THEOREM. *There exist an absolute constant c such that in the notation introduced in* (1) *for any* $a\in H$, $r\geqslant 0$, *integer* $n\geqslant 1$, *and* $\delta: 0\leqslant\delta<\frac{1}{9}$

$$\Delta_n(a,r)\leqslant c(\delta)(\sigma_6^{-3}\beta\sigma^{-3}n^{-1/2})^{1+\delta} + c\left(\prod_1^6\sigma_i^{-1}\right)\beta\sigma^{-3}(1+|a|^3)\,n^{-1/2}. \qquad (3)$$

Comparing this with V. V. Senatov's example (2) we see that (3) is an asymptotically precise estimate.

In what follows we will assume for simplicity that $\sigma=1$. The general case is reduced to this one if we replace X_j by $\sigma^{-1}X_j$, $j\geqslant 1$.

Proof. The theorem follows from Lemmas 1, 6, and 12 proved below if condition (9) is satisfied. When condition (9) is violated the theorem is obvious.

3. Auxiliary Lemmas

LEMMA 1. *Let* X_j, $j=1,2,\ldots$, S_n *be the same as in the theorem and let* χ_j *be the indicator function on* $\{|X_j|<n^{1/2}\}$, $S_n'=n^{-1/2}\sum_1^n X_j\chi_j$. *Then for any Borel set* $A\subset H$,

$$\Delta_1=|P(S_n\in A)-P(S_n'\in A)|\leqslant\beta n^{-1/2}.$$

Proof. We have (cf. (39) in [8, p. 95]),

$$\Delta_1 \leqslant |P(S_n \in A, |X_j| < n^{1/2}, j = \overline{1, n})$$
$$- P(S'_n \in A, |X_j| < n^{1/2}, j = \overline{1, n})| + \sum_1^n P(|X_j| \geqslant n^{1/2})$$
$$= nP(|X_1| \geqslant n^{1/2}) \leqslant \beta n^{-1/2}.$$

Let P be the distribution of X_1. Fix n and define

$$P_1(A) = P(A \cap B_{n^{1/2}}) + P(B^c_{n^{1/2}})\, \chi_A(0),$$
$$P_2(A) = P(A \cap B_R)/P(B_R),$$

assuming that $P(B_R) > 0$ (below R will be specified, but in Lemma 2 R is any number satisfying (5)).

Denote V_k the covariance operator of P_k and let $\sigma^2_{k_1} \geqslant \sigma^2_{k_2} \geqslant \cdots$ be its eigenvalues, $k = 1, 2$.

LEMMA 2. *We have*

$$\sigma^2_{1i} \leqslant \sigma^2_i, \qquad \sigma^2_{2i} \leqslant \rho\sigma^2_i, \qquad i = 1, 2, ..., \tag{4}$$

where $\rho = 1/P(B_R)$. *Moreover, if*

$$\int_{|x| \geqslant R} |x|^2\, P(dx) \leqslant \sigma^3_6/3, \tag{5}$$

then

$$\sigma^2_{2i} \geqslant (2/3)\, \sigma^2_i, \qquad i = \overline{1, 6}, \tag{6}$$

and for $n^{1/2} \geqslant R$,

$$\sigma^2_{1i} \geqslant (5/9)\, \sigma^2_i, \qquad i = \overline{1, 6}. \tag{7}$$

Proof. Inequalities for σ^2_{2i} are proved in Lemma 8 in [5]. As to σ^2_{1i} we have similarly to (9) in [5],

$$(V_1 y, y) = (Vy, y) - \int_{|x| \geqslant n^{1/2}} (x, y)^2\, P(dx) - \left(\int_{|x| \geqslant n^{1/2}} (x, y)\, P(dx)\right)^2$$
$$\geqslant (Vy, y) - (1 + q) \int_{|x| \geqslant n^{1/2}} (x, y)^2\, P(dx),$$

where $q = P(B^c_{n^{1/2}})$. Now the same reasoning as in the proof of Lemma 8 in [5] gives (4) for σ^2_{1i} and (7).

Put now

$$s = 3\beta\sigma_6^{-2}, \qquad R = (\beta n)^{1/3}, \tag{8}$$

and observe that if

$$\beta\sigma_6^{-3} n^{-1/2} \leqslant 3^{-3/2} \tag{9}$$

then $s \leqslant R \leqslant n^{1/2}$ and hence $P(|X_1| \geqslant R) \leqslant \frac{1}{2}$, since we always have $P(|X_1| \geqslant s) \leqslant \frac{1}{2}$. Thus if (9) is satisfied then since $P_1 \geqslant P_2/2$, $P_2 \geqslant P_2'/2$, where $P_2'(A) = P(A \cap B_s)/P(B_s)$ (cf. [8, p. 89]), we have

$$P_1 = (P_2 + P_3)/2, \qquad P_2 = (P_2' + P_3')/2, \tag{10}$$

P_3, P_3' being some probability measures.

In what follows, s, R will be always defined by (8) and condition (9) will be assumed to be satisfied (if not, the theorem is obviously true). Thus (5) is also fulfilled.

Let P_4 be the probability measure corresponding to the H-valued random variable $Z' = \xi Z'' + Y' + EZ$; where ξ is a bounded real random variable such that $E\xi = 0$, $E\xi^2 = \frac{1}{2}$, $E\xi^3 = 1$ (see [8, p. 84]), $Z'' = Z - EZ$, Z is distributed according to P_2, Y' has Gaussian distribution with parameters $(0, V_2/2)$, and ξ, Z, Y' are independent. Note that Z' has mean EZ, covariance operator V_2, and for any $h_1, h_2, h_3 \in H$,

$$E(Z', h_1)(Z', h_2)(Z', h_3) = E(Z, h_1)(Z, h_2)(Z, h_3)$$

(cf. [8, p. 85]); i.e., Z and Z' have the same first, second, and third moments. Finally, put $P_5 = (P_4 + P_3)/2$.

Lemma 3. *Let $X_j^{(k)}$, $j = 1, 2, \ldots$ be independent H-valued random variables with distribution P_k, $k = \overline{1, 5}$. For any $p \geqslant 0$, $n_1 : 1 \leqslant n_1 \leqslant n$,*

$$E\left|n^{-1/2} \sum_{j=1}^{n_1} X_j^{(k)}\right|^p \leqslant c(p), \qquad k = \overline{1, 5}. \tag{11}$$

Moreover, if Z is a Gaussian H-valued random variable then

$$E|Z|^p \leqslant c(p)(E|Z|^2)^{p/2}. \tag{12}$$

Proof. For $p \geqslant 2$, we have

$$E\left|n^{-1/2} \sum_{j=1}^{n_1} X_j^{(k)}\right|^p \leqslant c(p)\, n^{-p/2}((n_1 |EX_1^{(k)}|)^p + (n_1 E|X_1^{(k)}|^2)^{p/2} + n_1 E|X_1^{(k)}|^p) \tag{13}$$

(see, e.g., [10]). Obviously $|X_j^{(k)}| \leqslant n^{1/2}$ a.s., $E\,|X_j^{(k)}|^2 \leqslant 2$ for $k=\overline{1,3}$, so that

$$E\,|X_1^{(k)}|^p \leqslant 2n^{p/2-1}, \qquad k=\overline{1,3}, p \geqslant 2. \tag{14}$$

Furthermore,

$$|EX_j^{(1)}| = \left| \int_{|x| \geqslant n^{1/2}} xP(dx) \right| \leqslant n^{-1/2},$$

$$|EX_j^{(2)}| = P(B_k)^{-1} \left| \int_{|x| \geqslant R} xP(dx) \right| \leqslant 2\beta R^{-2} \leqslant n^{-1/2},$$

and, by (10),

$$|EX_j^{(3)}| \leqslant 2\,|EX_j^{(1)}| + |EX_j^{(2)}| \leqslant 3n^{-1/2},$$

so that

$$|EX_j^{(k)}| \leqslant 3n^{-1/2}, \qquad k=\overline{1,3}. \tag{15}$$

Inequalities (13)–(15) imply (11) for $k=\overline{1,3}$, $p \geqslant 2$. Inequality (12) when Z has mean zero is proved, e.g., in [8, pp. 85–86]. The general case follows easily.

Now represent $X_j^{(4)}$ as

$$X_j^{(4)} = \xi_j(X_j^{(2)} - EX_j^{(2)}) + Y_j + EX_j^{(2)}, \tag{16}$$

where ξ_j, $X_j^{(2)}$, Y_j are independent and ξ_j, Y_j are distributed as ξ, Y' (see the definition of p_4 above). We have

$$E\left| n^{-1/2} \sum_1^{n_1} X_j^{(4)} \right|^p \leqslant c(p) \left(E\left| n^{-1/2} \sum_1^{n_1} \xi_j(X_j^{(2)} - EX_j^{(2)}) \right|^p \right.$$
$$\left. + E\left| n^{-1/2} \sum_1^{n_1} Y_j \right|^p + (n_1\, n^{-1/2}\, |EX_j^{(2)}|)^p \right).$$

Using, as before, the inequality from [10] and observing that ξ_j are bounded by an absolute constant and ξ_j are independent of $X_j^{(2)}$, we obtain $E\,|n^{-1/2} \sum_1^{n_1} \xi_j\, X_j^{(2)}|^p \leqslant c(p)$.

Now since $n^{-1/2} \sum_1^{n_1} Y_j$ is $(0, (n_1/2n)\, V_2)$ Gaussian, by (12),

$$E\left| n^{-1/2} \sum_1^{n_1} Y_j \right|^p \leqslant c(p)((n_1/n)\, E\,|X_1^{(2)}|^2)^{p/2} \leqslant c(p).$$

Finally, as we observed above, $n^{1/2}\,|EX_j^{(2)}| \leqslant 1$. Hence (11) is true for $k=4$, $p \geqslant 2$.

If $0 \leqslant P < 2$, $k = \overline{1, 4}$ the lemma follows now from the well-known moment inequalities.

Denoting $\bar{P}_k(\cdot) = P_k(n^{1/2} \cdot)$ and using (11) with $k = 3, 4$ we have

$$E\left|n^{-1/2} \sum_1^{n_1} X_j^{(5)}\right|^p = \int |x|^p \bar{P}_5^{n_1}(dx)$$

$$= 2^{-n_1} \sum_{i=0}^{n_1} \binom{n_1}{i} \int |x|^p \bar{P}_4^i * \bar{P}_3^{n_1-i}(dx)$$

$$\leqslant c(p)\, 2^{-n_1} \sum_{i=0}^{n_1} \binom{n_1}{i} \int (|x|^p + |y|^p)\, \bar{P}_4^i(dx)\, \bar{P}_3^{n_1-i}(dy)$$

$$\leqslant c(p).$$

The lemma is proved.

Remark. For future use note that while proving (11) with $k = 4$, we also proved that

$$E\left|n^{-1/2} \sum_1^{n_1} (\xi_j(X_j^{(2)} - EX_j^{(2)}) + EX_j^{(2)})\right|^p \leqslant c(p).$$

LEMMA 4. *Let Z_j, $j = 1, 2, \ldots$, be H-valued independent random variables with the same distribution Q such that $Q = (Q_1 + Q_2)/2$, where Q_1, Q_2 are probability measures, $Q_1(B_L) = 1$ for some $L > 0$, and the covariance operator V' of Q_1 has trace $\operatorname{tr} V' \leqslant 2$. Let $Y_j, j = 1, 2, \ldots$, be independent $(0, V)$ Gaussian random variables and let Z_0 be an H-valued random variable independent of Y_j, Z_j, $j = 1, 2, \ldots$. Finally let l, m, n_1, n be positive integers satysfying $l \leqslant m$, $l + m \leqslant n$, $b_1 \leqslant n$. Put*

$$U_1 = n^{-1/2} \sum_1^{l+m} Z_j, \qquad U_2 = n^{-1/2} \sum_1^{n_1} Y_j.$$

Then for any $A > 0$, even $k \geqslant 0$, integer $k' \geqslant 0$, $k_q \geqslant 0$, $q = \overline{1, k'}$, and any t, $x_j \in H$, $j = \overline{1, k}$, if $l \leqslant L^2$ or if $l > L^2$ and

$$|t| \leqslant c(A)\, L^{-1} n (l \ln(L^{-2} l))^{-1/2}$$

$$I = \left|E \exp\{it\,|U_1 + Z_0|^2\}\, |U_1|^k \prod_{q=1}^{k'} (x_q, U_1)^{k_q}\right|$$

$$\leqslant K \bar{c}_1 \prod_{q=1}^{k'} |x_q|^{k_q} (\exp\{-\bar{c}_2 l\} + \bar{c}'(L^2/l)^A + h^{1/2}(\bar{c}'' t_1^2\, lm/n^2, V')), \quad (17)$$

where

$$K = \sup_{1 \leqslant l \leqslant k} E\left|n^{-1/2} \sum_1^l Z_j\right|^k,$$

$\bar{k}=k+\sum_1^{k'} k_q$, $\bar{c}_j$ *are functions of* $\bar{k}$; $\bar{c}'$, $\bar{c}''$ *are functions of* A *and* $\bar{k}$,

$$t_1=\min\{|t|, L^{-1}n(m\ln(m/L^2))^{-1/2}\},$$

$$h(s, V_1')=\prod_{j=1}^{\infty}(1+2s(\sigma_j')^4)^{-1/2},$$

and $(\sigma_1')^2 \geqslant (\sigma_2')^2 \geqslant \cdots$ *are eigenvalues of* V'.

Moreover, for any t,

$$I \leqslant E|Z_1|^{k}\prod_{q=1}^{k'}|x_q|^{k_q},$$

$$\left|E\exp\{it|U_2+Z_0|^2\}|U_2|^k\prod_{q=1}^{k'}(x_q, U_2)^{k_q}\right|$$

$$\leqslant \bar{c}_3\prod_{q=1}^{k'}(Vx_q, x_q)^{k_q/2}h^{1/2}(\bar{c}_4t^2n_1^2/n^2, V). \tag{18}$$

We omit the proof of this lemma since it is basically the same as the proof of Lemma 11 in [5] which it is a generalisation of.

LEMMA 5. *Let* $X_j^{(k)}$ *be the same as in Lemma* 3. *Then for any* $b\in H$, $n'\leqslant n$, $0\leqslant\delta<1/9$,

$$\Delta'=\left|P\left(\left|n^{-1/2}\sum_1^{n'}X_j^{(2)}-b\right|<r\right)-P\left(\left|n^{-1/2}\sum_1^{n'}X_j^{(4)}-b\right|<r\right)\right|$$

$$\leqslant\alpha^{-3}\left(c(\delta)\,\beta^{1+\delta}\sigma_6^{-3-2\delta}n^{-(1+\delta)/2}+cs_1(b)\left(\prod_1^6\sigma_j^{-1}\right)n^{-1/2}\right),$$

where $\alpha=n'/n$,

$$s_1(b)=(\beta^{4/3}+n^{-1/3}E(X_1^{(2)}, b)^4)^{3/4}.$$

Proof. From the structure of $X_j^{(4)}$ (see (16)) it follows that $n^{-1/2}\sum_1^{n'}X_j^{(4)}$ can be represented as a sum of two independent H-valued random variables Z_1 and Y_1, where Y_1 is $(0, (\alpha/2)\,V_2)$ Gaussian. Hence, since (9) implies (5), by (28) (see below) and Lemma 2, the density function of $|Y_1+Z_1-b|^2$ which is equal to $\int p(b-v, r)\,P_{Z_1}(dv)$, where $p(b-v, r)$ is the density function of $|Y_1+v-b|^2$, is not greater then $c\alpha^{-1}\sigma_1^{-1}\sigma_2^{-1}$. Applying Esseen's inequality (see, e.g., [11, Theorem 2, Section 1, Chap. V]), we have

$$\Delta'\leqslant c\left(\int_{|t|\leqslant T}g_n(t)\,|t|^{-1}\,dt+\alpha^{-1}\sigma_1^{-1}\sigma_2^{-1}T^{-1}\right), \tag{19}$$

where

$$g_n(t) = |f_{2n}(t) - f_{4n}(t)|,$$

$$f_{2n}(t) = E\exp\left\{it\left|n^{-1/2}\sum_1^{n'} X_j^{(2)} - b\right|^2\right\},$$

$$f_{4n}(t) = E\exp\left\{it\left|n^{-1/2}\sum_1^{n'} X_j^{(4)} - b\right|^2\right\}.$$

Choose

$$T = c(3/2)\,\alpha^{-1/2}(s^{-1}n^{1/2})^{10/9}\,(\ln(\alpha(s^{-2}n)^{8/9}))^{-1/2}, \qquad s = 3\beta\sigma_6^{-2},$$

where $c(3/2)$ is from Lemma 4. Also put

$$T_1 = (n/s_1^2(b))^{1/6}, \qquad T_2 = c(3/2)(ns^{-2}/\ln(ns^{-2}))^{1/2}.$$

First we estimate $I_1 = \int_{T_2 \leqslant |t| \leqslant T} g_n(t)\,|t|^{-1}\,dt$. Obviously,

$$I_1 \leqslant \int_{T_2 \leqslant |t| \leqslant T} (|f_{2n}(t)| + |f_{4n}(t)|)\,|t|^{-1}\,dt = I_{11} + I_{12}. \tag{20}$$

Using the representation $n^{-1/2}\sum_1^{n'} X_j^{(4)} = Y_1 + Z_1$ (see above), Lemma 1 in [8, p. 82] and Lemma 2 we have

$$\begin{aligned}
|f_{4n}(t)| &= |E\exp\{it\,|Y_1 + Z_1 - b|^2\}| \\
&\leqslant \prod_1^{\infty} (1 + t^2\alpha^2\sigma_{2j}^4)^{-1/4} \\
&\leqslant \prod_1^{3} (1 + (4/9)\,t^2\alpha^2\sigma_j^4)^{-1/4} \\
&\leqslant \prod_1^{2} (1 + (4/9)\,t^2\alpha^2\sigma_j^4)^{-1/4}\,(1 + (4/9)\,t^2\alpha^2\sigma_3^4)^{-1/8}.
\end{aligned} \tag{21}$$

It follows, since $\sigma_6^{-2}(\sigma_1\sigma_2\sigma_3^{1/2})^{-1} \leqslant \prod_1^6 \sigma_j^{-1} + \sigma_6^{-3}$, that if $0 \leqslant \delta < 1/9$,

$$\begin{aligned}
I_{12} &\leqslant c\alpha^{-5/4}(\sigma_1\sigma_2\sigma_3^{1/2})^{-1}\,T_2^{-5/4} \\
&\leqslant c\alpha^{-5/4}\,\beta\sigma_6^{-2}(\sigma_1\sigma_2\sigma_3^{1/2})^{-1}\,n^{-1/2}(s^2n^{-1})^{1/8}\,(\ln(s^{-2}n))^{5/8} \\
&\leqslant \alpha^{-5/4}\left(c(\delta)\,\sigma_6^{-3-2\delta}\beta^{1+\delta}n^{-(1+\delta)/2} + c\left(\prod_1^6 \sigma_i^{-1}\right)\beta\,n^{-1/2}\right).
\end{aligned} \tag{22}$$

To estimate I_{11} we apply Lemma 4 with $Q=P_2$, $Q_1=P_2'$ (see (10), $L=s$, $A=\frac{3}{2}$, $\bar{k}=0$, $l\sim n'(s^2/n)^{1/9}$, $m=n'-l$. We obtain

$$|f_{2n}(t)| \leqslant c(\alpha^{-3/2}(s^2n^{-1})^{4/3} + \prod_{i=1}^{2} (1+c\alpha^2 T_2^2 (s^2n^{-1})^{1/9}\,\sigma_3^4)^{-1/4}) \tag{23}$$

for all $T_2 \leqslant |t| \leqslant T$. Hence if $\delta\colon 0\leqslant\delta<1/9$

$$\begin{aligned} I_{11} &\leqslant c(\alpha^{-3/2}s^2n^{-1} + \alpha^{-5/4}(\sigma_1\sigma_2\sigma_3^{1/2})^{-1}\,T_2^{-5/4}(s^2n^{-1})^{-5/72})\ln T \\ &\leqslant c\alpha^{-3/2}(s^2n^{-1} + \beta\sigma_6^{-2}(\sigma_1\sigma_2\sigma_3^{1/2})^{-1}\,n^{-1/2}\,(s^2n^{-1})^{\delta/2})\ln T \\ &\leqslant \alpha^{-3/2}\left(c(\delta)\,\sigma_6^{-3-2\delta}\beta^{1+\delta}n^{-(1+\delta)/2} + c\left(\prod_1^6 \sigma_i^{-1}\right)\beta\,n^{-1/2}\right). \end{aligned} \tag{24}$$

Next we estimate $I_2=\int_{T_1\leqslant|t|\leqslant T_2} g_n(t)\,|t|^{-1}\,dt$ assuming that $T_1\leqslant T_2$. We have $I_2=I_{21}+I_{22}$, where (see (21))

$$\begin{aligned} I_{22} &= \int_{T_1\leqslant|t|\leqslant T_2} |f_{4n}(t)|\;|t|^{-1}\,dt \\ &\leqslant 2\int_{T_1}^{\infty} |t|^{-1}\prod_{i=1}^{6}(1+(4/9)\,t^2\alpha^2\sigma_i^4)^{-1/4}\,dt \\ &\leqslant c\alpha^{-1}\left(\prod_1^6 \sigma_j^{-1}\right)s_1(b)\,n^{-1/2} \end{aligned} \tag{25}$$

and by Lemma 4, with the same parameters as in (23) except that now $l\sim n'/2$, $m=n'-l$,

$$\begin{aligned} I_{21} &= \int_{T_1\leqslant|t|\leqslant T_2} |f_{2n}(t)|\;|t|^{-1}\,dt \\ &\leqslant c\int_{T_1}^{T_2}|t|^{-1}\,(\alpha^{-3/2}(s^2n^{-1})^{3/2} + \prod_{j=1}^{6}(1+ct^2\alpha^2\sigma_j^4)^{-1/4})\,dt \\ &\leqslant c\alpha^{-3}\left(s^2n^{-1} + \left(\prod_1^6\sigma_j^{-1}\right)s_1(b)\,n^{-1/2}\right). \end{aligned} \tag{26}$$

Finally, estimate $I_3=\int_{|t|\leqslant T'} g_n(t)\,|t|^{-1}\,dt$, where $T'=\min(T_1,T_2)$. Denote $\bar{P}_2$ (resp. $\bar{P}_4$) the distribution of $X_1^{(2)}n^{-1/2}$ (resp. $X_1^{(4)}n^{-1/2}$). We have

$$I_3 \leqslant \int_{|t| \leqslant T'} |t|^{-1} \left| \int \exp\{it\,|x-b|^2\} (\bar{P}_2^{n'} - \bar{P}_4^{n'})(dx) \right| dt$$

$$\leqslant \sum_{m=0}^{n'-1} \int_{|t| \leqslant T'} |t|^{-1} \left| \iint \exp\{it\,|x+y-b|^2\} \right.$$

$$\left. \times \bar{P}_2^m * \bar{P}_4^{n'-m-1}(dx)(\bar{P}_2 - \bar{P}_4)(dy) \right| dt = \sum_{m=0}^{n'-1} I_{3m}.$$

Note that $\bar{P}_2$ and $\bar{P}_4$ have the same moments of the first three orders. Thus expanding $f(\lambda) = \exp\{it\,|x+\lambda y - b|^2\}$ by Taylor's formula we may write

$$I_{3m} = \int_{|t| \leqslant T'} |t|^{-1} \left| \int_0^1 \iint f^{(4)}(\lambda)\, \bar{P}_2^m * \bar{P}_4^{n'-m-1}(dx)(\bar{P}_2 - \bar{P}_4)(dy)(1-\lambda)^3\, d\lambda \right| dt,$$

where

$$f^{(4)}(\lambda) = f(\lambda)((2it)^4\,(x+\lambda y - b, y)^4$$
$$+ 6(2it)^3\,(x+\lambda y - b, y)^2\,|y|^2 + 3(2it)^2\,|y|^4).$$

To estimate the inner integral of the type

$$I = \int f(\lambda)(x, y)^k\, \bar{P}_2^m * \bar{P}_4^{n'-m-1}(dx)$$

$$= \iint \exp\{it\,|x_1 + x_2 + \lambda y - b|^2\}(x_1 + x_2, y)^k\, \bar{P}_2^m(dx_1)\, \bar{P}_4^{n'-m-1}(dx_2),$$

we may apply Lemma 4 as above, assuming without loss of generality that n' is large enough, say $n' \geqslant 5$. Namely if $m \geqslant (n'-1)/2$ we use (17) with $A = 3/2$, $l \sim m/2$, $m \sim m/2$, and if $n' - m - 1 > (n'-1)/2$ we use (18). Also applying Lemma 3 we thus obtain for all $|t| \leqslant T_2$,

$$|I| \leqslant c(k)\,|y|^k \left(\alpha^{-3/2}(s^2 n^{-1})^{3/2} + \prod_{j=1}^{6} (1 + ct^2\alpha^2\sigma_j^4)^{-1/4} \right).$$

Observing that for $k \geqslant 4$,

$$\int |y|^k\, \bar{P}_2(dy) \leqslant n^{-2} E\,|X_1^{(2)}|^4$$

and, by (12),

$$\int |y|^k\, \bar{P}_4(dy) \leqslant c(k)\, n^{-2} E |X_1^{(2)}|^4,$$

we find

$$I_{3m} \leqslant cn^{-2} \int_{|t| \leqslant T'} (|t|^3 + |t|)(\alpha^{-3/2}(s^2/n)^{3/2} + \prod_{j=1}^{6} (1 + ct^2\alpha^2\sigma_j^4)^{-1/4})\, dt (E(X_1^{(2)}, b)^4 + E(X_1^{(4)}, b)^4 + E\,|X_1^{(2)}|^4).$$

Furthermore, using (9) we obtain

$$\int_{|t| \leqslant T'} (|t|^3 + |t|) \left(\alpha^{-3/2}(s^2/n)^{3/2} + \prod_{j=1}^{6} (1 + ct^2\alpha^2\sigma_j^4)^{-1/4} \right) dt \leqslant c\alpha^{-3}\, T_1 \prod_{1}^{6} \sigma_j^{-1}.$$

Moreover, for Y_1 from the representation (see (16))

$$X_1^{(4)} = \xi_1(X_1^{(2)} - EX_1^{(2)}) + Y_1 + EX_1^{(2)},$$

we have

$$E(Y_1, b)^4 \leqslant 3(\tfrac{1}{2} V_2 b, b)^2 \leqslant \tfrac{3}{4} E(X_2^{(2)}, b)^4,$$

so that $E(X_1^{(4)}, b)^4 \leqslant cE(X_1^{(2)}, b)^4$ and $E\,|X_1^{(2)}|^4 \leqslant RE\,|X_1^{(2)}|^3 \leqslant 2n^{1/3}\,\beta^{4/3}$. Hence,

$$I_{3m} \leqslant c\alpha^{-3} \left(\prod_{1}^{6} \sigma_j^{-1} \right) (s_1(b))^{-1/3} (\beta^{4/3} + n^{-1/3}\, E(X_1^{(2)}, b)^4)\, n^{-3/2}$$

$$\leqslant c\alpha^{-3} \left(\prod_{1}^{6} \sigma_j^{-1} \right) s_1(b)\, n^{-3/2}. \tag{27}$$

Combining (22), (24)–(27), we obtain for $0 \leqslant \delta < 1/9$,

$$\left| \int_{|t| \leqslant T} g_n(t)\, t^{-1}\, dt \right| \leqslant \alpha^{-3} \left(c(\delta)\, \sigma_6^{-3-2\delta} \beta^{1+\delta} n^{-(1+\delta)/2} + c \left(\prod_{1}^{6} \sigma_j^{-1} \right) s_1(b)\, n^{-1/2} \right).$$

Now this relation, together with (19) and the obvious inequality

$$(\sigma_1\sigma_2\sigma_6^2)^{-1} \leqslant \left(\prod_{1}^{6} \sigma_j^{-1} + \sigma_6^{-3} \right) \Big/ 2,$$

implies the lemma.

LEMMA 6. *Let* $X_j^{(k)}, p_k, k=\overline{1,5}, j=1,2,\ldots,$ *be the same as in Lemma* 3. *Define* S_n' *as in Lemma* 1 *and put* $S_n''=n^{-1/2}\sum_1^n X_j^{(5)}$. *Then for any* $\delta: 0\leqslant\delta<1/9$,

$$\Delta_2=|P(|S_n'-a|<r)-P(|S_n''-a|<r)|$$

$$\leqslant c(\delta)\,\sigma_6^{-3-2\delta}\beta^{1+\delta}n^{-(1+\delta)/2}+c\left(\prod_1^6\sigma_j^{-1}\right)s_1(a)\,n^{-1/2},$$

where $s_1(\cdot)$ *is the same as in Lemma* 5.

Proof. Let as above $\bar{P}_k(\cdot)=P_k(n^{1/2}\cdot)$. By (10) the distribution of S_n' may be written as $((\bar{P}_2+\bar{P}_3)/2)^n$. Similarly, the distribution of S_n'' may be written as $((\bar{P}_4+\bar{P}_3)/2)^n$. Consequently,

$$\Delta_2=|(((\bar{P}_2+\bar{P}_3)/2)^n-((\bar{P}_4+\bar{P}_3)^n)(B_r(a))|$$

$$\leqslant 2^{-n}\left(\sum_1+\sum_2\right)\binom{n}{m}|(\bar{P}_2^m-\bar{P}_4^m)*\bar{P}_3^{n-m}(B_r(a))|$$

$$=I_1+I_2,$$

where $\sum_1$ is the summation over all integers m such that $|m-n/2|<n/4$ and $\sum_2$ is the summation over the remaining m, $0\leqslant m\leqslant n$. By the exponential inequality for the binomial distribution (see, e.g., [11]), $2^{-n}\sum_2\binom{n}{m}\leqslant 2\exp(-n/8)$. Hence $I_2\leqslant cn^{-1/2}$. Furthermore,

$$(\bar{P}_2^m-\bar{P}_4^m)*\bar{P}_3^{n-m}(B_r(a))$$

$$=\int(\bar{P}_2^m-\bar{P}_4^m)(B_r(a-x))\,\bar{P}_3^{n-m}(dx)$$

and, by Lemma 5, if $|m-n/2|<n/4$,

$$|(\bar{P}_2^m-\bar{P}_4^m)(B_r(a-x))|$$

$$=\left|P\left(\left|n^{-1/2}\sum_1^m X_j^{(2)}-a+x\right|<r\right)\right.$$

$$\left.-P\left(\left|n^{-1/2}\sum_1^m X_j^{(4)}-a+x\right|<r\right)\right|$$

$$\leqslant c(\delta)\;\sigma_6^{-3-2\delta}\beta^{1+\delta}n^{-(1+\delta)/2}+c\left(\prod_1^6\sigma_j^{-1}\right)s_1(a-x)\,n^{-1/2}.$$

Note also that, since

$$E(X_1^{(2)},a-x)^4\leqslant c(E(X_1^{(2)},a)^4+E\,|X_1^{(2)}|^4\,|x|^4)$$

$$\leqslant c(E(X_1^{(2)},a)^4+\beta^{4/3}\,n^{1/3}\,|x|^4),$$

we have $s_1(a-x) \leqslant c(s_1(a)+\beta\,|x|^3)$ and, by Lemma 3,

$$\int |x|^3\,\bar{P}_3^{n-m}(dx) = E\left|n^{-1/2}\sum_1^{n-m} X_j^{(3)}\right|^3 \leqslant c.$$

Hence,

$$I_1 \leqslant c(\delta)\,\sigma_6^{-3-2\delta}\beta^{1+\delta}n^{-(1+\delta)/2} + c\left(\prod_1^6 \sigma_j^{-1}\right) s_1(a)\,n^{-1/2}.$$

The lemma is proved.

LEMMA 7. *Let $F_1(u)$, $F_2(u)$ be real continuous functions defined on $[0,\infty)$ such that there exist continuous derivatives $F_1'(u)$, $F_2'(u)$ on $(0,\infty)$ and*

$$|F_1'(u)| + |F_2'(u)| \leqslant c\max(1, u^{-1/2}).$$

Define $F_3(u) = \int_0^u F_1(u-v)\,F_2(v)\,dv$. Then for any $u \geqslant 0$,

$$F_3'(u) = F_1(0)\,F_2(u) + \int_0^u F_1'(u-v)\,F_2(v)\,dv$$

$$= F_1(u)\,F_2(0) + \int_0^u F_1(u-v)\,F_2'(v)\,dv$$

and for $u > 0$,

$$F_3''(u) = F_1(0)\,F_2'(u) + F_1'(u)\,F_2(0) + \int_0^u F_1'(u-v)\,F_2'(v)\,dv.$$

The proof of the lemma employs standard reasoning used in analysis and is omitted.

LEMMA 8. *Let Y_i be $(0,\sigma_i^2)$ Gaussian real random variables $i=\overline{1,6}$, $1 \geqslant \sigma_1^2 \geqslant \cdots \geqslant \sigma_6^2$. Assume that Y_i are independent and denote $p_{m)} = p_{m)}(x,u)$ (resp. $\bar{p}_{m)} = \bar{p}_{m)}(x,u)$), $x=(x_1,\ldots,x_m)\in R^m$, the density function of $\sum_1^m (Y_i+x_i)^2$ (resp. $\sum_1^m (\sqrt{2}\,Y_i+x_i)^2$). Then for $m=4, 6$, $u>0$*

$$|p_{m)}| \leqslant c\sigma_1^{-1}\sigma_2^{-1} \tag{28}$$

$$\left|\frac{\partial p_{m)}}{\partial u}\right| \leqslant c\prod_1^4 \sigma_i^{-1}, \qquad \left|\frac{\partial^2 p_{m)}}{\partial x_j\,\partial u}\right| \leqslant c\sigma_m^{-1}\prod_1^4 \sigma_i^{-1} \tag{29}$$

$$\left|\frac{\partial^2 p_{4)}}{\partial u^2}\right| \leqslant c(u^{-1/2}+\sigma_2^{-1})\,\sigma_4^{-1}\prod_1^4 \sigma_i^{-1}, \qquad \left|\frac{\partial^2 p_{6)}}{\partial u^2}\right| \leqslant c\prod_1^6 \sigma_i^{-1} \tag{30}$$

$$\left|\frac{\partial^{\delta_1+\delta_2+\delta_3}}{\partial x_i^{\delta_1}\,\partial x_j^{\delta_2}\,\partial x_k^{\delta_3}\,p_{m)}}\right| \leqslant c\sigma_i^{-\delta_1}\sigma_j^{-\delta_2}\sigma_k^{-\delta_3}\,\bar{p}_{m)} \tag{31}$$

where $i,j,k=\overline{1,6}$ and $\delta_1+\delta_2+\delta_3 \leqslant 3$.

Proof. The density function $p_i(u) = p_i(\sigma_i, x_i, u)$ of $(Y_i + x_i)^2$ is equal to

$$p_i(u) = (2\pi u)^{-1/2} \sigma_i^{-1} d_i(u),$$

where

$$d_i(u) = d_i(\sigma_i, x_i, u) = \frac{1}{2}\left(\exp\left\{-\frac{(u^{1/2} - x_i)^2}{2\sigma_i^2}\right\} + \exp\left\{-\frac{(u^{1/2} + x_i)^2}{2\sigma_i^2}\right\}\right),$$

the density function of $(\sqrt{2}\, Y_i + x_i)^2$ is $\bar{p}_i(u) = (\pi u)^{-1/2}(2\sigma_i)^{-1}\, \bar{d}_i(u)$, where $\bar{d}_i(u) = d_i(\sqrt{2}\,\sigma_i, x_i, u)$. Denote $p_A(u) = p_A(x, u)$ (resp. $\bar{p}_A(u)$) the density function of $\sum_{i \in A} (Y_i + x_i)^2$ (resp. $\sum_{i \in A} (\sqrt{2}\, Y_i + x_i)^2$). We have

$$\begin{aligned} p_{ij}(u) &= \int_0^u p_i(u - v)\, p_j(v)\, dv \\ &= (2\pi\sigma_i\sigma_j)^{-1} \int_0^1 \frac{d_i(u(1-v))\, d_j(uv)}{((1-v)\, v)^{1/2}}\, dv. \end{aligned} \tag{32}$$

Hence $p_{ij}(u) \leqslant c\sigma_i^{-1}\sigma_j^{-1}$ and

$$p_{m)}(u) = \int_0^u p_{12}(u - v)\, p_{\overline{3,m}}(v)\, dv \leqslant c\sigma_1^{-1}\sigma_2^{-1};$$

i.e., (28) is true. Furthermore, by Lemma 5 in [12], for all $u \geqslant 0$,

$$p_{ij}(u) \leqslant c \min\{(\sigma_i^{-1} + \sigma_j^{-1})\, u^{-1/2}, \sigma_i^{-1}\sigma_j^{-1}\}. \tag{33}$$

From representation (32) it is easy to deduce that $p_{ij}(u)$ is continuous at all $u \geqslant 0$ on $[0, \infty)$ and $p'_{ij}(u)$ exists and is continuous at all $u > 0$. Moreover, since $|x|^\alpha \exp(-x^2) \leqslant c(\alpha)$, $\alpha \geqslant 0$, we have for $s \geqslant 0$

$$\left|\frac{\partial d_i(su)}{\partial u}\right| \leqslant c\sigma_i^{-1}(s/u)^{1/2}\, \bar{d}_i(su), \tag{34}$$

$$\left|\frac{\partial^l d_i(su)}{\partial x_i^l}\right| \leqslant c\sigma_i^{-l}\bar{d}_i(su), \qquad l = \overline{0, 3}. \tag{35}$$

Note also that $p_i(u) \leqslant \sqrt{2}\, \bar{p}_i(u)$ and by induction it is easy to see that for any $A \subset \{1, ..., 6\}$,

$$p_A(u) \leqslant 2^{a/2} \bar{p}_A(u), \tag{36}$$

where a is the number of elements in A.

If $i<j$ we have by (34)

$$|p'_{ij}(u)| \leqslant c\sigma_j^{-1} u^{-1/2} \bar{p}_{ij}(u) \leqslant c\sigma_i^{-1} \sigma_j^{-2} u^{-1/2}. \tag{37}$$

Furthermore, applying Lemma 7 to p_{12} and p_{34} we can write

$$p'_{4)}(u) = p_{12}(0)\, p_{34}(u) + \int_0^u p'_{12}(u-v)\, p_{34}(v)\, dv, \; u \geqslant 0, \tag{38}$$

and, by (33), (37)

$$|p'_{4)}(u)| \leqslant c\left(\prod_1^4 \sigma_i^{-1} + \int_0^u \sigma_1^{-1} \sigma_2^{-2} (u-v)^{-1/2}\, \sigma_4^{-1} v^{-1/2}\, dv\right)$$

$$\leqslant c \prod_1^4 \sigma_i^{-1}; \tag{39}$$

moreover, since $p_{4)}(0)=0$, by Lemma 7 applied to $p_{4)}(u)$ and $p_{5,6}(u)$

$$p'_{6)}(u) = \int_0^u p'_{4)}(u-v)\, p_{5,6}(v)\, dv. \tag{40}$$

Hence

$$|p'_{6)}(u)| \leqslant \int_0^u |p'_{4)}(u-v)|\, p_{5,6}(v)\, dv \leqslant c \prod_1^4 \sigma_i^{-1}.$$

This proves the first inequality in (29).

From (32) and (35) we have if $i<j$

$$\left|\frac{\partial p_{ij}(x,u)}{\partial x_i}\right|, \left|\frac{\partial p_{ij}(x,u)}{\partial x_j}\right| \leqslant c\sigma_j^{-1} \bar{p}_{ij}(x,u) \leqslant c\sigma_i^{-1} \sigma_j^{-2}. \tag{41}$$

By (33), (37), (41) it follows from (38) that if $i=3, 4$

$$\left|\frac{\partial^2 p_{4)}(x,u)}{\partial x_i\, \partial u}\right| \leqslant c\sigma_4^{-1} \prod_1^4 \sigma_i^{-1}. \tag{42}$$

Using the relation

$$p'_{4)}(u) = p_{12}(u)\, p_{34}(0) + \int_0^u p_{12}(v)\, p'_{34}(u-v)\, dv$$

instead of (38), we get (42) if $i=1, 2$. Differentiating (40) now we obtain, by (39), (41), and (42),

$$\left|\frac{\partial^2 p_{6)}}{\partial x_i\, \partial u}\right| \leqslant c\sigma_6^{-1} \prod_1^4 \sigma_i^{-1}.$$

By Lemma 7 applied to p_{12}, p_{34} we also have

$$p''_{4)}(u) = p_{12}(0)\, p'_{34}(u) + p'_{12}(u)\, p_{34}(0) + \int_0^u p'_{12}(u-v)\, p'_{34}(v)\, dv.$$

Together with (37) this implies the first inequality in (30). Applying Lemma 7 to $p'_{4)}$ and $p_{5,6}$ we deduce from (40)

$$p''_{6)}(u) = p'_{4)}(0)\, p_{5,6}(u) + \int_0^u p''_{4)}(u-v)\, p_{5,6}(v)\, dv.$$

Using now (33), (39), and the first inequality in (30) we get

$$\begin{aligned} |p''_{6)}(u)| &\leqslant c\left(\prod_1^6 \sigma_i^{-1} + \sigma_4^{-1}\sigma_6^{-1}\left(\prod_1^4 \sigma_i^{-1}\right)\right. \\ &\quad \left.\times \int_0^u (u-v)^{-1/2}\, v^{-1/2}\, dv + \sigma_2^{-1}\sigma_4^{-1}\prod_1^4 \sigma_i^{-1}\right) \\ &\leqslant c\prod_1^6 \sigma_i^{-1}. \end{aligned}$$

Finally for any different i, j, k we have

$$\begin{aligned} p_{ijk}(x, u) = (2\pi)^{-3/2}\, \sigma_i^{-1}\sigma_j^{-1}\sigma_k^{-1} u^{1/2} \int_0^1 \int_0^1 \frac{d_i(u(1-w)(1-v))}{((1-v)\, vw)^{1/2}} \\ \times d_j(u(1-w)v)\, d_k(uw)\, dv\, dw. \end{aligned}$$

Together with (35), (36) this implies (31). The lemma is proved.

LEMMA 9. *Let $f_1(u)$ be a real continuous function on R such that $f_1(0)=0$ and $f'_1(u)$ exists everywhere except, possibly, at 0 and $|f'_1(u)| \leqslant c$. Let $f_2(u)$ be a continuously differentiable function such that $f_2(u)=0$ if $|u| \geqslant A$. Then $f(u)=\int_{-\infty}^u f_1(u-v)\, f_2(v)\, dv$ is a continuously differentiable function and $f'(u)=\int_{-\infty}^u f'_1(u-v)\, f_2(v)\, dv$.*

The proof is elementary and we omit it.

LEMMA 10. *Let Y_i, $i=\overline{1,6}$, $p_{m)}$ be the same as in Lemma 8. Let ξ be an independent of Y_i, $i=\overline{1,6}$, real random variable with continuously differentiable density function $f_\xi(u)$ such that $f_\xi(u)=0$ if $|u| \geqslant 1$. For a $T>0$ denote $\tilde{p}_{m)}(u)=\tilde{p}_{m)}(x, u)$ the density function of $\sum_1^m (Y_i+x_i)^2+\xi T^{-1}$,*

$x=(x_1, ..., x_m) \in R^m$. *Then* $\tilde{p}_{4)}(u)$ *is continuously differentiable,* $\tilde{p}_{6)}(u)$ *is twice continuously differentiable, and for* $m=4, 6$,

$$|\tilde{p}_{m)}| \leqslant c\sigma_1^{-1}\sigma_2^{-1}, \qquad \left|\frac{\partial \tilde{p}_{m)}}{\partial u}\right| \leqslant c\prod_1^4 \sigma_i^{-1} \tag{43}$$

$$\left|\frac{\partial^2 \tilde{p}_{6)}}{\partial u^2}\right| \leqslant c\prod_1^6 \sigma_i^{-1}, \qquad \left|\frac{\partial^2 \tilde{p}_{m)}}{\partial x_j\, \partial u}\right| \leqslant c\sigma_m^{-1}\prod_1^4 \sigma_i^{-1} \tag{44}$$

$$\left|\frac{\partial^{\delta_1+\delta_2+\delta_3}}{\partial x_i^{\delta_1}\,\partial x_j^{\delta_2}\,\partial x_k^{\delta_3}}\tilde{p}_{m)}\right| \leqslant c\sigma_i^{-\delta_1}\sigma_j^{-\delta_2}\sigma_k^{-\delta_3}\bar{\tilde{p}}_{m)}, \tag{45}$$

where $i, j, k=\overline{1, 6}$, $\delta_1+\delta_2+\delta_3 \leqslant 3$, *and* $\bar{\tilde{p}}_{m)}$ *is the density function of* $\sum_1^m (\sqrt{2}\, Y_i + x_i)^2 + \xi T^{-1}$.

Proof. We have obviously

$$\tilde{p}_{m)}(u) = \int p_{m)}(u-v)\, Tf_\xi(Tv)\, dv$$

and the lemma follows easily from Lemmas 8 and 9.

LEMMA 11. *Let Z be an H-valued* (0, *W*) *Gaussian random variable with* tr $W \leqslant 2$. *Let* $\tau_1^2 \geqslant \tau_2^2 \geqslant \cdots$ *denote the eigenvalues and* $e_1, e_2, ...$ *the corresponding eigenvectors of W. Let* P_5 *be the same as in Lemma* 3, *G be* (0, *V*) *Gaussian, and* $R=\bar{P}_5-\bar{G}$, *where* $\bar{P}_5(\cdot)=P_5(n^{1/2}\cdot)$, $\bar{G}(\cdot)=G(n^{1/2}\cdot)$. *Then for any* $r \geqslant 0$, $b \in H$,

$$\delta' = \left|\int P(|Z+x-b|<r)\, R(dx)\right|$$
$$= c\left(\prod_1^6 \tau_i^{-1}\right)\left(\beta + \sum_{i=2}^3 E|(X_1, \bar{b}_j)|^3\right) n^{-3/2},$$

where $\bar{b}_2 = \sum_{i=5}^6 (b, e_i)\, e_i$, $\bar{b}_3 = \sum_{i=7}^\infty (b, e_i)\, e_i$.

Proof. We will show that if ξ is a real random variable, independent of Z, with the continuously differentiable density function $f_\xi(u)$ such that $f_\xi(u)=0$ if $|u| \geqslant 1$, then for any $T>0$,

$$\delta'_T = \left|\int P(|Z+x-b|^2+\xi T^{-1}<r^2)\, R(dx)\right|$$
$$\leqslant c\left(\prod_1^6 \tau_i^{-1}\right)\left(\beta + \sum_{j=2}^3 E|(X_1, \bar{b}_j)|^3\right) n^{-3/2}, \tag{46}$$

where c is an absolute constant. Letting $T\to\infty$, we obtain the lemma since $\delta'_T\to\delta'$ as $T\to\infty$.

For any $h\in H$ define $h_i=(h,e_i)$, $\bar h_1=\sum_1^4 h_ie_i$, $\bar h_2=\sum_5^6 h_ie_i$, $\bar h_3=\sum_7^\infty h_ie_i$ and put $\xi_i(\lambda)=|\bar Z_i+\lambda\bar x_i-\bar b_i|^2$, $i=\overline{1,3}$. Letting $\tilde q_6$ denote the density function of $\xi_1(1)+\xi_2(1)+\xi T^{-1}$ we can write

$$Q=P(|Z+x-b|^2+\xi T^{-1}<r^2)$$

$$=E\int_{-T^{-1}}^{\eta_3(1)}\tilde q_6(u)\,du,$$

where $\eta_3(\lambda)=r^2-\xi_3(\lambda)$. Now expand the function $F_3(\lambda)=\int_{-T^{-1}}^{\eta_3(\lambda)}\tilde q_6(u)\,du$ by Taylor's formula up to the term of the third order. We have $Q=\sum_0^3(j!)^{-1}Q_j$, where

$$Q_0=E\int_{-T^{-1}}^{\eta_3(0)}\tilde q_6(u)\,du,$$

$$Q_1=-2E(\bar Z_3-\bar b_3,\bar x_3)\,\tilde q_6(\eta_3(0))$$

$$Q_2=-2E\left[|\bar x_3|^2\,\tilde q_6(\eta_3(0))-2(\bar Z_3-\bar b_3,\bar x_3)^2\frac{\partial}{\partial u}\tilde q_6(\eta_3(0))\right]$$

$$Q_3=12E(\bar Z_3+\theta\bar x_3-\bar b_3,\bar x_3)|\bar x_3|^2\frac{\partial}{\partial u}\tilde q_6(\eta_3(\theta))$$

$$-8E(\bar Z_3+\theta\bar x_3-\bar b_3,\bar x_3)^3\frac{\partial^2}{\partial u^2}\tilde q_6(\eta_3(\theta)),\qquad 0\leqslant\theta\leqslant 1,$$

and $\delta'_T=\sum_0^3(j!)^{-1}\int Q_jR(dx)=\sum_0^3 I_j$.

To estimate I_j, $j=\overline{0,3}$, we first observe that since $P_1=(P_2+P_3)/2$ and $P_5=(P_4+P_3)/2$ have the same first and second moments (see the paragraph preceding Lemma 3) and the same is true for P and G, we have

$$\left|\int(g,x)R(dx)\right|=n^{-1/2}\left|\int_{|x|>n^{1/2}}(g,x)P(dx)\right|$$

$$\leqslant c(\beta+E|(X_1,g)|^3)\,n^{-3/2}\tag{47}$$

and, similarly,

$$\left|\int(g,x)(h,x)R(dx)\right|\leqslant c(\beta+E|(X_1,g)|^3+E|(X_1,h)|^3)\,n^{-3/2}$$

$$\left|\int|\bar x_j|^2R(dx)\right|\leqslant\beta n^{-3/2},\qquad j=\overline{1,3}.\tag{48}$$

We also will need estimates of

$$J_1 = \int |(h, x)|^3 (\bar{P}_5 + \bar{G})(dx), \qquad J_2 = \int |x|^k (\bar{P}_5 + \bar{G})(dx),$$

where $k \geqslant 3$ is an integer. We obviously have

$$\int |(h, x)|^3 G(dx) = E|(Y, h)|^3 \leqslant c(Vh, h)^{3/2}$$
$$= c(E(X_1, h)^2)^{3/2} \leqslant cE|(X_1, h)|^3$$

and (see (12))

$$\int |x|^k G(dx) \leqslant c(k).$$

To estimate integrals with respect to $P_5 = (P_4 + P_3)/2$ we first observe that since $P_3 \leqslant 2P_1$

$$\int |(h, x)|^3 P_3(dx) \leqslant 2E|(X_1, h)|^3$$

$$\int |x|^k P_3(dx) \leqslant 2\beta n^{(k-3)/2}.$$

It remains to estimate only integrals with respect to probability measure P_4 corresponding to $\xi_1 X_1^{(2)} + Y_1 + (1 - \xi_1) EX_1^{(2)}$ (see (16)). Since Y_1 is $(0, V_2/2)$ Gaussian and $P_2 \leqslant 2P_1$ we have

$$E|(Y_1, h)|^3 \leqslant c(E(Y_1, h)^2)^{3/2} \leqslant c(E(X_1^{(2)}, h)^2)^{3/2}$$
$$\leqslant c(E(X_1, h)^2)^{3/2} \leqslant cE|(X_1, h)|^3$$

and consequently

$$\int |(h, x)|^3 P_4(dx) \leqslant c(E|(X_1^{(2)}, h)|^3 + E|(Y_1, h)|^3)$$
$$\leqslant cE|(X_1, h)|^3.$$

Finally

$$\int |x|^k P_4(dx) \leqslant c(k)(E|X_1^{(2)}|^k + E|Y_1|^k)$$
$$\leqslant c(k)(E|X_1^{(1)}|^k + (E|Y_1|^2)^{k/2})$$
$$\leqslant c(k)\, \beta n^{(k-3)/2}.$$

Thus,

$$J_1 \leqslant cE|(X_1, h)|^3 n^{-3/2}, \qquad J_2 \leqslant c(k)\, \beta n^{-3/2}. \tag{49}$$

From (12), (43), (44), and (49) it follows that

$$|I_3| \leqslant c \left(\prod_1^6 \tau_i^{-1} \right) (\beta + E|(X_1, \bar{b}_3)|^3)\, n^{-3/2} \tag{50}$$

To estimate I_2 we use Taylor's expansion of $\tilde{q} = \tilde{q}_6$ and $\tilde{q} = (\partial/\partial u)\, \tilde{q}_6$ as a function of $x_1, ..., x_6$:

$$\tilde{q}(x_1, ..., x_6, u) = \tilde{q}(0, ..., 0, u) + \sum_{i=1}^{6} x_i \frac{\partial}{\partial x_i} \tilde{q}(\theta x_1, ..., \theta x_6, u), \qquad 0 \leqslant \theta \leqslant 1.$$

Applying (43)–(45), (48), (49), we obtain

$$|I_2| \leqslant c\tau_6^{-1} \left(\prod_1^4 \tau_i^{-1} \right) (\beta + E|(X_1, \bar{b}_3)|^3)\, n^{-3/2}. \tag{51}$$

Consider now I_1. Using Taylor's expansions, represent $\tilde{q}_6(x_1, ..., x_6, u)$ as

$$\tilde{q}_6(x_1, ..., x_6, u) = \tilde{q}_6(0, ..., 0, x_5, x_6, u) + S_1(u), \tag{52}$$

where

$$\begin{aligned} S_1(u) = \sum_{i=1}^{4} x_i \Bigg[& \frac{\partial}{\partial x_i} \tilde{q}_6(0, ..., 0, u) \\ & + \left(\sum_{j=5}^{6} x_j \frac{\partial}{\partial x_j} \right) \frac{\partial}{\partial x_i} \tilde{q}_6(0, ..., 0, \theta_i x_5, \theta_i x_6, u) \Bigg] \\ & + \frac{1}{2} \left(\sum_{i=1}^{4} x_i \frac{\partial}{\partial x_i} \right)^2 \tilde{q}_6(\theta x_1, ..., \theta x_4, x_5, x_6, u) \end{aligned}$$

and $0 \leqslant \theta,\ \theta_i \leqslant 1$. Put $Q_{11} = -2E(\bar{Z}_3 - \bar{b}_3, \bar{x}_3)\, \tilde{q}_6(0, ..., 0, x_5, x_6, \eta_3(0))$, $Q_{12} = -2E(\bar{Z}_3 - \bar{b}_3, \bar{x}_3)\, S_1(\eta_3(0))$. Denoting $\tilde{q}_4$ the density function of $\xi_1(1) + \xi T^{-1}$, we have

$$\tilde{q}_6(0, ..., 0, x_5, x_6, v) = E\tilde{q}_4(0, ..., 0, v - \xi_2(1))$$

and expanding $\tilde{q}_4(0, ..., 0, v - \xi_2(\lambda))$ as a function of λ, we may write

$$\begin{aligned} \tilde{q}_4(0, ..., 0, v - \xi_2(1)) = {} & \tilde{q}_4(0, ..., 0, v - \xi_2(0)) \\ & - 2 \int_0^1 (\bar{Z}_2 + \theta \bar{x}_2 - \bar{b}_2, \bar{x}_2) \frac{\partial \tilde{q}_4}{\partial u} (0, ..., 0, v - \xi_2(\theta))\, d\theta. \end{aligned}$$

Thus

$$Q_{11} = -2E(\bar{Z}_3 - \bar{b}_3, \bar{x}_3)\, E_{\bar{Z}_3}\tilde{q}_4(0, ..., 0, \eta_3(0) - \xi_2(1))$$

$$= -2E(\bar{Z}_3 - \bar{b}_3, \bar{x}_3)\, \tilde{q}_4(0, ..., 0, \eta_3(0) - \xi_2(0))$$

$$+ 4 \int_0^1 E(\bar{Z}_3 - \bar{b}_3, \bar{x}_3)|\bar{x}_2|^2 \frac{\partial \tilde{q}_4}{\partial u}(0, ..., 0, \eta_3(0) - \xi_2(\theta))\, \theta\, d\theta$$

$$+ 4 \int_0^1 E(\bar{Z}_3 - \bar{b}_3, \bar{x}_3)(\bar{Z}_2 - \bar{b}_2, \bar{x}_2) \frac{\partial \tilde{q}_4}{\partial u}(0, ..., 0, \eta_3(0) - \xi_2(\theta))\, d\theta$$

$$= Q_{111} + Q_{112} + Q_{113}.$$

Denote $I_{1i} = \int Q_{1i} R(dx)$, $i = 1, 2$, $I_{11j} = \int Q_{11j} R(dx)$, $j = \overline{1, 3}$. All I_{12}, I_{111}, I_{112} are estimated by applying (47)–(49) and Lemma 10. We have

$$|I_{12}|,\ |I_{111}|,\ |I_{112}| \leqslant c\tau_1^{-1}\tau_2^{-1}\tau_4^{-1}\tau_6^{-1}(\beta + E|(X_1, \bar{b}_3)|^3)\, n^{-3/2}.$$

To estimate I_{113} we observe that if $\tilde{q}(u)$ is $\tilde{q}_4(0, ..., 0, u)$ or $(\partial/\partial u)\, \tilde{q}_4(0, ..., 0, u)$ then for any θ, $0 \leqslant \theta \leqslant 1$, $k = \overline{1, 3}$,

$$E\tilde{q}(v - \xi_{2k}(\theta)) = \int \tilde{q}(v - s)\, p(s)\, ds, \tag{53}$$

where $\xi_{2k}(\theta) = |\bar{Z}_2/\sqrt{k} + \theta \bar{x}_2 - b_k|^2$ and $p(s) = p(x_5, x_6, s)$ is its density function, $b_k \in H$. On the other hand,

$$p(x_5, x_6, s) = p(0, 0, s) + \sum_{i=5}^{6} x_i \frac{\partial p}{\partial x_i}(\theta' x_5, \theta' x_6, s), \tag{54}$$

where $0 \leqslant \theta' \leqslant 1$, and exactly as in (41) we have for $i = 5, 6$,

$$\left| \frac{\partial p}{\partial x_i} \right| \leqslant c\tau_6^{-1} \bar{p}, \tag{55}$$

where $\bar{p}$ is the density function of $|\sqrt{2/k}\, \bar{Z}_2 + \theta \bar{x}_2 - b_k|^2$.

Now let Z_{21}, Z_{22} be independent and distributed as $\bar{Z}_2/\sqrt{2}$. Then from (53)–(55), (43) we deduce

$$E(\bar{Z}_2 - \bar{b}_2, \bar{x}_2)\, \tilde{q}(v - \xi_2(\theta))$$

$$= E(\bar{Z}_{21} - \bar{b}_2/2, \bar{x}_2)\, \tilde{q}(v - |Z_{22} + \theta \bar{x}_2 - \bar{b}_2 + Z_{21}|^2)$$

$$+ E(Z_{22} - \bar{b}_2/2, \bar{x}_2)\, \tilde{q}(v - |Z_{21} + \theta \bar{x}_2 - \bar{b}_2 + \bar{Z}_{22}|^2)$$

$$= \int (y - \bar{b}_2/2, \bar{x}_2) \int \tilde{q}(v - s)\, p_1(0, 0, s)\, ds\, P_{Z_{21}}(dy)$$

$$+ \int (y - \bar{b}_2/2, \bar{x}_2) \int \tilde{q}(v - s)\, p_2(0, 0, s)\, ds\, P_{Z_{22}}(dy) + S_2, \tag{56}$$

where $p_1(0, 0, s)$ is the density function of $|Z_{22} - \bar{b}_2 + y|^2$, $p_2(0, 0, s)$ is the density function of $|Z_{21} - \bar{b}_2 + y|^2$, and

$$|S_2| \leqslant c(E|(\bar{Z}_2, \bar{x}_2)| + |(\bar{b}_2, \bar{x}_2)|)(|x_5| + |x_6|)\, \tau_6^{-1} \prod_1^4 \tau_i^{-1}. \tag{57}$$

Hence denoting $r^2 - |z - \bar{b}_3|^2$ by z_1 we have by (43), (48), (49),

$$\begin{aligned}
|I_{113}| &= 4\left|\int_0^1 \iint (z - \bar{b}_3, \bar{x}_3) E(\bar{Z}_2 - \bar{b}_2, \bar{x}_2) \right. \\
&\qquad \left. \times \frac{\partial \tilde{q}_4}{\partial u}(z_1 - \xi_2(\theta))\, P_{\bar{Z}_3}(dz) R(dx)\, d\theta \right| \\
&\leqslant 4\left|\int_0^1 \iiiint (z - \bar{b}_3, \bar{x}_3)(y - \bar{b}_2/2, \bar{x}_2) R(dx) \right. \\
&\qquad \left. \times \frac{\partial \tilde{q}_4}{\partial u}(z_1 - s)\; p_1(0, 0, s)\, ds\, P_{Z_{21}}(dy)\; P_{\bar{Z}_3}(dz)\, d\theta \right| \\
&\quad + 4\left|\int_0^1 \iiiint (z - \bar{b}_3, \bar{x}_3)(y - \bar{b}_2/2, \bar{x}_2) R(dx) \right. \\
&\qquad \left. \times \frac{\partial \tilde{q}_4}{\partial u}(z_1 - s)\; p_2(0, 0, s)\, ds\, P_{Z_{22}}(dy)\; P_{\bar{Z}_3}(dz)\, d\theta \right| \\
&\quad + \left|\int_0^1 E(\bar{Z}_3 - \bar{b}_3, \bar{x}_3)\, S_2 R(dx)\, d\theta\right| \\
&\leqslant c\tau_6^{-1} \prod_1^4 \tau_i^{-1} \left(\beta + \sum_{i=2}^3 E|(X_1, \bar{b}_j)|^3\right) n^{-3/2}.
\end{aligned} \tag{58}$$

Thus

$$|I_1| \leqslant c\tau_6^{-1} \prod_1^4 \tau_i^{-1} \left(\beta + \sum_{j=2}^3 E|(X_1, \bar{b}_j)|^3\right) n^{-3/2}. \tag{59}$$

To estimate I_0 we represent Q_0 in the form

$$\begin{aligned}
Q_0 &= P(\xi_1(1) + \xi_2(1) + \xi_3(0) + \xi T^{-1} < r^2) \\
&= E\int_{-T^{-1}}^{\eta_2(1)} \tilde{q}_4(u)\, du,
\end{aligned}$$

where $\tilde{q}_4$ as above is the density function of $\xi_1(1) + \xi T^{-1}$ and $\eta_2(\lambda) = r^2 - \xi_2(\lambda) - \xi_3(0)$. Expanding $F_2(\lambda) = \int_{-T^{-1}}^{\eta_2(\lambda)} \tilde{q}_4(u)\, du$ by Taylor's formula up to the term of the second order we have $Q_0 = \sum_0^2 Q_{0j}$, where

$$Q_{00}=E\int_{-T^{-1}}^{\eta_2(0)}\tilde{q}_4(u)\,du,$$

$$Q_{01}=-2E(\bar{Z}_2-\bar{b}_2,\bar{x}_2)\,\tilde{q}_4(\eta_2(0))$$

$$Q_{02}=-2\int_0^1 E\left[|\bar{x}_2|^2\tilde{q}_4(\eta_2(\theta))-2(\bar{Z}_2+\theta\bar{x}_2-\bar{b}_2,\bar{x}_2)^2\frac{\partial}{\partial u}\tilde{q}_4(\eta_2(\theta))\right](1-\theta)\,d\theta.$$

Thus we have $I_0=\sum_0^2 I_{0j}$, $I_{0j}=\int Q_{0j}R(dx)$.

The term I_{00} is estimated by expanding $\tilde{q}_4(u)=\tilde{q}_4(x_1,...,x_4,u)$ as a function of $x_1,...,x_4$:

$$\tilde{q}_4(x_1,...,x_4,u)=\sum_{i=0}^{2}(i!)^{-1}\left(\sum_{j=1}^{4}x_j\frac{\partial}{\partial x_j}\right)^i\tilde{q}_4(0,...,0,u)$$
$$+\frac{1}{6}\left(\sum_{j=1}^{4}x_j\frac{\partial}{\partial x_j}\right)^3\tilde{q}_4(\theta x_1,...,\theta x_4,u),$$

where $0\leqslant\theta\leqslant 1$. Applying (47)–(49), (45) we obtain

$$|I_{00}|\leqslant c\tau_4^{-3}\beta n^{-3/2}.$$

To estimate I_{01} we use Taylor's expansion of $\tilde{q}_4(x_1,...,x_4,u)$ as a function of $x_1,...,x_4$ up to the terms of the second order and obtain similarly, using (47)–(49), (43), (45),

$$|I_{01}|\leqslant c\tau_1^{-1}\tau_2^{-1}\tau_4^{-2}(\beta+E|(X_1,\bar{b}_2)|^3)\,n^{-3/2}.$$

Finally, to estimate I_{02} we use Taylor's expansions of $\tilde{q}_4$ and $\partial\tilde{q}_4/\partial u$ as functions of $x_1,...,x_4$ up to the terms of the first order and write $Q_{02}=\sum_1^3 q_{02i}$, where

$$Q_{021}=-2\int_0^1|\bar{x}_2|^2\,E\tilde{q}_4(0,...,0,\eta_3(0)-\xi_2(\theta))(1-\theta)\,d\theta$$

$$Q_{022}=4\int_0^1 E(\bar{Z}_2-\bar{b}_2,\bar{x}_2)^2\frac{\partial}{\partial u}\tilde{q}_4(0,...,0,\eta_3(0)-\xi_2(\theta))(1-\theta)\,d\theta$$

$$Q_{023}=\int_0^1\left[-2|\bar{x}_2|^2\,E\left(\sum_{i=1}^4 x_i\frac{\partial\tilde{q}_4}{\partial x_i}(\theta_1x_1,...,\theta_1x_4,\eta_2(\theta))\right.\right.$$
$$+4E(\bar{Z}_2+\theta\bar{x}_2-\bar{b}_2,\bar{x}_2)^2\left(\sum_{i=1}^4 x_i\frac{\partial^2\tilde{q}_4}{\partial x_i\,\partial u}(\theta_2x_1,...,\theta_2x_4,\eta_2(\theta))\right.$$
$$\left.+4E(2\theta(\bar{Z}_2-\bar{b}_2,\bar{x}_2)|\bar{x}_2|^2+\theta^2|\bar{x}_2|^4)\frac{\partial}{\partial u}\tilde{q}_4(0,...,0,\eta_2(\theta))\right](1-\theta)\,d\theta,$$

$0 \leqslant \theta_1, \ \theta_2 \leqslant 1$. Then $I_{02} = \sum_1^3 I_{02i}$, $I_{02i} = \int Q_{02i} R(dx)$. Applying (49) and Lemma 10, we obtain

$$|I_{023}| \leqslant c\tau_4^{-1} \left(\prod_1^4 \tau_i^{-1} \right) (\beta + E|(X_1, \bar{b}_2)|^3)\, n^{-3/2}.$$

Furthermore using (53)–(55), (48), (49), and (43) we find

$$|I_{021}| \leqslant c\tau_1^{-1} \tau_2^{-1} \tau_6^{-1} \beta n^{-3/2}.$$

It remains only to estimate I_{022}. To this aim we represent $\bar{Z}_2$ as $\sum_1^3 Z'_{2j}$, where Z'_{2j}, $j = \overline{1, 3}$ are independent and distributed as $\bar{Z}_2/\sqrt{3}$. Then Q_{022} is a finite sum of terms

$$4 \int_0^1 E(Z'_{2i} - \bar{b}_2/3, \bar{x}_2)(Z'_{2j} - \bar{b}_2/3, \bar{x}_2) \frac{\partial}{\partial u} \tilde{q}_4 \Bigg(0, ..., 0, \eta_3(0) - \Bigg| Z'_{2k} + \theta \bar{x}_2 - \bar{b}_2 + \sum_{r \neq k} Z'_{2r} \Bigg|^2 \Bigg) (1 - \theta)\, d\theta,$$

where $k \neq i$, $k \neq j$, $i, j, k = \overline{1, 3}$. Using (53)–(55) and reasoning as in (56)–(58) we obtain

$$|I_{022}| \leqslant c\tau_6^{-1} \left(\prod_1^4 \tau_i^{-1} \right) (\beta + E|(X_1, \bar{b}_2)|^3)\, n^{-3/2}.$$

Combinig the above estimates we find

$$|I_0| \leqslant c \left(\prod_1^6 \tau_i^{-1} \right) (\beta + E|(X_1, \bar{b}_2)|^3)\, n^{-3/2}. \tag{60}$$

Relations (50), (51), (59), and (60) imply (46). This proves the lemma.

LEMMA 12. *In the notation of the theorem and Lemmas* 6 *and* 11,

$$\begin{aligned} \Delta_3 &= |P(|S''_n - a| < r) - P(|Y - a| < r) \\ &\leqslant c \left(\prod_1^6 \sigma_i^{-1} \right) \left(\beta + \sum_{j=2}^3 (E|(X_1, \bar{a}_j)|^3 + E|(X_1, \bar{a}'_j)|^3 \right) n^{-1/2}, \end{aligned}$$

where $\bar{a}_j$ is a constructed according to $W = V$ and $\bar{a}'_j$ according to $W = V_2$.

Proof. Without loss of generality we will assume that $n \geqslant 2$. In the notation of Lemma 11 we may write

$$\Delta_3 = |(\bar{P}_5^n - \bar{G}^n)(B_r(a))| \leqslant \sum_{m=0}^{n-1} I_m,$$

where

$$I_m = |\bar{P}_5^m * \bar{G}^{n-m-1} * R(B_r(a))|.$$

If $m \leqslant (n-1)/2$ we will write I_m as

$$I_m = \left| \iint \bar{G}^{n-m-1}(B_r(a) - x - y) R(dx) \bar{P}_5^m(dy) \right|$$

and observe that $\bar{G}^{n-m-1}$ is $(0, ((n-m-1)/n) V)$ Gaussian with the covariance operator having eigenvalues $((n-m-1)/n) \sigma_i^2 \geqslant \sigma_i^2/4$. Applying Lemma 11, we obtain

$$\left| \int \bar{G}^{n-m-1}(B_r(a) - x - y) R(dx) \right|$$

$$\leqslant c \left(\prod_1^6 \sigma_i^{-1} \right) \left(\beta + \sum_{j=2}^{3} E|(X_1, \bar{a}_j - \bar{y}_j)|^3 \right) n^{-3/2}.$$

But $E|(X_1, \bar{a}_j - \bar{y}_j)|^3 \leqslant c(E|(X_1, \bar{a}_j)|^3 + \beta |y|^3)$, $j = 2, 3$, and by Lemma 3 $\int |y|^3 \bar{P}_5^m(dy) \leqslant c$. Thus if $m \leqslant (n-1)/2$,

$$I_m \leqslant c \left(\prod_1^6 \sigma_i^{-1} \right) \left(\beta + \sum_{j=2}^{3} E|(X_1, \bar{a}_j)|^3 \right) n^{-3/2}. \tag{61}$$

If $m > (n-1)/2$ we write I_m as

$$I_m = \left| \iint \bar{P}_5^m(B_r(a) - x - y) R(dx) \bar{G}^{n-m-1}(dy) \right|. \tag{62}$$

We have

$$\bar{P}_5^m = ((\bar{P}_4 + \bar{P}_3)/2)^m = (\Sigma_1 + \Sigma_2) 2^{-m} \binom{m}{k} \bar{P}_4^k * \bar{P}_3^{m-k},$$

where Σ_1 is the summation over all integers k such that $|k - m/2| < m/4$ and Σ_2 is the summation over all remaining k from $\overline{0, m}$. By the exponential inequality for the binomial distribution (see, e.g., [11]) $2^{-m} \Sigma_2 \binom{m}{k} \leqslant 2 \exp(-m/8)$ and in our case $\exp(-m/8) \leqslant cn^{-3/2}$. Hence

$$I_m \leqslant 2^{-m} \Sigma_1 \binom{m}{k} I_{mk} + cn^{-3/2}, \tag{63}$$

where

$$I_{mk} = \left| \iiint \bar{P}_4^k(B_r(a) - x - y - z) R(dx) \bar{G}^{n-m-1}(dy) \bar{P}_3^{m-k}(dz) \right|.$$

Furthermore, $\bar{P}_4^k = G_k * \tilde{P}_k$, where G_k is $(0, (k/2n) V_2)$ Gaussian and $\tilde{P}_k$ is the distribution of $n^{-1/2} \sum_1^k (\xi_j(X_j^{(2)} - EX_j^{(2)}) + EX_j^{(2)})$. Thus

$$I_{mk} = \left| \iiiint G_k(B_r(a) - x - y - z - u) R(dx)\, \bar{G}^{n-m-1}(dy)\, \bar{P}_3^{m-k}(dz)\, \tilde{P}_k(du) \right|.$$

Applying Lemma 11 we have by Lemma 2 and the remarks after it

$$\left| \int G_k(B_r(a) - x - y - z - u) R(dx) \right|$$

$$\leqslant c \left(\prod_1^6 \sigma_i^{-1} \right) \left(\beta + \sum_{j=2}^3 E |(X_1, \bar{a}_j' - \bar{y}_j' - \bar{z}_j' - \bar{u}_j')|^3 \right) n^{-3/2},$$

since $k/n > (n-1)/8n \geqslant 1/16$. But

$$E|(X_1, \bar{a}_j' - y_j' - \bar{z}_j' - \bar{u}_j')|^3 \leqslant c(E|(X_1, \bar{a}_j')|^3 + \beta(|y|^3 + |z|^3 + |u|^3),$$

$j = 2, 3$, and by Lemma 3 and the remark following it,

$$\iiint (|y|^3 + |z|^3 + |u|^3)\, \bar{G}^{n-m-1}(dy)\, \bar{P}_3^{m-k}(dz)\, \tilde{P}_k(du) \leqslant c.$$

Thus

$$I_{mk} \leqslant c \left(\prod_1^6 \sigma_i^{-1} \right) \left(\beta + \sum_{j=2}^3 E|(X_1, \bar{a}_j')|^3 \right) n^{-3/2}. \tag{64}$$

Combining (61)–(64) we obtain the lemma.

References

[1] Yurinskii, V. V. (1982). On the precision of the Gaussian approximation for the probability of hitting a ball. *Teor. Veroyatnost. i Primenen.* **27** 270–278. [Russian]

[2] Götze, F. (1979). Asymptotic expansions for bivariate von Mises functionals. *Z. Wahrsch. Verw. Gebiete* **50** 333–355.

[3] Nagaev, S. V. (1983). On the accuracy of normal approximation for distributions of sums of independent Hilbert space-valued random variables. In *Probability Theory and Mathematical Statistics. Fourth USSR-Japan Symposium Proceedings, 1982.* Lecture Notes in Mathematics Vol. 1021, pp. 461–473. Springer-Verlag, Berlin/Heidelberg/New York.

[4] Sazonov, V. V., Ulyanov, V. V., and Zalesskii, B. A. (1987). On normal approximation in Hilbert space. In *Probability Theory and Mathematical Statistics. Proceedings of the Fourth Vilnius Conference, 1985*, Vol. II, pp. 561–580. VNU Sci. Press, Utrecht.

[5] Sazonov, V. V., Ulyanov, V. V., and Zalesskii, B. A. (1988). Normal approximation in Hilbert space I, II, *Teor. Veroyatnost. i Primenen.* **33** 225–245, 508–521. [Russian]

[6] SENATOV, V. V. (1985). On the dependency of estimates of the convergence rate in the central limit theorem on the covariance operator of the summands. *Teor. Veroyatnost. i Primenen.* **30** 354–357. [Russian]

[7] SENATOV, V. V. (1985). Four examples of lower estimates in the multidimensional central limit theorem, *Teor. Veroyatnost. i Primenen.* **30** 750–758. [Russian].

[8] SAZONOV, V. V. (1981). Normal approximation—Some recent advances. In *Lecture Notes in Mathematics Vol. 879.* Springer–Verlag, Berlin/Heidelberg/New York.

[9] NAGAEV, S. V., AND CHEBOTAREV, V. I. (1978). On the estimates of the convergence rate in the central limit theorem for random l_2-valued vectors. In *Mathematical Analysis and Allied Problems in Mathematics*, pp. 153–182, Nauka, Novosibirsk. [Russian]

[10] PINELIS, I. F. (1980). Estimates of moments of infinite dimensional martingales. *Mat. Zamyetki* **27** 953–958. [Russian]

[11] PETROV, V. V. (1987). *Limit Theorems for Sums of Independent Random Variables.* Nauka, Moscow. [Russian]

[12] NAGAEV, S. V. (1985). On the rate of convergence to normal law in Hilbert space. *Teor. Veroyatnost. i Primenen.* **30** 19–32. [Russian]

The Estimation of the Bispectral Density Function and the Detection of Periodicities in a Signal

T. SUBBA RAO AND M. M. GABR

University of Manchester, Manchester, England

In a recent paper Subba Rao and Gabr (*J. Time Ser. Anal.* (1987), in press) considered the estimation of the spectrum and the inverse spectrum based on the method by Pisarenko (*Geophys. J. Roy. Astronom. Soc.* **28** (1972), 511–531). The asymptotic properties of these estimates were studied using the properties of Wishart matrices. In this paper we show how the method can be extended to the estimation of the bispectral density function, an important tool in the study of non-Gaussian time series. All these methods of estimation are illustrated with simulated examples. In the illustrations considered, the emphasis is on the detection of periodicities in the "signal" (possibly in the presence of noise). We also considered an example based on real data. These data arise in the study of the earth's magnetic reversals and the detection of periodicities.

1. INTRODUCTION

The second-order spectrum plays an important role in Gaussian time series analysis and in signal processing. In view of its importance several techniques have been proposed for estimating the spectral density function given sample data from a time series. The methods of estimation proposed so far can be grouped into two categories, viz. (i) nonparametric methods, (ii) parametric methods.

The parametric methods are based on model fitting (usually of the AR type) while the standard nonparametric method is based on "smoothing" the periodogram by a suitable weight function or a "spectral window" (see, e.g., [12]). There are, however, two special nonparametric approaches which have attracted considerable attention in the engineering literature. They are (a) Pisarenko's method [9, 10] and (b) Capon's method [2].

Multivariate Statistics and Probability
ISBN 0-12-580205-6

Reprinted from *J. Mult. Anal.* **27**(2).

The second-order spectra will not adequately characterise the series, (unless it is Gaussian) and hence there is a need for higher order spectral analysis. The simplest type of higher order spectral analysis is *bispectral analysis.* In recent years the bispectrum has been used in a number of investigations, for example, testing linearity [16] and deconvolution of seismic signals [6]. There are two widely used methods of estimation of the bispectrum and they are: (i) using fast Fourier transforms and (ii) smoothing the third-order periodogram (see [17]). However, in this study we concentrate on generalising Pisarenko's method to the bispectral case.

In Section 2, the spectral and bispectral properties of various models are discussed. The "truncated bispectrum" is defined in Section 3, and its estimation is considered in Section 4. This method of estimation is a generalisation of Pisarenko's method given for the estimation of the second-order spectrum [18]. The estimation of spectrum and bispectrum of simulated data is considered in Section 5. The detection of periodicities via the spectrum and bispectrum is considered in Section 6 and is illustrated with simulated examples. The methods are further illustrated with application to real data in Section 7.

2. Spectral and Bispectral Density Functions

Let $\{X(t)\}$ be a real-valued discrete parameter third-order stationary time series with $\mu = E(X(t))$, $R(s) = E(X(t)-\mu)\,(X(t+s)-\mu)$, $c(s_1, s_2) = E(X(t)-\mu)\,(X(t+s_1)-\mu)\,(X(t+s_2)-\mu)$. Since $X(t)$ is real valued we have the obvious symmetry relations,

$$R(s) = R(-s) \quad \text{and} \quad c(s_1, s_2) = c(s_2, s_1) = c(-s_1, s_2 - s_1) = c(s_1 - s_2, -s_2).$$

The spectral and the bispectral density functions are defined respectively by

$$\begin{aligned} h(w) &= \frac{1}{2\pi} \sum_{-\infty}^{\infty} R(s)\, e^{-is\omega}, \qquad |\omega| \leqslant \pi, \\ h(\omega_1, \omega_2) &= \frac{1}{(2\pi)^2} \sum_{\tau_1} \sum_{\tau_2} c(\tau_1, \tau_2)\, e^{-i\tau_1\omega_1 - i\tau_2\omega_2}, \qquad -\pi \leqslant \omega_1, \omega_2 \leqslant \pi. \end{aligned} \tag{2.1}$$

In view of the symmetry of the third-order covariances, we have

$$\begin{aligned} h(\omega_1, \omega_2) &= h(\omega_2, \omega_1) = h(-\omega_1, -\omega_1 - \omega_2) \\ &= h(-\omega_1 - \omega_2, \omega_2) = h^*(-\omega_1, -\omega_2) \end{aligned} \tag{2.2}$$

(where $h^*(\omega_1, \omega_2)$ denotes the complex conjugate of $h(\omega_1, \omega_2)$).

The bispectral density function $h(\omega_1, \omega_2)$ is usually complex and can sometimes be explicitly evaluated from a given model. For example, let $X(t) = \sum_0^\infty g(u)\, e(t-u)$, where $\{e(t)\}$ are mutually independent with $E(e(t)) = 0$, $E(e^2(t)) = \sigma_e^2$, $E(e^3(t)) = \mu_3$. Then the relations $h(\omega) = \sigma_e^2 (2\pi)^{-1}\, |H(\omega)|^2$, $h(\omega_1, \omega_2) = \mu_3 (2\pi)^{-2}\, H(-\omega_1 - \omega_2)\, H(\omega_1)\, H(\omega_2)$, where $H(\omega) = \sum_u g(u)\, e^{-iu\omega}$ can easily be obtained.

In many practical situations, $X(t)$ may correspond to a "signal," but one observes a contaminated version of the signal, say, $Z(t)$. Let us assume, for each t, we can write $Z(t) = X(t) + Y(t)$, where the "noise" $Y(t)$ is assumed to be a zero mean stationary (up to third order) process. Further, we assume that $X(t)$ and $Y(t)$ are independent. Then we have $h_Z(\omega) = h_X(\omega) + h_y(\omega)$, $h_Z(\omega_1, \omega_2) = h_X(\omega_1, \omega_2) + h_y(\omega_1, \omega_2)$. An important problem in signal processing is the estimation of the parameters (say, frequencies) of the signal $X(t)$ when we observe $\{Z(t)\}$. We notice from these relations the estimation depends heavily on the behaviour of $h_y(\omega)$ at the "natural frequencies" of $\{X(t)\}$, even if $\{Y(t)\}$ is Gaussian.

However, if $\{Y(t)\}$ is Gaussian (or has *any* symmetric distribution), then $h_Z(\omega_1, \omega_2) = h_X(\omega_1, \omega_2)$, for all ω_1 and ω_2. This shows that the evaluation (and estimation) of the bispectrum can be an extremely important part of signal processing, and we will illustrate this usefulness in later sections.

In an earlier paper [18], we considered the estimation of the "truncated spectral density function" and its relationship with the Pisarenko estimate. In the following section we define a "truncated bispectrum" and then consider its estimation.

3. Truncated Bispectrum

Let $(X(1), X(2), \ldots, X(n))$ be a sample from the series $\{X(t)\}$ and let $\bar{X} = (1/n) \sum X(t)$. We evaluate the finite Fourier transform, $J_x(\omega) = \sum (X(t) - \bar{X})\, e^{-it\omega}$, and the third-order periodogram $I_n(\omega_1, \omega_2, \omega_3)$ by

$$I_n(\omega_1, \omega_2, \omega_3) = \frac{1}{(2\pi)^2 n} J_x(\omega_1)\, J_x(\omega_2)\, J_x(\omega_3). \tag{3.1}$$

Then we can show that (provided $\omega_1 + \omega_2 + \omega_3 = 0 \pmod{2\pi}$)

$$\begin{aligned} &E(I_n(\omega_1, \omega_2, \omega_3)) \\ &\quad = \frac{1}{(2\pi)^2 n} \sum_{t_1} \sum_{t_2} \sum_{t_3} c(t_1 - t_3, t_2 - t_3)\, e^{-i(t_1 - t_3) w_1 - i(t_2 - t_3)\omega_2} \\ &\quad = h_n(\omega_1, \omega_2), \qquad \text{say.} \end{aligned} \tag{3.2}$$

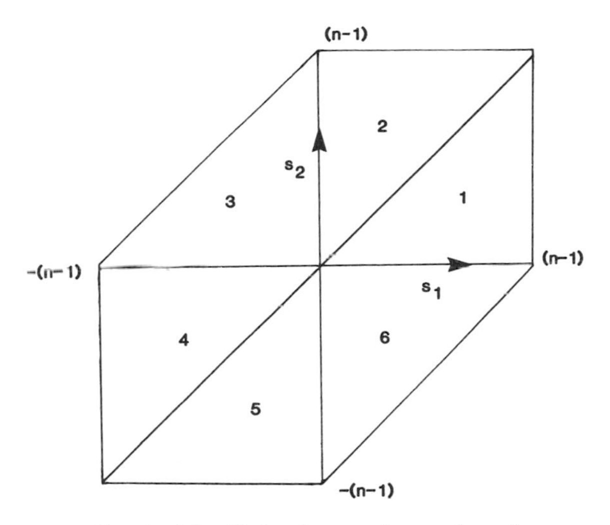

FIG. 1. The third-order covariances $c(s_1, s_2)$.

We will call $h_n(\omega_1, \omega_2)$ the *truncated bispectral density function*, and we now consider its estimation. (Note that $h_n(\omega_1, \omega_2)$ tends to $h(\omega_1, \omega_2)$ as $n \to \infty$).

In view of the symmetry relations, $c(\tau_1, \tau_2)$ is calculated only in one sector of Fig. 1, say sector (2). To simplify the triple summations in (3.2), we proceed as follows: Let

$$c^*(s_1, s_2) = \begin{cases} (n - s_1)\, c(s_1, s_2) & \text{if } 0 \leqslant s_2 \leqslant s_1 \leqslant n-1 \text{ (i.e.) sector (1)} \\ c^*(s_2, s_1) & \text{if } 0 \leqslant s_1 \leqslant s_2 \leqslant n-1 \text{ (i.e.) sector (2)} \\ c^*(s_2 - s_1, -s_1) & \text{if } (s_1, s_2) \text{ lies in sector (3)} \\ c^*(s_1 - s_2, -s_2) & \text{if } (s_1, s_2) \text{ lies in sector (4)} \\ c^*(-s_2, s_1 - s_2) & \text{if } (s_1, s_2) \text{ lies in sector (5)} \\ c^*(-s_1, s_2 - s_1) & \text{if } (s_1, s_2) \text{ lies in sector (6).} \end{cases}$$

We can now write (3.2) as

$$h_n(\omega_1, \omega_2) = \frac{1}{(2\pi)^2 n} \sum_{s_1=1}^{2n-1} \sum_{s_2=1}^{2n-1} c^*(s_1 - n, s_2 - n)\, e^{-i(s_1 - n)\,\omega_1 - i(s_2 - n)\,\omega_2}. \tag{3.3}$$

As in the case of a truncated spectrum (see [18]) we can write (3.3) in terms of eigenvalues and eigenvectors of a symmetric matrix $\mathbf{c}^*$ of order $(2n-1)\times(2n-1)$, given by

$$\mathbf{c}^* = \begin{bmatrix} c^*(n-1,n-1) & c^*(n-1,n-2) & \cdots & c^*(n-1,0) & 0 \quad 0 & \cdots & 0 \\ c^*(n-2,n-1) & c^*(n-2,n-2) & \cdots & c^*(n-2,0) & c^*(n-2,-1) & \cdots & 0 \\ c^*(0,n-1) & c^*(0,n-2) & \cdots & c^*(0,0) & c^*(0,-1) & \cdots & c^*(0,-n+1) \\ 0 & 0 & & & & & \\ 0 & 0 & \cdots & c^*(-n+1,0) & c^*(-n+1,-1) & \cdots & c^*(-n+1,-n+1) \end{bmatrix}.$$

Let $\{\mu_{n,j},\ j=-(n-1), ..., 0, 1, ..., (n-1)\}$ be the eigenvalues of $\mathbf{c}^*$ and $\mathbf{A}_{n,-(n-1)}, \mathbf{A}_{n,-(n-2)}, ..., \mathbf{A}_{n,(n-1)}$ the corresponding normalised eigenvectors. Since $\mathbf{c}^*$ is symmetric, we have $\mathbf{c}^* = \sum_{j=-(n-1)}^{n-1} \mu_{n,j}\, \mathbf{A}_{n,j}\, \mathbf{A}'_{n,j}$, where $\mathbf{A}'_{n,j} = (a_j(-n+1), a_j(-n+2), ..., a_j(n-1))$. Hence we obtain

$$h_n(\omega_1, \omega_2) = \frac{1}{(2\pi)^2 n} \sum_{j=-(n-1)}^{n-1} \mu_{n,j}\, \mathbf{A}^*_{n,j}(\omega_1)\, A^*_{n,j}(\omega_2), \tag{3.4}$$

where $A^*_{n,j}(\omega) = \sum_{s=-(n-1)}^{n-1} a_j(s)\, e^{is\omega}$.

At this stage it may be noted that the eigenvalues $\mu_{n,j}$ cannot be in any way related to $h_n(\omega_1, \omega_2)$, unlike the case of spectral density functions [18]. However, there is an advantage in writing (3.3) in terms of the eigenvalues and vectors, and when we consider the estimation of (3.4), the reason for doing so will become clear.

4. Estimation of the Truncated Bispectral Density Function $h_n(\omega_1, \omega_2)$

Given a sample $(X(1), X(2), ..., X(n))$ from $\{X(t)\}$, let $n = Mk$, where M and k are integers. Divide the data into M groups, where each group consists of k observations. Let the observations in the lth group $(l = 1, 2, ..., M)$ be denoted by the vector $\mathbf{X}_l$, where $\mathbf{X}_l = (X((l-1)k+1), X((l-1)k+2), ..., X(lk))$ $(l = 1, 2, ..., M)$. Let $\bar{X}_j = (1/M)\sum_{l=1}^{M} X_l(j)$, where $X_l(j) = X((l-1)k+j)$,

$$\hat{c}_j(s_1, s_2) = \frac{1}{M}\sum_{l=1}^{M} (X_l(j) - \bar{X}_j)\,(X_l(j+s_1) - \bar{X}_{j+s_1})\,(X_l(j+s_2) - \bar{X}_{j+s_2}),$$

$$\hat{c}^*(s_1, s_2) = (1/k)\sum_{j=1}^{k-|\tau|} \hat{c}_j(s_1, s_2),\ (s_1 = 0, \pm 1, \pm 2, ..., \pm(k-1),$$

$$s_2 = 0, \pm 1, +2, ..., \pm(k-1)),$$

where $\tau = \max(s_1, s_2)$. We now define a symmetric matrix $\hat{\mathbf{c}}^*$ of order $(2k-1) \times (2k-1)$ similar to $\mathbf{c}^*$ where, for example, we replace $c^*(n-1, n-1)$ by $\hat{c}^*(k-1, k-1)$, etc. Let $\{\hat{\mu}_{k,j}, j=0, \pm 1, ..., \pm(k-1)\}$, $\{\hat{A}_{k,j}, j=0, \pm 1, ..., \pm(k-1)\}$ be the eigenvalues and normalised eigenvectors of $\hat{\mathbf{c}}^*$. Consider the estimate

$$\hat{h}_k(\omega_1, \omega_2) = \frac{1}{(2\pi)^2 k} \sum_{j=-(k-1)}^{k-1} \hat{\mu}_{k,j} \hat{A}^*_{k,j}(\omega_1) \hat{A}^*_{k,j}(\omega_2), \tag{4.1}$$

where $\hat{\mathbf{A}}^*_{k,j}(\omega) = \sum_{s=-(k-1)}^{k-1} \hat{a}_{k,j}(s) e^{-is\omega}$.

In order to study the asymptotic sampling properties of the estimate (4.1), we need to know the sampling properties of $\{\hat{\mu}_{k,j}\}$, and $\{\hat{a}_{k,j}(s)\}$; and at present these are not known (since $\hat{\mathbf{c}}^*$ is not a Wishart matrix). However, it is reasonable to conjecture that for fixed k and as $M \to \infty$, $\hat{h}_k(\omega_1, \omega_2)$ will be a consistent estimate of $h_k(\omega_1, \omega_2)$. We now discuss the advantages of using the expression (4.1) for estimating $h_n(\omega_1, \omega_2)$.

The choice of k, in relation to n, is quite important, and in a way is similar to the choice of the truncation point in the estimation of spectral density functions. One way of choosing k is to plot $\hat{c}^*(s_1, s_2)$ against s_1, s_2, and see whether $\hat{c}^*(k, k)$, where $k = \max(s_1, s_2)$, decays to zero beyond some value, k_0, say. If it does, we can choose $k = k_0$. This is consistent with the assumption that $\sum \sum |c(s_1, s_2)| < \infty$. Though in theory it is possible to find k_0 in this way, we see that k_0 must be found from a 3-dimensional plot and this can be quite difficult. This is where the representation (4.1) in terms of the eigenvalues $\{\mu_{k,j}\}$ can be extremely useful. Since the eigenvalues contain most of the information contained in the matrix $\hat{\mathbf{c}}^*$, an examination of $\{\hat{\mu}_{k,j}\}$ for some values of k, will clearly indicate the choice of k_0. Besides, the modulus of the bispectral estimate computed from (4.1) is usually very smooth.

5. Numerical Illustrations

In the following section we illustrate the methods of estimation of spectrum and bispectrum. The theoretical forms of the estimates are given below (for details see [18]).

Let $(X(1), X(2), ..., X(n))$ be a sample from the zero mean third-order stationary time series $\{X(t)\}$. Let $R(t, s) = \text{cov}(X(t), X(s))$. Define the Toeplitz matrix $\mathbf{R}_n$ of order $n \times n$, where the element corresponding to the tth row, sth column $(t, s = 1, 2, ..., n)$ is $R(t, s)$. Let $\lambda_{n,0}, \lambda_{n,1}, ..., \lambda_{n,n-1}$ be the eigenvalues of $\mathbf{R}_n$ and let $\mathbf{b}_{n,0}, \mathbf{b}_{n,1}, ..., \mathbf{b}_{n,n-1}$ be the corresponding normalised eigenvectors. Further, let $\mathbf{b}_{n,j} = (\mathbf{b}_{n,j}(0), b_{n,j}(1), \ldots, b_{n,j}(n-1))$ $(j=0, 1, \ldots, n-1)$. We define the truncated spectral density $h_n(\omega)$ and the theoretical form of the Capon's estimator, as $h_{n,\,\text{cap}}(\omega)$. They are

$$h_n(\omega) = \frac{1}{4\pi} \sum_{j=0}^{n-1} \lambda_{n,j} B_{n,j}(\omega), \tag{5.1}$$

$$h_{n,\,\text{cap}}(\omega) = \frac{1}{\pi} \left[\sum_{j=0}^{n-1} \lambda_{n,j}^{-1} B_{n,j}(\omega) \right]^{-1}, \tag{5.2}$$

where $B_{n,j}(\omega) = (2/n) | \sum_{t=0}^{n-1} b_{n,j}(t) e^{it\omega} |^2$. We proceed as in Section 4 and form the sample variance–covariance matrix $\hat{\mathbf{R}}_k = M^{-1} \sum_{j=1}^{M} \mathbf{X}_j \mathbf{X}_j$. Let $\hat{\lambda}_{k,j}$ $(j=0, 1, 2, \ldots, k-1)$ be the eigenvalues of the matrix $\hat{\mathbf{R}}_k$ and let the corresponding normalised eigenvectors be $\hat{\mathbf{b}}_{k,j}$ $(j=0, 1, 2, \ldots, k-1)$, where $\hat{\mathbf{b}}_{k,j} = (\hat{b}_{k,j}(0), \hat{b}_{k,j}(1), \ldots, \hat{b}_{k,j}(k-1))$. Then the estimates of $h_n(\omega)$, $h_{n,\,\text{cap}}(\omega)$ are obtained by $\hat{h}_k(\omega)$, $\hat{h}_{k,\,\text{cap}}(\omega)$, respectively. These estimates are

$$\hat{h}_k(\omega) = \frac{1}{4\pi} \sum_{j=0}^{k-1} \hat{\lambda}_{k,j} \hat{B}_{k,j}(\omega), \tag{5.3}$$

$$\hat{h}_{k,\,\text{cap}}(\omega) = \frac{1}{\pi} \left[\sum_{j=0}^{k-1} \hat{\lambda}_{k,j}^{-1} \hat{B}_{k,j}(\omega) \right]^{-1}, \tag{5.4}$$

where $\hat{B}_{k,j}(\omega) = (2/k) | \sum_{t=0}^{k-1} \hat{b}_{k,j}(t) e^{it\omega} |^2$. The examples considered for illustration are as follows:

EXAMPLE 1. Let the time series $\{X(t)\}$ satisfy the equation

$$X(t) - 0.4X(t-1) + 0.7X(t-2) = e(t), \tag{5.5}$$

where $\{e(t)\}$ are independent, identically distributed normal variables with mean zero and variance unity. The theoretical spectral density function $h(\omega)$, for the above model is given by $h(\omega) = (2\pi)^{-1} | 1 - 0.4e^{-i\omega} + 0.7e^{-2i\omega} |^{-2}$. The spectrum $h(\omega)$ has a maximum at $\omega = 0.4\pi$.

Two time series of lengths $n = 2800$ and $n = 3600$ are generated from the model (5.5). The above estimates (at the frequencies $\omega_j = j\pi$, $j = 0(0.1)\,1$) are computed in three cases. They are: (i) $n = 2800$, $k = 20$, $M = 140$; (ii) $n = 3360$, $k = 24$, $M = 140$; and (iii) $n = 3360$, $k = 28$, $M = 120$. The graphs of the theoretical spectrum $h(\omega)$ and the estimates $\hat{h}_k(\omega)$ and $\hat{h}_{k,\,\text{cap}}(\omega)$ for

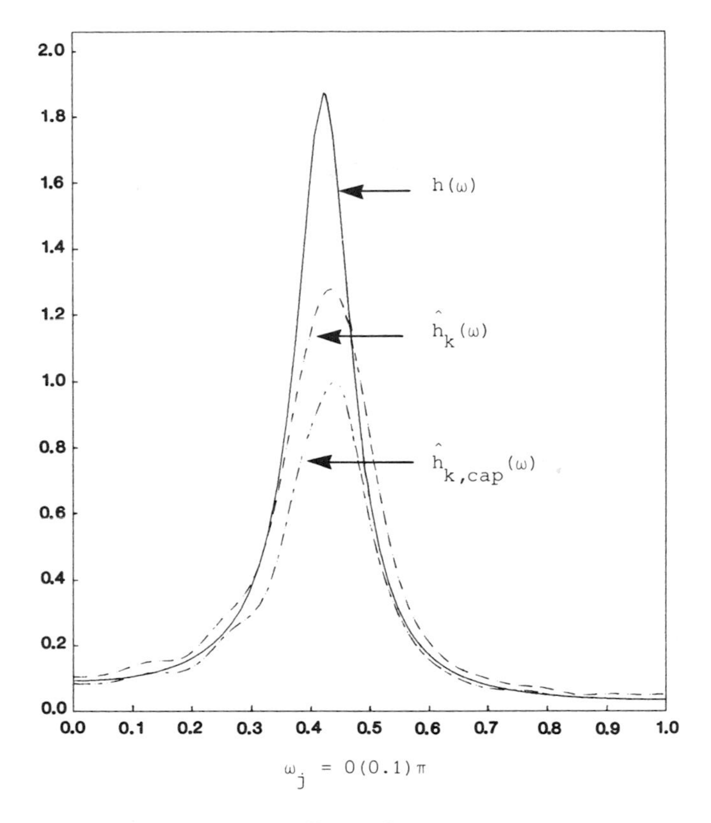

FIGURE 2

case (i) are given in Fig. 2; for case (ii) in Fig. 3; for case (iii) in Fig. 4. In each of these cases, there is a clear peak at $\omega = 0.4\pi$ in the estimated spectrum. From the simulations we have performed we found that Capon's high resolution estimate is good when the ratio k/M is small. Since the series $\{X(t)\}$ is Gaussian, the bispectral density function is zero.

EXAMPLE 2. We now consider a time series $\{X(t)\}$ generated from the model

$$X(t) - 0.4X(t-1) + 0.7X(t-2) = \eta(t), \tag{5.6}$$

$\eta(t) = e^2(t) - 1$, where $\{e(t)\}$ are as in Example 1. Since the variance of $\eta(t)$ is 2, the spectral density function of $X(t)$ is $2h(\omega)$, where $h(\omega)$ is given in

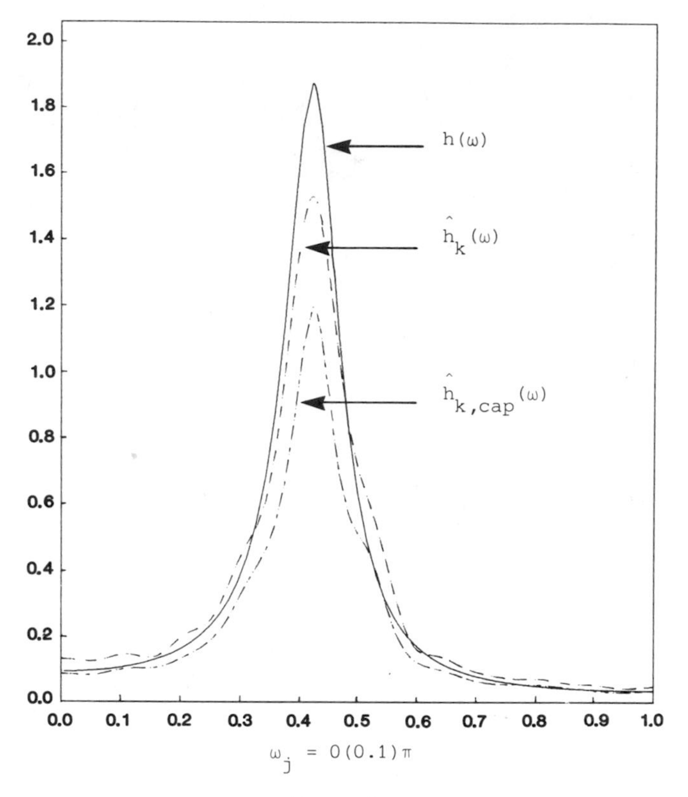

FIGURE 3

Example 1. In this case $X(t)$ is non-Gaussian (though linear). The bispectral density function is given by

$$h(\omega_1, \omega_2) = \frac{8}{(2\pi)^2} G(e^{-i\omega_1}) G(e^{-i\omega_2}) G(e^{+i(\omega_1+\omega_2)}), \tag{5.7}$$

where $G(e^{-i\omega}) = (1 - 0.4e^{-i\omega} + 0.7e^{-2i\omega})^{-1}$. The modulus of $h(\omega_1, \omega_2)$ is given in Fig. 5, and the estimated modulus is given in Fig. 6 (here $n = 2000$, $k = 20$, $M = 100$). We see clear peaks at $\omega_1 = 0$, $\omega_2 = 0.4\pi$; $\omega_1 = 0.4\pi$, $\omega_2 = 0$; and $\omega_1 = \omega_2 = 0.4\pi$ confirming that the bispectrum can be used to detect pseudo periods the time series may have.

EXAMPLE 3. A nonlinear (and non-Gaussian) time series $\{X(t)\}$ is generated from the model $X(t) = 0.7X(t-4)\, e(t-4) + e(t)$. The theoretical spectrum, the estimated spectrum and the high resolution estimate are

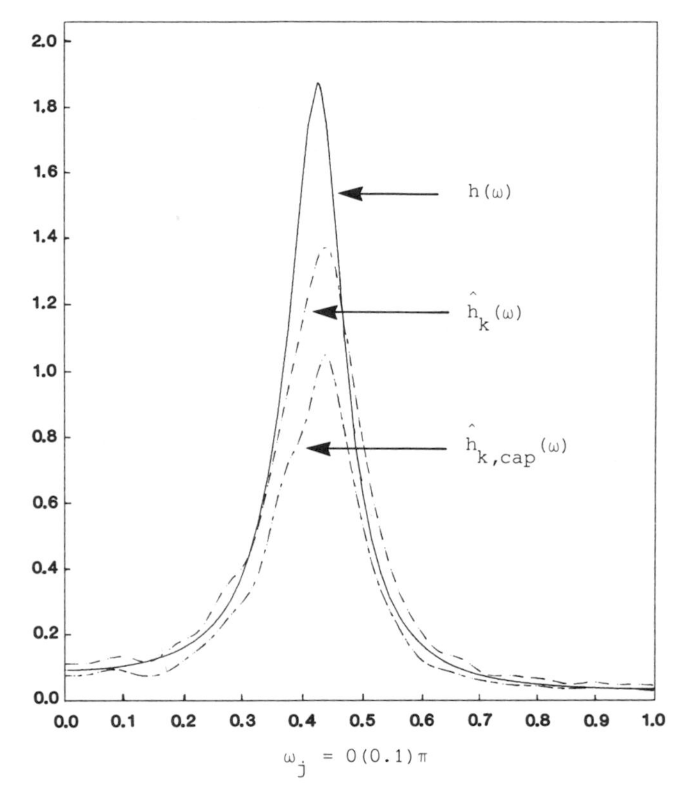

FIGURE 4

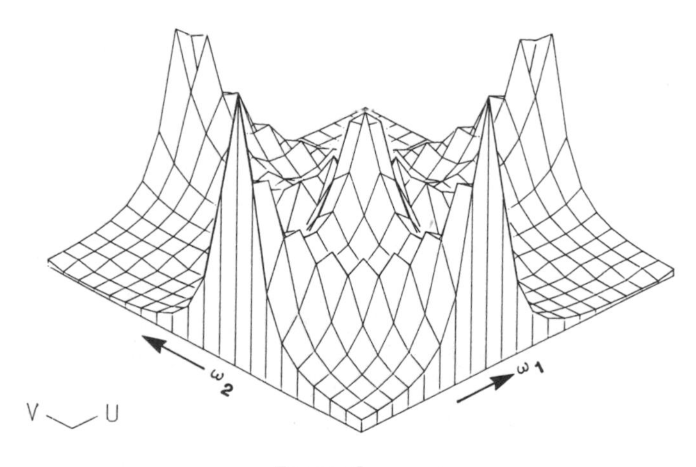

FIGURE 5

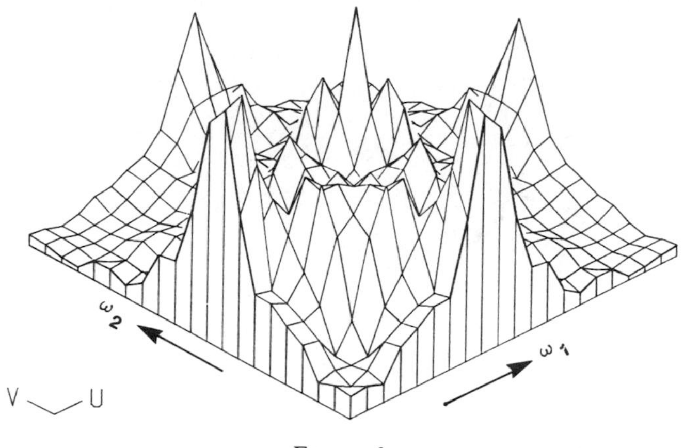

FIGURE 6

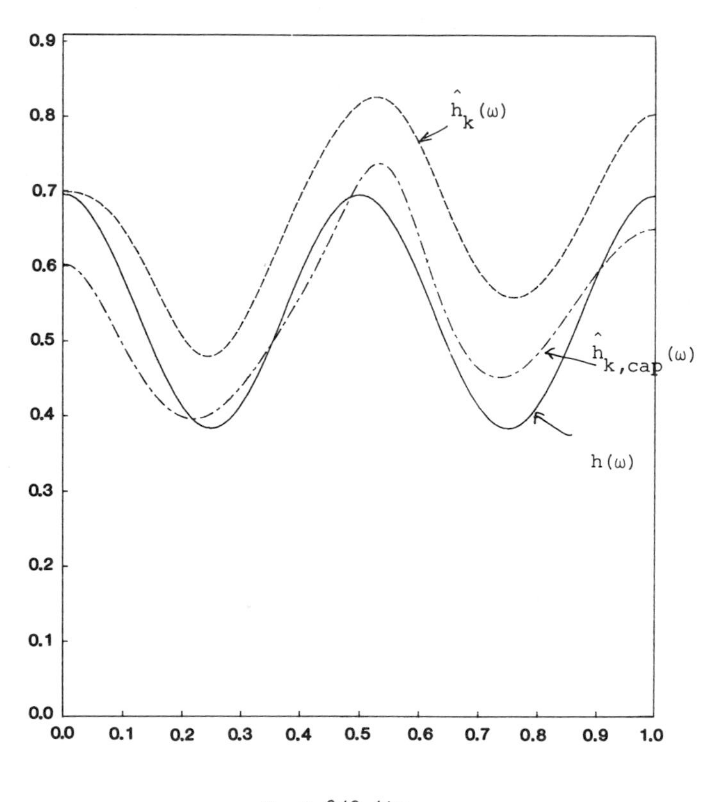

FIGURE 7

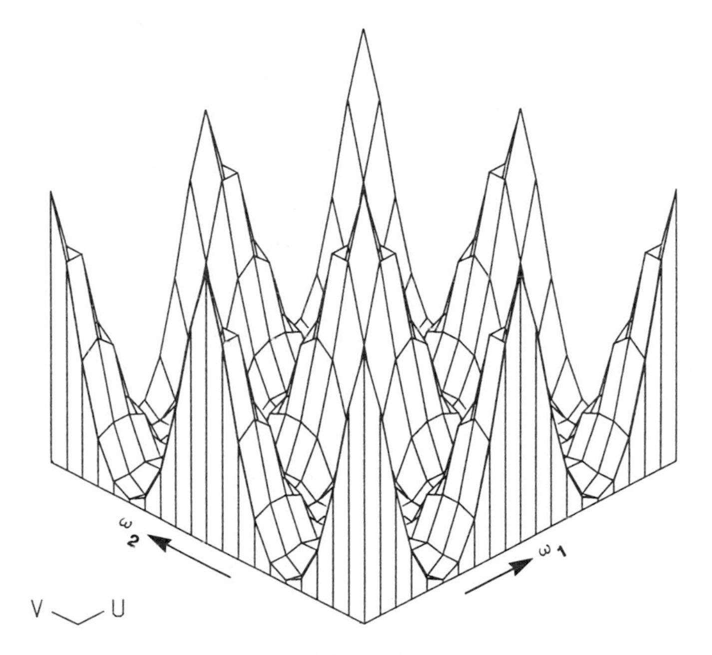

FIGURE 8

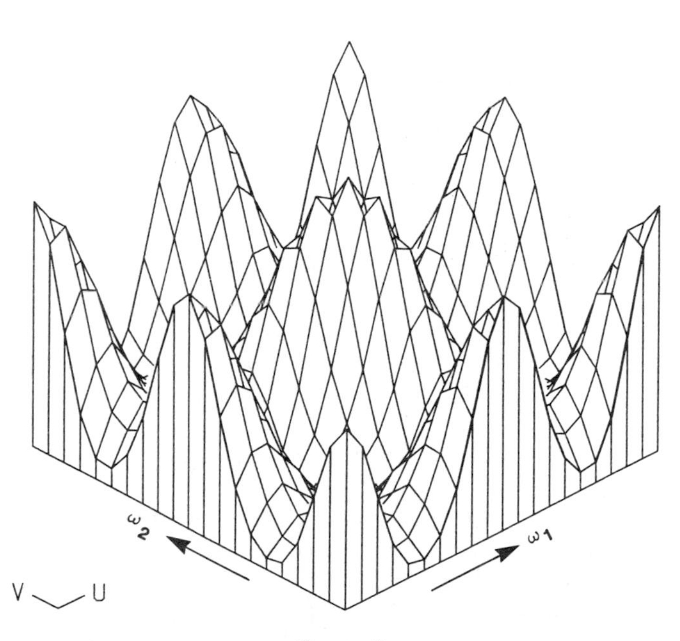

FIGURE 9

given in Fig. 7; and the theoretical (modulus) bispectrum and its estimate (where $n = 1000$, $k = 10$, $M = 100$) are given in Figs. 8 and 9, respectively. It is interesting to observe that the estimates reproduced the periodic features of the spectrum and bispectrum.

6. The Retrieval of Harmonics via Spectrum and Bispectrum

Parametric estimates of spectrum, such as AR estimates and ARMA estimates, are widely used to detect periodicities of signals which are represented by harmonic processes. To show how AR spectral estimates can be used to detect the periodicity, let $X(t) = A\,\mathrm{Sin}(\omega t + \psi)$. Since $\mathrm{Sin}(\omega t + \psi) = 2\cos\omega\,\mathrm{Sin}(\omega(t-1) + \psi) - \mathrm{Sin}(\omega(t-2) + \psi)$, we have the difference equation $X(t) - 2\cos\omega\, X(t-1) + X(t-2) = 0$. In other words, harmonic process with a single freqency ω can be written as an AR(2) process with the input term $e(t)$ identically zero. The associated characteristic polynomial $Z^2 - 2\cos\omega Z + 1$, has roots $Z = e^{i\omega}, e^{-i\omega}$. Therefore, if the AR(2) spectrum is computed for the series $\{X(t)\}$, one observes a peak at the frequency ω corresponding to the frequency of the signal $\{X(t)\}$. This extends to the case of several harmonic terms (see Chan, Lavoie, and Plant [3]). Let $X(t) = \sum_{j=1}^{m} A_j\,\mathrm{Sin}(\omega_j t + \psi_j)$; then $X(t)$ satisfies the equation $X(t) = \sum_{j=1}^{2m} a_j X(t-j)$. Suppose now that instead of observing the signal $\{X(t)\}$, one observes a contaminated version, say, $Z(t) = X(t) + Y(t)$, where $Y(t)$ is noise (see Section 2). Then the above model for $X(t)$ can be written as an ARMA $(2m, 2m)$ of the form

$$Z(t) - \sum_{j=1}^{2m} a_j Z(t-j) = Y(t) - \sum_{j=1}^{2m} a_j Y(t-j). \tag{6.1}$$

Therefore, if one wants to extract the harmonics of a signal contaminated by noise, an ARMA spectrum has to be computed, and not just an AR spectrum. Alternatively, as shown by Ulrych and Clayton [19] (see also Subba Rao, [15]), one can perform principal component analysis on the variance matrix of $(Z(t), Z(t-1), \ldots, Z(t-2m))$ and obtain the estimate of the variance of $Y(t)$ and the parameters $(a_1 . a_2, \ldots, a_{2m})$, and these in turn give the harmonic components because they correspond to the roots of the polynomial $Z^{2m} - a_1 Z^{2m-1} - a_2 Z^{2m-2} \cdots - a_{2m}$ (see [5]). As shown by Ulrych and Clayton [19], this is the basis of Pisarenko's algorithm (Pisarenko, [10]) for estimating the parameters of the signal $X(t)$, when $X(t)$ satisfies an AR model and the noise $\{Y(t)\}$ is independent of $X(t)$. The method proposed by Subba Rao [15], was based on canonical factor analysis, which here reduces to principal component analysis.

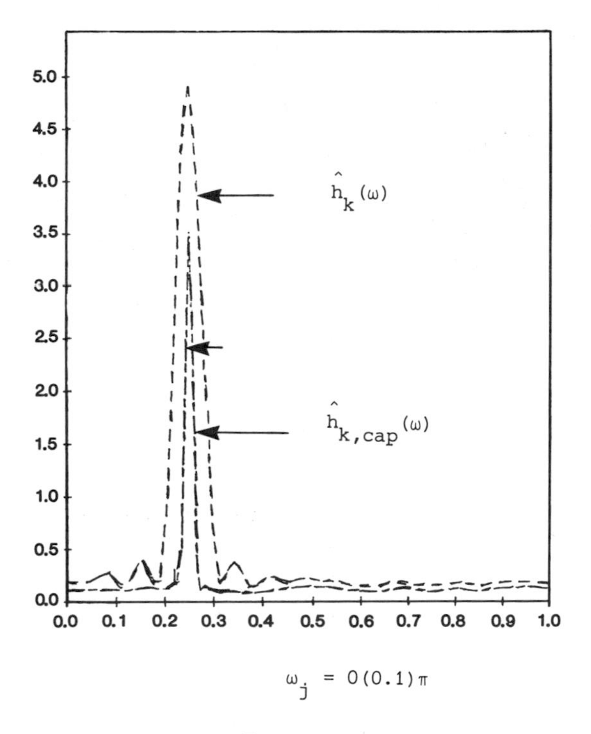

FIGURE 10

The above approaches depend on spectral analysis (or covariance analysis) for detecting the periodicities of the signal in the presence of noise. As pointed out in Section 2, an alternative would be to estimate the bispectrum of $\{Z(t)\}$ and this would be equal to the bispectrum of the signal (assuming the signal is non-Gaussian and the noise is Gaussian) and the following examples illustrate clearly its usefulness.

EXAMPLE 4. A time series $\{X(t)\}$ is generated from the model $X(t) = 2\ \mathrm{Sin}(0.25\pi)t + e(t)$ $(t = 1, 2, \ldots, n)$, where $\{e(t)\}$ are as before. $n = 3000$, $k = 30$, $M = 100$, $\hat{h}_k(\omega)$, $\hat{h}_{k,\mathrm{cap}}(\omega)$, and the bispectral modulus are calculated. The spectral estimates are given in Fig. 10 and the bispectral modulus is given in Fig. 11. In the spectral estimate there is a clear peak at $\omega = 0.25\pi$ and in the bispectrum at $\omega_1 = \omega_2 = 0.25\pi$.

EXAMPLE 5. A time series $\{X(t)\}$ is generated from the model

$$X(t) = 4\ \mathrm{Sin}(0.15\pi)t + 4\ \mathrm{Sin}(0.35\pi)t + e(t).$$

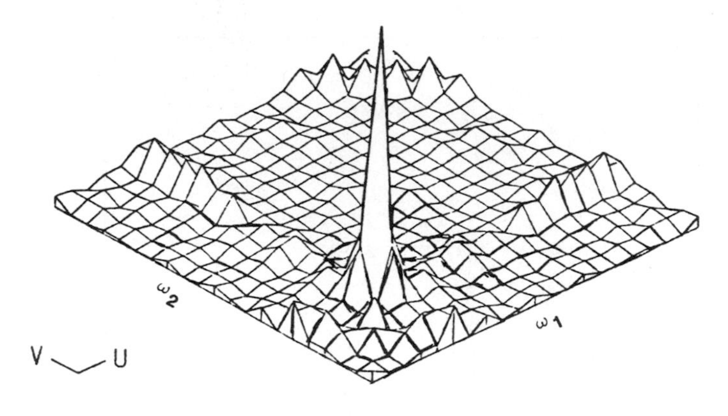

FIGURE 11

For the estimation of the spectrum, we have chosen $n = 5000$, $k = 50$, $M = 100$, and for the bispectrum, $n = 3000$, $k = 30$, $M = 100$. The graphs of the spectral estimates and the modulus of the bispectral estimate are given in Figs. 12 and 13, respectively. The peaks at $\omega_1 = 0.15\pi$ and $\omega_2 = 0.35\pi$ in

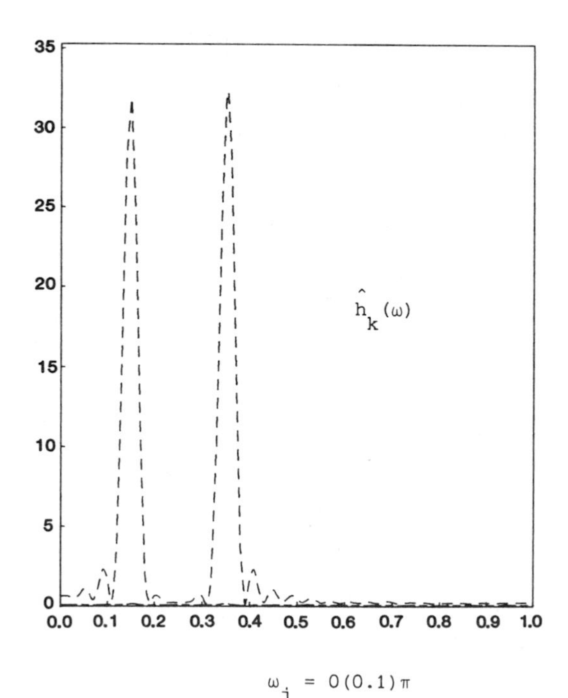

FIGURE 12

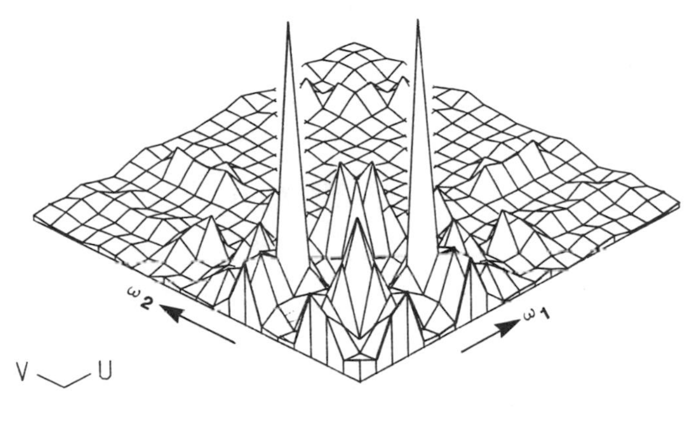

FIGURE 13

$\hat{h}_k(\omega)$ stood out clearly, but it is not the case in $\hat{h}_{k,\,\mathrm{cap}}(\omega)$. In the bispectrum there are clear peaks at $\omega_1=\omega_2=0.15\pi$, $\omega_1=0.15\pi$, $\omega_2=0.35\pi$, and $\omega_1=\omega_2=0.35\pi$. At other frequencies, the values of the modulus are very small.

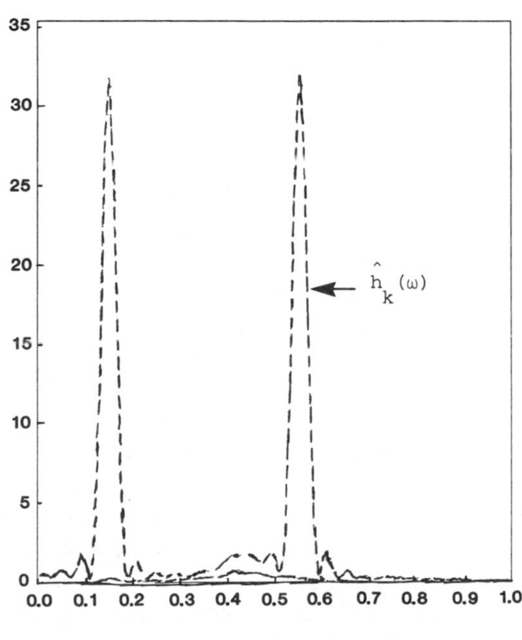

FIGURE 14

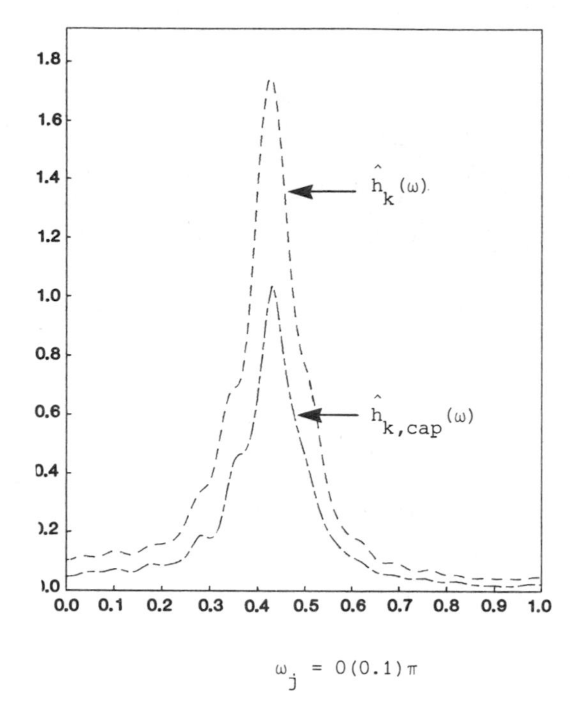

FIGURE 15

EXAMPLE 6. As our final illustration, a time series $\{Z(t)\}$ is generated from the model $Z(t) = 4\,\mathrm{Sin}(0.15\pi)t + 4\,\mathrm{Sin}(0.55\pi)\,t + Y(t)$, where $Y(t)$ is a coloured Gaussian noise generated from the model $Y(t) - 0.4Y(t-1) + 0.7Y(t-2) = e(t)$, where $\{e(t)\}$ is defined earlier. We note that the spectrum of $\{Y(t)\}$ has a peak at $\omega = 0.4\pi$, which is in between the frequencies of the signal $X(t)$, and this complicates the "identification" procedure. The estimates $\hat{h}_k(\omega)$ and $\hat{h}_{k,\,\mathrm{cap}}(\omega)$ are calculated using: (i) $n = 5000$, $k = 50$, $M = 100$; (ii) $n = 4000$, $k = 40$, $M = 100$. The graphs of these are given in Figs. 14 and 15. When $k = 50$ and $M = 100$, there are clear peaks in $\hat{h}_k(\omega)$ (see Fig. 14) at the frequencies $\omega_1 = 0.15\pi$ and $\omega_2 = 0.55\pi$, which are frequencies of the signal. When $k = 40$ and $M = 100$ there are no visible peaks (see Fig. 15) at these frequencies; instead, we observe a peak at $\omega = 0.4\pi$ which corresponds to "pseudo" periodicity of the noise. In order to understand why this happened, we note that $R_Z(s) = R_X(s) + R_y(s)$, and $R_y(s) \to 0$ as $|s| \to \infty$. Therefore, unless we include terms of very high-order lagged autocovariances, the periodicity of the signal will not be visible in the estimate. This is in fact similar to the observation made by Priestley [11], in his analysis of mixed spectra and the construction of his $P(\lambda)$ test.

FIGURE 16

Let us now look at the bispectral estimate. The bispectral estimate of $\{Z(t)\}$ is estimated using $n=3000$, $k=30$, and $M=100$, and the modulus is plotted in Fig. 16. We see clear peaks at $\omega_1=\omega_2=0.15\pi$ and $\omega_1=\omega_2=0.55\pi$, and smaller peaks at $\omega_1=0.15\pi$ and $\omega_2=0.55\pi$. This example clearly demonstrates the usefulness of evaluating the bispectrum, in addition to the spectrum.

7. The Periodicity of the Earth's Magnetic Reversals

We now illustrate the above methods of estimation with a real example which has received considerable attention in geophysics literature. The problem is to detect the periodicity in the earth's magnetic reversals. The theoretical results postulate long term periodicity in magnetic stratigraphy with reversal periods of 285, 114, 64, 47, and 34 million years. Recently several authors [19, 8, 13, 7] have analysed this data. Negi and Tiwari [8], have come to the conclusion that the spectral peaks at around 285, 114, 64, 47, and 34 million years seem to be very significant. However, the data sets analysed by various authors seem to be different. Stothers [14] considered the 296 magnetic reversals over the past 165 million years, the dates (intervals) of these reversals are given by Harland [4]. The data analysed by Stothers [14] corresponds to the number of reversals over 4 million year intervals. For our illustration we considered the number of reversals during the first 124 million years as given by Harland [4]; the data corresponding to the number of reversals over 2 million year intervals. Thus we have 62 observations, the spacing between observations being 2 million years. The spectrum and the bispectrum are estimated using $k=6$

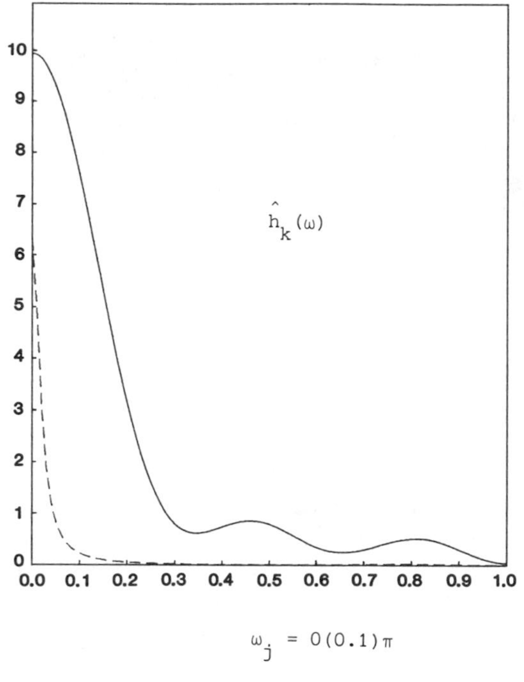

FIGURE 17

and $M = 10$, the spectrum is plotted in Fig. 17 and the values of the modulus of the bispectrum are given in Table I. No clear peaks in the low frequency of the spectrum are seen, but we observe two small peaks at the frequencies $\omega = 0.45\pi$ and $\omega = 0.8\pi$. The peak of $\omega = 0.45\pi$ corresponds to approximately 9m years, and the peak of $\omega = 0.8\pi$ correspond to 5m years. It is instructive to examine the values of the bispectrum (modulus) given in Table I. We see that the values are very large in the low frequency range, conforming that this might be due to a long periodicity. The value at $\omega_1 = \omega_2 = 0.05\pi$ is very significant. Though this does not correspond to a peak, we see that there is a sudden drop in magnitude at the next frequency. This frequency, $\omega_1 = \omega_2 = 0.05\pi$, corresponds to 80 million years, and Negi and Tiwari [8] pointed out that this may correspond to the variational period of the sun being perpendicular to the galactic plane which is 85m years. There are other peaks at $\omega_1 = 0$, $\omega_2 = 0.45\pi$ corresponding to, approximately, 9 million years and the peak at $\omega_1 = 0$, $\omega_2 = 0.8\pi$ corresponds to 5 million years. These peaks are also observed by Stothers [14] and others. The data set we have analysed is not large enough to draw any valid conclusions. However, the above preliminary bispectral analysis clearly shows that the above data is non-Gaussian.

TABLE I

Values of $|h(\omega_1, \omega_2)| \to j(\omega_j = j\pi)$

ω_j \ ω_j	0.00	0.05	0.10	0.15	0.20	0.25	0.30	0.35	0.40	0.45	0.50	0.55	0.60	0.65
1.00	0.094													
0.95	0.323	0.138	0.212											
0.90	0.861	0.494	0.181	0.270	0.259									
0.85	1.364	1.025	0.519	0.165	0.192	0.054	0.250							
0.80	1.524	1.362	0.907	0.391	0.096	0.032	0.208	0.515	0.685					
0.75	1.283	1.310	1.027	0.578	0.185	0.012	0.111	0.362	0.577	0.611	0.441			
0.70	0.875	0.960	0.830	0.538	0.220	0.019	0.044	0.157	0.326	0.410	0.347	0.174	0.033	
0.65	0.676	0.635	0.520	0.347	0.168	0.055	0.051	0.051	0.102	0.151	0.143	0.072	0.023	0.090
0.60	0.937	0.654	0.394	0.252	0.181	0.101	0.032	0.020	0.019	0.083	0.127	0.120	0.069	
0.55	1.583	1.108	0.586	0.284	0.223	0.124	0.126	0.201	0.150	0.032	0.157	0.201		
0.50	2.234	1.775	1.080	0.483	0.225	0.127	0.229	0.412	0.425	0.251	0.051			
0.45	2.446	2.222	1.585	0.859	0.359	0.165	0.267	0.520	0.637	0.522				
0.40	2.049	2.127	1.746	1.122	0.560	0.242	0.217	0.456	0.649					
0.35	1.376	1.576	1.484	1.135	0.712	0.374	0.169	0.264						
0.30	1.201	1.145	1.200	1.147	0.941	0.644	0.348							
0.25	2.382	1.504	1.266	1.382	1.313	1.012								
0.20	5.371	3.374	1.851	1.516	1.564									
0.15	9.856	6.983	3.998	1.997										
0.10	14.743	11.607	7.596											
0.05	18.5391	15.928												
0.00	19.970													

ACKNOWLEDGMENT

We are very thankful to Professor M. B. Priestley for going through the manuscript and making many helpful suggestions.

REFERENCES

[1] BRILLINGER, D. R., AND ROSENBLATT, M. (1967). Asymptotic theory of estimates of kth order spectra. *Spectral Analysis of Time Series* (B. Harris, Ed.), pp. 153–188. Wiley, New York.

[2] CAPON, J. (1969). High resolution frequency—Wave number spectrum analysis. *Proc. IEEE* **57** 1408–1418.

[3] CHAN, Y. Y., LAVOIE, J. M. M., AND PLANT, J. B. (1981). A parametric estimation approach to estimation of frequencies of sinusoids. *IEEE Trans. Acoust. Speech Signal Process.* **ASSP-29**, No. 2. 214–219.

[4] HARLAND, W. B. (1982). *A Geological Time Scale*. Caambridge Univ. Press, London.

[5] KAY, S. M., AND MARPLE, S. L. (1981). Spectrum analysis: A modern perspective. *Proc. IEEE* **69**, No. 11 1380–1419.

[6] LII, K. S., AND ROSENBLATT, M. (1982). Deconvolution and estimation of transfer function, phase and coefficients for non Gaussian linear processes. *Ann. Statist.* **10** 1195–1208.

[7] LUTZ, T. M. (1985). The magnetic reversal record is not periodic. *Nature* **317** 404–407.

[8] NEGI, J. G., AND TIWARI, R. K. (1983). Matching long term periodicities of geomagnetic reversals and gelactic motions of the solar system. *Geophys. Res. Lett.* **10**, No. 8 713–716.

[9] PISARENKO, V. F. (1972). On the estimation of spectra by means of nonlinear functions of the covariance matrix. *Geophys. J. Roy. Astronom. Soc.* **28** 511–531.

[10] PISARENKO, V. F. (1973). The retrieval of harmonics from a covariance function. *Geophys. J. Roy. Astronom. Soc.* **33** 347–366.

[11] PRIESTLEY, M. B. (1962). The analysis of stationary processes with mixed spectra, I. *J. Roy. Statist. Soc. B* **24** 215–233.

[12] PRIESTLEY, M. B. (1981). *Spectral Analysis and Time Series*. Academic Press, New York/London.

[13] RAUP, D. M. (1985). Rise and fall of periodicity. *Nature* **317** 384–385.

[14] STOTHERS, R. B. (1986). Periodicity of the earth's magnetic reversals. *Nature* **318** 444–446.

[15] SUBRA RAO, T. (1976). Canonical factor analysis and stationary time series models. *Sankyā Ser. B* **38** 256–271.

[16] SUBBA RAO, T., AND GABR, M. M. (1980). A test for linearity of stationary time series. *J. Time Ser. Anal.* **1** 145–158.

[17] SUBBA RAO, T., AND GABR, M. M. (1984). An introduction to bispectral analysis and bilinear time series models. *Lecture notes in Statistics* Vol. 24. Springer-Verlag, New York/Berlin.

[18] SUBBA RAO, T., AND GABR, M. M. (1987). Estimation of spectrum, inverse spectrum and bispectrum of a stationary time series. *J. Time Ser. Anal.*, in press.

[19] ULRYCH, T. J., AND CLAYTON, R. W. (1976). Time series modelling and maximum entropy. *Phys. Earth Planetary Interiors* **12** 188–200.

Analysis of Odds Ratios in $2 \times n$ Ordinal Contingency Tables*

K. SUBRAMANYAM

University of Pittsburgh

AND

M. BHASKARA RAO

North Dakota State University

The set of all bivariate probability distributions with support contained in $\{(i, j);\ i=1, 2 \text{ and } j=1, 2, ..., n\}$ which are totally positive of order two is shown to be a convex set under some conditions on one of the marginal distributions. The extreme points of this compact convex set are explicitly enumerated. Using the structure of this convex set, we show that the power function of any test for testing the hypothesis of independence against the hypothesis of strict total positivity of order two in $2 \times n$ ordinal contingency tables has a simple form in terms of the extreme points. A numerical illustration is provided.

1. INTRODUCTION

Let X and Y be two random variables each taking a finite number of values. For simplicity, assume that X takes values 1, 2, ..., m and Y takes values 1, 2, ..., n. Let $p_{ij} = \Pr(X = i, Y = j)$, $i = 1$ to m, and $j = 1$ to n. In

* This work is supported by Contract N00014-85-K-0292 of the Office of Naval Research and Contract F49620-35-C-0008 of the Air Force Office of Scientific Research. The U.S. Government's right to retain a nonexlusive royalty-free license in and to the copyright covering this paper, for governmental purposes, is acknowledged.

Multivariate Statistics and Probability
ISBN 0-12-580205-6

Reprinted from *J. Mult. Anal.* **27**(2).

order to describe *local association* between X and Y, $(m-1)(n-1)$ odds ratios defined by

$$\theta_{ij} = \frac{p_{ij} p_{i+1,j+1}}{p_{i+1,j} p_{i,j+1}}, \quad i = 1, 2, ..., m-1, \; j = 1, 2, ..., n-1$$

have been commonly used in the literature. See Agresti [1]. In practice, the joint distribution of X and Y will be unknown and one would like to test the hypothesis

$$H_0\colon X \text{ and } Y \text{ are independent}$$

against certain ordered alternatives involving the odds ratios θ_{ij}'s based on a random sample of size N on (X, Y). See Grove [7, 8], Patefield [13], Barlow, Bartholomew, Bremner, and Brunk [2], and Bartholomew [4], among others. One such alternative hypothesis is given by

$$H_1\colon \theta_{ij} \geqslant 1, \quad i = 1, 2, ..., m-1; j = 1, 2, ..., n-1. \tag{1.1}$$

The condition imposed by H_1 is also stated in the form

$$p_{ij} p_{i+1,j+1} \geqslant p_{i,j+1} p_{i+1,j}, \quad i = 1, 2, ..., m-1; j = 1, 2, ..., n-1, \tag{1.2}$$

or, equivalently, in the form that the determinants

$$\begin{vmatrix} p_{ij} & p_{i,j+1} \\ p_{i+1,j} & p_{i+1,j+1} \end{vmatrix} \geqslant 0, \quad i = 1, 2, ..., m-1; j = 1, 2, ..., n-1. \tag{1.3}$$

Using induction, one can show that (1.3) is equivalent to

$$\begin{vmatrix} p_{i_1 j_1} & p_{i_1 j_2} \\ p_{i_2 j_1} & p_{i_2 j_2} \end{vmatrix} \geqslant 0 \tag{1.4}$$

for all $1 \leqslant i_1 < i_2 \leqslant m$ and $1 \leqslant j_1 < j_2 \leqslant n$. Condition (1.4) is precisely the condition that the matrix $P = (p_{ij})$ is totally positive of order two (TP_2) or the joint distribution of X and Y is totally positive of order two. See Karlin [10, p. 18]. For this definition and its ramifications, see Barlow and Proschan [3, p. 143]. In the literature, this notion also goes by the name *positive likelihood ratio dependence*. See Lehmann [11, p. 1150].

There are various tests available in the literature for testing H_0 against H_1 given above. In the context of $2 \times n$ bivariate distributions, Grove [7] derived the likelihood ratio test for H_0 versus an alternative which is slightly weaker than H_1 given above. Patefield [13] and Hirotsu [9] worked within the framework of H_0 and H_1 given above. One of the major stumbling blocks on a critical examination of the tests used in this connection is the lack of a suitable apparatus by which one can compute the power at any given distribution in the alternative. Comparison of the performance of the tests is also fraught with similar difficulties.

In this paper, by looking at the notion of total positivity of order two from a global point of view, we show that some of the difficulties mentioned above can be overcome under some conditions. Let $M(\mathrm{TP}_2)$ denote the collection of all bivariate distributions with support contained in $\{(i, j); 1 \leqslant i \leqslant m, 1 \leqslant j \leqslant n\}$. Any member of $M(\mathrm{TP}_2)$ can be regarded as a matrix $P = (p_{ij})_{1 \leqslant i \leqslant m, 1 \leqslant j \leqslant n}$ such that each p_{ij} is nonnegative, $\Sigma_i \Sigma_j p_{ij} = 1$ and all the second-order determinants of the type mentioned above are nonnegative. In Section 2, we examine the convexity properties of the set $M(\mathrm{TP}_2)$. Using the structure of the convex sets described in Section 2, we give a simple formula for evaluating the power function of any test proposed to test independence of X and Y against the alternative hypothesis of strict total positivity of order two for X and Y in Section 3. This formula is useful in evaluating the exact size and power of any test proposed. The mechanism of the formula is explained with the help of a particular example. Section 4 is concerned with extensions of the results of Section 2.

2. Convexity Properties

In this section we assume that $m = 2$. Let $q_1, q_2, ..., q_n$ be n positive numbers such that $q_1 + q_2 + \cdots + q_n = 1$. Let $M_q(\mathrm{TP}_2)$ be the collection of all bivariate distributions of total positivity of order two and whose second marginal distribution is $q_1, q_2, ..., q_n$, where $q = (q_1, q_2, ..., q_n)$. More precisely

$$M_q(\mathrm{TP}_2) = \{P = (p_{ij}) \in M(\mathrm{TP}_2); p_{1j} + p_{2j} = q_j, j = 1 \text{ to } n\}.$$

The following result gives the structure of the above set.

Theorem 1. *The set $M_q(\mathrm{TP}_2)$ is a compact convex set. It has exactly $(n+1)$ extreme points given by*

$$P_0 = \begin{bmatrix} 0 & 0 & \cdots & 0 \\ q_1 & q_2 & \cdots & q_n \end{bmatrix}$$

$$P_i = \begin{bmatrix} q_1 & q_2 & \cdots & q_i & 0 & 0 & \cdots & 0 \\ 0 & 0 & \cdots & 0 & q_{i+1} & q_{i+2} & \cdots & q_n \end{bmatrix}, \qquad 1 \leqslant i \leqslant n.$$

Proof. It is clear that $M_q(\mathrm{TP}_2)$ is bounded and closed. We prove the convexity of $M_q(\mathrm{TP}_2)$. Let $P=(p_{ij})_{1\leqslant i\leqslant 2, 1\leqslant j\leqslant n}$ be a given matrix. Then $P\in M_q(\mathrm{TP}_2)$ if and only if

(i) $p_{ij}\geqslant 0$ for all i and j,

(ii) $p_{1j}+p_{2j}=q_j$ for all j, and

(iii) $p_{1j_1}q_{j_2}-p_{1j_2}q_{j_1}\geqslant 0$ for all $1\leqslant j_1<j_2\leqslant n$.

Let $P=(p_{ij})$ and $Q=(q_{ij})$ belong to $M_q(\mathrm{TP}_2)$ and $0\leqslant\lambda\leqslant 1$. Then for $1\leqslant j_1<j_2\leqslant n$,

$$\begin{vmatrix}\lambda p_{1j_1}+(1-\lambda)\,q_{1j_1} & \lambda p_{1j_2}+(1-\lambda)\,q_{1j_2}\\ \lambda p_{2j_1}+(1-\lambda)\,q_{2j_1} & \lambda p_{2j_2}+(1-\lambda)\,q_{2j_2}\end{vmatrix}$$
$$=\lambda[p_{1j_1}q_{j_2}-p_{1j_2}q_{j_1}]+(1-\lambda)[q_{1j_1}q_{j_2}-q_{1j_2}q_{j_1}]$$
$$\geqslant 0 \quad \text{in view of property (iii) above.}$$

Consequently, $\lambda P+(1-\lambda)Q\in M_q(\mathrm{TP}_2)$. This proves that $M_q(\mathrm{TP}_2)$ is a convex set.

It is obvious that each $P_i\in M_q(\mathrm{TP}_2)$ and is also an extreme point of $M_q(\mathrm{TP}_2)$. In order to show that these are the only extreme points of $M_q(\mathrm{TP}_2)$, it suffices to show that every member of $M_q(\mathrm{TP}_2)$ is a convex combination of these P_i's. Let $P=(p_{ij})\in M_q(\mathrm{TP}_2)$ be given. Let $\alpha_0=1-p_{11}/q_1$, $\alpha_i=p_{1i}/q_i-p_{1i+1}/q_{i+1}$, $i=1, 2, ..., n-1$, and $\alpha_n=p_{1n}/q_n$. One can check that $\alpha_0+\alpha_1+\cdots+\alpha_n=1$, $\alpha_i\geqslant 0$ for $i=1, 2, ..., n-1$ from property (iii) above, and $\alpha_n\geqslant 0$ and $\alpha_0\geqslant 0$ from $p_{11}+p_{21}=q_1$. Further,

$$P=\alpha_0P_0+\alpha_1P_1+\cdots+\alpha_nP_n.$$

This completes the proof.

Thus we see that every distribution P in $M_q(\mathrm{TP}_2)$ is a mixture of a fixed finite number of *special* distributions in $M_q(\mathrm{TP}_2)$. Is the representation of P in terms of P_0, P_1, ..., P_n given above unique? In the parlance of *identifiability of mixtures* the above question translates into whether the family of distributions in $M_q(\mathrm{TP}_2)$ is identifiable with respect to $\{P_0, P_1, ..., P_n\}$. See Teicher [14, p. 244]. This is indeed the case. This follows from the fact that the vectors $(q_1, 0, 0, ..., 0)$, $(q_1, q_2, 0, 0, ..., 0)$, $(q_1, q_2, q_3, 0, 0, ..., 0)$, ..., $(q_1, q_2, ..., q_n)$ are a (Hamel) basis for the n-dimensional Euclidean space R^n. Thus we have the following result.

THEOREM 2. *The family of distributions $M_q(\mathrm{TP}_2)$ is identifiable with respect to $\{P_0, P_1, ..., P_n\}$.*

Remarks. 1. A close look at the extreme points of $M_q(\mathrm{TP}_2)$ reveals the following information. Under each of P_0 and P_n, X and Y are independently distributed. Under each of $P_1, P_2, \ldots, P_{n-1}$, X and Y are not independently distributed.

2. The set $M(\mathrm{TP}_2)$ is not convex. For example, take $n=2$ and look at the following two bivariate distributions.

$$P=\begin{bmatrix} \frac{1}{4} & \frac{1}{4} \\ \frac{1}{4} & \frac{1}{4} \end{bmatrix}; \qquad Q=\begin{bmatrix} \frac{2}{9} & \frac{1}{9} \\ \frac{4}{9} & \frac{2}{9} \end{bmatrix}.$$

Each of P and Q is TP_2 but not $\frac{1}{2}P+\frac{1}{2}Q$.

3. An examination of Theorem 1 provides the following information on the position of zeros of any bivariate distribution $P=(p_{ij})$ in $M_q(\mathrm{TP}_2)$. The matrix P is one of the following types:

A. $P=P_i$ for some $i=0, 1, 2, \ldots, n$.

B. Every entry in P is positive.

C. P can be partitioned as

$$\begin{bmatrix} P_{11} & P_{12} \\ 0 & P_{22} \end{bmatrix}$$

in which every entry in the submatrices P_{11}, P_{12}, and P_{22} is positive.

D. P can be partitioned as

$$\begin{bmatrix} P_{11} & 0 \\ P_{21} & P_{22} \end{bmatrix}$$

in which every entry in the submatrices P_{11}, P_{21}, and P_{22} is positive.

E. P can be partitioned as

$$\begin{bmatrix} P_{11} & P_{12} & 0 \\ 0 & P_{22} & P_{23} \end{bmatrix}$$

in which every entry in the submatrices P_{11}, P_{12}, P_{22}, and P_{23} is positive.

4. If X and Y are independent under P, then P is a convex combination of P_0 and P_n.

3. An Application

Theorem 1 is useful in computing the size and power function of any given test under the following setting. Let (X, Y) be a random vector with some probability law $P=(p_{ij})_{1 \leqslant i \leqslant 2, 1 \leqslant j \leqslant n}$. The only information we have about P is that the marginal distribution $q=(q_1, q_2, ..., q_n)$ of Y is known and that $P \in M_q(\mathrm{TP}_2)$. Suppose we wish to test the null hypothesis H_0 that X and Y are independent against the alternative H_1, that X and Y are strictly totally positive of order two, i.e., X and Y are totally positive of order 2 but not independent, based on N independent realizations (X_1, Y_1), (X_2, Y_2), ..., (X_N, Y_N) of (X, Y). Note that both the hypotheses are composite. Suppose $T=T((X_1, Y_1), (X_2, Y_2), ..., (X_N, Y_N))$ is a test statistic proposed and C is the critical region of the test based on T to discriminate the hypotheses H_0 and H_1. Let $\beta_T(\cdot)$ be the power function of the test based on T, i.e.,

$$\beta_T(P)=\Pr\{T \in C/P\}, \qquad P \in M_q(\mathrm{TP}_2).$$

The computations of $\beta_T(P)$ for P in $M_q(\mathrm{TP}_2)$ can be simplified by using Theorem 1. For a given P in $M_q(\mathrm{TP}_2)$ we can find nonnegative numbers $\alpha_0, \alpha_1, ..., \alpha_n$ with sum equal to unity such that $P=\alpha_0 P_0+\alpha_1 P_1+\cdots+\alpha_n P_n$. The joint distribution of (X_1, Y_1), (X_2, Y_2), ..., (X_N, Y_N) is given by the product probability measure

$$\begin{aligned} P^N &= P \otimes P \otimes \cdots \otimes P \\ &= \Sigma_{i_1} \Sigma_{i_2} \cdots \Sigma_{i_N} \alpha_{i_1} \alpha_{i_2} \cdots \alpha_{i_N} (P_{i_1} \otimes P_{i_2} \otimes \cdots \otimes P_{i_N}), \end{aligned}$$

where each $i_j \in \{0, 1, 2, ..., n\}$, $j=1, 2, ..., N$. Assume that T is a symmetric function of (X_1, Y_1), (X_2, Y_2), ..., (X_N, Y_N). It is not difficult to see that

$$\begin{aligned} \beta_T(P) &= \Sigma_{i_1} \Sigma_{i_2} \cdots \Sigma_{i_N} \alpha_{i_1} \alpha_{i_2} \cdots \alpha_{i_N} \\ &\quad \times \beta_T(P_{i_1} \otimes P_{i_2} \otimes \cdots \otimes P_{i_N}) \\ &= \Sigma(N!/r_0!\, r_1! \cdots r_n!)\, \alpha_0^{r_0} \alpha_1^{r_1} \cdots \alpha_n^{r_n} \\ &\quad \times \beta_T(P_0^{r_0} \otimes P_1^{r_1} \otimes \cdots \otimes P_n^{r_n}), \end{aligned} \tag{3.1}$$

where the summation is taken over all nonnegative integers, $r_0, r_1, ..., r_n$ subject to the condition that $r_0+r_1+\cdots+r_n=N$.

The above formula expresses the power of the test T evaluated at P as a convex combination of the powers of the test T evaluated at the distributions $P_0^{r_0} \otimes P_1^{r_1} \otimes \cdots \otimes P_n^{r_n}$ with $r_0+r_1+\cdots+r_n=N$ with the coefficients in the convex combination coming from the multinomial distribution $(N; \alpha_0, \alpha_1, ..., \alpha_n)$. The precise meaning of $\beta_T(P_0^{r_0} \otimes P_1^{r_1} \otimes \cdots \otimes P_n^{r_n})$

is given by $\Pr(T \text{ rejects } H_0 \mid (X_1, Y_1), \ldots, (X_{r_0}, Y_{r_0})$ has distribution P_0; $(X_{r_0+1}, Y_{r_0+1}), \ldots, (X_{r_0+r_1}, Y_{r_0+r_1})$ has distribution $P_1, \ldots,$ and $(X_{r_0+r_1+\cdots+r_{n-1}+1}, Y_{r_0+r_1+\cdots+r_{n-1}+1}), \ldots, (X_N, Y_N)$ has distribution P_n). For moderate values of N, the above formula can be used effectively to evaluate the exact power of the test T at any distribution P in $M_q(\mathrm{TP}_2)$.

We can also give a simple formula to evaluate the size α of the test T. Let $M_{I,q}(\mathrm{TP}_2)$ be the family of all distributions in $M_q(\mathrm{TP}_2)$ under which X and Y are independent. $M_{I,q}(\mathrm{TP}_2)$ is precisely the family of all distributions specified by the null hypothesis H_0. $M_{I,q}(\mathrm{TP}_2)$ is a compact convex sct with extreme points P_0 and P_n. This can be seen as follows. Let

$$P = \begin{bmatrix} p_{11} & p_{12} & \cdots & p_{1n} \\ p_{21} & p_{22} & \cdots & p_{2n} \end{bmatrix} \in M_{I,q}(\mathrm{TP}_2).$$

Let $p_{11} + p_{12} + \cdots + p_{1n} = p_1$ and $p_{21} + p_{22} + \cdots + p_{2n} = p_2$. Then $p_{ij} = p_i q_j$ for all i and j, and also

$$P = p_2 P_0 + p_1 P_n.$$

Consequently,

$$\beta_T(P) = \sum_{r=0}^{N} \binom{N}{r} p_2^r p_1^{N-r} \beta_T(P_0^r \otimes P_n^{N-r}),$$

and the size of the test T is given by

$$\alpha = \sup_{0 \leqslant p_1 \leqslant 1} \sum_{r=0}^{N} \binom{N}{r} (1-p_1)^r p_1^{N-r} \beta_T(P_0^r \otimes P_n^{N-r}). \tag{3.2}$$

Note that the numbers $\beta_T(P_0^r \otimes P_n^{N-r})$ depend on r and $q_1, q_2, \ldots, q_n$ only.

We illustrate the foregoing ideas by an example. At this juncture, some comments on Goodman–Kruskal's gamma ratio Γ are in order. For any bivariate distribution $P = (p_{ij})_{1 \leqslant i \leqslant 2, 1 \leqslant j \leqslant n}$, the Goodman–Kruskal gamma ratio $\Gamma(P)$ is defined by

$$\Gamma(P) = (\pi_c - \pi_d)/(\pi_c + \pi_d),$$

where

$$\pi_c = p_{11}(p_{22} + p_{23} + \cdots + p_{2n}) + p_{12}(p_{23} + p_{24} + \cdots + p_{2n}) + \cdots + p_{1n-1} p_{2n}$$

and

$$\pi_d = p_{1n}(p_{21} + p_{22} + \cdots + p_{2n-1}) + p_{1n-1}(p_{21} + p_{22} + \cdots + p_{2n-2}) + \cdots + p_{12} p_{21}.$$

It is easy to verify that $0 \leqslant \Gamma(P) \leqslant 1$ for every P in $M_q(\mathrm{TP}_2)$. Further, $\Gamma(P_i) = 1$ for $i = 1, 2, ..., n-1$. The gamma ratio also characterizes independence as explained in the following result.

THEOREM 3. *For any bivariate distribution* $P = (p_{ij})_{1 \leqslant i \leqslant 2, 1 \leqslant j \leqslant n}$ *in* $M_q(\mathrm{TP}_2)$, $\Gamma(P) = 0$ *if and only if* X *and* Y *are independent under* P.

Proof. If X and Y are independent under P, it is obvious that $\Gamma(P) = 0$. Suppose $\Gamma(P) = 0$. This implies that $\pi_c - \pi_d = 0$ and also

$$\begin{vmatrix} p_{1j_1} & p_{1j_2} \\ p_{2j_1} & p_{2j_2} \end{vmatrix} = 0$$

for every $1 \leqslant j_1 < j_2 \leqslant n$. We distinguish two cases.

Case 1. None of the column marginal totals is zero. Then we can write

$$\begin{bmatrix} p_{1j} \\ p_{2j} \end{bmatrix} = c_j \begin{bmatrix} p_{11} \\ p_{21} \end{bmatrix}$$

for some constants $c_2, c_3, ..., c_n$. Let p_1, p_2 be the row marginal totals and $q_1, q_2, ..., q_n$ the column marginal totals. Then $p_{11} + p_{12} + \cdots + p_{1n} = p_1 = (1 + c_2 + c_3 + \cdots + c_n) p_{11}$ and $q_j = c_j q_1$ for $j = 2, 3, ..., n$. Consequently, $1 = q_1 + q_2 + \cdots + q_n = (1 + c_2 + c_3 + \cdots + c_n) q_1$ and $1 + c_2 + c_3 + \cdots + c_n = 1/q_1$. This implies that $p_{11} = p_1 q_1$. Using a similar argument, one can show that $p_{ij} = p_i q_j$ for all i and j.

Case 2. Some of the column marginal totals are each equal to zero. Ignoring these columns and dealing with the reduced matrix, one can establish independence by adapting the argument given in Case 1.

The foregoing discussion indicates that it is reasonable to construct a test based on the gamma ratio. An estimator $\hat{\Gamma}$ of Γ is built as follows. Let N_{ij} = number of (X_r, Y_r)'s with $X_r = i$ and $Y_r = j$, $i = 1, 2$; $j = 1, 2, ..., n$. The data $\{(X_r, Y_r);\ r = 1, 2, ..., N\}$ can be summarized in the form of a contingency table:

$$\begin{bmatrix} N_{11} & N_{12} & \cdots & N_{1n} \\ N_{21} & N_{22} & \cdots & N_{2n} \end{bmatrix}.$$

The estimator of $\Gamma = \hat{\Gamma} = (C - D)/(C + D)$, where

$$\begin{aligned} C = {} & N_{11}(N_{22} + N_{23} + \cdots + N_{2n}) \\ & + N_{12}(N_{23} + N_{24} + \cdots + N_{2n}) + \cdots + N_{1n-1} N_{2n} \end{aligned}$$

and

$$D = N_{1n}(N_{21} + N_{22} + \cdots + N_{2n-1}) + N_{1n-1}(N_{21} + N_{22} + \cdots + N_{2n-2}) + \cdots + N_{12}N_{21}.$$

$\hat{\Gamma}$ is obviously a symmetric function of (X_1, Y_1), (X_2, Y_2), ..., (X_N, Y_N). One can build a test T based on $\hat{\Gamma}$:

Test T: Reject H_0 if and only if $\hat{\Gamma} \geqslant c$ for some fixed $0 < c < 1$.

A NUMERICAL ILLUSTRATION. Let $n = 2$ and $N = 6$. The extreme points of $M_q(\mathrm{TP}_2)$ are

$$P_0 = \begin{bmatrix} 0 & 0 \\ q_1 & q_2 \end{bmatrix}, \quad P_1 = \begin{bmatrix} q_1 & 0 \\ 0 & q_2 \end{bmatrix}, \quad P_2 = \begin{bmatrix} q_1 & q_2 \\ 0 & 0 \end{bmatrix}.$$

Random Sample: (X_1, Y_1), (X_2, Y_2), ..., (X_6, Y_6).

If the joint distribution of the random sample is $P_0^{r_0} \otimes P_1^{r_1} \otimes P_2^{r_2}$ with $r_0 + r_1 + r_2 = 6$, i.e., each of (X_1, Y_1), (X_2, Y_2), ..., (X_{r_0}, Y_{r_0}) has distribution P_0; each of (X_{r_0+1}, Y_{r_0+1}), (X_{r_0+2}, Y_{r_0+2}), ..., $(X_{r_0+r_1}, Y_{r_0+r_1})$ has distribution P_1; and each of the remaining (X_i, Y_i)'s has distribution P_2, we denote this joint distribution by (r_0, r_1, r_2).

One can check that the estimator $\hat{\Gamma}$ can take any one of the seven values -1, $-\frac{3}{5}$, $-\frac{1}{2}$, 0, $\frac{1}{2}$, $\frac{3}{5}$, and 1. The probability that $\hat{\Gamma} = d$ under any given joint distribution of the sample is of the form $d_1 q_1^6 + d_2 q_1^5 q_2 + d_3 q_1^4 q_2^2 + d_4 q_1^3 q_2^3 + d_5 q_1^2 q_2^4 + d_6 q_1 q_2^5 + d_7 q_2^6$ for some nonnegative integers d_1, d_2, ..., d_7 which depend on the joint distribution and the value d. We denote this probability by the vector $(d_1, d_2, ..., d_7)$ under the joint distribution and the value d. If each $d_i = 0$, we denote the corresponding vector by $\tilde{0}$. The distribution of $\hat{\Gamma}$ is listed in Table I under each of the 28 possible joint distributions of the sample.

Using the distribution of $\hat{\Gamma}$, we can compute the size of any test based on $\hat{\Gamma}$, and also its power function. We calculate the size of the following three tests under different values of q_1.

Test	Critical region
T_1	Reject H_0 if $\hat{\Gamma} \geqslant 1$
T_2	Reject H_0 if $\hat{\Gamma} \geqslant c$ for any fixed c satisfying $\frac{3}{5} \leqslant c < 1$
T_3	Reject H_0 if $\hat{\Gamma} \geqslant c$ for any fixed c satisfying $\frac{1}{2} \leqslant c < \frac{3}{5}$

TABLE I

Distribution of $\hat{\Gamma}$ under Various Joint Distributions of the Sample

Joint distribution	$\hat{\Gamma}$: -1	$-\frac{3}{5}$	$-\frac{1}{2}$	0	$\frac{1}{2}$	$\frac{3}{5}$	1
1. (6, 0, 0)	$\tilde{0}$	$\tilde{0}$	$\tilde{0}$	(1, 6, 15, 20, 15, 6, 1)	$\tilde{0}$	$\tilde{0}$	$\tilde{0}$
2. (5, 0, 1)	(0, 1, 5, 10, 10, 5, 0)	$\tilde{0}$	$\tilde{0}$	(1, 0, 0, 0, 0, 0, 1)	$\tilde{0}$	$\tilde{0}$	(0, 5, 10, 10, 5, 1, 0)
3. (4, 0, 2)	(0, 2, 1, 4, 6, 4, 0)	$\tilde{0}$	(0, 0, 8, 0, 0, 0, 0)	(1, 0, 0, 12, 0, 0, 1)	(0, 0, 0, 0, 8, 0, 0)	$\tilde{0}$	(0, 4, 6, 4, 1, 2, 0)
4. (3, 0, 3)	(0, 3, 3, 1, 3, 3, 0)	(0, 0, 0, 9, 0, 0, 0)	$\tilde{0}$	(1, 0, 9, 0, 9, 0, 1)	$\tilde{0}$	(0, 0, 0, 9, 0, 0, 0)	(0, 3, 3, 1, 3, 3, 0)
5. (2, 0, 4)	(0, 4, 6, 4, 1, 2, 0)	$\tilde{0}$	(0, 0, 0, 0, 8, 0, 0)	(1, 0, 0, 12, 0, 0, 1)	(0, 0, 8, 0, 0, 0, 0)	$\tilde{0}$	(0, 2, 1, 4, 6, 4, 0)
6. (1, 0, 5)	(0, 5, 10, 10, 5, 1, 0)	$\tilde{0}$	$\tilde{0}$	(1, 0, 0, 0, 0, 0, 1)	$\tilde{0}$	$\tilde{0}$	(0, 1, 5, 10, 10, 5, 0)
7. (0, 0, 6)	$\tilde{0}$	$\tilde{0}$	$\tilde{0}$	(1, 6, 15, 20, 15, 6, 1)	$\tilde{0}$	$\tilde{0}$	$\tilde{0}$
8. (4, 1, 1)	(0, 1, 1, 4, 6, 4, 0)	$\tilde{0}$	(0, 0, 4, 0, 0, 0, 0)	(1, 0, 0, 6, 0, 0, 1)	(0, 0, 0, 0, 4, 0, 0)	$\tilde{0}$	(0, 5, 10, 10, 5, 2, 0)
9. (3, 1, 2)	(0, 2, 1, 1, 3, 3, 0)	(0, 0, 0, 3, 0, 0, 0)	(0, 0, 2, 0, 0, 0, 0)	(1, 0, 6, 6, 3, 0, 1)	(0, 0, 0, 0, 6, 0, 0)	(0, 0, 0, 6, 0, 0, 0)	(0, 4, 6, 4, 3, 3, 0)
10. (2, 1, 3)	(0, 3, 3, 1, 1, 2, 0)	(0, 0, 0, 3, 0, 0, 0)	(0, 0, 0, 0, 2, 0, 0)	(1, 0, 3, 6, 6, 0, 1)	(0, 0, 6, 0, 0, 0, 0)	(0, 0, 0, 6, 0, 0, 0)	(0, 3, 3, 4, 6, 4, 0)
11. (1, 1, 4)	(0, 4, 6, 4, 1, 1, 0)	$\tilde{0}$	(0, 0, 0, 0, 4, 0, 0)	(1, 0, 0, 6, 0, 0, 1)	(0, 0, 4, 0, 0, 0, 0)	$\tilde{0}$	(0, 2, 5, 10, 10, 5, 0)
12. (3, 2, 1)	(0, 1, 0, 1, 3, 3, 0)	$\tilde{0}$	(0, 0, 2, 0, 0, 0, 0)	(1, 0, 3, 6, 0, 0, 1)	(0, 0, 0, 0, 6, 0, 0)	(0, 0, 0, 3, 0, 0, 0)	(0, 5, 10, 10, 6, 3, 0)
13. (2, 2, 2)	(0, 2, 1, 0, 1, 2, 0)	(0, 0, 0, 2, 0, 0, 0)	$\tilde{0}$	(1, 0, 4, 4, 4, 0, 1)	(0, 0, 4, 0, 4, 0, 0)	(0, 0, 0, 8, 0, 0, 0)	(0, 4, 6, 6, 6, 4, 0)
14. (1, 2, 3)	(0, 3, 3, 1, 0, 1, 0)	$\tilde{0}$	(0, 0, 0, 0, 2, 0, 0)	(1, 0, 0, 6, 3, 0, 1)	(0, 0, 6, 0, 0, 0, 0)	(0, 0, 0, 3, 0, 0, 0)	(0, 3, 6, 10, 10, 5, 0)
15. (2, 3, 1)	(0, 1, 0, 0, 1, 2, 0)	$\tilde{0}$	$\tilde{0}$	(1, 0, 3, 3, 0, 0, 1)	(0, 0, 2, 0, 6, 0, 0)	(0, 0, 0, 6, 0, 0, 0)	(0, 5, 10, 11, 8, 4, 0)
16. (1, 3, 2)	(0, 2, 1, 0, 0, 1, 0)	$\tilde{0}$	$\tilde{0}$	(1, 0, 0, 3, 3, 0, 1)	(0, 0, 6, 0, 2, 0, 0)	(0, 0, 0, 6, 0, 0, 0)	(0, 4, 8, 11. 10, 5, 0)
17. (1, 4, 1)	(0, 1, 0, 0, 0, 1, 0)	$\tilde{0}$	$\tilde{0}$	(1, 0, 0, 0, 0, 0, 1)	(0, 0, 4, 0, 4, 0, 0)	(0, 0, 0, 6, 0, 0, 0)	(0, 5, 11, 14, 11, 5, 0)
18. (5, 1, 0)	$\tilde{0}$	$\tilde{0}$	$\tilde{0}$	(1, 1, 5, 10, 10, 5, 1)	$\tilde{0}$	$\tilde{0}$	(0, 5, 10, 10, 5, 1, 0)
19. (0, 1, 5)	$\tilde{0}$	$\tilde{0}$	$\tilde{0}$	(1, 5, 10, 5, 5, 1, 1)	$\tilde{0}$	$\tilde{0}$	(0, 1, 5, 15, 10, 5, 0)
20. (4, 2, 0)	$\tilde{0}$	$\tilde{0}$	$\tilde{0}$	(1, 0, 1, 4, 6, 4, 1)	$\tilde{0}$	$\tilde{0}$	(0, 6, 14, 16, 9, 2, 0)
21. (0, 2, 4)	$\tilde{0}$	$\tilde{0}$	$\tilde{0}$	(1, 4, 6, 4, 1, 0, 1)	$\tilde{0}$	$\tilde{0}$	(0, 2, 9, 16, 14, 6, 0)
22. (3, 3, 0)	$\tilde{0}$	$\tilde{0}$	$\tilde{0}$	(1, 0, 0, 1, 3, 3, 1)	$\tilde{0}$	$\tilde{0}$	(0, 6, 15, 19, 12, 3, 0)
23. (0, 3, 3)	$\tilde{0}$	$\tilde{0}$	$\tilde{0}$	(1, 3, 3, 1, 0, 0, 1)	$\tilde{0}$	$\tilde{0}$	(0, 3, 12, 19, 15, 6, 0)
24. (2, 4, 0)	$\tilde{0}$	$\tilde{0}$	$\tilde{0}$	(1, 0, 0, 0, 1, 2, 1)	$\tilde{0}$	$\tilde{0}$	(0, 6, 15, 20, 14, 4, 0)
25. (0, 4, 2)	$\tilde{0}$	$\tilde{0}$	$\tilde{0}$	(1, 2, 1, 0, 0, 0, 1)	$\tilde{0}$	$\tilde{0}$	(0, 4, 14, 20, 15, 6, 0)
26. (1, 5, 0)	$\tilde{0}$	$\tilde{0}$	$\tilde{0}$	(1, 0, 0, 0, 0, 1, 1)	$\tilde{0}$	$\tilde{0}$	(0, 6, 15, 20, 15, 5, 0)
27. (0, 5, 1)	$\tilde{0}$	$\tilde{0}$	$\tilde{0}$	(1, 0, 0, 0, 0, 0, 1)	$\tilde{0}$	$\tilde{0}$	(0, 6, 15, 20, 15, 6, 0)
28. (0, 6, 0)	$\tilde{0}$	$\tilde{0}$	$\tilde{0}$	(1, 0, 0, 0, 0, 0, 1)	$\tilde{0}$	$\tilde{0}$	(0, 6, 15, 20, 15, 6, 0)

TABLE II

Size of the Tests T_1, T_2, and T_3

Test \ q_1	0.1	0.2	0.3	0.4	0.5	0.6	0.7	0.8	0.9
T_1	0.247	0.346	0.359	0.332	0.291	0.332	0.359	0.346	0.247
T_2	0.248	0.352	0.373	0.353	0.323	0.353	0.373	0.352	0.248
T_3	0.252	0.362	0.394	0.391	0.381	0.391	0.394	0.362	0.252

Note that

$$\text{Size of } T_i = \max_{0 \leqslant p_1 \leqslant 1} \sum_{r=0}^{6} \binom{6}{r} (1-p_1)^r p_1^{6-r} \times \beta_{T_i}(P_0^r \otimes P_2^{6-r}), \qquad i = 1, 2, 3.$$

Comments on Table II. It appears that the size depends so little on the critical region chosen. Since the sample size N is small, the range of values that $\hat{\Gamma}$ takes is very limited, and the probabilities $P(\hat{\Gamma} \geqslant 1)$, $P(\hat{\Gamma} \geqslant \frac{3}{5})$, and $(\hat{\Gamma} \geqslant \frac{1}{2})$ are not all that different under each of the joint distributions $P_0^r \otimes P_2^{6-r}$, $r = 0, 1, 2, ..., 6$, of the sample (X_1, Y_1), (X_2, Y_2), ..., (X_6, Y_6). From Table I, the following information can be gleaned for $q_1 = q_2 = \frac{1}{2}$, for the tail probabilities of $\hat{\Gamma}$

Joint distribution	$P(\hat{\Gamma} \geqslant 1)$	$P(\hat{\Gamma} \geqslant \frac{3}{5})$	$P(\hat{\Gamma} \geqslant \frac{1}{2})$
P_2^6	0	0	0
$P_0 \otimes P_2^5$	$31(\frac{1}{2})^6$	$31(\frac{1}{2})^6$	$31(\frac{1}{2})^6$
$P_0^2 \otimes P_2^4$	$17(\frac{1}{2})^6$	$17(\frac{1}{2})^6$	$25(\frac{1}{2})^6$
$P_0^3 \otimes P_2^3$	$13(\frac{1}{2})^6$	$22(\frac{1}{2})^6$	$22(\frac{1}{2})^6$
$P_0^4 \otimes P_2^2$	$17(\frac{1}{2})^6$	$17(\frac{1}{2})^6$	$25(\frac{1}{2})^6$
$P_0^5 \otimes P_2$	$31(\frac{1}{2})^6$	$31(\frac{1}{2})^6$	$31(\frac{1}{2})^6$
P_0^6	0	0	0

The sizes of T_1, T_2, and T_3 work out to be

$$T_1 = (\tfrac{1}{2})^6 \max_{0 \leqslant p \leqslant 1} \left\{ 31 \binom{6}{1} (1-p) p^5 + 17 \binom{6}{2} (1-p)^2 p^4 + 13 \binom{6}{3} (1-p)^3 p^3 + 17 \binom{6}{4} (1-p)^4 p^2 + 31 \binom{6}{5} (1-p)^5 p \right\},$$

$$T_2 = (\tfrac{1}{2})^6 \max_{0 \leqslant p \leqslant 1} \left\{ 31 \binom{6}{1} (1-p) p^5 + 17 \binom{6}{2} (1-p)^2 p^4 \right.$$
$$+ 22 \binom{6}{3} (1-p)^3 p^3 + 17 \binom{6}{4} (1-p)^4 p^2$$
$$\left. + 31 \binom{6}{5} (1-p)^5 p \right\},$$

$$T_3 = (\tfrac{1}{2})^6 \max_{0 \leqslant p \leqslant 1} \left\{ 31 \binom{6}{1} (1-p) p^5 + 25 \binom{6}{2} (1-p)^2 p^4 \right.$$
$$+ 22 \binom{6}{3} (1-p)^3 p^3 + 25 \binom{6}{4} (1-p)^4 p^2$$
$$\left. + 31 \binom{6}{5} (1-p)^5 p \right\}.$$

From these expressions, it is clear that one cannot expect substantial differences between the sizes.

As q_1 moves away from $\frac{1}{2}$, the three columns of probabilities in the above table tend to be closer leading to very small differences between the sizes.

Power Function. The power of each of the above three tests has been evaluated under each of the following joint distributions of X and Y figuring in H_1:

1. $(0.2)\, P_0 + (0.2)\, P_1 + (0.6)\, P_2$
2. $(0.2)\, P_0 + (0.4)\, P_1 + (0.4)\, P_2$
3. $(0.2)\, P_0 + (0.6)\, P_1 + (0.2)\, P_2$
4. $(0.4)\, P_0 + (0.2)\, P_1 + (0.4)\, P_2$
5. $(0.4)\, P_0 + (0.4)\, P_1 + (0.2)\, P_2$
6. $(0.6)\, P_0 + (0.2)\, P_1 + (0.2)\, P_2$.

Let α_0, α_1, and α_2 be the generic symbols for the coefficients of P_0, P_1, and P_2, respectively, in the above. In Table III the joint distribution of X and Y is denoted by $(\alpha_0, \alpha_1, \alpha_2)$. The power function of a test T, in this case, works out explicitly as

$$\beta_T(\alpha_0, \alpha_1, \alpha_2) = \sum_{r_0} \sum_{r_1} \sum_{r_2} \frac{6!}{r_0!\, r_1!\, r_2!} \alpha_0^{r_0} \alpha_1^{r_1}$$
$$\times \alpha_2^{r_2} \beta_T(P_0^{r_0} \otimes P_1^{r_1} \otimes P_2^{r_2}),$$

where the summation is taken over all r_0, r_1, $r_2 \geqslant 0$ with $r_0 + r_1 + r_2 = 6$, and $\beta_T(P_0^{r_0} \otimes P_1^{r_1} \otimes P_2^{r_2}) = \Pr(T$ rejects $H_0 |$ the joint distribution of

$(X_1, Y_1), ..., (X_6, Y_6)$ is $P_0^{r_0} \otimes P_1^{r_1} \otimes P_2^{r_2}$). The computations are summarized in Table III.

General Case. In the case of $2 \times n$ tables with sample size N, the complexity of the calculations involved in the exact evaluation of power and size of tests increase as N increases. One needs to compute the powers $\beta_T(P_0^{r_0} \otimes P_1^{r_1} \otimes \cdots \otimes P_n^{r_n})$ for all partitions $r_0 + r_1 + r_2 + \cdots + r_n = N$ of N (see (3.1) and (3.2)). But the number of partitions is enormous even for moderate values of N. The evaluation of the probability $\beta_T(P_0^{r_0} \otimes P_1^{r_1} \otimes \cdots \otimes P_n^{r_n})$ involves the determination of the exact distribution of the test statistic on which the test T is based under the joint distribution $P_0^{r_0} \otimes P_1^{r_1} \otimes \cdots \otimes P_n^{r_n}$ of the sample. If the sample size N is small this may not be difficult. It is now clear that the formulas (3.1) and (3.2) are useful from a practical point of view for evaluation of exact size and power of tests when N is small. For large N, one may have to take recourse to asymptotics to evaluate size and power of tests approximately.

4. Some Generalizations

As has been pointed out in Remark 2 in Section 2, the set $M(\text{TP}_2)$ is not convex in general. Even if we fix both the marginal distributions, the set is not convex. More specifically, let $p = (p_1, p_2, ..., p_m)$ and $q = (q_1, q_2, ..., q_n)$ be two fixed probability vectors. Let $M_{p,q}(\text{TP}_2)$ be the collection of all bivariate distributions with support contained in $\{(i, j);\ i = 1$ to m and $j = 1$ to $n\}$, the first marginal p and the second marginal distribution q. This set is not convex. As an example, let $p = (\frac{1}{3}, \frac{1}{3}, \frac{1}{3}) = q$ and look at the following two bivariate distributions.

$$P_1 = \begin{bmatrix} \frac{1}{9} & \frac{1}{9} & \frac{1}{9} \\ \frac{1}{9} & \frac{1}{9} & \frac{1}{9} \\ \frac{1}{9} & \frac{1}{9} & \frac{1}{9} \end{bmatrix}; \qquad P_2 = \begin{bmatrix} \frac{1}{3} & 0 & 0 \\ 0 & \frac{1}{3} & 0 \\ 0 & 0 & \frac{1}{3} \end{bmatrix}.$$

Each of P_1 and P_2 is TP_2 but $(\frac{2}{3})\, P_1 + (\frac{1}{3})\, P_2$ is not TP_2.

However, under certain special circumstances certain convex combinations of TP_2 distributions turn out to be TP_2. Let $M_q(\text{TP}_2)$ be the collection of all TP_2 bivariate distributions with support contained in $\{(i, j):\ i = 1$ to m and $j = 1$ to $n\}$ and the second marginal distribution being q. If P_1 and P_2 are two bivariate distributions in $M_q(\text{TP}_2)$ under each of which X and Y are independent, then $\alpha P_1 + (1 - \alpha)\, P_2$ is also TP_2 for every $0 \leqslant \alpha \leqslant 1$. We simply note that under $\alpha P_1 + (1 - \alpha)\, P_2$, X and Y are independent.

TABLE III

Power Function of the Tests T_1, T_2, and T_3

q_1 Joint Distribution	Power 0.1	0.2	0.3	0.4	0.5	0.6	0.7	0.8	0.9
Test 1									
(0.2 0.2 0.6)	0.324	0.471	0.512	0.497	0.478	0.410	0.358	0.290	0.184
(0.2 0.4 0.4)	0.355	0.532	0.602	0.615	0.599	0.568	0.519	0.433	0.278
(0.2 0.6 0.2)	0.371	0.579	0.685	0.734	0.748	0.736	0.690	0.585	0.377
(0.4 0.2 0.4)	0.257	0.372	0.410	0.417	0.417	0.417	0.410	0.372	0.257
(0.4 0.4 0.2)	0.277	0.432	0.517	0.567	0.599	0.615	0.603	0.533	0.356
(0.6 0.2 0.2)	0.184	0.290	0.358	0.410	0.458	0.497	0.512	0.471	0.324
Test 2									
(0.2 0.2 0.6)	0.325	0.480	0.532	0.528	0.492	0.441	0.378	0.299	0.185
(0.2 0.4 0.4)	0.358	0.546	0.633	0.661	0.651	0.614	0.549	0.447	0.280
(0.2 0.6 0.2)	0.373	0.591	0.713	0.774	0.794	0.777	0.717	0.597	0.379
(0.4 0.2 0.4)	0.260	0.387	0.445	0.469	0.475	0.469	0.445	0.387	0.260
(0.4 0.4 0.2)	0.279	0.445	0.548	0.613	0.651	0.661	0.634	0.546	0.358
(0.6 0.2 0.2)	0.185	0.299	0.378	0.441	0.492	0.528	0.532	0.480	0.325
Test 3									
(0.2 0.2 0.6)	0.330	0.495	0.560	0.570	0.550	0.509	0.445	0.348	0.205
(0.2 0.4 0.4)	0.369	0.576	0.681	0.723	0.723	0.690	0.620	0.497	0.300
(0.2 0.6 0.2)	0.386	0.625	0.763	0.833	0.855	0.836	0.767	0.631	0.392
(0.4 0.2 0.4)	0.276	0.430	0.509	0.543	0.553	0.543	0.509	0.430	0.276
(0.4 0.4 0.2)	0.298	0.495	0.618	0.690	0.723	0.723	0.681	0.577	0.369
(0.6 0.2 0.2)	0.205	0.348	0.445	0.509	0.550	0.570	0.560	0.495	0.330

5. Concluding Remarks

In the context of $2 \times n$ bivariate distributions, extreme point methods have been used to provide explicit formulas for the evaluation of size and power of any test one proposes for testing

H_0: X and Y are independent against

H_1: All odds ratios are $\geqslant 1$ with at least one strict inequality, based on a random sample of size N on (X, Y). These formulas are also helpful in comparing the performance of two competing tests. If N and n are small, one can find the exact distribution of the test statistic involved and the computation of size and power becomes practically feasible.

Bhaskara Rao, Krishnaiah, and Subramanyam [5] examined the problem of testing H_0 against H_1: X and Y are strictly positive quadrant dependent. The notion of positive quadrant dependence is weaker than total positivity of order two. Nguyen and Sampson [12] examined the convexity property of the set of all discrete bivariate positive quadrant dependent distributions. Cochran [6] presents a 2×5 contingency table which seems to conform to the pattern described by H_1 above.

Acknowledgment

The authors are grateful to the referee for his comments which are responsible for an improved version of the original manuscript.

References

[1] Agresti, A. (1984). *Analysis of Ordinal Categorical Data.* Wiley, New York.

[2] Barlow, R. E., Bartholomew, D. J., Bremner, J. M., and Brunk, H. D. (1972). *Statistical Inference under Order Restrictions.* Wiley, New York.

[3] Barlow, R. E., and Proschan, F. (1981). *Statistical Theory of Reliability and Life Testing: Probability Models.* Holt, Reinhart & Winston, Silver Spring, MD.

[4] Bartholomew, D. J. (1959). A test of homogeneity for ordered alternatives. *Biometrika* **16** 36–48.

[5] Bhaskara Rao, M., Krishnaiah, P. R., and Subramanyam, K. (1987). A structure theorem on bivariate positive quadrant dependent distributions and tests for independence in two-way contingency tables. *J. Multivariate Anal.* **23** 96–119.

[6] Cochran, W. G. (1954). Some methods for strengthening the common χ^2 tests. *Biometrics* **10** 417–451.

[7] Grove, D. H. (1980). A test of independence against a class of ordered alternatives in a $2 \times C$ contingency table. *J. Amer. Statist. Assoc.* **75** 454–459.

[8] Grove, D. M. (1984). Positive association in a two-way contingency table. *Comm. Statist.—Theory Methods* **13** 931–945.

[9] HIROTSU, C. (1982). Use of cumulative efficient scores for testing ordered alternatives in discrete models. *Biometrika* **69** 567–577.

[10] KARLIN, S. (1968). *Total Positivity*. Stanford Univ. Press, Stanford, CA.

[11] LEHMANN, E. L. (1966). Some concepts of dependence. *Ann. Math. Statist.* **37** 1137–1153.

[12] NGUYEN, T. T., AND SAMPSON, A. (1985). The geometry of certain fixed marginal distributions. *Linear Algebra Appl.* **70** 73–87.

[13] PATEFIELD, W. M. (1982). Exact tests for trends in ordered contingency tables. *J. R. Statist. Soc. C* **31** 32–43.

[14] TEICHER, H. (1961). Identifiability of mixtures. *Ann. Math. Stat.* **32** 244–248.

Asymptotic Expansions of the Distributions of Some Test Statistics for Gaussian ARMA Processes

Masanobu Taniguchi

Hiroshima University, Hiroshima, Japan

Let $\{X_t\}$ be a Gaussian ARMA process with spectral density $f_\theta(\lambda)$, where θ is an unknown parameter. The problem considered is that of testing a simple hypothesis $H: \theta = \theta_0$ against the alternative $A: \theta \neq \theta_0$. For this problem we propose a class of tests $\mathscr{S}$, which contains the likelihood ratio (LR), Wald (W), modified Wald (MW) and Rao (R) tests as special cases. Then we derive the χ^2 type asymptotic expansion of the distribution of $T \in \mathscr{S}$ up to order n^{-1}, where n is the sample size. Also we derive the χ^2 type asymptotic expansion of the distribution of T under the sequence of alternatives $A_n: \theta = \theta_0 + \varepsilon/\sqrt{n}$, $\varepsilon > 0$. Then we compare the local powers of the LR, W, MW, and R tests on the basis of their asymptotic expansions.

1. Introduction

In multivariate analysis, the asymptotic expansions of the distributions of various test statistics have been investigated in detail (e.g., Peer [4], Hayakawa [1, 2], Hayakawa and Puri [3]). On the other hand, in time series analysis, the first systematic study was tried by Whittle [12]. For an autoregressive process or a moving average process, he gave the limiting distribution of a test statistic of likelihood ratio type, and indicated a method to give its Edgeworth expansion. Recently Phillips [5] gave the Edgeworth expansion of the t-ratio test statistic in the estimation of the coefficient of a first-order autoregressive process (AR(1)). For an AR(1) process, Tanaka [6] gave the higher order approximations for the distributions of the likelihood ratio, Wald and Lagrange multiplier tests under both the null and alternative hypotheses. Also Taniguchi [8] derived the asymptotic expansion for the distribution of the likelihood ratio criterion

Multivariate Statistics and Probability
ISBN 0-12-580205-6

Reprinted from *J. Mult. Anal.* **27**(2).

for a Gaussian autoregressive moving average (ARMA) process under a sequence of local alternatives.

In this paper we consider a Gaussian ARMA process with the spectral density $f_\theta(\lambda)$ which depends on an unknown parameter θ. We assume that θ is scalar in order to avoid unnecessarily complex notations and formulas. The problem considered is that of testing a simple hypothesis $H:\theta=\theta_0$ against the alternative $A:\theta\neq\theta_0$. For this problem we propose a class of tests $\mathscr{S}$, which contains the likelihood ratio (LR), Wald (W), modified Wald (MW) and Rao (R) tests as special cases. Then we derive the χ^2 type asymptotic expansion of the distribution of $T\in\mathscr{S}$ up to order $1/n$, where n is the sample size.

In Section 4 we investigate a correction factor ρ which makes the term of order $1/n$ in the asymptotic expansion of the distribution of $(1+\rho/n)T$ vanish (i.e., Bartlett's adjustment) and give the necessary and sufficient condition for $T\in\mathscr{S}$ such that T is adjustable in the sense of Bartlett.

In Section 5 we derive the χ^2 type asymptotic expansion of the distribution of $S\in\mathscr{S}$ under the sequence of alternatives $A_n:\theta=\theta_0+\varepsilon/\sqrt{n}$, $\varepsilon>0$. Using the asymptotic expansion for S, we compare the local powers of the LR, W, MW, and R tests on the basis of their asymptotic expansions. Then it is shown that none of the above tests is uniformly superior.

2. Preliminaries

We introduce $\mathscr{D}$ and $\mathscr{D}_{\text{ARMA}}$, the spaces of functions on $[-\pi, \pi]$,

$$\mathscr{D}=\Big\{f:f(\lambda)=\sum_{u=-\infty}^{\infty} a(u)\exp(-iu\lambda),\ a(u)=a(-u),$$

$$\sum_{u=-\infty}^{\infty}(1+|u|)\,|a(u)|<d, \text{ for some } d<\infty\Big\},$$

$$\mathscr{D}_{\text{ARMA}}=\Big\{f:f(\lambda)=\frac{\sigma^2}{2\pi}\frac{|\sum_{j=0}^{q} a_j e^{ij\lambda}|^2}{|\sum_{j=0}^{p} b_j e^{ij\lambda}|^2},\ (\sigma^2>0),$$

$$\underline{c}\leqslant\frac{|\sum_{j=0}^{q} a_j z^j|^2}{|\sum_{j=0}^{p} b_j z^j|^2}\leqslant\bar{c}, \text{ for } |z|\leqslant 1,\ 0<\underline{c}<\bar{c}<\infty\Big\}.$$

We set down the following assumptions.

Assumption 1. $\{X_t; t=0, \pm 1, ...\}$ is a Gaussian stationary process with the spectral density $f_{\theta_0}(\lambda)\in\mathscr{D}_{\text{ARMA}}$, $\theta_0\in C\subset\Theta\subset R^1$, and mean 0. Here Θ is an open set of R^1 and C is a cmpact subset of Θ.

ASSUMPTION 2. The spectral density $f_\theta(\lambda)$ is continuously five times differentiable with respect to $\theta \in \Theta$, and the derivatives $\partial f_\theta/\partial\theta$, $\partial^2 f_\theta/\partial\theta^2$, $\partial^3 f_\theta/\partial\theta^3$, $\partial^4 f_\theta/\partial\theta^4$, and $\partial^5 f_\theta/\partial\theta^5$ belong to $\mathscr{D}$.

ASSUMPTION 3. There exists $d_1 > 0$ such that

$$I(\theta) = \frac{1}{4\pi}\int_{-\pi}^{\pi} \left\{\frac{\partial}{\partial\theta}\log f_\theta(\lambda)\right\}^2 d\lambda \geqslant d_1 > 0, \qquad \text{for all} \quad \theta \in \Theta.$$

Suppose that a stretch, $\mathbf{X}_n = (X_1, ..., X_n)'$ of the series $\{X_t\}$ is available. Let Σ_n be the covariance matrix of $\mathbf{X}_n$. The likelihood function based on $\mathbf{X}_n$ is given by

$$L_n(\theta) = (2\pi)^{-n/2}\, |\Sigma_n|^{-1/2} \exp\{-\tfrac{1}{2}\mathbf{X}_n'\Sigma_n^{-1}\mathbf{X}_n\}.$$

Let

$$Z_1(\theta) = \frac{1}{\sqrt{n}}\frac{\partial}{\partial\theta}\log L_n(\theta),$$

$$Z_2(\theta) = \frac{1}{\sqrt{n}}\left\{\frac{\partial^2}{\partial\theta^2}\log L_n(\theta) - E_\theta \frac{\partial^2}{\partial\theta^2}\log L_n(\theta)\right\},$$

and

$$Z_3(\theta) = \frac{1}{\sqrt{n}}\left\{\frac{\partial^3}{\partial\theta^3}\log L_n(\theta) - E_\theta \frac{\partial^3}{\partial\theta^3}\log L_n(\theta)\right\}.$$

The asymptotic moments (cumulants) of $Z_1(\theta)$, $Z_2(\theta)$, and $Z_3(\theta)$ are evaluated by Taniguchi [9] as follows.

LEMMA 1. *Under Assumptions* 1–3, *we have*

$$E_\theta\{Z_1(\theta)^2\} = I(\theta) + O(n^{-1}),$$

$$E_\theta\{Z_1(\theta)\, Z_2(\theta)\} = J(\theta) + O(n^{-1}),$$

$$E_\theta\{Z_1(\theta)^3\} = \frac{1}{\sqrt{n}} K(\theta) + O(n^{-3/2}),$$

$$E_\theta\{Z_1(\theta)\, Z_3(\theta)\} = L(\theta) + O(n^{-1}),$$

$$\mathrm{Var}_\theta\{Z_2(\theta)\} = M(\theta) + O(n^{-1}),$$

$$E_\theta\{Z_1(\theta)^2 Z_2(\theta)\} = \frac{1}{\sqrt{n}} N(\theta) + O(n^{-3/2}),$$

$$\mathrm{cum}_\theta\{Z_1(\theta), Z_1(\theta), Z_1(\theta), Z_1(\theta)\} = \frac{1}{n} H(\theta) + O(n^{-2}),$$

where

$$J(\theta) = -\frac{1}{2\pi}\int_{-\pi}^{\pi} \left\{\frac{\partial}{\partial\theta} f_\theta(\lambda)\right\}^3 f_\theta(\lambda)^{-3}\, d\lambda$$
$$+\frac{1}{4\pi}\int_{-\pi}^{\pi} \left\{\frac{\partial^2}{\partial\theta^2} f_\theta(\lambda)\right\}\left\{\frac{\partial}{\partial\theta} f_\theta(\lambda)\right\} f_\theta(\lambda)^{-2}\, d\lambda,$$

$$K(\theta) = \frac{1}{2\pi}\int_{-\pi}^{\pi} \left\{\frac{\partial}{\partial\theta} f_\theta(\lambda)\right\}^3 f_\theta(\lambda)^{-3}\, d\lambda,$$

$$L(\theta) = \frac{3}{2\pi}\int_{-\pi}^{\pi} \left\{\frac{\partial}{\partial\theta} f_0(\lambda)\right\}^4 f_\theta(\lambda)^{-4}\, d\lambda$$
$$-\frac{3}{2\pi}\int_{-\pi}^{\pi} \left\{\frac{\partial^2}{\partial\theta^2} f_\theta(\lambda)\right\}\left\{\frac{\partial}{\partial\theta} f_\theta(\lambda)\right\}^2 f_\theta(\lambda)^{-3}\, d\lambda$$
$$+\frac{1}{4\pi}\int_{-\pi}^{\pi} \left\{\frac{\partial^3}{\partial\theta^3} f_\theta(\lambda)\right\}\left\{\frac{\partial}{\partial\theta} f_\theta(\lambda)\right\} f_\theta(\lambda)^{-2}\, d\lambda,$$

$$M(\theta) = \frac{1}{\pi}\int_{-\pi}^{\pi} \left\{\frac{\partial}{\partial\theta} f_\theta(\lambda)\right\}^4 d\lambda$$
$$-\frac{1}{\pi}\int_{-\pi}^{\pi} \left\{\frac{\partial}{\partial\theta} f_\theta(\lambda)\right\}^2 \left\{\frac{\partial^2}{\partial\theta^2} f_\theta(\lambda)\right\} f_\theta(\lambda)^{-3}\, d\lambda$$
$$+\frac{1}{4\pi}\int_{-\pi}^{\pi} \left\{\frac{\partial^2}{\partial\theta^2} f_\theta(\lambda)\right\}^2 f_\theta(\lambda)^{-2}\, d\lambda,$$

$$N(\theta) = -\frac{1}{\pi}\int_{-\pi}^{\pi} \left\{\frac{\partial}{\partial\theta} f_\theta(\lambda)\right\}^4 f_\theta(\lambda)^{-4}\, d\lambda$$
$$+\frac{1}{2\pi}\int_{-\pi}^{\pi} \left\{\frac{\partial^2}{\partial\theta^2} f_\theta(\lambda)\right\}\left\{\frac{\partial}{\partial\theta} f_\theta(\lambda)\right\}^2 f_\theta(\lambda)^{-3}\, d\lambda,$$

$$H(\theta) = \frac{3}{2\pi}\int_{-\pi}^{\pi} \left\{\frac{\partial}{\partial\theta} f_\theta(\lambda)\right\}^4 f_\theta(\lambda)^{-4}\, d\lambda.$$

Henceforth, for simplicity, we sometimes use $Z_1, Z_2, Z_3, I, J, K,$ etc. instead of $Z_1(\theta)$, $Z_2(\theta)$, $Z_3(\theta)$, $I(\theta)$, $J(\theta)$, $K(\theta)$, etc., respectively.

Now we consider the equation

$$\frac{\partial}{\partial\theta} l_n(\theta) = 0, \qquad \theta \in \Theta, \tag{2.1}$$

where $l_n(\theta) = \log L_n(\theta)$. The maximum likelihood estimator $\hat{\theta}_n$ of θ_0 is defined by a value of θ that satisfies Eq. (2.1). The following lemma is due to Taniguchi [10].

LEMMA 2. *Assume that Assumptions* 1–3 *hold. Let* α *be an arbitrary fixed number such that* $0<\alpha<\frac{3}{8}$.

(1) *There exists a statistic* $\hat{\theta}_n$ *which solves* (2.1) *such that for some* $d_2>0$,

$$P^n_{\theta_0}[|\hat{\theta}_n-\theta_0|<d_2 n^{\alpha-1/2}]=1-o(n^{-1}), \tag{2.2}$$

uniformly for $\theta_0 \in C$.

(2) *For* $\{\hat{\theta}_n\}$ *satisfying* (2.2), *we have the stochastic expansion*

$$\begin{aligned}\sqrt{n}(\hat{\theta}_n-\theta_0)=&\frac{Z_1}{I_n}+\frac{Z_1 Z_2}{I^2\sqrt{n}}-\frac{3J+K}{2I^3\sqrt{n}}Z_1^2\\&+\frac{1}{I^3 n}\left\{Z_1 Z_2^2+\frac{1}{2}Z_1^2 Z_3-\frac{9J+3K}{2I}Z_1^2 Z_2+\frac{(3J+K)^2}{2I^2}Z_1^3\right.\\&\left.-\frac{4L+3M+6N+H}{6I}Z_1^3\right\}+o_p(n^{-1}),\end{aligned} \tag{2.3}$$

where $I_n=E(Z_1^2)$.

3. ASYMPTOTIC EXPANSIONS FOR THE NULL DISTRIBUTIONS

Consider the transformation

$$\begin{aligned}W_1&=Z_1/\sqrt{I},\\W_2&=Z_2-J\cdot I^{-1}Z_1,\\W_3&=Z_3-L\cdot I^{-1}Z_1.\end{aligned}$$

For the testing problem $H:\theta=\theta_0$ against $A:\theta\neq\theta_0$, we introduce the following class of tests:

$$\begin{aligned}\mathscr{S}_H=\Bigg\{T\,|\,T=&W_1^2+\frac{1}{\sqrt{n}}(a_1 W_1^2 W_2+a_2 W_1^3)\\&+\frac{1}{n}(b_1 W_1^2+b_2 W_1^2 W_2^2+b_3 W_1^4+b_4 W_1^3 W_2+b_5 W_1^3 W_3)+o_p(n^{-1}),\\&\text{under } H, \text{ where } a_i\ (i=1,2) \text{ and } b_i\ (i=1,\ldots,5)\\&\text{are nonrandom constants}\Bigg\}.\end{aligned}$$

This class $\mathscr{S}_H$ is a very natural one.

(i) The likelihood ratio test $\mathrm{LR} = 2[l_n(\hat{\theta}_n) - l_n(\theta_0)]$ belongs to $\mathscr{S}_H$. In fact, expanding LR in a Taylor series at $\theta = \theta_0$, and noting Lemma 1 and (2.2), we obtain

$$\begin{aligned}
\mathrm{LR} &= 2(\hat{\theta}_n - \theta_0)\frac{\partial}{\partial\theta} l_n(\theta_0) + (\hat{\theta}_n - \theta_0)^2 \frac{\partial^2}{\partial\theta^2} l_n(\theta_0) \\
&\quad + \frac{1}{3}(\hat{\theta}_n - \theta_0)^3 \frac{\partial^3}{\partial\theta^3} l_n(\theta_0) + \frac{1}{12}(\hat{\theta}_n - \theta_0)^4 \frac{\partial^4}{\partial\theta^4} l_n(\theta_0) + o_p(n^{-1}) \\
&= 2\sqrt{n}(\hat{\theta}_n - \theta_n)\, Z_1(\theta_0) - \{\sqrt{n}(\hat{\theta}_n - \theta_0)\}^2 \{I(\theta_0) + \Delta(\theta_0)/n\} \\
&\quad + \frac{1}{\sqrt{n}} Z_2(\theta_0)\{\sqrt{n}(\hat{\theta}_n - \theta_0)\}^2 \\
&\quad + \frac{1}{3\sqrt{n}} \{\sqrt{n}(\hat{\theta}_n - \theta_0)\}^3 \left\{E \frac{1}{n}\frac{\partial^3}{\partial\theta^3} l_n(\theta_0)\right\} \\
&\quad + \frac{1}{3n} \{\sqrt{n}(\hat{\theta}_n - \theta_0)\}^3 Z_3(\theta_0) \\
&\quad + \frac{1}{12n} \{\sqrt{n}(\hat{\theta}_n - \theta_0)\}^4 \left\{E \frac{1}{n}\frac{\partial^4}{\partial\theta^4} l_n(\theta_0)\right\} + o_p(n^{-1}),
\end{aligned} \tag{3.1}$$

where $EZ_1(\theta)^2 = I(\theta) + \Delta(\theta)/n + o(n^{-1})$. Notice that

$$E\left[\frac{1}{n}\frac{\partial^3}{\partial\theta^3} l_n(\theta_0)\right] = -3J - K + O(n^{-1}), \tag{3.2}$$

$$E\left[\frac{1}{n}\frac{\partial^4}{\partial\theta^4} l_n(\theta_0)\right] = -4L - 3M - 6N - H + O(n^{-1}) \tag{3.3}$$

(see Taniguchi [9]). Substituting (2.3), (3.2), and (3.3) in (3.1) we have

$$\begin{aligned}
\mathrm{LR} &= W_1^2 + \frac{1}{3\sqrt{n}\, I^{3/2}}(3I^{1/2} W_1^2 W_2 - K W_1^3) + \frac{1}{12 n I^3}[-12 I^2 \Delta W_1^2 + 12 I W_1^2 W_2^2 \\
&\quad + \{3(J+K)^2 - I(3M + 6N + H)\} W_1^4 - 12 I^{1/2}(J+K) W_1^3 W_2 \\
&\quad + 4 I^{3/2} W_1^3 W_3] + o_p(n^{-1}),
\end{aligned}$$

which implies that LR belongs to $\mathscr{S}_H$.

Similarly, we can get results (ii)–(iv):

(ii) Wald's test $\mathrm{W} = n(\hat{\theta}_n - \theta_0)^2 I(\hat{\theta}_n)$ belongs to $\mathscr{S}_H$ with the coefficients $a_1 = 2/I$, $a_2 = J/I^{3/2}$, $b_1 = -2\Delta/I$, $b_2 = 3/I^2$, $b_3 = -(3J^2 + 4JK + K^2)/4I^3 + (4L + 3N + H)/6I^2$, $b_4 = -K/I^{5/2}$, and $b_5 = 1/I^{3/2}$.

(iii) A modified Wald's test $\mathrm{MW}=n(\hat{\theta}_n-\theta_0)^2 I(\theta_0)$ belongs to $\mathscr{S}_H$ with the coefficients $a_1=2/I$, $a_2=-(J+K)/I^{3/2}$, $b_1=-2\Delta/I$, $b_2=3/I^2$, $b_3=(9J^2+14JK+5K^2)/4I^3-(L+3M+6N+H)/3I^2$, $b_4=-(6J+4K)/I^{5/2}$, and $b_5=1/I^{3/2}$.

(iv) Rao's test $\mathrm{R}=Z_1(\theta_0)^2 I(\theta_0)^{-1}$ belongs to $\mathscr{S}_H$ with the coefficients $a_1=a_2=b_1=b_2=b_3=b_4=b_5=0$.

To derive the asymptotic expansion of the distribution of $T\in\mathscr{S}_H$, we need the following lemma (see Taniguchi [10]).

LEMMA 3. *Uner Assumptions* 1–3, $\mathbf{W}=(W_1, W_2, W_3)'$ *has the following Edgeworth expansion*:

$$\begin{aligned}
P_{\theta_0}^n[\mathbf{W}\in B] &= \int_B f_1(w_1)\, f_2(w_2, w_3)\Bigg[1+\frac{1}{6\sqrt{n}}\sum_{j,k,l=1}^{3} c_{jkl}^{(1)} H_{jkl}(\mathbf{w}) \\
&\quad +\frac{1}{2n}\sum_{j,k=1}^{3} c_{jk}^{(3)} H_{jk}(\mathbf{w})+\frac{1}{24n}\sum_{j,k,l,m=1}^{3} c_{jklm}^{(1)} H_{jklm}(\mathbf{w}) \\
&\quad +\frac{1}{72n}\sum_{j,k,l,j',k',l'=1}^{3} c_{jkl}^{(1)} c_{j'k'l'}^{(1)} H_{jklj'k'l'}(\mathbf{w})\Bigg]\, d\mathbf{w}+o(n^{-1}) \\
&= \int_B q_n(\mathbf{w})\, d\mathbf{w}+o(n^{-1}) \qquad \textit{say},
\end{aligned}$$

where B is a Borel set of R^3, $\mathbf{w}'=(w_1, w_2, w_3)$, $f_1(w_1)=(2\pi)^{-1/2}e^{-w_1^2/2}$, $f_2(w_2, w_3)=(2\pi)^{-1}\,|\Omega_2|^{-1/2}\exp-\frac{1}{2}(w_2, w_3)\,\Omega_2^{-1}(w_2, w_3)'$, and $H_{j_1,\ldots,j_s}(\mathbf{w})$ are the Hermite polynomials. Here the above coefficients $c_{\cdots}^{(\cdot)}$ and the matrix Ω_2 can be expressed by using the spectral density.

For $T\in\mathscr{S}_H$, define $c_T(t)=E[e^{itT}]$. By Lemma 3 we have

$$\begin{aligned}
c_T(t) &= \iiint \exp it\left\{w_1^2+\frac{1}{\sqrt{n}}(a_1 w_1^2 w_2+a_2 w_1^3)\right. \\
&\quad \left.+\frac{1}{n}(b_1 w_1^2+b_2 w_1^2 w_2^2+b_3 w_1^4+b_4 w_1^3 w_2+b_5 w_1^3 w_3\right\} q_n(\mathbf{w})\, d\mathbf{w}+o(n^{-1}) \\
&= \iiint \exp(itw_1^2)\times\Big[1+\frac{it}{\sqrt{n}}(a_1 w_1^2 w_2+a_2 w_1^3) \\
&\quad +\frac{it}{n}(b_1 w_1^2+b_2 w_1^2 w_2^2+b_3 w_1^4+b_4 w_1^3 w_2+b_5 w_1^3 w_3) \\
&\quad +\frac{(it)^2}{2n}(a_1 w_1^2 w_2+a_2 w_1^3)^2\Big]\, q_n(\mathbf{w})\, d\mathbf{w}+o(n^{-1}).
\end{aligned}$$

In the first place we calculate the above integral with respect to w_2 and w_3. Second, integrating it with respect to w_1, it is not difficult to show the following lemma.

LEMMA 4. *Under Assumptions* 1–3, *the characteristic function* $c_T(t)$ *has the following asymptotic expansion*:

$$c_T(t) = (1-2it)^{-1/2}\left[1 + n^{-1}\sum_{j=0}^{3} A_j^{(T)}(1-2it)^{-j}\right] + o(n^{-1}),$$

where

$$\begin{aligned}
A_0^{(T)} &= \{9I^2(IM-J^2)\,a_1^2 + 6I(IN-JK)a_1 - 12I^3b_1 - 12I^2(IM-J^2)b_2 \\
&\quad - 12I^2\,\Delta + 3IH - 5K^2\}/24I^3, \\
A_1^{(T)} &= \{-6I^2(IM-J^2)a_1^2 - 8I(IN-JK)a_1 + 15I^3a_2^2 + 6I^{3/2}Ka_2 \\
&\quad + 4I^3b_1 + 4I^2(IM-J^2)b_2 - 12I^3b_3 + 4I^2\Delta - 2IH + 5K^2\}/8I^3, \\
A_2^{(T)} &= \{3(I^3M - I^2J^2)\,a_1^2 + 6I(IN-JK)a_1 - 30I^3a_2^2 - 16KI^{3/2}a_2 \\
&\quad + 12I^3b_3 + IH - 5K^2\}/8I^3, \\
A_3^{(T)} &= 5(3I^{3/2}a_2 + K)^2/24I^3.
\end{aligned}$$

From the above lemma we have

THEOREM 1. *Under Assumptions* 1–3, *the asymptotic expansion of the distribution of* $T \in \mathscr{S}_H$ *is given by*

$$P_{\theta_0}^n[T \leqslant x] = P[\chi_1^2 \leqslant x] + n^{-1}\sum_{j=0}^{3} A_j^{(T)}P[\chi_{1+2j}^2 \leqslant x] + o(n^{-1}). \quad (3.4)$$

For concrete spectral models we can give the coefficients $A_j^{(T)}$ in (3.4) for the four tests T = LR, W, MW, and R in simple forms (cf. Taniguchi [9]).

EXAMPLE 1. For the autoregressive spectral density

$$f_{\theta_0}(\lambda) = \frac{\sigma^2}{2\pi}\,|1-\alpha e^{i\lambda}|^{-2} \qquad (\theta_0 = \alpha),$$

we can show that

(i) for T = LR (likelihood ratio test),

$$A_0^{(LR)} = 1, \qquad A_1^{(LR)} = -1, \qquad A_2^{(LR)} = A_3^{(LR)} = 0;$$

(ii) for T = W (Wald's test),

$$A_0^{(W)} = \frac{5\alpha^2 - 1}{4(1-\alpha^2)}, \qquad A_1^{(W)} = -\frac{\alpha^2 + 1}{2(1-\alpha^2)}, \qquad A_2^{(W)} = \frac{3}{4}, \qquad A_3^{(W)} = 0;$$

(iii) for T = MW (modified Wald's test),

$$A_0^{(MW)} = \frac{5\alpha^2 - 1}{4(1-\alpha^2)}, \qquad A_1^{(MW)} = \frac{2 - \alpha^2}{2(1-\alpha^2)},$$

$$A_2^{(MW)} = \frac{-33\alpha^2 - 3}{4(1-\alpha^2)}, \qquad A_3^{(MW)} = \frac{15\alpha^2}{2(1-\alpha^2)};$$

(iv) for T = R (Rao's test),

$$A_0^{(R)} = \frac{11 - 15\alpha^2}{4(1-\alpha^2)}, \qquad A_1^{(R)} = \frac{27\alpha^2 - 10}{2(1-\alpha^2)},$$

$$A_2^{(R)} = \frac{9 - 69\alpha^2}{4(1-\alpha^2)}, \qquad A_3^{(R)} = \frac{15\alpha^2}{2(1-\alpha^2)}.$$

EXAMPLE 2. For the moving average spectral density

$$f_{\theta_0}(\lambda) = \frac{\sigma^2}{2\pi} |1 - \beta e^{i\lambda}|^2 \qquad (\theta_0 = \beta),$$

we can show that

(i) for T = LR,

$$A_0^{(LR)} = -\frac{1 + 2\beta^2}{2(1-\beta^2)}, \qquad A_1^{(LR)} = \frac{1 + 2\beta^2}{2(1-\beta^2)}, \qquad A_2^{(LR)} = A_3^{(LR)} = 0;$$

(ii) for T = W,

$$A_0^{(W)} = \frac{-9 - 7\beta^2}{4(1-\beta^2)}, \qquad A_1^{(W)} = \frac{5\beta^2 - 3}{2(1-\beta^2)},$$

$$A_2^{(W)} = \frac{15 - 33\beta^2}{4(1-\beta^2)}, \qquad A_3^{(W)} = \frac{15\beta^2}{2(1-\beta^2)};$$

(iii) for T = MW,

$$A_0^{(MW)} = \frac{-9 - 7\beta^2}{4(1-\beta^2)}, \qquad A_1^{(MW)} = \frac{5\beta^2}{2(1-\beta^2)},$$

$$A_2^{(MW)} = \frac{-3\beta^2 + 9}{4(1-\beta^2)}, \qquad A_3^{(MW)} = 0;$$

(iv) for $\mathrm{T}=\mathrm{R}$,

$$A_0^{(\mathrm{R})}=\frac{11-3\beta^2}{4(1-\beta^2)}, \qquad A_1^{(\mathrm{R})}=\frac{21\beta^2-10}{2(1-\beta^2)},$$

$$A_2^{(\mathrm{R})}=\frac{3(3-23\beta^2)}{4(1-\beta^2)}, \qquad A_3^{(\mathrm{R})}=\frac{15\beta^2}{2(1-\beta^2)}.$$

4. Bartlett's Adjustment

In this section we illuminate Bartlett's adjustment for $T\in\mathscr{S}_H$. Since $T\in\mathscr{S}_H$, it is easy to show that

$$E(\mathrm{T})=1-\rho/n+o(n^{-1}),$$

where

$$\rho=-\{I^2\Delta+I^3b_1+I^2(IM-J^2)b_2+3I^3b_3+Ia_1(IN-JK)+I^{3/2}Ka_2\}/I^3.$$

Thus we have

$$\mathrm{T}/E(\mathrm{T})=\left(1+\frac{\rho}{n}\right)\mathrm{T}+o_p(n^{-1}).$$

The above ρ is called Bartlett's adjustment factor. If the terms of order n^{-1} in the asymptotic expansion of the distribution of $\mathrm{T}^*=(1+\rho/n)\mathrm{T}$ vanish (i.e., $P^n_{\theta_0}[\mathrm{T}^*\leqslant x]=P[\chi_1^2\leqslant x]+o(n^{-1})$), we say that T is adjustable in the sense of Bartlett.

Denoting $c_{\mathrm{T}^*}(t)=Ee^{it\mathrm{T}^*}$, we have

$$\begin{aligned}
c_{\mathrm{T}^*}(t)&=c_{\mathrm{T}}(t)+E\left\{e^{it\mathrm{W}_1^2}\cdot\frac{it\rho\mathrm{W}_1^2}{n}\right\}+o(n^{-1})\\
&=c_{\mathrm{T}}(t)+(1-2it)^{-1/2}\left\{\frac{\rho}{2n}\left(\frac{1}{1-2it}-1\right)\right\}+o(n^{-1})\\
&=(1-2it)^{-1/2}\left[1+n^{-1}\left\{A_0^{(\mathrm{T})}-\frac{\rho}{2}+\left(A_1^{(\mathrm{T})}+\frac{\rho}{2}\right)(1-2it)^{-1}\right.\right.\\
&\qquad\left.\left.+A_2^{(\mathrm{T})}(1-2it)^{-2}+A_3^{(\mathrm{T})}(1-2it)^{-3}\right\}\right]+o(n^{-1}).
\end{aligned}\tag{4.1}$$

In (4.1), putting $A_0^{(T)} - \rho/2 = 0$, $A_1^{(T)} + \rho/2 = 0$, $A_2^{(T)} = 0$, and $A_3^{(T)} = 0$, we have the following theorem.

THEOREM 2. *The test statistic* $\mathrm{T} \in \mathcal{S}_H$ *is adjustable in the sense of Bartlett if and only if the coefficients* $\{a_j\}$ *and* $\{b_j\}$ *satisfy the relations* (i) *and* (ii):

(i) $a_2 = -K/3I^{3/2}$,

(ii) $3I^2(IM - J^2)a_1^2 + 6I(IN - JK)a_1 + 12I^3 b_3 + IH - 3K^2 = 0.$

Among the four tests LR, W, MW, and R, the LR test is the only one which is adjustable in the sense of Bartlett.

For the LR test, Bartlett's adjustment factor $\rho = \rho_{\mathrm{LR}}(\theta_0)$ is given by

$$\rho_{\mathrm{LR}}(\theta_0) = \frac{-M + 2N + H}{4I^2} + \frac{3J^2 - 6JK - 5K^2}{12I^3}.$$

In particular, for the ARMA spectral density

$$f_{\theta_0}(\lambda) = \frac{\sigma^2}{2\pi} \frac{|1 - \beta e^{i\lambda}|^2}{|1 - \alpha e^{i\lambda}|^2}$$

the Bartlett's adjustment factors are given by

$$\rho_{\mathrm{LR}}(\sigma^2) = -1/3, \qquad \text{for} \quad \theta_0 = \sigma^2,$$

$$\rho_{\mathrm{LR}}(\alpha) = 2, \qquad \text{for} \quad \theta_0 = \alpha,$$

$$\rho_{\mathrm{LR}}(\beta) = \frac{-1 - 2\beta^2}{1 - \beta^2}, \qquad \text{for} \quad \theta_0 = \beta.$$

5. Asymptotic Expansions for the Nonnull Distributions

Here we introduce a class $\mathcal{S}_A$ of tests and derive the χ^2 type asymptotic expansion of the distribution of $S \in \mathcal{S}_A$ under the sequence of alternatives $A_n: \theta = \theta_0 + \varepsilon/\sqrt{n}$, $\varepsilon > 0$. Consider the transformation

$$U_1(\theta) = Z_1(\theta)/\sqrt{I(\theta)}, \tag{5.1}$$

$$U_2(\theta) = (Z_2(\theta) - J(\theta)\, I(\theta)^{-1} Z_1(\theta))/(\gamma_\theta I(\theta)), \tag{5.2}$$

where $\gamma_\theta = (M(\theta)\, I(\theta) - J(\theta)^2)^{1/2}/I(\theta)^{3/2}$. In this section, for simplicity, we use $U_1, U_2, Z_1, Z_2, I, J, K, \gamma$, instead of $U_1(\theta)$, $U_2(\theta)$, $Z_1(\theta)$, $Z_2(\theta)$, $I(\theta)$,

$J(\theta)$, $K(\theta)$, γ_θ, respectively, if they are evaluated at $\theta=\theta_0+\varepsilon/\sqrt{n}$. Define the following class of tests:

$$\mathscr{S}_A=\{S\,|\,S=\{U_1+I(\theta_0)^{1/2}\varepsilon\}^2+\frac{1}{\sqrt{n}}[c_1U_1^3+c_2U_1^2U_2$$
$$+\{c_3U_1^2+c_4U_1U_2\}\varepsilon+\{c_5U_1+c_6U_2\}\varepsilon^2+c_7\varepsilon^3]+o_p(n^{-1/2}),$$
$$\text{under } A_n, \text{ where } c_7=I^{3/2}c_1-Ic_3+I^{1/2}c_5\}.$$

This class $\mathscr{S}_A$ is also very natural:

(i) The likelihood ratio test $\mathrm{LR}=2[l_n(\hat{\theta}_n)-l_n(\theta_0)]$ belongs to $\mathscr{S}_A$. In fact, expanding LR in a Taylor series at $\theta=\hat{\theta}_n$, we obtain

$$\begin{aligned}\mathrm{LR}&=-(\theta_0-\hat{\theta}_n)^2\frac{\partial^2}{\partial\theta^2}l_n(\hat{\theta}_n)+\frac{1}{3}(\hat{\theta}_n-\theta_0)^3\frac{\partial^3}{\partial\theta^3}l_n(\hat{\theta}_n)+o_p(n^{-1/2})\\
&=-(\hat{\theta}_n-\theta+\theta-\theta_0)^2\left\{\frac{\partial^2}{\partial\theta^2}l_n(\theta)+(\hat{\theta}_n-\theta)\frac{\partial^3}{\partial\theta^3}l_n(\theta)\right\}\\
&\quad+\frac{1}{3}(\hat{\theta}_n-\theta+\theta-\theta_0)^3\frac{\partial^3}{\partial\theta^3}l_n(\theta)+o_p(n^{-1/2})\\
&=-\frac{1}{n}\frac{\partial^2}{\partial\theta^2}l_n(\theta)(v^2+2v\varepsilon+\varepsilon^2)\\
&\quad-\frac{1}{3\sqrt{n}}\left\{\frac{1}{n}\frac{\partial^3}{\partial\theta^3}l_n(\theta)\right\}(2v^3+3v^2\varepsilon-\varepsilon^3)+o_p(n^{-1/2}),\end{aligned}\tag{5.3}$$

where $v=\sqrt{n}(\hat{\theta}_n-\theta)$. Substituting

$$v=Z_1/I+\left\{Z_1Z_2-\frac{3J+K}{2I}Z_1^2\right\}\Big/\{I^2\sqrt{n}\}+o_p(n^{-1/2})\qquad(\text{cf. (2.3)}),$$

$$E\left\{\frac{1}{n}\frac{\partial^3}{\partial\theta^3}l_n(\theta)\right\}=-3J-K+O(n^{-1})\qquad(\text{cf. (3.2)}),$$

$$\frac{1}{n}\frac{\partial^2}{\partial\theta^2}l_n(\theta)=-I+\frac{1}{\sqrt{n}}Z_2+O(n^{-1}),$$

for (5.3) we have

$$\mathrm{LR}=\{U_1+I(\theta_0)^{1/2}\varepsilon\}^2+\frac{1}{\sqrt{n}}\left[-\frac{K}{3I^{3/2}}U_1^3+\gamma U_1^2U_2\right.$$
$$\left.+\left\{\frac{J+K}{I^{1/2}}U_1-\gamma IU_2\right\}\varepsilon^2+\frac{3J+2K}{3}\varepsilon^3\right]+o_p(n^{-1/2}).$$

Similarly we can get the results (ii)–(iv):

(ii) Wald's test $\mathrm{W}=n(\hat{\theta}_n-\theta_0)^2 I(\hat{\theta}_n)$ belongs to $\mathscr{S}_A$ with the coefficients $c_1=J/I^{3/2}$, $c_2=2\gamma$, $c_3=(3J+K)/I$, $c_4=2\gamma I^{1/2}$, $c_5=2(2J+K)/I^{1/2}$, $c_6=0$, and $c_7=(2J+K)$.

(iii) The modified Wald's test $\mathrm{MW}=n(\hat{\theta}_n-\theta_0)^2 I(\theta_0)$ belongs to $\mathscr{S}_A$ with the coefficients $c_1=-(J+K)/I^{3/2}$, $c_2=2\gamma$, $c_3=-(3J+2K)/I$, $c_4=2\gamma I^{1/2}$, $c_5=-(2J+K)/I^{1/2}$, and $c_6=c_7=0$.

(iv) Rao's test $\mathrm{R}=Z_1(\theta_0)^2 I(\theta_0)^{-1}$ belongs to $\mathscr{S}_A$ with the coefficients $c_1=c_2=0$, $c_3=K/I$, $c_4=-2\gamma I^{1/2}$, $c_5=(J+2K)/I^{1/2}$, $c_6=-2\gamma I$, and $c_7=J+K$.

The following lemma is essentially due to Taniguchi [10].

LEMMA 5. *Under Assumptions* 1–3,

$$
\begin{aligned}
&P^n_{\theta_0+\varepsilon/\sqrt{n}}[U_1<y_1, U_2<y_2] \\
&=\int_{-\infty}^{y_1}\int_{-\infty}^{y_2}\phi(u_1)\,\phi(u_2) \\
&\times\left[1+\frac{1}{6\sqrt{n}}\left\{\frac{K(\theta_0)}{I(\theta_0)^{3/2}}(u_1^3-3u_1)+3c_{112}(u_1^2u_2-u_2)\right.\right. \\
&\left.\left.+3c_{122}(u_1u_2^2-u_1)+c_{222}(u_2^3-3u_2)\right\}\right]du_1\,du_2+o(n^{-1/2}) \\
&=\int_{-\infty}^{y_1}\int_{-\infty}^{y_2} f(u_1,u_2)\,du_1\,du_2+o(n^{-1/2}) \qquad \textit{say},
\end{aligned}
$$

where $\phi(u)=(1/\sqrt{2\pi})\exp(-u^2/2)$, *and the coefficients* c_{112}, c_{122}, *and* c_{222} *are expressed by the spectral density* (see Taniguchi [10]).

Using Lemma 5 we can evaluate the characteristic function $c_{\mathrm{S}}(t)$ of $\mathrm{S}\in\mathscr{S}_A$, under A_n. In fact,

$$
\begin{aligned}
c_{\mathrm{S}}(t)&=E_{\theta_0+\varepsilon/\sqrt{n}}\{e^{it\mathrm{S}}\} \\
&=\iint f(u_1,u_2)\exp[it\{u_1+I(\theta_0)^{1/2}\varepsilon\}^2] \\
&\quad\times\left[1+\frac{it}{\sqrt{n}}\{c_1u_1^3+c_2u_1^2u_2\right. \\
&\quad+(c_3u_1^2+c_4u_1u_2)\varepsilon+(c_5u_1+c_6u_2)\varepsilon^2 \\
&\quad\left.+c_7\varepsilon^3\}\right]du_1\,du_2+o(n^{-1/2})
\end{aligned}
$$

$$= \iint \phi(u_1)\,\phi(u_2)\exp[it\{u_1 + I(\theta_0)^{1/2}\varepsilon\}^2]$$

$$\times\Bigg[1 + \frac{it}{\sqrt{n}}\{c_1u_1^3 + c_2u_1^2u_2$$

$$+ (c_3u_1^2 + c_4u_1u_2)\varepsilon + (c_5u_1 + c_6u_2)\varepsilon^2 + c_7\varepsilon^3\}$$

$$+ \frac{1}{6\sqrt{n}}\Bigg\{\frac{K(\theta_0)}{I(\theta_0)^{3/2}}(u_1^3 - 3u_1)$$

$$+ 3c_{112}(u_1^2u_2 - u_2) + 3c_{122}(u_1u_2^2 - u_1)$$

$$+ c_{222}(u_2^3 - 3u_2)\Bigg\}\Bigg]\,du_1\,du_2 + o(n^{-1/2}). \tag{5.4}$$

Integration of (5.4) with respect to u_2 yields

$$c_S(t) = \exp\left\{\frac{itI(\theta_0)\varepsilon^2}{1 - 2it}\right\}\cdot(1 - 2it)^{-1/2}\int(2\pi)^{-1/2}(1 - 2it)^{1/2}$$

$$\times\exp\left[-\frac{1 - 2it}{2}\left\{u_1 - \frac{2\varepsilon itI(\theta_0)^{1/2}}{1 - 2it}\right\}^2\right]$$

$$\times\Bigg[1 + \frac{it}{\sqrt{n}}\{c_1u_1^3 + c_3u_1^2\varepsilon + c_5u_1\varepsilon^2 + c_7\varepsilon^3\}$$

$$+ \frac{K(\theta_0)}{6\sqrt{n}I(\theta_0)^{3/2}}(u_1^3 - 3u_1)\Bigg]\,du_1 + o(n^{-1/2}).$$

Calculation of the above integral leads to

LEMMA 6. *Under Assumptions* 1–3, *the characteristic function* $c_S(t)$ *of* $S \in \mathscr{S}_A$ *under* $\theta = \theta_0 + \varepsilon/\sqrt{n}$ *has the asymptotic expansion*

$$c_S(t) = \exp\left\{\frac{itI(\theta_0)\varepsilon^2}{1 - 2it}\right\}\times(1 - 2it)^{-1/2}$$

$$\times\left[1 + n^{-1/2}\sum_{j=0}^{3}B_j^{(S)}(1 - 2it)^{-j}\right] + o(n^{-1/2}),$$

where

$$B_0^{(S)} = \tfrac{1}{6}[\{-9I(\theta_0)^{3/2}c_1 + 6I(\theta_0)c_3 - 3I(\theta_0)^{1/2}c_5 - K(\theta_0)\}\varepsilon^3$$

$$+ \{9I(\theta_0)^{1/2}c_1 - 3c_3 + 3K(\theta_0)/I(\theta_0)\}\varepsilon],$$

$$B_1^{(S)} = \tfrac{1}{2}[\{6I(\theta_0)^{3/2}c_1 - 3I(\theta_0)c_3 + I(\theta_0)^{1/2}c_5 + K(\theta_0)\}\varepsilon^3$$

$$+ \{c_3 - 6I(\theta_0)^{1/2}c_1 - 2K(\theta_0)\,I(\theta_0)^{-1}\}\varepsilon],$$

$$B_2^{(S)} = \tfrac{1}{2}[\{I(\theta_0)c_3 - 4I(\theta_0)^{3/2}c_1 - K(\theta_0)\}\varepsilon^3 + \{3I(\theta_0)^{1/2}c_1 + K(\theta_0)\,I(\theta_0)^{-1}\}\varepsilon],$$

$$B_3^{(S)} = \tfrac{1}{6}\{3I(\theta_0)^{3/2}c_1 + K(\theta_0)\}\,\varepsilon^3.$$

This lemma implies,

THEOREM 3. *Under Assumptions* 1–3, *the distribution function of* $\mathrm{S} \in \mathscr{S}_A$ *for* $\theta = \theta_0 + \varepsilon/\sqrt{n}$ *has the asymptotic expansion*

$$P^n_{\theta_0 + \varepsilon/\sqrt{n}}[\mathrm{S} \leqslant x] = P[\chi_1^2(\delta) \leqslant x] + n^{-1/2} \sum_{j=0}^{3} B_j^{(S)} P[\chi^2_{1+2j}(\delta) \leqslant x] + o(n^{-1/2}),$$

where $\delta^2 = I(\theta_0)\varepsilon^2/2$, *and* $\chi_j^2(\delta)$ *is a noncentral* χ^2 *random variable with* j *degrees of freedom and noncentrality parameter* δ^2.

For the four tests S = LR, W, MW, and R, we can give more explicit expressions for the coefficients $B_j^{(S)}$ in Theorem 3.

EXAMPLE 3. (i) S = LR (likelihood ratio test)

$$B_0^{(LR)} = -(3J(\theta_0) + K(\theta_0))\varepsilon^3/6, \qquad B_1^{(LR)} = J(\theta_0)\varepsilon^3/2,$$

$$B_2^{(LR)} = K(\theta_0)\varepsilon^3/6, \qquad B_3^{(LR)} = 0,$$

(ii) S = W (Wald's test)

$$B_0^{(W)} = -(K(\theta_0) + 3J(\theta_0))\varepsilon^3/6,$$

$$B_1^{(W)} = \{J(\theta_0)\varepsilon^3 - (3J(\theta_0) + K(\theta_0))\varepsilon/I(\theta_0)\}/2,$$

$$B_2^{(W)} = \{-J(\theta_0)\varepsilon^3 + (3J(\theta_0) + K(\theta_0))\varepsilon/I(\theta_0)\}/2,$$

$$B_3^{(W)} = (K(\theta_0) + 3J(\theta_0))\varepsilon^3/6,$$

(iii) S = MW (modified Wald's test)

$$B_0^{(MW)} = -(K(\theta_0) + 3J(\theta_0))\varepsilon^3/6,$$

$$B_1^{(MW)} = \{J(\theta_0)\varepsilon^3 + (3J(\theta_0) + 2K(\theta_0))\varepsilon/I(\theta_0)\}/2,$$

$$B_2^{(MW)} = \{(K(\theta_0) + J(\theta_0))\varepsilon^3 - (3J(\theta_0) + 2K(\theta_0))\varepsilon/I(\theta_0)\}/2,$$

$$B_3^{(MW)} = -(2K(\theta_0) + 3J(\theta_0))\varepsilon^3/6,$$

(iv) S = R (Rao's test)

$$B_0^{(R)} = -(K(\theta_0) + 3J(\theta_0))\varepsilon^3/6, \qquad B_1^{(R)} = (J(\theta_0)\varepsilon^3 - K(\theta_0)\varepsilon/I(\theta_0))/2,$$

$$B_2^{(R)} = K(\theta_0)\varepsilon/\{2I(\theta_0)\}, \qquad B_3^{(R)} = K(\theta_0)\varepsilon^3/6.$$

6. Power Comparisons between the Test Criteria

In view of Theorem 3 we can investigate the local power properties in the class $\mathscr{S}_A$. By Theorem 3 and Example 3, it is not difficult to show that for $S \in \mathscr{S}_A$,

$$\begin{aligned}
&P^n_{\theta_0+\varepsilon/\sqrt{n}}[S > x] - P^n_{\theta_0+\varepsilon/\sqrt{n}}[LR > x] \\
&\quad = \frac{1}{\sqrt{n}}\Bigg[\frac{1}{2}\{P(\chi_7^2(\delta) > x) - P(\chi_5^2(\delta) > x)\}\, Q_3^{(S)}(\theta_0) \\
&\qquad + \frac{1}{2}\{P(\chi_5^2(\delta) > x) - P(\chi_3^2(\delta) > x)\}\, Q_2^{(S)}(\theta_0) \\
&\qquad + \frac{1}{2}\{P(\chi_3^2(\delta) > x) - P(\chi_1^2(\delta) > x)\}\, Q_1^{(S)}(\theta_0)\Bigg] + o(n^{-1/2}), \qquad (6.1)
\end{aligned}$$

where

$$\begin{aligned}
Q_1^{(S)}(\theta_0) &= \{3I(\theta_0)^{3/2}c_1 - 2I(\theta_0)c_3 + I(\theta_0)^{1/2}c_5 - J(\theta_0)\}\varepsilon^3 \\
&\quad + \{c_3 - 3I(\theta_0)^{1/2}c_1 - K(\theta_0)/I(\theta_0)\}\varepsilon, \\
Q_2^{(S)}(\theta_0) &= \{I(\hat{\theta}_0)c_3 - 3I(\theta_0)^{3/2}c_1 - K(\theta_0)\}\varepsilon^3 + \{3I(\theta_0)^{1/2}c_1 + K(\theta_0)/I(\theta_0)\}\varepsilon, \\
Q_3^{(S)}(\theta_0) &= \{3I(\theta_0)^{3/2}c_1 + K(\theta_0)\}\varepsilon^3/3.
\end{aligned}$$

The following relation is well known,

$$P[\chi_{j+2}^2(\delta) > x] - P[\chi_j^2(\delta) > x] = 2p_{j+2}(x; \delta), \qquad (6.2)$$

where $p_j(x; \delta)$ is the probability density function of $\chi_j^2(\delta)$. (6.1) and (6.2) above imply

Theorem 4. *Under Assumptions* 1–3,

$$\begin{aligned}
&P^n_{\theta_0+\varepsilon/\sqrt{n}}[S > x] - P^n_{\theta_0+\varepsilon/\sqrt{n}}[LR > x] \\
&\quad = \frac{1}{\sqrt{n}}[Q_3^{(S)}(\theta_0)\, p_7(x; \delta) + Q_2^{(S)}(\theta_0)\, p_5(x; \delta) \\
&\qquad + Q_1^{(S)}(\theta_0)\, p_3(x; \delta)] + o(n^{-1/2}),
\end{aligned}$$

for $S \in \mathscr{S}_A$.

By Theorem 4, for an ARMA process, we can compare the local power properties among the four tests LR, W, MW, and R.

Consider the following ARMA (p, q) spectral density

$$f_{\theta_0}(\lambda)=\frac{\sigma^2}{2\pi}\frac{\prod_{k=1}^{q}(1-\psi_k e^{i\lambda})(1-\psi_k e^{-i\lambda})}{\prod_{k=1}^{p}(1-\rho_k e^{i\lambda})(1-\rho_k e^{-i\lambda})}, \tag{6.3}$$

where $\psi_1, ..., \psi_q, \rho_1, ..., \rho_p$ are real numbers such that $|\psi_j|<1$, $j=1, ..., q$, $|\rho_j|<1$, $j=1, ..., p$. For the spectral density (6.3) we can get the following local power comparisons.

EXAMPLE 4. W versus LR under A_n,

$$P^n_{\theta_0+\varepsilon/\sqrt{n}}[\mathrm{W}>x]-P^n_{\theta_0+\varepsilon/\sqrt{n}}[\mathrm{LR}>x]$$
$$=\frac{1}{\sqrt{n}}\{3J(\theta_0)+K(\theta_0)\}\left\{\frac{\varepsilon^3}{3}p_7(x;\delta)+\frac{\varepsilon}{I(\theta_0)}p_5(x;\delta)\right\}+o(n^{-1/2}).$$

(i) If $\theta_0=\sigma^2$, then $3J(\theta_0)+K(\theta_0)=-2/\sigma^6<0$, which implies that LR is more powerful than W.

(ii) If $\theta_0=\psi_k$, then $3J(\theta_0)+K(\theta_0)=6\psi_k/(1-\psi_k^2)^2$, which implies that W is more powerful than LR if $\psi_k>0$ and vice versa.

(iii) If $\theta_0=\rho_k$, then $3J(\theta_0)+K(\theta_0)=0$, which implies that LR and W have identical local powers.

EXAMPLE 5. MW versus LR under A_n,

$$P^n_{\theta_0+\varepsilon/\sqrt{n}}[\mathrm{MW}>x]-P^n_{\theta_0+\varepsilon/\sqrt{n}}[\mathrm{LR}>x]$$
$$=\frac{1}{\sqrt{n}}\{-3J(\theta_0)-2K(\theta_0)\}\left\{\frac{\varepsilon^3}{3}p_7(x;\delta)+\frac{\varepsilon}{I(\theta_0)}p_5(x;\delta)\right\}+o(n^{-1/2}).$$

(i) If $\theta_0=\sigma^2$, then $-3J(\theta_0)-2K(\theta_0)=1/\sigma^6>0$, which implies that MW is more powerful than LR.

(ii) If $\theta_0=\psi_k$, then $-3J(\theta_0)-2K(\theta_0)=0$, which implies that MW and LR have identical local powers.

(iii) If $\theta_0=\rho_k$, then $-3J(\theta_0)-2K(\theta_0)=-6\rho_k/(1-\rho_k^2)^2$, which implies that LR is more powerful than MW if $\rho_k>0$ and vice versa.

EXAMPLE 6. R versus LR under A_n,

$$P^n_{\theta_0+\varepsilon/\sqrt{n}}[\mathrm{R}>x]-P^n_{\theta_0+\varepsilon/\sqrt{n}}[\mathrm{LR}>x]$$
$$=\frac{K(\theta_0)}{\sqrt{n}}\left\{\frac{\varepsilon^3}{3}p_7(x;\delta)+\frac{\varepsilon}{I(\theta_0)}p_5(x;\delta)\right\}+o(n^{-1/2}).$$

(i) If $\theta_0 = \sigma^2$, then $K(\theta_0) = 1/\sigma^6 > 0$, which implies that R is more powerful than LR.

(ii) If $\theta_0 = \psi_k$, then $K(\theta_0) = -6\psi_k/(1-\psi_k^2)^2$, which implies that R is more powerful than LR if $\psi_k < 0$ and vice versa.

(iii) If $\theta_0 = \rho_k$, then $K(\theta_0) = 6\rho_k/(1-\rho_k^2)^2$, which implies that LR is more powerful than R if $\rho_k < 0$ and vice versa.

These examples show that none of the LR, W, MW, and R tests is uniformly superior.

References

[1] Hayakawa, T. (1975). The likelihood ratio criterion for a composite hypothesis under a local alternative. *Biometrika* **62** 451–460.

[2] Hayakawa, T. (1977). The likelihood ratio criterion and the asymptotic expansion of its distribution. *Ann. Inst. Statist. Math.* **29** 359–378.

[3] Hayakawa, T., and Puri, M. L. (1985). Asymptotic expansions of the distributions of some test statistics. *Ann. Inst. Statist. Math.* **37** 95–108.

[4] Peers, H. W. (1971). Likelihood ratio and associated test criteria. *Biometrika* **58** 577–587.

[5] Phillips, P. C. B. (1977). Approximations to some finite sample distributions associated with a first order stochastic difference equations. *Econometrica* **45** 463–486.

[6] Tanaka, K. (1982). *Chi-square Approximations to the Distributions of the Wald, Likelihood Ratio and Lagrange Multiplier Test Statistics in Time Series Regression*. Tech. Rep. 82. Kanazawa University.

[7] Taniguchi, M. (1983). On the second order asymptotic efficiency of estimators of Gaussian ARMA processes. *Ann. Statist.* **11** 157–169.

[8] Taniguchi, M. (1985). An asymptotic expansion for the distribution of the likelihood ratio criterion for a Gaussian autoregressive moving average process under a local alternative. *Econom. Theory* **1** 73–84.

[9] Taniguchi, M. (1986). Third order asymptotic properties of maximum likelihood estimators for Gaussian ARMA processes. *J. Multivariate Anal.* **18** 1–31.

[10] Taniguchi, M. (1987). Validity of Edgeworth expansions of minimum contrast estimators for Gaussian ARMA processes. *J. Multivariate Anal.* **21** 1–28.

[11] Wakaki, H. (1986). *Comparison of Powers of a Class of Tests for Covariance Matrices*. Tech. Rep. 183. Hiroshima Statistical Research Group, Hiroshima University.

[12] Whittle, P. (1951). *Hypothesis Testing in Time Series Analysis*. Uppsala.

Estimating Multiple Rater Agreement for a Rare Diagnosis*

JOSEPH S. VERDUCCI

Ohio State University

MICHAEL E. MACK

Stuart Pharmaceutical Division, ICI Americas Inc.

AND

MORRIS H. DEGROOT

Carnegie-Mellon University

This paper addresses the problem of estimating the population coefficient of agreement kappa (κ) among a set of raters who independently classify a randomly selected subject into one of two categories. Of the many possible probability models for these classifications, only mixtures of binomial models incorporate random rater effects, although limiting forms of additive and multiplicative (log-linear) models may themselves be represented as mixtures of binomials. Mixture models also motivate a simple new estimator $\tilde{\kappa}_x$ of κ that is appropriate in the important situation where one of the categories is rare. In the case of a rare category, simulations under multiplicative and mixture models demonstrate the substantially smaller mean squared error of $\tilde{\kappa}_x$ compared to its more popular competitor. An example of psychiatric classification illustrates the plausibility of a simple mixture model as well as sizable discrepancies among estimators of κ.

1. INTRODUCTION

1.1. *Motivation*

Suppose that a diagnostic procedure is established to classify subjects into a fixed set of categories. Various types of inter-rater reliability have

* This research was supported in part by the National Science Foundation under Grant DMS–8701770 and the National Institute of Mental Health under Grant 5–T32–MH15758.

Multivariate Statistics and Probability
ISBN 0-12-580205-6

Reprinted from *J. Mult. Anal.* **27**(2).

been proposed to measure the agreement among raters who independently apply the diagnostic procedure to the same set of subjects. Several such indices are reviewed in Landis and Koch [18]. In the case where subjects are thought to represent a population of interest, a population κ coefficient takes the form

$$\kappa = (p_{\text{agree}} - p_{\text{chance}})/(1 - p_{\text{chance}}), \tag{1.1}$$

where p_{agree} is the probability that two raters will agree about the classification of a randomly chosen subject, and p_{chance} is the probability that the two raters will agree if they independently choose a category with probability given by a fixed marginal distribution. If the raters themselves are chosen at random from a population of raters, then for a positive–negative dichotomous categorization

$$p_{\text{agree}} = E[p_i^2 + (1 - p_i)^2],$$

where p_i is the proportion of raters categorizing subject i as positive, and the expectation is taken over the population of subjects. With this notation, $\mu = E(p_i)$ is the overall proportion of positive responses among all raters and all subjects, and a simple measure of chance agreement is

$$p_{\text{chance}} = \mu^2 + (1 - \mu)^2. \tag{1.2}$$

Kraemer [17] motivates this choice of p_{chance} by showing that the increase in sample size needed to compensate for errors in categorization is a function of κ defined in this way.

Alternatively, let μ_a and μ_b represent the proportions of subjects given positive categorization by raters a and b, respectively. If each rater independently chooses the positive category at random according to his own propensity for assigning that category, then the probability that raters a and b will agree is $\mu_a\mu_b + (1 - \mu_a)(1 - \mu_b)$. Thus

$$p_{\text{chance}} = E^*[\mu_a\mu_b + (1 - \mu_a)(1 - \mu_b)], \tag{1.3}$$

where the expectation E^* is now taken over all pairs of different raters.

For any finite population of R raters, p_{chance} given by (1.3) is smaller than that given by (1.2) by a factor proportional to the variance of the μ_a. If the population of raters is large, there is little difference between (1.2) and (1.3). For a fixed value of p_{agree}, κ is a decreasing function of p_{chance}, and so Kraemer's κ defined by (1.1) and (1.2) cannot be greater than κ defined by (1.1) and (1.3). Thus for two diagnostic procedures giving the same p_{agree}, Kraemer's κ penalizes the procedure that has the greatest variability among raters. Presumably, the diagonostic procedure will ultimately be employed by individual raters, since that is the type of reliability that is being measured. Kraemer's κ thus offers some protection against the potentially disasterous situations when diagnoses are made by outlying raters.

This feature of penalizing rater variability is a property of intraclass correlations extolled by Bartko [4]. Fleiss and Cohen [26] first demonstrated that their coefficient of agreement, the sample analog of Kraemer's κ, is indeed an intraclass correlation and argued forcibly for its use. Nevertheless, Tanner and Young [23] advocate adjusting for differences in the marginal distributions of raters before measuring agreement.

In this paper we have chosen to adopt Kraemer's [17] definition of κ because of its importance in distinguishing good diagnostic procedures and because of its potential use in designing experiments that allow for errors in the diagnostic classification. Our goal is to identify good estimators of κ under various conditions, especially when the probability μ of a positive response is small

1.2. *Basic Model, Literature Review, and Summary of Results*

Consider the situation where a sample of n subjects is selected at random from a population, and the ith subject receives R_i independent ratings as to whether or not a certain characteristic is present. Let $X_{ij}=1$ if the jth rating of subject i is positive, and $X_{ij}=0$ otherwise ($j=1, ..., R_i$; $i=1, ..., n$). All of the models in this paper are special cases of the basic model in which the set of R_i variables $\{X_{ij} \mid j=1, ..., R_i\}$ are exchangeable with $P(X_{ij}=1)=p_i$. Loosely speaking, we refer to this assumption as the finite exchangeability of raters. If, in addition, we assume that the finite sequence $\{X_{ij} \mid j=1, ..., R_i\}$ is part of an infinite sequence $\{X_{ij} \mid j=1, 2, ...\}$ of exchangeable random variables, we may invoke deFinetti's famous theorem to conclude that:

For $i=1, ..., n$, the variables $\{X_{ij} \mid j=1, ..., R_i\}$ are conditionally independent given $p_i=P(X_{ij}=1)$. (1.4)

Because subjects are sampled, the p_i represent i.i.d. random effects rather than fixed numbers. Further interpretation of model (1.4) depends on its application. Consider the following two situations.

In the first situation, different sets of raters are selected at random from a population of qualified raters. The probabilities p_i may then be interpreted as the proportion of raters in the population who would judge on the basis of a particular examination that subject i has the characteristic in question. Variation of $\{X_{ij} \mid j=1, ..., R_i\}$ might then be due to raters focusing on different aspects of the examination, or raters having different beliefs about the association of these aspects with the characteristic in question.

In the second situation, each subject is repeatedly evaluated by a nonintrusive, memoryless mechanism for detection of a stable characteristic. Here p_i is the long-run proportion of times that subject i displays evidence of the characteristic in question. In this case variation of $\{X_{ij} | j=1, ..., R_i\}$ reflects random variation in the behavior of subject i under repeated examinations.

The first situation involves measuring inter-rater reliability; the second involves measuring test-retest reliability. In this paper we adopt language appropriate for the first situation, but this language easily translates to cover the type of test–retest reliability described in the second situation. The only way to distinguish the two situations mathematically is to identify fixed rater effects in the first situation. Given our motivation we do not undertake such an analysis here, but see Landis and Koch [19] for an analysis with fixed rater effects in an additive model, or Tanner and Young [23] for a similar analysis with multiplicative models.

Under model (1.4) with $\mu = E(p_i)$, Kraemer's κ measure of reliability may be expressed as

$$\begin{aligned}\kappa &= \{E[p_i^2 + (1-p_i)^2] - [\mu^2 + (1-\mu)^2]\}/\{1 - [\mu^2 + (1-\mu)^2]\} \\ &= \operatorname{var}(p_i)/\mu(1-\mu).\end{aligned}$$

If all raters simply give positive ratings at random (without examining subjects) with probability μ, then $p_i = \mu$ for each i, and thus $\kappa = 0$. At the opposite extreme, the maximal variance of the p_i is obtained when the proportion μ of subjects have $p_i = 1$ and the remaining proportion $1-\mu$ of subjects have $p_i = 0$. In this case $\operatorname{var}(p_i) = \mu(1-\mu)$ and $\kappa = 1$.

Model (1.4) permits a reduction of the data to the statistics $Y_i = \Sigma_j X_{ij}$, since these statistics are sufficient for estimating the random effects p_i. That is, under model (1.4) the likelihood of the effects p_i given the observations $\{x_{ij}\}$ is expressed as

$$L(\{p_i\} | \{x_{ij}\}) = \prod_{i=1}^{n} \binom{R_i}{y_i} (p_i)^{y_i} (1-p_i)^{R_i - y_i}. \tag{1.5}$$

Any further reduction depends upon assumptions about the distribution of the p_i. In particular, we shall be interested in distributional forms under which the statistical information for κ is large when μ is small. Estimators that are optimal under such conditions should be generally efficient whenever μ is small; that is, whenever the diagonosis is rare.

Historically, an index of multiple rater agreement first appeared in Fleiss [11]. In the case when $R_1 = \cdots = R_n = R$, Fleiss proposed the following statistic, which he described as a generalization of Cohen's [7] kappa:

$$\hat{\kappa}_F = \{T - [\bar{x}^2 + (1-\bar{x})^2]\}/\{1 - [\bar{x}^2 + (1-\bar{x})^2]\}, \tag{1.6}$$

where

$$T = n^{-1} \sum_{i=1}^{n} [2/R(R-1)] \sum_{j=1}^{R-1} \sum_{k=j+1}^{R} [X_{ij}X_{ik} + (1 - X_{ij})(1 - X_{ik})]$$

$$= \left\{ \sum_{i=1}^{n} \left[\binom{y_i}{2} + \binom{R - y_i}{2} \right] \right\} \Big/ n \binom{R}{2}$$

is the observed proportion of pairs of raters who agree, and

$$\bar{x} = (nR)^{-1} \sum_{i=1}^{n} \sum_{j=1}^{R} X_{ij} = (nR)^{-1} \sum_{i=1}^{n} Y_i$$

is the observed proportion of positive ratings. Fleiss and Cohen [26] related (1.6) to the intraclass correlation coefficient of reliability studied by Bartko [4]. Fleiss [30] also showed how correcting for chance agreement reduces many other indices of association to (1.6). When the sample is the entire population, Kraemer [17] showed that $\hat{\kappa}_F$ in (1.6) is the same as κ defined by (1.1) and (1.2). It is easy to prove that, under model (1.4), $\hat{\kappa}_F$ converges almost surely to κ as n becomes large.

In the special case where $R_1 = \cdots = R_n = R$, an alternative formulation to (1.4) is to think of $X_i = (X_{i1}, ..., X_{iR})$ as an observation in a 2^R contingency table. The assumption of the finite exchangeability of raters then corresponds to the assumption of homogeneous marginal distributions for that table. Landis and Koch [18] took this approach to test the assumption of marginal homogeneity as well as various other hypotheses about agreement, using the additive models of Grizzle, Starmer, and Koch [15] for contingency tables. Fleiss, Nee, and Landis [12] showed that Fleiss' $\hat{\kappa}_F$ in (1.6) is closely related to the weighted least squares estimates of κ obtained in Landis and Koch [19].

Under chance agreement the statistics Y_i come from the same binomial distribution. Altham [1] offers two generalizations to the binomial distribution: the multiplicative model which is the distribution induced upon the Y_i when the $\{X_{ij}\}$ follow a marginally homogeneous log-linear model (cf. Bishop, Fienberg, and Holland [5]); and the "additive" model for Y_i, which corresponds to a Lancaster additive model for $\{X_{ij}\}$. For a comparison of log-linear, additive (in the sense of Grizzle, Starmer, and Koch [15], and Lancaster additive models for contingency tables, see Darroch [27] or Darroch and Speed [28].

In Section 2, we adopt a log-linear model for contingency tables, and show that $\hat{\kappa}_F$ is the maximum likelihood estimator of κ under the assumption of homogeneous two-way marginal distributions and no higher order interactions. Our treatment improves upon that of Altham in that our parameterization allows us to compute explicit maximum likelihood estimates, whereas she employed an iterative method. Our investigation of $\hat{\kappa}_F$ under the multiplicative model roughly parallels that of Landis and Koch [18] who determined its asymptotic distribution under an additive model. We derive the asymptotic distribution of $\hat{\kappa}_F$ under the multiplicative model, and examine its small sample behavior in a simulation study. Finally, in situations where raters really are representatives from a population of raters, we note a logical inconsistency inherent in any finitely exchangeable model, such as the additive and multiplicative contingency table models, in which the raters are not infinitely exchangeable.

Infinite exchangeability and its consequential form (1.4) are assumed throughout Section 3. Using deFinetti's theorem, we prove that only such induced distributions on the $\{X_{ij}\}$ are compatible when different numbers of raters rate each subject. After considering general mixing distributions for the p_i, we examine two particular distributions in detail, namely the beta distribution and a distribution concentrated on two points, one of which is 0. This latter distribution leads to a simple case of Kraemer's [17] "true dichotomy" model, with a somewhat different interpretation, and allows for precise estimation of κ when μ is small. We do not propose this model itself as being realistic; but it does afford a simple rationale for constructing a simple estimator that may improve upon $\hat{\kappa}_F$ in the case of a rare diagnosis.

Crowder [8] was the first to examine the role of the incidental parameter μ in estimating κ (his σ^2) under the assumption that the Y_i follow a beta-binomial distribution. Although stable over most of the range of μ, the variance of the maximum likelihood estimator (m.l.e.) of κ grows rapidly as μ decreases below a value of about 0.1. In contrast, the m.l.e. for κ under the special mixing model remains relatively well behaved. This result suggests using this last estimator, or a simple approximation to it, whenever μ is small. In Subsection 3.4 we simulate the small sample behavior of several estimators under the multiplicative and special mixing models when $\mu = 0.1$. The simple approximation to the m.l.e. under special mixing performs remarkably better than $\hat{\kappa}_F$ in these situations. Finally, in Section 4 we compare the various models and estimators on a set of psychiatric ratings obtained from Fleiss [11].

The new results in this paper are (1) the characterization of $\hat{\kappa}_F$ as the m.l.e. for κ in the multiplicative interaction model and the derivation of its asymptotic distribution under this model; (2) the conclusion that mixing models of the form (1.4) are the only models logically consistent with randomly selected raters; and (3) identification of a simple estimator, $\tilde{\kappa}_x$ defined in (3.7), that may serve as a better index of reliability than $\hat{\kappa}_F$ in the case of a rare diagnosis whether or not raters are randomly selected from a population of raters.

2. Multiplicative Model

Throughout this section, we treat a set of R ratings on a subject i as an observation $X_i = (X_{il}, ..., X_{iR})$ of 0's and 1's in a 2^R contingency table, and we adopt a log-linear model in order to obtain some parsimony in explaining the distribution of X_i over subjects. In particular, we examine the simplest such model of interest, that of no second or higher order interaction. This first-order log-linear model is the multiplicative interaction counterpart of the additive interaction one-way ANOVA model in Landis and Koch [19]. The multiplicative interaction model introduced in Subsection 2.1 allows an explicit m.l.e. for κ and also a simple method for computing its asymptotic variance, both given in Subsection 2.2. In Subsection 2.3 the asymptotic variance is compared to the exact small sample variance under several instances of the model. Finally in Subsection 2.4 we explain the source of difficulty in extending the multiplicative model to situations where different numbers of raters rate each subject, which motivates the mixing models of Section 3.

2.1. *Derivation of the Multiplicative Interaction Model from the Symmetric Log-Linear Model*

In the notation of Bishop, Fienberg, and Holland [5], the first-order log-linear model gives the probability of observing a particular vector x of ratings as

$$P(X = x) = \exp\left[u_0 + \sum_{j=1}^{R} u_j(x_j) + \sum_{j=1}^{R-1} \sum_{k=j+1}^{R} u_{jk}(x_j, x_k)\right], \qquad (2.1)$$

where u_j measures the main effect of the judgment of rater j; u_{jk} measures the interaction between a pair of raters; and u_0 scales the function $P(\cdot)$ so that it is a *bona fide* probability mass function.

By the basic assumption in Section 1, the ratings $(X_{i1}, \;\, , X_{iR})$ are assumed to be exchangeable random variables. This assumption implies that the functions $u_j(\cdot)$ must be identical for all $j = 1, ..., R$, as must be the functions $u_{jk}(x_j, x_k)$ for $j = 1, ..., R-1$ and $k = j+1, ..., R$. This distributional symmetry leaves just two unconstrained parameters that we can identify using the following system:

$$u_j(1) = u,$$
$$u_j(0) = 0,$$
$$u_{jk}(0, 0) = u_{jk}(1, 1) = v, \text{ and}$$
$$u_{jk}(1, 0) = u_{jk}(0, 1) = 0.$$

Under this system model (2.1) induces the following distribution upon Y, the number of positive ratings that a randomly chosen subject receives:

$$P(Y=y)=\left[\binom{R}{y}\phi(u,v;y)\right]\Big/\psi(u,v), \tag{2.2}$$

where

$$\phi(u,v;y)=\exp\left\{yu+\left[\binom{y}{2}+\binom{R-y}{2}\right]v\right\}$$

and

$$\psi(u,v)=\sum_{y=0}^{R}\binom{R}{y}\phi(u,v;y).$$

Moreover, Y is sufficient for the parameters u and v under this symmetry version of (2.1). To see how the parameters u and v relate to κ, first note that for randomly sampled subjects

$$\mu=P(X_{ij}=1)=\sum_{y=1}^{R}\binom{R-1}{y-1}\phi(u,v;y)/\psi(u,v),$$

is the probability of a positive diagonosis; and

$$p_{11}=P(X_{ij}=1, X_{ik}=1)=\sum_{y=2}^{R}\binom{R-2}{y-2}\phi(u,v;y)/\psi(u,v),$$

for each $j\neq k$, is the probability that two raters give a positive diagonosis to the same subject. With some simple algebra, Kraemer's [17] κ defined by (1.1) and (1.2) becomes

$$\kappa=(p_{11}-\mu^2)/(\mu-\mu^2). \tag{2.3}$$

The range of the parameter space $\{(\mu,\kappa)\}$ is $0<\mu<1$, and $\lambda(\mu)\leqslant\kappa\leqslant 1$, where $\lambda(\mu)$ is a symmetric function about $\mu=\frac{1}{2}$, approaching its upper bound 0 as $\mu\to 1$, and achieving a lower bound of $-(R-1)^{-1}$ for $\mu=\frac{1}{2}$ when R is even and for $\mu=(R\pm 1)/2R$ when R is odd. Although the m.l.e.'s of u and v generally do not have a closed form, those for μ and κ do, as we now prove.

2.2. *Maximum Likelihood Estimates*

THEOREM 2.1. *Let $\{y_i \mid i=1,\ldots,n\}$ be the observed numbers of positive ratings received by a random sample of n subjects, where each random* *variable Y_i is distributed according to the multiplicative model (2.2). Then the maximum likelihood estimators (m.l.e's) of μ and κ are*

$$\hat{\mu} = S/(nR) \tag{2.4}$$

and

$$\hat{\kappa}_F = 1 - (1 - T)/[2\hat{\mu}(1 - \hat{\mu})], \tag{2.5}$$

where

$$S = \sum_{i=1}^{n} y_i$$

is the total number of positive ratings, and

$$T = \left\{ \sum_{i=1}^{n} \left[\binom{y_i}{2} + \binom{R - y_i}{2} \right] \right\} \Big/ n \binom{R}{2} \tag{2.6}$$

is the proportion of pairs of raters who agree.

Proof. The log-likelihood of observing $\{y_i \mid i = 1, ..., n\}$ under (2.2) is

$$LL(u, v; \{y_i\}) = Su + n \binom{R}{2} Tv - n \log[\psi(u, v)], \tag{2.7}$$

where $\psi(u, v)$ is defined as in (2.2).

Setting the partial derivatives of (2.7) with respect to u and v equal to zero produces the equations

$$S = nE(Y) = nR\mu$$

and

$$n \binom{R}{2} T = nE \left[\binom{y_i}{2} + \binom{R - y_i}{2} \right] = n \binom{R}{2} q,$$

where $q = p_{\text{agree}}$ is the probability that any two raters agree on the diagnosis of a subject. Because the log-likelihood is strictly concave, the m.l.e.'s of μ and q are uniquely determined by (2.5) and (2.7), respectively.

Again some simple algebra shows that Kraemer's κ can be written in the form

$$\kappa = 1 - \{(1 - q)/[2\mu(1 - \mu)]\}. \tag{2.8}$$

Substituting the m.l.e.'s for μ and q into (2.8) gives the m.l.e. for κ. ∎

The estimator $\hat{\kappa}_F$ in Eq. (2.5) is exactly the index of interrater agreement proposed by Fleiss [11]. Theorem 1 characterizes this index as the m.l.e. of

Kraemer's κ under the multiplicative model (2.2). This formulation makes explicit the assumptions underlying the use of $\hat{\kappa}_F$ as an efficient estimator:

1. Raters are finitely exchangeable.
2. No second or higher order multiplicative interaction arises among the R diagnoses given to a subject.

Under these conditions κ is interpretable as the common (intraclass) correlation between pairs of raters.

A further benefit of this formulation is that the asymptotic distribution of $\hat{\kappa}_F$ can be determined for arbitrary values of the parameter κ. The variance of the asymptotic normal distribution of $\hat{\kappa}_F$ is given by

$$\operatorname{var}(\hat{\kappa}_F) = n^{-1}\Delta'\Sigma^{-1}\Delta, \tag{2.9}$$

where

$$\Delta' = (\partial\kappa/\partial u,\ \partial\kappa/\partial v)$$

and

$$\Sigma = \begin{pmatrix} \partial^2\psi/\partial u^2 & \partial^2\psi/\partial u\partial v \\ \partial^2\psi/\partial u\partial v & \partial^2\psi/\partial v^2 \end{pmatrix}.$$

This result follows directly from the theory of exponential families (see, for example, Barndorff-Nielsen [3, Chap. 8]). The partial derivatives are easily calculated from expressions (2.2) and (2.3).

Because the variance (2.9) of $\hat{\kappa}_F$ depends on the unknown parameters u and v, estimates of u and v must be substituted into that expression to obtain a numerical estimate for $\operatorname{var}(\hat{\kappa}_F)$. The m.l.e.'s for u and v may be obtained numerically from the log-likelihood (2.7). In fact, the log-likelihood provides a device for calculating the parameter pair (u, v) as a function of the (μ, κ) pair of parameters. For any admissible values (μ, κ), substitute $E(S) = nR\mu$ and $E(T) = 1 - [2\mu(1-\mu)(1-\kappa)]$ for S and T, respectively, in Eq. (2.7), and solve for the maximizing values of u and v. This technique is used in the next section to examine the exact small sample mean and variance of $\hat{\kappa}_F$.

2.3. *Small Sample Behavior of* $\hat{\kappa}_F$

In this section, the exact small sample expectation and standard deviation of $\hat{\kappa}_F$ are computed for various values of μ and κ in the multiplicative model. To do this, we first generate all $\binom{n+R}{R}$ possible samples $\{Y_i \mid i = 1, ..., n\}$ of fixed size n (see Feller [10, p. 52]). For each set $\{Y_i\}$ we compute $\hat{\kappa}_F$ (set equal to 1 when $\hat{\mu} = 0$ or 1) as well as the probability of observing $\{Y_i\}$ under independent sampling from the multiplicative model (2.2) with parameters $u = u(\mu, \kappa)$ and $v = v(\mu, \kappa)$.

Table I lists the expected value and standard deviation of $\hat{\kappa}_F$ for $R = 3$;

TABLE I

Small Sample Expectation and Standard Deviation of Fleiss' Kappa under the Multiplicative Model

	κ	0.1		0.5		0.9	
Sample size	μ	0.1	0.5	0.1	0.5	0.1	0.5
1		0.629 0.644	−0.004 0.704	0.800 0.510	0.435 0.724	0.956 0.248	0.883 0.398
2		0.491 0.591	−0.066 0.455	0.729 0.485	0.339 0.564	0.938 0.250	0.847 0.368
3		0.375 0.555	−0.027 0.363	0.663 0.487	0.351 0.458	0.921 0.266	0.841 0.327
4		0.288 0.511	0.004 0.315	0.609 0.485	0.376 0.389	0.906 0.280	0.842 0.292
5		0.223 0.466	0.026 0.282	0.565 0.477	0.397 0.341	0.892 0.292	0.844 0.266
6		0.176 0.423	0.040 0.257	0.531 0.466	0.411 0.307	0.880 0.300	0.846 0.249
7		0.142 0.383	0.050 0.238	0.504 0.453	0.419 0.282	0.870 0.305	0.846 0.239
8		0.117 0.347	0.057 0.222	0.484 0.439	0.425 0.265	0.862 0.309	0.844 0.235
9		0.099 0.316	0.063 0.209	0.468 0.425	0.428 0.252	0.854 0.312	0.842 0.235
10		0.086 0.289	0.067 0.198	0.457 0.412	0.429 0.242	0.848 0.313	0.839 0.237
10[a]		0.261	0.197	0.367	0.204	0.192	0.111

[a] Asymptotic approximation of the standard deviation.

Note. The first entry of each cell is the expectation and the second entry is the standard deviation.

$n = 1\ (1)\ 10$: $\kappa = 0.1, 0.5, 0.9$; and $\mu = 0.1, 0.5$. The last line of the table contains the asymptotic approximation obtained from (2.9) to the standard deviation of $\hat{\kappa}_F$ when $n = 10$. Several features of this table are noteworthy:

(1) s.d. $(\hat{\kappa}_F)$ is larger when $\mu = 0.1$ than when $\mu = 0.5$;

(2) s.d. $(\hat{\kappa}_F)$ is largest when $\kappa = 0.5$ rather than 0.1 or 0.9;

(3) $\text{var}(\hat{\kappa}_F) > \text{bias}^2(\hat{\kappa}_F)$;

(4) formula (2.9) underestimates $\text{var}(\hat{\kappa}_F)$ when $n = 10$;

(5) $\text{bias}(\hat{\kappa}_F)$ appears to be getting worse as n increases in this range!

Features (1) and (2) indicate a weakness of $\hat{\kappa}_F$; namely it performs worst over that portion of the range of (μ, κ) of greatest interest—infrequent diagnosis and moderate agreement. In Section 3 we develop an estimator that performs better in this region of interest and not much worse elsewhere. Feature (3) indicates that most gains are to be had by trying to reduce the variance of $\hat{\kappa}_F$ rather than its bias. Features (4) and (5) should remind us not to rely too heavily on asymptotic formulas.

We are also interested in how well $\hat{\kappa}_F$ performs when the underlying distribution is not in the multiplicative family. If $\kappa = \kappa(F)$ is expressed as a function of the distribution F of Y, then $\hat{\kappa}_F = \kappa(\hat{F})$, where $\hat{F}$ is the empirical distribution function, and the function $\kappa(\cdot)$ is continuous in the weak topology. Thus the almost sure convergence of the empirical distribution $\hat{F}$ to F implies that $\hat{\kappa}_F \to \kappa$ (a.s.), and $\hat{\kappa}_F$ is a consistent estimator of κ under any distribution F. In fact, $\hat{\kappa}_F$ is Fisher consistent (cf., Rao [22, p.345]), a much stronger property. Although this type of consistency is a desirable property, some estimators of κ may be more efficient at estimating κ than $\hat{\kappa}_F$ is, especially if the distribution fuction F is naturally restricted to a parametric family other than the multiplicative one.

We now offer some reasons why the multiplicative model in particular, and the log-linear model in general, may not naturally correspond to the ratings of randomly chosen judges for each subject in a representative sample of subjects. The basic difficulty is that interactions among the random variables $\{X_{ij} \mid j = 1, ..., R\}$, although associated with the variability of the random effects p_i, depend on the number R of raters involved. In particular, the multiplicative model, which assumes no interactions of order higher than two, makes different assumptions about the distribution of the sum of a fixed number k of ratings on a subject, as the total number R of raters varies. This problem and a possible solution are discussed in the next section.

2.4. *Varying Numbers of Raters*

Altham [1] noted that if a 2^R contingency table corresponding to a multiplicative model is collapsed over one of its dimensions, the resulting 2^{R-1} table does not correspond to a multiplicative model. (This is true for $R \geqslant 4$ and $\kappa \neq 0$.) Thus the assumption of no second or higher order interaction among R raters is different from the assumption of no second or higher order interaction among $R-1$ raters $(R \geqslant 4)$. Why should the distribution of the ratings given by any three raters depend in any way on the presence or absence of a fourth?

Even if we were willing to accept different assumptions about the distribution of positive diagnoses among groups of subjects who are assessed by different numbers of raters, there are practical problems with maximum likelihood estimation in this situation, because the parameters u and v

depend on R. Thus the log-likelihood (2.7) for each group cannot be simply added together. The parameters μ and κ do remain the same when the tables are collapsed, and these common values may still be estimated under the combined multiplicative models, although with some difficulty.

One way to mend the problem of incompatibility of the assumptions of no interaction is to define a sequence of probability functions for different numbers of raters in such a way as to ensure compatibility.

DEFINITION. A sequence of probability functions $f^{(R)}$ $(R=1, 2, ...,)$ defined on $(0, 1, ..., R)$ is *symmetrically marginally compatible* (SMC) if

$$f^{(R-1)}(y)=[(R-y)/R]f^{(R)}(y)+[(1+y)/R]f^{(R)}(1+y) \qquad (2.10)$$

for each $y=0, 1, ..., R-1$ and $R=2, 3, ...$.

It is possible to construct a sequence of SMC probability functions whose members are all multiplictive models with the same μ and κ. The construction proceeds according to the following theorem whose proof is an immediate consequence of the convergence of the iterative proportional fitting algorithm (cf. Andersen [2, Theorem 6] or Bishop, Fienberg, and Holland [5, Theorem 3.5-1]).

THEOREM 2.2. *Let $f^{(1)}(y)$ be Bernoulli (μ) and let $f^{(R)}(y)=P(y)$ be the multiplicative model defined by* (2.2) *for $R=2, 3$. Corresponding to each of these $f^{(R)}$, let M_R be the symmetric log-linear model on a 2^R table $(R=1, 2, 3)$. That is, under M_R*

$$P(X=x)=[f^{(R)}(y)]\Big/\binom{R}{y} \qquad (R=1, 2, 3),$$

where $y=\sum x_j$. For $R\geqslant 4$ let M_R be the symmetric log-linear model on a 2^R table with fixed 2^{R-1} marginal probabilities given by M_{R-1} and no Rth order interaction; and let $f^{(R)}(y)$ be the multiplicative model associated with M_R. Then the sequence $f^{(R)}$ $(R=1, 2, ...,)$ is SMC.

In the next section we show that $\{f^{(R)}\}$ is SMC if and only if the infinite sequence $X_1, X_2, ...,$ of ratings given by the judges to any one subject is exchangeable. In that context, Theorem 2.2 gives a method for constructing an infinite exchangeable sequence with each finite subsequence following a log-linear model.

3. MIXING MODELS

We return to the basic model (1.4) where the probability p of receiving a positive diagnosis varies from subject to subject and where, for each subject, raters make their diagnoses independently. Thus each subject has his own personal probability p of being given a positive diagnosis by any randomly selected rater and has probability

$$\binom{R}{y} p^y(1-p)^{R-y}$$

of receiving exactly y positive diagnoses from any group of R raters. The probability p is a random variable that has a distribution function ξ over the population of subjects. It follows that the probability distribution of the number of positive diagnoses Y that a randomly chosen subject will receive from R raters is

$$P(Y=y)=\int_0^1 \binom{R}{y} p^y(1-p)^{R-y}\, d\xi(p) \qquad y=0, \ldots, R. \tag{3.1}$$

This formulation allows us to interpret agreement among raters in terms of properties of the distribution ξ. In particular, the inter-rater reliability of the diagnostic procedure is related to the dispersion of the mixing distribution ξ. Among all distributions with a given μ, extremes of dispersion are given by the distributions ξ^0 and ξ_0 defined as

$$P(p=0 \mid \xi^0)=1-\mu, \qquad P(p=1 \mid \xi^0)=\mu,$$

and

$$P(p=\mu \mid \xi_0)=1.$$

Under ξ^0 all the mass is concentrated at $p=0$ and $p=1$; hence all raters agree on every subject. Under ξ_0 all the mass is concentrated at a single point; hence raters randomly assign positive diagnoses with probability μ to all subjects.

Let Z_j be the response of the jth rater to a randomly chosen subject. The mixing model (3.1) implies that

$$E(Z_1 Z_2)=\int_0^1 p^2\, d\xi(p)=E(p^2 \mid \xi)$$

and

$$E(Z_1)=E(Z_2)=\mu=E(p \mid \xi),$$

so that

$$\operatorname{cov}(Z_1, Z_2)=E(p^2 \mid \xi)-\mu^2=\operatorname{var}(p \mid \xi)$$

and

$$\text{var}(Z_1) = \text{var}(Z_2) = \mu - \mu^2.$$

Thus

$$\kappa = \text{corr}(Z_1, Z_2) = [E(p^2 \mid \xi) - \mu^2]/[\mu - \mu^2] = \text{var}(p \mid \xi)/\text{var}(p \mid \xi^0). \tag{3.2}$$

That is, the common correlation between pairs of raters is measured by the ratio of the variance of the mixing distribution to the largest possible variance for mixing distributions with the same mean.

When each subject in a random sample is rated by the same number R of raters, the moment estimator of $\text{var}(p \mid \xi)$ is simply the sample variance of Y/R. In this case a consistent estimator of κ is

$$\tilde{\kappa}_m = \left[\sum (y_i/R - \hat{\mu})^2\right] \Big/ [(n-1)\,\hat{\mu}(1-\hat{\mu})], \tag{3.3}$$

where y_i is the number of positive ratings given to subject i ($i = 1, ..., n$) and $\hat{\mu} = (\sum y_i)/nR$. If the number R_i of raters judging subject i differs for some subjects, then some weighting of the Y_i according to R_i may be appropriate. In general, optimal weights depend on ξ.

3.1. *Compatibility*

The following theorem shows that mixing models and only mixing models are compatible over varying numbers of raters.

THEOREM 3.1. *Let* $\{f^{(R)} \mid R = 1, 2, ...,\}$ *be a sequence of probability functions with* $f^{(R)}$ *defined on* $\{0, ..., R\}$. *Then* $\{f^{(R)}\}$ *is SMC if and only if* $f^{(R)}$ *has the form* (3.1) *with the same* ξ *for each* R.

Proof. First suppose that each $f^{(R)}$ has the form (3.1) with the same ξ. Then

$$[(R-y)/R]\,f^{(R)}(y) = \binom{R-1}{y} \int_0^1 (1-p)\,p^y(1-p)^{(R-y)-1}\,d\xi(p) \tag{3.4}$$

and

$$[(1+y)/R]\,f^{(R)}(1+y) = \binom{R-1}{y} \int_0^1 p^{y+1}(1-p)^{(R-y)-1}\,d\xi(p). \tag{3.5}$$

Adding (3.4) and (3.5) shows that $f^{(R)}$ is SMC.

To prove the converse, it suffices to show that each $f^{(R)}$ is the probability function for the finite sum

$$Y^{(R)} = \sum_{j=1}^{R} Z_j$$

of an infinite exchangeable sequence of Bernoulli random variables. Let $\{Z_j\}$ be such an exchangeable sequence with moments defined by

$$E\left(\prod_{j=1}^{k} Z_j\right)=f^{(k)}(k), \qquad k=1, 2, \ldots.$$

It remains only to show that

$$f^{(R)}(y)=P(Y^{(R)}=y)$$

for each $y=0, \ldots, R$ and $R=1, 2, \ldots$. This equality is a consequence of the following three facts:

(1) Each $f^{(R)}(y)$ is determined by the probabilites $\{f^{(k)} \mid k=1, \ldots, R\}$ and the recurrence relation defined by the SMC property.

(2) $f^{(k)}(k)=P(Z_1=\cdots=Z_k=1)=P(Y^{(k)}=k)$, $k=1, 2, \ldots$.

(3) The probability functions of $Y^{(k)}$ are SMC. ▮

Under the following parametric model estimation of κ is simple, even with varying numbers of raters.

3.2. *Beta-Binomial Model*

When the mixing distribution ξ in (3.1) is a beta distribution, then the distribution of Y is beta-binomial. If each subject i is independently judged by R_i raters ($i=1, \ldots, n$) then the log-likelihood from these beta-binomial models can be written explicitly as a function of μ and κ, namely

$$\begin{aligned} LL(\mu, \kappa) = \text{constant} + \sum_{i=1}^{n} \Bigg\{ & \sum_{j=1}^{y_i} \log\left[\mu+(j-1)\,\kappa/(1-\kappa)\right] \\ & + \sum_{j=1}^{R_i-y_i} \log\left[(1+\mu)+(j-1)\,\kappa/(1-\kappa)\right] \\ & - \sum_{j=1}^{R_i} \log\left[1+(j-1)\,\kappa/(1-\kappa)\right]\Bigg\}. \end{aligned}$$

The m.l.e. $\hat{\kappa}_B$ for κ under the beta-binomial model comes directly from numerically maximizing this expression.

Although the beta-binomial distribution has been applied to many different areas of statistics (cf. Griffiths [14], for refeences), Plackett and Paul [21] seem to be the first to consider this distribution, as a special case of the Dirichlet-multinomial distribution, for modeling observer agreement. Kraemer [17] also mentions its use in this context. Since the publication of these papers some further technical work on the beta-binomial distribution has appeared. We briefly connect these newer results to the context of diagnostic agreement.

A major technical problem with the beta-binomial model, not shared by the multiplicative model, is that the conditional distribution of $\{Y_i \mid \hat{\mu} = (nR)^{-1} \sum y_i\}$ still depends on the nuisance parameter μ. Tarone [24] uses the $C(\alpha)$ procedure of Neyman [20] to surmount this problem when testing $H_0: \kappa = 0$. He further shows that the asymptotically optimal tests of this null hypothesis versus beta-binomial and multiplicative alternatives are not equivalent. Thus model selection does play a role in testing for inter-rater agreement. One shortcoming of the $C(\alpha)$ method is that its assumptions are not valid for testing $H_0: \kappa = \kappa_0$ when $\kappa_0 \neq 0$ in the beta-binomial model.

Crowder [8] carefully examined the likelihood function for κ conditional on the observed value μ and concluded that it is a fairly constant function of μ, except when μ is close to 0 or 1. Thus, although the beta-binomial model may be convenient for estimating inter-rater agreement for a prevalent diagnostic characteristic, it is difficult to form precise inference about κ when the diagnosis is rare. One possibility is to form conservative tests as, for example, in Potthoff and Whittinghill [29]. Another is to adopt a more appropriate model, like the special mixing model.

3.3. *Special Mixture of Two Binomials*

Suppose that the population of subjects consists of two subpopulations, those that possess the characteristic in question and those that do not. For simplicity, assume that those which do not possess the characteristic are never misdiagnosed, but that any subject which possesses the characteristic has fixed probability π of receiving a positive diagnosis. The key assumption here is that ξ has a mass point at $p = 0$. If the prevalence of the characteristic in the population is 1-ζ, then the distribution of positive responses Y is

$$P(Y = y) = \begin{cases} \zeta + (1 - \zeta)(1 - \pi)^R & \text{for} \quad y = 0, \\ (1 - \zeta)\binom{R}{y} \pi^y (1 - \pi)^{R-y} & \text{for} \quad y = 1, ..., R. \end{cases} \tag{3.6}$$

The following theorem shows how to find the m.l.e. $\hat{\kappa}_x$ for κ under model (3.6). Its proof appears in the appendix.

THEOREM 3.2. *Let* $\{y_i \mid i = 1, ..., n\}$ *be a random sample from model* (3.6), *and suppose that* $0 < S = \sum y_i < nR$, *so that the m.l.e's* $\hat{\pi}$, $\hat{\zeta}$, *and* $\hat{\kappa}_x$ *are well defined. Let* $A = \sum I(y_i = 0)$, *where* $I(Y = 0)$ *is an indicator of the event*

$\{Y=0\}$. *If* $A/n \leqslant (1-S/nR)^R$, *then* $\hat{\pi}=S/nR$, $\hat{\zeta}=0$, *and* $\hat{\kappa}_x=0$. *If* $A/n > (1-S/nR)^R$, *then* $\hat{\pi}$ *is the unique root of*

$$f(\pi)=S(1-\pi)^R+(n-A)\,R\pi-S=0$$

in the range $0<\pi<1$; $\hat{\zeta}=(A-n\hat{q})/(n-n\hat{q})$, *where* $\hat{q}=(1-\hat{\pi})^R$; *and* $\hat{\kappa}_x=\hat{\pi}\hat{\zeta}/[(1-\hat{\pi})+\hat{\pi}\hat{\zeta}]$.

The above theorem separates the cases where the observed number of perfect negative agreements A is more or less than an estimate of its expected value under chance guessing. If the underlying κ is close to 1, then the observed A will almost always be grater than its chance expected value. In this case, as well as in the situation where R is large, a good approximation to $\hat{\kappa}_x$ is

$$\tilde{\kappa}_x=[A/(n-A)][S/(nR-S)], \tag{3.7}$$

which is obtained from the simplified estimators $\tilde{\zeta}=A/n$ and $\tilde{\pi}=S/(n-A)\,R$.

The estimator $\tilde{\kappa}_x$ is easy to compute and performs well in the situation of most interest: rare but moderately reliable diagnosis (μ small, κ large). We now offer a convenient formula for its variance, before comparing its performance against that of $\hat{\kappa}_F$.

The asymptotic variance of $\tilde{\kappa}_x$ may be approximated by assuming that $A\sim\text{Binomial}(n,\zeta)$ and $S\sim\text{Binomial}(nR(1-\zeta),\pi)$ are independent. Because the approximation depends only on the moments of A and S it is not necessary for $nR(1-\zeta)$ to be an integer. Letting $B=A/(n-A)$ and $C=S/(nR-S)$ gives

$$\text{var}(\tilde{\kappa}_x)=\text{var}(B)\,\text{var}(C)+E(B)^2\,\text{var}(C)+E(C)^2\,\text{var}(B). \tag{3.8}$$

Using the δ-method (cf. Bishop, Fienberg, and Holland [5, p. 481]), we compute the asymptotic means

$$E(B)=\zeta/(1-\zeta)+O(1/n)$$

and

$$E(C)=[(1-\zeta)\,\pi]/[1-(1-\zeta)\,\pi]+O(1/n);$$

and the asymptotic variances

$$\text{var}(B)=\zeta/n(1-\zeta)+o(1/n)$$

and

$$\text{var}(C)=[(1-\zeta)\,\pi]/nR[1-(1-\zeta)\,\pi]+o(1/n).$$

Substituting these and the sample values $\zeta = A/n$ and $\pi = S/(n-A)R$ into (3.8) leads to the approximation

$$\mathrm{var}(\tilde{\kappa}_x) = \tilde{\kappa}_x[(B/R) + C]/n. \tag{3.9}$$

3.4. *Small Sample Comparison of* $\tilde{\kappa}_x$ *and* $\hat{\kappa}_F$

Using the same sample-generating method as in Section 2, we now compare the exact small sample expectation, standard deviation, and mean squared error of $\tilde{\kappa}_x$ and $\hat{\kappa}_F$ under both the multiplicative and special mixing models. For ease of computation we fix $R=3$ and focus on the case $\kappa=0.7$, $\mu=0.1$ ($\pi=0.73$, $\zeta=\frac{63}{73}$) as a typical point in the interesting region of the parameter space.

As Table II shows, the values for $\tilde{\kappa}_x$ are very close to those for $\hat{\kappa}_x$. Moreover, the mean squared error (mse) of $\tilde{\kappa}_x$ is consistently smaller than that of $\hat{\kappa}_F$ under either model, due mainly to the smaller standard deviation of $\tilde{\kappa}_x$. In fact when $n=10$ the mse of $\tilde{\kappa}_x$ is less than half that of $\hat{\kappa}_F$.

Table II makes the point that it is possible to improve substantially on $\hat{\kappa}_F$ as an estimator of κ under certain circumstances (models like the multiplicative or mixing models; rare but moderately reliable diagnoses). We now compare the estimators on actual data to see if such circumstances are realized.

4. EXAMPLE

An example in Fleiss' [11] paper concerns the reliability of psychiatric diagnoses of 30 patients who are each judged by 6 different raters. Here we fit the multiplicative, special mixing, and beta-binomial models to the observed distribution of positive ratings for the diagnosis of schizophrenia. Table III displays the data along with the estimated expected values based on the maximum likelihood estimates we have derived under each of the three models. The relative squared error (χ^2 goodness of fit statistic) is computed for each model. Although the small expected counts suggest that the distribution of this statistic is not well approximated by the χ^2 distribution, the relative squared error is still a reasonable gauge for comparing these models, because each model has the same number of estimated parameters.

Table III shows the similarity between the fits of the multiplicative and beta-binomial models. Both fit the distribution of number of positive diagnoses of schizophrenia poorly. The reason is that the distribution is bimodal, with the second mode not too close to $Y=R$. Diagnosis of schizophrenia is better modeled by the special mixing distribution.

TABLE II

Small Sample Expectation, Standard Deviation, and Mean Squared Error of $\hat{\kappa}$, $\hat{\kappa}_x$, and $\tilde{\kappa}_x$ for 3 Raters under the Multiplicative and Special Mixing Distributions with $\kappa = 0.7$ and $\mu = 0.1$

Model / Estimator	Multiplicative			Mixing		
Sample size	$\hat{\kappa}_F$	$\hat{\kappa}_x$	$\tilde{\kappa}_x$	$\hat{\kappa}_F$	$\hat{\kappa}_x$	$\tilde{\kappa}_x$
1	0.877	0.901	0.918	0.878	0.908	0.918
	0.410	0.332	0.273	0.409	0.309	0.272
	0.199	0.151	0.122	0.199	0.139	0.122
2	0.833	0.874	0.884	0.859	0.898	0.903
	0.397	0.301	0.276	0.340	0.250	0.235
	0.175	0.121	0.106	0.141	0.102	0.096
3	0.791	0.845	0.852	0.830	0.878	0.882
	0.408	0.298	0.282	0.334	0.239	0.230
	0.175	0.110	0.103	0.128	0.089	0.086
4	0.755	0.818	0.824	0.802	0.858	0.861
	0.417	0.301	0.288	0.337	0.239	0.232
	0.177	0.105	0.098	0.124	0.082	0.075
5	0.725	0.794	0.800	0.777	0.840	0.843
	0.421	0.301	0.291	0.338	0.239	0.233
	0.178	0.099	0.095	0.120	0.077	0.075
6	0.701	0.773	0.779	0.756	0.824	0.826
	0.420	0.300	0.291	0.338	0.238	0.234
	0.176	0.095	0.091	0.117	0.072	0.071
7	0.682	0.755	0.760	0.739	0.810	0.812
	0.417	0.297	0.289	0.335	0.236	0.232
	0.174	0.091	0.087	0.110	0.064	0.063
8	0.666	0.740	0.745	0.724	0.797	0.799
	0.412	0.293	0.286	0.330	0.234	0.230
	0.171	0.087	0.084	0.109	0.064	0.063
9	0.654	0.727	0.731	0.712	0.786	0.789
	0.406	0.287	0.281	0.325	0.230	0.227
	0.167	0.083	0.080	0.106	0.060	0.059
10	0.645	0.716	0.720	0.702	0.777	0.779
	0.398	0.282	0.276	0.319	0.226	0.223
	0.161	0.080	0.077	0.102	0.057	0.056

Note. The first entry of each cell is the expectation, the second entry is the standard deviation, and the last entry is the mean squared error.

The next logical step is to see how the choice of model affects the estimation of the parameter κ. Table IV compares all the estimators that we have discussed, namely Fleiss' $\hat{\kappa}_F$ (eq. (1.5)), the moment estimator $\tilde{\kappa}_m$

TABLE III

Distribution of Positive Ratings for the Diagnosis of Schizophrenia

	Number of positive ratings							
	0	1	2	3	4	5	6	RSE[a]
Observed data	22	0	1	2	3	2	0	–
Multiplicative model	18.6	5.2	1.9	1.0	0.8	1.0	1.7	16.0
Beta-binomial model	16.3	3.7	2.4	1.9	1.7	1.7	2.1	9.7
Special mixing model	22.1	0.2	0.9	2.1	2.6	1.7	0.5	0.8

[a] RSE is relative squared error $= \sum$ (observed-expected)^2/expected.

(Eq. (3.3)), the m.l.e. under the beta-binomial $\hat{\kappa}_B$ (Section 3.2), the m.l.e. under special mixing $\hat{\kappa}_x$ (Theorem 3.2), and its approximation $\tilde{\kappa}_x$ (Eq. (3.7)), Standard errors were computed according to the asymptotic formulas for $\hat{\kappa}_F$, $\hat{\kappa}_B$, and $\hat{\kappa}_x$, formula (3.9) for $\tilde{\kappa}_x$, and the bootstrap method (Efron [9]) for $\tilde{\kappa}_m$. As a comparison, the bootstrap standard error for $\hat{\kappa}_F$ was 0.105 versus the asymptotic approximation 0.133; the bootstrap standard error for $\tilde{\kappa}_x$ was 0.073 versus the approximation 0.110 given by Eq. (3.9).

The table demonstrates sizable differences in the estimates of κ, from 0.517 for $\hat{\kappa}_F$ to 0.620 for $\tilde{\kappa}_m$. The value of $\tilde{\kappa}_x = 0.550$ represents a rough compromise between these. Greater relative differences are seen in the estimated standard errors. The standard error of $\hat{\kappa}_F$ is estimated to be between 21 to 35% larger than that of $\tilde{\kappa}_x$, for the asymptotic and bootstrap estimates, respectively. Thus all indications are that $\tilde{\kappa}_x$ is more stable in this region of the parameter space.

TABLE IV

Estimates of Kappa for the Diagnosis of Schizophrenia

	Estimator				
	$\hat{\kappa}_F$	$\tilde{\kappa}_m$	$\hat{\kappa}_B$	$\hat{\kappa}_x$	$\tilde{\kappa}_x$
Estimate	0.517	0.620	0.552	0.549	0.550
Asymptotic SE	0.133		0.117	0.078	0.110
Bootstrap SE	0.105	0.065			0.073

5. Concluding Remarks

This paper is meant not just as a concept paper, linking the log-linear and mixing model approaches, but also as a demonstration that very practical estimators such as $\tilde{\kappa}_x$ arise out of the mixing distribution approach to measuring and estimating diagnostic agreement. In future work we hope to develop the mixing distribution approach for multiple categories.

APPENDIX: Proof of Theorem 3.2

The log-likelihood $\mathrm{LL}(\pi, \zeta)$ for a sample $\{y_i \mid i=1, ..., n\}$ from the distribution (3.6) is

$$\mathrm{LL}(\pi, \zeta) = A \log[\zeta + (1-\zeta)\, q] + (n-A) \log[(1-\zeta)\, q] + S \log[\pi/(1-\pi)] + \sum \log \binom{R}{y_i}$$

for $0 \leqslant \zeta \leqslant 1$ and $0 < \pi < 1$, where $q = (1-\pi)^R$, $A = \sum I\{y_i = 0\}$, and I is an indicator function. The partial derivative of the log-likelihood with respect to ζ is

$$\partial\, \mathrm{LL}/\partial\zeta = [(A-nq) - (n-nq)\,\zeta]/\{[q+(1-q)\,\zeta](1-\zeta)\}.$$

Thus $\mathrm{LL}(\pi, \zeta)$ is strictly decreasing in ζ when $q > A/n$, and in the range $0 < q \leqslant A/n$ the log-likelihood has a unique maximum at $\zeta = (A-nq)/(n-nq)$. We can therefore restrict the search for the maximum likelihood point (π, ζ) to the path $[\pi, \zeta(\pi)]$ defined by

$$\zeta(\pi) = \begin{cases} 0 & \text{if } 0 < \pi \leqslant \pi_0, \\ (A-nq)/(n-nq) & \text{if } \pi_0 < \pi < 1, \end{cases}$$

where $\pi_0 = 1 - (A/n)^{1/R}$. Define

$$g(\pi) = \begin{cases} nR \log(1-\pi) + S \log[\pi/(1-\pi)] & \text{for } 0 < \pi \leqslant \pi_0, \\ A \log[A/(n-A)] + n \log(1 - A/n) & \\ \quad + (n-A) \log[q/(1-q)] + S \log[\pi/(1-\pi)] & \text{for } \pi_0 < \pi < 1, \end{cases}$$

so that $g(\pi)$ is a monotone function of $\mathrm{LL}[\pi, \zeta(\pi)]$ along the path $[\pi, \zeta(\pi)]$. The problem reduces to maximizing $g(\pi)$ over $0 < \pi < 1$. Notice that both $g(\cdot)$ and its derivative

$$g'(\pi) = \begin{cases} (S - nR\pi)/[\pi(1-\pi)] & \text{for } 0 < \pi \leqslant \pi_0, \\ [S(1-q) - (n-A)\, R\pi]/[\pi(1-\pi)(1-q)] & \text{for } \pi_0 < \pi < 1, \end{cases}$$

are continuous throughout the range $0 < \pi < 1$. Over the course of the range $0 < \pi \leqslant \pi_0$, $g(\pi)$ is either strictly increasing or has a unique maximum at $\pi = S/nR$ depending on whether or not $\pi_0 < S/nR$. Let $h(\pi) = S(1-q) - (n-A)\,R\pi$. Then $h(1) = S - S(n-A)\,R \leqslant 0$, and $h'(\pi) = RS(1-\pi)^{R-1} - R(n-A)$ is non-increasing. It follows that $g(\cdot)$ is either strictly decreasing or has a unique maximum in the range $\pi_0 < \pi < 1$ depending on whether or not $h(\pi_0) < 0$. Since $q = A/n$ when $\pi = \pi_0$, $h(\pi_0) = 0$ if and only if $\pi_0 = S/nR$. Thus if $\pi_0 \geqslant S/nR$ then $\hat{\pi} = S/nR$ and $\hat{\zeta} = 0$; if $\pi_0 < S/nR$ then $\hat{\pi}$ is the unique root of $h(\pi)$ in the range $\pi_0 < \pi < 1$ and $\hat{\zeta} = (A - n\hat{q})/(n - n\hat{q})$, where $\hat{q} = (1-\hat{\pi})^R$. The result for $\hat{\kappa}_x$ comes from formula (3.2) and the invariance property of maximum likelihood estimators.

References

[1] Altham, P. M. E. (1978). Two generalizations of the binomial distribution. *Appl. Statist.* **27** 162–167.

[2] Andersen, A. H. (1974). Multidimensional contingency tables. *Scand. J. Statist.* **1** 115–127.

[3] Barndorff-Nielsen, O. (1978). *Information and Exponential Families in Statistical Theory.* Wiley, New York.

[4] Bartko, J. J. (1966). The intraclass correlation coefficient as a measure of reliability. *Psychol. Rep.* **19** 3–11.

[5] Bishop, Y., Fienberg, S. E., and Holland, P. (1975). *Discrete Multivariate Analysis.* MIT Press, Cambridge, MA.

[6] Chatfield, C., and Goodhart, G. J. (1970). The beta-binomial model for consumer purchasing behavior. *Appl. Stastist.* **19** 240–250.

[7] Cohen, J. (1960). A coefficient of agreement for nominal scales, *Educ. Psychol. Meas.* **20** 37–46.

[8] Crowder, M. J. (1979). Inference about the intraclass correlation coefficient in the beta binomial ANOVA for proportions. *J. Roy. Stastist. Soc. Ser. B* **41** 230–234.

[9] Efron, B. (1979). Bootstrap methods; Another look at the jackknife. *Ann. Statist.* **7** 1–26.

[10] Feller, W. (1968). *Probability Theory and Its Applications*, Vol. 1, 3rd ed. Wiley, New York.

[11] Fleiss, J. L. (1971). Measuring nominal scale agreement among many raters. *Psychol. Bull.* **76**, 378–382.

[12] Fleiss, J. L., Nee, J. C. M., and Landis, J. R. (1979). The large sample variance of kappa in the case of different sets of raters. *Psychol. Bull.* **86** 974–977.

[13] Fleiss, J. L. (1981). *Statistical Methods for Rates and Proportions*, 2nd ed. Wiley, New York.

[14] Griffiths, D. A. (1973). Maximum likelihood estimation for the beta-binomial distribution and an application to the household distribution of the total number of cases of a disease. *Biometrics* **29** 637–648.

[15] Grizzle, J. E., Starmer, C. F., and Koch, G. G. (1969). Analysis of categorical data by linear models. *Biometrics* **25** 489–504.

[16] Haberman, S. J. (1974). *The Analysis of Frequency Data.* Univ. of Chicago Press, Chicago.

[17] Kraemer, H. C. (1979). Ramifications of a population model for κ as a coefficient of reliability. *Psychometrika* **44** 461–472.

[18] Landis, J. R., and Koch, G. G. (1977). The measure of observer agreement for categorical data. *Biometrics* **33** 159–174.

[19] Landis, J. R., and Koch, G. G. (1977). A one-way components of variance model for categorical data. *Biometrics* **33** 671–679.

[20] Neyman, J. (1959). Optimal asymptotic tests of composite statistical hypotheses. In *Probability and Statistics* (U. Grenander, Ed.), pp. 213–234. Wiley, New York.

[21] Plackett, R. L., and Paul, S. R. (1978). Dirichlet models for square contingency tables. *Comm. Statist. A—Theory Methods* **7** 939–952.

[22] Rao, C. R. (1973). *Linear Statistical Inference and Its Applications*, 2nd ed. Wiley, New York.

[23] Tanner, M. A., and Young, M. A. (1985). Modeling agreement among raters. *J. Amer. Statist. Assoc.* **80** 175–180.

[24] Tarone, R. E. (1979). Testing the goodness of fit of the binomial distribution. *Biometrika* **66** 585–590.

[25] Williams, D. A. (1975). The analysis of binary responses from toxicological experiments involving reproduction and teratogenicity. *Biometrics* **31** 949–952.

[26] Fleiss, J. L., and Cohen, J. (1973). The equivalence of weighted kappa and the intraclass correlation coefficient as measures of reliability. *Educ. Psychol. Meas.* **33** 613–619.

[27] Darroch J. N. (1974). Multiplicative and additive interactions in contingency tables. *Biometrika* **61** 207–214.

[28] Darroch, J. N., and Speed, T. P. (1983). Additive and multiplicative models and interactions. *Ann. Statist.* **11** 724–738.

[29] Potthoff, R. F., and Whittinghill, M. (1966). Testing for homogeneity. I. The binomial and multinomial distributions; II. The Poisson distribution. *Biometrika* **53** 167–182; 183–190.

[30] Fleiss, J. L. (1975). Measuring agreement between two judges on the presence or absence of a trait. *Biometrics* **31** 651–659.

Author Index

Subject Index

A

B

C

D

E